12TH INTERNATIONAL CONFERENCE ON VIBRATIONS IN ROTATING MACHINERY

PROCEEDINGS OF THE INTERNATIONAL CONFERENCE ON VIBRATIONS IN ROTATING MACHINERY (ONLINE, UK, 2020)

12TH INTERNATIONAL CONFERENCE ON VIBRATIONS IN ROTATING MACHINERY

Editor

Institution of Mechanical Engineers

CRC Press
Taylor & Francis Group
Boca Raton   London   New York

CRC Press is an imprint of the
Taylor & Francis Group, an **informa** business

A BALKEMA BOOK

Front cover image credit: Rolls-Royce

CRC Press/Balkema is an imprint of the Taylor & Francis Group, an informa business

© 2020 Taylor & Francis Group, London, UK

Typeset by Integra Software Services Pvt. Ltd., Pondicherry, India

Library of Congress Cataloging-in-Publication Data

Applied for

Published by: CRC Press/Balkema
        Schipholweg 107C, 2316XC Leiden, The Netherlands
        e-mail: Pub.NL@taylorandfrancis.com
        www.routledge.com – www.taylorandfrancis.com

ISBN: 978-0-367-67742-8 (Hbk)
ISBN: 978-0-367-68335-1 (pbk)
ISBN: 978-1-003-13263-9 (eBook)
DOI: 10.1201/9781003132639
https://doi.org/10.1201/9781003132639

Table of Contents

Organising Committee

Tribology Group, Mechatronics, Informatics & Control Group and Institution of Mechanical Engineers

Member Credits:

Patrick Keogh, Professor of Machine Systems, University of Bath
Andrew Rix, Associate Fellow – Rotor Dynamics, Rolls-Royce plc
Christoph Schwingshackl, Senior Lecturer in Structural Dynamics, Imperial College London
Arthur Lees, Professor Emeritus (Engineering), Swansea University
Georges Jacquet-Richardet, Professor in the Dynamics and Control of Structures, INSA Lyon
David Ewins, Distinguished Research Fellow, Imperial College London
Robert Herbert, Convenor of ISO/TC 108/SC 2/WG 31 Balancing, ex RWE npower
Seamus Garvey, Professor of Dynamics, University of Nottingham
Izhak Bucher, Professor of Dynamics and Mechatronics, Israel Institute of Technology
Philip Bonello, Reader in Engineering Dynamics, University of Manchester

International Scientific Committee:

Member Credits:

Mihai Arghir, Université de Poitiers, France
Nicolo Bachschmid, Politecnico di Milano, Italy
Katia Lucchesi Cavalca, Universidade Estadual de Campinas, Brazil
Aly El-Shafei, RITEC, Egypt
Roger Fittro, University of Virginia, USA
Eric Hahn, University of New South Wales, Australia
Timo Holopainen, ABB, Finland
Rainer Nordmann, Fraunhofer-Institut für Betriebsfestigkeit und Systemzuverlässigkeit LBF, Germany

Nevzat Ozgüven, Middle East Technical University, Turkey
Paolo Pennacchi, Politecnico di Milano, Italy
Ilmar Santos, Technical University of Denmark, Denmark
Damian Vogt, University of Stuttgart, Germany
Fulei Chu, Tsinghua University, China

Identification of misaligned additive forces and moments of coupling in turbo-generator system integrated with an active magnetic bearing

R. Siva Srinivas[1], Rajiv Tiwari[1]*, Ch. Kanna Babux[2]

[1]Department of Mechanical Engineering, Indian Institute of Technology Guwahati, Guwahati, Indiax
[2]Aero Engine Research and Design Centre, Hindustan Aeronautics Limited, Bangalore, India

ABSTRACT

Turbo generators run up to multiple stages and are interconnected by couplings.In the present paper, additive coupling stiffness (ACS), which is time varying in nature, is used to model the misalignment that exists between bearing centers of two rotor systems integrated with an active magnetic bearing (AMB). While the static stiffness of coupling is ever present, ACS exists only in the presence of misalignment. Presence of misalignment in/about a direction generates harmonics of additive forces and moments in the corresponding direction. These are identified from a linear least-squares problem using amplitude and phase of harmonics from full spectrum of rotor vibration and AMB current.

1   INTRODUCTION

Misalignment is one of the common faults encountered in rotating machinery. Various symptoms of the fault are loops in orbits [1, 2], high second order vibration [3], high axial vibration (4) and multiple harmonics in full spectrum [4]. Unconventional techniques such as thermal imaging [5], acoustic emission [6], and wavelet transforms [7] too have been used for detection. In the presence of misalignment, there is a time-varying stiffness component of coupling in addition to its static stiffness [8]. When it comes to measurement techniques, the effect of misalignment can be directly measured by measuring forces/moments using load cells/torque sensors [4] and measuring orbital motion with proximity probes [3]. The expressions for forces and moments generated due to parallel and/or angular misalignment have been given by various authors [9, 10]. A few authors have estimated the forces and moments in misaligned rigid and flexible coupled rotors with inverse problem [11].Inverse problem fundamentally has to do with identifying the various unknown parameters (so called difficult to theoretically model) and faults of rotor system in question from known test vibration data. Vibration data is acquired from locations, which are readily accessible and DOFs which are practically measurable. When the identifiable parameters/faults are linked to rotational DOFs (for example, the rotational additive stiffness of a crack, angular stiffness of a coupling) the corresponding DOFs cannot be grouped as slaves in dynamic condensation. At the same time, it would be difficult to measure them practically. Though today there are instruments that measure vibration of rotational DOFs (i.e., the slope of shaft), measurement of translational vibration is still more widely used in data acquisition owing to its

---

* Corresponding author: wydeek@gmail.com[1], rtiwari@iitg.ac.in[2], chkannababu@gmail.com[3]

simplicity. Hence, modifying the mathematical model in terms of translational DOFs becomes imperative. This is achieved by performing the next level of reduction, which is called hybrid/high-frequency condensation.

AMBs are widely used for levitating industrial rotors, suppression of vibrations (synchronous, asynchronous, and seismic) and identification of cracks, faults in pumps and bearings [12]. Feasibility studies were performed on integrating AMB that runs on PID control law with seven stage turbo generator [13]. Some authors have used AMBs as auxiliary supports which perform the dual role of suppressing vibration and identifying faults [14].

The present paper attempts to identify the time-varying component of coupling stiffness in the presence of misalignment from an inverse problem, which uses rotor vibration and AMB current data. The rotor system considered in this work is similar to two-stage test rig set-up used by [15] to study the vibratory behavior of coupled rotating shafts containing a transverse crack. An AMB is integrated with the rotor system for the purpose of vibration attenuation. The direct problem/mathematical model is first developed by finite element method using Timoshenko beam elements. The actual vibratory behavior of coupling in the presence of misalignment is introduced in the model by using an appropriate steering function. An identification algorithm, based on the inverse problem, is developed to identify additive stiffness parameters of the coupling. The uniqueness of the work lies in the (i) development of coupling model in the presence of misalignment and integrating with standard FE model and (ii) using data of the harmonics of full spectrum in inverse problem. From the ACS matrix one can quantify both severity and direction of misalignment.

2 TURBO GENERATOR MATHEMATICAL MODEL

Figure 1(a) shows the situation when the bearing centers are perfectly aligned. In such case there is only static coupling stiffness $\mathbf{K}_c$ as shown in Eqn. (1) .

$$\mathbf{f}_c = \boldsymbol{K}_c\{\boldsymbol{\eta}_0 + \boldsymbol{\eta}(t)\} \tag{1}$$

Figure 1(b) shows the situation where there is misalignment between bearing centers. Then additive component of coupling stiffness $\Delta\mathbf{K}_c(t)$ comes into play in addition to the static stiffness K_c as shown in Eqn.(2).

$$\begin{aligned}\mathbf{f}_c &= [\mathbf{K}_c + \Delta\mathbf{K}_c(t)]\{\boldsymbol{\eta}_0 + \boldsymbol{\eta}(t)\} \\ &= \underbrace{\mathbf{K}_c\boldsymbol{\eta}}_{\mathbf{f}_{SCS}} + \underbrace{\Delta\mathbf{K}_c(t)\boldsymbol{\eta}_0 + \Delta\mathbf{K}_c(t)\boldsymbol{\eta}}_{\mathbf{f}_{ACS}}\end{aligned} \tag{2}$$

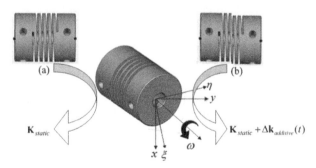

Figure 1. Coupling stiffness (a) with no misalignment (b) with misalignment.

2

2.1 Static stiffness force of coupling

The static stiffness force of the coupling in *real* form is given below [17]

$$
\mathbf{f}^r_{scs} = \mathbf{K_c}\eta =
\begin{bmatrix}
k_{xx} & k_{xy} & 0 & 0 & -k_{xx} & -k_{xy} & 0 & 0 \\
k_{yx} & k_{yy} & 0 & 0 & -k_{yx} & -k_{yy} & 0 & 0 \\
0 & 0 & k_{\varphi_y\varphi_y} & 0 & 0 & 0 & -k_{\varphi_y\varphi_y} & 0 \\
0 & 0 & 0 & k_{\varphi_x\varphi_x} & 0 & 0 & 0 & -k_{\varphi_x\varphi_x} \\
-k_{xx} & -k_{xy} & 0 & 0 & k_{xx} & k_{xy} & 0 & 0 \\
-k_{yx} & -k_{yy} & 0 & 0 & k_{yx} & k_{yy} & 0 & 0 \\
0 & 0 & -k_{\varphi_y\varphi_y} & 0 & 0 & 0 & k_{\varphi_y\varphi_y} & 0 \\
0 & 0 & 0 & -k_{\varphi_x\varphi_x} & 0 & 0 & 0 & k_{\varphi_x\varphi_x}
\end{bmatrix}
\begin{Bmatrix}
x_c \\
y_c \\
\varphi_{y_c} \\
\varphi_{x_c} \\
x_{c+1} \\
y_{c+1} \\
\varphi_{y_{c+1}} \\
\varphi_{x_{c+1}}
\end{Bmatrix}
\tag{3}
$$

In complex form, Eqn. (3) can be rewritten as

$$
\mathbf{f}^c_{scs} = 0.5
\begin{Bmatrix}
k_{xx}\{(u_c+\bar{u}_c)-(u_{c+1}+\bar{u}_{c+1})\}-jk_{xy}\{(u_c-\bar{u}_c)-(u_{c+1}-\bar{u}_{c+1})\} \\
+j\big[k_{yx}\{(u_c+\bar{u}_c)-(u_{c+1}+\bar{u}_{c+1})\}-jk_{yy}\{(u_c-\bar{u}_c)-(u_{c+1}-\bar{u}_{c+1})\}\big] \\
k_{\varphi_y\varphi_y}(\varphi_c+\bar{\varphi}_c)-k_{\varphi_y\varphi_y}(\varphi_{c+1}+\bar{\varphi}_{c+1})+k_{\varphi_x\varphi_x}(\varphi_c-\bar{\varphi}_c)-k_{\varphi_x\varphi_x}(\varphi_{c+1}-\bar{\varphi}_{c+1}) \\
k_{xx}\{(u_{c+1}+\bar{u}_{c+1})-(u_c+\bar{u}_c)\}-jk_{xy}\{(u_{c+1}-\bar{u}_{c+1})-(u_c-\bar{u}_c)\} \\
+j\big[k_{yx}\{(u_{c+1}+\bar{u}_{c+1})-(u_c+\bar{u}_c)\}-jk_{yy}\{(u_{c+1}-\bar{u}_{c+1})-(u_c-\bar{u}_c)\}\big] \\
-k_{\varphi_y\varphi_y}(\varphi_c+\bar{\varphi}_c)+k_{\varphi_y\varphi_y}(\varphi_{c+1}+\bar{\varphi}_{c+1})+k_{\varphi_x\varphi_x}(\varphi_{c+1}-\bar{\varphi}_{c+1})-k_{\varphi_x\varphi_x}(\varphi_c-\bar{\varphi}_c)
\end{Bmatrix}
\tag{4}
$$

If cross coupled coupling coefficients are not considered then $k_{xy} = k_{yx} = 0$. For a symmetric coupling $k_{xx} = k_{yy} = k_{rad}$, and $k_{\varphi_{xx}} = k_{\varphi_{yy}} = k_{\varphi rad}$. The final form of static coupling force is then given by

$$
\mathbf{f}^c_{scs} =
\begin{bmatrix}
k_{rad} & 0 & -k_{rad} & 0 \\
0 & k_{ang} & 0 & -k_{ang} \\
-k_{rad} & 0 & k_{rad} & 0 \\
0 & -k_{ang} & 0 & k_{ang}
\end{bmatrix}
\begin{Bmatrix}
u_c \\
\varphi_c \\
u_{c+1} \\
\varphi_{c+1}
\end{Bmatrix}
\tag{5}
$$

2.2 Additive stiffness force of coupling

Additive stiffness force of coupling $\mathbf{f}_{acs}$ in Eqn. (2) can be expanded as

$$
\mathbf{f}_{acs} = \Delta\mathbf{K}_c(t)\left(\eta_{0c}+\bar{\eta}_c e^{j\omega t}\right)
\tag{6}
$$

Due to the very large weight of turbo generators going up to few tens of tones and comparatively negligible weight of the coupling, it can be assumed that vibratory displacement at coupling nodes is less than static deflection due to the self-weight of rotor at the coupling location, i.e.

$$
\bar{\eta}_c << \eta_{0_c}
\tag{7}
$$

Then

$$
\mathbf{f}_{acs} = \Delta\mathbf{K}_c(t)\eta_{0_c}
\tag{8}
$$

With

3

$$\Delta\mathbf{K}_{\mathbf{c}_{\text{rot}}}(t) = s(t)\begin{bmatrix} \Delta k_{\xi\xi} & 0 & 0 & 0 & -\Delta k_{\xi\xi} & 0 & 0 & 0 \\ 0 & \Delta k_{\eta\eta} & 0 & 0 & 0 & -\Delta k_{\eta\eta} & 0 & 0 \\ 0 & 0 & \Delta k_{\varphi_\eta\varphi_\eta} & 0 & 0 & 0 & -\Delta k_{\varphi_\eta\varphi_\eta} & 0 \\ 0 & 0 & 0 & \Delta k_{\varphi_\xi\varphi_\xi} & 0 & 0 & 0 & -\Delta k_{\varphi_\xi\varphi_\xi} \\ -\Delta k_{\xi\xi} & 0 & 0 & 0 & \Delta k_{\xi\xi} & 0 & 0 & 0 \\ 0 & -\Delta k_{\eta\eta} & 0 & 0 & 0 & \Delta k_{\eta\eta} & 0 & 0 \\ 0 & 0 & -\Delta k_{\varphi_\eta\varphi_\eta} & 0 & 0 & 0 & \Delta k_{\varphi_\eta\varphi_\eta} & 0 \\ 0 & 0 & 0 & -\Delta k_{\varphi_\xi\varphi_\xi} & 0 & 0 & 0 & \Delta k_{\varphi_\xi\varphi_\xi} \end{bmatrix} \tag{9}$$

$$\eta_{0_c} = \left\{ \delta x_c \quad \delta y_c \quad \delta\varphi_{y_c} \quad \delta\varphi_{x_c} \quad \delta x_{c+1} \quad \delta y_{c+1} \quad \delta\varphi_{y_{c+1}} \quad \delta\varphi_{x_{c+1}} \right\}^T \tag{10}$$

Parallel misalignment generates time dependent additive coupling stiffness coefficients Δk_{T_ξ} and Δk_{T_η} in the rotating frame of reference. Likewise, Δk_{R_η} and Δk_{R_ξ} are additive coupling stiffness coefficients generated due to the angular misalignment. Subscript 'c' denotes the number of the coupling node, $s(t)$ is the steering function, which simulates the multi-harmonic behavior of misalignment.

In real rotor systems, unequal amounts of misalignment exists in $x-y$ and $z-x$ planes, and therefore the coefficients are assumed to be unequal, $\Delta k_{T_\xi} \neq \Delta k_{T_\eta}$ and $\Delta k_{R_\xi} \neq \Delta k_{R_\eta}$, which makes the total number of direct ACS coefficients four. All cross-coupling coefficients are ignored in this work. The matrix $\Delta\mathbf{K}_c(t)$ is in the rotating frame of reference. We convert Eqn. (9) from rotating to stationary frame of reference using the transformation shown in Eqn. (11)

$$\Delta\mathbf{K}_{c_{\text{stat}}} = \mathbf{T}^T \Delta\mathbf{K}_{c_{\text{rot}}}(t)\mathbf{T} = s(t)\left(\mathbf{T}^T \Delta\mathbf{K}_{c_{\text{rot}}}\mathbf{T}\right) \tag{11}$$

Where transformation matrix $\mathbf{T}$ is given by

$$T = \begin{bmatrix} \mathbf{R} & \mathbf{0} & \mathbf{0} & \mathbf{0} \\ \mathbf{0} & \mathbf{R} & \mathbf{0} & \mathbf{0} \\ \mathbf{0} & \mathbf{0} & \mathbf{R} & \mathbf{0} \\ \mathbf{0} & \mathbf{0} & \mathbf{0} & \mathbf{R} \end{bmatrix} \text{ with } \mathbf{R} = \begin{bmatrix} \cos\omega t & \sin\omega t \\ -\sin\omega t & \cos\omega t \end{bmatrix} \tag{12}$$

Table 1. Correlation between misalignment and coupling additive stiffness coefficients.

Misalignment type	Direction	$\Delta k_{\xi\xi}$	$\Delta k_{\eta\eta}$	$\Delta k_{\varphi_\eta\varphi_\eta}$	$\Delta k_{\varphi_\xi\varphi_\xi}$
Parallel	x	✓	✗	✗	✗
	y	✗	✓	✗	✗
Angular	φ_x	✗	✗	✓	✗
	φ_y	✗	✗	✗	✓
Combined	x, y	✓	✓	✗	✗
	φ_x, φ_y	✗	✗	✓	✓

The final form of the ACS force in the complex form (from Eqns. (8), (9) and (10)) is given by

$$\mathbf{f}^c_{acs} = 0.5s(t)\begin{Bmatrix} (\delta x_c - \delta x_{c+1})\left\{\Delta k_{T\xi}(1 + e^{j2\omega t}) + \Delta k_{T\eta}(1 - e^{j2\omega t})\right\} \\ (\delta\varphi_{y_c} - \delta\varphi_{y_{c+1}})\left\{\Delta k_{R\xi}(1 + e^{j2\omega t}) + \Delta k_{R\eta}(1 - e^{j2\omega t})\right\} \\ (\delta x_{c+1} - \delta x_c)\left\{\Delta k_{T\xi}(1 + e^{j2\omega t}) + \Delta k_{T\eta}(1 - e^{j2\omega t})\right\} \\ (\delta\varphi_{y_{c+1}} - \delta\varphi_{y_c})\left\{\Delta k_{R\xi}(1 + e^{j2\omega t}) + \Delta k_{R\eta}(1 - e^{j2\omega t})\right\} \end{Bmatrix} \tag{13}$$

where the steering function is assumed to be square wave with 40% duty cycle as given in Eqn. (14). The Fourier expansion of the wave is as follows

$$
\begin{aligned}
s(t) ={}& 0.5 + 0.6055\cos(\omega t) + 0.1871\cos(2\omega t) - 0.1247\cos(3\omega t) - 0.1514\cos(4\omega t) \\
&+ 0.1009\cos(6\omega t) + 0.0535\cos(7\omega t) - 0.0468\cos(8\omega t) - 0.0673\cos(9\omega t) \\
&+ 0.055\cos(11\omega t) + 0.0312\cos(12\omega t) - 0.0288\cos(13\omega t) - 0.0432\cos(14\omega t) \\
&+ 0.0378\cos(16\omega t) + 0.022\cos(17\omega t)
\end{aligned}
\tag{14}
$$

Eqn. (13) is rearranged as below by taking the steering function $s(t)$ inside and multiplying with exponential terms.

$$\mathbf{f}^c_{acs}(t) = 0.5\begin{Bmatrix} (\delta x_c - \delta x_{c+1})\left(\Delta k_{T\xi}\sum_{i=-n}^{i+n} p_i e^{ij\omega t} + \Delta k_{T\eta}\sum_{i=-n}^{i+n} q_i e^{ij\omega t}\right) \\ (\delta\varphi_{y_c} - \delta\varphi_{y_{c+1}})\left(\Delta k_{R\xi}\sum_{i=-n}^{i+n} p_i e^{ij\omega t} + \Delta k_{R\eta}\sum_{i=-n}^{i+n} q_i e^{ij\omega t}\right) \\ (\delta x_{c+1} - \delta x_c)\left(\Delta k_{T\xi}\sum_{i=-n}^{i+n} p_i e^{ij\omega t} + \Delta k_{T\eta}\sum_{i=-n}^{i+n} q_i e^{ij\omega t}\right) \\ (\delta\varphi_{y_{c+1}} - \delta\varphi_{y_c})\left(\Delta k_{R\xi}\sum_{i=-n}^{i+n} p_i e^{ij\omega t} + \Delta k_{R\eta}\sum_{i=-n}^{i+n} q_i e^{ij\omega t}\right) \end{Bmatrix} \tag{15}$$

The harmonics generated by coupling force defined by Eqn.(15) are similar to the experimental spectra shown in [4]. This requirement is fulfilled by the suitable steering function in the form of rectangular function) defined by Eqn. (14)). Presence of additive stiffness coefficients in a particular direction would physically represent the existence of misalignment in the corresponding direction. By estimating the values of $\Delta k_{T\xi}p_i$, $\Delta k_{T\eta}q_i$, $\Delta k_{R\xi}p_i$, $\Delta k_{R\eta}q_i$ in Eqn. (15) the fluctuating forces/moments that arise due to misalignment can be estimated. As the severity of misalignment increases the number of harmonics in the full spectrum also increase and the index i increases likewise. The misalignment in the coupling is known to generate multiple harmonics [4, 11] and to mimic the same, the concept of time-varying additive stiffness coefficients of coupling is introduced. The product of time-varying additive stiffness coefficients of coupling and the static deflections at coupling nodes produces multi-harmonic coupling forces. This behavior is also noticed in rotors with breathing cracks [25].

2.3 Global EOM
The global EOM of turbo generator system integrated with AMB is given by

$$(\mathbf{M}_{sh} + \mathbf{M}_d)\ddot{\mathbf{u}} + [\mathbf{C}_b + \mathbf{C}_c + \mathbf{C}_{sh} - j\omega(\mathbf{G}_{sh} + \mathbf{G}_d)]\dot{\mathbf{u}} + (\mathbf{K}_b + \mathbf{K}_{scs} + \mathbf{K}_{sh})\mathbf{u} + s(t)\Delta K_c \eta_{0_c} = \mathbf{f}_u + \mathbf{f}_{amb} \tag{16}$$

Equation (16) is rewritten as

$$\mathbf{M}_g\ddot{\mathbf{u}} + \left(\mathbf{C}_g - j\omega\mathbf{G}_g\right)\dot{\mathbf{u}} + \mathbf{K}_g\mathbf{u} = \mathbf{f}_{unb} + \mathbf{f}_{amb} - \mathbf{f}_{acs} \tag{17}$$

With $\mathbf{f}_{u_i} = m_{d_i}e_{d_i}\omega^2 e^{j\omega t}e^{j\beta_i}$; $\mathbf{f}_{amb} = k_s\mathbf{u}_{amb} - k_i\mathbf{i}_c$

$\mathbf{M}_g = \mathbf{M}_{sh} + \mathbf{M}_d$; $\mathbf{C}_g = \mathbf{C}_b + \mathbf{C}_c + \mathbf{C}_{sh}$; $\mathbf{G}_g = \mathbf{G}_{sh} + \mathbf{G}_d$; $\mathbf{K}_g = \mathbf{K}_b + \mathbf{K}_{scs} + \mathbf{K}_{sh}$

The elemental matrices in Ref. [16], which have real terms, are used in the present work.The nodal displacements of x-z plane are arranged in real domain and those of y-z plane in imaginary domain. Eqn. (17) is the global EOM of the rotor system expressed in terms of *all* translational and rotational DOFs.The next section describes the procedure to reduce the number of DOFs be means of condensation technique.

3 DYNAMIC CONDENSATION CONSIDERING GYROSCOPIC EFFECT OF SHAFT AND DISC

The nodes where the external forces act on the rotor system are usually chosen as master DOFs. In the present case, the master DOFs correspond to the nodal locations of *all identifiable parameters.* They include the translational DOFs at the nodal locations of bearings, discs, coupling, AMBs and rotational DOFs at coupling nodes. In a turbo-generator system having n_b isotropic bearings, n_d discs, n_{amb} AMBs and n_c interconnecting couplings there are $(n_b + n_d + n_{amb} + 4n_c)$ master DOFs.

$$\eta_m = \left\{ u_b^1 \quad u_d^1 \quad u_{amb}^1 \quad \cdots \quad u_c^1 \quad \varphi_c^1 \quad u_c^2 \quad \varphi_c^2 \quad \cdots \quad u_d^i \quad u_{amb}^i \quad \cdots \quad u_c^p \quad \varphi_c^p \quad \cdots \quad u_d^i \quad u_b^i \right\} \quad (18)$$

Rest of the DOFs are grouped as slave DOFs. The transformation matrix $\mathbf{T}^d$ links master DOFs to the total DOFs, as

$$\eta = \left\{ \begin{matrix} \eta_m^d \\ \eta_s^d \end{matrix} \right\} = \mathbf{T}^d \eta_m^d \quad (19)$$

The final form of $\mathbf{T}^d$ is given by

$$\mathbf{T}^d = \left\{ \begin{matrix} \mathbf{I} \\ -[\mathbf{K}_{ss} - \omega_0^2 \mathbf{M}_{ss} - j\omega_0^2 \mathbf{G}_{ss}]^{-1} [\mathbf{K}_{sm} - \omega_0^2 \mathbf{M}_{sm} - j\mathbf{G}_{sm}] \end{matrix} \right\} \quad (20)$$

As tabulated in Table 1, the gyroscopic matrix is considered in the derivation of transformation matrix. The EOM in terms of reduced coordinates is given by

$$\mathbf{M}^d \ddot{\eta}_m^d + (\mathbf{C}^d - j\omega \mathbf{G}^d) \dot{\eta}_m^d + \mathbf{K}^d \eta_m^d = \mathbf{f}_{unb}^d + \mathbf{f}_{amb}^d - \mathbf{f}_{acs}^d \quad (21)$$

with $\mathbf{M}^d = (\mathbf{T}^d)^T \mathbf{M}_g \mathbf{T}^d$; $\mathbf{K}^d = (\mathbf{T}^d)^T \mathbf{K}_g \mathbf{T}^d$; $\mathbf{C}^d = (\mathbf{T}^d)^T \mathbf{C}_g \mathbf{T}^d$; $\mathbf{G}^d = (\mathbf{T}^d)^T \mathbf{G}_g \mathbf{T}^d$

$$\mathbf{f}_{unb}^d = (\mathbf{T}^d)^T \mathbf{f}_{unb}; \mathbf{f}_{amb}^d = (\mathbf{T}^d)^T \mathbf{f}_{amb}; \mathbf{f}_{acs}^d = (\mathbf{T}^d)^T \mathbf{f}_{acs}$$

By substituting $\eta_m = \bar{\eta}_m e^{j\omega t}$, the EOM in time domain are converted to frequency domain, as

$$\left(-\mathbf{M}^d (i\omega)^2 + j(i\omega)(\mathbf{C}^d - j\omega \mathbf{G}^d) + \mathbf{K}^d \right) \bar{\eta}_m^d e^{ij\omega t} = \mathbf{f}_{unb}^d + \mathbf{f}_{amb}^d - \mathbf{f}_{acs}^d \quad (22)$$

Eqn. (22) expresses the EOM of rotor system in terms of master DOFs alone. All rotational DOFs except those at locations of couplings are grouped as slave DOFs. The next section describes the procedure to remove rotational DOFs at coupling location from the EOMs.

4 HYBRID CONDENSATION

Out of $(n_b + n_d + n_{amb} + 4n_c)$ master DOFs, $(n_b + n_d + n_{amb} + 2n_c)$ are translational DOFs and $2n_c$ are rotational DOFs.Rotational DOFs $\left\{ \varphi_c^1 \quad \varphi_c^2 \quad \cdots \quad \varphi_c^{n_c} \quad \varphi_c^{n_c+1} \right\}$ are linked to the static and additive stiffness parameters of couplings k_{ang}, Δk_{R_ξ} and Δk_{R_η}.

It would be desirable to remove these rotational DOFs owing to the difficulty of measurement in experiments. We therefore perform one more step wherein the $2n_c$ rotational DOFs are made as *hybrid* slave DOFs and the rest of $(n_b + n_d + n_a + 2n_c)$ translational DOFs are grouped as *hybrid* master DOFs. The coordinates η_m^h and η_m^d are linked by the transformation matrix as shown below

$$\eta_m^d = \left\{ \begin{matrix} \boldsymbol{\eta}_m^h \\ \boldsymbol{\eta}_s^h \end{matrix} \right\} = \mathbf{T}^h \boldsymbol{\eta}_m^h \tag{23}$$

With

$$\mathbf{T}^h = \left\{ \begin{matrix} \mathbf{I} \\ -[\mathbf{K}_{ss}^D - \omega_0^2 \mathbf{M}_{ss}^D - \mathrm{j}\omega_0^2 \mathbf{G}_{ss}^D]^{-1}[\mathbf{K}_{sm}^D - \omega_0^2 \mathbf{M}_{sm}^D - \mathrm{j}\mathbf{G}_{sm}^D] \end{matrix} \right\} \tag{24}$$

The final EOM in frequency domain that utilizes data from *hybrid master DOFs alone* is given by

$$\left(-\mathbf{M}^h (i\omega)^2 + j(i\omega)\left(\mathbf{C}^h - j\omega\mathbf{G}^h\right) + \mathbf{K}^h \right) \bar{\boldsymbol{\eta}}_m^h e^{ij\omega t} = \mathbf{f}_{unb}^h + \mathbf{f}_{amb}^h - \mathbf{f}_{acs}^h \tag{25}$$

The next section describes the development of identification algorithm used for estimating the unknown system parameters.

Table 2. Various rotor parameters and faults identified using hybrid condensation.

		Condensation technique					
		Dynamic			Hybrid		
	Estimated parameters	M	K	G	M	K	G
[18]	$[C_c]^*$	✓	✗	✗	-NA-		
[19]	$[C_c]$	✓	✓	✗	✓	✗	✗
[20]	$[C_c]$	✓	✓	✗	✓	✓	✗
[21]	$[K_b], [K_c], [C_b], [C_c], [U^{re}], [U^{im}]$	✓	✓	✗	✓	✗	✗
[22]	$k_{s_{x1}}^{amb}, k_{s_{x2}}^{amb}, k_{s_{y1}}^{amb}, k_{s_{y2}}^{amb},$	✓	✓	✗	-NA-		
[23]	$[U^{re}], [U^{im}], [k_s^{amb}], [k_i^{amb}]$	✓	✓	✓	-NA-		
[24]	$[K_c], [C_b], [U^{re}], [U^{im}], k_s^{amb}, k_i^{amb}$	✓	✓	✗	✓	✗	✗
Present work	$[\varDelta K_c], [K_c], [K_b], [U^{re}], [U^{im}], k_s^{amb}, K_i^{amb}$	✓	✓	✓	✓	✓	✓
$*[C_{cr}]$: Crack flexibility matrix							

5   IDENTIFICATION ALGORITHM USING LEAST-SQUARES METHOD

Eqn. (25) can be rearranged as shown in Eqn. (26) so that all unknown identifiable parameters are on the left side and all known parameters are on the right side.

$$
\mathbf{f}_{unb}^h + \mathbf{f}_{amb}^h - \mathbf{f}_{acs}^h - \mathbf{K}_{scs}^h \bar{\eta}_m^h - \mathbf{K}_b^h \bar{\eta}_m^h =
$$
$$
\left(-(\mathbf{M}_{sh}^h + \mathbf{M}_d^h)(i\omega)^2 + \mathrm{j}(i\omega)(\mathbf{C}_b^h + \mathbf{C}_{sh}^h - \mathrm{j}\omega(\mathbf{G}_{sh} + \mathbf{G}_d)) + \mathbf{K}_{sh}^h \right) \bar{\eta}_m^h e^{ij\omega t}
\tag{26}
$$

The above equation can be arranged in the form

$$
\mathbf{Ax} = \mathbf{b}
\tag{27}
$$

which is again written as

$$
\left\{ \begin{array}{c} \mathbf{A}^{re} \\ \mathbf{A}^{im} \end{array} \right\} \mathbf{x} = \left\{ \begin{array}{c} \mathbf{b}^{re} \\ \mathbf{b}^{im} \end{array} \right\}
\tag{28}
$$

The vector of unknowns **x** comprises of unbalance eccentricity, phase, AMB stiffness constants, bearing stiffness coefficients, coupling static and additive stiffness coefficients. The inputs to matrix **A**, vector **b** are the real and imaginary parts of rotor vibration and AMB current in frequency domain. xcan be solved by linear least-squares method.

6   RESULTS AND DISCUSSION

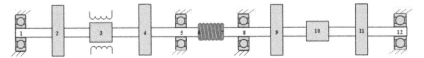

Figure 2. Coupled rotor-bearing-AMB System.

The representative turbo-generator system having two coupled rotors with AMB integrated on rotor-1 is shown in Figure 2 .Rotors 1 and 2 each carry two discs and are supported on two isotropic ball bearings. An intermediate coupling connects the rotor-1 and rotor-2. The drive end coupling of the rotor system is not considered in the FE model. An auxiliary AMB is located between two discs and it supports rotor-1. The mass of AMB core that is attached to rotor 1(at node-3 location) and damping of the coupling are ignored in the present problem. The various parameters of the rotor system shown in Figure 2 are listed in Table 2.

Critical speeds of rotor system are identified at 80 rad/s and 190 rad/s from Hilbert envelope of vibration response at nodes. The identification is performed at a speed of 39 Hz (245 rad/s) away from turbo generator critical speeds. In the present problem, five harmonics have been considered on the positive and negative side of full spectrum. The number of harmonics in real rotors would depend on the amount of misalignment present in between bearing centers. To improve the condition number of the matrix, vibration and current data at hybrid master DOFs from clockwise and anticlockwise rotations have been considered.

Table 2. Rotor-bearing-coupling-AMB system specifications.

Disc					
Mass, kg	**Moment of inertia,kg-m^2**		**Unbalance**		
			Amplitude,kg-m		**Phase, rad**
m_{d_1} 2.25	I_{d_1} .0024	I_{p_1} .0048	e_1 4e-5		$\beta_1 \pi/10$
m_{d_2} 1.75	I_{d_2} .0035	I_{p_2} .007	e_2 3e-5		$\beta_2 \pi/4$
m_{d_3} 2.5	I_{d_3} .0045	I_{p_3} .009	e_3 4e-5		$\beta_3 \pi/15$
m_{d_4} 1.6	I_{d_4} .004	I_{p_4} .008	e_4 3e-5		$\beta_4 \pi/12$
Bearing					
Stiffness, N/m			Damping, Ns/m		
k_{b1} 500000			c_{b1} 1350		
k_{b2} 650000			c_{b2} 1300		
k_{b3} 575000			c_{b3} 1300		
k_{b4} 595000			c_{b1} 1250		
Coupling Stiffness					
Radial, N/m			Angular, N-m/rad		
k_{rad} 225000			k_{ang} 200000		
Δk_{T_ξ} 60000			Δk_{R_ξ} 75000		
Δk_{T_η} 40000			Δk_{R_η} 100000		
AMB					
$k_p = 12200$ A/m			$k_s^{amb} = 105210$ N/m		
$k_i = 2000$ A/m-s			$k_i^{amb} = 42.1$ A/m		
$k_d = 3$ A-s/m					
Shaft					
d, m	l_e, m	a_0	a_1		
0.17	0.25	0.154	1e-5		

The results of the inverse problem and the effect of considering **K**, **M** and **G** in the derivation of transformation matrix on the accuracy of parameter estimation are tabulated in Table 3. Case 1 is an illustration of when **M** alone is considered in the derivation of $\mathbf{T}^h$ matrix. Only 3 out of 20 parameters show an error within 10%. Rest of the parameters deviate by a large value. Case 2 shows that by considering both **M** and **K** in $\mathbf{T}^h$ there is a significant improvement in the estimates. The largest deviation is noticed in k_{ang}(9.65%), k_s^{amb}(-8.78%). Other parameters are estimated within 5% of their assumed values. In Case 3, **G** is considered along with **K** and **M** in the derivation of $\mathbf{T}^h$. This results in further improvement of the estimates with the highest deviation noticed in Δk_{R_η}(4.56%), k_{ang}(3.73%). To simulate experimental conditions, random nose is added to the numerically simulated vibration and current output. Figure 3 shows the sensitivity of algorithm to signal noise for three cases (0%, 1% and 2%).The estimates are reasonably close with the highest deviation being 8% shown by Δk_{T_η}.

Figure 4 shows the robustness of algorithm against modelling error in the static deflection at coupling nodes$(\delta x_c, \delta x_{c+1}, \delta \varphi_{y_c}$ and $\delta \varphi_{y_{c+1}})$. Three cases (0%, 1% and 2%) have been considered and the highest error in estimates is less than 10%. Substituting the coefficients of ΔK in Eqn. (15) yields fluctuating forces and moments of the coupling due to misalignment. For the case of pure angular misalignment, the fluctuating moments of coupling due to additive stiffness is shown in Figure 5. In real rotors the thickness of shims would decide the amount of angular/parallel misalignment present which in turn would affect the magnitude of fluctuating forces & moments.

Table 3. Comparison of various condensation methods employed for parameter estimation.

Parameter	Assumed value	% error in estimated value		
		Matrices considered in condensation		
		Case 1: DC: K,M,G HC: M	Case 2: DC: K,M,G HC: K,M	Case 3: DC: K,M,G HC: K,M,G
U_1^{re}	3.80E-05	24.58	0.00	0.00
U_1^{im}	2.12E-05	178.24	0.01	0.02
U_2^{re}	3.91E-05	164.81	-0.04	-0.05
U_2^{im}	2.90E-05	786.55	-0.07	-0.07
U_3^{re}	1.24E-05	9.58	0.25	0.25
U_3^{im}	2.12E-05	7.32	-0.08	-0.08
U_4^{re}	8.32E-06	121.37	-0.51	-0.51
U_4^{im}	7.76E-06	537.21	-0.51	-0.51
k_{b1}	500000	7.70	-0.01	-0.01
k_{b2}	650000	85.51	0.14	0.20
k_{b3}	575000	1103.75	1.70	1.62
k_{b4}	595000	-349.34	0.41	0.40
k_s^{amb}	105210	-8475.66	-8.78	-3.78
k_i^{amb}	42.10	-1672.80	-1.75	-0.75
k_{rad}	225000	-2314.29	0.05	0.06
k_{ang} (linked to φ_c^1, φ_c^2)	200000	97.89	9.65	3.73
$\Delta k_{T\xi}$	60000	-2246.67	0.33	0.32
$\Delta k_{T\eta}$	40000	-4472.69	3.24	3.20
$\Delta k_{R\xi}$ (linked to φ_c^1, φ_c^2)	75000	101.13	3.80	3.95
$\Delta k_{R\eta}$ (linked to φ_c^1, φ_c^2)	100000	101.14	4.40	4.56

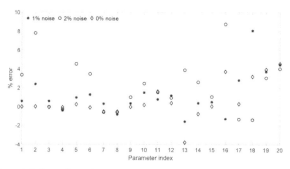

Figure 3. Sensitivity of estimates to noise in vibration and current data.

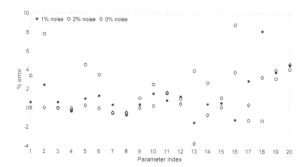

Figure 4. Sensitivity of estimates to modelling error.

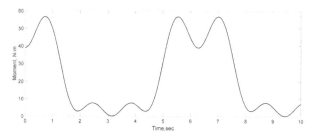

Figure 5. Fluctuating moment of coupling due to angular misalignment.

7 CONCLUSIONS

The mathematical model of coupled rotor systems integrated with an auxiliary AMB is developed using Timoshenko beam elements. Coupling misalignment is modelled in the form of additive stiffness and plugged into the rotor model. The simulated responses in both time and frequency domains are similar to the experimental responses reported in literature. Hybrid condensation is used to express the rotor EOM purely in terms of translational DOFs without a need for data from rotational DOFs. The identification algorithm derived from the direct problem is

11

a linear least-squares regression problem. The inputs to the algorithm are amplitude and phase of the harmonics of full spectrum FFT.It has been shown that considering **M**, **K** and **G** in the derivation of transformation matrix results in good estimates of parameters. The algorithm is found to be robust to noisy signals and to errors in modelling. The algorithm can be experimentally tested on coupled rotor test rig with integrated AMB set up by the following procedure: (i) the real and imaginary parts (i.e. the amplitude and phase or displacement signals and a common reference signal of the rotor vibration and AMB current must be captured at the physical locations on the rotor corresponding to the nodes of the FEM model or condensation scheme could be used to reduce DOFs of the model) at any speed away from critical speeds (ii) This data must be input to the identification algorithm that is built from the mathematical model of the rotor-bearing-coupling-AMB system. Parametric studies investigating the additive coupling stiffness by treating coupling types, shim thickness and speed as variables would be the logical extension to the present work.

Nomenclature

K Stiffness matrix	f force vector
M Mass matrix	u_c complex translational displacement
G Gyroscopic matrix	φ_c complex rotational displacement
C Damping matrix	i index
T Transformation matrix	$j \sqrt{-1}$
$\Delta \mathbf{K_c(t)}$: ACS matrix	ω angular speed
η_0: Static deflection vector	t time
η: Vibratory displacement vector	λ constant
Subscripts	
acs additive coupling stiffness	*sh* shaft
scs static coupling stiffness	*g* global
rad radial	*d* disc
ang angular	*m* master
amb active magnetic bearing	*s* slave
c coupling	*b* bearing
Superscripts	
d dynamic	*c* complex
h hybrid	*T* transpose

REFERENCES

[1] Monte, M., Verbelen, F., Vervisch, B., Detection of Coupling Misalignment by Extended orbits, Experimental Techniques, Rotating Machinery, and Acoustics, 8, Proceedings of the Society for Experimental Mechanics Series, pp 243–250, February, 2015.

[2] Monte M, Verbelen, F, Vervisch, B., The Use of Orbitals and Full Spectra to Identify Misalignment, Structural Health Monitoring, 5, pp 215–222, February, 2014.

[3] Rao, J.S., Sreenivas, R., Chawla, A., Experimental Investigation of Misaligned Rotors, Proceedings of ASME Turbo Expo, June 4–7, 7, Louisiana, USA, 2001.

[4] Patel, T., Darpe, A.K., Experimental Investigations on Vibration Response of Misaligned rotors, Mechanical Systems and Signal Processing, 23, pp 2236–2252, 2009.

[5] Mohanty, A.R., Fatima, S, Shaft misalignment detection by thermal imaging of support bearings, IFAC-Papers Online, 48, 21, pp 554–559, 2015.

[6] Chacon, J.L.F., Andicoberry, E.A., Kappatos, V., Asfis, G., Gan, T.H., Balachandran, W, Shaft Angular Misalignment Detection using Acoustic Emission, Applied Acoustics,85, pp 12–22, 2014.

[7] Chandra, N. H., Sekhar, A, S., Fault Detection in Rotor Bearing Systems using Time Frequency Techniques, Mechanical Systems and Signal Processing, 72–73, pp 105–133, May, 2016.

[8] Lees, A.W., Misalignment in Rigidly Coupled Rotors, Journal of Sound and Vibration, 305, pp 261–271, 2007.

[9] Gibbons, C.B., Coupling Misalignment Forces, Proceedings of the Fifth Turbomachinery Symposium, pp 111–116, 1976.

[10] Sekhar, A.S., Prabhu, B.S., Effects of Coupling Misalignment on Vibrations of Rotating Machinery, Journal of Sound and Vibration, 185, 4, pp 655–671, 1995.

[11] Lal, M., Tiwari, R., Multiple Fault Identification in Simple Rotor-Bearing-Coupling Systems based on Forced Response Measurements, Mechanism and Machine Theory, 51, pp 87–109, 2012.

[12] Siva Srinivas, R., Tiwari, R., Kannababu, Ch., Application of Active Magnetic Bearings in Flexible Rotordynamic Systems - A State-of-The-Art Review, Mechanical Systems and Signal Processing, 106, pp 537–572, 2018.

[13] Pilotto, R., Nordmann, R., Atzrodt, H., Herold, S., Use of Magnetic Bearings in Vibration Control of a Steam Turbine with Oil Film Bearings", 24th International Congress on Sound and Vibration, pp 23–27 July, London, UK, 2017.

[14] Siva Srinivas, R., Tiwari, R., Kannababu, Ch., Model Based Analysis and Identification of Multiple Fault Parameters in Coupled Rotor Systems with Offset Discs in the Presence of Misalignment and Integrated With an Active Magnetic Bearing, Journal of Sound and Vibration, 450, pp 109–140, 2019.

[15] Mayes, I.W., Davies, W.G.R., A Method of Calculating the Vibrational Behaviour of Coupled Rotating Shafts Containing a Transverse Crack, 2nd international conference, Vibrations in Rotating Machinery, pp 17–28, September, Cambridge, UK, 1980.

[16] Chen, W. J., A Note on Computational Rotor Dynamics, Journal of Vibration and Acoustics, 120, pp 228–233, January, 1998.

[17] Tiwari, R., Rotor Systems: Analysis and Identification, 1st edition, 2017, Taylor & Francis Group, CRC Press, Boca Raton, USA, November, 2017.

[18] Dharmaraju, N., Tiwari, R., Talukdar, S., Identification of an Open Crack Model in a Beam based on Force–Response Measurements, Composites and Structures, 82, pp 167–179, 2004.

[19] Dharmaraju, N., Tiwari, R., Talukdar, S., Development of a Novel Hybrid Reduction Scheme for Identification of an Open Crack Model in a Beam, Mechanical Systems and Signal Processing, 19, pp 633–657, 2005.

[20] Kartikeyan, M., Talukdar, S., Development of a Novel Algorithm for a Crack Detection, Localization, and Sizing in a Beam Based on Forced Response Measurements, Journal of Vibration and Acoustics, 130, pp 021002–1 - 021002–14, 2008.

[21] Lal, M., Tiwari, R.,Quantification of Multiple Fault Parameters in Flexible Turbo-Generator Systems with Incomplete Rundown Vibration Data, Mechanical Systems and Signal Processing, 41, pp 546–563, 2013.

[22] Tiwari, R., Chougale, A., Identification of Bearing Dynamic Parameters and Unbalance States in a Flexible Rotor System Fully Levitated on Active Magnetic Bearings, Mechatronics, 24, pp 274–286, 2014.

[23] Prasad, V., Tiwari, R., Identification of Speed-Dependent Active Magnetic Bearing Parameters and Rotor Balancing -60in High-Speed Rotor Systems, Journal of Dynamic Systems Measurement and Control, 141, 4, 041013, 11, 2019.

[24] Sampath, K. K and Lal, M., Dual Flexible Rotor System with Active Magnetic Bearings for Unbalance and Coupling Misalignment Faults Analysis, Sadhana, 44:188, 16, 2019.

[25] Mayes, I. W., and Davies, W.G.R.: The Vibrational Behaviour of a Rotating Shaft System Containing a Transverse Crack. Conference on Vibrations in Rotating Machinery, University of Cambridge, pp 53–64, 1976.

12th International Conference on Vibrations in Rotating Machinery -
Institution of Mechanical Engineers, ISBN 978-0-367-67742-8

On the analysis of a rotor system subjected to rub using a continuous model

A.G Mereles, K.L Cavalca

Faculty of Mechanical Engineering, University of Campinas, Brazil

ABSTRACT

Rub related phenomena are some of the most researched topics in rotordynamics and they are yet not fully comprehended. Thus, this work aims at studying the different outcomes possible for different configurations of a rotor system consisting of a simply-supported shaft with an off-centred disk. For this purpose, a continuous model is used, which takes into account the inertia and gyroscopic effects of both the shaft and the rigid disk. The governing partial differential equations are discretized by means of the modal expansion method. A finite element model is also presented and its results compared with the continuous model.

1 INTRODUCTION

Rubbing occurs when there is physical contact between rotating and stationary parts in a rotating machine, which is a serious malfunction that may lead to a complete failure of the machine. Some of the effects of rubbing in rotating machinery are: rotor stiffening effect, changes in response amplitudes and phase, appearance of high frequency components, wear, noise, thermal bow and effects on fluid dynamic forces (1). The complexity of rubbing arises due to the several physical phenomena involved, such as friction, physical contacts, and stiffness changes (1). This makes the theoretical study of rubbing very challenging, since it is difficult to establish a complete mathematical model due to the several parameters involved.

The behaviour of rotor systems subjected to rubbing appears to have either a forward or a backward whirl, depending on several parameters such as damping, stator stiffness and friction. Also, during forward whirl, the response can further be divided in forward annular rub (FWAR) and forward partial rub (FWPR). In the former, the rotor is continuously in contact with the stator, whirling within the clearance. This behaviour is seen mainly when the friction between the rotor and stator is very low. However, when friction is high enough, the rotor enters a bouncing state, or in FWPR, with successive impacts during the whirl motion. This state has been shown to possess rich dynamics characteristics from periodic to chaotic (2-4). In the backward whirl cases, the rotor slides or rolls continuously throughout the whirl motion at a nonsynchronous speed. The configuration of rolling without sliding is known as dry whirl while the sliding motion is called dry whip (5,6); being the latter much more dangerous to the rotor system.

Most of the studies on rubbing in rotor systems use lumped mass models (2, 7-9). While these models can be used to give insights on the dynamics of the systems, the results are mainly qualitative. Since the rubbing often excites a wide spectrum band, more detailed models are required. A common approach is to use the Finite Element Method (FEM), as done in (10,11). However, FEM based models for non-linear problems often require a highly discretized mesh, which may create a high numerical cost.

In this work, a continuous model of a rotor-disk system subjected to rub is presented. The system consists of a simply-supported shaft with an off-centred disk. The partial differential equation that govern the motion of the system is discretised by the eigenfunctions obtained directly from the equations of motion. The eigenfunctions obtained considers the gyroscopic effects of the disk and the shaft, thus it is speed dependent. Furthermore, a finite element model is established to a primary validation the results given by the continuous model.

2   MATHEMATICAL MODEL

The rotor-disk system studied consists of a continuous shaft with a disk located at $x = a$ and simply supported at both ends, as depicted in Figure 1. The vertical and horizontal displacements are given as $w(x, t)$ and $v(x, t)$, respectively. A massless stator is positioned at $x = b$ with a radial clearance of d_c. The contact is modeled by means of a nonlinear spring-damping system. The shaft is considered to be homogeneous and its rotary inertia is considered in the equations. The equation of motion of a Rayleigh shaft with a disk can be written in complex form as (4),

$$\rho A \frac{\partial^2 u}{\partial t^2} - \rho A r_0^2 \left(\frac{\partial^4 u}{\partial x^2 \partial t^2} - 2j\Omega \frac{\partial^3 u}{\partial x^2 \partial t} \right) + EI \frac{\partial^4 u}{\partial x^4} + M_d \frac{\partial^2 u}{\partial t^2} \delta_d(x - a) + 2j\Omega I_D \frac{\partial^2 u}{\partial x \partial t} \delta_d'(x - a) = p(x, t) \tag{1}$$

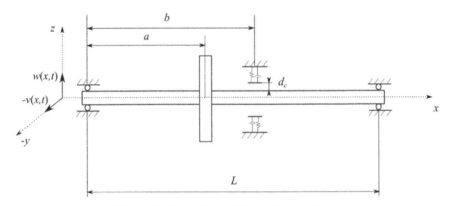

Figure 1. Depiction of the rotor system.

being $u(x, t) = v(x, t) + jw(x, t)$ the rotor's displacement in complex form, ρ the mass density, A the cross-sectional area, r_0 the radius of gyration, E the modulus of elasticity, I the area moment of inertia, Ω the rotational speed, M_d the mass of the disk, I_D the diametral mass moment of inertia of the disk, δ_d is the Dirac delta and $\delta_d' = d(\delta_d)/dx$ its derivative with respect to x; and $p(x, t) = p_y(x, t) + jp_z(x, t)$ is the external load and its components in the y and z direction. The external forces considered in this work consist of the unbalance, damping and the contact force due to the stator rubbing. Also, the second order effect of the diametral mass moment of inertia of the thin disk is not considered in the model. The boundary conditions are simply supported, thus,

$$u(0, t) = u(L, t) = \frac{\partial^2 u(0, t)}{\partial x^2} = \frac{\partial^2 u(L, t)}{\partial x^2} = 0 \tag{2}$$

16

The discretization of Equation (1) can be done in several ways. The most commonly approach is by means of the modal expansion or assumed modes method, where the following form for the displacement is assumed,

$$u(x,t) = \sum_{n=1}^{\infty} \phi_n(x) q_n(t) \tag{3}$$

being ϕ_n the basis functions or mode shapes and q_n the modal coordinates. The mode shapes ϕ_n can be obtained by means of the eigenfunctions (satisfy the free, undamped governing equations and geometric and natural boundary conditions) or comparison functions (satisfy the geometric and natural boundary conditions) or admissible functions (only satisfy the geometric boundary conditions) (12). The approach followed here used the eigenfunctions, which are obtained directly from the free and undamped equation of motion, Equation (1) with $p = 0$.

Due to the gyroscopic effect, the mode shapes ϕ_n depend on the rotational speed Ω. Also, for each mode the mode shape has two forms: a forward and a backward form. Each of which has its own whirl frequency that depends on the rotor's speed. In this way, Equation (3) is rewritten as,

$$u(x,t) = \sum_{n=1}^{\infty} \left(\phi_n^F(x) q_n^F(t) + \phi_n^B(x) q_n^B(t) \right) = \sum_{i=F,B} \sum_{n=1}^{\infty} \phi_n^i(x) q_n^i(t) \tag{4}$$

where F and B denote the forward and backward modes and i sums their both contributions. For rotor systems where the gyroscopic effects are not strong, Equation (3) can be used in discretizing the equations with no great errors associated with it. This was done by Azeez and Vakakis (4), where comparison functions obtained from a non-spinning shaft was used to discretize the partial differential equations. However, for rotor systems with large diameter disks, or that operate in high speeds, the use of the correct mode shapes is necessary.

The main problem that arises when using the eigenfunctions taken from the free equation of motion, Equation (1) with $p = 0$; is that the resulting eigenvalue problem is non-self-adjoint, due to the gyroscopic force. Thus, the regular procedure used for non-rotating beams to uncouple the differential equations for the modal coordinates q_n is not applicable. In this work, the modal analysis method (13) is used to uncouple the modal equations of motion, which gives the following first-order uncoupled equations for the modal amplitudes (14,15),

$$\dot{q}_n^i(t) = \lambda_n^i q_n^i(t) + \int_0^L \bar{\psi}_n^i(x) p(x,t) dx \text{ for } n = 1,2,\ldots \text{and } i = F, B \tag{5}$$

where λ_n^i is the eigenvalue which for undamped gyroscopic systems are given as $\lambda_n^i = j\omega_n^i$, being ω_n^i the natural frequencies; and $\bar{\psi}_n^i$ are the adjoint eigenfunctions of ϕ_n^i. Equation (5) consist of two sets, forward and backward, of infinite first order uncoupled modal equations, which are in reality truncated at N modes. The solution of the equations gives the complex modal coordinates $q(t)$. The physical displacements are obtained from Equation (4) and the velocity can be obtained as,

$$\dot{u}(x,t) = \sum_{n=1}^{\infty} \left(\lambda_n^F \phi_n^F(x) q_n^F(t) + \lambda_n^B \phi_n^B(x) q_n^B(t) \right) \tag{6}$$

the horizontal $v(x,t)$ and vertical $w(x,t)$ components of the displacement and velocity are simply the real and imaginary parts of $u(x,t)$ and $\dot{u}(x,t)$, respectively.

2.1 Unbalance force

The unbalance force is considered acting only at the disk. The following relation was considered,

$$p_{un}(x,t) = me\Omega^2 e^{j\Omega t}\delta_d(x-a) \tag{7}$$

being me the unbalance amount and a the location of the disk along the shaft.

2.2 External damping

The damping in this work is considered as an external non-proportional force. The damping force is written as,

$$p_{damp}(x,t) = -c_a\dot{u}(x,t) = -c_a\sum_{n=1}^{\infty}\left(\lambda_n^F\phi_n^F(x)\dot{q}_n^F(t) + \lambda_n^B\phi_n^B(x)\dot{q}_n^B(t)\right) \tag{8}$$

being c_a the external damping coefficient. It's worth noting that the damping couple the forward and backward modes.

2.3 Contact force

The impact model considered is based on the Hunt and Crossley (16) model, which is given as,

$$F_c(t) = k_h\delta^n(t) + c_h\delta^n(t)\dot{\delta}^m(t) \tag{9}$$

being δ and $\dot{\delta}$ the indentation and its rate of change, k_h the stiffness coefficient; and c_h the damping coefficient. The exponents n and m are assumed as $n = m = 1$. The indentation and its rate of change are given as, respectively,

$$\delta(t) = |u(b,t)| - d_c = \sqrt{v(b,t)^2 + w(b,t)^2} - d_c \tag{10}$$

$$\dot{\delta}(t) = |\dot{u}(b,t)| = \frac{v(b,t)\dot{v}(b,t) + w(b,t)\dot{w}(b,t)}{|u(b,t)|} \tag{11}$$

being d_c the radial clearance and b the impact location along the shaft. The tangential force is modelled based on the Coulomb model, and the sign of the relative velocity at the contact point v_{rel} is taken into account. Thus, the coefficient of friction is assumed to be (17),

$$\mu = \mu(v_{rel}) = \mu_m\tanh\left(\frac{v_{rel}}{v_0}\right) \tag{12}$$

being μ_m the maximum coefficient of friction and v_0 a curve-fitting parameter, which was set to $v_0 = 10^{-3}$. The $\tanh(x)$ function is used instead of the $\text{sign}(x)$ function to avoid the discontinuity presented in the latter. The tangential force and relative velocity are given as, respectively

$$F_t = \mu(v_{rel})F_c \tag{13}$$

$$v_{rel} = \dot{\theta}(t)|u(b,t)| + \Omega R \tag{14}$$

where R is the shaft's radius and $\dot{\theta}$ is the whirl speed, which is given as,

$$\dot{\theta}(t) = \frac{d}{dt}\left[\tan^{-1}\left(\frac{w(b,t)}{v(b,t)}\right)\right] = \frac{v(b,t)\dot{w}(b,t) - w(b,t)\dot{v}(b,t)}{|u(b,t)|^2} \tag{15}$$

The complex form of the contact forces is expressed as,

$$p_{cont}(x,t) = -\zeta[F_c(t) + jF_t(t)]e^{j\theta}\delta_d(x-b) = -\zeta[1 + j\mu(v_{rel})]F_c(t)\frac{u(b,t)}{|u(b,t)|}\delta_d(x-b) \tag{16}$$

where ζ is defined as,

$$\zeta = \begin{cases} 1 \text{ for } |u(b,t)| \geq d_c \\ \\ 0 \text{ for } |u(b,t)| < d_c \end{cases} \tag{17}$$

2.4 Modal equations

The final modal equations can be obtained by substituting the unbalance, damping and contact forces, Equations (7), (8) and (16), into Equation (5), which gives,

$$\ddot{q}_n^i(t) = \lambda_n^i q_n^i(t) + \int_0^L \bar{\psi}_n^i(x)p_{un}(x,t)dx + \int_0^L \bar{\psi}_n^i(x)p_{damp}(x,t)dx + \int_0^L \bar{\psi}_n^i(x)p_{cont}(x,t)dx \tag{18}$$

for $n = 1, 2, \ldots$ and $i = F, B$

or, equivalently

$$\ddot{q}_n^i(t) - \lambda_n^i q_n^i(t) + c_a \sum_{m=1}^{\infty}\left(\lambda_m^F q_m^F(t)c_{nm}^{Fi} + \lambda_m^B q_m^B(t)c_{nm}^{Bi}\right) = me\Omega^2 e^{j\Omega t}\bar{\psi}_n^i(a) - \zeta[1 + j\mu(v_{rel})]F_c(t)\frac{u(b,t)}{|u(b,t)|}\bar{\psi}_n^i(b) \tag{19}$$

for $n = 1, 2, \ldots$ and $i = F, B$

where,

$$c_{nm}^{ik} = \int_0^L \bar{\psi}_n^i(x)\phi_m^k(x)dx \tag{20}$$

The modal equations (19), are now coupled due to the damping, and are nonlinear when $|u(b,t)| > d_c$, and linear otherwise.

3 FINITE ELEMENTS MODEL

For the purpose of comparison with the continuous model, a finite element (FE) model was also established. The model consists of $N_e = 9$ standard beam elements considering rotary inertia. Figure 2 shows the mesh used, which consisted of 7 elements of same length $l_1 = 100$ mm, and two elements of length $l_1 - l_2$ and l_2, where $l_2 = 40$ mm. This was done to change the impact location if needed by changing the length l_2. The disk is positioned at node 5 as shown. The Equations of motion of the FE model are given as (18),

$$[M_e]\{\ddot{X}_e(t)\} + [D_e]\{\dot{X}_e(t)\} + [K_e]\{X_e(t)\} = \{F_e(t)\} \tag{21}$$

where $\{X_e(t)\}$ are the node displacements, $[M_e]$ and $[K_e]$ are the mass and stiffness matrices, respectively; $[D_e]=[C_e]+\Omega[G_e]$ being $[C_e]$ and $[G_e]$ the damping and the gyroscopic matrices. $\{F_e(t)\}$ are the external forces applied on the rotor, which consist of the unbalance and contact forces. The damping matrix was considered proportional to the stiffness matrix, that is, $[C_e] = \beta[K_e]$, being β the proportionality constant. Also, the disk matrices were summed at the corresponding degrees of freedom of the disk node, and the diametral moment of inertia was not considered.

In the FE model, the same impact model given by Equation (9) was used, and applied at the corresponding nodes; which was the same location as the distance b in the continuous model. In addition, no reduction was used in the FE equations (21). Thus, they were integrated considering all the nodes of the FE mesh.

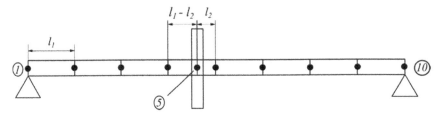

Figure 2. Mesh of the FE model.

4   MODE SHAPES AND NATURAL FREQUENCIES

A first comparison between the FE and the Continuous model was the mode shapes and critical speeds given by each one. The main parameters used in the simulations are listed in Table 1. These parameters where used in both continuous and FE models. Also, the damping was neglected in both models in this study.

Table 2 shows the critical speed comparison. It can be noted that the agreement between the models is very good, and they differ more as the mode considered increases. It's worth noting, however, that, if more elements are considered in the FE model, the agreement in the higher critical speeds between the results is better. Figure 3 shows the Campbell diagram obtained with both models, which showed a very close behaviour. The mode shapes for a speed of $\Omega = 10^4$ rpm are shown in Figure 4. It is seen that the results agree very well, showing that the continuous model can be used for the mode shape obtainment with great accuracy. The mode shapes of the continuous model were normalized in the figures using the biorthonormality conditions.

Here it is important to point out that most the works which used continuous models used the sine function, $A_n \sin(n\pi)$ with n being the node number and A_n a constant, to discretize the equations of motion. While this approach is reasonable for low speeds and for rotors with centred disks, it gives wrong results for higher rotational speeds, as the mode shapes of the rotor are strongly dependent on the speed. Therefore, the eigenfunctions used in this work gives more reliable results, since it depends on the rotor speed and gives similar results as the FE model regarding the rotor mode shapes and natural frequencies.

Table 1. Parameters used in the simulations.

Parameter	Symbol	Value
Length of the rotor	L	800 mm
Diameter of the rotor's shaft	$2R$	12 mm
Density of the material	ρ	7850 kg/m^3
Modulus of Elasticity	E	200 GPa
Diameter of the disk	D_d	200 mm
Length of the disk	h_d	9.32 mm
Mass of the disk	M_d	2.3 kg
Location of the disk	a	0.45L
Unbalance mass	me	7.4×10^{-5} kgm
Damping coefficient	c_a	7 Ns/m
Proportionality constant (FE model)	β	2×10^{-5}

Table 2. Comparison of the critical speeds obtained with both models.

Critical speed	Continuous model (rpm)	FE model (rpm)	Error (%)
First (BW)	817.90	817.86	0.004
Second (FW)	818.8	818.9	0.001
Third (BW)	4491	4491	0.002
Fourth (FW)	11100	11105	0.043
Fifth (BW)	12401	12407	0.043
Sixth (FW)	17087	17104	0.095

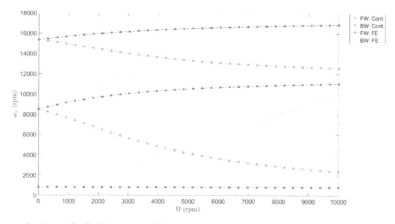

Figure 3. Campbell diagram of both models. The black dashed line is where $\Omega = \omega$.

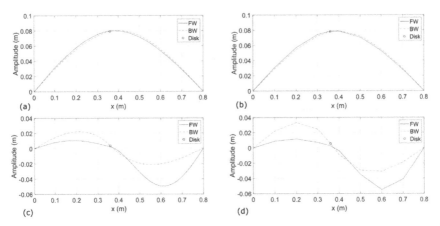

Figure 4. Results for the two firsts mode shapes of the rotor system at $\Omega = 10^4$ rpm using: (a) and (c) Continuous model and in (b) and (d) FE model.

5   NUMERICAL SETUP

Equation (19) consists of N complex coupled equations for the forward and backward modes. Because both forward and backward modes have to be taken into account in the obtainment of the physical coordinates v and w to determine if the rotor contacts the stator, they must be solved simultaneously. Therefore, a total of $2N$ first-order ordinary differential equations are solved; the equations are linear when no impact occurs and nonlinear otherwise. The numerical integration in the rub cases were performed using the variable-step size integrator *ode45* of the software Matlab, which is a six-stage, fifth-order, Runge-Kutta method (19). The FE equations were integrated using the integrator *ode15s*, since the *ode45* scheme was very slow for the FE model. The *ode15s* integrator is best suited for systems with high numerical stiffness.

In continuous systems, the number of modes used for discretizing the equations of motion is highly important for the reliability of the results, and they depend closely on the impact stiffness k_h, since the higher the latter, the higher the number of modes have to be considered in the discretization. Besides the intensity of the impact, the location where it occurs along the shaft and the external damping also affects the number of modes needed. When the impact occurs on the disk or a high damping is considered, a lower number of modes can be chosen with no great differences in the results. The mode number $N = 5$ was sufficient for reliable results for all cases studied, from mild to very intense rub. It is worth noting that in this case the higher frequency modes are not considered, since the analysis performed in this work is mainly close to the first forward critical speed. However, this need not to be the case, for example, for high speed rotor systems; where the low frequency modes may be discarded due to their little influence in the overall response. This offers a good advantage over the FE model, as in the latter, if no reduction is applied in the equations, all the modes are necessarily considered in the solution, which has a very high mode number for a highly refined mesh.

By using $N = 5$ the number of equations to be solved in the continuous model is 10, and the FE model, with 9 elements, is 36 (as 4 degrees of freedom are eliminated due to the simply-supported condition). Since the FE model has to be turned to a first order differential equation, the total number of equations integrated was 72. Although the equations of the continuous model are coupled, their numerical solution is much easier than the FE model, saving a great deal of computational time.

6   FORCED OSCILLATIONS

6.1. Continuous model results

In the forced oscillations study, the important parameters that affect the rubbing phenomena were varied and the results analysed. The rubbing was induced by the unbalanced force and the initial conditions were all set to zero. In the following simulations, the ratio of the shaft radius by the radial clearance was set to $R/d_c = 80$, thus $d_c = 0.075$ mm (75 μm) and disk position is $a = 0.45L$. The impact stiffness and damping were assumed as $k_h = 10^6$ N/m and $c_h = 0.5k_h$, respectively. The other parameters used are listed in Table 1.

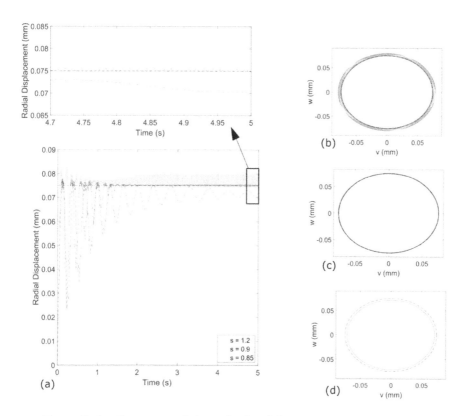

Figure 5. Continuous model results for different rotational speed with
$\mu_m = 0.025$ **and** $b = 0.5L$: **(a) radial displacement and rotor orbits in steady state for (b)** $s = 1.2$, **(c)** $s - 0.9$ **and (d)** $s = 0.85$. **The black dashed line represents the stator clearance.**

A first study performed is shown in Figure 5, where the friction coefficient was $\mu_m = 0.025$, and the impact location is set on the shaft at its mid-span, $b = 0.5L$. In the figure, $s = \Omega/\omega_{cr,1}^F$, being $\omega_{cr,1}^F$ the first forward critical speed; and the black dashed line represents the stator clearance. The orbits shown were taken after 500 cycles, being the last 20 cycles shown, and the displacements v and w correspond to the horizontal and vertical displacement at $x = b$. Three cases are outlined in Figure 5. When $s = 0.85$, the rotor contacts the stator several times but eventually reaches a no-rub steady state. This happens mainly because the maximum no-rub steady state displacement is less than the clearance. By increasing the friction in this case only makes the rotor disengage the stator sooner, thus there is no coexistence of two stable whirl solutions. This is true, however, for moderate friction, as with friction as high as $\mu_m = 0.5$ backward rub can occur even at very low speed as shown experimentally in (20). By increasing the speed to $s = 0.9$, a state of forward annular rub is seen (FWAR), where the rotor continuously contacts the stator throughout its whirl motion. This state is unavoidable, as the steady state displacement of the rotor at this speed is higher than the clearance. For $s = 1.2$, the rotor first enters in an annular rub state which later changes to a forward partial rub state (FWPR) due to friction. This state is characterized by a series of impacts throughout the whirl motion and possess rich dynamics characteristics which range from periodic to chaotic (2,3,4). Also, in FWPR cases, sub- and superharmonics are commonly seen in the frequency spectrum of the response (8,21).

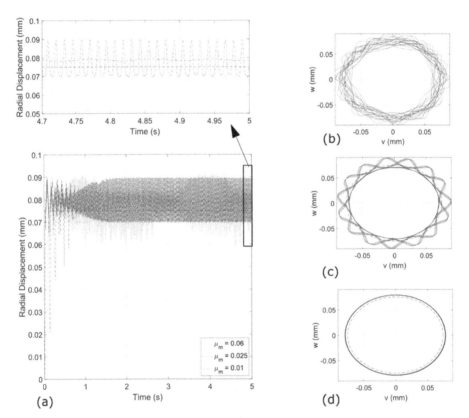

Figure 6. Continuous model results for different friction coefficients with $s = 1.6$ and $b = 0.5L$: (a) radial displacement and rotor orbits in steady state for (b) $\mu_m = 0.06$, (c) $\mu_m = 0.025$ and (d) $\mu_m = 0.01$.

Another study is shown in Figure 6, where the speed was held constant at $s = 1.6$ and the friction coefficient was varied. The friction cases correspond to high, moderate and low friction, respectively. Since the speed is not close from the first critical speed, the influence of the higher modes of vibration are stronger in this case. For a low friction of $\mu_m = 0.01$, the rotor enters in FWAR. This result is interesting because at this speed the no rub displacement of the rotor is much lower than the radial clearance. This means that two whirl solutions coexist, and they can be interchanged for different initial conditions. For higher friction coefficient of $\mu_m = 0.025$, the rotor first enters in a FWAR state which then changes to a FWPR, with several impacts over the whirl motion. It is worth noting that this result does not differ from ones obtained with Jeffcott models (7, 9). However, the continuous model gives the global response of the rotor, where the displacements at any point of the shaft can be obtained. This can be useful, for example, to asses if another point on shaft can also contact its surrounding stator or seal. By increasing the friction even more to $\mu_m = 0.025$, Figure 6 shows that the rotor enters in a stronger rub state, with more influence of the higher modes of vibration. By comparing the orbits shown in 6b and 6c, it is seen that the rotor transitioned from periodic to a quasi-periodic state with the increasing friction, as the orbit in 6b is not closed. This result shows that with high friction, more modes are excited by the contact force and thus the disregard of the higher modes of vibration may lead to very different outcomes.

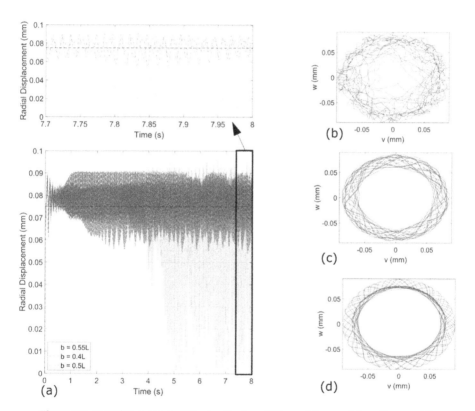

Figure 7. Continuous model results for different impact locations with $\mu_m = 0.04$ **and** $s = 1.6$**: (a) radial displacement and rotor orbit in steady state for (b)** $b = 0.55L$**, (c)** $b = 0.4L$ **and (d)** $b = 0.5L$**.**

In the last case studied, the position b where the impact occurred was varied. The friction coefficient was set to $\mu_m = 0.04$ and the speed $s = 1.6$. Figure 7 shows the results obtained. The impact positions were $b = 0.55L$, $b = 0.5L$ and $b = 0.4L$. Since the disk is placed at $a = 0.45L$, in this last case the impact is occurring before the disk (see Figure 1). The results when $b = 0.5L$ and $b = 0.4L$, were very similar, as the distance from the impact location to the disk position was the same. Since neither of the orbits in Figures 7c and 7d are closed, the dynamic behaviour was quasi-periodic. An interesting result is seen when the impact location is positioned farther from the disk with $b = 0.55L$. Figure 7b shows the orbit for this case, where the behaviour is seemingly chaotic. However, advanced numerical tools such as Lyapunov exponents must be used to firmly state the behaviour to be chaotic, which was not done in this work. What can be stated is that the rotor was very close to enter in a backward whirl motion, as a frequency analysis (not shown here) showed a high negative component in the response's spectrum. These results show that the position where the impact occurs is very important in the dynamic of the rotor system and the farther from the disk the contact occurs, the stronger the rubbing phenomenon is.

6.2. Comparison with FE results

In order to evaluate the results given by the continuous model, they were compared with the responses given by the FE model in the same conditions. For that matter, the parameters used in the FE and continuous models - including radial clearance, disk locations and properties, impact stiffness and damping, and so on - were the same. Since the first forward critical speed of both models are fairly the same (See Table 2), the parameter $s = \Omega/\omega_{cr,1}^F$ was also used to define the rotor speed in the FE model. In addition, the impact force used was the same and was added at the corresponding degrees of freedom of the midspan node of the FE mesh, node 6 (Figure 2).

The first comparison between the models is shown in Figure 8, where the friction coefficient was held at $\mu_m = 0.06$ and the speed was varied. The displacements v and w in the FE orbits correspond to the horizontal and vertical displacements at the impact node. The results given by both models are very similar. When the speed was $s = 0.9$, both models predicted a full annular rub with low indentation. At the speed of $s = 1.2$, the models predicted a FWPR, but their behavior were rather different. While the continuous results, Figure 8b, was closed, meaning a periodic solution, the FE orbit, Figure 8e, showed a quasi-periodic behaviour. Despite of this qualitative difference, both models showed very close indentations. For a higher speed, $s = 1.6$, the results were very similar, and the FE model also showed an influence of the higher modes of vibration.

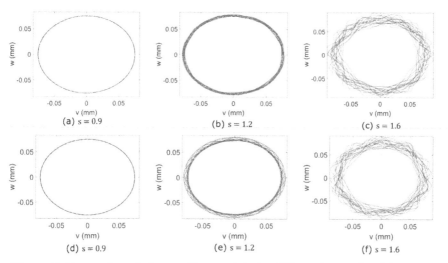

Figure 8. Comparison between the models for different speeds with $\mu_m = 0.06$ and $b = 0.5L$: (a), (b) and (c) continuous model and (d), (e) and (f) FE model.

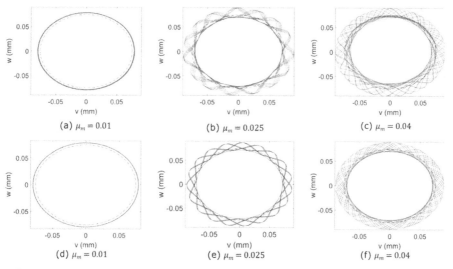

Figure 9. Comparison between the models for different friction coefficients with $s = 1.6$ and $b = 0.5L$: (a), (b) and (c) continuous model and (d), (e) and (f) FE model.

Figure 9 shows the comparison between the models with different friction coefficients, where the speed was held at $s = 1.6$. At first glance, it is seen that, in a qualitative point of view, the results given by the continuous and FE models were very close. For a low friction with $\mu_m = 0.01$, both models predicted the annular rubbing state, and the indentation was also very similar. For $\mu_m = 0.025$, a periodic solution is seen in the results of the models. Although the results look similar qualitatively, period numbers were slightly different and the indentation was higher in the continuous model. With a higher friction, $\mu_m = 0.04$, the orbit of the FE model, Figure 9f, is not closed, thus the behaviour is a quasi-periodic one; which is the same as the continuous model, Figure 9c. Given the results, it is fair to say that the models were in good agreement. However, the influence of the modes of vibration is stronger in the continuous model in comparison with the FE model.

7 CONCLUSIONS

This paper presents a continuous model for a rotor-disk system that rubs against a massless stator. The shaft was homogeneous and the Euler-Bernoulli beam theory was used. The gyroscopic effect of both the shaft and the rigid disk were considered in the equations of motion, which, in turn, were discretized by means of the modal expansion method. The eigenfunctions used in the discretization were obtained directly from the free and undamped equation of motion, resulting in a mode shape function with dependence upon the rotor speed and with forward and backward cases. Also, the equations of motion for the modal coordinates were obtained through the modal analysis method described in (13).

In order to evaluate the continuous model presented, a FE model was also established, which had 9 standard beam elements with rotary inertia. No reduction was considered in the FE model, and the simply supported condition was applied by directly removing the displacements degrees of freedom in the boundary nodes. In a first comparison between the models, the critical speeds, Campbell diagrams and mode shapes given by them were compared. The results showed that the continuous model can be used with great accuracy to obtain the natural frequencies and mode shapes, as the results were very close from the FE model.

By means of numerical simulations, the behaviour of the rotor system was studied in different operating conditions. The parameters varied were the rotational speed Ω, friction coefficient μ_m and impact location b. It was shown that by increasing the friction, the higher modes of vibration influence more the response of the system. Also, it was shown that depending where the impact is occurring, far from or close to the disk, the dynamic behaviour of the system is dramatically changed. These results from the continuous model were further compared with the FE model outcomes. The comparison showed that the models were in good agreement, as their dynamic behaviour did not differ much. Also, the comparisons prove the validity of the present continuous approach in modelling non-smooth systems, which was the rub forces in this case.

The continuous model offered some advantages over the FE regarding the numerical integration, as the number of equations needed to be integrated was lower. This saves a great deal of computational time and requires less numerical processing power; which can be very important in large rotor systems, that mostly require large processing units. Another important advantage of the continuous model is that it requires only the modes that will actually influence the response of the system, thus not only higher frequency modes, but the lower ones, can be discarded. It also worth pointing out that the continuous model can be applied to more complex geometries, where its advantages would be even greater.

REFERENCES

[1] Agnieszka Muszynska. *Rotordynamics*. CRC press, 2005.
[2] Paul Goldman and Agnes Muszynska. Chaotic behavior of rotor/stator systems with rubs. Journal of Engineering for Gas Turbines and Power, 116 (3):692–701, 1994.

[3] Fredric F Ehrich. Some observations of chaotic vibration phenomena in high-speed rotordynamics. Journal of vibration and acoustics, 113(1):50–57, 1991.

[4] Mohammed F Abdul Azeez and Alexander F Vakakis. Numerical and experimental analysis of a continuous overhung rotor undergoing vibro-impacts. International journal of non-linear mechanics, 34(3):415–435, 1999.

[5] Georges Jacquet-Richardet, Mohamed Torkhani, Patrice Cartraud, Fabrice Thouverez, T Nouri Baranger, Mathieu Herran, Claude Gibert, Sebastien Baguet, Patricio Almeida, and Loic Peletan. Rotor to stator contacts in turbomachines. review and application. Mechanical Systems and Signal Processing, 40(2):401–420, 2013.

[6] Dara W Childs and Avijit Bhattacharya. Prediction of dry-friction whirl and whip between a rotor and a stator. 2007.

[7] Jun Jiang and Heinz Ulbrich. The physical reason and the analytical condition for the onset of dry whip in rotor-to-stator contact systems. Journal of vibration and acoustics, 127(6):594–603, 2005.

[8] Fredric F Ehrich. High order subharmonic response of high speed rotors in bearing clearance. Journal of Vibration, Acoustics, Stress, and Reliability in Design, 110(1):9–16, 1988.

[9] Jie Hong, Pingchao Yu, Dayi Zhang, and Yanhong Ma Nonlinear dynamic analysis using the complex nonlinear modes for a rotor system with an additional constraint due to rub-impact. Mechanical Systems and Signal Processing, 116:443-461, 2019.

[10] Sébastien Roques, Mathias Legrand, Patrice Cartraud, Carlo Stoisser, and Christophe Pierre. Modeling of a rotor speed transient response with radial rubbing. Journal of Sound and Vibration, 329(5):527–546, 2010.

[11] Mehdi Behzad, Mehdi Alvandi, David Mba, and Jalil Jamali. A _nite element-based algorithm for rubbing induced vibration prediction in rotors. Journal of Sound and Vibration, 332(21):5523–5542, 2013.

[12] Ali H Nayfeh and P Frank Pai. Linear and nonlinear structural mechanics. John Wiley & Sons, 2008.

[13] Chong-Won Lee, R Katz, AG Ulsoy, and RA Scott. Modal analysis of a distributed parameter rotating shaft. Journal of Sound and Vibration, 122(1):119–130, 1988.

[14] Leonard Meirovitch. Computational methods in structural dynamics, volume 5. Springer Science & Business Media, 1980.

[15] Chong-Won Lee. Vibration analysis of rotors, volume 21. Springer Science & Business Media, 1993.

[16] Kenneth H Hunt and Frank R Erskine Crossley. Coefficient of restitution interpreted as damping in vibroimpact. Journal of applied mechanics, 42 (2):440–445, 1975.

[17] Elijah Chipato, AD Shaw, and MI Friswell. Frictional effects on the nonlinear dynamics of an overhung rotor. Communications in Nonlinear Science and Numerical Simulation, 104875, 2019.

[18] Yukio Ishida and Toshio Yamamoto. Linear and Nonlinear Rotordynamics: a modern treatment with applications. John Wiley & Sons, 2013.

[19] Lawrence F Shampine and Mark W Reichelt. The matlab ode suite. SIAM journal on scientific computing, 18(1):1–22, 1997.

[20] Ulrich Ehehalt, Oliver Alber, Richard Markert, and Georg Wegener. Experimental observations on rotor-to-stator contact. Journal of Sound and Vibration, 446:453-467, 2019.

[21] Fulei Chu and Wenxiu Lu. Experimental observation of nonlinear vibrations in a rub-impact rotor system. Journal of Sound and Vibration, 283(3-5):621–643, 2005.

Optimization of rotating machinery by BESO method

E.S. Carobino, R. Pavanello

Department of Computational Mechanics, Faculty of Mechanical Engineering, University of Campinas, Brazil

J. Mahfoud

University of Lyon, INSA-Lyon, CNRS UMR5259, LaMCoS, Villeurbanne, France

ABSTRACT

Rotating systems are the main components in turbomachinery, and are presents in several industrial applications, such as oil and gas or energy. The efficiency of these machines is directly dependent on the mechanical characteristics and their dynamic behaviours. Thus it is important to develop specific numerical methods and algorithms that can help to predict and to optimize the dynamic behaviour of these machines. The goal of this work is to implement a mathematical model to simulate and to optimize the dynamic behaviour of rotating machines and to develop an evolutionary optimization method in order to improve their response for a large frequency band, placing the nominal rotational speed as far as possible from resonance regions. A modal analysis was made considering the effects of mass, stiffness, damping and gyroscopic effects, in order to have an optimized reparation of the shaft diameters by using a finite element analysis, considering its components (shaft, disks and bearings). The optimized distribution of the shaft diameters was obtained to maximize the frequency span of two natural frequencies around the operating speed. Constraints related to the total shaft volume and to limit values for the diameters were considered. The optimization problem is solved by using the Bidirectional Evolutionary Structural Optimization (BESO) method. A MATLAB code was developed to simulate and to optimize the rotating systems. The Campbell diagram and the unbalance response of the rotor bearing system has been used to assess the performance of the optimized configuration.

1   INTRODUCTION

Turbomachines are widely presents in different industrial fields. The design of their rotating parts deserves special attention in order to optimize the performances and to avoid dynamical problems. In this context, many techniques of optimization have been employed to a large variety of mechanical structures, improving their topology, shape and material (1).

The finite element method can be used to model the rotor (2,3), and take into account specific considerations of the machine such as the positions of the bearings, disks and seals.

Among many possibilities, the topology optimization could be applied to change the eigenvalues of the system in order to increase the frequency span of two consecutive natural frequencies (4). A shape optimization of Euler-Bernoulli beams was executed for maximizing band gaps (5,6), frequency regions where the propagation of elastic waves is annulled.

Genetic algorithms were applied to optimize rotors intending to move the operational speed away from the closest resonance in some different forms, such as changing the diameters of the elements without varying the total weight or element lengths (7). In a more specific way, Immune-Genetic Algorithm (IGA) (8) and Multi-Objective Genetic Algorithm (MOGA) (9) were used with multi-objective functions. The former intending to minimize the weight and the force transmitted to the bearings, and the later one minimizing the total weight and placing the operational speed as far as possible from the closest resonance, both changing the diameters of the elements as well.

A gradient based algorithm was used with a multi-objective function to minimize the rotor weight and unbalance response, increasing the critical speed (10). Based on derivatives of the objective function, the evolutionary optimization methods express their importance, such as the Evolutionary Structural Optimization method (ESO) (11) and the bi-directional Evolutionary Structural Optimization method (BESO) (12). The first one removes elements from the structure, and the last one can remove or add elements, both methods work according to a sensitivity analysis. The ESO method was applied to a rotating machinery following a versatile objective function, intending to reduce the weight, the amplification factor (Q factor), and move the critical speeds as far as possible from the operational speed under dynamic constraints (13), by changing the lengths and diameters of the elements.

Therefore, the main objective of this work is to apply the BESO method focusing on developing the formulation of maximizing the separation margin by using a specific sensitivity analysis, in order to design a simple but representative optimized rotating machinery.

2   MECHANICAL MODEL AND OPTIMIZATION PROCEDURE

2.1   Mechanical model of the rotor
The rotor system is modelled by a finite element analysis with an assembly of discrete bearings and rotor segments with elements of concentrate mass, what means beam elements in bending with some disks connected. The Equation 1 shows the equation of motion of the complete system.

$$[\mathbf{M}]\{\ddot{\delta}\} + [\mathbf{c}(\Omega)]\{\dot{\delta}\} + [\mathbf{K}]\{\delta\} = \{\mathbf{F}(\Omega, t)\} \tag{1}$$

Where $[\mathbf{M}]$ is the global mass matrix, $[\mathbf{C}] = -\Omega[\mathbf{G}] + [\mathbf{C_b}]$ and $[\mathbf{K}]$ are the global matrices of damping and stiffness respectively. $[\mathbf{G}]$ is the gyroscopic matrix, $[\mathbf{C_b}]$ is the damping matrix of the bearings, $\{\mathbf{F}(\Omega, t)\}$ is the global load vector, $\{\delta\}$ is the vector of nodal displacements, and Ω is the rotational speed.

2.1.1 *Eigenvalue problem*
In order to solve the associated complex eigenvalue problem it is convenient to rewrite the Equation (1) in the space state form, neglecting the applied loads as shows the Equation (2).

$$[\mathbf{A}]\{\dot{\mathbf{q}}\} + [\mathbf{B}]\{\mathbf{q}\} = \mathbf{0} \tag{2}$$

Where

$$[\mathbf{A}] = \begin{bmatrix} [\mathbf{C}(\Omega)] & [\mathbf{M}] \\ [\mathbf{M}] & \mathbf{0} \end{bmatrix}, [\mathbf{B}] = \begin{bmatrix} [\mathbf{K}] & \mathbf{0} \\ \mathbf{0} & -[\mathbf{M}] \end{bmatrix}, \{\mathbf{q}\} = \begin{Bmatrix} \{\delta\} \\ \{\dot{\delta}\} \end{Bmatrix}$$

Supposing a harmonic solution $\{\mathbf{q}\} = \{\varphi_i\}e^{\lambda_i t}$, the associated eigenvalue problem can be obtained as shows the Equation (3).

31

$$([\mathbf{A}]\lambda_i+[\mathbf{B}])\{\varphi_i\}=\mathbf{0} \tag{3}$$

Where λ_i and $\{\varphi_i\}$ are respectively the eigenvalues and eigenvectors. The mentioned eigensystem must be evaluated for different values of rotational speeds Ω in order to calculate the eigenfrequencies and obtain the Campbell Diagram [2].

2.1.2 *Unbalance response in steady-state*
The mass unbalance force can be expressed as done in Equation (4), where $m_{unb}\varepsilon$ is the unbalance.

$$\mathbf{f(t)}=\mathbf{m_{unb}}\varepsilon\Omega^2\mathbf{e^{j\Omega t}} \tag{4}$$

Assuming the steady-state response in the form of the Equation (5), substituting the Equation (4) in (1), and considering the unbalance force is added to the global load vector, the steady-state unbalance response $\{p\}$ is obtained as shows the Equation (6), where $\{\mathbf{F_0(\Omega)}\}$ is the vector of amplitudes of $\{F(\Omega, t)\}$.

$$\{\delta\}=\{\mathbf{p}\}\mathbf{e^{j\Omega t}} \tag{5}$$

$$\{\mathbf{p(\Omega)}\}=\left[-[\mathbf{M}]\Omega^2+j[\mathbf{C}]\Omega+[\mathbf{K}]\right]^{-1}\{\mathbf{F_0(\Omega)}\} \tag{6}$$

2.2   Evolutionary optimization procedure
Among many optimization methods applicable to maximize the gaps between natural frequencies the BESO method was chosen, basing its evolutionary procedure in a sensitivity analysis that is made by direct differentiation of the objective function with respect to the design variable. In this work, the cross section area of each finite element is taken as design variable, and according to the BESO procedure these values can be increased or decreased for each iteration. The Figure 1 illustrates a hypothetical result obtained by using the proposed method in the case of circular cross-section shafts. The Figure 2 illustrates a flowchart of the implemented BESO method, where V^* is the final volume that will be reached at the end, AR and ER are the percentage of the total volume that is added and removed for each iteration respectively, dS is the changing rate of cross section areas, S_{min} and S_{max} are respectively the lower and upper design limits.

The algorithm works with the cross section areas of the elements as the design variable in order to avoid the non-linear dependencies and to obtain a smoother evolution (14). The changing rate of the cross section areas dS is defined at the beginning, thus the optimization deals only with discrete values. Shaft sections such as bladed stages, seals, impellers, bearings and others, remain constant and do not take part in the optimization.

Figure 1. Illustration of a topology optimization for a rotor.

The stop criterion is based on the comparison between the objective function values of the latest iterations as shows the Equation (7).

$$\frac{\sum_{j=1}^{5} f(x_i)_{k-j+1} - \sum_{j=1}^{5} f(x_i)_{k-j+4}}{\sum_{j=1}^{5} f(x_i)_{k-j+1}} \leq T \tag{7}$$

Where k is the current iteration, τ is the tolerance, and $f(x_i)$ is the objective function. Thus, the total shaft volume is reduced during the evolutionary process until the volume constraint V^* is reached and the stop criterion is satisfied as shows the flow-chart in the Figure 2.

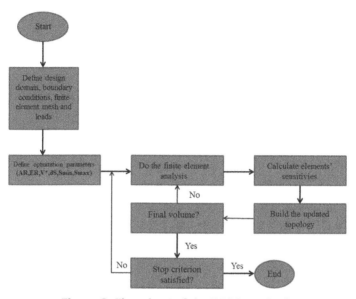

Figure 2. Flowchart of the BESO method.

2.2.1 *Maximization of natural frequencies separation*

The adopted objective function is the gap between two chosen natural frequencies. Due to gyroscopic effects, the natural frequencies are dependent on the rotational speeds, thus these frequencies must be related with their respective rotational speeds. Therefore, the statement of the optimization problem can be written as follows.

$$\begin{cases} \text{Maximize: } \left| \left| \lambda_{i,\Omega} \right| \right| - \left| \left| \lambda_{i',\Omega'} \right| \right| \\ \text{Subjectto: } V^* - \sum_{e=1}^{N} S_e L_e = 0 \\ S_{min} \leq S_e \leq S_{max} \end{cases} \tag{8}$$

Where i and i' are the ith modes, Ω and Ω' are the rotational speeds, S_{min} and S_{max} are the minimum and maximum values of the surfaces of the elements, V^* is the final volume to be reached, and L_e is the length of each element.

2.2.2 *Sensitivity analysis*

In order to carry out the optimization, it is necessary to calculate the sensitivity of each element, that can be obtained by differentiating the objective function with respect to the design parameter S_e of the jth element.

Taking the derivative of Equation (3) with respect to S_e leads to (9).

$$\left(\frac{\partial \lambda_i}{\partial S_{ej}}[A] + \lambda_i \frac{\partial [A]}{\partial S_{ej}} + \frac{\partial [B]}{\partial S_{ej}}\right)\{\varphi_i\} + (\lambda_i[A] + [B])\frac{\partial \{\varphi_i\}}{\partial S_{ej}} = 0 \tag{9}$$

Multiplying (9) by $\{\varphi\}_i^T$ yields the Equation 10.

$$\frac{\partial \lambda_i}{\partial S_{ej}}\{\varphi_i\}^T[A]\{\varphi_i\} + \lambda_i\{\varphi_i\}^T\frac{\partial [A]}{\partial S_{ej}}\{\varphi_i\} + \{\varphi_i\}^T\frac{\partial [B]}{\partial S_{ej}}\{\varphi_i\} + \{\varphi_i\}^T(\lambda_i[A] + [B])\frac{\partial \{\varphi_i\}}{\partial S_{ej}} = 0 \tag{10}$$

Then it is possible to solve (10) for the eigenvalue sensitivity such as shows the Equation (11).

$$\frac{\partial \lambda_i, \Omega}{\partial S_{ej}} = -\frac{\lambda_i P_i' + q_i'}{P_i} \tag{11}$$

Where

$$P_i' = \{\varphi_i\}^T\frac{\partial [A]}{\partial S_{ej}}\{\varphi_i\}, q_i' = \{\varphi_i\}^T\frac{\partial [B]}{\partial S_{ej}}\{\varphi_i\}, P_i = \{\varphi_i\}^T[A]\{\varphi_i\}$$

Considering a general eigenvalue with real and imaginary parts *Re* and *Im* respectively, its absolute value is done by (12).

$$||\lambda_{i,\Omega}|| = \sqrt{R_e^2 + I_m^2} \tag{12}$$

Applying the derivative with respect to the design variable in (12) to obtain the derivative of the module as function of the real and imaginary parts of the eigenvalue, it is used the fact that the derivatives of the mentioned parts are the respective real and imaginary parts of the derivative of the complete eigenvalue done by the Equation 11. Thus, the sensibility of the ith natural frequency is done by (13).

$$\frac{\partial ||\lambda_i, \Omega||}{\partial S_{ej}} = \frac{1}{\sqrt{R_e^2 + I_m^2}}\left[R_e\left(\text{real}\,\frac{\partial \lambda_i, \Omega}{\partial S_{ej}}\right) + I_m\left(\text{imag}\,\frac{\partial \lambda_i, \Omega}{\partial S_{ej}}\right)\right] \tag{13}$$

And the derivative of the objective function can be obtained as showed in the Equation 14.

$$\frac{\partial f(S_e)}{\partial S_{ej}} = \frac{\partial ||\lambda_i, \Omega||}{\partial S_{ej}} - \frac{\partial ||\lambda_{i'}, \Omega'||}{\partial S_{ej}} \tag{14}$$

3 NUMERICAL RESULTS

A Jeffcot rotor is used in this study (Figure 3) intending to increase the frequency span between the first two resonances. The disk and the shaft are made of steel, with Young Modulus $E=200,00$ GPa, Poisson's ratio $\nu = 0,30$, and density $\rho = 7800,00 kg/m^3$. The initial shaft diameters is $d=0,10$ m, the total length is $L = 1,30$m, and 78 equal length beam elements with two nodes and 4 degree of freedom per node are used.

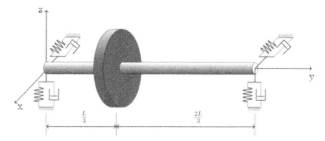

Figure 3. Finite element mesh of the initial rotor.

The Tables 1 and 2 show the properties of the disk and bearings. D_{disk} and th_{disk} are respectively the diameter and thickness of the disk, K_{xx}, K_{zz}, and C_{xx}, C_{zz} are respectively the stiffness and damping values for directions x and z.

Table 1. Properties of the disk.

Parameter	Value
$D_{disk}[m]$	0,40
$th_{disk}[m]$	0,05
Node	25

Table 2. Properties of the bearings.

Parameter	Value
Nodes	1 & 79
$K_{xx}\left[\frac{MN}{m}\right]$	70,00
$K_{zz}\left[\frac{MN}{m}\right]$	70,00
$C_{xx}\left[\frac{N}{ms}\right]$	500,00
$C_{zz}\left[\frac{N}{ms}\right]$	500,00

3.1 Optimization

The operating speed $\Omega_{op}=25000$RPM is situated between the second and the third resonances ($\Omega_1=16370$RPM and $\Omega_2=32830$RPM). The unbalance response is calculated. The unbalance of $m_{unb}\varepsilon=110,53$gmm was chosen to be of G 6 quality based on ISO 1940. As the damping effect is low the resonance frequencies are close to the natural frequencies, thus the norm of the respective eigenvalues are used for the optimization.

In order to avoid abrupt modifications throughout the optimization, the algorithm uses the parameters $AR=0,30\%$, $ER=1,00\%$, $V^*=80,00\%$, $S_{min}=0,0049$m^2, $S\_max=0,0089$m^2, $dS=0,0001$m^2, $\tau=0,20\%$, intending to obtain a soft evolution of the objective function. Figures 4 and 5 show the results obtained after 32 iterations that maximized the objective function.

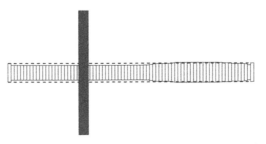

Figure 4. Finite element mesh of the optimized shaft.

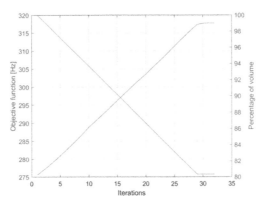

Figure 5. Evolution of the objective function and volume.

The Table 3 shows the resonance speeds before and after the optimization. Figure 6 shows the Campbell Diagrams for the initial and optimized systems. Before the optimization, the operational speed was *143,83 Hz* (*34,52%* of the operational speed) higher than the second resonance, and after the optimization it passes to *159,50 Hz* (*38,28%* of the operational speed). On the other hand, the operational speed was *130,50 Hz* (*31,32%* of the operational speed) lower than the third resonance, and it passes to *165,50 Hz* (*39,72%* of the operational speed) after the optimization.

Table 3. Critical speeds around the operational speed.

	Initial Rotor [RPM]	**Optimized Rotor [RPM]**
First critical speed	16370	15430
Second critical speed	32830	34930

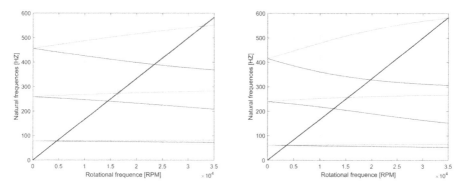

Figure 6. Campbell Diagrams before (left) and after (right) the optimization.

The response in the frequency domain for the 18th node is represented in the Figure 7, considering an unbalancing of $m_{unb}\varepsilon = 110,53gmm$.

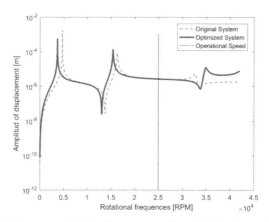

Figure 7. Response in the frequency domain for the 18th node.

4   CONCLUSION

The BESO method was applied in the topology optimization of a Jeffcot rotor, taking into account the main characteristics such as gyroscopic effects, stiffness and damping. The aim of the work was to increase the frequency span between the second and third resonances.

To deal with the mechanical elements (bearings, impellers and others) that can be connected to the shaft, the algorithm only changes the sections where the shaft is free of these elements.

An analytical sensibility analysis was made in order to evaluate the influence of each shaft finite element to the objective function, and a volume constraint was imposed besides the chosen optimization parameters to smooth the intermediary topologies as well as minimize the mass of the rotating machinery.

Until the moment this work is done, that is the first application of the BESO method to optimize the separation margin in the context of rotordynamics, and the representative Jeffcot rotor explored in this work can lead the way to the optimization of complex rotating machineries.

REFERENCES

(1) M. P. Bendsøe and O. Sigmund. Optimization of structural topology, shape, and material, volume 414. Springer, 1995.

(2) M. Lalanne and G. Ferraris. Rotordynamics prediction in engineering, volume 2. Wiley, 1998.

(3) R. D. Cook, D. S. Malkus, M. E. Plesha, and R. J. Witt. Concepts and applications of finite element analysis, volume 4. Wiley New York, 1974.

(4) Z.-D. Ma, H.-C. Cheng, and N. Kikuchi. Structural design for obtaining desired eigenfrequencies by using the topology and shape optimization method. Computing Systems in Engineering, 5(1):77–89, 1994.

(5) N. Olhoff, B. Niu, and G. Cheng. Optimum design of band-gap beam structures. International Journal of Solids and Structures, 49(22):3158–3169, 2012.

(6) L. Brillouin and M. Parodi. Propagation of waves in periodic structures. Foreign literature, 1953.

(7) B. G. Choi and B. S. Yang. Optimum shape design of rotor shafts using genetic algorithm. Journal of Vibration and Control, 6(2):207–222, 2000.

(8) B.-K. Choi and B.-S. Yang. Multiobjective optimum design of rotor-bearing systems with dynamic constraints using immune-genetic algorithm. Journal of engineering for gas turbines and power, 123(1):78–81, 2001.

(9) I. M'laouhi, N. B. Guedria, and H. Smaoui. Multi-objective discrete rotor design optimization. In Condition Monitoring of Machinery in Non-Stationary Operations, pages 193–200. Springer, 2012.

(10) F. Strauß, M. Inagaki, and J. Starke. Reduction of vibration level in rotordynamics by design optimization. Structural and Multidisciplinary Optimization, 34 (2):139–149, 2007.

(11) X. Huang and M. Xie. Evolutionary topology optimization of continuum structures: methods and applications. John Wiley & Sons, 2010.

(12) Huang, X., & Xie, Y. M. (2009). Bi-directional evolutionary topology optimization of continuum structures with one or multiple materials. *Computational Mechanics*, *43*(3), 393.

(13) Y.-H. Kim, A. Tan, B.-S. Yang, W.-C. Kim, B.-K. Choi, and Y.-S. An. Optimum shape design of rotating shaft by eso method. Journal of Mechanical Science and Technology, 21(7):1039–1047, 2007.

(14) Casas W.J.P., Pavanello R., Optimization of fluid-structure systems by eigenvalues gap separation with sensitivity analysis, Applied Mathematical Modelling, Volume 42, Pages 269–289, 2017.

(15) RAO, Singiresu S. Mechanical vibrations, 1995. *Addsion-Wesley, MA*, 2019.

Influence of thrust bearings in lateral vibrations of turbochargers under axial harmonic excitation

T.F. Peixoto, K.L. Cavalca

School of Mechanical Engineering, University of Campinas, Brazil

ABSTRACT

Turbocharger rotordynamic modelling usually neglects thrust bearing coupling effects on lateral oscillations. However, using thrust bearings in turbochargers is mandatory, due to the compressor and turbine gas flows axial force imbalances. Furthermore, approximating the bearings dynamic characteristics by linear coefficients is not suitable in this rotor-bearing system, because of the high rotational speeds achieved by a typical automotive turbocharger. This paper brings a model of the entire turbocharger, considering the nonlinear behaviour of both thrust and floating ring bearings, with thermal effects, analysing the turbocharger response to an axial harmonic excitation and the thrust bearing influence on lateral vibrations.

1 INTRODUCTION

The current practice of engine downsizing with a turbocharger (TC) requires a great knowledge of both the engine and turbocharger behaviours. A reliable model of the turbocharger is of primal importance for the development and optimization of newer designs. Given the low mass and extremely high rotational speeds achieved by a typical automotive turbocharger, high nonlinear phenomena can be observed in turbocharger oscillations. To support these high vibrations, the common practice is to utilize a floating ring bearing (FRB) to support the rotating shaft. The FRB has proved to increase the damping characteristics of radial bearings and it is an optimal choice to sustain a high-speed TC [1]. However, because of the two oil films arranged in series, with the floating ring working as a mass pedestal between both films, the phenomenon of fluid-induced instability is particularly strong in this type of bearing [2]. Either the inner film, the outer film, or even both films simultaneously, can induce lateral oscillations [3] and recent researches have focused on minimizing these sub-synchronous oscillations [4].

Several key components may influence the FRB instability, such as the inner or outer film clearances, the inner-to-outer film clearances ratio, the bearing lengths, the bearing length-to-diameter ratios, the ring size, the mass unbalance level, the injection pressure, among others [5]. Study the effect of each parameter in the system response is crucial to understand the overall TC dynamic characteristics. Recent research has suggested increasing the outer film clearance may reduce, or even totally suppress, some sub-synchronous components [6]. However, this is not an optimal solution, as other sub-synchronous components may become higher. Unfortunately, most works on turbocharger focus in the lateral vibrations, entirely neglecting the axial dynamics. Previous works completely neglect axial loads and the thrust bearing (TB) effect on the overall turbocharger behaviour [1]–[6]. Just recently, a few works have begun to include the TB on the complete turbocharger dynamic simulations. Novotný et al. approximated the TB by a linear spring with a constant stiffness [7], but improved the TB model to account for nonlinear stiffness and damping characteristics in a time-transient analysis [8]. The key factor to account for the TB nonlinear axial behaviour was the creation of a lookup table, a Database, storing the bearing load-capacity for different oil film thicknesses. This so called Database Method, term coined by [9], allows for a fast evaluation of the bearing characteristics. However, the most limiting assumption on [8] was neglecting the TB angular misalignment effects. It has been show [10]–[13] that the angular misalignment may influence the axial supported load of a typical turbocharger thrust bearing and induce restoring moments in the shaft, affecting the lateral dynamics of such systems. The influence of thrust bearings in lateral vibrations has already been verified in other types of rotating machines [14], [15], but just recently this concept has been applied to turbochargers. Chatzisavvas et al. [10] considered the angular misalignment and showed the thrust bearing influence on lateral oscillations, emphasizing its capacity on suppressing some sub-synchronous components and changing the onset of oil whirl in the lateral oscillations. This model was further investigated by Koutsovasilis [11], analysing the influence of the thrust bearing position along the axis, noticing the bearing size and position may have positive, negative or neutral impact on lateral oscillations. However, the modelling of [10], [11], although considering angular misalignments, neglected thermal variations, which has also been shown by Peixoto and Cavalca [12], [13] to affect the load-carrying capacity of turbocharger thrust bearings. The simplifications adopted either by [8], [11] are justifiable, in order to complete a time-transient analysis in a reasonable amount of computational time, given the high computational cost of solving the Reynolds equation at each time step altogether with the Energy equation. To overcome this time problem to analyse transients in turbocharger dynamics, but still considering shaft angular misalignment and lubricant thermal effects, the proposed Database Method has a great appeal. [7], [8] mention the creation of a database to use it in the time integration of the turbocharger dynamic equations, but the work focus on the development of the Reynolds equation, and some variations of it, to account for different phenomena on the bearings. The analysis admits some strong assumptions, that may not be suitable for all turbocharger applications, and little detail is given to the Database Method itself. Chasalevris and Louis [9] presented a detailed explanation of the Database Method, solving for the loads in isothermal floating ring bearings in a turbocharger. The study does not focus on the bearing modelling itself, using a simplified isothermal approach, but instead compares the Database Method Solution to a Direct Solution, solving the Reynolds equation at each time step. The results revealed a great agreement between both solutions, evidencing the advantages of the Database Method in terms of computational time and noticing the method could be extended to include any other desired effect. This work proposes to use the Database Method to model thrust bearings, including angular misalignment effects and temperature variations in the oil films. The study presents the thrust bearing modelling and the use of the Database Method in the analysis. It utilizes the method to study a turbocharger subject to a harmonic axial excitation and the coupling effect the thrust bearing may have on lateral oscillations.

2 METHODOLOGY

The turbocharger model consists of a Finite Element (FE) model of the rotating shaft with the compressor and turbine wheels modelled as rigid discs. The rotating shaft is supported by floating ring bearings and a double-acting thrust bearing, to support both the lateral and axial oscillations. Because the thrust collar, initially parallel to the thrust bearing, is attached to the shaft, any shaft rotation imposes a collar rotation, inducing restoring moments of the thrust bearing on the shaft, which may influence lateral oscillations. The next sections present the FE model of the rotating shaft and the bearing models. This entire model is detailed in [13]. We focus on the description of the thrust bearing database and the differences on the response compared to a direct solution.

2.1 Turbocharger model

Given the high nonlinear characteristics of a typical automotive turbocharger, approximating the bearing dynamics by its equivalent linear coefficients is not a suitable approach [4], [8], [11], [13]. In order to accurately model the hydrodynamic forces, a nonlinear approach must be sought. In order to accomplish it, the turbocharger equations of motion are written as:

$$M\ddot{q} + (C - \Omega G)\dot{q} + Kq = f_s + f_{ext} + f_h \tag{2.1}$$

wherein in the left-hand side of the equation, the mass, damping, gyroscopic and stiffness matrix are obtained through a FE modelling of the rotating shaft, compressor and turbine wheels [13], [16]. In the right-hand side of the equation, the excitation vector is separated in three terms. The first term relates to the static forces, the turbocharger weight. The second term relates to the external excitation forces. The third term represents the nonlinear hydrodynamic bearing forces. During the time integration of these equations of motion, the hydrodynamic forces are considered in its full form, instead of approximating it by equivalent stiffness and damping coefficients. Time integration is performed utilizing the Newmark integration scheme [13].

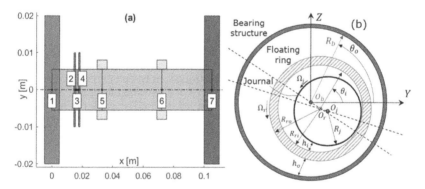

Figure 1. (a) TC FE model and (b) FRB variables.

The FE model utilized in this work is shown in Figure 1a. Nodes 1 and 7 comprise the compressor and turbine wheels, as rigid discs. Nodes 5 and 6 represent the floating ring bearings at compressor-side (CS) and turbine-side (TS). The double-acting thrust bearing is discretized in nodes 2 to 4. Each oil film will produce an axial force and restoring moments in the thrust collars in nodes 2 and 4, so the CS thrust bearing is located at node 2 and the TS thrust bearing is at node 4.

2.2 Floating ring bearing lubrication model

The floating ring bearing is utilized to increase the damping characteristics of a typical radial bearing. The outer film, a second layer of lubricant, provides a higher damping characteristic to this type of journal bearing [3], [4] and is especially useful to support the lateral oscillations in a high-speed turbocharger. A cross section of the floating ring bearing, with its main variables, is shown in Figure 1b. The shaft has a radius R_j, the floating-ring has an inner and outer radius R_{ri} and R_{ro}, respectively, and the bearing house has a radius R_b. The shaft rotates with an angular speed Ω_j, while the ring rotates with an angular speed Ω_r induced by fluid shear in the oil films. The inner and outer clearances are C_1 and C_2, respectively. Both oil films are assumed to be infinitely short, given its dimensions provide a length-to-diameter ratio close to 0.5. The complete derivation of the short bearing equations modelling the isothermal floating ring bearing is provided by [2], while the thermal effects were included and described by [13]. Essentially, the inner and outer film forces in both directions and the torques on the ring are given by:

$$F_{iy,z} = \mu_i\left(\Omega_j + \Omega_r\right)R_{ri}L_i\left(\frac{R_{ri}}{C_1}\right)^2\left(\frac{L_i}{D_{ri}}\right)^2 f_{iy,z} \quad T_i = 2\pi\frac{\mu_i R_{ri}^3 L_i\left(\Omega_j - \Omega_r\right)}{C_1\sqrt{1-\epsilon_i^2}} + \frac{\left(y_j F_{iz} - z_j F_{iy}\right)}{2}$$

$$F_{oy,z} = \mu_o\Omega_r R_{ro}L_o\left(\frac{R_{ro}}{C_2}\right)^2\left(\frac{L_o}{D_{ro}}\right)^2 f_{oy,z}, \quad T_o = 2\pi\frac{\mu_o R_{ro}^3 L_o\Omega_r}{C_2\sqrt{1-\epsilon_o^2}} - \frac{\left(Y_r F_{oz} - Z_r F_{oy}\right)}{2}$$

(2.2)

wherein the dimensionless forces $f_{i,o-y,z}$ are obtained from short bearing theory applied to each film, considering global thermal effects [13].

2.3   Thrust bearing lubrication model

The main geometric variables of a typical turbocharger TB pad are shown in Figures 2a and 2b. A TB pad has an inner and outer radii r_i and r_o, respectively, and an angular extent θ_0. The converging gap has an angular length θ_{ramp} and shoulder height s_h. The minimum oil film thickness is h_0. If the thrust collar is able to rotate around the Y and Z axes, as illustrated by Figure 2c-f, the oil film profile must consider these rotations. The TB modelling is the same as the one described by [12], [13], consisting of a full three-dimensional thermo-hydrodynamic model of the entire TB . Pressure distribution is governed by the generalized Reynolds Equation (2.3), while temperature is governed by the 3D Energy Equation (2.4). Pressure and temperature equations are compactly written using the ∇ operator. The integrals F_0, F_1 and F_2 in the Reynolds equation account for the viscosity variation throughout the oil film thickness, according to Eq. (2.5), and the oil film shape in the TB and its time derivative are described by Eqs. (2.6) and (2.7), considering the collar rotations. The pressure and temperature equations are solved simultaneously by the Finite Volume Method for all pads in the TB and the axial load and restoring moments of the TB are estimated integrating the converged pressure over each pad area, Eq. (2.8).

$$\nabla \cdot (F_2 \nabla p) = \Omega_j \partial(F_1/F_0)/\partial\theta + \partial h/\partial t$$

(2.3)

$$\rho c_p(u \cdot \nabla T) = \nabla \cdot (k\nabla T) + \mu\left((\partial v_r/\partial x)^2 + (\partial v_\theta/\partial x)^2\right)$$

(2.4)

$$F_0 = {}_0^h\frac{1}{\mu}dx, \quad F_1 = {}_0^h\frac{x}{\mu}dx, \quad F_2 = {}_0^h\frac{x^2}{\mu}dx - \frac{F_1^2}{F_0}$$

(2.5)

$$h = \begin{cases} h_0 + s_h\left(1 - \frac{\theta}{\theta_{ramp}}\right) + r\left(\phi_y \sin\theta - \phi_z \cos\theta\right), & \theta \leq \theta_{ramp} \\ h_0 + r\left(\phi_y \sin\theta - \phi_z \cos\theta\right), & \theta > \theta_{ramp} \end{cases}$$

(2.6)

$$rtialh/\partial t = \dot{h}_0 + r\left(\dot{\phi}_y \sin\theta - \dot{\phi}_z \cos\theta\right)$$

(2.7)

$$F_x = \int_A pr\,dr\,d\theta, \quad M_y = \int_A pr^2 \sin\theta\,dr\,d\theta, \quad M_z = - \int_A pr^2 \cos\theta\,dr\,d\theta$$

(2.8)

Details on the modelling and the implementation on the actual equations to estimate the thrust bearing load and restoring moments are given in [12].

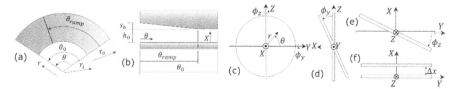

Figure 2. (a,b) TB variables, (c) allowable displacements, rotations around (d) Y axis and (e) Z axis and (f) displacements along X axis.

2.4   Thrust bearing database model

The actual lubrication model described by Eqs. (2.3)-(2.8) is computationally expensive to solve, due to the coupling between pressure and temperature equations. An iterative approach is sought in order to correctly estimate both the pressure and temperature fields in the oil film in each TB pad. Also, the analysis cannot be simplified to only one pad, because the collar rotation induces different pressure distributions on each pad. This greatly enhances the computational time to solve the TB governing equations. Further simplifications can drastically reduce the computational time to solve it, such as the isothermal approach, neglecting heat generation within the bearing. However, it has been shown this may lead to erroneous predictions [12], [13] and the best approach to accurately model the dynamical behaviour of turbochargers supported by thrust bearings must consider thermal effects. In a dynamical analysis, given the TB high nonlinearities, the Reynolds equation is solved on every step of the time integration scheme. To avoid the high computational cost of solving Eqs. (2.3)-(2.8) at every step, the Database Method described by [9] will be applied to thrust bearings, with correct adaptions. The Database Method consists in creating a lookup table, with the TB axial forces and restoring moments, for different input variables, such as the collar displacements and velocities, and, once the bearing loads are required, an interpolation in this database is performed. Different interpolation schemes can be utilized. In this work, we propose to utilize the linear interpolation scheme. Some important observations must be done before further developing the Database Method:

1. Different thrust bearings will have different geometry, which means a database should be constructed for each different thrust bearing considered in the simulations. Given the thrust bearing geometrical parameters r_i, r_o, θ_{ramp}, θ_0, s_h and N_{pad}, one could consider create one database for every desired combination of these parameters. This, however, is not recommended, as the database will have some terabytes in size and storage may be problematic.
2. The same observation is valid for different oil films, if thermal effects are considered. Isothermal analysis reduces the number of variables, solving the isothermal Reynolds equation in dimensionless form. To include thermal variations, however, it is necessary to specify the oil parameters ρ, c_p and k and the viscosity-temperature relation. Nondimensional analysis can characterize heat transfer for different lubricant flows, but still three independent parameters (the Péclet, Eckert and Reynolds numbers) must be defined [12] to create the database. Moreover, the dimensionless analysis relies on the definition of a reference temperature, usually taken equal to the replacement oil temperature, which raises the number of independent variables to at least four. The same storage problems may be encountered here if one tries to create a single database for different lubricants.
3. Given a TB and lubricant, the number of variables in the determination of the bearing characteristics may be six or seven, depending on whether thermal

effects or not will be accounted. If thermal effects will be considered, the rotational speed is an important parameter related to the shear dissipation in the oil film, so it should be considered as an independent input parameter. This dependency does not exist under the isothermal hypothesis, as the dimensionless Reynolds equation is solved.

Considering the three aforementioned observations, the strategy to develop a database accounting for thermal effects considers the fixed-geometry thrust bearings with known parameters and just one type of lubricant. With that in sight, observing the expressions for the bearing load, Eq. (2.8), and its dependence on the Reynolds and Energy Equations, (2.3) and (2.4), the oil film shape (2.6) and its derivative (2.7), the axial force and restoring moments can be written as a function of seven variables:

$$F_x, M_y, M_z = f_{1,2,3}\left(\Omega, h_0, \phi_y, \phi_z, \dot{h}_0, \dot{\phi}_y, \dot{\phi}_z\right) \tag{2.9}$$

After creating a 7D grid for different values of the rotational speed Ω, the minimum oil film thickness h_0, the rotations ϕ_y and ϕ_z, the squeezing velocity $\dot{h}_0$ and the rotation velocities $\dot{\phi}_y$ and $\dot{\phi}_z$, a database can be stored for each interested parameter and further evaluations of the axial load and restoring moments can be quickly performed by a 7D interpolation on the database. The database can be extended to store any desired parameter during the calculations, such as the average oil film temperature on the bearing, $T_{avg} = T_{avg}\left(\Omega, h_0, \phi_y, \phi_z, \dot{h}_0, \dot{\phi}_y, \dot{\phi}_z\right)$ or even one database for each bearing pad, such as $F_x^n = F_x^n\left(\Omega, h_0, \phi_y, \phi_z, \dot{h}_0, \dot{\phi}_y, \dot{\phi}_z\right)$, $n = 1 : N_{pad}$. In this work, we store the TB axial load and restoring moments and the overall average lubricant temperature, to further investigations on thermal effects. The creation of the database should be carefully planned. The evaluation of the thrust bearing thermal characteristics is highly cost, and the right choice for the input parameters is mandatory to create the database in a reasonable amount of time. The displacement variables h_0, ϕ_y and ϕ_z are restricted to geometrical limitations. Given that a double-acting thrust bearing supports the turbocharger, the maximum oil film thickness is the sum of both thrust bearing clearances, while, clearly, the minimum value should be zero. The zero value makes no physical sense, however, one can discretize this variable starting in a small, but nonzero, value. The rotations are limited to the restriction $|\phi_{y,z}r_o/h_0| < 0.4$, indicating that, with the rotation, the film thickness is always positive. The specified limit is smaller than 1.0 to assure that, in the case of simultaneous rotation ϕ_y and ϕ_z, the film thickness is always positive [12], [15]. The presented simulations suggest it is sufficient to consider the values $10 \cdot \mu m \leq h_0 \leq 40 \cdot \mu m$ and $-3 \cdot mrad \leq \phi_{y,z} \leq 3 \cdot mrad$ in the database creation. The geometrical limitations cannot be extended to restrict the velocities. Even with the nondimensionalization proposed by [12], whether the variables are the dimensional set $\dot{h}_0$, $\dot{\phi}_y$ and $\dot{\phi}_z$ or the nondimensional set $\dot{h}$, $\dot{\phi}_y$ and $\dot{\phi}_z$, both sets of variables have the domain $(-\infty, \infty)$, which makes the variable discretization unfeasible. Chasalevris and Louis [9] propose a second change of variables to the dimensionless Reynolds equation applied to radial bearings, dividing it by a factor Q and utilizing a second variable $\bar{x}_{1,2}$ to account for the squeeze term. The proposed change of variables, however, is not suitable for the thrust bearing, nor for thermo-hydrodynamic analysis. Novotný et al. [8] restricts the approximation velocity of both thrust bearing surfaces to a positive, and yet arbitrary, limit, $\dot{h}_0 \in (0, \dot{h}_0^{max})$, but does not mention any appropriate value for setting this limit. This approach also neglects strong cavitation effects that may appear in the separation of both thrust bearing surfaces (for negative velocities). In the absence of any reference value, we propose to create the database for the set of variables $(\Omega, h_0, \phi_y, \phi_z, H, \bar{\varphi}_y, \bar{\varphi}_z)$, wherein the nondimensional variables related to the velocity terms are defined as:

$$\bar{\dot{H}} = \bar{\dot{h}}/r_o\Omega, \ \bar{\dot{\varphi}}_y = \bar{\dot{\phi}}_y/\Omega, \ \bar{\dot{\varphi}}_z = \bar{\dot{\phi}}_z/\Omega \qquad (2.10)$$

and the range of each nondimensional variable is restricted to a maximum arbitrary value $\left|\bar{\dot{H}}\right| < \bar{\dot{H}}\max$, $\left|\bar{\dot{\varphi}}_y\right| < \bar{\dot{\varphi}}_y\max$ and $\left|\bar{\dot{\varphi}}_z\right| < \bar{\dot{\varphi}}_z\max$. The choice of this arbitrary value, however, is not completely random. If one knows, or at least has a good estimate, of the maximum operating velocities of the turbocharger, this values are promptly ready to use. However, if this information is not known beforehand, we recommend the adoption of a maximum estimated velocity following a (simplified) isothermal analysis. Given that the isothermal Reynolds equation is much faster to evaluate than the generalized Reynolds equation, several techniques can be employed to solve the equations-of-motion of the turbocharger. Aside from the already established Finite Difference solution, Koutsovasilis approach [11] uses a Galerkin form to get a fast approximated solution of the isothermal Reynolds equation by means of a truncated series of sine terms. Analytical equations are also available to some thrust bearing configurations [17]. Either way, after a time transient analysis with an isothermal thrust bearing, a good estimate of the maximum velocities will be known. Given that the load-carrying capacity of a turbocharger thrust bearing decreases with the inclusion of thermal effects [12], [13], these maximum velocities estimated by the isothermal modelling may be multiplied by a safety factor SF in order to get the range within the variables $\bar{\dot{H}}, \bar{\dot{\varphi}}_y, \bar{\dot{\varphi}}_z$ can be discretized. Our simulations suggest the values of $-2 \times 10^{-4} \leq \bar{\dot{H}} \leq 2 \times 10^{-4}$ and $-5 \times 10^{-4} \leq \bar{\dot{\varphi}}_{y,z} \leq 5 \times 10^{-4}$.

Creating the database requires solving the governing equations (2.3)-(2.8) for all combinations of the input parameters. This is the highly cost operation on the method. Its advantage is that the database is evaluated only once. Every further evaluation of the thrust bearing performance parameters is done interpolating the desired output from the database. Because 7 different input variables are necessary to create the database, a 7D multivariate interpolation is necessary. This study restricts to linear interpolation, but observe any interpolation scheme can be employed. For multilinear interpolation, because we have a function of 7 variables, the 7-linear interpolation will be the weighted average of $2^7 = 128$ neighbour values. Essentially, each parameter (the axial force F_x, the restoring moments $M_{y,z}$ and the fluid film mean temperature T_{avg}) will be evaluated as

$$\theta = \sum_{i=1}^{128} N_i\theta_i \qquad (2.11)$$

wherein each value θ_i is the calculated force/moment/temperature for an input value of the 7D grid and the coefficient N_i is the weight coefficient of the linear interpolation. This is simply an extension of the linear interpolation presented by [9] to a 7D interpolation. The great benefit of the Database Method is the low computational time to evaluate the interpolation. It is stated in [9] that this evaluation time is comparable to analytical solution evaluations. This greatly reduces the computational evaluation of the thrust bearing thermal performance. On average, the implemented routine to evaluate the pressure and temperature distributions of a thrust bearing takes 30 to 60 s to converge, for a known thrust collar rotational speed, displacements and velocities. If an excessive number of points on the fluid mesh discretization is utilized, this time can greatly increase. The interpolation scheme proposed is independent of the mesh size, as the evaluation relies simply on the database interpolation, and this interpolation time reduces to a few milliseconds, which reduces the total computational time to do a turbocharger time transient analysis.

45

2.5 External excitation

The external axial excitation considered in the turbocharger is a harmonic excitation on the turbine node. Lüddecke et al. [18] provided some experimental results in dimensionless forms of the thrust force in a turbocharger due to engine operation, indicating the thrust force from the engine is periodic. They observed that the thrust force has a cycle average (static) value directed from the turbine towards the compressor and the cycle resolved (dynamic) thrust load may vary depending on the operation point of turbocharger. In the absence of a better description of the harmonic excitation, and noticing that any periodic function can be expanded in a Fourier series, the external axial force, applied on node 7, considers only the first Fourier term in the series expansion, as described by Eq. (2.12).

$$F_x^{ext} = -A_0 - A_1 \sin \omega_1 t \qquad (2.12)$$

The negative sign in the equation defines the force direction from the turbine to the compressor. The cosine term is also neglected as it introduces only a phase difference in the external excitation. Lüddecke et al. [18] notices that if the static average value is much greater than the dynamic thrust force, almost no axial displacement is observed in the rotor. On the other hand, if the amplitude of the oscillating force is high, the axial rotor displacement is mainly influenced by the engine load conditions. They also provided all results in dimensionless form, dividing the axial thrust forces by the turbocharger own weight. Observing these points, in order to observe the axial displacement of the rotor, but to also consider the more realistic case of a nonzero cycle average thrust force, we admit the terms in Eq. (2.12) to be equal to ten times the rotor weight, i.e., $A_0 = A_1 = 10W_{TC} = 51.3N$. The frequency of excitation is composed of two pulses at a complete engine revolution [18], so that the excitation frequency is approximately $\omega_1 = 628$ rad/s; it is twice the engine speed, admitted for an engine running at 50 Hz (3,000 rpm).

3 RESULTS

3.1 Turbocharger model

The turbocharger model is the same one described by [13]. The FE model of the turbocharger shown in Figure 1, whose parameters are listed in Table 1, is composed of six beam elements, two rigid discs representing the compressor and turbine wheels, two equal rotating floating ring bearings and a double-acting thrust bearing. The oil circulating in the bearings is the Essolube X2 20W, whose temperature-viscosity relation is described by Eq. (3.1).

$$\mu(T) = 0.2023 / \left(1.000 - 0.0468T + 0.0029T^2\right) \qquad (3.1)$$

Table 1. Finite element model parameters.

Shaft parameters				
Density (kg/m^3)	Young modulus (GPa)	Poisson's ratio (–)	Diameter (mm)	Length (mm)
7860	200	0.30	11.0	15.5-1.00-1.00- 15.5-39.0-33.0

Rigid discs parameters			
	Mass (kg)	Polar MoI (10^{-6} kg.m^2)	Transverse MoI (10^{-6} kg.m^2)
Compressor Wheel	0.118	44.0	32.7
Turbine wheel	0.326	81.0	77.0

Floating ring bearing geometrical parameters					
	Radial clearance (µm)	Bore radius (mm)	Length (mm)	Ring mass (g)	Ring polar MoI (10^{-7} kg.m^2)
Inner film	35.0	5.535	6.5	7.5	3.504
Outer film	75.0	8.000	9.0		

Thrust bearing geometrical parameters					
Inner radius (mm)	Outer radius (mm)	Shoulder height (µm)	Pad angular extent (°)	Ramp angular extent (°)	Number of pads (–)
5.5	10.0	20.0	100	75	3

Oil physical parameters			
Density (kg/m^3)	Specific heat capacity (J/kg.°C)	Thermal conductivity (W/m.°C)	Inlet oil temperature (°C)
880	1950	0.130	100

3.2 Database method in time transient analysis

Given the rotor-bearing system model described in the previous section and the external axial harmonic excitation force discussed on Eq. (2.12), it is possible to simulate the system time response. In order to observe only the bearing damping characteristics, the damping matrix of the rotating shaft is assumed to be zero. In addition, in order to correctly ascertain the different phenomena in this rotating system, the unbalance forces will also be neglected. Finally, the engine excitations due to base motion will also be neglected. This assumption corresponds to the investigations of Tian et al. [2], who noticed that for rotor speeds higher than 45 krpm, the engine induced vibrations are suppressed by the dominant sub-synchronous oscillations due to oil whirl and whip in the FRBs. The simulations are performed for a rotational speed of 100 krpm, so it is expect little influence of the engine excitations on the lateral oscillations. With these assumptions, the work aims to investigate the fluid-induced instability presented in the FRBs and the TB coupling effect in lateral oscillations.

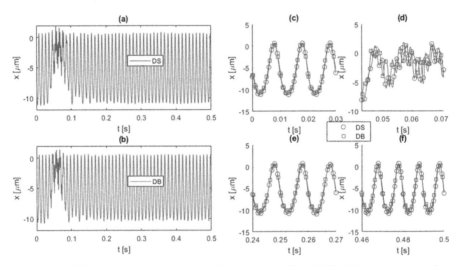

Figure 3. – TB axial displacement. (a) direct solution (DS), (b) database solution (DB), (c-f) comparison of both solutions at selected time spans.

Figure 3 presents the axial displacement of the thrust bearing midpoint (node 3 in the FE discretization), when the system is subjected to the axial harmonic excitation. Figures 3a and 3b present the response utilizing the Direct Solution (DS), solving the Reynolds equation at each time step, and the Database Method solution (DB), respectively. Figures 3c-f highlight different time intervals in order to observe the differences utilizing each approach. Figure 3c presents the solution at the very beginning of the simulation, Figure 3d, at the instability transition, further explained, Figure 3e, at some time after the instability transition and, finally, Figure 3f at the end of the simulation. Clearly, both approaches do not predict the exact same response, but they have an excellent agreement. The biggest differences are observed only during the instability transition, in Figure 3d.

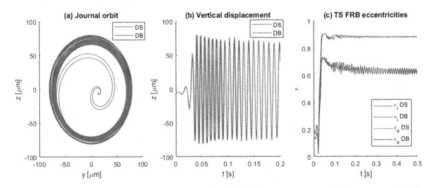

Figure 4. – TS FRB: (a) orbit, (b) vertical displacement and (c) eccentricities.

The initial conditions admitted for the simulations are the static equilibrium position of the rotor structure. Provided no external loads or perturbations interfere in the floating ring bearings, the system would remain in this static equilibrium position forever [13]. This static equilibrium position, however, is an unstable position. Given that the TB may induce bending moments on the shaft, the perturbation from the TB induces the FRB instability, as shown in Figure 4a. Figure 4a presents the journal orbit of the TS FRB. It is possible to identify the journal rapidly diverges from its initial position, but reaches a limit cycle smaller than the bearing clearance and remains within this cycle. This can also be observed on the eccentricities of the inner and outer films on Figure 4c. Figure 4b presents the vertical displacement of the journal as a function of time, during the first half of total time, in order to show the negligible differences of both models. A great agreement is observed between both curves. The small differences in the lateral response are due to the TB restoring moments estimated by the two different methods.

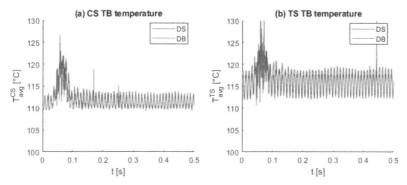

Figure 5. – Average oil temperature on CS TB (a) and TS TB (b).

Thermal effects can also be addressed by the database method. The direct solution is extremely cost, as it requires to solve the generalized Reynolds equation altogether with the energy equation at every time step during the time integration scheme, while the database method just interpolates the values from the lookup table. During the time integration scheme, the interpolation of the database method can also be used to analyse any other interested parameter. Storing the average temperature of the fluid film in each TB allows for the comparison of the temperature evolution of both schemes. Figure 5 illustrates this behaviour. In Figure 5a, the global average film temperature of CS TB is shown, as predicted by both methods, while Figure 5b shows the TS TB average film temperature. The oscillating predicted behaviour is captured by the Database solution. The biggest difference is observed on the TS TB, as the temperature amplitude oscillation predicted by the Database Method is a little higher than the direct solution. Both solutions are capable to show the temperature increase during the instability transition and the temperature oscillations due to axial and rotational motion of the thrust collars. Finally, another important aspect of the Database method not revealed in any results is the computational time to evaluate this time transient solution. The code for both solutions was written and compiled with Intel Visual Fortran 11.1.038 and both solutions were run in a personal computer with an Intel Core i7-5500 CPU @ 2.40 GHz processor and 8.00 GB of RAM. The direct solution took 3,368 minutes (56 hours) to run, while the database solution took only 2 minutes. That is an impressive gain in terms of computational time and shows the suitability of the Database Method to model thrust bearings and perform thermal transient analysis on turbochargers.

4   CONCLUSIONS

This work compared two methods for a time transient analysis of a turbocharger supported by floating ring and thrust bearings. The hydrodynamic thrust bearing loads and moments are estimated by the generalized Reynolds equation, as thermal effects are also considered. The two approaches in the time integration were to solve the Reynolds equation at each time step (the Direct Solution) or to interpolate it from a lookup table (the Database Method). Both approaches produce results in very good agreement, and the suitability of the Database Method over the Direct Solution concerns the computational time to evaluate both models. The Database Method is an extremely fast model and should pose as an excellent alternative to the reliable, but slower, Direct method, solving the governing equations every time step. The dynamic simulations of the turbocharger revealed the unstable characteristics of the floating ring bearings, as a small perturbation on the shaft lateral motion induces a diverging movement of the rotating shaft from its static equilibrium position, reaching a limit cycle (smaller than the bearing clearances) and remaining within this cycle. When the shaft is axially excited, the thrust collar rotations induce restoring moments on the shaft, that are great enough to perturb the lateral motion and induce the oil whirl/whip on the floating ring bearings. The axial thrust was modelled as a harmonic force, with a mean value directed from the turbine to the compressor. The axial response is a harmonic motion, and a perturbation on axial displacements is observed in the lateral oscillations instability transition, revealing the natural coupling of the axial and lateral dynamics. Thermal effects were also addressed and both methods could capture the overall behaviour. The temperature variations resemble the axial displacement: a harmonic variation of the average film temperature.

ACKNOWLEDGEMENTS

The authors would like to thank BorgWarner Brasil Ltda., SAE-Unicamp, CAPES, CNPq and FAPESP grant #2015/20363-6 for supporting this research.

REFERENCES

[1] C.-H. Li, "Dynamics of Rotor Bearing Systems Supported by Floating Ring Bearings," *Journal of Lubrication Technology*, vol. 104, no. 4, pp. 469–476, Oct. 1982, doi: 10.1115/1.3253258.

[2] L. Tian, W. J. Wang, and Z. J. Peng, "Dynamic behaviours of a full floating ring bearing supported turbocharger rotor with engine excitation," *Journal of Sound and Vibration*, vol. 330, no. 20, pp. 4851–4874, Sep. 2011, doi: 10.1016/j.jsv.2011.04.031.

[3] B. Schweizer, "Total instability of turbocharger rotors—Physical explanation of the dynamic failure of rotors with full-floating ring bearings," *Journal of Sound and Vibration*, vol. 328, no. 1–2, pp. 156–190, Nov. 2009, doi: 10.1016/j.jsv.2009.03.028.

[4] L. Tian, W. J. Wang, and Z. J. Peng, "Effects of bearing outer clearance on the dynamic behaviours of the full floating ring bearing supported turbocharger rotor," *Mechanical Systems and Signal Processing*, vol. 31, pp. 155–175, Aug. 2012, doi: 10.1016/j.ymssp.2012.03.017.

[5] R. J. Trippett and D. F. Li, "High-Speed Floating-Ring Bearing Test and Analysis," *A S L E Transactions*, vol. 27, no. 1, pp. 73–81, Jan. 1984, doi: 10.1080/05698198408981547.

[6] P. Koutsovasilis, N. Driot, D. Lu, and B. Schweizer, "Quantification of sub-synchronous vibrations for turbocharger rotors with full-floating ring

bearings," *Arch Appl Mech*, vol. 85, no. 4, pp. 481–502, Apr. 2015, doi: 10.1007/s00419-014-0924-0.

[7] P. Novotný, P. Škara, and J. Hliník, "The effective computational model of the hydrodynamics journal floating ring bearing for simulations of long transient regimes of turbocharger rotor dynamics," *International Journal of Mechanical Sciences*, vol. 148, pp. 611–619, Nov. 2018, doi: 10.1016/j.ijmecsci.2018.09.025.

[8] P. Novotný, J. Hrabovský, J. Juračka, J. Klíma, and V. Hort, "Effective thrust bearing model for simulations of transient rotor dynamics," *International Journal of Mechanical Sciences*, vol. 157–158, pp. 374–383, Jul. 2019, doi: 10.1016/j.ijmecsci.2019.04.057.

[9] Chasalevris and Louis, "Evaluation of Transient Response of Turbochargers and Turbines Using Database Method for the Nonlinear Forces of Journal Bearings," *Lubricants*, vol. 7, no. 9, p. 78, Sep. 2019, doi: 10.3390/lubricants7090078.

[10] I. Chatzisavvas, A. Boyaci, P. Koutsovasilis, and B. Schweizer, "Influence of hydrodynamic thrust bearings on the nonlinear oscillations of high-speed rotors," *Journal of Sound and Vibration*, vol. 380, pp. 224–241, Oct. 2016, doi: 10.1016/j.jsv.2016.05.026.

[11] P. Koutsovasilis, "Automotive turbocharger rotordynamics: Interaction of thrust and radial bearings in shaft motion simulation," *Journal of Sound and Vibration*, vol. 455, pp. 413–429, Sep. 2019, doi: 10.1016/j.jsv.2019.05.016.

[12] T. F. Peixoto and K. L. Cavalca, "Investigation on the angular displacements influence and nonlinear effects on thrust bearing dynamics," *Tribology International*, vol. 131, pp. 554–566, Mar. 2019, doi: 10.1016/j.triboint.2018.11.019.

[13] T. F. Peixoto and K. L. Cavalca, "Thrust bearing coupling effects on the lateral dynamics of turbochargers," *Tribology International*, p. 106166, Jan. 2020, doi: 10.1016/j.triboint.2020.106166.

[14] N. Mittwollen, T. Hegel, and J. Glienicke, "Effect of Hydrodynamic Thrust Bearings on Lateral Shaft Vibrations," *Journal of Tribology*, vol. 113, no. 4, p. 811, 1991, doi: 10.1115/1.2920697.

[15] L. San Andrés, "Effects of Misalignment on Turbulent Flow Hybrid Thrust Bearings," *Journal of Tribology*, vol. 124, no. 1, p. 212, 2002, doi: 10.1115/1.1400997.

[16] H. D. Nelson, "A Finite Rotating Shaft Element Using Timoshenko Beam Theory," *J. Mech. Des.*, vol. 102, no. 4, p. 793, 1980, doi: 10.1115/1.3254824.

[17] S. Liu and L. Mou, "Hydrodynamic Lubrication of Thrust Bearings with Rectangular Fixed-Incline-Pads," *Journal of Tribology*, vol. 134, no. 2, p. 024503, Apr. 2012, doi: 10.1115/1.4006022.

[18] B. Luddecke, P. Nitschke, M. Dietrich, D. Filsinger, and M. Bargende, "Unsteady Thrust Force Loading of a Turbocharger Rotor During Engine Operation," p. 10, 2015.

Cylindrical roller bearing under elastohydrodynamic lubrication with localized defects modeling

N.A.H. Tsuha, K.L. Cavalca

School of Mechanical Engineering, University of Campinas (Unicamp), Brazil

ABSTRACT

The main objective of this study is to model a cylindrical roller bearing under elastohydrodynamic lubrication (EHL) with localized surface defects. The dynamic model of the cylindrical roller bearing is applied to a rotor system modeled by the finite element method. The load-displacement relationship of the lubricated contact force model is based on an equivalent EHL stiffness and damping obtained by numerical simulation. Comparison results are presented of the frequency spectra in simulation with healthy and faulty bearings under EHL and Hertzian classical dry contact.

1   INTRODUCTION

One of the most frequent failure modes of rolling element bearings is due to surface defects in raceways and their elements, such as pitting, spalling and cracks. Fault prediction is one major actual challenge (1-3) and modeling of rolling element bearing defects is a way to make failure prediction methods more robust.

There is an effort to model surface defects in the cylindrical roller bearing, as shown in the works of Shao et al. (4), Wang et al. (5) and Liu et al. (6). However, aiming at more accuracy between the results of modeling and experiment, Wang et al. (5) suggested that other factors should be considered during modeling process, as skewing between bearing elements and lubrication.

Due to the reduced contact area and the high load applied to the rolling bearings, the most common type of lubrication is the elastohydrodynamic (EHL). One major difficulty in roller bearing failure modeling is combining the time-varying deflection caused by the localized defect and the effects of the lubricant oil film.

Thus, the objective of this study is to model a cylindrical roller bearing under elastohydrodynamic lubrication and to analyze the vibration response with localized surface defects. A reduced model of a two-degree-of-freedom bearing with EHL contact force between raceways and rollers is described with time-varying excitation generated by localized defects on raceway surfaces. The dynamic model of the cylindrical roller bearing is applied to a rotor system modeled by the finite element method. The load-displacement relationship of the lubricated contact force model is based on an equivalent EHL stiffness and damping. These parameters were obtained by numerical simulation of elastohydrodynamic system of equations: Reynolds equation and film thickness equation, considering the elastic deformation on contact area and force balance. Comparative results of the frequency spectra in simulation with healthy and faulty bearings under EHL and Hertzian classical dry contact.

2   EHL CONTACT MODEL

To model the elastohydrodynamic lubricated contact and calculate the equivalent stiffness and damping coefficients, firstly, it is necessary to solve numerically the EHL system of equations.

2.1   Elastohydrodynamic system of equations

The EHL system of equations is given by the Reynolds equation, the film thickness equation and the equation of motion at the lubricated contact. The variation of viscosity and density of the fluid in function of pressure are also accounted for due to the high pressures on nonconformal contacts.

The Reynolds equation for a bi-dimensional contact area is:

$$\frac{\partial}{\partial x}\left(\frac{\rho h^3}{\eta}\frac{\partial p}{\partial x}\right) + \frac{\partial}{\partial y}\left(\frac{\rho h^3}{\eta}\frac{\partial p}{\partial y}\right) = 6u_s\frac{\partial(\rho h)}{\partial x} + 12\frac{\partial(\rho h)}{\partial t} \tag{1}$$

where p is pressure, h is lubricant film thickness, η is viscosity of the lubricant oil, ρ is density of fluid, $u_s = u_1 + u_2$ is the sum of the velocities of both surfaces in contact and t is the time reference.

The viscosity-pressure and density-pressure relationships used in this work are (7, 8):

$$\eta(p) = \eta_0 \exp\left\{\frac{\alpha\, p_0}{z}\left[-1 + \left(1 + \frac{p}{p_0}\right)^z\right]\right\} \tag{2}$$

$$\rho(p) = \rho_0 \left(\frac{5.9 \cdot 10^8 + 1.34 \cdot p}{5.9 \cdot 10^8 + p}\right) \tag{3}$$

where p_0 is the atmospheric pressure, η_0 is the viscosity at p_0, α is the pressure-viscosity coefficient and z is the viscosity-pressure ratio.

In EHL finite line contacts, the film thickness equation can be written as:

$$h(x,y,t) = -\delta(t) + \frac{x^2}{2R} + \frac{2}{\pi E'}\iint \frac{p(x',y',t)}{\sqrt{(y-y')^2 + (x-x')^2}}dx'dy' \tag{4}$$

where δ is the mutual approach between both surfaces in contact, R is the equivalent radius of curvature of bodies 1 and 2, given by $R^{-1} = R_1^{-1} + R_2^{-1}$ and E' is the equivalent modulus of elasticity, namely, $E' = 2/\left[(1-\nu_1^2)/E_1 + (1-\nu_2^2)/E_2\right]$.

The equation of motion of the EHL contact dynamic surface motion is:

$$m_e\ddot{\delta}(t) + \int\int_\Omega p(x,y,t)dx\,dy = f(t) \tag{5}$$

where f is the external force applied to the contact, $m_e = \pi R^2 l \rho_c$ is the roller mass in the reduced contact, l is the effective length of roller and ρ_c is the material density of bodies and $\ddot{\delta}$ is the acceleration.

To solve the system of equations, the numerical methods widely used in EHL problems were applied (9): multigrid and MLMI (Multi-Level Multi-Integration). During the numerical convergence process, the pressure assumes only non-negative values due to cavitation condition accounted (Gümbel cavitation model). Once the numerical solution of the EHL finite line contact is evaluated, the reduced order force model of lubricated contact can be calculated.

2.2 EHL contact stiffness and damping

The reduced order model of EHL contact force between the rolling element and the raceways can be represented in terms of restitutive (stiffness) and dissipative (damping) forces – see Figure 1.

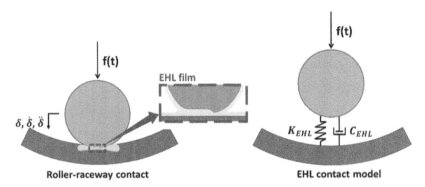

Figure 1. EHL contact between rolling element and raceway representation to a single degree of freedom system.

According to Tsuha and Cavalca (10, 11), the restitutive force f_k of the EHL finite line contact of non-profiled roller can be represented as:

$$f_k = k_{EHL}\delta_0 + \Delta F \qquad (6)$$

where k_{EHL} is the equivalent EHL stiffness, δ_0 is the mutual approach δ between contact bodies when $t = 0$ and ΔF is a surface separation EHL constant.

The EHL stiffness k_{EHL} and parameter ΔF are evaluated by curve fitting the mutual approach δ_0 to a range of static contact forces f_k, when $t = 0$, using the Levenberg-Marquardt method (12). The restitutive force model has the advantage of being an explicit force-displacement relation with independent parameters (k_{EHL} and ΔF).

In contrast with the classical Hertzian dry contact model, the lubricated contact force model also has a dissipative component based on a viscous damping c_{EHL}:

$$f_c = c_{EHL}\dot{\delta} \qquad (7)$$

To calculate the viscous damping c_{EHL}, the equation of motion must be solved numerically in the time domain under free vibration. Rewriting the equation of motion – Eq. (5) – as a function of reduced order force parameters:

$$m_e\ddot{\delta}(t) + c_{EHL}\dot{\delta}(t) + k_{EHL}\delta(t) + \Delta F = f(t) \qquad (8)$$

Thus, applying a disturbance in displacement as the initial condition of the free vibration problem and considering the principle of energy conservation, viscous damping c_{EHL} can be calculated for each rotation speed, according to (13):

$$c_{EHL} = \frac{\frac{1}{2}k_{EHL}[\Delta\delta(0)]^2}{\int_0^{t\to\infty}\left(\Delta\dot{\delta}\right)^2 dt} \qquad (9)$$

where $\Delta\delta = \delta - \delta_0$.

3   ROLLER BEARING WITH LOCALIZED DEFECT

Once the lubricated contact between the rolling element and raceways are modeled, the whole rolling element bearing vibration response can be investigated. In this work, the cylindrical roller bearing has two degrees of freedom: translations in y and z (see Figure 2).

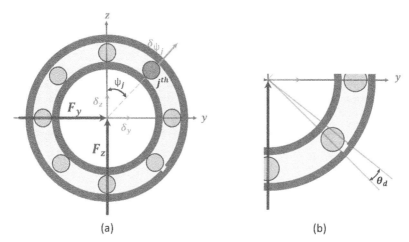

(a)   (b)

**Figure 2. (a) Cylindrical roller bearing with two degrees of freedom.
(b) Detail of a localized defect on the outer raceway.**

The EHL restitutive bearing forces are (14):

$$F_y = \sum_{j=1}^{Z} \left(k_t \, \delta_{\psi_j} \, sen\left(\psi_j\right)\right) + \sum_{j=1}^{Z} \Delta F \, sen\left(\psi_j\right) \tag{10}$$

$$F_z = \sum_{j=1}^{Z} \left(k_t \, \delta_{\psi_j} \, \cos\left(\psi_j\right)\right) + \sum_{j=1}^{Z} \Delta F \cos\left(\psi_j\right) \tag{11}$$

where k_t is the equivalent EHL stiffness between roller and both inner and outer raceways. The parameter ΔF_t is the equivalent surface separation EHL constant and parameter δ_{ψ_j} is the equivalent displacement, being ψ_j the azimuth angle of jth rolling element:

$$\psi_j = \psi_0 + \omega_c t + \frac{2\pi}{Z}(J-1) \tag{12}$$

where ψ_0 is the angular position of the reference roller, Z is the number of rolling elements and ω_c is the rotation speed of the cage:

$$\omega_c = \frac{\Omega_o - \Omega_i}{2d_m}(d_m - D) \tag{13}$$

In Equation (13), Ω_o is the angular velocity of the outer raceway, Ω_i the angular velocity of the inner race and d_m is the pitch diameter.

The equivalent displacement δ_{ψ_j} is then:

$$\delta_{\psi_j} = \delta_z \cos\left(\psi_j\right) + \delta_y \, sen\left(\psi_j\right) - \frac{P_d}{2} - H_d \tag{14}$$

where δ_y and δ_z are, respectively, the displacements in directions y and z (Figure 2). The parameter P_d is the diametral clearance and H_d is the defect depth. The localized defect considered in this work has a simple straight geometry given by:

$$H_d = \begin{cases} \Delta_d, & \theta_i \le mod(\theta, 2\pi) \le \theta_d + \theta_i \\ 0, & other \end{cases} \tag{15}$$

The detail of the localized defect can be seen in Figure 2. The parameter Δ_d is the maximal additional deflection of defect, θ is the angular position, θ_i is the initial angular position of defect from z-axis and θ_d is the length of the defect in tangential direction.

Analogous to the restitutive forces, the dissipative forces in the y- and z-axes are:

$$F_{\dot{y}} = \sum_{j=1}^{z} [c_t \dot{\delta}_{\psi_j}] sen\left(\psi_j\right) \tag{16}$$

$$F_{\dot{z}} = \sum_{j=1}^{z} [c_t \dot{\delta}_{\psi_j}] \cos\left(\psi_j\right) \tag{17}$$

where c_t is the equivalent viscous damping in both raceways and $\dot{\delta}_{\psi_j}$ is the velocity, or the first displacement δ_{ψ_j} derivative from Equation (14):

$$\dot{\delta}_{\psi_j} = \dot{\delta}_z \cos(\psi_j) - \delta_z \, sen(\psi_j)\, \dot{\psi}_j + \dot{\delta}_y \, sen(\psi_j) + \delta_y \cos(\psi_j)\, \dot{\psi}_j \tag{18}$$

From the explicit equations of the reduced order force, it is possible to insert the effects of the roller bearing and its defect on the rotor in time domain. The sum of restitutive and dissipative forces in both y and z directions can be easily included as external forces in the finite element model of the rotating system:

$$[M]\{\ddot{q}\} + ([C] + \Omega[G])\{\dot{q}\} + [K]\{q\} = \{F_e\} \tag{18}$$

where $\{q\}$ is the coordinate vector, $[M]$ is the mass matrix, $[K]$ is the rotor stiffness matrix, $[C]$ is the shaft proportional damping matrix, $[G]$ is the gyroscopic matrix, $\{F_e\}$ is the external force vector.

4   RESULTS AND DISCUSSION

For the case study, cylindrical roller bearings NJ 202 were selected and the geometric data used in numerical simulation, as in Table 1. The lubricant oil is ISO VG 32 at 27°C ($a = 2.23 \times 10^{-8}\, Pa^{-1}$, $\eta_0 = 4.879 \times 10^{-2} Pa$, $z = 0.68$) and the rotation speed is 40 Hz. Four levels with the most discretized grid of 193x193 points were used in numerical solution with multilevel methods. For the rotation speed of 40 Hz, the EHL contact parameters evaluated are: $k_t = 7.942 10^7 N/m$, $\Delta F_t = 62.08 N$ and $c_t = 4.77 N.s/m$.

Table 1. Geometry data of cylindrical roller bearing NJ 202.

Number of cylindrical rollers	10
Roller diameter [mm]	5.5
Effective length of roller [mm]	4.6
Bearing pitch diameter [mm]	24.8
Diametral clearance [mm]	0

In order to study the dynamic behavior due to the localized defect, the rolling element bearings were inserted in a rotor finite element model, using Timoshenko beam and disc elements. The rotating system is composed of two cylindrical roller bearings NJ 202, a shaft with 15 mm of diameter, magnetic actuator and disc. In Figure 3, the roller bearings are at nodes 2 and 10. The magnetic actuator is located at node 5 and it applies a constant horizontal force of 112 N. The disc is at node 8.

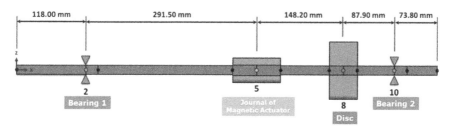

Figure 3. Rotor modeled by finite element method.

The proportional damping coefficient is $\beta = 7 \times 10^{-5}$, the modulus of elasticity of the shaft is $E = 2.1 \times 10^{11} Pa$ and its density is $\rho = 7860 kg/m^3$. The unbalance given by disc mass is $3.1 \times 10^{-4} kg.m$. First, the rotor is analyzed with healthy bearings, namely, without any defects (Figure 4). In sequence, the system is evaluated under bearings with a localized defect on the outer raceway with a defect depth of $\Delta_d = 100$, initial angular position of the defect $\theta_i = 0°$ and angular length of the defect $\theta_d = 1°$ (Figure 5). Finally, considering that the localized defect has enlarged, the rotor with bearings with $\theta_d = 5°$ and remaining $\Delta_d = 100$ and $\theta_i = 0°$ (Figure 6). To compare with the EHL contact results, the classical Hertzian dry contact model of (15, 16) was applied. From the time domain simulations, the DFT (Discrete Fourier Transform) of the cylindrical roller bearings parameters were calculated. In this work, there are only results of the roller bearing 1 (Figure 3), since both bearings present similar behavior. Tables 2 and 3 show the comparison of DFT amplitude evaluated for the rotation frequency (1x) and RPFO (Roller Passing Frequency of Outer) in y-direction and z-direction respectively. The rotation frequency (1x) is 40 Hz and the RPFO is 155.6 Hz.

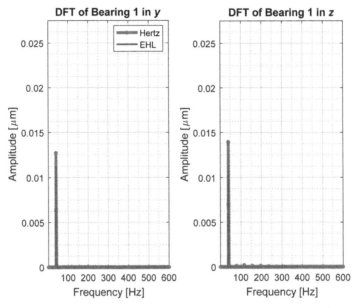

Figure 4. DFT of healthy cylindrical roller bearing displacement for Hertzian and EHL contacts.

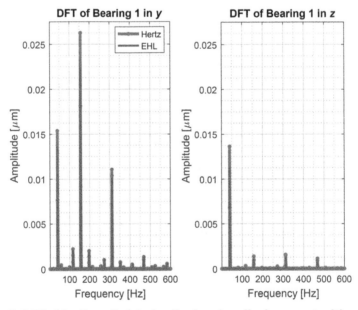

Figure 5. DFT of faulty cylindrical roller bearing displacement with angular length of the defect $\theta_d = 1°$.

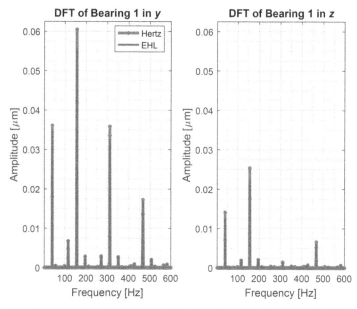

Figure 6. DFT of faulty cylindrical roller bearing displacement with angular length of the defect $\theta_d = 5°$.

Table 2. Comparison of DFT amplitudes for the rotation frequency (1x) and RPFO between Hertzian and EHL contacts in the y-direction.

Cases	Healthy bearing	Localized defect $\theta_d = 1°$	Localized defect $\theta_d = 5°$
Amplitude 1x – EHL $[\mu m]$	0.01120	0.01376	0.03463
Amplitude 1x – Hertz $[\mu m]$	0.01271	0.01538	0.03607
Difference of amplitude 1x between EHL and Hertz [%]	13.48	11.77	4.16
Amplitude RPFO – EHL $[\mu m]$	0	0.02235	0.05485
Amplitude RPFO – Hertz $[\mu m]$	0	0.02626	0.06049
Difference of amplitude of RPFO between EHL and Hertz [%]	0	17.49	10.28

Table 3. Comparison of DFT amplitudes for the rotation frequency (1x) and RPFO between Hertzian and EHL contacts in the z-direction.

Cases	Healthy bearing	Localized defect $\theta_d = 1°$	Localized defect $\theta_d = 5°$
Amplitude 1x – EHL $[\mu m]$	0.01120	0.01118	0.01124
Amplitude 1x – Hertz $[\mu m]$	0.01396	0.01363	0.01410
Difference of amplitude 1x between EHL and Hertz [%]	24.64	21.91	25.44
Amplitude RPFO – EHL $[\mu m]$	0	0.00097	0.02207
Amplitude RPFO – Hertz $[\mu m]$	0	0.00138	0.02539
Difference of amplitude of RPFO between EHL and Hertz [%]	0	42.27	15.04

For the case with roller bearings without localized defects (Figure 4), there is a predominance of the rotation speed of the shaft at 40 Hz. The RPFO does not appear significant here, since the EHL contact force model used is linear and the diametral clearance of bearing, parameter that creates non-linearities in the system, was considered null. When comparing the 1x frequency amplitudes between dry and lubricated contact conditions, there is are considerable differences in their values. Using the EHL contact as a reference for calculation of error between dry and lubricated contacts, the amplitude of Hertzian contact bearing is 13.48% higher than the EHL roller bearing in the y-direction. In the z-direction, the difference is more significant at 24.64%. In both directions, the amplitudes with the dry contact roller bearing conditions are higher than EHL case.

When the roller bearings are modeled with a localized defect in the outer raceway, the RPFO and their multiples appear in both directions. In this case, where there is a constant force applied in the y-direction by the magnetic actuator, the frequencies showed a higher amplitude on the y-axis. Comparing Figures 4 and 5, there is an increase in the displacement amplitude of the 1x rotation frequency for dry and lubricated contact with the defect of $\theta_d = 1°$, but the Hertzian problem continues to show higher amplitudes. For the $\theta_d = 1°$ (Figure 5), the values of RPFO amplitude for the dry and EHL contact models present significant differences: 17.49% in the y-direction and 42.27% in the z-direction.

To simulate a problem when the defect is enlarged, a case with $\theta_d = 5°$ was analyzed. The 1x frequency and RPFO with their multiple harmonics also are shown in the DFT of Figure 6. As the values of amplitude of this case are higher than the previous analysis, the scale needed to be changed. There is an increase in all amplitudes of frequencies for both dry and EHL contact in comparison with case with $\theta_d = 1°$, mainly in y-direction. The difference between the Hertzian and EHL models is still relevant, reaching more than 25% in the 1x amplitude and a 15.04% in the RPFO amplitude in the z-direction.

5   CONCLUSIONS

Rolling bearings are classic elements in machines and have wide applications in mechanical systems. When a localized defect on the roller bearing surface occurs, there is a significant change in the dynamics of bearing and, consequently, in the rotating system. These faults can, therefore, lead to unwanted vibration on the machines.

When inserting the defect on the outer race, an increase in amplitude was noticed in the frequency spectrum in both the 1x rotation frequency and the RPFO with their higher harmonics. When enlarging the defect, these amplitudes tend to increase even more.

When comparing the EHL contact model and the classical Hertzian dry contact model, it is noticed that both present the frequencies in the DFT. However, the amplitude value is different in frequency spectrum, which reiterates the importance of considering lubrication for problems that need precision. This work allows study of roller bearing defects more accurately by considering the effects of lubrication on roller and raceway contacts. Consequently, it is possible to detect bearing failure parameters more accurately.

ACKNOWLEDGMENTS

The authors thank Petrobras and CNPq for the support of this research.

REFERENCES

(1) Yang, Y., Liu, C., Jiang, D., Behdinan, K., "Nonlinear vibration signatures for localized fault of rolling element bearing in rotor-bearing-casing system", *International Journal of Mechanical Sciences*, 173, 2020, 05449. doi:10.1016/j.ijmecsci.2020.105449.

(2) 2 Randall, R. B., Antoni, J, "Rolling element bearing diagnostics-A tutorial", *Mechanical System and Signal Processing*, 25 (2), 2011, pp. 485–520. doi:10.1016/j.ymssp.2010.07.017.

(3) 3 Rai, A., Upadhyay, S. H., "A review on signal processing techniques utilized in the fault diagnosis of rolling element bearings", *Tribology International*, 96, 2016, pp. 289–306. doi:10.1016/j.triboint.2015.12.037.

(4) 4 Shao, Y., Liu, L., Ye, J., "A new method to model a localized surface defect in a cylindrical roller bearing dynamic simulation", *Proceedings of the Institution of Mechanical Engineers, Part J: Journal of Engineering Tribology*, 228(2), 2014, pp. 140–159. doi:10.1177/1350650113499745.

(5) 5 Wang, F., Jing, M., Yi, J., Dong, G., Liu, H., Ji, B. "Dynamic modelling for vibration analysis of a cylindrical roller bearing due to localized defects on raceways", *Proceedings of the Institution of Mechanical Engineers, Part K: Journal of Multi-Body Dynamics*, 229(1), 2015, pp.39–64. doi:10.1177/1464419314546539.

(6) 6 Liu, Y., Zhu, Y., Yan, K., Wang, F., Hong, F., "A novel method to model effects of natural defect on roller bearing", *Tribology International*, 122, 2018, pp. 169–178. doi:10.1016/j.triboint.2018.02.028.

(7) 7 Roelands, C. J. A. Correlational Aspects of the Viscosity-Temperature-Pressure Relationship of Lubricating Oils. Delft, The Netherlands: Technical University Delft, 1966. Thesis (PhD).

(8) 8 Dowson, D., Higginson, G. R., "Elasto-hydrodynamic Lubrication – SI Edition", *Pergamon Press*, 1st. ed, Great Britain, 1977.

(9) 9 Venner, C. H., Lubrecht, A. A., "Multilevel Methods in Lubrication", Elsevier Science, 2000.

(10) 10 Tsuha, N. A. H., Nonato, F., Cavalca, K. L., "Formulation of a reduced order model for the stiffness on elastohydrodynamic line contacts applied to cam-follower mechanism", *Mechanism and Machine Theory*, 113, 2017, pp. 22–39. doi:10.1016/j.mechmachtheory.2017.03.002.

(11) 11 Tsuha, N. A. H., Cavalca, K. L., "Finite line contact stiffness under elastohy-drodynamic lubrication considering linear and nonlinear force models", *Tribology International*, 146, 2020, 106219. doi:10.1016/j.triboint.2020.106219.

(12) 12 Marquardt, D. W., "An Algorithm for Least-Squares Estimation of Nonlinear Parameters", *Journal of the Society for Industrial and Applied Mathematics*, 11 (2), 1963, pp. 431–441. doi:10.1137/0111030.

(13) 13 Tsuha, N. A. H., Nonato, F., Cavalca, K. L., "Stiffness and Damping Reduced Model in EHD Line Contacts", in: K. Cavalca, H. Webe (Eds.), *Proceedings of the 10th International Conference on Rotor Dynamics – IFToMM*. IFToMM 2018. Mechanisms and Machine Science, 2019, pp. 43–55. doi:10.1007/978-3-319-99262-4_4.

(14) 14 Nonato, F., Cavalca, K. L., "An approach for including the stiffness and damping of elastohydrodynamic point contacts in deep groove ball bearing equilibrium models", *Journal of Sound and Vibration*, 333(25), 2014, pp. 6960–6978. doi:10.1016/j.jsv.2014.08.011.

(15) 15 Harris, T. A., "Rolling Bearing Analysis", *John Wiley & Sons*, New York, 1991.

(16) 16 Palmgren, A., "Ball and Roller Bearing Engineering", 3rd. ed, *S.H. Burbank*, Philadelphia, 1959.

12th International Conference on Vibrations in Rotating Machinery -
Institution of Mechanical Engineers, ISBN 978-0-367-67742-8

Analytical study of rotordynamic behaviour and rolling element bearing transient response in a high-speed race transmission

B. Friskney, S. Theodossiades, M. Mohammad-Pour

Wolfson School of Mechanical, Electrical and Manufacturing Engineering, Loughborough University, UK

ABSTRACT

In high-speed motorsport power transmission systems, rotordynamic effects are often significant. Where several gear pairs and flexible transmission shafts are present, gyroscopic action can lead to considerable speed-driven changes in the natural frequencies of coupled lateral-torsional vibration modes. When system excitations such as gear meshing are considered, these changes must be taken into account in order to confidently model the vibrational responses that occur at the system critical speeds. This study applies an analytical finite element method to the targeted problem in order to study the system's behaviour, providing a tool for design and diagnostics.

1 INTRODUCTION

An accurate understanding of the dynamics of high-speed drivetrain systems is crucial to both design and prediction of performance. In a system consisting of many different components, such as multiple shafts, gears and rolling element bearings, the consideration of dynamics phenomena is key to the undertaking of a full system-level analysis. In the present study, the natural frequencies and vibrational modes of the system are gained through the application of rotordynamics to a high-speed gearbox. Analysis of the forced vibration response of the system is included, considering non-linear representations of the gear mesh and rolling element bearings.

The gyroscopic effect is of great importance to accurately predicting the natural frequencies of a rotating system. This effect arises as a result of rocking motions as a laterally deflected rotor rotates. Where this rocking motion coincides with a component (e.g. gear) or a flexible shaft element passing through the centreline of the undeflected rotor, the varying angular momentum due to rocking induces a torque that is a function of the rate of change of the slope with respect to the undeflected centreline. Figure 1 illustrates the generation of the above torque T, where the dashed outline represents the deflected rotor of polar moment of inertia I, at an angle φ to the undeflected centreline. Naturally, torque T is also a function of the spin speed ω. The system sees this effect as a spring of variable stiffness (stiffening effect). As such, the natural frequency of a particular lateral vibration mode with rocking motion varies with spin speed, with effects pronounced at the highest speeds. It is therefore essential to include the above in the analysis of a high-speed rotating system.

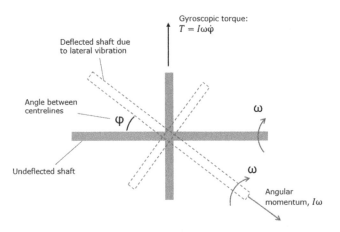

Figure 1. Generation of gyroscopic torque in a rotating shaft with a mounted disk.

Commonly, analytical methods for studying rotating systems consist of a finite element formulation in which the mass, stiffness, gyroscopic and damping matrices are derived for each component, then assembled into larger matrices which are applicable to the system as a whole. The process is described in detail by Rao (1), of which variations and developments are employed in several works (2–6). A useful summary of the various analytical techniques employed in rotordynamics is provided by Nelson (7).

Studies in the rotordynamics of dual-shaft geared systems are well documented. Rao et al. (8) investigated a turbine system with slender shaft elements and a spur gear pair, in which significant changes in the natural frequencies of coupled lateral-torsional modes were observed due to gyroscopic action. Coupling of lateral-torsional modes means that both lateral and torsional mode shapes can participate simultaneously at the same natural frequency. They are coupled by the action of the gear mesh (9,10). Chen et al. (11) and Hu et al. (12) studied finite element shaft-bearing systems connected by helical gear pairs and spur gear pairs respectively. Their applications are automotive, hence shafts are shorter and have greater diameter-to-length ratios than in the case of Rao et al. (8). Both studies employ nonlinear representations of gear mesh stiffness to investigate vibration response, whereas in the majority of the aforementioned studies, the gear mesh and bearings are represented as linear components with constant stiffness. More complex, i.e. nonlinear representations of these components tend to be confined to studies focused primarily on gear mesh and/or bearing dynamics (13–18). Shafts are often rigid (13,15,18) or treated as flexible (17,19). However, a recent study by Hu et al. (20) also applies flexible shaft formulations to a dual-shaft problem with nonlinear gear mesh terms. In this study, the bearing stiffnesses are linear. Detailed studies presenting frequency response functions of gear vibrations, including vibration response with an increased amplitude across an excitation frequency region near to the gear pair's natural frequency (nonlinear hardening/softening behaviour) are presented by several authors in computation (13,15,16) and experiment (15,21). The new aspect of the current work is to combine the methodologies of finite element rotordynamics and nonlinearities of both the gear pair and rolling element bearings into a single study. Additionally, the targeted problem involves high speeds associated with race applications, and greater geometrical complexity as a result of the presence of several gear pair and bearing components, and cross-sectional variations in the transmission shafts.

2   SYSTEM GEOMETRY AND COMPONENTS

The eight-speed drivetrain system used for the present study is shown in Figure 2. It consists of input and output hollow shafts supported by four rolling element bearings, with eight steel spur gear pairs mounted on the shafts. The engaged gear pair is shown in red. In order to generate the governing equations of motion, the system is de-constructed into its constituent parts, to which representations are applied. Equations for each component are then re-assembled to form the mass, stiffness, gyroscopic and damping matrices for the full system.

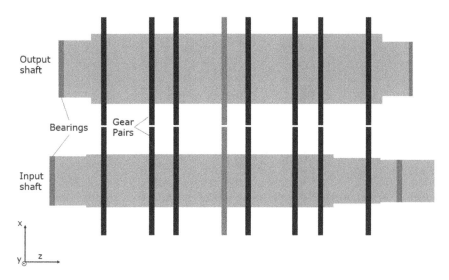

Figure 2. Layout of the drivetrain system.

Nodes are placed within the system at locations where a component (gear wheel or bearing) is encountered, or a shaft cross-section changes. They are numbered moving from left-to-right, starting with the first node at the left-most end of the input shaft. Upon reaching the right-most end, numbering continues at the left-most end of the output shaft. A global coordinate system is shown in Figure 2, in which rotations about the principal axes follow the right-hand rule. The motion of the ith node is described by the column vector:

$$\{q\}_i = \{v_i\ w_i\ \theta_i\ \varphi_i\ \alpha_i\}^T \tag{1}$$

In Eq. (1), v and w denote lateral motions, along the x- and y- axes respectively. Θ and φ are the angles of rotation (slopes) about the same axes. The angle a represents the angle of rotation about the z-axis, which is the sense of shaft spin. As there are to be no external sources imparting axial motion upon the system and the gears are spur, axial motion is expected to be negligible compared to the rotational and bending motions. As such, the translation along the z-axis is omitted.

Bearings are modelled as two-dimensional massless components, with principal axes coinciding with the x- and y-axes of their nodes (Figure 3). In the time-domain response problem, restoring forces are developed in the bearings in response to deflections v and w along these axes. These forces are a summation of the radial loads taken by each rolling element, resolved along the x- and y-axes. The forces are a function of the angular positions of the rollers, the roller-to-raceway contact stiffness, the rotation frequency of the retaining cage, the geometry of the bearing and the lateral deflections of the system at the bearing node. In the present study, the bearings have zero clearance. The reader is referred to Mohammad-Pour et al. (14) for the detailed derivation. For the purposes of the eigen problem, the bearings are simplified as parallel spring-dampers aligned to the x- and y-axes, with a constant linear stiffness value as the overall mean stiffness of the bearing.

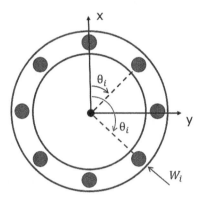

Figure 3. Principal axes of rolling element bearings. Θ_i = angular position of the ith roller, W_i = restoring force of the ith roller.

The gear wheels are represented by purely inertial rigid disks with zero thickness, therefore they make no contribution to the stiffness of the system and the total length of a shaft is divided amongst shaft elements. As both bearings and gear wheels coincide with single nodes, their associated matrices are of order 5. The individual terms constituting these matrices are taken from Rao et al. (8) and Chen et al. (11).

The lengths of the shaft connecting the nodes of the system are modelled as flexible elements, constructed using a Timoshenko beam formulation with distributed mass and gyroscopic terms. Such a formulation is necessary if inertial and gyroscopic effects are expected to be significant (22). The shafts have stiffness in bending and torsion. As the elements are constrained by two nodes, the matrices of a shaft element are of order 10. In order to join the two shafts together and couple them, terms for the gear mesh of the engaged gear pair must be added. The gear mesh stiffness is a function of the teeth geometry and applied load, which vary as the gear pair rotates. In the time-domain problem, the gear mesh stiffness k_m is expressed as a Fourier series:

$$k_m(a_p) = k_0 + \sum_{n=1}^{8} [a_n \cos(nz_p a_p) + b_n \sin(nz_p a_p)] \tag{2}$$

In Eq. (2), a_p is the angular displacement of the pinion, z_p is the number of teeth on the pinion, and k_0, a_n and b_n are Fourier coefficients calculated from Loaded Tooth Contact Analysis (LTCA). The dynamic transmission error (DTE), u, must be calculated, which is the difference between the ideal positions of the pinion and gear and their actual positions as aresult of tooth and shaft deflections under loading, commonly used for analysis of vibrations in geared systems. It is shown in Eq. (3), in which $r_{p,g}$ are the base radii of the pinion and gear.

$$u = v_g - v_p - (r_p \alpha_p + r_g \alpha_g) \tag{3}$$

The mesh stiffness is acting using a backlash switching function, which compares the magnitude of the DTE to the size of half of the backlash of the gear pair in order to determine whether there is contact on the tooth driving face, reverse face, or no contact between the interacting teeth (13,15). Similar to the bearings, the terms are linearised for the eigen problem, in which only the mean stiffness k_0 is used. The reader is referred to Rao et al (8) for the derivation of the gear mesh stiffness matrix, and Chen et al (11) for the damping matrix and method for locating the gear mesh terms within the stiffness and damping matrices of the full system.

3   EIGEN PROBLEM

The matrices of the full system are assembled by stepping from left-to-right through the system, and summing terms at nodes which have contributions to the same matrices by different components. A column vector is thus formed, containing all n degrees of freedom of the system:

$$\{q\} = \left\{ \{q\}_1{}^T \{q\}_2{}^T \{q\}_3{}^T \ldots \{q\}_{end}{}^T \right\}^T \tag{4}$$

All system matrices are of the order n. The assembled and coupled equations of motion of the system are:

$$[M]\{\ddot{q}\} + (\Omega_1[G] + [C])\{\dot{q}\} + [K]\{q\} = \{Q\} \tag{5}$$

In Eq. (5), $[M]$ represents the mass/inertia matrix, $[G]$ is the gyroscopic matrix, $[C]$ is the damping matrix, $[K]$ is the stiffness matrix and $\{Q\}$ is the vector of external forces. Ω_1 is the rotational speed of the input shaft. For the eigen problem, the linearised terms for the bearing and gear mesh stiffnesses are used as described in Section 2, and the vector $\{Q\}$ is filled with zeros. The damped modal analysis method and characteristic determinant method described by Meirovitch (23) are followed in order to solve for the natural frequencies. An undamped method is used to find the mode shapes (24).

4   VIBRATION RESPONSE

In order to produce the vibration response of the system, the gear mesh nonlinearities (time-varying mesh stiffness and backlash) are added to the $[K]$ matrix, rendering it time-varying. The bearing stiffness terms are removed from the $[K]$ matrix and the restoring forces are placed in vector $\{Q\}$ as reaction loads. The model is driven by the speed of the input shaft, which is given as akinematic input to the equations of motion at asingle node of the input shaft. Aresistive torque is applied at the final node of the output shaft. The equations of motion are solved for aconstant speed and torque using acommercially-available ODE solver to produce the displacement and velocity time-histories for all degrees of freedom. The coordinate system can be rotated about its z-axis such that the x- and y- axes align to the line of action (LOA) and off-line of action (OLOA) of the gear mesh, respectively. This way, the behaviour of the bearings is more easily studied, as the dominant force applied to the bearings is expected to align to the x-axis. The basic information of the system, including orders of magnitude of stiffness values, is presented in Table 1.

Table 1. System Parameters.

Parameter	Value	Units
Shaft material density	7900	kg/m^3
Shaft modulus of elasticity	200	GPa
Shaft Poisson's ratio	0.3	-
Bearing and gear mesh linear stiffnesses	O (10^8)	N/m
Total backlash of gear pair	24	μm
Number of teeth on pinion of engaged gear pair	23	-
Total inertia of input shaft assembly	0.00325	kgm^2
Total inertia of output shaft assembly	0.00419	kgm^2

5   RESULTS

5.1   Natural frequencies

The Campbell Diagram for the system is displayed in Figure 4. The horizontal lines represent the damped natural frequencies (modes) of the system, which vary as a function of the spin speed due to the gyroscopic effect. Modes with torsional participation are shown in blue. The mode with a natural frequency near 0 Hz is a torsional rigid body mode. No rigid body modes are found for the lateral motions, as these motions are constrained by the bearings. The red diagonal line is the Synchronous Whirl Line (SWL), which in this study is the meshing frequency of the engaged gear pair, calculated as the number of teeth of the pinion multiplied by the spin speed of the input shaft. This is chosen as the action of the gear mesh is a significant source of internal excitation. The spin speeds at which the SWL intersects the natural frequencies are known as the critical speeds of the system, at which the corresponding natural frequency is excited.

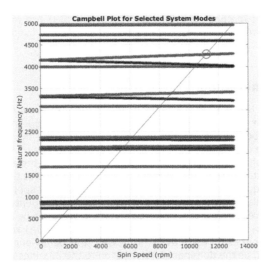

Figure 4. Campbell diagram for the examined system.

5.2   Resonances

In order to study the vibration response of the system, the analyses of two types of resonance are presented. In the first case, the natural frequency of the engaged gear pair according to the gear mesh stiffness and total inertia of the system is targeted to be excited. In the second, a flexible system natural frequency calculated from the eigen problem (Campbell Diagram) is targeted.

5.2.1   Case 1 – gear pair natural frequency

First, a natural frequency excitation case is considered which involves the action of the gear mesh alone, and does not account for the flexibilities of the shaft. Therefore, this case does not appear in the Campbell Diagram. To calculate the natural frequency of the gear pair, a characteristic mass and stiffness are required. For the stiffness, the constant term k_0 of the meshing stiffness Fourier series is used. An equivalent mass m is calculated from the total inertia of the components on the input and output shafts, and the base radii of the engaged pinion and gear (13). The natural frequency ω_0 is then:

$$\omega_0 = \sqrt{k_0/m} \tag{6}$$

For the engaged gear pair used to produce the Campbell Diagram, the spin speed at which this natural frequency coincides with the meshing frequency is low in comparison to the typical operating speeds of the system. Therefore, a different gear pair is selected in order to study this case, whose natural frequency fall within the duty cycle. The new gear pair has 13 teeth on the pinion and is located at the right-most end of the shafts (Figure 2). The frequency calculated using Eq. (6) is 1237 Hz, which matches the gear mesh frequency at an input shaft speed of 5709 rpm. However, excitation at this frequency is difficult to achieve due to the comparatively low torque transmitted by the system at this speed. Despite this, excitation at a greater speed corresponding to a greater duty torque may be possible, due to the broadband excitation behaviour that can be observed as a result of the nonlinearity of the gear mesh. In a linear system, the response at the resonant frequency is characterised by a single peak in the frequency response function, where the excitation frequency is equal to the natural frequency. With nonlinear stiffness terms, this peak becomes bent, broadening the range around the gear pair's natural frequency over which excitation is possible. As such, a higher speed can be chosen which is desired to fall within this range. Two simulations are performed with a duration of 1 s, in order to study the behaviour within and away from this region. Torque values are chosen from a real race duty cycle to correspond to the speeds in this gear.

1. Input shaft speed of 6628 rpm, with input shaft torque of 220 Nm (close to natural frequency of the gear pair).
2. Input shaft speed of 8002 rpm, with input shaft torque of 253 Nm (away from the natural frequency of the gear pair).

In order to ascertain the excitation of the gear pair's natural frequency, the magnitude of the DTE is compared between the two simulations. The time histories of the DTE for the final few cycles of motion are displayed in Figure 5. The size of oscillations in the first simulation is greater, with a peak-to-peak value of 14.2 μm compared to 13.2 μm in the second simulation. The difference in amplitudes indicates that the gear mesh excitation frequency corresponding to the input speed of 6628 rpm may lie within the frequency region of greater response amplitude in the frequency response function of the nonlinear system. The mean value of the DTE is higher in the second simulation, measuring 78.6 μm compared to 69.9 μm. This is due to the greater torque acting upon the system in the second simulation, as greater tooth deflection due to higher transmitted torque moves the DTE further away from the half-backlash limit, which is 12 μm in this case.

5.2.2   Case 2 – system natural frequency

To activate a flexible natural frequency of the system, the model must be driven at a critical speed. The crossover point at 11150 rpm between the SWL (gear meshing frequency) and a vibration mode of the system is selected and indicated by the green circle in Figure 4. A simulation is performed at constant input speed of 11150 rpm and high torque of 540 Nm, which is representative of race conditions. In order to determine the activation of the selected vibration mode, the mode shapes found from the Eigen problem are compared to the forced response. This physical shape is produced by plotting instantaneous lateral and torsional displacements of all the system nodes at their respective locations. The comparison is shown in Figure 6.

The eigen problem predicts a shape in the x-direction of the output shaft corresponding to a 2nd lateral mode, with negligible deflection in the input shaft in comparison. However, the simulation results show shapes of a 1st mode. This difference may be a result of modal damping content. Each mode in the Campbell Diagram (Figure 4) is associated with a particular damping ratio for vibration at that mode. The value of this ratio increases for higher modes. In this case, the higher order modes may be dampened, including the mode at which the mesh frequency is targeted here.

It is observed that the shape of the expected bent shape of the input shaft in the y-direction is produced in the time domain simulation, albeit mirrored about the plot's x-axis at the selected instant. As the LOA of the gear pair aligns to the x-direction of the system (Figure 2), the gear mesh excitation force is not applied in the perpendicular y-direction, hence the system is allowed to vibrate freely in this plane and the effects of modal damping with excitation are not significant. The torsional deflection of the input shaft traces a similar shape to that predicted by the mode shape, again mirrored about the plot's x-axis.

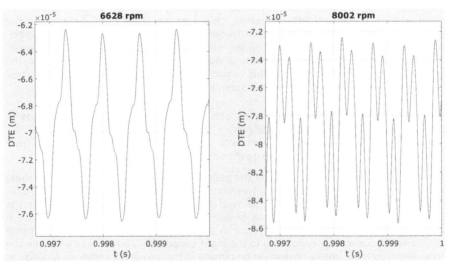

Figure 5. DTE time histories for two selected cases of speed and torque.

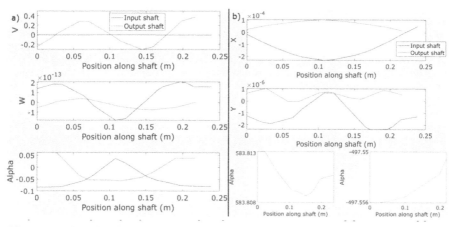

Figure 6. A) Mode shapes at the chosen system natural frequency, b) shapes of the deflected shaft (forced vibration response).

5.3 Rolling element bearing response

The behaviour of the rolling element bearings is briefly studied using the results of the first simulation in section 5.2.1. The radial deflection and its associated wavelet spectrum, as well as the loads in x- and y-directions of a single roller in the leftmost bearing of the input shaft (Figure 2) are shown in Figure 7. The timeframe used to produce the wavelet is longer than that used for the time histories, for the purposes of greater clarity.

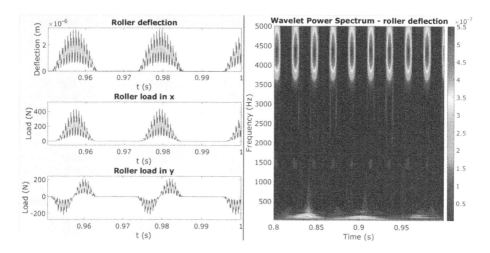

Figure 7. Time histories of bearing roller deflection and load.

The periodic nature of the deflection time history is consistent with the roller moving into and out of the loaded zone of the bearing, at the frequency of rotation of the retaining cage ($\sim$ 45.4 Hz). The higher-frequency components of this motion correspond to the gear meshing frequency ($\sim$ 1440 Hz) and its third harmonic ($\sim$ 4320 Hz).

The strength of these components in the roller deflection signal grows and disappears as the roller enters and leaves the loaded zone. The presence of these frequencies in the roller deflection spectrum agrees qualitatively with the results of Mohammad-Pour et al. (14). The alignment of the gear pair's LOA to the x-axis of the bearing leads to negligible deflection of the bearing centre in y-direction in comparison to that which occurs in x-direction. As such, a high load is taken by the roller in x-direction, the sign of which does not change. The reaction load of a single roller is radial and thus acts towards the centre of the bearing (Figure 3). Although there is negligible centre deflection in y-direction, a portion of this radial load is exerted in y-. The sign of this load changes as the roller passes through the centre of the bearing loaded zone. The sum of the loads in y-direction for all rollers in the loaded zone is close to zero, hence there is negligible resultant reaction load produced. This is consistent with the negligible deflection in the same direction that is imparted upon the bearing.

6   CONCLUSIONS

The following conclusions are drawn from the analysis presented.

1. The method of constructing the system's equations of motion combined with the damped modal analysis technique employed to solve the eigen problem provides a good picture of the system's rotordynamics in the frequency domain, including the gyroscopic effect. This is expounded by excitation of the system at a critical speed. Differences between the predicted mode shape and the shape of the deflected rotor during such excitation are likely due to the effects of modal damping in the simulation model under excitation with respect to the eigen problem.
2. Comparison between DTE results obtained from excitation near to and far from the gear pair natural frequency indicates that activation of response near to the natural frequency is achievable.
3. The rolling element bearings respond in the expected manner, with strong contributions from the gear mesh frequency, its harmonics and the cage rotation frequency to the response histories of a single roller.

The presented study of a full gearbox arrangement provides a tool through which previously established finite element rotordynamics methodologies can be applied alongside nonlinear gear mesh and rolling element bearing representations to high speed problems with complex geometry. This can act as a driver for system design and a source of boundary conditions or pre-diagnostics for bearing vibrational, misalignment and tribo-dynamic studies.

Further work would aim to complete the picture surrounding the frequency response curve near to the natural frequency of the gear pair, by simulating more speed cases and studying how the magnitude of the DTE changes between each. Experimental measurements are required in order to provide an independent validation of both the results of the eigen problem and the forced vibration response of the system.

REFERENCES

[1] Rao JS. Rotor Dynamics. 3rd ed. Delhi: New Age International Ltd.; 1996.
[2] Ozguven HN. A Non-Linear Mathematical Model for Dynamic Analysis of Spur Gears Including Shaft and Bearing Dynamics. J ofSound Vib. 1991;145 (2):239–60.
[3] Kubur M, Kahraman A, Zini DM, Kienzle K. Dynamic Analysis of a Multi-Shaft Helical Gear Transmission by Finite Elements: Model and Experiment. J Vib Acoust. 2004;126(3):398.

[4] Kahraman A, Ozguven HN, Houser DR, Zakrajsek JJ. Dynamic Analysis of Geared Rotors by Finite Elements. J Mech Des. 2008;114(3):507.

[5] Özgüven HN, Özkan ZL. Whirl Speeds and Unbalance Response of Multibearing Rotors Using Finite Elements. In: Journal of Vibration Acoustics Stress and Reliability in Design. 1984. p. 72.

[6] Zorzi ES, Nelson HD. Finite Element Simulation of Rotor-Bearing Systems With Internal Damping. J Eng Power. 1977;99(1):71–6.

[7] Nelson HD. Rotordynamic Modelling and Analysis Procedures: A Review. JSME Int J [Internet]. 1998;41(1). Available from: http://www.mendeley.com/research/geology-volcanic-history-eruptive-style-yakedake-volcano-group-central-japan/

[8] Rao JS, Shiau TN, Chang JR. Theoretical analysis of lateral response due to torsional excitation of geared rotors. Mech Mach Theory. 1998;33(6):761–83.

[9] Choi S-T, Mau S-Y. Dynamic Analysis of Geared Rotor-Bearing Systems by the Transfer Matrix Method. J Mech Des. 2002;123(4):562.

[10] Iida H, Tamura A, Kikuch K, Agata H. Coupled Torsional-Flexural Vibration of a Shaft in a Geared System of Rotors (1st Report). Bull JSME [Internet]. 1980;23(186):2111. Available from: http://www.mendeley.com/research/geology-volcanic-history-eruptive-style-yakedake-volcano-group-central-japan/

[11] Chen S, Tang J, Li Y, Hu Z. Rotordynamics analysis of a double-helical gear transmission system. Meccanica. 2016;51(1):251–68.

[12] Hu Z, Tang J, Chen S. Analysis of coupled lateral-torsional vibration response of a geared shaft rotor system with and without gyroscopic effect. Proc Inst Mech Eng Part C J Mech Eng Sci. 2018;232(24):4550–63.

[13] Theodossiades S, Natsiavas S. Periodic and chaotic dynamics of motor-driven gear-pair systems with backlash. Chaos, Solitons and Fractals. 2001;12 (13):2427–40.

[14] Mohammadpour M, Johns-Rahnejat PM, Rahnejat H. Roller bearing dynamics under transient thermal-mixed non-Newtonian elastohydrodynamic regime of lubrication. Proc Inst Mech Eng Part K J Multi-body Dyn. 2015;229(4):407–23.

[15] Kahraman A, Singh R. Non-Linear Dynamics of a Spur Gear Pair. J Sound Vib. 1990;142(1):49–75.

[16] Özgüven HN, Houser DR. Dynamic Analysis of High Speed Gears by Using Loaded Static Transmission Error. J Sound Vib. 1988;125(1):71–83.

[17] De-Juan A, Viadero F, Garcia P, Diez-Ibarbia A, Iglesias M, Fernandez-del-Rincon A. Enhanced model of gear transmission dynamics for condition monitoring applications: Effects of torque, friction and bearing clearance. Mech Syst Signal Process [Internet]. 2016;85:445–67. Available from: http://dx.doi.org/10.1016/j.ymssp.2016.08.031

[18] Viadero F, Fernandez Del Rincon A, Sancibrian R, Garcia Fernandez P, De Juan A. A model of spur gears supported by ball bearings. WIT Trans Modelling Simul. 2007;46:711–22.

[19] Theodossiades S, Natsiavas S. On geared rotordynamic systems with oil journal bearings. J Sound Vib. 2001;243(4):721–45.

[20] Hu B, Zheng M, Zhou C. Tribo-dynamics model of a spur gear pair with gyroscopic effect and flexible shaft. Forsch Im Ingenieurwes [Internet]. 2019 Sep;83 (3):367–77. Available from: https://doi.org/10.1007/s10010-019-00371-4

[21] Kang MR, Kahraman A. Measurement of vibratory motions of gears supported by compliant shafts. Mech Syst Signal Process [Internet]. 2012;29:391–403. Available from: http://dx.doi.org/10.1016/j.ymssp.2011.11.007

[22] Nelson HD. A Finite Rotating Shaft Element Using Timoshenko Beam Theory. J Mech Des. 1980;102(4):793.

[23] Meirovitch L. Analytical Methods in Vibrations. New York: The Macmillan Company; 1967.

[24] Rao SS. Mechanical Vibrations. 5th ed. Upper Saddle River, NJ: Pearson Education, Inc.; 2011.

Nonlinear analysis of hydrodynamic forces for multi-lobe bearings

C.A.A. Viana, D.S. Alves, T.H. Machado

School of Mechanical Engineering, UNICAMP, Brazil

ABSTRACT

In rotordynamics, hydrodynamic bearings are vital components. Throughout history, several types of bearing geometries have been developed since the classical cylindrical bearings are susceptible to instability at high rotating speeds and/or low loads. However, it was found that preloaded segments in lobed geometries could postpone the instability threshold. Among the bearings with fixed geometry, elliptical and three-lobe bearings are the most used. The elliptical bearing consists of two circular arcs, being these centres located in the same line. Instead, the three-lobed bearing consists of three eccentric lobes, in which the centre of each lobe is equally spaced, often producing three wedges of hydrodynamic pressure. An important parameter that characterizes these bearings is the preload, defined by the ratio between the distance of the lobe's centre of curvature and the bearing radial clearance. Another critical feature of hydrodynamic bearings is that, depending on the rotor operating conditions, the bearings may exhibit nonlinear behavior. In these cases, the dynamics of the oil film can no longer be represented by the classical linear theory. Therefore, there is a need to use nonlinear forces derived directly from the solution of Reynolds equation to model the bearings, resulting in a high computational effort, since this equation must be solved at each time instant. In this context, the goal of the present paper is to evaluate the influence of preload on elliptical and three-lobed bearings regarding the linearization of the hydrodynamic forces, defining when the linear model can, or cannot, satisfactorily represent the system. In addition, different rotor rotating speeds are evaluated to verify the influence of this operational parameter on the liner or nonlinear character of the hydrodynamic forces.

1   INTRODUCTION

In Rotordynamics, hydrodynamic bearings are vital components since they accommodate the shaft in its interior and transmit forces between rotating parts, being indicated in cases with high loads and high rotating speeds. Therefore, they are the intermediary components between the support structure and rotor that transmit vibrations from one part to another. Thus, the dynamic characteristics of the rotor are highly influenced by bearings properties, even causing instability problems that arise at high rotating speeds, especially when cylindrical geometry bearings are used. Several types of bearing geometries have been developed since classical cylindrical bearings are susceptible to instability at high rotating speeds and/or low loads. However, it was found that preloaded segments in lobed geometries could postpone the instability threshold. Among all types of bearings with fixed geometry, the elliptical and three-lobed are the most used.

The behaviour of the oil film (inside the bearing) is governed by the Reynolds equation, which consists in a classical approach for hydrodynamic lubrication problems. Several attempts to solve the Reynolds equation have been made throughout history, since it is a differential equation with no complete analytical solution. Onward progress of computer science, mainly after mid-twentieth century, made complete numerical solution becomes viable. Pinkus (1), (2) and (3) are examples of contributions towards the correct analysis, with proper boundary conditions, obtaining the pressure distribution in elliptical, partial arc and three-lobed bearings, respectively.

One of the most traditional ways for representing the dynamics of hydrodynamic bearings is through using equivalent stiffness and damping coefficients. This methodology was originally developed by Lund (4,5). To obtain these coefficients, it is necessary to linearize the hydrodynamic forces around the equilibrium position of the shaft inside the bearing radial clearance. Since then, linear coefficients became more used to represent the influence of bearings on rotating systems responses, as can be seen in (6-9).

However, hydrodynamic bearings may have a highly nonlinear nature, depending on the operating condition, in which new phenomena related to the presence of subharmonics, superharmonics or chaotic vibrations can develop. Under these conditions, the use of linear stiffness and damping coefficients to calculate the rotor's dynamic response can generate results that do not match experimental conditions revealed by real machines, since the bearings nonlinear behaviour influences the equation of motion of the entire rotating system. Many studies have been published in order to improve the understanding of nonlinearities in journal bearings, both with regard to conditions of instability that the machine can present (10,11), as well as in operational and more realistic conditions (12-17).

Still regarding the studies of nonlinear effects in hydrodynamic bearings, analysing different approaches for calculating hydrodynamic forces, Zhao et al. (18) propose three models to describe hydrodynamic forces through nonlinear coefficients. The study concluded that under conditions of high amplitude of excitation, in which linear model is no longer valid, all models were consistent, bringing results close to each other. In particular, using a model with 24 coefficients, the reliability of the calculated hydrodynamic forces was improved. Yang et al. (19) compared the hydrodynamic forces calculated analytically via dynamic coefficients. The results showed that when more orders of dynamic coefficients are included in a representative model, the approximated forces are more consistent with analytical solutions. In the same way, Alves (20) developed a model including nonlinear coefficients for calculating the hydrodynamic forces, obtained through expansion of these forces in Taylor series up to the 5th order. It was verified, through experimental validation, that the response obtained through nonlinear coefficients in situations without the presence of external controls in the system, were satisfactory, with slight results discrepancies due to effects such as misalignment and bow. Zhang et al. (21) managed to perform an analytical manipulation, by separation of variables, of the particular and homogeneous solutions of Reynolds Equation in additive and multiplicative terms, obtaining three linear differential equations. A semi-analytical model for nonlinear forces of oil films was proposed for hydrodynamic bearings with 2 axial grooves. The comparative results of forces were shown to be in agreement with the results obtained by finite difference method.

In this context, the present paper investigates differences in linear and nonlinear models for hydrodynamic forces present in elliptical and three-lobed journal bearings, with system operating under high loading conditions. Several numerical simulations were performed varying the preload parameter and the rotating speed of the system. The presented analyses are of great importance for mapping the influence of nonlinearities in different types of lobed journal bearings with fixed geometry, through assessing the influences of preload parameter on the system's dynamic response. This study is important since there are few published papers dealing with the influence of nonlinearities of hydrodynamic bearings on the behaviour of rotating systems under operating conditions.

It is also important to highlight that the results presented here can be used for imme-diate application focuses on the time domain simulation of rotors supported by lobed hydrodynamic bearings operating near nonlinear conditions. Computational model-based analysis such as fault detection and vibration control of rotating machines can be significantly more efficient by correctly identifying the presence of high nonlineari-ties, improving the quality of simulations and bringing direct benefits, specially, to the industry.

2 MATERIALS AND METHODS

2.1 Reynolds Equation and Multi-lobe Journal Bearings

Reynolds equation is considered the basis of the classical theory of hydrodynamic lubrication. This equation is formulated assuming a viscous fluid, applying together the conservation of momentum and continuity equation. The solution of the Reynolds equation, for hydrodynamic bearings, provides the fluid pressure distribution, infor-mation necessary to analyse the vast majority of basic problems in this matter. Its origin comes from the Navier-Stokes equations, adopting some simplifying hypoth-eses in the model. The lubricant is considered as a Newtonian fluid and its viscosity is constant over the entire length of the oil film, which flow is considered laminar. Fur-thermore, the radius of curvature of the bearing is much greater than the oil film thick-ness, enabling to ignore the effects of the fluid curvature. Another important simplifying hypothesis is that over the oil film thickness, the pressure is considered constant, since thickness is very small in relation to the shaft radius. Eq. (1) shows the Reynolds equation in its isothermal form, considering a dynamic loading in the journal bearing.

$$\frac{\partial}{\partial \theta}\left(h^3 \frac{\partial P}{\partial \theta}\right) + \frac{\partial}{\partial z}\left(h^3 \frac{\partial P}{\partial z}\right) = 6\mu U \frac{\partial h}{\partial \theta} + 12\mu \frac{\partial h}{\partial t} \tag{1}$$

where θ is the coordinate in circumferential direction, z the coordinate in axial direc-tion, h the oil film thickness, U the relative velocity in tangential direction between journal and bearing's surface, p is the pressure and μ the fluid viscosity. It is important to mention that in this work the cavitation model known as Swift-stieber Condition or Reynolds Condition was used.

In order to obtain the pressure field, it is necessary to calculate the oil film thick-ness in advance. This calculation differs for each type of bearing, being essential the analysis of the geometries with the load W referring to rotor weight, as can be seen in Figure (1). The distance e connecting the centres of bearing and shaft is called "eccentricity". With the radial clearance C_r, it is possible to define the "eccentricity ratio" as $n = e/C_r$. The angle ϕ between the Y direction and the line of centres is called "attitude angle".

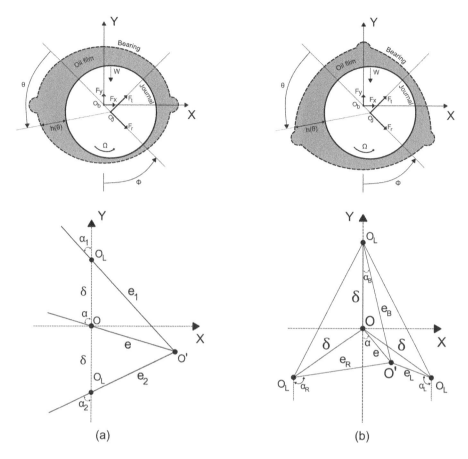

Figure 1. Basic geometry for multi-lobe journal bearings: (a) elliptical; (b) three-lobed.

Elliptical bearings consist of two circular arcs, with centres equidistant from bearing's centre by the "ellipticity" parameter δ. Defining the "ellipticity ratio" as $= \varepsilon/\delta$, it is possible to write for each lobe the "eccentricity ratio" and the "equilibrium angles" for elliptical bearings as can be seen in Eq. (2) and Eq. (3) respectively. These two parameters are necessary to calculate the oil film thickness and consequently the pressure field.

$$n_1 = \sqrt{n^2 + m^2 + 2nm\cos\phi}; \quad n_2 = \sqrt{n^2 + m^2 - 2nm\cos\phi} \tag{2}$$

$$\alpha_1 = \arcsin\left(\frac{n\,\text{sen}\,\phi}{n_1}\right); \quad \alpha_2 = \arcsin\left(\frac{n\,\text{sen}\,\phi}{n_2}\right) \tag{3}$$

The tri-lobed bearing, on the other hand, consists of the intersection of three circular arcs with centres equally spaced by the ellipticity. Equations (4) and (5) contain, respectively, "eccentricity ratio" and the "equilibrium angles" for this type of noncircular journal bearing. As well as the elliptical bearing, these parameters are necessary to calculate the oil film thickness and consequently the pressure field.

$$n_B = \sqrt{n^2 + m^2 + 2nm\cos\phi}; \quad n_R = \sqrt{n^2 + m^2 - 2nm\cos\left(\frac{\pi}{3} + \phi\right)};$$

$$n_L = \sqrt{n^2 + m^2 - 2nm\cos\left(\frac{\pi}{3} - \phi\right)} \tag{4}$$

$$\alpha_B = \arcsin\left(\frac{n\operatorname{sen}\phi}{n_B}\right); \quad \alpha_R = \frac{2\pi}{3} - \arcsin\left[\frac{n\sin\left(\frac{\pi}{3} + \phi\right)}{n_R}\right];$$

$$\alpha_L = \frac{2\pi}{3} - \arcsin\left[\frac{n\sin\left(\frac{\pi}{3} - \phi\right)}{n_L}\right] \tag{5}$$

2.2 Hydrodynamic Forces

The forces resulting from the pressure distribution of the oil film, inside the bearing, can be obtained directly by numerical integration, calculating the components of the hydrodynamic force in the local frame (radial and tangential directions), as shown in Eq. (6). Having the "attitude angle" ϕ it is possible to recalculate the hydrodynamic forces in the reference frame (x and y directions), as can be seen in Eq. (7). This methodology results in the nonlinear calculation of the forces, since it takes into account the direct integration of the pressure distribution along the lubricant.

$$\begin{Bmatrix} F_r \\ F_t \end{Bmatrix} = \int_0^{2\pi} \int_{-\frac{L}{2}}^{\frac{L}{2}} p \begin{Bmatrix} \cos\theta \\ \sin\theta \end{Bmatrix} dz R d\theta \tag{6}$$

$$\begin{Bmatrix} F_x \\ F_y \end{Bmatrix} = \begin{Bmatrix} -F_r\sin\phi + F_t\cos\phi \\ F_r\cos\phi + F_t\sin\phi \end{Bmatrix} \tag{7}$$

For calculating the linear forces, the finite disturbances technique, around the equilibrium position, proposed by Lund (4,5) is used. First, the equilibrium position is calculated, then the dynamic stiffness and damping coefficients of the bearing are determined by calculating partial derivative of the forces around this equilibrium position (Eq. (8)). Eq. (9) shows the linearized forces through a first-order Taylor series expansion, being F_{x0} and F_{y0} the forces at journal equilibrium position. $(\Delta x, \Delta y)$ and $(\Delta\dot{x}, \Delta\dot{y})$ are the disturbances in position and velocity and $(i,j) = (x \text{ or } y)$.

$$K_{ij} = \left(\frac{\partial F_i}{\partial\Delta_j}\right)_0; \quad C_{ij} = \left(\frac{\partial F_i}{\partial\Delta_j}\right)_0 \tag{8}$$

$$F_x = F_{x0} + K_{xx}\Delta x + K_{xy}\Delta y + C_{xx}\Delta\dot{x} + + C_{xy}\Delta\dot{y}$$
$$F_y = F_{y0} + K_{yx}\Delta x + K_{yy}\Delta y + C_{yx}\Delta\dot{x} + + C_{yy}\Delta\dot{y} \tag{9}$$

2.3 Solution methods

The results presented in this paper compare the forces obtained with the linear and nonlinear models for two different configurations of bearings. Thus, it is important to present the steps used in the calculation algorithms for each approach. The process starts at a certain rotating speed. For the nonlinear analysis, the procedure is given by calculating the pair eccentricity/attitude angle for each position. Once calculated these parameters, it is possible to compute the pressure field and then perform a numerical integration to obtain the nonlinear hydrodynamic forces. For the linear analysis, the procedure is somewhat different, since it is necessary to carry out the calculation of the equilibrium position for the analysed rotating speed. With the equilibrium pair (eccentricity/attitude angle), the procedure of finite perturbations is performed to obtain the dynamic coefficients matrices (stiffness and damping). Finally, the hydro-dynamic forces calculation is given as Eq. (9) in both directions. Figure (2) shows the flowchart summarizing the two model types.

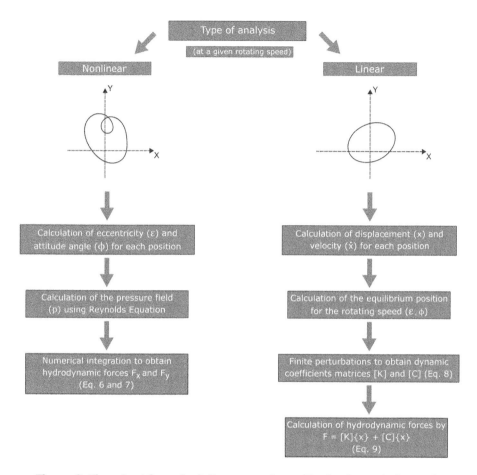

Figure 2. Flowchart for calculating procedure of hydrodynamic forces in linear and nonlinear models.

3   RESULTS AND DISCUSSION

This section presents the hydrodynamic forces resulting from the analysis considering a Laval rotor. This rotor consists of 19 nodes (18 beam elements) with two journal bearings located at nodes 3 and 17, far approximately 580 mm from each other, lubri-cated by ISO VG 32 oil at a constant temperature of 40º C (average viscosity of $\mu = 2.7622 \times 10^{-2}$ *Pa.s*). The bearings have 20 mm length, 31 mm at inner diameter with 90 μm of radial clearance. The shaft has a decentralized coupled disk placed at node 7, distant 200 m from bearing 1, with an unbalanced mass $m_0 = 3g$ located 37 mm from the disk's centre. Besides this unbalance force, it is considered an external force applied at node 7 with magnitude of 200 N, in order to simulate a severe load condition in the system. A sketch of the rotating system just described is shown in Figure (3).

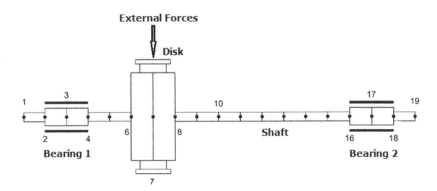

Figure 3. Configuration of the used rotor (out of scale for better visualization).

Since the system has an asymmetrical configuration, shaft's displacements inside the two journal bearings are different due to differences in loading and gyroscopic effect. Therefore, in this work, just the forces calculated in bearing 1 (closer to disk) are shown, since they contain the information at the most critical loading situation. A previous analysis of the system was carried out and the first critical speed of the simulated rotor was calculated at about 37 Hz. In this way, a set of 3 rotating speeds is chosen [15, 35, 60 Hz], in order to verify the influence of this parameter on the operational behaviour of the rotor in situations below, at the vicinity and above the first critical speed.

Two types of lobed bearings are used to generate the results – elliptical and three-lobed. In addition to the bearing type, it was picked two preload levels, being 0.25 and 0.8, simulating high and low eccentricities conditions respectively. Thus, the objective is to analyse the behaviour over time of the hydrodynamic forces in four different bearing configurations, calculated using linear and nonlinear models as described in Figure (2). The forces were calculated with a time step of 10^{-4} s, saving values for F_x and F_y every 10 loops due to their proximity. Simulations were run until the final time in which orbits reach steady state. To remove the transient regime, only the last 0.5 s of the simulation were selected.

Figure (4) shows the hydrodynamic forces for elliptical bearing with preload 0.25, in both x and y directions, calculated with linear and nonlinear models. Figure (4a) contains the analysis at 15 Hz being zoomed in Figs. (4d) and (4e) making it easy to view potential differences. The results showed no difference, considering the curves superimposed on each other. For 35 Hz, it can be seen in Figure (4b) slightly discrepancies between the curves (especially for the x direction forces), because near the critical speed region the bearing, in general, presents greater nonlinear behaviour. For 60 Hz, Figure (4c) also does not show a differentiation between linear and nonlinear models, since at this rotating speed the orbits are smaller due to the shaft self-centring effect which means that the bearing behaves mostly linearly.

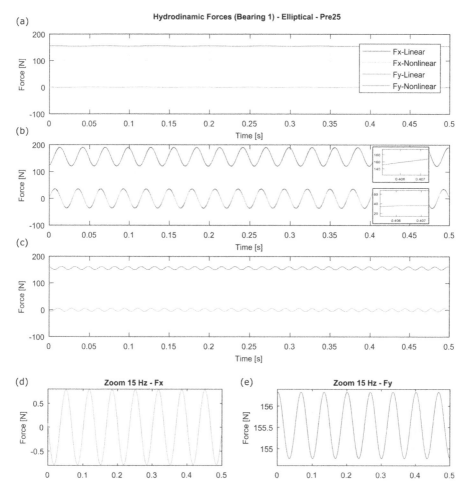

Figure 4. Hydrodynamic forces for elliptical bearing 1 (preload 0.25) – (a) 15 Hz, (b) 35 Hz, (c) 60 Hz, (d) zoom 15 Hz in x direction, (e) zoom 15 Hz in y direction.

For the elliptical bearing with preload 0.8, it is possible to notice, looking at Figure (5), that the differences are much less evident, being able to consider the results identical, since the margin of numerical error would be greater than the differences observed with the values. This is verified for all rotating speeds analysed in this bearing configuration (Figures 5a, 5b, 5c), even in the region close to the resonance frequency that presents greater amplitudes of vibration (35 Hz). With a higher preload value, the shaft goes closer to the bearing wall, which makes the oil film stiffer and, consequently, the orbits are smaller, making the bearing more linear.

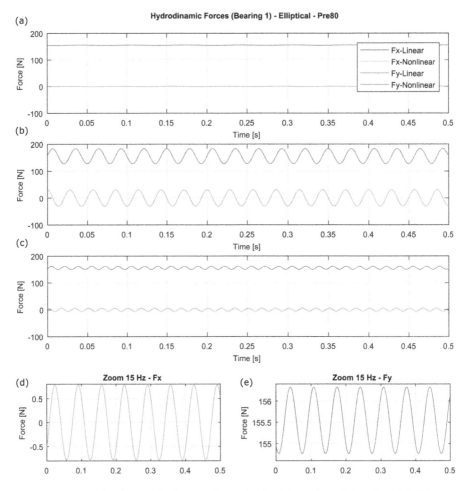

Figure 5. Hydrodynamic forces for elliptical bearing 1 (preload 0.8) – (a) 15 Hz, (b) 35 Hz, (c) 60 Hz, (d) zoom 15 Hz in x direction, (e) zoom 15 Hz in y direction.

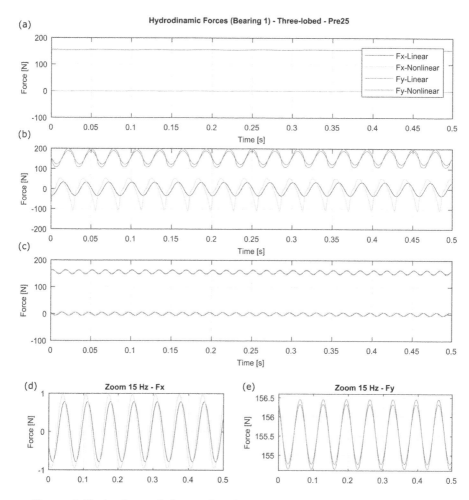

Figure 6. Hydrodynamic forces for three-lobe bearing 1 (preload 0.25) – (a) 15 Hz, (b) 35 Hz, (c) 60 Hz, (d) zoom 15 Hz in x direction, (e) zoom 15 Hz in y direction.

Changing the bearing type, Figure (6) presents the results for the three-lobed bearing with preload 0.25. In this case, differences between linear and nonlinear models appear in all compared rotating speeds. For 15 Hz, as can be seen in Figure (6a), zooming in Figs. (6d), the nonlinear forces in x direction have higher amplitude values than linear ones as well as a small lag in the nonlinear vector behaviour. In y direction, Figure (6e) it is noticed that differences in amplitude are a little smaller, but also evident. Increasing the rotating speed for 35 Hz, at first critical region, it can be observed the biggest discrepancies, with evident increase in amplitudes in nonlinear cases, both x and y directions. For nonlinear force F_x, beyond amplitude, there is either a large deformation in curve's shape. For 60 Hz the same pattern is repeated, but with similar shape of curves and minor differences in amplitude and phase, since the rotor overcame the critical region and centralizes itself, as can be seen in Figure (6c).

Finally, increasing preload for 0.8 in the three-lobed bearing the models also showed differences for this case (Figure (7)). Figures (7a, 7d and 7e) show the results for the 15 Hz rotation, in which a similar behaviour is observed to the previous case with pre-load 0.25, however the nonlinear forces in the x direction present a greater lag than in the previous case. In the other two rotations the same behaviour can be found. In Figure (7b) at 35 Hz the amplitude of the force does not increase considerably, even in the critical region, as well as at 60 Hz in figure (7c), however, the signal lag is more evident in all rotations as mentioned.

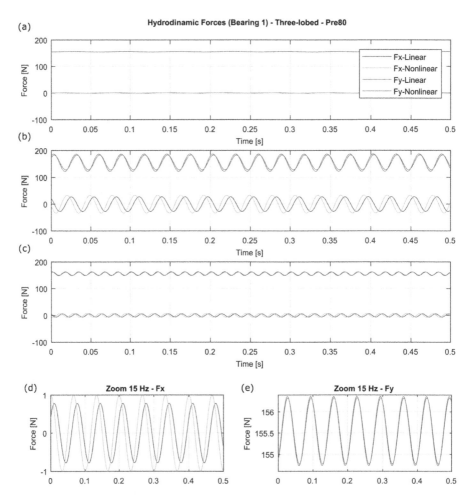

Figure 7. Hydrodynamic forces for three-lobe bearing 1 (preload 0.8) – (a) 15 Hz, (b) 35 Hz, (c) 60 Hz, (d) zoom 15 Hz in x direction, (e) zoom 15 Hz in y direction.

Finally, making a comparative analysis of the four cases analysed, it can be noticed that, with respect to geometry, the three-lobe bearing was more sensitive to the effects of nonlinearities compared to the elliptical bearing, presenting a greater discrepancy of linear and nonlinear model for hydrodynamic forces. Regarding the preload, as higher as its value, more linear is the behaviour of both bearings, mainly due to the increase in oil film stiffness to higher preload values.

4   ADDITIONAL COMMENTS

To complement the analyses carried out in the present work, new investigations are under development, and their outcome will be published in journals of the subject.

One of the aforementioned studies concerns the analysis of the response in the time and frequency domains. Regarding the orbits (temporal response), the initial analyses show that in addition to the variation in amplitude and shape of the responses, comparing the linear and nonlinear models, there is also a variation in the slope of the orbits. This variation probably explains the phase shift clearly seen on the figures presented in this paper for the three-lobe bearings between linear and nonlinear forces. Regarding frequency analysis, the calculated time series at the bearing positions are processed through DFT Full Spectrum, being able to observe the signal composition in harmonics - $\pm$ 1x, $\pm$ 2x and $\pm$ 3x. With different bearings configurations, only the +1x and -1x components were present in the linear model signal. However, when the nonlinear model was used, higher order harmonics ($\pm$ 2x, $\pm$ 3x) appeared, especially in situations with high nonlinear behaviour, such as the vicinity of critical speeds, changing orbits amplitude and shape.

Another ongoing investigation is the analysis of the rotor's behaviour when submitted to run-up. It was possible to observe a change in the fluid-induced instability threshold, which occurs significantly earlier for the nonlinear model.

5   CONCLUSION

The paper presents an investigation about the nonlinear behaviour of hydrodynamic forces in elliptical and three-lobed bearings under high static load and two different pre-load conditions for different rotating speeds. Simulations considering the direct integration of the Reynolds equation and the use of linear equivalent coefficients to represent the forces are shown.

It was possible to observe that elliptical bearings present linear behaviour except for cases close to the critical speed. When the pre-load is increased, the bearing seems to become even more linear, being both models presenting similar responses. In this case it is possible to say that the linear model can well represent the behaviour of elliptical bearings.

However, the linear model is not suitable to be used in three-lobed bearings, since the nonlinear hydrodynamic forces are considerably different from the linear one. In these cases, not even the amplitude of the forces differ, but also its phase, what will change the dynamics of the machine. Thus, for vibration control or fault detection on machines using three –lobed bearings the nonlinear bearing model is recommended.

ACKNOWLEDGEMENTS

The authors would like to thank Coordination of Superior Level Staff Improvement – CAPES, National Council for Scientific and Technological Development – CNPq, grant # 424899/2018-3, and grants # 2017/07454-8 and # 2018/21581-5 from the São Paulo Research Foundation (FAPESP) for the financial support to this research.

REFERENCES

[1] Pinkus, O.: Analysis of Elliptical Bearings. Transactions of ASME, v. 78, p. 965–973, 1956.

[2] Pinkus, O.: Solution of Reynolds Equation for Finite Journal Bearings. Transactions of ASME, v. 80, p. 858–864, 1958.

[3] Pinkus, O.: Analysis and Characteristics of Three-lobe Bearings. Journal of Basic Engineering, p. 49–55, 1959.

[4] Lund, J. W., Thomsen, K. K.: A Calculation Method and Data for the Dynamic Coefficients of Oil-Lubricated Journal Bearings. Topics in Fluid Bearing and Rotor Bearing System Design and Optimization, ASME, p. 11–28, 1978.

[5] Lund, J. W.: Review of the Concept of Dynamic Coefficients for Fluid Film Journal Bearings. ASME Journal of Tribology, v. 109, p. 37–41, 1987.

[6] Qiu, Z. L., Tieu, A. K.: The Effect of Perturbation Amplitudes on Eight Force Coefficients of Journal Bearing. Tribology Transactions, v. 39(2), p. 469–475, 1996.

[7] Müller-Karger, C. M., Granados, A. L.: Derivation of Hydrodynamic Bearing Coefficients Using the Minimum Square Method. Transactions of the ASME, v. 119, p. 802–807, 1997.

[8] Zhao, S. X., Zhou, H., Meng, G., Zhu, J.: Experimental Identification of Linear Oil-Film Coefficients using Least-Mean-Square Method in Time Domain. Journal of Sound and Vibration, v. 287, p. 809–825, 2005.

[9] Li, K., et al., Identification of oil-film coefficients for a rotor-journal bearing system based on equivalent load reconstruction. Tribology International, v.104, p. 285–293, 2016.

[10] Hattori, H.: Dynamic Analysis of a Rotor-Journal Bearing System with Large Dynamic Loads (Stiffness and Damping Coefficients Variation in Bearing Oil Films). JSME International Journal, Series C, v. 36 (2),p. 251–257, 1993.S.

[11] Hu, A., Hou, L., Xiang, L.: Dynamic Simulation and Experimental Study of an Asymmetric Double-Disk Rotor-Bearing System with Rub-Impact and Oil-Film Instability". Nonlinear Dynamics, v. 84, p. 641–659, 2016.

[12] Dakel, M., et al., Dynamic analysis of a harmonically excited on-board rotor-bearing system. 10th IMechE International Conference on Vibrations in Rotating Machinery (VIRM10), Sep 2012, London, United Kingdom. pp.C1326/024.S.

[13] Smolík, L. et al., Threshold stability curves for a nonlinear rotor-bearing system. Journal of Sound and Vibration, v. 442, p. 698–713, 2019.

[14] Machado, T., et al., Discussion about nonlinear boundaries for hydrodynamic forces in journal bearing. Nonlinear Dynamics, v. 92, p. 2005–2022, 2018.

[15] Li, C., She, H., Tang, Q.: "The Effect of Blade Vibration on the Nonlinear Characteristics of Rotor-Bearing System Supported by Nonlinear Suspension. Nonlinear Dynamics, v. 89, p. 987–1010, 2017.

[16] Arumugam, P., Swarnamani, S., Prabhu, B.S.: Effects of Journal Misalignment on the Performance Characteristics of Three-Lobe Bearings. Wear, v. 206 (1-2), p. 122–129, 1997.

[17] Prabhakaran Nair, K., Sukumaran Nair, V. P., Jayadas, N. H.: Static and Dynamic Analysis of Elastohydrodynamic Elliptical Journal Bearing with Micropolar Lubricant. Tribology International, v. 40(2), p. 297–305, 2007.

[18] Zhao, S. X., et al., An experimental study of nonlinear oil-film forces of a journal bearing. Journal of Sound and Vibration, v. 287, p. 827–843, 2005.

[19] Yang, L., Wang, W., Yu, L., Nonlinear dynamic oil-film forces in infinite-short journal bearings. International Joint Tribology Conference (2011), pp.61045/119-121.

[20] Alves, D. S., Wu, M. F., Cavalca, K. L., Application of gain-schedule vibration control to nonlinear journal-bearing supported rotor. Journal of Sound and Vibration, v. 442, p. 714–737, 2019.

[21] Zhang, Y., et al., A semianalytical approach to nonlinear fluid film forces of a hydrodynamic journal bearing with two axial grooves. Applied Mathematical Modelling, v. 65, p. 318–332, 2019.

Some further reflections on misalignment

A.W. Lees

College of Engineering, Swansea University, Bay Campus, Crymlyn Burrows, Swansea, UK

ABSTRACT

The work here develops the concepts outlined by the author at the conference of this series in Manchester. A study is presented of an investigation into the generation of harmonics of shaft rotation speed in the vibration of misaligned multi-bearing rotor systems. The author challenges the conventional explanation that these harmonics arise principally from the bearing non-linearity. Whilst this may give rise to part of the observed behaviour, a more fundamental source of the harmonic terms arises from the properties of the coupling joining the rotors. The operation of a flange coupling is analysed with particular emphasis on the link between transmitted torque and lateral vibration. A calculation route is presented together with an example of the resulting response.

Keywords: Rotors, Machines, Faults, Misalignment.

1   INTRODUCTION

It is surprising that, despite extensive developments in the understanding of rotor behaviour over recent decades, the behaviour of misaligned rotors remains an area where clear analysis is somewhat lacking. This is not merely an interesting academic point but one of considerable industrial importance as rotor misalignment is generally agreed to be the second most important fault in rotating machinery; only rotor imbalance has greater significance. But whereas the effects of rotor imbalance are now clearly appreciated, the same is not true for misalignment. This point was discussed by Jalan and Mohanty [1]. To directly quote from their paper 'In defiance of its importance, the study of shaft misalignment is still inadequate.'

There has been considerable confusion in the literature, and in industry, on the general features of a misaligned rotor reflecting the fact that it remains incompletely understood. This is partially a result of the complex nature of the problem. Whilst many authors have associated the presence of harmonics of shaft speed (i.e. the generation of vibration signals at multiples of shaft rotation speed) with misalignment, [2-7], Al-Hussain and Redmond [3] did not find evidence in their model and were driven to the conclusion that non-linearities in either the rotor or the bearings/support structure must be the origin of the phenomena. The source of these harmonics is not entirely clear but most authors have associated the phenomena with the non-linearity of the bearings. Redmond [7] has addressed the problem again more recently but still views the bearings as the main source on harmonics generation. Lees and Penny [8] presented a calculation method based on the analysis of Adiletta et al [9] but such studies do not fully represent the complexity of the situation. This is not a completely satisfactory explanation.

More recently Dal and Karkacay [10] considered a system with pressurised bearings but their analysis relies on the bearing non-linearity to generate harmonics. In a real machine, undoubtedly non-linearities in the bearing will make some contribution, but the objective of the present paper is to illustrate that there are more fundamental origins of the behaviour. Pennacchi *et al.*[11] considered a system with non-linear bearings and cyclic properties in the coupling. It is worth noting, however, that the analysis does not consider the transmitted torque. Similarly, Li et al [12] made no mention of the transmitted torque. Sekkar and Prabbu [13], however, did include the effect of transmitted torque in their calculations building on the work of Gibbons [14]. This work does indeed generate harmonic response at twice shaft speed, but this is the only harmonic. Nembhard and Sinha [15] have used a similar basis for the fault diagnosis studies. Rig trials by Patel and Darpe have shown the three times shaft speed term to be higher [16]. It is interesting to consider the importance of transmitted torque; recently monitoring torque has been proposed by Reddy and Sekkar [17] as a detection means for misalignment. An alternative strategy has been followed by Lal and Tiwari [18] who presented a method for the identification of coupling forces and moments from plant data.

In 2007, Lees [19] offered a different analysis. Using a rather idealised representation of a rigid coupling, the paper studied the Lagrange equations of the system and showed how the overall stiffness matrix becomes time dependent, hence generating a range of higher frequency components in the vibration spectrum. Whilst this study was restricted is scope, it did offer a means of considering the wider implications. Many authors report the generation of harmonic terms in the vibration signal but most have reported the dominance of the twice per revolution component, contradicting the results of Patel and Darpe [16]. These variations in themselves suggest that the phenomena are rather complex and this paper seeks to explore the generation of the harmonics.

In this paper, the factors influencing the dynamics of machines with flange couplings are examined and a complex pattern of behaviour emerges. This complicated performance and the significant number of factors which influence it may partially explain why this area of machine dynamics has remained obscure for so long. Some of the ideas have been reported in a simplified form previously (Lees, [20,21]) but here the models are extended using more realistic assumptions. Lees [21] proposed a model in which torque became a combination of moments and forces owing to the non-uniform distribution of friction across the face of the coupling. In [21] it was assumed that friction was proportional to the normal force, which of course, is only a limiting case. At the end of that paper an example was presented with limited friction, but the general case requires a somewhat different approach. This introduces a fundamental element into the study in that the response to misalignment becomes dependent on the transmitted torque, not only in amplitude, but also in spectral content. This is the subject of the current paper.

2 FLANGE COUPLINGS

2.1 Operation

Clearly the couplings play a crucial role in determining the dynamic behaviour of a misaligned machine. Ideally, they will transmit only torque but no forces or bending moment – but often this will not be the case. A simplified view of a flange coupling is shown in Figure 1: In this sketch only two bolts are shown but on a real machine there will be many, perhaps 16 or 32, but their essential role is to maintain an axial load between the mating surfaces of the coupling.

During operation, torque will be transmitted partly through friction between the mating faces and, at high torque levels, partly via shear forces in the bolts. Lees [19] considers the case in which torque is transmitted predominantly through the bolts and it was shown using Lagrange's Equation that the effective stiffness of the system becomes time dependent. At lower torques, where the friction terms predominate, [20,21] gave a basic analysis of the situation and the purpose of the current paper is to refine and extend this study.

Clearly the performance of the coupling is dependent on the interfacial pressure, P, between the mating surfaces and this is determined by the tension in the bolts and the coupling geometry. Given this uniform pressure distribution across the face, it is straightforward to calculate the maximum torque that can be transmitted. This is given by

$$T_c = \frac{2}{3}\mu P \pi r^2 \tag{1}$$

where P is the pressure, μ is the friction coefficient and r is the radius at which the coupling bolts are positioned.

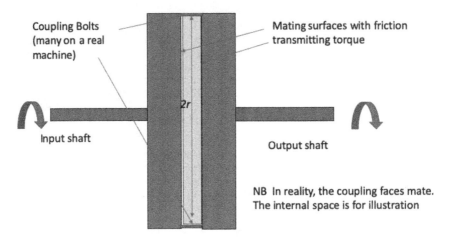

Figure 1. Idealised flange coupling.

In any real rotor system there will be bending moments orthogonal to the rotor which will distort the pressure profile from uniformity and reduce the torque which can be transmitted by friction. Generally, there will be a bending moment acting, one arising from imbalance and this will rotate with the shaft, the other (if present) arises from misalignment and will be fixed in space. These two bending moments are designated Γ_u and Γ_s respectively and using a finite element model of the rotor, they can be readily calculated using the procedure outlined in the appendix. Note that the misalignment moment will be subject only to a minor variation with rotor speed (arising from bearing properties) whereas the imbalance term varies substantially and in a complex manner. Hence at any time instant the rotor is at orientation $\theta = \omega t$, the resultant bending moment is

$$\Gamma = \left\{ \begin{array}{l} \Gamma_{sx} + \Gamma_{ux}cos(\theta - \varphi) \\ \Gamma_{sy} + \Gamma_{uy}sin(\theta - \varphi) \end{array} \right\} \tag{2}$$

where the phase shift φ is determined by the dynamics of the system. Hence the magnitude of the bending moment acting on the coupling face is given by

$$|\Gamma| = \sqrt{\left(\Gamma_{sx} + \Gamma_{ux}cos(\theta - \varphi)\right)^2 + \left(\Gamma_{sy} + \Gamma_{uy}sin(\theta - \varphi)\right)^2} \tag{3}$$

and the phase is

$$\phi = tan^{-1}\left(\frac{\Gamma_{sy} + \Gamma_{uy}sin(\theta - \varphi)}{\Gamma_{sx} + \Gamma_{ux}cos(\theta - \varphi)}\right) \tag{4}$$

The crucial point to make is that this time varying bending moment imposes a linear variation on the interfacial pressure profile in the direction of the resultant moment. Hence this resultant is constantly changing both magnitude and direction.

In Lees, [20, 21] the basic argument was presented with the simplifying assumption that friction is proportional to the pressure whereas a more realistic assumption makes the calculation a little more involved, but the principle is unchanged. In the present study, friction per unit area at any point of the mating surface is taken to be less than or equal to the coefficient of friction multiplied by the pressure. Hence, using a local frame of reference in which the y-axis is aligned with the maximum bending moment, it may be assumed that the profile of the friction is as shown in Figure 2, that is constant for $x < 1 - \delta$ and reducing linearly for $x > 1 - \delta$ (so long as it stays positive). As illustrated in Figure 2, δ is the radius at which the pressure begins to decrease and this is one of the parameters to be determined. This distribution of the friction would appear reasonable but it is a point that should at some stage be checked experimentally, however this may require some careful design of the facility.

At this stage, the transition level, $x = \delta$, between constant friction and linear variation is unknown but it clear that at this point, the friction force per unit area is

$$F(\delta) = \mu P(\delta) \tag{5}$$

Given a linear pressure variation due to the bending moments, it is convenient to express this as

$$P(x) = P_0\left(1 - \frac{|\Gamma|}{\Gamma_0}x\right) \tag{6}$$

where the moment scaling factor Γ_0 is the bending moment which reduces the pressure to zero at the outer edge of the coupling face and where x is measured along the (local) direction orthogonal to the bending moment. Letting $\alpha = \frac{\Gamma}{\Gamma_0}$ this is re-written as

$$P(x) = P_0(1 - \alpha x) \tag{6a}$$

Γ_0 can be calculated as

$$\Gamma_0 = \int_0^\pi P_0(1 - r'sin\phi)r'sin\phi cosdr'd\phi = \frac{4}{9}P_0 r^3 \tag{7}$$

In [21] a specific example was given of a reduced friction study, but in general a more detailed discussion of the determination of the transition point is needed. Note that δ is measured from the centre of the disc and hence $\delta = 1$ means there is no transition.

Across the full face of the coupling, the interface pressure is given by equation 6a. What does vary in the two regions (i.e. $x < \delta$ and $x > \delta$) is the friction which transmits the torque, provided this is below a critical value. The force per unit area takes the form

$$F(x) = F_0 \qquad x < \delta$$
$$F(x) = F_0(1 - a[x - \delta]) \quad x > \delta \tag{8}$$

The factor $F_0 \leq \mu P_0$ takes on the values required to transmit the torque. In this expression the coordinate system is local and instantaneous, so that the overall bending moment is about the (local) x -axis. This seemingly

complex procedure reduces the problem to the determination of the two parameters F_0, δ, since a is determined by the imposed bending moment. Here the direction of the gradient is given by ϕ (Equation 4). Note, however, that this non-uniform pressure profile will reduce the torque capacity of the coupling, in which case load would be transferred to the bolts. This capacity reduction factor is denoted R_c given by

$$R_c(a, \delta) = \int\limits_{-1}^{1} \int\limits_{-\sqrt{1-y^2}}^{\sqrt{1-y^2}} \frac{F(y)}{F_0} \frac{y^2}{\sqrt{x^2 + y^2}} dx dy \tag{9}$$

For any position of the transition point, δ, and the normalised bending moment, a, is readily calculated by integrating the frictional forces across the face. For any instant in the motion, the bending moment is calculated and using this the appropriate values of a, δ are evaluated using the approach described in the following section. Once these two parameters are established, however, several features follow as a consequence.

If $0 < \delta < 1$ then the centre of torque (i.e. the point in the cross section at which there is no net lateral force) will deviate from the centre of gravity. The forces of friction will not be symmetrical about the centre line and there will be a net force in the direction of the resultant bending moment, which varies, but not in a uniform manner.

$R_c(a, \delta)$ is shown in Figure 3 and this function can be tabulated prior to the main calculations. For clarity, this figure shows only a reduced resolution of the actual tabulation. Since this (or any other credible) pattern of friction is not symmetrical about the (local) x-axis, the frictional forces will exert a resultant force on the geometric centre of the coupling, meaning that a force will act on the shaft. The local force (at right angles to the direction of the pressure gradient) may be tabulated as

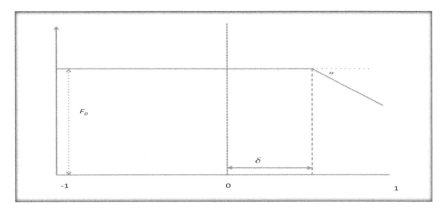

Figure 2. Friction Profile.

92

$$f_{loc}(\alpha,\delta) = \int\limits_{\delta}^{1} \int\limits_{-\sqrt{1-y^2}}^{\sqrt{1-y^2}} F_o \alpha \frac{y^2}{\sqrt{x^2+y^2}} dxdy \tag{10}$$

2.2 Determination of the parameters a, δ

The forces acting at the coupling are completely determined by the three parameters F_0, a, δ. Of these, the bending moment applied, a is known but the other two parameters must now be established.

The force level F_0 is readily calculated as

$$F_0 = \frac{T}{T_c R_c(\alpha,\delta)} \mu P \tag{11}$$

But at the transition point, the interfacial pressure will be $P_0(1 - \alpha\delta)$ and as friction here will be limiting

$$F_o = \frac{T}{T_c R_c(\alpha,\delta)} \mu P_0 = \mu P_0(1 - \alpha\delta) \tag{12}$$

Hence

$$\left(1 - \frac{T}{T_c R_c(\alpha,\delta)}\right) = \alpha\delta \tag{13}$$

Using interpolation, this non-linear equation is readily solved to yield the transition point δ. Once this is established the resulting forces are completely determined. Note that solutions for δ exist only for cases in which the coupling is not overloaded, i.e.

$$\frac{T}{T_c R_c(\alpha,\delta)} \leq 1 \tag{14}$$

If overloading occurs, some load will be carried by the bolts and this in itself can give rise to harmonic excitation (Lees [7]). A comprehensive method to include this effect has yet to be developed. In the present study, the 'excess' torque is neglected. There is considerable uncertainty in this case and a detailed three-dimensional model together with rig trial may be necessary to resolve this issue.

3   THE CALCULATION
As a prelude to the main calculations, three basic functions are tabulated for use in interpolations. These functions are;

    a) the torque capacity as a function of the bending moment, a, and the transmitted torque, or rather the remaining margin, R_c. This is shown in Figure 4.
    b) The local force as a function of the bending moment, α, and the transition point, δ.

      This is also shown in Figure 3.

    c) The position of the transition point, δ, as a function of the applied bending moment and the torque margin available.

All of these functions depend strongly on the physical parameters of the coupling.

At each rotor speed considered, a sequence of calculation steps is followed.

1) The static and rotating bending moments U_s and U_r are calculated using the approach given in the appendix. Note that the static term may only show a minor variation with speed arising from the variation in bearing stiffness. In the example studied here, U_s has been taken as constant. Both bending moments are now scaled using Equation 7.

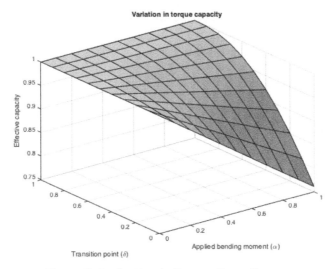

Figure 3. Reduction in Torque Capacity, R_c.

2) For a number of time steps within 1 revolution, calculate the magnitude of the resultant bending moment, a, and its direction, φ. In this study, ten time steps were used.
3) Using this value of a, determine the transition point, δ, at this instant. This is basically solving equation 13 and interpolated the capacity function.
4) The instantaneous force is now interpolated from the predetermined force function, calculated from Equation 14.
5) The force is now transformed into the global coordinates as

$$\begin{Bmatrix} F_x \\ F_y \end{Bmatrix} = P_0 \frac{T}{T_c R_c(a, \delta)} \times r^2 \times f_{loc}(a, \delta) \begin{Bmatrix} cos\phi \\ sin\phi \end{Bmatrix} \tag{15}$$

6) Having completed these calculations at each step in the cycle, a number (say 10) identical records are concatenated in order to improve the resolution of the ensuing Fourier Transform. The assembled signal then goes through FFT to yield the harmonics of the particular rotor speed.

4   AN EXAMPLE

The machine shown in Figure 5 is considered to examine the effects of the process out-lined above. For the sake of convenience, attention is restricted to a four-bearing machine comprising a single stage turbine driving an alternator. The overall length of the machine is 17m, and the bearings are plain journals with a clearance of 60μm. The mass of the two rotors is 24 and 50 tonnes respectively. A range of transmitted torques were examined corresponding to 60,70, 80 and 90% of the capacity as given by Equation 1. In practice, of course, a turbine generating 125 MW would have more than a single stage, but this model is taken for the sake of simplicity. Imbalance is imposed on both rotors: the resulting response is as shown over the running speed in Figure 6.

Clearly, the parameters of the coupling is of crucial importance here and the parameters chosen were Number of bolts 16 Pitch diameter 0.8m Bolt diameter 40mm Bolt tightness 90% yield Coefficient of friction 0.6

The bearings in this example are set in a straight line. Clearly this is inappropriate as self-weight bending imposes moments at the coupling, which lies between nodes 9 and 10 in the model. This model is rather stiffer than a typical machine and consequently the catenary of the alternator section is reduced. This is suitable to demonstrate the principles as the couplings can be connected in a simple straight-line datum. In a real machine, bearings heights must be adjusted to enable couplings to be connected. However, alignment errors can, and do, occur due to thermal movements.

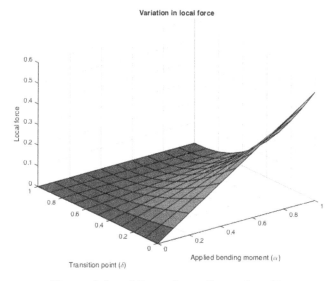

Figure 4. Local Force (non-dimensional).

(Reduced mesh for clarity)

The rotordynamic calculations were completed using the MATLAB toolbox given by Friswell *et al* [22]. The model uses Timoshenko beam elements and linearised representations of the oil journal bearings, the loading of which was determined from the rotor masses. Some variation in this loading is an obvious next step.

With the rotors set out in a horizontal alignment, the bending moment at the coupling is given by $U_s = \{\, -4.87 \quad 1.53\,\}^T \times 10^4$Nm., and this is converted to non-dimensional terms using equation 12. Note that on a real machine this will vary slightly with speed owing to the varying bearing coefficients. In the present study, however this fixed value has been used.

Using the approach described in the preceding section the magnitude of the forces acting on the coupling faces are calculated and these are as a function of rotor speed and are shown in Figure 7. Recognising space limitations, only the vertical forces at bearing 3 are presented: being close to the coupling, it is likely that these are representative. It is straightforward to evaluate force at all bearings. Note that there is a strong dependence on transmitted torque. The reasons for this behaviour become clear in Figure 8 which shows the manner in which the transition point varies in time and with torque. In this figure, the ratio mentioned is just $T/T_c\backslash$.

Table 1. Main parameters of model.

Bearing Locations	1	1 m
	2	4 m
	3	5.5 m
	4	15.5 m
Rotor 1 details		
Main body diameter		0.8 m
Main body length		2 m
Stub diameter		0.6 m
Rotor 2 details		
Main body diameter		1..2m
Main body length		6 m
Stub diameter		0.6 m
Bearing Clearance		0.5 mm
Oil viscosity		0.04 Ns/m^2

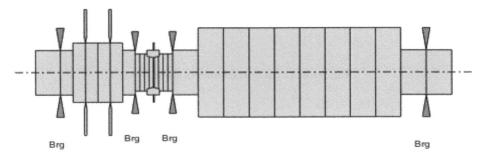

Brg   Brg   Brg   Brg

Figure 5. Rotor Model.

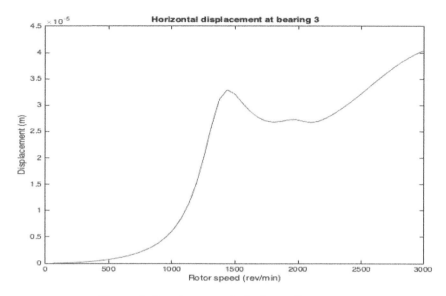

Figure 6. Synchronous Imbalance Response.

5   DISCUSSION

Rotor misalignment has been an important industrial consideration for many years and it has long been associated with the generation of harmonics of rotor speed. A full explanation of the phenomena has not yet emerged but the majority of authors have ascribed the harmonics to the non-linearity of journal bearings [12]. There is little doubt that the bearings will play a part but in this study the bearing non-linearity has been omitted to emphasize the presence of other factors.

It is shown here that flange couplings (and, no doubt, other types) show non-linear behaviour that can give rise to excitation at various multiple of shaft rotation fre-quency. To analyse this situation fully would require a complete three-dimensional finite element model of the coupling which may be of limited applicability in practice. To overcome this, a simplified analysis is presented here aimed at clarifying the phys-ics of the situation. Further work is needed in two areas firstly to confirm or modify the friction distribution across the face of the coupling (Figure 7) and secondly to combine this work with non-linear bearings. A third activity may also be added aimed at incorp-orating the analysis of the high torque case discussed by Lees [19]). An interesting question arises as to the possibility of incorporating the complete range without the need for a fully three-dimensional analysis.

The friction profile shown in Figure 2 is an assumption. A significant problem is that on real plant there is little detailed knowledge of how components will behave. This indi-cates the pressing need for some rig testing on this topic an it is one of the author's objective here to highlight the feasible mechanism and support laboratory work in this area. The uncertainty of the profile leads to doubts over the numerical values pre-dicted, but these are less important than the two key features of the model predictions namely a) the amplitude and frequency composition of the response of a misaligned system is dependent on the transmitted torque, and b) a range of harmonics develops. Whilst the numerical values will be model dependent, any credible friction profile will lead to a similar conclusion.

It may be argued that a full three-dimensional model is needed to fully investigate the phenomena. This would be particularly advantageous at higher torques when the torsional load is shared with the bolts. But the model has to be constructed with appropriate inclusion of the effects discussed in this paper (a mixture of bolt shear and friction as torque transmission mechanisms). However, detailed models can be slow in operation and simplified models, as presented here, offer an effective means of illustrating principles and exploring sensitivities, even if being less precise in numerical predictions. Two requirements for further work are clear: firstly, a detailed model (but note, the difficulty in specifying the form of the friction remains) and secondly, rig testing, including a study of the dependence of harmonics with transmitted torque.

The study presented here gives an account of how the performance of the coupling itself can give rise to harmonic forces and it offers a number of interesting insights. From this study it is clear why the ideal situation is one in which there is zero shear force and bending moments at the coupling. Furthermore, it shows that the phenomena are multi-factorial depending on misalignment, imbalance, coupling parameters and transmitted torque. This complexity probably explains why there has been a lack of agreement in the discussions of misalignment. It is far more complex than imbalance, which is simply dependent on the mass eccentricity.

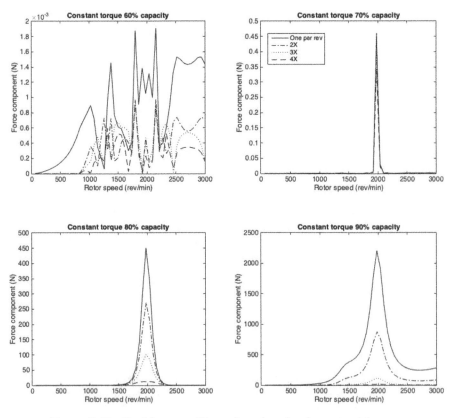

Figure 7. Vertical forces with various levels of constant torque.

Whilst any direct diagnostic technique for machines appears to be some way off (if at all possible), the ideas presented here form a useful basis. For example, the variation of the second harmonic with transmitted torque may provide important information to a machine operator. The study also provides some insights into some design requirements of the couplings used.

6   CONCLUSIONS

The study has sought to clarify some of the confusion in the literature on the phenomena arising in rigidly coupled, multi- rotor machines. In particular

a) Harmonics of shaft speed can be generated by the coupling
b) These forces show a complicated variation with respect to rotor speed.
c) The form of the frequency content depends on the interaction of misalignment, imbalance and torque.
d) Other strong influences are coupling geometry and tension in the coupling bolts.
e) The complex behaviour may explain the lack of clarity in the literature to date.

A number of areas in which further work is needed have been identified.

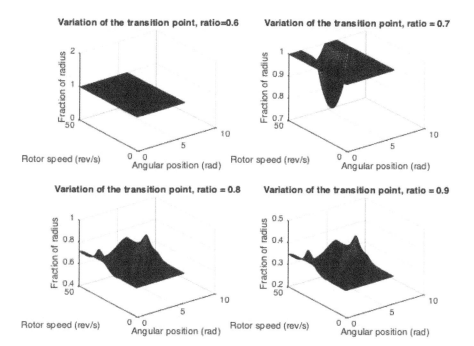

Figure 8. Transition point at differing torque levels.

APPENDIX

Evaluation of Bending Moments

The catenary of the rotor is easily calculated as

$$\mathbf{Y_c} = \mathbf{K}^{-1}\mathbf{MG} \tag{A1}$$

where $\mathbf{K}$ and $\mathbf{M}$ are the stiffness and mass matrices. $\mathbf{G}$ is a vector having the value g (9.81m/s) in vertical components and zero elsewhere. Having established this deflection, the static bending moment is calculated following the procedure described below. The imbalance response at speed Ω is given by

$$\mathbf{x} = \left[\mathbf{K} + j\Omega\mathbf{C} - \Omega^2\mathbf{M}\right]^{-1}\mathbf{M}\Omega^2\mathbf{e} \tag{A2}$$

where $\mathbf{e}$ is the mass eccentricity vector and C is the damping matric. Usually there will only be measurements of the rotor motion available at or close to the bearings, but the model may be used to 'recover' estimates of the motion at all the other nodal points, subject of course to all the usual concerns on model accuracy and other uncertainties. Let us say the coupling being studied is at node m of the model, then the four degrees of freedom at the coupling will be $[4m-3, 4m-2, 4m-1, 4m]$ and to calculate the bending moment the motion of nodes m-1 and m+1 must be considered.

For the sake of clarity the Euler beam formulation is used here. The displacements within the chosen elements are given

$$\begin{aligned} u(z_e) &= a_0 + a_1 z_e + a_2 z_e^2 + a_3 z_e^3 \\ v(z_e) &= b_0 + b_1 z_e + b_2 z_e^2 + b_3 z_e^3 \end{aligned} \tag{A3}$$

z_e is the axial position within the element. Now the bending moments in the two orthogonal directions are given by

$$\Gamma_x = EI\left(\frac{\partial^2 v}{\partial z_e^2}\right) = EI(2b_2 + 6b_3 z_e) \qquad \Gamma_y = EI\left(\frac{\partial^2 u}{\partial z_e^2}\right) = EI(2a_2 + 6a_3 z_e) \tag{A4}$$

So the problem now is to determine the parameters a and b and to do this it is convenient to consider the two directions separately. Taking the first direction, knowing the displacements at the two nodes of the element, the element parameters may be expressed as

$$\begin{Bmatrix} a_0 \\ a_1 \\ a_2 \\ a_3 \end{Bmatrix} = \begin{bmatrix} 1 & 0 & 0 & 0 \\ 0 & 1 & 0 & 0 \\ 1 & z_e & z_e^2 & z_e^3 \\ 0 & 1 & 2z_e & 3z_e^2 \end{bmatrix}^{-1} \begin{Bmatrix} u_{4m-7} \\ u_{4m-4} \\ u_{4m-3} \\ u_{4m} \end{Bmatrix} \tag{A5}$$

Similarly, the b parameters can be written in terms of u 4m-6, 4m-5, 4m-2 and 4m-1. With these results the bending moments in the element immediately before the coupling is defined and using the same approach those in the element after the coupling can also be evaluated. Note that since only displacement and slope are continuous across element boundaries, different estimates for bending moments and forces are obtained in the two elements. The optimum estimate is simply the mean of the corresponding values.

Hence, given an imbalance on the rotor, there will be a rotating bending moment vector at the coupling, or to be precise, at the flanges of the coupling.

REFERENCES

[1] A.K. Jalan, A.R. Mohanty, Model based fault diagnosis of a rotor–bearing system for misalignment and unbalance under steady-state condition, J. Sound Vib., 327 (3–5) (2009) 604–622.

[2] Dewell, D.L., Mitchell, L.D., "Detection of misaligned disk coupling using spectrum analysis", *Journal of Vibration, Acoustics, Stress & Reliability in Design*, 106, 1984,pp. 9–16.

[3] Al-Husain, K.M., Redmond, I., "Dynamic response of two rotors connected by rigid coupling with parallel misalignment", *Journal of Sound and Vibration*, 249 (3), 2002, pp. 483–498.

[4] Krodkiewski, J.M., Ding, J., "Theory and experiment on a method for on-site identification of configuration of multi-bearing Systems", *Journal of Sound and Vibration*, 164, 1993, pp. 281–293.

[5] Xu, M. Maragoni, R.D, "Vibration analysis of a motor-flexible coupling rotor system subject to misalignment and unbalance –Part 1: Theoretical Model Analysis", *Journal of Sound and Vibration*, 176, 1994, pp. 663–379.

[6] Xu, M. Maragoni, R.D, "Vibration analysis of a motor-flexible coupling rotor system subject to misalignment and unbalance –Part 2: Experimental Validation", *Journal of Sound and Vibration*, 176, 1994, pp. 681–691.

[7] Redmond I, "Study of a misaligned flexibly coupled shaft system having nonlinear bearings and cyclic coupling stiffness – Theoretical model and analysis", *Journal of Sound and Vibration*, 329, 2010, pp. 700–720.

[8] Lees, A.W. and Penny, J.E.T., "The Development of Harmonics in Rotor Misalignment", I.Mech.E. Conference "Vibrations in Rotating Machinery", Exeter, September 2008.

[9] Adiletta, G., Guido, A.R., and Rossi, C., "Chaotic motions of a rigid rotor in short journal bearings", *Non-linear Dynamics*, 10, pp251–269, 1996.

[10] Dal, A. and Karacay, T, "Effects of angular misalignment on the performance of rotor-bearing systems supported by externally pressurized air bearing", Tribology International, 111, pp 276–288,2017.

[11] Pennacchi,P., Vania, A. and Chaterton, S., "Nonlinear effects caused by coupling misalignment in rotors equipped with journal bearings", *Mechanical Systems and Signal Processing*, 30, pp.306–322, 2012.

[12] Li, H.L.. Chen, Y.S., Hou, L. and Zhang, Z.Y., "Periodic response analysis of a misaligned rotor system by harmonic balance method with alternating frequency/time domain technique", Science China-Technological Sciences, 59 (11), pp.1717–1729, 2016.

[13] Sekhar, A.S. and Prabhu, B.S., "Effects of coupling misalignment on vibrations of rotating machinery", Journal of Sound & Vibration, 185, pp. 655–671, 1995.

[14] Gibbons, C.B., "Coupling misalignment forces", *Proceedings of the 5th Turbomachinery Symposium*, Turbomachinery Laboratory, Texas A&M University, College Station, TX, October 1976.

[15] Nembhard, A. and Sinha, J.K., "Comparison of experimental observations in rotating machines with simple mathematical simulations" *Measurement*, 89, pp.120–136, 2016.

[16] Patel, T.J. and, Darpe, A.K., "Vibration Response of Misaligned Rotors", *Journal of Sound and Vibration*, 325, 2009, pp. 609–628.

[17] Reddy, M.C.S. and Sekkar, A.S., "Detection and monitoring of coupling misalignment in rotors using torque measurements", *Measurement*, 61, pp.111–122, 2015.

[18] Lal, M. and Tiwari, R. "Quantification of multiple fault parameters in flexible turbo-generator systems with incomplete rundown vibration data", *Mechanical Systems and Signal Processing*, 41, pp.546–563, 2013.

[19] Lees, A.W., "Misalignment in Rigidly Coupled Rotors", *Journal of Sound and Vibration*, 303, 2007, pp. 261–271.

[20] Lees, A.W., 'Vibration Problems in Machines: Diagnosis and Resolution', *CRC Press*, 2016.

[21] Lees, A.W., "Rotor Misalignment: Some Novel Insights into an Old Problem", I. Mech.E. Conference "Vibrations in Rotating Machinery", Manchester, September 2016.

[22] Friswell, M.I., Penny, J.E.T., Garvey, S.D. and Lees, A.W., 2010, "Dynamics of Rotating Machines", *Cambridge University Press*, NY.

Attenuating influence of time-delay on stability of rotors supported on active magnetic bearings

Tukesh Soni

UIET, Panjab University, Chandigarh, India

J.K. Dutt

Mechanical Engineering, IIT Delhi, New Delhi, India

A.S. Das

Mechanical Engineering, Jadavpur University, Kolkata, India

ABSTRACT

Active Magnetic Bearings (AMBs) are being increasingly employed for supporting high-speed rotors because of their contact-less operation leading to almost no friction, and the consequent heating, noise and energy loss. The other inspiration is that these bearings are active elements, or their properties can be changed if required. For high-speed applications, involving rotor speeds greater than 10,000 rpm such as high speed milling of aluminium and precision grinding of small parts, AMBs may be preferred over conventional rolling element or hydrodynamic bearings, as these become either unstable or very inefficient beyond a certain rotor spin-speed limit. However, the AMBs may also face stability issues in high speed rotor applications due to the inevitable time-delay involved in sensing and transporting the rotor displacement signal to the controller and then applying the control action on the rotor shaft. Therefore, the present work addresses the issue of delay on stability of rotors supported on AMBs. Control law (the transfer function between displacement of rotor in the air gap and the force applied on it by the bearing) has a pronounced effect on the stability of rotor supported by an AMB. Conventionally, PD and PID control laws are widely used for designing AMBs. This paper has proposed some novel control laws based on the constitutive relationship of viscoelastic semisolids and compare the stability for the rotor AMB system, controlled with conventional control laws, and novel control laws, when the delay is taken into consideration. To this end, a rigid rotor with centrally placed rotor disk supported on AMB has been modelled as a single degree of freedom system, to keep the work simple. The effect of various parameters on the performance of AMB in supporting the rotor shaft with time delay is analysed. The novel control laws have been shown to hold immense potential to attenuate the influence of delay on stability of high-speed rotors.

1   INTRODUCTION

The operating spin speed of rotors, among other factors, is limited by the suspension system used for supporting such rotors. Active Magnetic Bearings (AMBs) offer contact-less suspension for such rotors, thereby, increasing the possible rotor spin speed for such systems without increasing the noise levels, heat generated in the bearings and energy loss due to friction. The dual advantage of AMBs is that they can also be used for active control of lateral vibrations in such systems (1). However, usage of AMBs involves the issue of time-delay between the sensed rotor shaft displacement and the actual build-up of control current in the actuator which determines the control force generated in the actuator. The time-delay between the sensing of the rotor shaft displacement and actual build-up of control current can be due to (a) the delay in sensing the shaft displacement, (b) transportation delay from sensor to the controller (c) processing of the feedback sensed shaft displacement based on the control law transfer function and (d) delay due to time required in building up of the control current in the actuator coils. Now, it may be easily perceived that the direction and magnitude of the control force should be applied on the rotor at an appropriate instant of time, else, the actuator may cause instability rather than stabilizing the system. This work therefore attempts to analyse the effect of time-delay on the stability of the AMB-rotor system with conventional control laws. Novel control laws based on the constitutive relationship of the viscoelastic materials have recently shown promise in improving stability of parametrically excited AMB system (2) and in improving performance of an actuator in controlling vibration levels in an overhung rotor shaft system (3). In this work the stability of the rotor-AMB system with conventional control law such as Proportional-Integral-Derivative (PID) and with novel control laws based on the viscoelastic materials is compared.

Ji (4) analyzed the effect of time delay on the stability of a simple rotor supported by an AMB. Presence of Hopf bifurcation for certain values of time delay was reported. The results were verified using numerical simulation. Ji (5) studied the dynamics of a Jeffcott rotor with an AMB acting over the rotor disk with time delay. PD control law was used to decide the control current in the AMB. Linear stability analysis was carried out and the characteristics of a single Hopf bifurcation was determined using a central manifold. The analytical results were experimentally validated. The influence of a time delay on the stability of a rotor supported on an AMB was considered by Ji and Hansen (6). The analysis included geometric coordinate coupling and thus a set of coupled two-degree-of-freedom nonlinear differential equations were obtained. A critical value of time delay was found, beyond which the equilibrium position of the rotor became unstable and Hopf bifurcation was observed with two different kinds of periodic motion.

Wang and Liu (7) performed linear stability for a rotor shaft suspended on AMB and studied the bifurcation diagram in an appropriate parameter plane. It was shown that beyond a sequence of critical time delay values, the Hopf bifurcation occurred in the system. The authors also presented an algorithm for finding the direction of Hopf bifurcation using theory of normal form and center manifold. The results were verified using numerical simulations. Wang and Jiang (8) investigated multiple stabilities in an AMB system with time delay. Li et al. (9) investigated the effect of time delay on the dynamics of a single degree of freedom AMB system with velocity feedback control. The critical value of the time delay was found and the direction of and stability of the Hopf bifurcation was discussed. The effect of time delay on the forced vibration of the system was also analyzed using numerical simulations. The study concluded that the effect of time delay cannot be ignored in the design of controller for the AMB system. Jiang et al. (10) analyzed AMB system with time delay and studied the eigenvalue problem for the linearized system at the system equilibrium.

A robust fuzzy logic-based controller for a nonlinear AMB system with time delay has been proposed by Zheng et al. (11). The nonlinear AMB system was represented using a Takagi-Sugeno fuzzy model and a fuzzy-based Parallel Distributed Compensation (PDC) controller was designed with delay-dependent stabilization criterion. Wenjun Su et al. (12) studied the influence of time delay in AMB system and presented explicit and numerical solutions to find the maximum delay time before instability commences in the system for a single-DOF system. Inoue (13) considered the effect of second-order delay of the electric current and the first-order delay of the magnetic flux in the AMB system. Magnetic flux of the AMB was represented by power series function of electric current and shaft displacement, and the nonlinear analytical analysis was carried out for a rigid rotor system which was also experimentally verified.

A fuzzy logic-based controller for a non-linear AMB system with time delay was designed by Zheng et al. (14). Liu et al. (15) considered the problem of non-linear motion of flexible rotor with drill-bit and analysed the effect of stick-slip and time delay. Zhang et al. (16) investigated the effect of double time delay in rotors supported by the high-speed self-acting gas-lubricated bearings. The authors analysed the influence of controller gains on the dynamic response of the rotor-bearing non-linear system with time delay. Yoon et al. (17) developed an unbalance compensation method for an AMB with delay in feedback control. The proposed method was experimentally validated using an AMB test rig.

Based on the literature survey presented above, it can be concluded that the analysis of rotor AMB system with time delay with the AMB utilizing the conventional PID control law, which is the most prevalent control law in the industry, has not been carried out. The probable reason for this is that with PID control law the governing equation for the rotor AMB system results in an integro-differential equation. This has formed the motivation for the present work. This work compares the stability regions for the rotor-AMB system with conventional PID control law and the recently introduced control law based on constitutive relation for a viscoelastic semisolid (2), (3) called the Four Element (FE) control law for the system with time delay. A single degree of freedom model for the rotor-AMB system is analysed and the stability regions with regards to various system parameters, controller gains and the actuator constants is presented and analysed.

2   RIGID ROTOR MODEL AND ACTIVE MAGNETIC BEARING

In the present work a rotor system with a central rotor disk mounted on an AMB is modelled as a single degree of freedom system. A schematic of the rigid rotor shaft with centrally placed rotor disk is shown in *Figure 1*. The Active Magnetic Bearings (AMBs) work on the principle of feedback control and basically comprises of two pairs of electromagnets, proximity sensor, controller, signal amplifier, and power supply. The proximity sensor senses the rotor shaft displacement at the AMB location and sends this information to the controller. The controller based on the control law and the feedback signal decides the value of control current required in the electromagnetic actuator, thus the electromagnets apply the force on the rotor shaft in opposite direction of the ensuing shaft vibration. A schematic of the AMBs depicting the working principle is shown in *Figure 2*.

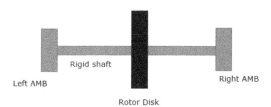

Rigid shaft

Left AMB

Right AMB

Rotor Disk

Figure 1. Rigid rotor shaft mounted on AMB.

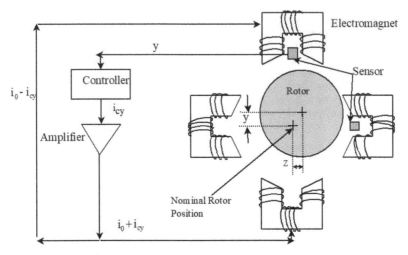

Figure 2. Schematic showing the working of an AMB.

For the given rotor shaft displacement in positive y direction the controller decides the value of control current i_{cy} which is subtracted from the upper electromagnets and added to the lower electromagnets so as to apply a downward force on the rotor shaft (see **Figure 2**). Similar, mechanism takes place for the horizontal pair of electromagnets as shown in. The rigid rotor model with centrally placed disk can be modelled as a single degree of freedom system with electromagnets applying the control force on the mass element of the system as shown in **Figure 3**.

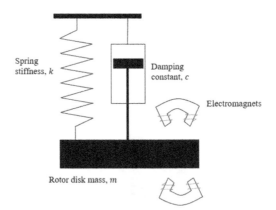

Figure 3. Single degree of freedom system with electromagnets.

2.1 AMB Force on the rotor shaft

The force applied by the AMB on the rotor shaft depends upon control current in the electromagnets and the radial air-gap between the electromagnet and the rotor shaft. The force applied by a pair of electromagnets on the on the rotor shaft is given as (1),

$$F_{AMB}(t) = K_{mag}\left(\frac{(i_0 + i_{cy})^2}{(s_0 - y)^2} - \frac{(i_0 - i_{cy})^2}{(s_0 + y)^2}\right) \qquad (1)$$

where, i_0 is the bias current value, i_{cy} is the control current in the vertical electromagnets, y is the vertical displacement of the rotor shaft at the magnetic bearing location and $K_{mag} = \frac{1}{4}\mu_0 n^2 A \cos\theta$ is a constant for a pair of electromagnets with μ_0 being the permissibility of free air, n is the number of coil turns in the electromagnet and A is the projected are of the pole face on the rotor shaft and θ is the equal to $22.5°$ for the case of radial magnetic bearings.

Eq. (1) models the electromagnetic force imposed on the rotor shaft by a pair of magnetic bearing electromagnets. However, the equation is nonlinear and therefore the same cannot be easily used for controller implementation and simulation. Eq. (1) can be linearized using Taylor's series expansion assuming that (a) the rotor displacement in vertical and horizontal directions at the magnetic bearings is small as compared to the rotor stator air gap, and (b) the control current in both the coils is small as compared to the bias current value. Mathematically, these assumptions can be written as,

$$y, z \ll s_0$$
$$i_{cy}, i_{cz} \ll i_0 \qquad (2)$$

Based on these assumptions and using Taylor's series expansion the Eq. (1) can be linearized and the same is given below for vertical and horizontal electromagnets:

$$(F_{AMB})_y = k_i i_{cy} - k_s y$$
$$(F_{AMB})_z = k_i i_{cz} - k_s z \qquad (3)$$

where, $k_i = \frac{4K_{mag}i_0}{s_0^2}$ is the force-current factor and $k_s = -\frac{4K_{mag}i_0^2}{s_0^3}$ is the force-displacement factor and is constant for an electromagnetic actuator.

2.2 Equations of motion for the single degree of freedom system with AMB

The equations of motion for the system shown in **Figure 3** can be derived using the Newton's law of motion and the same is given as,

$$m\ddot{x} + c\dot{x} + kx = f_{ext} + f_{AMB} \qquad (4)$$

where, m is the mass in kg, c is the damping coefficient in $\frac{Ns}{m}$ and k is the spring stiffness in N/m. f_{ext} is the external force on the mass and f_{AMB} is the force due to electromagnet pair acting on the single degree of freedom system. The external force on the mass can be due to unbalance in the rotor, misalignment of the rotor shaft, hydrodynamic force due to fluid flow etc. From equation (3), the force due to electromagnetic pole pair is given as,

$$f_{AMB} = k_i i_c - k_s x \qquad (5)$$

The AMB force largely depends upon the control current i_c which is decided based upon the control law used in the controller.

2.3 Control law and Pade's approximation

Control law governs the relationship between the control current and sensed displacement of the rotor shaft or the mass element of the single degree of freedom system. The conventional PID control law is given as,

$$i_c = -\left(K_P + \frac{K_I}{\mathcal{D}} + K_D\mathcal{D}\right)x_d \tag{6}$$

where, K_P is called the proportional gain, K_I is called the integral gain and K_D is called the derivative gain of the PID control law. $\mathcal{D}$ is the differentiation operator and x_d is the time-delayed value of the feedback signal reaching the controller and is mathematically given as,

$$x_d(t) = x(t - t_d) \tag{7}$$

Recently, a novel control law, based on the constitutive modelling of the viscoelastic material, called the 'Four-Element' or 'FE' control law has been introduced by Soni et al. (2), (3) and the same is given as,

$$i_c = -\left(\frac{k_1k_2 + (c_2(k_1 + k_2) + c_1k_2)\mathcal{D} + c_1c_2\mathcal{D}^2}{k_2 + c_2\mathcal{D}}\right)x_d \tag{8}$$

where, k_1, k_2, c_1 and c_2 are the four stiffness and damping elements of the control law. This control law has been shown to have better parametric stability than the conventional control laws, however, the same has not been tested and compared for the case of rotor system with time-delay.

Pade's approximants (18) are a family of approximations with different accuracy and associated complexity and are based on frequency response comparisons. The first order Pade's approximation is given as,

$$e^{-st_d} = \frac{1 - \frac{t_d s}{2}}{1 + \frac{t_d s}{2}} = \frac{2 - t_d s}{2 + t_d s} \tag{9}$$

Taking Laplace transform on both sides of the equation (7), one gets,

$$L(x_d(t)) = L(x(t - t_d))$$
$$X_d(s) = e^{-st_d}X(s) \tag{10}$$

Using Pade's first order approximation from equation (9), one gets,

$$X_d(s) = \frac{2 - t_d s}{2 + t_d s}X(s) \tag{11}$$

Therefore, equation (11) provides an approximation to the delayed feedback signal in terms of actual feedback displacement signal in frequency domain.

2.4 Assembled equations of motion

Using the equations (1)-(11), one gets the final assembled equations in the time domain as,

$$m\ddot{x} + c\dot{x} + kx - k_i i_c + k_s x = f_{ext} \tag{12}$$

Now for PID control law, using equation (6), the governing equations are

$$m\ddot{x} + c\dot{x} + kx + k_i\left(K_p + \frac{K_I}{D} + K_D D\right)x_d + k_s x = f_{ext} \tag{13}$$

For the case of FE control law using equation (4) and (8), the equations of motion are given as,

$$m\ddot{x} + c\dot{x} + kx + k_i\left(\frac{k_1 k_2 + (c_2(k_1+k_2)+c_1 k_2)D + c_1 c_2 D^2}{k_2 + c_2 D}\right)x_d + k_s x = f_{ext} \tag{14}$$

3   TIME DELAY ANALYSIS OF SYSTEM IN FREQUENCY DOMAIN

The relationship between the delayed feedback displacement signal, $x_d(t, t_d)$ and the actual state variable, $x(t)$ may be difficult to approximate in the time domain, therefore, in order to avoid the complication, the equations of motion derived in (13) and (14) are transformed to the frequency domain so that the Pade's approximation can be exploited. To this end, taking Laplace transform on both sides of equation (13) and (14), for the case of PID control law, as given below:

3.1   Pid control law

$$\left(ms^2 + cs + k + k_s\right)X(s) + \left(k_i K_P + \frac{k_i K_I}{s} + k_i K_D s\right)X_d(s) = F_{ext}(s) \tag{15}$$

Dividing both sides of the equation by mass, m

$$\left(s^2 + \frac{c}{m}s + \frac{k}{m} + \frac{k_s}{m}\right)X(s) + \left(\frac{k_i K_P}{m} + \frac{k_i K_I}{ms} + \frac{k_i K_D s}{m}\right)X_d(s) = \frac{F_{ext}(s)}{m} \tag{16}$$

Now, using non-dimensional parameters such as natural frequency, $\omega_n = \sqrt{\frac{k}{m}}$, damping ratio $\xi = \frac{c}{2m\omega_n}$, $\gamma = \frac{k_s}{k}$, $\alpha = \frac{k_i K_P}{k}$, $\beta = \frac{k_i K_D}{c}$, $\eta = \frac{k_i K_I}{\omega_n k}$, in equation (16) one gets,

$$\left(s^2 + 2\xi\omega_n s + \omega_n^2(1+\gamma)\right)X(s) + \left(\alpha\omega_n^2 + \frac{\eta\omega_n^3}{s} + 2\xi\omega_n\beta s\right)X_d(s) = \frac{F_{ext}(s)}{m} \tag{17}$$

Simplifying, and using Pade's approximation from equation (11),

$$\left(s^3 + 2\xi\omega_n s^2 + \omega_n^2(1+\gamma)s\right)X(s) + \left(2\xi\omega_n\beta s^2 + \alpha\omega_n^2 s + \eta\omega_n^3\right)\frac{2 - t_d s}{2 + t_d s}X(s) = \frac{sF_{ext}(s)}{m} \tag{18}$$

Further simplifying,

$$\begin{aligned}(t_d s^4 + 2\xi\omega_n t_d(1-\beta) + 2)s^3 + \left(\omega_n^2(1+\gamma-\alpha)t_d + 4\xi\omega_n(1+\beta)\right)s^2 \\ + \left(2\omega_n^2(1+\gamma+\alpha) - \eta\omega_n^3 t_d\right)s + 2\eta\omega_n^3 X(s) = \frac{(2+t_d s)sF_{ext}(s)}{m}\end{aligned} \tag{19}$$

Choosing, $F_{ext}(s) = \frac{1}{s}$, i.e. a step input signal,

Finally,

$$(X(s))_{PID} = \frac{1}{m} \frac{(2 + t_d s)}{D(s)} \tag{20}$$

where, $D(s) = (t_d s^4 + (2\xi\omega_n t_d (1 - \beta) + 2)s^3 + (\omega_n^2(1 + \gamma - \alpha)t_d + 4\xi\omega_n(1 + \beta))s^2$
$+ (2\omega_n^2(1 + \gamma + \alpha) - \eta\omega_n^3 t_d)s + 2\eta\omega_n^3)$

3.2 FE control law

Similarly, for the FE control law,

$$(s^2 + 2\xi\omega_n s + \omega_n^2(1 + \gamma))X(s) + k_i \left(\frac{k_1 k_2 + (c_2(k_1 + k_2) + c_1 k_2)s + c_1 c_2 s^2}{(k_2 + c_2 s)m} \right) X_d(s) = \frac{F_{ext}(s)}{m} \tag{21}$$

$$(s^2 + 2\xi\omega_n s + \omega_n^2(1 + \gamma))X(s) + k_i \left(\frac{k_1 + \left(c_2\left(\frac{k_1}{k_2} + 1\right) + c_1\right)\mathbf{s} + \frac{c_1 c_2}{k_2}s^2}{\left(1 + \frac{c_2}{k_2}s\right)m} \right) X_d(s) = \frac{F_{ext}(s)}{m} \tag{22}$$

Now, defining non-dimensional parameters as, $\zeta = \frac{\omega_n c_2}{k_2}$, $\psi = \frac{c_2}{c_1}$, $\alpha = \frac{k_i k_1}{k}$, $\beta = \frac{k_i c_1}{c}$

$$(s^2 + 2\xi\omega_n s + \omega_n^2(1 + \gamma))X(s) + \frac{(\alpha\omega_n^2 + (\alpha\omega_n\zeta + 2(1 + \psi)\beta\xi\omega_n)s + 2\xi\zeta\beta s^2)}{1 + \frac{\zeta}{\omega_n}\mathbf{s}} X_d(s) = \frac{F_{ext}(s)}{m} \tag{23}$$

Simplifying, and using Pade's approximation,

$$\left(s^2 + 2\xi\omega_n s + \omega_n^2(1 + \gamma) + \frac{\zeta}{\omega_n}s^3 + 2\xi\zeta s^2 + \omega_n\zeta(1 + \gamma)s\right)X(s) + (\alpha\omega_n^2 + (\alpha\omega_n\zeta$$
$$+ 2(1 + \psi)\beta\xi\omega_n)s + 2\xi\zeta\beta s^2)\frac{2 - t_d s}{2 + t_d s}X(s) = \left(1 + \frac{\zeta}{\omega_n}\right)\frac{sF_{ext}(s)}{m} \tag{24}$$

$$(2 + t_d s)\left(s^2 + 2\xi\omega_n s + \omega_n^2(1 + \gamma) + \frac{\zeta}{\omega_n}s^3 + 2\xi\zeta s^2 + \omega_n\zeta(1 + \gamma)s\right)$$
$$X(s) + (2 - t_d s)\left(\alpha\omega_n^2 + (\alpha\omega_n\zeta + 2(1 + \psi)\beta\xi\omega_n)s + 2\xi\zeta\beta s^2\right) \tag{25}$$
$$X(s) = \left(1 + \frac{\zeta}{\omega_n}\right)\frac{sF_{ext}(s)}{m}$$

$$\left(\frac{\zeta}{\omega_n}t_d s^4 + \left(\frac{2\zeta}{\omega_n} + t_d + 2\xi\zeta t_d(1 - \beta)\right)s^3 +\right.$$
$$(2 + 4\xi\zeta(1 + \beta) + 2\xi\omega_n t_d(1 + (1 + \psi)\beta) + \omega_n\zeta(1 + \gamma - \alpha)t_d)s^2$$
$$+ (4\xi\omega_n(1 + (1 + \psi)\beta) + 2\omega_n\zeta(1 + \gamma + \alpha) + \omega_n^2 t_d(1 + \gamma - \alpha))s + (2\omega_n^2(1 + \gamma + \alpha)) \tag{26}$$
$$X(s) = \left(1 + \frac{\zeta}{\omega_n}\right)\frac{1}{m}$$

Therefore, for the FE control law, one gets,

$$(X(s))_{FE} = \frac{(\omega_n + \zeta)}{m\omega_n N(s)} \tag{27}$$

where, $N(s) = (\frac{\zeta}{\omega_n} t_d s^4 + (\frac{2\zeta}{\omega_n} + t_d + 2\xi\zeta t_d(1-\beta))s^3 + (2 + 4\xi\zeta(1+\beta) + 2\xi\omega_n t_d(1 + (1+\psi)\beta)$
$+\omega_n\zeta(1+\gamma-\alpha)t_d)s^2 + (4\xi\omega_n(1 + (1+\psi)\beta) + 2\omega_n\zeta(1+\gamma+\alpha) + \omega_n^2 t_d(1+\gamma-\alpha))s + (2\omega_n^2(1+\gamma+\alpha)))$

4 STABILITY RESULTS AND DISCUSSION

The stability for the system can be ascertained by analysing the denominator of the transfer function $(X(s))_{PID, FE}$ which gives the characteristic polynomial for the system. The roots of the characteristic polynomial which are the poles of the system decide the stability of the rotor AMB system with time delay. The condition for stability can be mathematically stated as,

$$max.(Re(\lambda)) < 0 \tag{28}$$

where, λ is the root of the characteristic polynomial for the system.

A test system is first characterised with the help of non-dimensional parameters, the same is given in **Table I**.

The effect of controller gains on the stability of the system is analysed first. The non-dimensional proportional gain for both PID and FE control laws is $\alpha = \frac{k_i K_P}{k} = \frac{k_i k_1}{k}$, the plot of maximum real part of the roots of the characteristic polynomial for various values of the time delay and non-dimensional proportional gain α is shown in **Figure 4** (a) and (b) for PID and FE control law respectively.

Table 1. Non-dimensional parameters of the rotor-AMB system.

Non-dimensional parameter	Value
ω_n	100
ξ	0.2
α	150
β	10
γ	-100
ζ	0.1
η	1
ψ	1

111

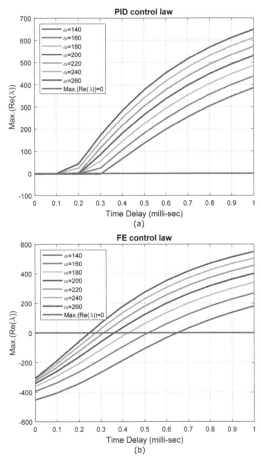

Figure 4. Effect of non-dimensional proportional gain, α on time delay stability of the system.

It can be seen from **Figure 4** (a) and (b) that the rotor AMB system with FE control law is more tolerant to the changes in gain values with regards to system stability. For example, at $\alpha = 140$, the time delay value at which the system becomes unstable is 0.65 milli-sec, whereas, the limiting time-delay value for PID control is 0.3 milli-sec. Similar trend is observed for higher values of α.

The non-dimensional damping gain for both the controllers has been defined as, $\beta = \frac{k_i K_D}{c} = \frac{k_i c_1}{c}$. The plot of the maximum real part of the roots of the characteristic polynomial versus the time delay value and the non-dimensional damping gain β is shown in **Figure 5** (a) and (b) for PID and FE control law respectively. As opposed to the non-dimensional proportional gain α, the higher values of the non-dimensional damping gain β, improves the stability of the rotor-AMB system. Again, the range of values of time delay for which the system is stable is much higher for the FE control law as compared to the PID control law for all values of β, as is evident from **Figure 5** (a) and (b).

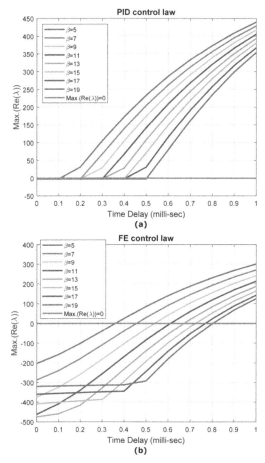

Figure 5. Effect of non-dimensional damping gain, β on the time delay stability of the rotor-AMB system.

The effect of the rotor-AMB system properties such as the natural frequency of the system ω_n, and the system damping ξ on the stability of the system is analysed using the plots shown in **Figure 6** and **Figure 7**. The higher values of system natural frequency reduce the range of time-delay values for which the system remains stable. However, again, for the FE control law the range of time-delay values for which the rotor-AMB system is stable is higher than that compared to the conventional PID control law as seen in **Figure 6** (a) and (b). For example, for a system natural frequency of 75 rad/s the limiting value of time delay for PID control law is 0.4 milli-sec while that for FE control law is 0.75 milli-sec (**Figure 6** (a) and (b)), i.e., $\cong$90% higher as compared to PID control law.

Figure 7 shows the plot of maximum of real part of roots of the characteristics polynomial versus the various possible system damping ξ and time delay values. As expected the system damping helps in maintaining stability of the rotor AMB system for higher time delay values. The FE control law proves to be more delay tolerant with regards to system stability as compared to the PID control law especially for the case of low system damping values. For example, for a system with lower value of damping ratio

ξ, of 0.1, the time delay value beyond which the system becomes unstable for FE control law is 0.35 milli-sec while that for the case of PID control law is 0.1 mill-sec only.

Apart from the above parameters which are common to both the control laws, some non-dimensional parameters which are unique to controllers may also affect the system stability. In order to ascertain the effect of such controller parameters, the variation of maximum of real part of roots of the characteristic polynomial is plotted and analysed for non-dimensional integral gain of PID control law, i.e. $\eta = \frac{k_i K_I}{\omega_n k'}$, and the non-dimensional ratio of secondary stiffness (k_2) and secondary damping (c_2) of the FE control law, i.e. $\zeta = \frac{\omega_n c_2}{k_2}$.

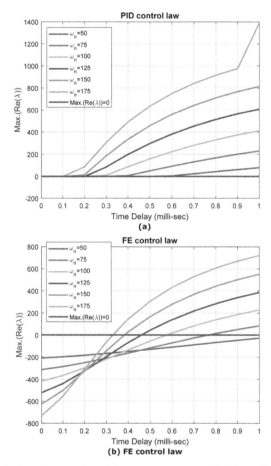

Figure 6. Effect of system natural frequency, ω_n on the time delay stability of the system.

The plot of the maximum value of the real part of the roots of the characteristic polynomial for different values of the non-dimensional integral gain for the PID control law, η is shown in **Figure 8**. It can be seen that the stable range of time delay values are insensitive to the integral gain values of the PID control law, as for different value of the η, the time delay value at which the maximum real value of λ becomes positive is almost same.

For the case of FE control law, the non-dimensional parameter ζ represents the ratio of the secondary damping and secondary stiffness of the FE controller. The plot of the maximum value of the real part of the roots of the characteristic polynomial for different values of ζ is shown in **Figure 9**. It is evident from **Figure 9** that a higher value of secondary damping as compared to secondary stiffness will reduce the stable range of the time delay values. From this one may conclude that the stable range of the time delay values can be further increased in the plots from **Figure 4** to **Figure 7** for lower values of the controller parameter $\zeta\zeta$. This is not possible with PID control law, as the stable region is seen to be insensitive to the non-dimensional integral gain, η.

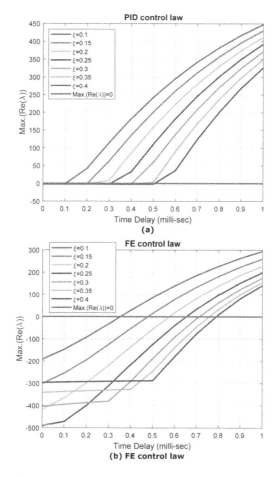

Figure 7. Effect of non-dimensional system damping ration on the time-delay stability of the rotor AMB system.

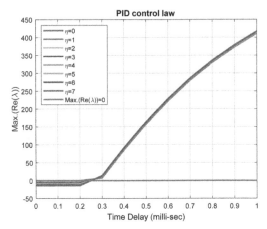

Figure 8. Effect on non-dimensional integral gain of PID control law on stability of the system.

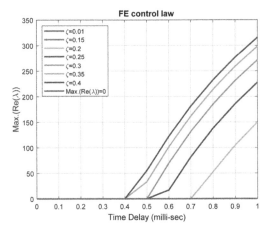

Figure 9. Effect of the non-dimensional ratio of secondary stiffness and damping of the FE control law on the stability of the system.

5   CONCLUSION

This work dealt with the stability analysis of a rigid rotor AMB system with time-delay for the case of AMB with conventional PID control law and a recently introduced Four Element (FE) control law. The system was modelled as a single degree of freedom system with an AMB and the stability analysis was carried out by transforming the system to frequency domain so as to exploit the Pade's approximation in relating the delayed feedback signal with the rotor shaft displacement. The stability of the system with time delay was ascertained by value of the roots of the characteristic polynomial of the system for different values of the non-dimensional parameters pertaining to the rotor shaft system and the controller. Stability regions with respect to system parameters such as the natural frequency, system damping and controller parameters such as the non-dimensional proportional gain, non-dimensional damping gain were drawn, which revealed that the system with FE control law is much more delay tolerant as compared to the conventional PID control law. Moreover, the tuning of the controller parameter unique to FE control law, namely the ratio of secondary damping to secondary stiffness of the FE controller would further increase the stable region for the rotor AMB system. This is not possible for the PID control law, as the stability region is seen to be insensitive to variations in the non-dimensional integral gain of the PID control law. From the findings of this work, it can be concluded that the FE control law outperforms the PID control law in terms of tolerating the inevitable time delay in the rotor AMB system.

REFERENCES

[1] Schweitzer G, Maslen EH, Bleuler H, Cole M, Keogh P, Larsonneur R, et al. Magnetic Bearings: Theory, Design and Application to Rotating Machinery. Springer-Verlag; 2009.

[2] Soni T, Dutt JK, Das AS. Parametric Stability Analysis of Active Magnetic Bearing Supported Rotor System With a Novel Control Law Subject to Periodic Base Motion. IEEE Trans Ind Electron. 2020 Feb;67(2):1160–70.

[3] Soni T, Das AS, Dutt JK. Active vibration control of ship mounted flexible rotor-shaft-bearing system during seakeeping. Journal of Sound and Vibration. 2020 Feb 17;467:115046.

[4] Ji JC. STABILITY AND HOPF BIFURCATION OF A MAGNETIC BEARING SYSTEM WITH TIME DELAYS. Journal of Sound and Vibration. 2003 Jan 23;259 (4):845–56.

[5] Ji JC. Dynamics of a Jeffcott rotor-magnetic bearing system with time delays. International Journal of Non-Linear Mechanics. 2003 Nov 1;38(9):1387–401.

[6] Ji JC, Hansen CH. Hopf Bifurcation of a Magnetic Bearing System with Time Delay. J Vib Acoust. 2005 Aug 1;127(4):362–9.

[7] Wang H, Liu J. Stability and bifurcation analysis in a magnetic bearing system with time delays. Chaos, Solitons & Fractals. 2005 Nov 1;26(3):813–25.

[8] Wang H, Jiang W. Multiple stabilities analysis in a magnetic bearing system with time delays. Chaos, Solitons & Fractals. 2006 Feb 1;27(3):789–99.

[9] Li Huiguang, Heng L, Lie Y. Effect of Time Delay in Velocity Feedback Loop on the Dynamic Behaviors of Magnetic Bearing System. In: 2007 International Conference on Mechatronics and Automation. 2007. p. 2911–6.

[10] Jiang W, Wang H, Wei J. A study of singularities for magnetic bearing systems with time delays. Chaos, Solitons & Fractals. 2008 May 1;36(3):715–9.

[11] Kai Zheng, Heng Liu, Lie Yu. Robust Fuzzy Control of a Nonlinear Magnetic Bearing System with Computing Time Delay. In: 2008 IEEE/ASME International Conference on Advanced Intelligent Mechatronics. 2008. p. 839–43.

[12] Wenjun Su, Kai Zheng, Heng Liu, Lie Yu. Time delay effects on AMB systems. In: 2009 International Conference on Mechatronics and Automation. 2009. p. 4682–6.

[13] Inoue T, Sugawara Y, Sugiyama M. Modeling and Nonlinear Vibration Analysis of a Rigid Rotor System Supported by the Magnetic Bearing (Effects of Delays of Both Electric Current and Magnetic Flux). J Appl Mech [Internet]. 2010 Jan 1 [cited 2019 Aug 22];77(1). Available from:/appliedmechanics/article/77/1/011005/426766/Modeling-and-Nonlinear-Vibration-Analysis-of-a

[14] Zheng K, Liu H, Yu L. Fuzzy Modelling and Output Feedback Stabilization of an Active Magnetic Bearing with Delayed Feedback: Proceedings of the Institution of Mechanical Engineers, Part J: Journal of Engineering Tribology [Internet]. 2011 Feb 1 [cited 2019 Aug 22]; Available from: https://journals.sagepub.com/doi/pdf/10.1177/2041305X10394407

[15] Liu X, Vlajic N, Long X, Meng G, Balachandran B. Nonlinear motions of a flexible rotor with a drill bit: stick-slip and delay effects. Nonlinear Dyn. 2013 Apr 1;72 (1):61–77.

[16] Zhang Y, Zhang S, Liu F, Zhou C, Lu Y, Müller N. Motion analysis of a rotor supported by self-acting axial groove gas bearing system with double time delays. Proceedings of the Institution of Mechanical Engineers, Part C: Journal of Mechanical Engineering Science. 2014 Nov 1;228(16):2888–99.

[17] Yoon SY, Di L, Lin Z. Unbalance compensation for AMB systems with input delay: An output regulation approach. Control Engineering Practice. 2016 Jan 1;46:166–75.

[18] Baker GAJ. Essentials of Padé Approximants. Elsevier; 1975. 319 p.

Calculation procedure to derive the threshold of vibration stability of soft mounted induction motors with elastic rotors and sleeve bearings fixed on active motor foot mounts for arbitrary controller structures

U. Werner

Faculty EFI, Nuremberg Tech, Germany

ABSTRACT

In the paper an iterative calculation procedure is presented for deriving the threshold of vibration stability of induction motors with flexible shafts and sleeve bearings, mounted on soft steel frame foundations with active motor foot mounts and arbitrary controller structures. The mathematical model considers the electromagnetic influence, stiffness and internal damping of the rotor, stiffness and damping of the bearing housings with end shields, of the foundation and of the oil film of the sleeve bearings, as well as the stiffness and damping of the motor foot mounts and the controlled forces, which are applied by the motor foot mounts.

1   INTRODUCTION

In rotating machinery not only forced vibrations have to be considered, but also self-excited vibrations, which may lead to vibration instability, often caused by the characteristic of the oil film in sleeve bearings and internal material damping of the rotor (rotating damping) (1-3). Large induction motors are often mounted directly on elastic steel frame foundations, which sometimes lead to vibration problems. The idea is to enlarge the system to a special mechatronic system by implementing active motor foot mounts – actuators – between the motor feet and foundation, sensors at the motor feet and a separate controller for each actuator (Figure 1).

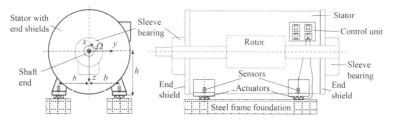

Figure 1. Induction motor enlarged to a special mechatronic system.

In general, implementing active vibration control is an appropriate method to reduce vibrations, which is used in different technical applications (4-7). The concept described here was basically investigated in (8), but the novelty of this paper here is the mathematical derivation of the calculation of the poles in the Laplace-domain and an example of stability analysis of a 2-pole induction motors with flexible rotor in combination with arbitrary controller structures.

2   VIBRATION MODEL

The vibration model is a simplified plane multibody model (Figure 2), containing two main mass – the rotor mass m_w and the stator mass m_s, with the moment of inertia θ_{sx} – which is basically described in (8).

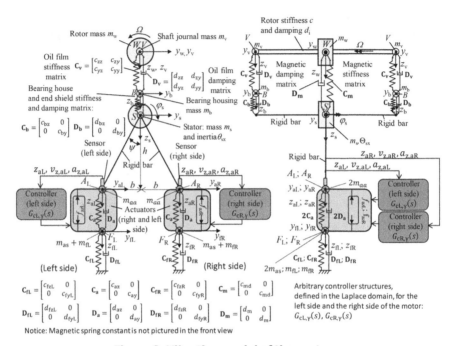

Figure 2. Vibration model of the system.

Additional mass are added for each shaft journal m_v, for each bearing housing m_b, for each actuator – m_{as} (stator) and m_{aa} (armature) – and for the foundation m_{fL} and m_{fR}. The rotor has the stiffness c and the internal damping (rotating damping) d_i and rotates with the rotary angular frequency Ω. The shaft journals of the rotor are linked to the sleeve bearing housings by the stiffness and the damping matrices $\mathbf{C_v}$ and $\mathbf{D_v}$ of the oil film. The stiffness and damping matrices $\mathbf{C_b}$ and $\mathbf{D_b}$ of the sleeve bearing housings with end shields connect the bearing housings to the stator. The stator – which is assumed to be rigid – is connected to the rotor in the air gap by the electromagnetic spring and damper matrices $\mathbf{C_m}$ and $\mathbf{D_m}$, considering simplified the electromagnetism of the induction motor, based on (9-12). The actuator stiffness and damping matrix $\mathbf{C_a}$ and $\mathbf{D_a}$ link the motor feet to the foundation, which has the stiffness matrices $\mathbf{C_{fL}}$ and $\mathbf{C_{fR}}$ and the damping matrices $\mathbf{D_{fL}}$ and $\mathbf{D_{fR}}$. The actuator forces are described by f_{azL} and f_{azR}. All values of the actuators are related on one motor side. Because of the planarity of the model, only one sensor for each motor side is necessary, measuring the vertical motor feet vibrations – displacements (z_{aL}, z_{aR}) or velocities (v_{aL}, v_{aR}) or accelerations (a_{aL}, a_{aR}), depending on the feedback strategy – and transmitting the signals to the controllers. The controllers have arbitrary controller structures, described in the Laplace-domain by the transfer functions $G_{cL,\gamma}(s)$ and $G_{cR,\gamma}(s)$:

$$G_{\mathrm{cL},\gamma}(s) = \frac{\sum\limits_{\mu L=0}^{mL} b_{\mu\mathrm{L},\gamma} \cdot s^{\mu L}}{\sum\limits_{\nu L=0}^{nL} a_{\nu\mathrm{L},\gamma} \cdot s^{\nu L}}; \quad G_{\mathrm{cR},\gamma}(s) = \frac{\sum\limits_{\mu R=0}^{mR} b_{\mu\mathrm{R},\gamma} \cdot s^{\mu R}}{\sum\limits_{\nu R=0}^{nR} a_{\nu\mathrm{R},\gamma} \cdot s^{\nu R}} \tag{1}$$

The coefficients $b_{\mu\mathrm{L},\gamma}$, $b_{\mu\mathrm{R},\gamma}$, $a_{\mu\mathrm{L},\gamma}$ and $a_{\mu\mathrm{R},\gamma}$ are the constants of the polynomial functions. The index γ depends on the chosen feedback strategy:

$$\gamma = \begin{cases} 0 & : \mathrm{No\,feedback\,(open\,control\,loops)} \\ z & : \mathrm{Feedback\,of\,the\,vertical\,motor\,feet\,displacements\ } z_{\mathrm{aL}}, z_{\mathrm{aR}} \\ v & : \mathrm{Feedback\,of\,the\,vertical\,motor\,feet\,velocities\ } v_{z,\mathrm{aL}}, v_{z,\mathrm{aR}} \\ a & : \mathrm{Feedback\,of\,the\,vertial\,motor\,feet\,accelerations\ } a_{z,\mathrm{aL}}, a_{z,\mathrm{aR}} \end{cases} \tag{2}$$

Referring to (13-15), the oil film stiffness and damping coefficients $c_{\mathrm{ij}} = c_{\mathrm{ij}}(\Omega)$ and $d_{\mathrm{ij}} = d_{\mathrm{ij}}(\Omega)$ can be calculated. The electromagnetic stiffness coefficient $c_{\mathrm{md}}(\Omega, \omega_{\mathrm{F}})$ and damping coefficient $d_{\mathrm{m}}(\Omega, \omega_{\mathrm{F}})$, depending additionally on the whirling angular frequency ω_{F}, can be calculated referring to (9-12). The mechanical damping coefficients d_{n} (d_{i}, d_{bz}, d_{by}, d_{fzL}, d_{fyL}, d_{fzR}, d_{fyR}, d_{az}, d_{ay}) can be derived by the corresponding mechanical loss factor $\tan\delta_{\mathrm{n}}$, the corresponding stiffness c_{n} and by the whirling frequency ω_{F}, referring to (3,8,12):

$$d_{\mathrm{n}} = \frac{c_{\mathrm{n}} \cdot \tan\delta_{\mathrm{n}}}{\omega_{\mathrm{F}}}\, with\, n = i, bz, by, fzL, fyL, fzR, fyR, az, ay;\, with\, c_{\mathrm{i}} = c \tag{3}$$

3   MATHEMATICAL DESCRIPTION

Based on (8), the following differential equation can be described:

$$\mathbf{M} \cdot \ddot{\mathbf{q}} \cdot \mathbf{D} \cdot \dot{\mathbf{q}} + \mathbf{C} \cdot \mathbf{q} = \mathbf{f_a} \tag{4}$$

$$\mathrm{with}: \mathbf{q}(t) = [z_s; z_w; y_s; y_w; \varphi_s; z_v; z_b; z_{fL}; z_{fR}; y_v; y_b; y_{fL}; y_{fR}]^T \tag{5}$$

including the linearization for the motor feet displacements, because of small displacements:-

$$z_{\mathrm{aL}} = z_s - \varphi_s \cdot b;\, z_{\mathrm{aR}} = z_s + \varphi_s \cdot b;\, y_{\mathrm{aL}} = y_{\mathrm{aR}} = y_s - \varphi_s \cdot h \tag{6}$$

The mass matrix is described by:

$$M = \begin{bmatrix}
m_s + 2m_{aa} & 0 & 0 & 0 & 0 & 0 & 0 & 0 & 0 & 0 & 0 & 0 & 0 \\
0 & m_w & 0 & 0 & 0 & 0 & 0 & 0 & 0 & 0 & 0 & 0 & 0 \\
0 & 0 & m_s + 2m_{aa} & 0 & -2m_{aa} \cdot h & 0 & 0 & 0 & 0 & 0 & 0 & 0 & 0 \\
0 & 0 & 0 & m_w & 0 & 0 & 0 & 0 & 0 & 0 & 0 & 0 & 0 \\
0 & 0 & -2m_{aa} \cdot h & 0 & \theta_{sx} + 2m_{aa}\left(b^2 + h^2\right) & 0 & 0 & 0 & 0 & 0 & 0 & 0 & 0 \\
0 & 0 & 0 & 0 & 0 & 2m_v & 0 & 0 & 0 & 0 & 0 & 0 & 0 \\
0 & 0 & 0 & 0 & 0 & 0 & 2m_b & 0 & 0 & 0 & 0 & 0 & 0 \\
0 & 0 & 0 & 0 & 0 & 0 & 0 & m_{as} + m_{fL} & 0 & 0 & 0 & 0 & 0 \\
0 & 0 & 0 & 0 & 0 & 0 & 0 & 0 & m_{as} + m_{fR} & 0 & 0 & 0 & 0 \\
0 & 0 & 0 & 0 & 0 & 0 & 0 & 0 & 0 & 2m_v & 0 & 0 & 0 \\
0 & 0 & 0 & 0 & 0 & 0 & 0 & 0 & 0 & 0 & 2m_b & 0 & 0 \\
0 & 0 & 0 & 0 & 0 & 0 & 0 & 0 & 0 & 0 & 0 & m_{as} + m_{fL} & 0 \\
0 & 0 & 0 & 0 & 0 & 0 & 0 & 0 & 0 & 0 & 0 & 0 & m_{as} + m_{fR}
\end{bmatrix} \tag{7}$$

The damping matrix is described by:

$$D = \begin{bmatrix}
2(d_{az} + d_{bz}) + d_m & -d_m & 0 & 0 & 0 \\
-d_m & d_m + d_i & 0 & 0 & 0 \\
0 & 0 & 2(d_{ay} + d_{by}) + d_m & -d_m & -2d_{ay}h \\
0 & 0 & -d_m & d_m + d_i & 0 \\
0 & 0 & -2d_{ay}h & 0 & 2(d_{ay}h^2 + d_{az}b^2) \\
0 & -d_i & 0 & 0 & 0 \\
-2d_{bz} & 0 & 0 & 0 & 0 \\
-d_{az} & 0 & 0 & 0 & d_{az}b \\
-d_{az} & 0 & 0 & 0 & -d_{az}b \\
0 & 0 & 0 & -d_i & 0 \\
0 & 0 & -2d_{by} & 0 & 0 \\
0 & 0 & -d_{ay} & 0 & d_{ay}h \\
0 & 0 & -d_{ay} & 0 & d_{ay}h
\end{bmatrix}$$

$$\begin{bmatrix}
0 & -2d_{bz} & -d_{az} & -d_{az} & 0 & 0 & 0 & 0 \\
-d_i & 0 & 0 & 0 & 0 & 0 & 0 & 0 \\
0 & 0 & 0 & 0 & 0 & -2d_{by} & -d_{ay} & -d_{ay} \\
0 & 0 & 0 & 0 & -d_i & 0 & 0 & 0 \\
0 & 0 & d_{az}b & -d_{az}b & 0 & 0 & d_{ay}h & d_{ay}h \\
2d_{zz}+d_i & -2d_{zz} & 0 & 0 & 2d_{zy} & -2d_{zy} & 0 & 0 \\
-2d_{zz} & 2(d_{zz}+d_{bz}) & 0 & 0 & -2d_{zy} & 2d_{zy} & 0 & 0 \\
0 & 0 & d_{az}+d_{fzL} & 0 & 0 & 0 & 0 & 0 \\
0 & 0 & 0 & d_{az}+d_{fzR} & 0 & 0 & 0 & 0 \\
2d_{yz} & -2d_{yz} & 0 & 0 & 2d_{yy}+d_i & -2d_{yy} & 0 & 0 \\
-2d_{yz} & 2d_{yz} & 0 & 0 & -2d_{yy} & 2(d_{yy}+d_{by}) & 0 & 0 \\
0 & 0 & 0 & 0 & 0 & 0 & d_{ay}+d_{fyL} & 0 \\
0 & 0 & 0 & 0 & 0 & 0 & 0 & d_{ay}+d_{fyR}
\end{bmatrix} \tag{8}$$

The stiffness matrix is described by:

$$\mathbf{C} = \begin{bmatrix}
2(c_{az}+c_{bz})-c_{md} & c_{md} & 0 & 0 & 0 \\
c_{md} & c-c_{md} & 0 & d_i\Omega & 0 \\
0 & 0 & 2(c_{ay}+c_{by})-c_{md} & c_{md} & -2c_{ay}h \\
0 & -d_i\Omega & c_{md} & c-c_{md} & 0 \\
0 & 0 & -2c_{ay}h & 0 & 2(c_{ay}h^2+c_{az}b^2) \\
-2c_{bz} & 0 & 0 & 0 & 0 \\
-c_{az} & 0 & 0 & 0 & c_{az}b \\
-c_{az} & 0 & 0 & 0 & -c_{az}b \\
0 & d_i\Omega & 0 & -c & 0 \\
0 & 0 & -2c_{by} & 0 & 0 \\
0 & 0 & -c_{ay} & 0 & c_{ay}h \\
0 & 0 & -c_{ay} & 0 & c_{ay}h
\end{bmatrix}$$

$$\begin{bmatrix}
0 & -2c_{bz} & -c_{az} & -c_{az} & 0 & 0 & 0 & 0 \\
-c & 0 & 0 & 0 & -d_i\Omega & 0 & 0 & 0 \\
0 & 0 & 0 & 0 & 0 & -2c_{by} & -c_{ay} & -c_{ay} \\
d_i\Omega & 0 & 0 & 0 & -c & 0 & 0 & 0 \\
0 & 0 & c_{az}b & -c_{az}b & 0 & 0 & c_{ay}h & c_{ay}h \\
2c_{zz}+c & -2c_{zz} & 0 & 0 & 2c_{zy}+d_i\Omega & -2c_{zy} & 0 & 0 \\
-2c_{zz} & 2(c_{zz}+c_{bz}) & 0 & 0 & -2c_{zy} & 2c_{zy} & 0 & 0 \\
0 & 0 & c_{az}+c_{fzL} & 0 & 0 & 0 & 0 & 0 \\
0 & 0 & 0 & c_{az}+c_{fzR} & 0 & 0 & 0 & 0 \\
2c_{yz}-d_i\Omega & -2c_{yz} & 0 & 0 & 2c_{yy}+c & -2c_{yy} & 0 & 0 \\
-2c_{yz} & 2c_{yz} & 0 & 0 & -2c_{yy} & 2(c_{yy}+c_{by}) & 0 & 0 \\
0 & 0 & 0 & 0 & 0 & 0 & c_{ay}+c_{fyL} & 0 \\
0 & 0 & 0 & 0 & 0 & 0 & 0 & c_{ay}+c_{fyR}
\end{bmatrix} \tag{9}$$

The actuator force vector $\mathbf{f_a}(t)$ can be split into the actuator force vector on the left side $\mathbf{f_{azL}}(t)$ and on the right side $\mathbf{f_{azR}}(t)$ of the motor:

$$\mathbf{f_a}(t) = \mathbf{f_{azL}}(t) + \mathbf{f_{azR}}(t) = \mathbf{P_{azL}} \cdot f_{azL}(t) + \mathbf{P_{azR}} \cdot f_{azR}(t) \text{ with :} \tag{10}$$

$$\mathbf{P_{azL}} = [1; 0; 0; 0; -b; 0; 0; -1; 0; 0; 0; 0; 0]^T \tag{11}$$

$$\mathbf{P_{azR}} = [1; 0; 0; 0; b; 0; 0; 0; -1; 0; 0; 0; 0]^T \tag{12}$$

Now the derivation of the calculation of the poles in the Laplace-domain can be performed. Therefore, the system has to be switched from time-domain to Laplace-domain. When transferring the differential equation in the Laplace-domain with initial conditions are zero, follows:

$$\mathbf{M} \cdot \mathbf{Q}(s) \cdot s^2 + \mathbf{D} \cdot \mathbf{Q}(s) \cdot s + \mathbf{C} \cdot \mathbf{Q}(s) = \mathbf{F_a}(s) \tag{13}$$

With the substitutions $\mathbf{X}_1(s) = \mathbf{Q}(s)$ and $\mathbf{X}_2(s) = \mathbf{Q}(s) \cdot s$ follows the state equation:

$$\underbrace{\begin{bmatrix} \mathbf{X}_1(s) \\ \mathbf{X}_2(s) \end{bmatrix}}_{\mathbf{X}(s)} \cdot s = \underbrace{\begin{bmatrix} \mathbf{0}_{13} & \mathbf{I}_{13} \\ -\mathbf{M}^{-1} \cdot \mathbf{C} & -\mathbf{M}^{-1} \cdot \mathbf{D} \end{bmatrix}}_{\mathbf{A}_{st}} \cdot \underbrace{\begin{bmatrix} \mathbf{X}_1(s) \\ \mathbf{X}_2(s) \end{bmatrix}}_{\mathbf{X}(s)} + \underbrace{\begin{bmatrix} \mathbf{0}_{13} \\ \mathbf{M}^{-1} \end{bmatrix}}_{\mathbf{B}_{st}} \cdot \mathbf{F_a}(s) \tag{14}$$

and the output equation:

$$\underbrace{\begin{bmatrix} \mathbf{X}_1(s) \\ \mathbf{X}_2(s) \\ \mathbf{X}_2(s) \cdot s \end{bmatrix}}_{\mathbf{Y}(s)} = \underbrace{\begin{bmatrix} \mathbf{I}_{13} & \mathbf{0}_{13} \\ \mathbf{0}_{13} & \mathbf{I}_{13} \\ -\mathbf{M}^{-1} \cdot \mathbf{C} & -\mathbf{M}^{-1} \cdot \mathbf{D} \end{bmatrix}}_{\mathbf{C}_{st}} \cdot \underbrace{\begin{bmatrix} \mathbf{X}_1(s) \\ \mathbf{X}_2(s) \end{bmatrix}}_{\mathbf{X}(s)} + \underbrace{\begin{bmatrix} \mathbf{0}_{13} \\ \mathbf{0}_{13} \\ \mathbf{M}^{-1} \end{bmatrix}}_{\mathbf{D}_{st}} \cdot \mathbf{F_a}(s) \tag{15}$$

with the zero-matrix $\mathbf{0}_{13} \in \mathbb{R}^{13 \times 13}$ and the unit-matrix $\mathbf{I}_{13} \in \mathbb{R}^{13 \times 13}$. Now the actuator force vector $\mathbf{F_a}(s)$ is described in the Laplace-domain by:

$$\mathbf{F_a}(s) = \mathbf{F_{azL}}(s) + \mathbf{F_{azR}}(s) = \mathbf{P_{azL}} \cdot F_{azL}(s) + \mathbf{P_{azR}} \cdot F_{azR}(s) \text{ with :} \tag{16}$$

$$F_{azL}(s) = \begin{cases} 0 & \text{for } \gamma = 0 \\ -Z_{aL}(s) \cdot G_{cL,z}(s) = -(Z_s(s) - \Phi_s(s) \cdot b) \cdot G_{cL,z}(s) & \text{for } \gamma = z \\ -V_{z,aL}(s) \cdot G_{cL,v}(s) = -(V_{z,s}(s) - V_{\varphi,s}(s) \cdot b) \cdot G_{cL,v}(s) & \text{for } \gamma = v \\ -A_{z,aL}(s) \cdot G_{cL,a}(s) = -(A_{z,s}(s) - A_{\varphi,s}(s) \cdot b) \cdot G_{cL,a}(s) & \text{for } \gamma = a \end{cases} \tag{17}$$

$$F_{azR}(s) = \begin{cases} 0 & \text{for } \gamma = 0 \\ -Z_{aR}(s) \cdot G_{cR,z}(s) = -(Z_s(s) + \Phi_s(s) \cdot b) \cdot G_{cR,z}(s) & \text{for } \gamma = z \\ -V_{z,aR}(s) \cdot G_{cR,v}(s) = -(V_{z,s}(s) + V_{\varphi,s}(s) \cdot b) \cdot G_{cR,v}(s) & \text{for } \gamma = v \\ -A_{z,aR}(s) \cdot G_{cR,a}(s) = -(A_{z,s}(s) + A_{\varphi,s}(s) \cdot b) \cdot G_{cR,a}(s) & \text{for } \gamma = a \end{cases} \tag{18}$$

The variable $Z_s(s)$ is the Laplace-transformed vertical displacement of the stator centre point S and $\Phi_s(s)$ is the Laplace-transformed angular displacement of S at the x-axis. The variable $V_{z,s}(s)$ is the Laplace-transformed vertical velocity of S and $V_{\varphi,s}(s)$ is the Laplace-transformed angular velocities of S at the x-axis. The variable $A_{z,s}(s)$ is the Laplace-transformed vertical acceleration of S and $A_{\varphi,s}(s)$ is the Laplace-transformed angular accelerations of S at the x-axis. Now a controller transfer matrix $\mathbf{T}_{st,\gamma}(s)$ can be derived, considering that the output vector $\mathbf{Y}(s)$ is lead back.

$$\mathbf{F_a}(s) = \mathbf{P_{azL}} \cdot F_{azL}(s) + \mathbf{P_{azR}} \cdot F_{azR}(s) = -\mathbf{T}_{st,\gamma}(s) \cdot \mathbf{Y}(s) \tag{19}$$

The controller transfer matrix $\mathbf{T}_{st,\gamma}(s)$ can now be described by:

$$\mathbf{T}_{st,\gamma}(s) = \begin{cases} \begin{bmatrix} \mathbf{0}_{13} & \mathbf{0}_{13} & \mathbf{0}_{13} \end{bmatrix} & \text{for } \gamma = 0 \\ \begin{bmatrix} \mathbf{T_z} & \mathbf{0}_{13} & \mathbf{0}_{13} \end{bmatrix} & \text{for } \gamma = z \\ \begin{bmatrix} \mathbf{0}_{13} & \mathbf{T_v} & \mathbf{0}_{13} \end{bmatrix} & \text{for } \gamma = v \\ \begin{bmatrix} \mathbf{0}_{13} & \mathbf{0}_{13} & \mathbf{T_a} \end{bmatrix} & \text{for } \gamma = a \end{cases} \tag{20}$$

with:

$$T_\gamma(s) = \begin{bmatrix} G_{cL,\gamma}(s) + G_{cR,\gamma}(s) & 0 & 0 & 0 & b\big[G_{cR,\gamma}(s) - G_{cL,\gamma}(s)\big] & 0 & 0 & 0 & 0 & 0 & 0 & 0 & 0 \\ 0 & 0 & 0 & 0 & 0 & 0 & 0 & 0 & 0 & 0 & 0 & 0 & 0 \\ 0 & 0 & 0 & 0 & 0 & 0 & 0 & 0 & 0 & 0 & 0 & 0 & 0 \\ 0 & 0 & 0 & 0 & 0 & 0 & 0 & 0 & 0 & 0 & 0 & 0 & 0 \\ b\big[G_{cR,\gamma}(s) - G_{cL,\gamma}(s)\big] & 0 & 0 & 0 & b^2\big[G_{cL,\gamma}(s) - G_{cR,\gamma}(s)\big] & 0 & 0 & 0 & 0 & 0 & 0 & 0 & 0 \\ 0 & 0 & 0 & 0 & 0 & 0 & 0 & 0 & 0 & 0 & 0 & 0 & 0 \\ 0 & 0 & 0 & 0 & 0 & 0 & 0 & 0 & 0 & 0 & 0 & 0 & 0 \\ -G_{cL,\gamma}(s) & 0 & 0 & 0 & b \cdot G_{cL,\gamma}(s) & 0 & 0 & 0 & 0 & 0 & 0 & 0 & 0 \\ -G_{cR,\gamma}(s) & 0 & 0 & 0 & -b \cdot G_{cR,\gamma}(s) & 0 & 0 & 0 & 0 & 0 & 0 & 0 & 0 \\ 0 & 0 & 0 & 0 & 0 & 0 & 0 & 0 & 0 & 0 & 0 & 0 & 0 \\ 0 & 0 & 0 & 0 & 0 & 0 & 0 & 0 & 0 & 0 & 0 & 0 & 0 \\ 0 & 0 & 0 & 0 & 0 & 0 & 0 & 0 & 0 & 0 & 0 & 0 & 0 \\ 0 & 0 & 0 & 0 & 0 & 0 & 0 & 0 & 0 & 0 & 0 & 0 & 0 \end{bmatrix}$$

with $T_z(s), T_v(s), T_a(s) \in \mathbb{C}^{13 \times 13}$ and $T_0(s) = 0_{13} \in \mathbb{R}^{13 \times 13}$. This can be pictured in the state space model in Figure 3. Index "*st*" is used for the matrices of the state space to avoid mix-up with the stiffness matrix $\mathbf{C}$ and damping matrix $\mathbf{D}$.

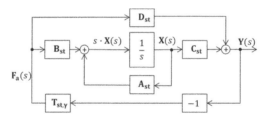

Figure 3. State space model for vibration control regarding stability analysis with negative feedback of the output vector.

Now the system can be described by:

$$\mathbf{X}(s) \cdot s = \mathbf{A}_{st} \cdot \mathbf{X}(s) + \mathbf{B}_{st} \cdot \mathbf{F_a}(s) \tag{21}$$

$$\mathbf{Y}(s) = \mathbf{C}_{st} \cdot \mathbf{X}(s) + \mathbf{D}_{st} \cdot \mathbf{F_a}(s) \tag{22}$$

With $\mathbf{F_a}(s) - \mathbf{T}_{st,\gamma}(s) \cdot \mathbf{Y}(s)$ follows:

$$\mathbf{X}(s) \cdot s = \mathbf{A}_{st} \cdot \mathbf{X}(s) - \mathbf{B}_{st} \cdot \mathbf{T}_{st,\gamma}(s) \cdot \mathbf{Y}(s) \tag{23}$$

$$\mathbf{Y}(s) = \big[\mathbf{I}_{39} + \mathbf{D}_{st} \cdot \mathbf{T}_{st,\gamma}(s)\big]^{-1} \cdot \mathbf{C}_{st} \cdot \mathbf{X}(s) \tag{24}$$

After inserting (24) in (23) follows:

$$\Big[\mathbf{I}_{26} \cdot s - \mathbf{A}_{st} + \mathbf{B}_{st} \cdot \mathbf{T}_{st,\gamma}(s) \cdot \big[\mathbf{I}_{39} + \mathbf{D}_{st} \cdot \mathbf{T}_{st,\gamma}(s)\big]^{-1} \cdot \mathbf{C}_{st}\Big] \cdot \mathbf{X}(s) = \mathbf{0} \tag{25}$$

with the unit-matrices $\mathbf{I}_{26} \in \mathbb{R}^{26 \times 26}$ and $\mathbf{I}_{39} \in \mathbb{R}^{39 \times 39}$. Therefore, the poles of the system can now be calculated by:

$$det\left[\mathbf{I}_{26} \cdot s - \mathbf{A}_{st} + \mathbf{B}_{st} \cdot \mathbf{T}_{st,\gamma}(s) \cdot \left[\mathbf{I}_{39} + \mathbf{D}_{st} \cdot \mathbf{T}_{st,\gamma}(s)\right]^{-1} \cdot \mathbf{C}_{st}\right] = 0 \qquad (26)$$

This direct procedure is only possible if the matrices $\mathbf{A}_{st}$ and $\mathbf{C}_{st}$ are independent of the whirling frequency ω_F, corresponding here to the natural angular frequency. However, this is not the case here. The rotary angular frequency Ω will be increased here until the real part of the pole, which will lead to instability, reaches zero, then the threshold of stability is reached. In this case, the rotary angular frequency Ω becomes Ω_{stab} and the critical pole becomes s_{stab}.

$$\Omega = \Omega_{stab}; s_{stab} = \pm j \cdot \omega_{stab} \qquad (27)$$

Therefore, the critical mode vibrates at Ω_{stab} with the natural angular frequency ω_{stab}, without decaying, and the whirling angular frequency ω_F is equal to the natural angular frequency ω_{stab} for the considered mode ($\omega_F = \omega_{stab}$). The oil film stiffness coefficients c_{ij} and the oil film damping coefficients d_{ij} are still functions of Ω, but the mechanical damping coefficients d_n are functions of the whirling angular frequency ω_F and so functions of the natural angular frequency ω_{stab}. The magnetic stiffness coefficient c_{md} and the magnetic damping coefficient d_m also depend on the rotary angular frequency Ω and on the natural angular frequency ω_{stab}. Therefore, the stiffness matrix $\mathbf{C}$ and the damping matrix $\mathbf{D}$ as well as the matrices $\mathbf{A}_{st}$ and $\mathbf{C}_{st}$ become depended on Ω and on ω_{stab}.

$$c_{ij}(\Omega), d_{ij}(\Omega), d_n(\omega_F), c_{md}(\Omega, \omega_F), d_m(\Omega, \omega_F) \rightarrow \mathbf{C}(\Omega, \omega_F), \mathbf{D}(\Omega, \omega_F) \rightarrow \mathbf{A}_{st}(\Omega, \omega_F), \mathbf{C}_{st}(\Omega, \omega_F) \quad (28)$$

Therefore, an iterative solution is necessary to derive the threshold of stability, which is described in Figure 4.

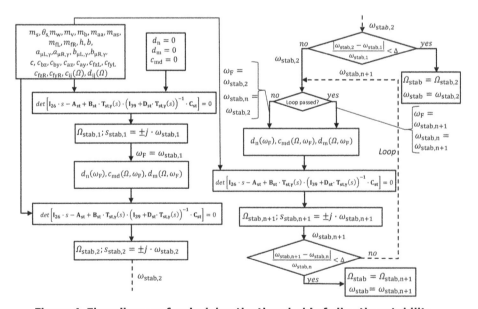

Figure 4. Flow diagram for deriving the threshold of vibration stability.

126

4 NUMERICAL EXAMPLE

In this section the threshold of stability is analysed for a 2-pole induction motor (Table 1), with sleeve bearings and flexible rotor. The induction motor is mounted on a soft foundation with active motor mounts between the motor feet and foundation and is driven by a converter with constant magnetization in the whole speed rang, operating without load $(s = 0)$.

Table 1. Data of the two-pole induction motor, actuators and foundation.

Data of the motor:	
- Mass of the stator	$m_s = 7040 \text{ kg}$
- Mass inertia of the stator at the x-axis	$\theta_{sx} = 1550 \text{ kgm}^2$
- Mass of the rotor	$m_w = 1900 \text{ kg}$
- Mass of the rotor shaft journal	$m_v = 10 \text{ kg}$
- Mass of the bearing housing	$m_b = 80 \text{ kg}$
- Stiffness of the rotor	$c = 1.0 \cdot 10^8 \text{ kg/s}^2$
- Undamped magnetic spring constant	$c_m = 7.0 \cdot 10^6 \text{ kg/s}^2$
- Height of the centre of gravity S	$h = 560 \text{ mm}$
- Distance between motor feet	$2b = 1060 \text{ mm}$
- Horizontal stiffness of bearing housing and end shield	$c_{by} = 4.8 \cdot 10^8 \text{ kg/s}^2$
- Vertical stiffness of bearing housing and end shield	$c_{bz} = 5.7 \cdot 10^8 \text{ kg/s}^2$
- Mechanical loss factor of the bearing housing and end shield	$\tan \delta_b = 0.04$
- Mechanical loss factor of the rotor	$\tan \delta_i = 0.03$
Data of the sleeve bearings:	
- Bearing shell	Cylindrical
- Lubricant viscosity grade	ISO VG 32
- Nominal bore diameter/Bearing width	$d_b = 110 \text{ mm}/b_b = 81.4 \text{ mm}$
- Ambient temperature/Supply oil temperature	$T_{amb} = 20°C/T_{in} = 40°C$
- Mean relative bearing clearance (DIN 31698)	$\Psi_m = 1.6 \text{ ‰}$
Data of the actuators (for each motor side):	
- Mass of the stator	$m_{as} = 10 \text{ kg}$
- Mass of the armature	$m_{aa} = 3 \text{ kg}$
- Vertical stiffness for each motor side	$c_{az} = 1.2 \cdot 10^8 \text{ kg/s}^2$
- Horizontal stiffness for each motor side	$c_{ay} = 3.0 \cdot 10^8 \text{ kg/s}^2$
- Mechanical loss factor of the actuators	$\tan \delta_a = 0.04$

(Continued)

Table 1. (*Continued*)

Data of the motor:	
Data of the foundation (for each motor side):	
- Mass left side	$m_{fL} = 30 \text{kg}$
- Mass right side	$m_{fR} = 30 \text{kg}$
- Vertical stiffness for each motor side	$c_{fzL} = c_{fzR} = 1.5 \cdot 10^8 \text{ kg/s}^2$
- Horizontal stiffness for each motor side	$c_{fyL} = c_{fyL} = 1.0 \cdot 10^8 \text{ kg/s}^2$
- Mechanical loss factor of the foundation	$\tan \delta_f = 0.04$

The electromagnetic spring value c_{md} and the electromagnetic damper value d_m are calculated according to (9-12), depending on rotor angular frequency Ω and on whirling angular frequency ω_F. For the control system, a feedback of the vertical motor feet accelerations is chosen, so that the index γ becomes a. The transfer function for both controllers – which is supposed to be identical for the left side and for the right side of the motor – is arbitrarily chosen here as polynomial functions of 3nd degree for the numerator and for the denominator, defined in the Laplace-domain:

$$G_{cL,a}(s) = G_{cR,a}(s) = \frac{b_{3,a} \cdot s^3 + b_{2,a} \cdot s^2 + b_{1,a} \cdot s + b_{0,a}}{a_{3,a} \cdot s^3 + a_{2,a} \cdot s^2 + a_{1,a} \cdot s + a_{0,a}} \tag{29}$$

The coefficients are described in Table 2.

Table 2. Coefficients of the transfer function of the controllers for feedback of the motor feet accelerations.

Coefficients of the numerator	Coefficients of the denominator
$b_{0,a} = 1 \cdot 10^8 \text{ [kg/s]}$	$a_{0,a} = 0 \text{ [1/s]}$
$b_{1,a} = 1 \cdot 10^6 \text{ [kg]}$	$a_{1,a} = 100 \text{ [-]}$
$b_{2,a} = 1 \cdot 10^6 \text{ [kgs]}$	$a_{2,a} = 10 \text{ [s]}$
$b_{3,a} = 0 \text{ [kgs}^2\text{]}$	$a_{3,a} = 1 \text{ [s}^2\text{]}$

The coefficients are only chosen roughly. The reason is that it is not the aim of the paper to find the optimal justification of the controllers for this example, but to show the fundamental influence of the control system. Now the threshold of stability is calculated for different cases (Table 3). Table 3 shows, that if the motor is mounted directly on a rigid foundation (case 1) the threshold of stability is lowest (3415 rpm). If the motor is now directly mounted on the soft foundation (case 2), the threshold of instability increases strongly by 40.7% to 4805 rpm. When now inserting the actuators between the soft foundation and the motor feet and operating with open control loops, so that the actuators only act passively (case 3), the threshold of stability decrease again to 4190 rpm (-12.8%). If now the control loops are closed (case 4), the threshold of stability increases to 6820 rpm (+62.8%).

Therefore this example shows, that here active vibration control is very effective, increasing the threshold of stability at about 2015 rpm (+41.9%) from 4805 rpm (case 2) to 6820 rpm (case 4), if the motor shall be operated on the soft foundation.

Table 3. Threshold of stability n_{stab} and whirling angular frequency $\omega_F = \omega_{stab}$ of the critical mode for different cases.

Case	n_{stab} [rpm]	$\omega_F = \omega_{stab}$ [rad/s]
1) Motor mounted directly on a rigid foundation, without actuators	3415	190.4
2) Motor mounted directly on the soft foundation, without actuators	4805	223.0
3) Motor mounted on the soft foundation with actuators, but with open control loops (actuators are only operating passively)	4190	208.6
4) Motor mounted on the soft foundation with actuators and with closed control loops (actuators are operating actively)	6820	225.5

When considering that the whirling angular frequency ω_F corresponds to the angular natural frequency of the critical mode, which gets instable – neglecting the boundary condition that this is strictly speaking only valid at the threshold of stability for the calculation process – the modal damping values of the critical mode can be calculated for the different cases (Figure 5).

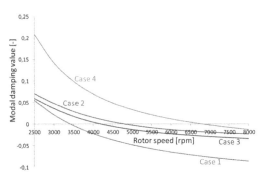

Figure 5. Modal damping values of the critical mode for the different cases.

5   CONCLUSION

In the paper an iterative calculation procedure is presented for deriving the threshold of vibration stability of induction motors with flexible shafts and sleeve bearings, mounted on soft steel frame foundations with active motor foot mounts for arbitrary controller structures. The mathematical model considers the electromagnetic influence, stiffness and internal damping of the rotor, stiffness and damping of the bearing housings with end shields, of the foundation and of the oil film of the sleeve bearings, as well as the stiffness and damping of the motor foot mounts and the controlled forces, which are applied by the motor foot mounts. Beside the mathematical description a numerical example was presented, where the effectiveness of this concept of active vibration control was shown. Currently experimental tests are running in the laboratory, based on a small 2-pole induction motor (11kW). However, tests based on large induction motors (>1MW) are planed in the future.

REFERENCES

[1] Genta G. (2005) *Dynamics of Rotating Systems*, Springer Science & Business Media.

[2] Rao, J. S. (1996) *Rotor Dynamics*, John Wiley & Sons, New York

[3] Gasch, R., Nordmann, R., Pfützner, H. (2002) *Rotordynamik*, Springer-Verlag, Berlin-Heidelberg.

[4] Preumont, A. (2011) *Vibration Control of Active Structures: An Introduction*, Springer.

[5] Janschek K. (2012) *Mechatronic Systems Design: Methods, Models, Concepts*, Springer.

[6] Ulbrich H. (1994) *A comparison of different actuator concepts for applications in rotating machinery*, International Journal of Rotating Machinery, Volume 1, 1, pp. 61–71.

[7] Chen H. M., Lewis P., Donald S., Wilson S. (1998) *Active mounts*, Journal of the Acoustical Society of America, Volume 91, 4.

[8] Werner U. (2019) *State space model for vibration control with arbitrary controller structures for soft mounted induction motors with sleeve bearings fixed on active motor foot mounts*, ZAMM, doi: 10.1002/zamm.201900126.

[9] Belmans R., Vandenput A., Geysen W. (1987) *Influence of unbalanced magnetic pull on the radial stability of flexible-shaft induction machines*, Iee proceedings-b electric power applications, 134 (2), pp. 101–109.

[10] Seinsch, H.-O. (1992) *Oberfelderscheinungen in Drehfeldmaschinen*, Teubner-Verlag, Stuttgart.

[11] Früchtenicht J., Jordan H., Seinsch H.O. (1982) *Exzentrizitätsfelder als Ursache von Laufinstabilitäten bei Asynchronmaschinen*. Teil 1, Seite 271-281, Teil 2, Seite 283-292, *Archiv für Elektrotechnik*, Bd. 65.

[12] Werner U. (2017) *Influence of electromagnetic field damping on the vibration stability of soft mounted induction motors with sleeve bearings, based on a multibody model*, SIRM-12th International Conference on Vibrations in Rotating Machines, Graz, Austria.

[13] Tondl A. (1965) *Some Problems of Rotordynamics*, Chapman and Hall, London.

[14] Glienicke J. (1966) *Feder- und Dämpfungskonstanten von Gleitlagern für Turbomaschinen und deren Einfluss auf das Schwingungsverhalten eines einfachen Rotors*, Dissertation TH Karlsruhe.

[15] Lund J., Thomsen K. (1978) *A calculation method and data for the dynamics of oil lubricated journal bearings in fluid film bearings and rotor bearings system design and optimization*, ASME, New York, pp.1–28.

12th International Conference on Vibrations in Rotating Machinery -
Institution of Mechanical Engineers, ISBN 978-0-367-67742-8

Experimental investigation on the static and dynamic characteristics of partially textured journal bearings

H. Taura

Nagaoka University of Technology, Japan

ABSTRACT

When a surface texture is formed on the bearing surface of a journal bearing, the lubrication characteristics change due to changes in the fluid film thickness. According to previous studies, when the entire bearing surface is textured, the load-carrying capacity of oil film decreases due to the increases in apparent film thickness. Besides, Yamada et al. demonstrated that the hydrodynamic force in the circumferential direction decreased, and the stability of the rotating shaft improved. Our numerical studies showed that maintaining the load-carrying capacity and improving the stability of a rotor supported in the bearing can be achieved at the same time if the texture is provided in an appropriate region, but the results have not been confirmed by the experiments. In the study, to investigate the effect of the texture position of the bearing surface on the static and dynamic characteristics of a partially textured journal bearing, experiments were conducted with a test bearing of which 90° texture is attached to the loaded or unloaded sides of the bearing surface. The experiments confirm that the load-carrying capacity depends on the circumferential position of the texture. When the texture is placed on the unloaded side of the bearing surface, the load-carrying capacity is more significant than that of the bearing with the texture placed on the loaded side. On the other hand, the reduction of the cross-coupled stiffness coefficient, which is the cause of the self-excited vibration, is slightly higher when the texure is placed on the loaded side.

1   INTRODUCTION

Surface texturing has evolved along with the development of micro-fabrication techniques and has been applied to various tribological elements, such as piston rings [1], mechanical seals [2], and thrust bearings [3]. Some experimental studies on the frictional characteristics of textured surfaces have demonstrated that surface texturing extends the operational conditions under a hydrodynamic lubrication regime [4], reduces their friction coefficient under a boundary-lubrication regime [5], and improves their anti-seizure performance [6].

This technique has also attracted attention in the field of the journal bearings. Many research groups have investigated the effect of surface texturing on the static characteristics of the bearings [7-10], and they obtained some valuable information. When a surface texture is formed on the bearing surface of a journal bearing, the lubrication characteristics change due to changes in the fluid film thickness. In general, the load-carrying capacity decreases because the apparent film thickness increases due to the surface texture. However, Tala-Ighil et al. [9] and Brizmer [10] have revealed that the load-carrying capacity was increased somewhat by manufacturing surface texture on an appropriate area of the bearing surface. On the other hand, the dynamic characteristics have only been investigated experimentally by Dadouche et al. [11], and they demonstrated that the direct stiffness and damping coefficients of textured bearings could become more substantial than those of smooth bearings under some geometrical parameters of textures and operating conditions. Yamada [12, 13] have also analysed numerically and experimentally the load-carrying capacity and dynamic coefficients of journal bearings with square dimples on the whole of the bearing surface, and they also demonstrated that the fully textured bearings improve the stability of a rotor supported in the bearings, but decrease the load-carrying capacity.

As mentioned above, fully textured bearings can improve the stability of the rotor supported in the bearings, but the load-carrying capacity decreases. To suppress the reduction of the load-carrying capacity, we focus on the partially textured bearings, of which the texturing area is limited on the bearing surface in the circumferential direction, and performed some numerical calculations. Some results showed that the partially textured bearing could improve the dynamic and stability characteristics at the same time while maintaining the load-carrying capacity if the textured region is placed on the unloaded side of the bearing surface [14]. Although these results have been obtained, they have not been confirmed experimentally.

In the study, we focus on the partially textured bearings, of which the texturing area is limited on the bearing surface in the circumferential direction, and numerically investigate the effect of the texturing region on the static and dynamic characteristics of the bearings.

2   EXPERIMENTS

2.1   Experimental setup
Figures 1 (a) & (b) show an overview and a photo of the experimental setup for the measurement of static and dynamic characteristics of journal bearings used in the present study. This experimental setup was used in our previous studies [14], and hence we described briefly here.

The test section is enclosed by a rigid frame (300×300×360mm) which is fixed on a rigid base. A test journal bearing is floating on a rotating shaft of 50mm in diameter supported at both ends by two angular bearings. The shaft is driven by an electric motor through a pulley-belt arrangement at a rotational speed up to 3000 rpm. A welded metal bellows can apply a static load downward in the vertical direction to the test bearing by injecting compressed air from a compressor. Four piezo-actuators as shakers are mounted to the test bearing at 45° to the vertical axis at the both bearing ends and can apply both static and dynamic loads to the test bearing via a parallel leaf spring block. These actuators can also adjust the axial alignment of bearing with the additional static load by the actuator.

The static and dynamic loads are measured by load cells attached to the test bearing, and the relative displacements between the test bearing and journal are measured with two pairs of four eddy current displacement sensors (range: 2mm, resolution: 0.3μm) installed at each end of the test bearing in the directions of dynamic forces (ξ and η directions in Figure 1(b)). The lubrication oil is supplied into the bearing clearance through axial grooves via oil supply holes. The exhausting oil temperature is measured with two K-type thermocouples at the oil-drain port. The signals of the displacement sensors, load cells and thermocouples are transmitted to a personal computer via A/D convertor.

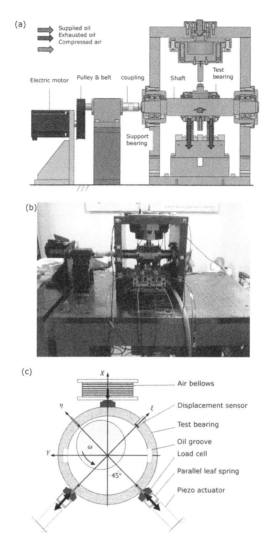

Figure 1. Experimental setup (a) Overview, (b) Photo (c) Schematic of the loading system.

2.2   Test bearing

Figure 2 shows a partially textured bearing we employed in the study (indicated by symbols "PTX"). The basic shape of it is a circular journal bearing with two axial oil grooves. Square dimples are attached to the quartered area of the bearing surface in the circumferential direction. As the shape of the test bearing excepting the dimples is symmetrical with respect to the oil grooves, the texture range in the circumferential direction can be arranged on the loaded side, or the unloaded side by changing the load direction for the bearing as shown in Figure 3.

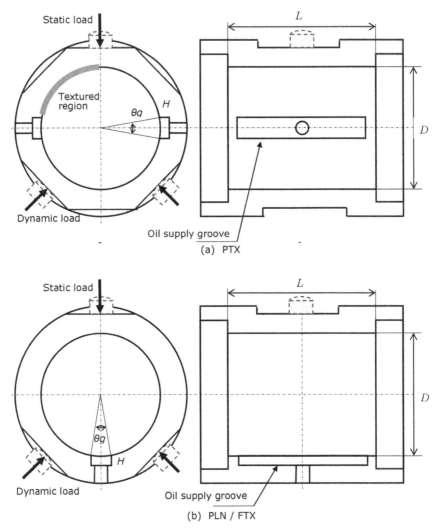

Figure 2. Test bearings (a) Partially textured bearing, (b) smooth bearing/ fully textured bearing.

To compare the experimental results, we also prepared a smooth journal bearing (indicated by symbols "PLN") and a fully textured journal bearing (indicated by symbols "FTX") as shown in Figure 2(b). These bearings were used for the previous studies. It should be noted that they have one axial oil groove. Table 1 shows a specification of the bearing.

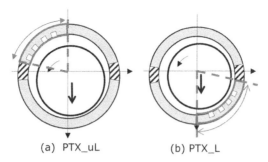

(a) PTX_uL (b) PTX_L

Figure 3. Load direction and test Bearings.

Table 1 . Specification of the test bearing.

Bearing type			PTX	FTX	PLN
Material			Brass (JIS C2801)		
Inner diameter	D	mm	50.05		
Length	L	mm	50.00		
Mean radial clearance	C	mm	26		
Mass	m	kg	2.8		

Figure 4(a) shows a schematic of the square dimples used in the present study, and the specification of the dimples is shown in Table 2. The dimples are textured uniformly on the bearing surface by electrical discharge machining with an accuracy of ± 0.02 mm for the dimple width. The appearance and the surface profile are shown in Figure 4(b) & (c).

Table 2 . Specification of the dimples.

Bearing type			PTX	FTX	PLN
Dimple width	$l_{\theta 1} = l_{z1}$	mm	0.65		—
Dimple pitch	$l_{\theta 2} = l_{z2}$	mm	1.30		—
Number of dimples	$N_t = N_\theta \times N_z$	-	988(26x38)	4294(113x38)	—
Dimple depth	h_t	μm	22	30	—
Total dimple area	$l_{\theta 1} \times l_{z1} \times N_t$	mm^2	417	1815	—

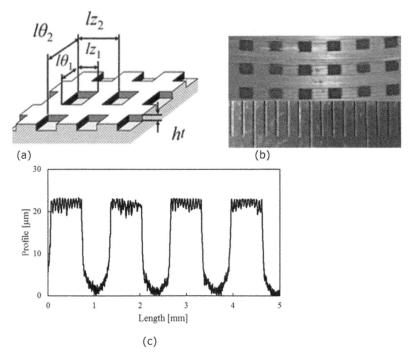

(a)   (b)

(c)

Figure 4. Detail of the dimples (a) Schematic of dimples, (b) Photo, (c) Surface profile of the bearing surface.

2. 3 Measurement of dynamic characteristics of journal bearings

When small displacements are applied to the rotor, Δx and Δy, and small velocities, $\Delta \dot{x}$ and $\Delta \dot{y}$, in vertical and horizontal directions around its equilibrium position, the linearized oil-film reaction forces f_x and f_y can be expressed with stiffness coefficients $k_{ij} c_{ij}$ and damping coefficients $(i, j = x, y)$ as follows:

$$f_x = f_{x0} + k_{xx}\Delta x + k_{xy}\Delta y + c_{xx}\Delta \dot{x} + c_{xy}\Delta \dot{y}$$
$$f_y = f_{y0} + k_{yx}\Delta x + k_{yy}\Delta y + c_{yx}\Delta \dot{x} + c_{yy}\Delta \dot{y}$$

(1)

where f_{x0}, f_{y0} are static oil film reaction forces in the x and y direction respectively.

In the present study, the sinusoidal-excitation method was employed to determine k_{ij} and c_{ij}. The sinusoidal dynamic force was applied to the test bearing, and the relative displacement of the bearing motion was measured simultaneously. Once the dynamic coefficients were obtained in the $\xi - \eta$ coordinate system, then they were transformed into the $x - y$ coordinate system using the rotation matrix of 45°.

In the results, we express the dimensionless stiffness and damping coefficients K_{ij}, C_{ij} defined as follows:

$$K_{ij} = \frac{C}{W} k_{ij}, C_{ij} = \frac{C\omega}{W} c_{ij} (i, j = x, y)$$

(2)

where ω is the angular speed of the shaft, and W is the static load.

2.4 Experimental conditions

The experimental conditions are listed in Table 3. In the study, the experiments were conducted for a wide range of Sommerfeld number S, which is defined by

$$S = \frac{N\mu DL}{W}\frac{R^2}{C^2} \tag{3}$$

where N is the shaft speed, and μ is the viscosity of the lubricating oil. The oil used in the study was ISO VG8 grade oil. S was varied from 0.2 to 1 by altering the value of W while keeping N constant. For the dynamic test, a sinusoidally oscillating force with a frequency of 2Hz was applied to the test bearing separately from two perpendicular directions of the ξ and η coordinates, and its amplitude was adjusted such that the magnitude of relative displacement between the bearing and the journal corresponded to one-tenth of the mean radial clearance C.

Table 3 . Experimental condition.

	Symbol	Unit	Value
Sommerfeld number	S	[-]	0.2-1.0
Static load	W	[N]	150-750
Shaft speed	N	$[s^{-1}]$	11 (660rpm)
Supply oil temperature	T_s	[°C]	30±3
Flow rate of oil supply	Q_s	[ml/min]	100
Viscosity of oil (@25°C)	μ	$[mPa \cdot s]$	14 (VG8)
Excitation frequency	f_v	[Hz]	2

In addition, an uncertainty analysis of the dynamic coefficients was performed to check the accuracy of the obtained experimental results in accordance with the ANSI/ASME standard on measurement uncertainty[15]. Although uncertainties varied with S, the average uncertainty values for all experimental results over a tested range of S are presented as follows:

$$K_{xx}, K_{yy} \cong 19\%, K_{xy}, K_{yx} \cong 11\%, C_{xx}, C_{yy} \cong 13\%, C_{xy}, C_{yx} \cong 20\% \tag{4}$$

3 RESULTS AND DISCUSSION

Figure 5(a) represents the journal centre loci measured in the experiments. In the figure, coloured circles represent the measured results for the partially textured bearing (Red: PTX_L or Blue: PTX_UL), and the black and white circles the smooth bearing (PLN) and the fully textured bearing (FTX) respectively. The solid line depicts the numerical results for PLN. The journal centre loci of PTX_L and PTX_UL move "inward" compared to those for PLN and FTX, i.e., it gets close to the vertical axis. This is due to the difference in the shape of the oil supply grooves.

Figure 5(b) shows a variation of the eccentricity ratio with the Sommerfeld number S. For the same value of S, the eccentricity of PTX_L and FTX is greater than that of PLN and PTX_UL. The load-carrying capacity is reduced due to the increase of the apparent bearing clearance on the loaded region of the bearing surface. These results suggest that the partially textured bearing can maintain the load-carrying capacity if the dimples are placed on the unloaded region of the bearing surface.

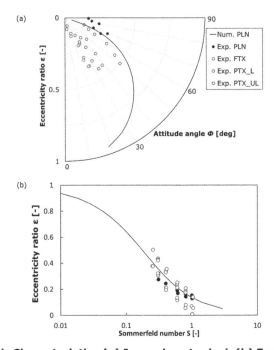

Figure 5. Static Characteristics (a) Journal centre loci, (b) Eccentricity ratio.

Figure 6 shows the results of four stiffness and four damping coefficients with S. The numerical results shown in the figures are calculated by using the perturbation method. K_{yx}, K_{yy}, C_{yx}, C_{yy} for both the partially textured bearings show a different trend with S compared to PLN and FTX. In particular, the force generated in the y direction is smaller than that in the x direction, especially in the range of high S. This may be mainly due to the difference in the oil grooves.

As for the texture position, the stiffness and damping coefficients for both the partially textured bearings vary in the qualitatively similar way with S. Besides, the magnitude of the cross-coupled stiffness coefficients, which is a cause of the oil whip or oil whirl, are slightly smaller when the texture is placed on the loaded side. These results suggest that the texture on the load side is more effective in reducing the destabilizing force[13].

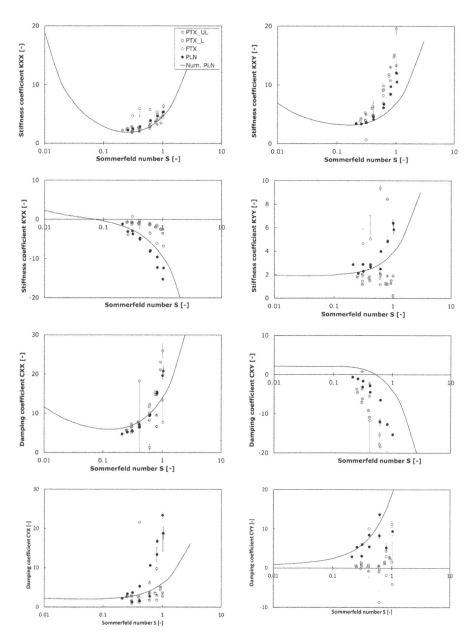

Figure 6. Stiffness and damping coefficients.

4 CONCLUSIONS

In this study, we investigate an effect of the texture position of the bearing surface on the static and dynamic characteristics of a partially textured journal bearing. Experiments were conducted with a test bearing of which 90° texture is attached on the loaded or unloaded sides of the bearing surface. The following conclusions are obtained;

(1) The experiments confirm that the loading capacity depends on the circumferential position of the texture. When the texture is placed on the unloaded side of the bearing surface, the load-carrying capacity is more significant than that of the bearing with the texture placed on the loaded side.
(2) When the texture is placed on the loaded side of the bearing surface, the reduction of the cross-coupled stiffness coefficient, which is the cause of the self-excited vibration, is slightly higher than that on the unloaded side.

REFERENCES

[1] Ryk, G., Kligerman, Y., and Etsion, I., 2002, "Experimental Investigation of Laser Surface Texturing for Reciprocating Automotive Components," Tribology Transactions, Vol. 45(4), pp.444–449.

[2] Etsion, I., and Halperin, G., 2002, "A Laser Surface Textured Hydrostatic Mechanical Seal," Tribology Transactions, Vol. 45(3), pp.430–434.

[3] Etsion, I., Halperin, G., Brizmer, V., and Kligerman, Y., 2004, "Experimental Investigation of Laser Surface Textured Parallel Thrust Bearings", Tribology Letters, Vol. 17(2),pp.295–300.

[4] Kovalchenkoa, A., Ajayi, O., Erdemir, A., Fenske, G., and Etsion, I., 2005, "The Effect of Laser Surface Texturing on Transitions in Lubrication Regimes during Unidirectional Sliding Contact," Tribology International, Vol. 38, pp.219–225.

[5] Podgornik, B., and Sedlacek, M., 2012, "Performance, Characterization and Design of Textured Surfaces," Transactions of the ASME, Journal of Tribology, Vol. 134, 041701.

[6] Kuroiwa, Y., Amanov, A., Tsuboi, R., Sasaki, S., and Kato, S., 2013, "Effectiveness of Surface Texturing for Improving the Anti-Seizure Property of Copper Alloy," Procedia Engineering, Vol. 68, pp.600–606.

[7] Ausas, R., Ragot, P., Leiva. J., Jai, M., Bayada, G., and Buscaglia, G.C., 2007, "The Impact of the Cavitation Model in the Analysis of Microtextured Lubricated Journal Bearings," Transactions of the ASME, Journal of Tribology, Vol. 129(4), pp.868–875.

[8] Cupillard, S., Cervantes, M., and Glavatskih, S., 2008, "A CFD Study of a Finite Textured Journal Bearing," IAHR 24th Symposium on Hydraulic Machinery and Systems, Brazil, Oct. 27–31.

[9] Tala-Ighil, N., Fillon, M., and Maspeyrot, P., 2011, "Effect of Textured Area on the Performances of a Hydrodynamic Journal Bearing", Tribology International, Vol. 44(3), pp.211–219.

[10] Brizmer, V., and Kligerman, Y., 2012, "A Laser Surface Textured Journal Bearing," Transactions of the ASME, Journal of Tribology, Vol. 134(3), 031702.

[11] Dadouche, A., Conlon, M. J., Dmochowski, W., Koszela, W., Galda, L., and Pawlus, P., 2011, "Effect of Surface Texturing on the Steady- State Properties and Dynamic coefficients of a Plain Journal Bearing: Experimental Study," Proceedings of ASME Turbo Expo 2011: Turbine Technical Conference and Exposition, Vol. 6, pp.695–704.

[12] Yamada, H., Taura, H., and Kaneko, S.: Static Characteristics of Journal Bearings with Square Dimples. Transactions of the ASME, Journal of Tribology, 139 (5):051703-051703-11(2017).

[13] Yamada, H., Taura, H., and Kaneko, S., 2017, "Numerical and Experimental Analyses of the Dynamic Characteristics of Journal Bearings With Square Dimples," Journal of Tribology, Vol. 140 (1),011703-011703-13.

[14] Taura, H., 2019, "Effect of Texture Region on the Static and Dynamic Characteristic of Partially Textured Journal Bearings," Proceedings of the 10th International Conference on Rotor Dynamics – IFToMM, Vol. 1, pp. 422–436.

[15] Measurement Uncertainty, ANSI/ASME PTC 19.1-1985 Part 1, 1986 (reaffirmed 1990).

12th International Conference on Vibrations in Rotating Machinery -
Institution of Mechanical Engineers, ISBN 978-0-367-67742-8

Rotating machines featuring new rotor topology and internal actuation for vibration mitigation

G.A. Fieux, N.Y. Bailey, P.S. Keogh

Department of Mechanical Engineering, University of Bath, UK

ABSTRACT

In the majority of applications, vibrations in a rotor system must be kept at a minimum to ensure optimal performance, minimised environmental transmission, and safe operation. Many approaches have been used to reduce their effects, including balancing, active bearing support, or rotor alterations, but few offer very compact and high-speed active control. This paper examines a novel active internal topology for hollow rotors. Low frequency actuation in the rotating frame of the rotor is thus equivalent to near synchronous actuation in a fixed, non-rotating, frame. Control strategies have been simulated and shown to decrease by 99% the runout and by 50% the reaction forces when crossing a critical speed.

1   INTRODUCTION

Rotating machinery plays a crucial role in the modern industrial world, but it is prone to generate different level of vibrations, which very often will have a detrimental effect on the performance of a system. An uncontrolled high level of vibrations will induce stresses in components, reducing their lifetime, generating noise, and causing rubs if clearances are exceeded. These disturbances come largely from unbalance, but also from thermal bends along the rotor, or from the environment around the machine.

The most straightforward approach to control vibrations is to balance the rotor, unbalance being the primary source of disturbances. Passive balancing is well known and described in the ISO standards [1] and is often a necessary step before any other type of vibration control is considered. However, the balancing condition can change when the rotor is in operation and when passive balancing is no longer accessible. Active balancing, where masses are moved around at different angles on the shaft can then be an option [2, 3, 4, 5, 6] to cater for variable balanced state, and an extensive review of these techniques can be found in [7]. However, this still enables only the control of synchronous disturbances. To control other types of vibrations, the support of the rotor can be altered, with for instance active magnetic bearings [8, 9], or ball bearings on piezo-electric actuators [10, 11, 12, 13]. However, the support is not always accessible to be modified, and some solutions like active magnetic bearings come with hardware to fit in the vicinity of the support. When space is very limited the rotor itself can be altered with alternative manufacturing techniques with, for instance, tie bolt assemblies [14] or composite shells [15, 16]. To improve further vibration reduction with a small footprint, some research has been carried out to apply direct bending on the rotor, for instance via functionally graded composites [17, 18], or piezo electric patches [19, 20, 21]. Although efficient at low speed, piezo patches are limited in bending moment whereas functionally graded material can be challenging to manufacture and service.

This paper introduces a novel internal topology for hollow rotor vibration mitigation, aimed at overcoming the force limitation of actuated rotors, whilst keeping the design modular and with a minimal footprint. The concept relies on miniaturised actuators applying controlled forces on disks perpendicular to the axis of rotation leading to a bend of the active section. This gives capability to the rotor to apply a counter bending moment and remain straight. This applies without the need for high frequency actuation since the actuator is located in the rotating frame.

Firstly, a description of the novel rotor topology and general dimensions based on an existing test rig is given. Its modelling via finite element analysis and a state space formulation with unbalance response follows. Then, the results and the effect of the proposed controlled method is presented. Finally, the results are discussed and ongoing and future work is described.

2   ROTOR MODELLING

2.1   General rotor dimensions

The dimensions of the rotor used for the analysis are shown on Figure 1, along with the concept of the active system to be controlled. To enable straightforward manufacturing, the rotor is divided into three sections joined together via rigid couplings. The two outer sections are simple tubes, which are mounted on bearings. They are hollow in order to carry the cables that the actuators will need for power and communication with the controller. The middle section is a hollow cylinder of a larger diameter and actuation to apply an equivalent bending moment at points where the diameter changes. A set of flanges with multiple holes at the couplings and in the middle allow balancing of the rotor and the mounting of trial masses. The material used for the active section is considered to be high strength aluminium 7075 to circumvent manufacturing constraints, while the passive sections are made of mild steel. A motor with a flexible coupling rotates the whole assembly up to 10,000 RPM. Between the actuator control force and the moment applied on the rotor, a force multiplication architecture as well as a lever arm mechanism will ensure a maximum bending moment estimated to go up to 190 N.m for an actuator force of 220 N.

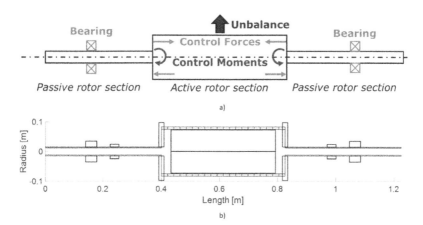

a)

b)

Figure 1. a) Rotor Concept b) Model dimensions.

2.2 Finite element modelling of the system

The rotor is modelled via a decomposition into Timoshenko beams finite elements with each node having two translational $\{x_1, y_1\}$ and two rotational $\{\theta_1, \psi_1\}$ degrees of freedom in the radial direction. Once decomposed into the N-1 elements, the generalised displacement vector can be described as

$$\delta = \{x_1 y_1 \theta_1 \psi_1, \ \ldots, \ x_i y_i \theta_i \psi_i, \ \ldots, x_N y_N \theta_N \psi_N\}^T \tag{1}$$

The generalised displacement vector and the shape functions of the Timoshenko beam formulation may be used to compute the strain energy U and kinetic energy T of the system, and equated to the external load F using the Lagrange equation:

$$\frac{d}{dt}\left(\frac{\partial T}{\partial \dot{\delta}}\right) - \frac{\partial T}{\partial \delta} + \frac{\partial U}{\partial \delta} = F \tag{2}$$

Equation (2) can then be rearranged as a function of the generalised displacement δ and its derivatives to reveal the traditional mass, damping, and stiffness matrices M, C and K, as well as the rotational speed of the rotor Ω:

$$[M]\ddot{\delta} + \Omega[C]\dot{\delta} + [K]\delta = F \tag{3}$$

This is the general equation for the unsupported rotor. In order to integrate the effect of bearing supports, stiffness and gyroscopic coefficients can be superimposed, respectively, on the K and C matrices at the nodes corresponding of the bearing locations. The control and unbalance forces will be added as parts of F on the appropriate nodes and degrees of freedom to simulate the effect of unbalance and of the control.

2.3 Unbalance response

In order to study the passive regime response of the system with very low computation time, the displacement and the external forces can be assumed to vary sinusoidally with speed:

$$\begin{cases} \{\delta\} = \{\Delta\} * e^{i\Omega t} \\ \{F\} = \{N_F\} * |F| * e^{i(\Omega t + \varphi)} \end{cases} \tag{4}$$

where $\{\Delta\}$ the amplitude vector of the nodal degrees of freedom, t is time, N_F the location of application of the external force or moment, $|F|$ its magnitude and φ is the phase difference. When substituting the displacement and force from equation (4) in equation (3) it is possible to obtain the expression of the amplitude of the generalised displacements as

$$\{\Delta\} = \left(-\Omega^2[M] + i\Omega[C] + [K]\right)^{-1}\left(\{N_{FU}\}F_u * e^{i\varphi_{Fu}} + \{N_{MC}\}M_C * e^{i\varphi_{MC}}\right) \tag{5}$$

where, respectively, F_U and M_C are the magnitudes of the unbalance force and control moment, and φ_{FU} and φ_{MC} are their phase differences. N_{FU} and N_{MC} are the location vectors of the unbalance force and control moments, with a 1 and a -i, respectively on the x and y axes of the node where they are applied:

$$\begin{cases} \{N_{FU}\} = \{0 \ldots 0 \ 1 - i \ 0 \ldots 0\} \\ \{N_{MC}\} = \{0 \ldots 0 \ 1 - i \ 0 \ldots 0\} \end{cases} \tag{6}$$

x_u y_u ϑ_{MC} ψ_{MC}

144

The magnitudes of the displacement and slopes at a given node can be found by taking the real part of the vector Δ from equation (5).

2.4   State space simulation
Alternatively, the system can be set up in a state space formulation by setting

$$\{q\} = \left\{ \begin{array}{c} \delta \\ \dot{\delta} \end{array} \right\} \tag{7}$$

rearranging equation (3) into

$$\left\{ \begin{array}{l} \{\dot{q}\} = [A] * \{q\} + [B] * \{u\} \\ \{p\} = [E] * \{q\} + [F] * \{u\} \end{array} \right. \tag{8}$$

where $\{p\}$ is the desired output, $\{u\}$ the external load, and A, B, E and F of equation (8) are defined as

$$\left\{ \begin{array}{c} [A] = \left[\begin{array}{cc} [0] & [I] \\ -[M]^{-1}[K] & -\Omega[M]^{-1}[C] \end{array} \right] \\ [B] = \left[\begin{array}{c} [0] \\ [M]^{-1} \end{array} \right] \\ [E] = [I] \\ [F] = [0] \end{array} \right. \tag{9}$$

The system can be then configured in a block diagram with software such as Simulink, with the external forces $\{u\}$ used as the input, of the system, and the generalised displacements and their derivatives $\{q\}$ coming as output and used to compute the external forces of the next time step.

Another use of the state space matrices is to facilitate the computation of the natural frequencies ω_n of the rotor system using the matrix $[A]$ defined in equation (9) through the diagonal eigenvalue matrix $[\lambda]$:

$$diag(\omega_n) = |diag[\lambda]| = |eig([A])| \tag{10}$$

From here, the corresponding eigenvectors $\{V\}$ can be found as they satisfy:

$$[A]\{V\} = \{V\}[\lambda] \tag{11}$$

The eigenvectors, when sorted in ascending order by their eigenvalues, will represent the deformed shape of the rotors when passing through the resonance frequencies or critical speeds, and are of interest to estimate the vibration signature of the rotor when running up in speed.

3   SIMULATION RESULTS

3.1   Modal analysis
Using equations (10) and (11), the eigenvalues and eigenvectors of the active rotor modelled can be estimated, in order to determine whether or not it will be operating in the flexible or rigid regime. Figure 2 displays the mode shapes, as well as the resonance frequencies. From these results, only the first bending mode will be crossed within the speed range of the motor of the test setup, rated for a maximum of 10,000 RPM. The second mode is seen at the maximal speed, depending where the unbalance is located.

145

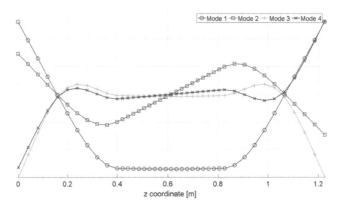

Figure 2. Rotor natural frequencies and eigen vectors with a bearing stiffness of 10^8 N/m (Mode 1: 3,601 RPM/Mode 2: 10,655 RPM/Mode 3: 35,753 RPM/ Mode 4: 37,353 RPM).

Figure 3 shows a speed sweep from 0 to 100,000 RPM with a G6 unbalance split between the middle, z = 600 mm, and the node at z = 360, which is where the highest displacement of the second bending mode is located. The displacement is then monitored at these two locations and the four first resonance frequencies appear clearly.

This configuration in terms of natural frequencies demonstrates how the rotor would operate in the flexible domain above the first bending mode natural frequency. The aim of the active system is to pass the first bending frequency and to mitigate its effect on the runout of the rotor.

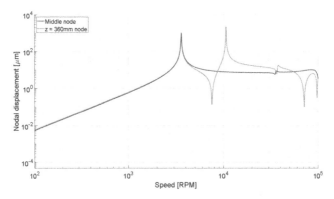

Figure 3. Rotor frequency sweep.

3.2 Control optimisation

Knowing the actuation capabilities, it is possible to vary the force at a given speed and monitor several parameters to find the optimal bending moment that the actuation system should apply to minimise a vibration metric. Among the parameters of

146

interest, the maximum displacement of the rotor is of the utmost importance in applications where clearances must be respected. Another interesting parameter to minimise is the average displacement, as this will influence the unbalance force. Finally, the reaction force of the bearings can be computed and brought to the lowest level possible, to enable a very small amount of force transmission to the stator.

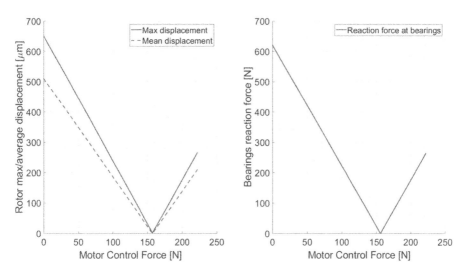

Figure 4. Control force optimisation at resonance (3,600 RPM) - Displacement (left) and bearing force (right) minimisation.

Figure 4 displays the results of the control force sweep between 0 and 220 N, which is predicted to be the amplitude available for the control of the prototype system. The force sweep is undertaken at a constant speed of 3,600 RPM, which is the first bending mode. A clear minimum appears at 157 N, which makes it the optimal control force, predictably minimising at the same time the displacements as well as the reaction forces at the bearing locations.

3.3   Controlled unbalance response

Using the force results of section 3.2, the first objective of the active rotor section will be to counter the bend the rotor when it passes through the first bending mode. In order to do this, the control forces are applied in the plane of the bending, resulting in a perpendicular control moment.

Figure 5 displays the effectiveness of the control while crossing resonance, as the deflection between the extremes of the rotor is reduced from 1,600 µm down to 14 µm, reducing the peak deformation by more than 99 %. This is done with a 157 N force when the maximal force of the motor is predicted to be around 220 N, leaving the potential for stronger control and a margin for losses that will inevitably occur between the actuation motor and where the control moment is applied.

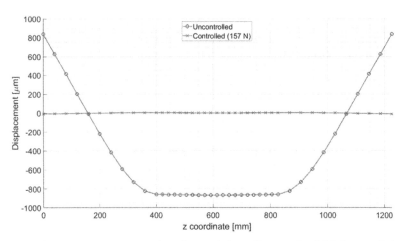

Figure 5. Resonance suppression through bending control (deformed shape at 3,600 RPM with G6 unbalance and a motor applying 157 N at 60 mm).

What is observed is that when subjected to the control action, the shaft will "snap through" the neutral position and start bending to the opposite direction if the force keeps increasing, giving an optimal force for each rotation speed that will minimise the runout as observed in the previous sections.

Other simulations have focused on softer bearing technologies with lower radial stiffness coefficients, such as journal bearings or active magnetic bearings. It was found that the shaft will straighten into a "w" shape as it cannot react so heavily against the bearings to transmit the forces to the stator. Thus, its average displacement will be harder to bring close to the rotation axis of the rotor, but it can be considered as externally "rigid" due to internal actuation.

3.4 Reduction of transmitted forces

Another use of on-board control via bending is to reduce the forces at the bearing location, hence mitigating the vibrations transmitted to the stator of the system. In the same way as for the displacement optimisation, the force can be optimised for each speed in order to obtain the best results in terms of vibration transmission.

This is particularly useful when passing through the resonance, as strong vibrations will be transmitted when crossing 3,600 RPM, in the case of the rotor used for the demonstration. Figure 6 displays the reaction forces at a bearing location, and a clear improvement is seen as the shaft is controlled with 157 N. Although the resonance peak is not eliminated, its maximal amplitude is reduced by 50% from 10,000 to 5,000 N. Another noticeable effect is that the increase of amplitude around the peak is reduced to a very narrow zone around the peak, reducing the effect of the resonance even when getting close to it.

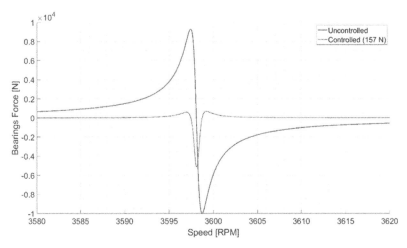

Figure 6. Bearing force reduction with bending control when passing through resonance.

4 CONCLUSIONS & FURTHER WORK

With the aim of designing a novel internal rotor topology for active control of rotor vibrations, a finite element model of a prototype rotor has been developed. A steady state complex formulation for unbalance response and state space formulation of the problem were introduced as a cost-efficient simulation strategy, in order to assess the efficiency of the novel rotor topology. The proposed rotor geometry appears to have a resonant frequency at 3,600 RPM which is within its speed range of 0 to 10,000 RPM, associated with an arc shape bending mode when supported by bearings. The actuation system applies control via internal bending moments. The optimisation of the control force appeared to validate the efficiency of the controller by reducing the force transmitted to the stator and the runout of the shaft.

Future work will include the manufacturing of a prototype actuated rotor in order to validate the model and the feasibility, as well as implementation of time dependent controller for the actuator.

REFERENCES

[1] ISO, "Mechanical vibration —Balance quality requirements for rotors in a constant (rigid) state — Part 1: Specification and verification of balance tolerances," Geneva, 2003.

[2] C. Alauze, J. Der Hagopian, L. Gaudiller and P. Voinis, "Active balancing of turbomachinery: application to large shaft lines," *JVC/Journal of Vibration and Control*, vol. 7, pp. 249–278, 2001.

[3] H. Fan, M. Jing, R. Wang, H. Liu and J. Zhi, "New electromagnetic ring balancer for active imbalance compensation of rotating machinery," *Journal of Sound and Vibration*, pp. 3837–3858, 2014.

[4] D. N. Pardivala, S. W. Dyer and C. D. Bailey, "Design modifications and active balancing on an integrally forged steam turbine rotor to solve serious reliability problems.," in *Turbomachinery and Pump Symposia*, 1998.

[5] J. Van de Vegte, "Balancing of flexible rotors during operation," *Journal of Mechanical Engineering Science*, vol. 23, pp. 257–261, 1981.

[6] J. Van de Vegte and R. T. Lake, "Balancing of rotating systems during operation," *Journal of Sound and Vibration*, vol. 57, pp. 225–235, 1978.

[7] S. Zhou and J. Shi, "Active balancing and vibration control of rotating machinery: A survey," *Shock and Vibration Digest*, vol. 33, pp. 361–371, 2001.

[8] E. Maslen, P. Hermann, M. Scott and R. R. Humphris, "Practical limits to the performance of magnetic bearings: Peak force, slew rate, and displacement sensitivity," *Journal of Tribology*, vol. 111, pp. 331–336, 1989.

[9] N. Dagnaes-Hansen and I. F. Santos, "Magnetic bearings for non-static flywheel energy storage systems (FESS)," vol. 60, 2019, pp. 116–131.

[10] T. S. Barrett, A. B. Palazzolo and A. F. Kascak, "Active vibration control of rotating machinery using piezoelectric actuators incorporating flexible casing effects," *Journal of Engineering for Gas Turbines and Power*, vol. 117, pp. 176–187, 1995.

[11] F. B. Becker, S. Heindel and S. Rinderknecht, "Active vibration isolation of a flexible rotor being subject to unbalance excitation and gyroscopic effect using H-optimal control," *Mechanisms and Machine Science*, vol. 21, pp. 1727–1739, 2015.

[12] S. Heindel, P. C. Muller and S. Rinderknecht, "Unbalance and resonance elimination with active bearings on general rotors," *Journal of Sound and Vibration*, vol. 431, pp. 422–440, 2018.

[13] R. Rong Lin, A. Palazzolo, A. F. Kascak and G. Montague, "Use of piezoelectric actuators in active vibration control of rotating machinery," 1990.

[14] J. J. Moore and A. H. Lerche, "Rotordynamic comparison of built-up versus solid rotor construction," Orlando, FL, United states, 2009.

[15] H. L. Wettergren, "The influence of imperfections on the eigenfrequencies of a rotating composite shaft," *Journal of Sound and Vibration*, vol. 204, pp. 99–116, 1997.

[16] R. Sino, T. N. Baranger, E. Chatelet and G. Jacquet, "Dynamic analysis of a rotating composite shaft," *Composites Science and Technology*, vol. 68, pp. 337–345, 2008.

[17] P. M. Przybyłowicz, "Stability of actively controlled rotating shaft made of functionally graded material," *Journal of theoretical and applied mechanics*, vol. 43, pp. 609–630, 2005.

[18] D. K. Rao and T. Roy, "Vibration Analysis of Functionally Graded Rotating Shaft System," Guwahati, 2016.

[19] P. M. Przybylowicz, "Near-critical behaviour of a rotating shaft actively stabilised by piezoelectric elements," *Systems Analysis Modelling Simulation*, vol. 42, pp. 527–537, 2002.

[20] P. J. Sloetjes and A. De Boer, "Vibration reduction and power generation with piezoceramic sheets mounted to a flexible shaft," *Journal of Intelligent Material Systems and Structures*, vol. 19, pp. 25–34, 2008.

[21] H.-G. Horst and H. P. Wolfel, "Active vibration control of a high speed rotor using PZT patches on the shaft surface," *Journal of Intelligent Material Systems and Structures*, vol. 15, pp. 721–728, 2004.

Effects of unbalance and AMB misalignment in a rigid rotor with an offset disc levitated by active magnetic bearings: A numerical investigation

Prabhat Kumar[1, 2], Rajiv Tiwari[2]

[1]Department of Mechanical Engineering, National Institute of Technology Manipur,
 Imphal West, Manipur, India
[2]Department of Mechanical Engineering, Indian Institute of Technology Guwahati,
 Guwahati, Assam, India

ABSTRACT

Rotor unbalance and misalignment are the most influential faults in rotating machines. Nowadays, the high-speed machines are employed with active magnetic bearings (AMBs) for rotor supporting and fault identification purposes. In the present paper, the dynamic analysis of an unbalanced and misaligned rigid rotor with a disc at the offset position supported on AMBs, has been numerically investigated. Non-collinearity between the rotating axis of rotor and the axis of supported AMBs is the main cause of misalignment fault. The prime intention of the paper is to study the rotor dynamic behaviour under the influence of disc eccentricity and AMB misalignment ratios.

1 INTRODUCTION

Rotating elements are well known and very common in modern manufacturing and production industries for various advantageous applications. Usually for the supporting purpose, the rotor in the rotating machines require bearings. For many years, rolling element bearings have been used to support rotors via physical contact, however the latest trend is towards utilizing active magnetic bearings which support the rotor without any contact due to electromagnetic forces induced in the rotating conductor. Moreover, in this bearing support, the rotor does not experience wear or frictional resistance during rotation. This overcomes various operating limitations of rolling element bearings. Active magnetic bearing force helps to rotate the rotor at high speed. One of the major advantages of this bearing system is that it is associated with a controller, which allows the rotor to operate stably for different system parameters and at multiple higher speeds [1].

Generally, the rotating elements may experience several faults, such as unbalance, crack, misalignment, bow in rotor, etc. These faults are extremely harmful for the rotor system as they may lead to dangerous accidents and low and ineffective production in industries. Unbalance is the most common fault which may cause high amplitude of vibration in the rotating machines, especially while approaching the critical speed. This fault is extremely unsafe and leads to noise, primarily caused due to uneven mass distribution around the rotational axis. In mathematical form, it is represented as the multiplication of mass of rotor, rotor eccentricity (i.e., the distance between the rotor centre of gravity and rotational centre) and square of the rotor spin speed. Thus, this fault is hazardous at high spin speeds (near critical speeds) of the rotor and even a small unbalance can be of high potential impact at higher speed. For exploring the dynamic analysis under unbalance fault and its identification, a brief survey on vibration signal-based health monitoring of various rotating machines was presented by the researchers [1] and [2]. De Queiroz [3] proposed a creative method using a robust vibration control mechanism, for the identification of unbalance fault parameters in a Jeffcott rotor. Markert et al. [4] and Platz et al. [5] as well as Sudhakar and Sekhar [6] discussed a model based on least-squares fitting technique, for presenting the dynamic vibrational behaviour and identification in an unbalanced simple rotor system supported by two conventional bearings. An unbalance identification technique, with a combination of modal expansion and optimization algorithm was proposed by Yao et al. [7] in a rotor and conventional bearing system associated with one disc. This technique was very robust and effective, and had also been validated with experimental results. Afterwards, utilizing a joint-input state estimation method, Shrivastava and Mohanty [8] conducted both numerical and experimental investigation for the unbalance identification in a rigid rotor supported by conventional bearing system.

Misalignment in a rotor system is also a severe fault, which may decrease the efficiency of the machine and cause failure due to unnecessary vibration. The cause of misalignment may be due to the non-collinearity between axes of two coupled shafts or non-collinearity between the supported bearing axis and rotating axis of shaft. Dewell and Mitchell [9] experimentally validated the developed mathematical expression for the moments in misaligned disc-coupling and found that as the misalignment increases, the second and fourth harmonics amplitudes in the response spectrum also increases. Vibrational behaviour of a misaligned coupled rotor supported on two conventional bearings and the misalignment fault identification was presented by [10, 11]. Lal and Tiwari [12] identified the coupling misalignment fault by developing a model-based identification algorithm in rigid and flexible rotor-conventional bearing-coupling systems. Later, AMBs were incorporated into the flexible turbo-generator system by Kuppa and Lal [13] and estimated the parameters related to coupling, AMBs and unbalance fault, etc.

Srinivas et al. [14, 15] described the working principles and components of AMBs, and their applications in a flexible rotating machines as well as identified the coupling misalignment using a steering function in a misaligned rotor-coupling-train system integrated with an AMB. Further, Tuckmantel and Cavalca [16] presented the dynamic nature of a coupled rotor-AMB system under the influence of unbalance and angular coupling misalignment faults. Kumar and Tiwari [17, 18] presented the vibrational behaviour of two-degrees-of-freedom rigid rotor system mounted on misaligned active magnetic bearings and identified the unbalance and residual AMB misalignment faults quantitatively. Recently, Zhao et al. [19] developed a fault detection and diagnosis method on the basis of multi-input convolutional neural network (MI-CNN) for differentiating the misalignment fault in coupled shaft and crack in the shaft. They noticed that the method was effective and robust.

Several researches have performed rotating system diagnosis as well as fault parameter identification in a rotating machine supported on conventional bearings. However, the investigation on dynamic analysis in active magnetic bearing-rotor systems associated with unbalance and the AMB lateral and angular misalignment faults has not been presented till now, which is very crucial. Due to large gap in AMB, the rotor will be allowed to operate in the presence of misalignment. But, an extra force will be generated at the locations of AMBs due to this misalignment, which will increase the amplitude of vibration and noise in the system as well as shorten the life span of the equipment. Moreover, an additional control current will be further needed to remove or reduce this force using AMB. This additional current unnecessarily will consume more electric power. Hence, to avoid unnecessary power consumption, the proper alignment is required. It is also worth mentioning that although the industrial manufacturers may have precise manufacturing tolerances on housing for rigid rotors in the initial time, the misalignment can occur due to the strain through continuous operations of the equipment, from thermal distortion of the supporting frame due to uneven thermal expansion, etc. [20]. Sensor measurement errors may also cause the rotor misalignment with the supported active magnetic bearings. In a fully floated rotor-AMB system, the center position of AMB in the vertical direction is located using the change in voltage detected from eddy current proximity sensor, by placing the core at the lower pole of AMB and then at upper pole of AMB. Similarly, the change in voltage perceived from sensor, by placing the core at the extreme backside pole of AMB and subsequently at the extreme front side pole of AMB, is used to detect the center position of AMB in the horizontal direction. However, it is difficult to exactly locate the center position of AMB in both the vertical and horizontal directions through the non-contact sensor due to measurement error, arising from the defective positioning of displacement sensors. This inability in finding the exact location of AMB centre will lead to misalignment of the axes of AMBs with respect to the absolute reference of operating axis of rotor. Similarly, the unbalance may be present in the rotor system during its operation due to deposition of any foreign particle, corrosion, and wear, distortion due to stress relief, or thermal distortion, in the rotor system. Therefore, it is mandatory to understand the dynamic nature of the unbalanced rigid rotor system supported on misaligned active magnetic bearings. This paper discusses the development of a mathematical model for the dynamic behaviour of an unbalanced and misaligned four degrees-of-freedom rigid rotor system levitated by two active magnetic bearings at both the ends of rotor.

In the present work, a rigid rotor with an unbalanced offset disc mounted on two active magnetic bearings has been mathematically modelled. The rigid rotor is assumed to be in a combination of parallel and angular misalignments with supported AMBs. Force due to the misaligned AMBs in the linearized form have been obtained and observed to contain additional constant forces along with the forces assisted with the AMB force-displacement and force-current constants. The equations of motion with inclusion of gyroscopic effect, unbalance force, inertia force and misaligned AMB force have been derived. Further, the equations have been non-dimensionalized with respect to the dimensionless system and faults parameters, and solved by exploring a SIMULINK^{TM} model, to generate the dimensionless rotor displacement and AMB current responses at both AMBs locations. The main purpose of the present work is to study the dynamic analysis of the rotor-AMB system for multiple ranges of disc eccentricity and AMB lateral and angular misalignment ratios. In this work, the rotor has been considered rigid due to some of advantages over the flexible rotor in experimentation. In comparison to the rigid rotor system, more number of eddy current proximity sensors are required to measure transverse displacement responses at different locations of flexible shaft. There is a difficulty in measuring the rotational displacements of the flexible shaft. Balancing of the flexible rotor is more complex than rigid rotor balancing. The flexible rotor continuously changes its elastic configuration as more critical speeds are encountered. Since it is recognized that rotor-bending operating deflections modify the resulting forces from residual unbalance, flexible rotor need to be balanced at high speed. Moreover, at higher speeds of rotor, the aligning of supported AMBs may be difficult due to large amplitude of vibration. Thus, it is better to first align the system at low speeds (where the rotor is considered rigid) by identifying the residual misalignment amount. After aligning, the system can cross its critical speeds with less vibration (i.e., for the case of flexible rotor).

2 ROTOR-AMB SYSTEM CONFIGURATION AND MATHEMATICAL MODELLING

This section explains the assumptions and details involved in modelling the rotor-AMB system. For studying and examining the dynamic behaviour, a system comprising of a rigid rotor with an offset disc misaligned with both the supported active magnetic bearings has been considered, which is illustrated in Figure 1. In the view of industrial applications, the disc in the system may be represented as a flywheel, rotary vanes of turbines and impellers of pumps, etc. Misalignment of both AMBs in the transverse directions are assumed to be different and represented by δ_{x1} and δ_{x2} in the x-direction and δ_{y1} and δ_{y2} in the y-direction. AMBs at the end positions are assumed to be isotropic with different force-displacement and force-current stiffness parameters.

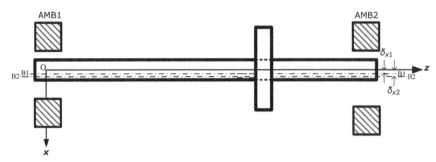

Figure 1. A rigid rotor system with an offset disc supported on two active magnetic bearings in *x-z* plane.

Owing to the rigid behaviour of rotor, the displacement of rotor centre of gravity can be expressed as functions of the displacement of rotor at AMB1 and AMB2 locations, and distance of centre of gravity from both AMBs. Figure 2 can be followed to determine this displacement, in which the left and right AMBs are shown by A1 and A2, the rotor centre of gravity and disc location are symbolized by G and D, respectively. The distances a_1 and a_2 are between AMB1 and AMB2 from the rotor centre of gravity and l_1 is the distance between disc and the rigid rotor centre of gravity. The displacement of rotor at the centre of gravity location can be written as

$$u_x = \bar{a}_2 u_{x1} + \bar{a}_1 u_{x2}; \; u_y = \bar{a}_2 u_{y1} + \bar{a}_1 u_{y2}; \; \varphi_y = (-u_{x1} + u_{x2})/l; \; \varphi_x = (u_{y1} - u_{y2})/l \tag{1}$$

with

$$\bar{a}_1 = \frac{a_1}{l}; \; \bar{a}_2 = \frac{a_2}{l}$$

where u_x and u_y are the x and y directional rigid rotor centre of gravity and the angular displacements of rotor in the y-z and x-z planes are represented by φ_y and φ_x, respectively. The rotor vibrational displacements at AMB1 and AMB2 positions in the x and y directions are given by (u_{x1}, u_{y1}) and (u_{x2}, u_{y2}).

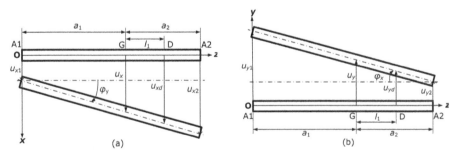

Figure 2. Rigid rotor translational and angular displacements at different locations in (a) *x-z* plane (b) *y-z* plane.

2.1 Unbalance force model

The exciting force due to disc unbalance fault on the rotor system in x and y directions are expressed as

$$f_{unbx} = m_d e \omega^2 \cos(\omega t + \beta); \; f_{unby} = m_d e \omega^2 \sin(\omega t + \beta) \tag{2}$$

where the disc mass, unbalance eccentricity, phase of unbalance and the rotor spin speed are symbolized by m_d, e, β and ω respectively.

2.2 Misaligned AMB force model

Equation (3) presents the force on rotor in x direction due to active magnetic bearings when the rotor axis is in alignment with the AMB axis [21], as shown in Figure 3(a).

$$f_x = k \left\{ \frac{(i_0 + i_x)^2}{(s_0 + u_x)^2} - \frac{(i_0 - i_x)^2}{(s_0 - u_x)^2} \right\} \tag{3}$$

where the constant k, which depends upon the structural parameters of actuator such as number of coils turned around poles N_a, angle between two consecutive poles a, cross sectional area of poles A_a, etc. and vacuum permeability μ_0 which is equal to $4\pi \times 10^{-7}$ H/m, is

$$k = \frac{1}{4} \mu_0 N^2 A_a \cos \frac{a}{2} \tag{4}$$

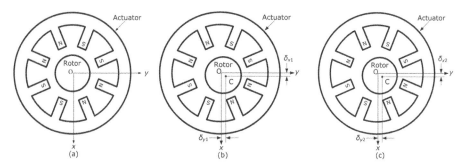

Figure 3. Side view of actuator in x-y plane (a) when it is perfectly aligned with the rotor at both AMBs position (b) when it is misaligned with rotor at AMB1 position (c) when it is misaligned with rotor at AMB2 position.

The x-directional rotor displacement and AMB controlling current are represented by u_x and i_x. The bias current and the AMB air gap are symbolized by i_0 and s_0, respectively. Further, simplification of equation (3) based on assumption $u_x \ll s_0$ and neglecting the higher order terms, i.e. u_x^2, i_x^2, $u_x i_x$ and $u_x i_x^2$, the linearized form for equation (3) can be expressed as

$$f_x = k_a u_x + k_i i_x \tag{5}$$

where the force-displacement and force-current stiffness constants of AMB are denoted by k_a and k_i, which is written as

$$k_a = \frac{4ki_0^2}{s_0^3}; \quad k_i = \frac{4ki_0}{s_0^2} \tag{6}$$

Similarly, in the perfectly aligned case, the force on rotor due to AMBs in y-direction can be expressed as

$$f_y = k_a u_y + k_i i_y \tag{7}$$

where u_y and i_y are the y-directional rotor vibrational displacement and AMB controlling current, respectively. However, in an actual rotor system, it is very difficult to find a perfectly aligned rotor axis with supported AMB axes. The misalignment between them may occur due to mechanical machining and rotor system assembling errors [22] as well as sensor measurement error while locating the centre of AMB as discussed in Section 1. Figures 3(b) and 3(c) depict the side view of actuators of AMB1 and AMB2 that is misaligned with the rotor by δ_{x1} and δ_{x2} in the x-direction and δ_{y1} and δ_{y2} in the y-direction. This non-coincidence of the axes may affect the performance of AMB for uneven air gap distribution. Therefore, the air gap between the rotor and an AMB is altered between the upper and lower poles. For the case of AMB1 misalignment in the x-direction, the modified lower and upper air gaps are $(s_0-\delta_{x1})$ and $(s_0+\delta_{x1})$, respectively. Thus, the misaligned AMB1 force on rotor due to this uneven air gap in the x-direction can be written as

$$f_{mx1} = k \left\{ \frac{(i_0 + i_{x1})^2}{(s_0 - \delta_{x1} + u_{x1})^2} - \frac{(i_0 - i_{x1})^2}{(s_0 + \delta_{x1} - u_{x1})^2} \right\} \tag{8}$$

The assumption of $(u_{x1} \ll s_0-\delta_{x1})$ and $(u_{x1} \ll s_0+\delta_{x1})$, and neglegence of higher order terms give the linearized form of the force f_{mx1} as

$$f_{mx1} = k_{ma1} u_{x1} + k_{mi1} i_{x1} + f_1 \tag{9}$$

with

$$k_{ma1} = \frac{k_{a1}}{\left(1 - \bar{\delta}_1^2\right)^2}; k_{mi1} = \frac{k_{i1}\left(1 + \bar{\delta}_1^2\right)}{\left(1 - \bar{\delta}_1^2\right)^2}; f_1 = \frac{f_{a1}\bar{\delta}_1}{\left(1 - \bar{\delta}_1^2\right)^2}; f_{a1} = \frac{4k_1 i_0^2}{s_0^2}; \bar{\delta}_1 = \frac{\delta_{x1}}{s_0} \tag{10}$$

Similarly, the misaligned AMB1 force in the y-direction can be expressed as

$$f_{my1} = k_{ma2} u_{x1} + k_{mi2} i_{x1} + f_2 \tag{11}$$

with

$$k_{ma2} = \frac{k_{a1}}{\left(1 - \bar{\delta}_2^2\right)^2}; k_{mi2} = \frac{k_{i1}\left(1 + \bar{\delta}_2^2\right)}{\left(1 - \bar{\delta}_2^2\right)^2}; f_2 = \frac{f_{a1}\bar{\delta}_2}{\left(1 - \bar{\delta}_2^2\right)^2}; f_{a1} = \frac{4k_1 i_0^2}{s_0^2}; \bar{\delta}_2 = \frac{\delta_{y1}}{s_0} \tag{12}$$

Following this concept, the x- and y-directional misaligned AMB2 forces on the rotor can be written as

$$f_{mx2} = k_{ma3}u_{x2} + k_{mi3}i_{x2} + f_3 \tag{13}$$

$$f_{my2} = k_{ma4}u_{x2} + k_{mi4}i_{x2} + f_4 \tag{14}$$

with

$$
\begin{aligned}
&k_{ma3} = \frac{k_{a2}}{\left(1 - \bar{\delta}_3^2\right)^2}; \; k_{ma4} = \frac{k_{a2}}{\left(1 - \bar{\delta}_4^2\right)^2}; \; k_{mi3} = \frac{k_{i2}\left(1 + \bar{\delta}_3^2\right)}{\left(1 - \bar{\delta}_3^2\right)^2}; \; k_{mi4} = \frac{k_{i2}\left(1 + \bar{\delta}_4^2\right)}{\left(1 - \bar{\delta}_4^2\right)^2}; \\
&f_3 = \frac{f_{a2}\bar{\delta}_3}{\left(1 - \bar{\delta}_3^2\right)^2}; \; f_4 = \frac{f_{a2}\bar{\delta}_4}{\left(1 - \bar{\delta}_4^2\right)^2}; \; f_{a1} = \frac{4k_1 i_0^2}{s_0^2}; \; f_{a2} = \frac{4k_2 i_0^2}{s_0^2}; \; \bar{\delta}_3 = \frac{\delta_{x2}}{s_0}; \; \bar{\delta}_4 = \frac{\delta_{y2}}{s_0}
\end{aligned}
\tag{15}
$$

The x- and y-directional current output of PID controller [23] at AMB1 and AMB2 locations can be given as

$$i_{x1} = -\left(k_P u_{x1} + k_I \int u_{x1} dt + k_D \dot{u}_{x1}\right); \; i_{y1} = -\left(k_P u_{y1} + k_I \int u_{y1} dt + k_D \dot{u}_{y1}\right) \tag{16}$$

$$i_{x2} = -\left(k_P u_{x2} + k_I \int u_{x2} dt + k_D \dot{u}_{x2}\right); \; i_{y2} = -\left(k_P u_{y2} + k_I \int u_{y2} dt + k_D \dot{u}_{y2}\right) \tag{17}$$

where k_P, k_D and k_I are the PID controller proportional, derivative and integral gains, respectively.

2.3   Non-dimensional equations of motion of rotor-AMB system

On the basis of the moment equilibrium method, considering the moment due to inertia force, misaligned AMB force, unbalance force and the gyroscopic effect in the x-z and y-z planes about AMB1 and AMB2 locations, the dynamic equations of motion of the rigid rotor-AMB system have been derived. The equations of motion of the rotor system can be expressed as

$$m\ddot{u}_x \frac{a_2}{l} - \frac{I_d}{l}\ddot{\varphi}_y + \frac{I_p}{l}\dot{\varphi}_x \omega - f_{unbx}\left(\frac{a_2}{l} - \frac{l_1}{l}\right) - f_{mx1} = 0 \text{ and}$$

$$m\ddot{u}_y \frac{a_2}{l} + \frac{I_d}{l}\ddot{\varphi}_x - \frac{I_p}{l}\dot{\varphi}_y \omega - f_{unby}\left(\frac{a_2}{l} - \frac{l_1}{l}\right) - f_{my1} = 0 \tag{18}$$

$$m\ddot{u}_x \frac{a_1}{l} + \frac{I_d}{l}\ddot{\varphi}_y - \frac{I_p}{l}\dot{\varphi}_x \omega - f_{unbx}\left(\frac{a_1}{l} + \frac{l_1}{l}\right) - f_{mx2} = 0 \text{ and}$$

$$m\ddot{u}_y \frac{a_1}{l} - \frac{I_d}{l}\ddot{\varphi}_x + \frac{I_p}{l}\dot{\varphi}_y \omega - f_{unby}\left(\frac{a_1}{l} + \frac{l_1}{l}\right) - f_{my2} = 0 \tag{19}$$

where m is the rotor (shaft and disc) mass, I_d and I_p are the diametral mass moment of inertia of the rigid rotor system and the disc polar mass moment of inertia, respectively. After substituting equation (1) into the equations (18) and (19), and further writing all the equations in non-dimensional form in terms of several dimensionless rotor-AMB system and faults parameters, we get the compact matrix form as

$$\bar{\mathbf{M}}\ddot{\bar{\mathbf{q}}}-\bar{\mathbf{G}}\dot{\bar{\mathbf{q}}}=\bar{\mathbf{f}}_{unb}+\bar{\mathbf{f}}_{mAMB} \qquad (20)$$

where the dimensionless mass and gyroscopic matrices are

$$\bar{\mathbf{M}}=\begin{bmatrix} (\bar{m}\bar{a}_2^2+\bar{i}_d) & 0 & (\bar{m}\bar{a}_1\bar{a}_2-\bar{i}_d) & 0 \\ 0 & (\bar{m}\bar{a}_2^2+\bar{i}_d) & 0 & (\bar{m}\bar{a}_1\bar{a}_2-\bar{i}_d) \\ (\bar{m}\bar{a}_1\bar{a}_2-\bar{i}_d) & 0 & (\bar{m}\bar{a}_1^2+\bar{i}_d) & 0 \\ 0 & (\bar{m}\bar{a}_1\bar{a}_2-\bar{i}_d) & 0 & (\bar{m}\bar{a}_1^2+\bar{i}_d) \end{bmatrix}; \bar{\mathbf{G}}=\begin{bmatrix} 0 & \bar{i}_p & 0 & -\bar{i}_p \\ \bar{i}_p & 0 & -\bar{i}_p & 0 \\ 0 & -\bar{i}_p & 0 & i_p \\ -\bar{i}_p & 0 & \bar{i}_p & 0 \end{bmatrix}$$

$$(21)$$

With $\bar{m}=\frac{ms_0\omega_{nf}^2}{W}; \bar{i}_d=\frac{I_d s_0\omega_{nf}^2}{Wl^2}; \bar{i}_p=\frac{I_p s_0\omega_{nf}^2}{Wl^2}; \bar{l}_1=\frac{l_1}{l}$

The dimensionless vibrational displacement vector is

$$\bar{\mathbf{q}}=\left\{\begin{matrix} \bar{u}_{x1} & \bar{u}_{y1} & \bar{u}_{x2} & \bar{u}_{y2} \end{matrix}\right\}^T \qquad (22)$$

The dimensionless unbalance force vector is

$$\bar{\mathbf{f}}_{unb}=\bar{m}_d\bar{e}\left\{\begin{matrix} \cos(\tau+\beta)(\bar{a}_2-\bar{l}_1) & \sin(\tau+\beta)(\bar{a}_2-\bar{l}_1) & \cos(\tau+\beta)(\bar{a}_1+\bar{l}_1) & \sin(\tau+\beta)(\bar{a}_1+\bar{l}_1) \end{matrix}\right\}^T \quad (23)$$

With $\bar{m}_d=\frac{m_d s_0\omega_{nf}^2}{W}; \bar{e}=\frac{e}{s_0}; \tau=\omega t; \bar{\omega}=\frac{\omega}{\omega_{nf}}$

The non-dimensional misaligned AMBs force vector is

$$\bar{\mathbf{f}}_{mAMB}=\left\{\begin{matrix} \bar{f}_{mx1} & \bar{f}_{my1} & \bar{f}_{mx2} & \bar{f}_{my2} \end{matrix}\right\}^T \qquad (24)$$

With $\bar{f}_{mx1}=\frac{f_{mx1}}{W}; \bar{f}_{my1}=\frac{f_{my1}}{W}; \bar{f}_{mx2}=\frac{f_{mx2}}{W}; \bar{f}_{my2}=\frac{f_{my2}}{W}$

On putting equations (9), (11), (13) and (14) into equation (24), we get

$$\bar{\mathbf{f}}_{mAMB}=\bar{\mathbf{K}}_{ma}\bar{\mathbf{q}}+\bar{\mathbf{K}}_{mi}\bar{\mathbf{i}}+\bar{\mathbf{f}}_c \qquad (25)$$

where the dimensionless current output vector of PID controller is

$$\bar{\mathbf{i}}=-\left\{\bar{k}_P\bar{\mathbf{q}}+\bar{k}_I\int\bar{\mathbf{q}}d\tau+\bar{k}_D\dot{\bar{\mathbf{q}}}\right\} \qquad (26)$$

with

$$\bar{\mathbf{i}}=\left\{\begin{matrix} \bar{i}_{x1} & \bar{i}_{y1} & \bar{i}_{x2} & \bar{i}_{y2} \end{matrix}\right\}^T \qquad (27)$$

$$\bar{\mathbf{K}}_{ma}=\begin{bmatrix} \bar{k}_{ma1} & 0 & 0 & 0 \\ 0 & \bar{k}_{ma2} & 0 & 0 \\ 0 & 0 & \bar{k}_{ma3} & 0 \\ 0 & 0 & 0 & \bar{k}_{ma4} \end{bmatrix}; \bar{\mathbf{K}}_{mi}=\begin{bmatrix} \bar{k}_{mi1} & 0 & 0 & 0 \\ 0 & \bar{k}_{mi2} & 0 & 0 \\ 0 & 0 & \bar{k}_{mi3} & 0 \\ 0 & 0 & 0 & \bar{k}_{mi4} \end{bmatrix}; \bar{\mathbf{f}}_c=\left\{\begin{matrix} \bar{f}_1 \\ \bar{f}_2 \\ \bar{f}_3 \\ \bar{f}_4 \end{matrix}\right\}$$

The non-dimensional misaligned AMB parameters and force constants in x and y directions for both AMBs are

158

$$\bar{k}_{ma1} = \frac{k_{ma1}s_0}{W}; \ \bar{k}_{ma2} = \frac{k_{ma2}s_0}{W}; \ \bar{k}_{ma3} = \frac{k_{ma3}s_0}{W}; \ \bar{k}_{ma4} = \frac{k_{ma4}s_0}{W}; \ \bar{k}_{mi1} = \frac{k_{mi1}i_0}{W}; \ \bar{k}_{mi2}$$

$$= \frac{k_{mi2}i_0}{W}; \ \bar{k}_{mi3} = \frac{k_{mi3}i_0}{W}; \quad\quad\quad\quad\quad (28)$$

$$\bar{k}_{mi4} = \frac{k_{mi4}i_0}{W}; \ \bar{f}_1 = \frac{f_1}{W}; \ \bar{f}_2 = \frac{f_2}{W}; \ \bar{f}_3 = \frac{f_3}{W}; \ \bar{f}_4 = \frac{f_4}{W}; \ \bar{k}_P = \frac{k_P s_0}{i_0}; \bar{k}_I = \frac{k_I s_0}{i_0 \omega}; \bar{k}_D = \frac{k_D \omega s_0}{i_0}$$

On substituting equations (10), (12) and (15), we can write the dimensionless constants associated with both misaligned AMBs as

$$\bar{k}_{ma1} = \frac{k_{a1}s_0}{W\left(1-\bar{\delta}_1^2\right)^2}; \bar{k}_{ma2} = \frac{k_{a1}s_0}{W\left(1-\bar{\delta}_2^2\right)^2}; \bar{k}_{ma3} = \frac{k_{a2}s_0}{W\left(1-\bar{\delta}_3^2\right)^2}; \bar{k}_{ma4} = \frac{k_{a2}s_0}{W\left(1-\bar{\delta}_4^2\right)^2}; \bar{\delta}_1 = \frac{\delta_{x1}}{s_0};$$

$$\bar{k}_{mi1} = \frac{k_{i1}i_0\left(1+\bar{\delta}_1^2\right)}{W\left(1-\bar{\delta}_1^2\right)^2}; \bar{k}_{mi2} = \frac{k_{i1}i_0\left(1+\bar{\delta}_2^2\right)}{W\left(1-\bar{\delta}_2^2\right)^2}; \bar{k}_{mi3} = \frac{k_{i2}i_0\left(1+\bar{\delta}_3^2\right)}{W\left(1-\bar{\delta}_3^2\right)^2}; \bar{k}_{mi4} = \frac{k_{i2}i_0\left(1+\bar{\delta}_4^2\right)}{W\left(1-\bar{\delta}_4^2\right)^2}; \bar{\delta}_2 = \frac{\delta_{y1}}{s_0}; \quad (29)$$

$$\bar{f}_1 = \frac{f_{a1}\bar{\delta}_1}{W\left(1-\bar{\delta}_1^2\right)^2}; \bar{f}_2 = \frac{f_{a1}\bar{\delta}_2}{W\left(1-\bar{\delta}_2^2\right)^2}; \bar{f}_3 = \frac{f_{a2}\bar{\delta}_3}{W\left(1-\bar{\delta}_3^2\right)^2}; \bar{f}_4 = \frac{f_{a2}\bar{\delta}_4}{W\left(1-\bar{\delta}_4^2\right)^2}; \bar{\delta}_3 = \frac{\delta_{x2}}{s_0}; \bar{\delta}_4 = \frac{\delta_{y2}}{s_0}$$

Following equation (20), it can be stated that the equations of motion of the four degrees-of-freedom (DOFs) rigid rotor system are in non-dimensional form, which contain several dimensionless rotor and AMB parameters (i.e., $\bar{m}$ $\bar{m}_d$, $\bar{i}_d$, $\bar{i}_p$, $\bar{k}_{ma}$, $\bar{k}_{mi}$ and $\bar{f}$) and dimensionless unbalance ($\bar{e}$) as well as misalignment ($\bar{\delta}_1$, $\bar{\delta}_2$, $\bar{\delta}_3$, $\bar{\delta}_4$) fault parameters. The next section describes the dynamic effect on dimensionless displacement and current responses (obtained from solution equation (20)) in the presence of disc unbalance eccentricity and AMB misalignment fault parameters.

3 NUMERICAL RESULTS AND DISCUSSION

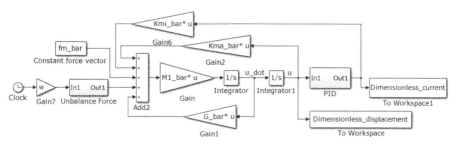

Figure 4. Simulink model of the problem.

Using the SIMULINK™ model is displayed in Figure 4, equation (20) has been solved to generate dimensionless displacement and controlling current responses of the rotor-AMB system at both AMBs positions, in the x- and y- directions.

Table 1. Dimensionless rotor-AMB parameters for numerical investigation.

Parameters	Assumed Values	Parameters	Assumed Values
Nondimensiona rotor mass, $\bar{m}$	11.6933	Dimensionless AMB1 force-displacement stiff-ness, $\bar{k}_{a1}$	0.8986
Nondimensional disc mass, $\bar{m}_d$	8.5532	Dimensionless AMB2 force-displacement stiff-ness, $\bar{k}_{a2}$	3.6402
Disc eccentricity phase, β (deg)	30.00	Dimensionless AMB1 force- current stiffness, $\bar{k}_{i1}$	0.8986
Dimensionless diam-etral moment of inertia of the rotor, $\bar{i}_d$	0.4705	Dimensionless AMB2 force- current stiffness, $\bar{k}_{i2}$	3.6402
Dimensionless polar moment of inertia of the disc, $\bar{i}_p$	0.1503	Dimensionless distance of AMB1 from rotor centre of gravity, $\bar{a}_1$	0.5925
Dimensionless propor-tional gain, $\bar{k}_P$	1.2000	Dimensionless distance of AMB2 from rotor centre of gravity, $\bar{a}_2$	0.4075
Dimensionless deriva-tive gain, $\bar{k}_D$	0.3213	Dimensionless distance of disc from rotor centre of gravity, $\bar{l}_l$	0.0325
Dimensionless integral gain, $\bar{k}_I$	0.0011	Dimensionless rotor spin Speed, $\bar{\omega}$	0.5866

Figure 4 reveals the presence of multiple blocks, which execute different functions. The time taken in the simulation is performed by the clock block. For the multiplication of constant or variable numbers and matrices, the triangular gain block has been used. The add block executes the addition of vectors associated with the unbalance force and forces due to gyroscopic couple effect and misaligned AMB force. The con-trolling current output of the system has been obtained utilizing the PID block, which takes rotor displacement as input. This block contains the matrix terms of propor-tional, derivative and integral factors of PID controller. Finally, the workspace block saves the dimensionless displacement and current signal data for plotting several fre-quency responses, at different values of disc eccentricity and misalignment ratios.

160

Table 2. Assumed values of disc eccentricity as well as x and y directions misalignment ratios along with the generated peak values of frequency response of dimensionless displacement at both AMBs locations.

Disc eccentricity ratio ($\bar{e}$)	Misalignment ratios				Peak values of frequency response of dimensionless displacement at AMB1	Peak values of frequency response of dimensionless displacement at AMB2
	($\bar{\delta}_1$)	($\bar{\delta}_2$)	($\bar{\delta}_3$)	($\bar{\delta}_4$)	($\bar{R}_1$)	($\bar{R}_2$)
0.125	0.250	0.300	0.275	0.2625	0.2194	0.0561
0.150	0.275	0.325	0.300	0.2875	0.2682	0.0663
0.175	0.300	0.350	0.325	0.3125	0.3201	0.0780
0.200	0.325	0.375	0.350	0.3375	0.3764	0.0921
0.225	0.350	0.400	0.375	0.3625	0.4388	0.1096
0.250	0.375	0.425	0.400	0.3875	0.5095	0.1317

Table 3. Assumed values of disc eccentricity as well as x and y directions misalignment ratios along with the generated peak values of frequency response of dimensionless controlling current at both AMBs locations.

Disc eccentricity ratio ($\bar{e}$)	Misalignment ratios				Peak values of frequency response of dimensionless current at AMB1	Peak values of frequency response of dimensionless current at AMB2
	($\bar{\delta}_1$)	($\bar{\delta}_2$)	($\bar{\delta}_3$)	($\bar{\delta}_4$)	($\bar{I}_1$)	($\bar{I}_2$)
0.125	0.250	0.300	0.275	0.2625	0.2670	0.0683
0.150	0.275	0.325	0.300	0.2875	0.3265	0.0807
0.175	0.300	0.350	0.325	0.3125	0.3897	0.0949
0.200	0.325	0.375	0.350	0.3375	0.4582	0.1121
0.225	0.350	0.400	0.375	0.3625	0.5342	0.1335
0.250	0.375	0.425	0.400	0.3875	0.6202	0.1603

The numerical simulation for the unbalanced and misaligned rigid rotor-active magnetic bearing system was undertaken for 5 s with the help of dimensionless parameters given in Table 1. The dimensionless displacement and current responses at both AMB positions have been obtained using a fourth-order Runge-Kutta differential solver with 0.0001 s fixed time step size. The chosen dimensionless rotor speed for the numerical investigation purpose is 0.5866, which is lower than the first critical speed (0.8896) of the rotor-AMB system. This is essentially required for exhibiting rigid rotor behaviour during operation. Tables 2 and 3 display the assumed values of the disc eccentricity ratio ($\bar{e}$) and misalignment ratios ($\bar{\delta}_1, \bar{\delta}_2, \bar{\delta}_3$ and $\bar{\delta}_4$) in the x- and y-directions of both AMBs. The simulation has been run individually for different values of fault parameters to generate the non-dimensional responses with respect to frequency ratio ($\bar{\omega}$), which is shown in Figure 5. Figures 5(a) and (b) represent the frequency response of dimensionless rotor displacement at the locations of AMB1 and AMB2, respectively. The frequency response of dimensionless controlling current at AMB1 and AMB2 positions are presented in Figure 5(c) and (d). The maximum values of frequency responses captured from Figure 5, for every assumed fault parameters at both AMBs is also presented in Table 2 and 3. Values of disc eccentricities and misalignment ratios expressed in these Tables are in the ascending order. On the basis of this, it would be noticeable that the percentage change in the responses for the individual values are also in the ascending order. The rise in percentage of the peak values of displacement at AMB1 from Table 2 are 22.24%, 45.90%, 71.56%, 100% and 132.22%, respectively, relative to the captured maximum displacement value of 0.2194 for ($\bar{e}$=0.125) and misalignment ratios ($\bar{\delta}_1$=0.25, $\bar{\delta}_2$=0.30, $\bar{\delta}_3$=0.275 and $\bar{\delta}_4$=0.2625). Similarly, with respect to the displacement value of 0.0561 at initial assumed values of fault parameters, the percentage increase in frequency response of dimensionless displacement at AMB2 are 18.18%, 39.04%, 64.17%, 95.36% and 134.76%, respectively.

Using Table 3, the rise in maximum magnitude in the percentage form have also been calculated for the frequency response of dimensionless controlling current at AMB1 and AMB2, respectively. Relative to the initial peak values of current, i.e. 0.2670 and 0.0683 at AMB1 and AMB2 locations, the percentage change in peak values are also found to be in the increasing order. The increase in percentages are 22.28%, 45.95%, 71.61%, 100.07% and 132.28% at AMB1 and 18.15%, 38.94%, 64.13%, 95.46% and 134.70% at the position of AMB2. Some of the major observations can be made from these results are that there is a rapid rise in the peak values of frequency response of non-dimensional displacement and current with the slight increment in the dimensionless fault parameters values. It can also be noticed the percentage increase in the displacement and current peak values at AMB1 are in the same order. In the similar way, at AMB2 position, the rise in maximum values of dimensionless displacement and current are almost equal. It is obvious because the current output of the PID controller directly depends upon the input rotor displacement at both AMBs, following equation (26). Moreover, from Figure 5 (a-d), it is observable that the size of curves of frequency response increases with increase in the dimensionless values of system fault parameters of disc eccentricity and misalignment ratios.

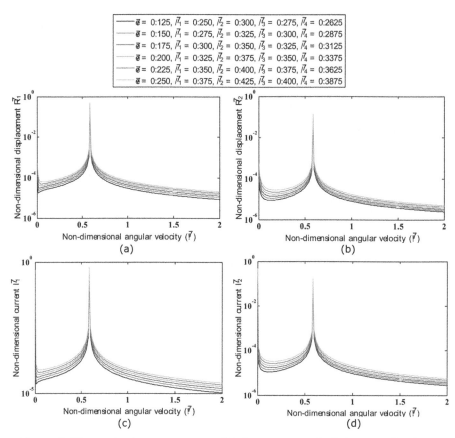

Figure 5. Dimensionless response with frequency ratio (a) Rotor displacement at AMB1 (b) Rotor displacement at AMB2 (c) Controlling current at AMB1 (d) Controlling current at AMB2.

Further, the analysis on the dynamic nature of rotor-AMB system under the effects of different ranges of fault parameters have also been explored, for combinations of each value of disc eccentricity ratios and every misalignment ratio. The obtained peak values of frequency response of non-dimensional displacement and current at AMB1 and AMB2 versus the disc eccentricity ratios are shown in Figure 6 (a-d). The values of maximum displacement and current at AMB1 position for the combination of every fault parameters are found to be in the range of 0.2194 to 0.5095 and 0.2670 to 0.6203, respectively. Similarly, the peak values of dimensionless displacement and current at AMB2, respectively, are within the range of 0.0561 to 0.1317 and 0.0683 to 0.1603. The main conclusion can also be drawn from Figure 6 that the magnitude of frequency response of dimensionless rotor displacement and controlling current at both AMBs locations rise rapidly with the increasing order of all value combination of assumed set of dimensionless parameters associated with unbalance and AMB misalignment faults. Moreover, the numerical simulation has also been performed to investigate the influence of AMB misalignment on the considered rotor system.

Figure 7 Displays the frequency response of dimensionless displacement and current at AMB1 location for without and with AMB misalignment. Unbalance fault due to disc is present in both the study cases. To explore the misalignment effect at the peak part of frequency response plots, the related portion has been magnified in Figure 7 (a) and (b). This figure has been plotted for disc eccentricity ratio ($\bar{e}$) of 0.25 and AMB misalignment ratios of (δ_1=0.375, δ_2=0.425, δ_3=0.400 and δ_4=0.3875). The peak values of dimensionless displacement as a function of frequency ratio, without and with misalignment are found to be 0.3916 and 0.5095, respectively.

In the same line, the current peaks values are 0.4767 and 0.6203, respectively, for unbalance fault only and integrated unbalance and AMB misalignment faults. Misalignment of both AMBs in the rotor system causes enhancement of the displacement and current maximum values by 30.11% and 30.12%, respectively, as compared to the case of perfect alignment of rotor and supported AMBs axes. These increments in response magnitudes represent the severity of misalignment fault.

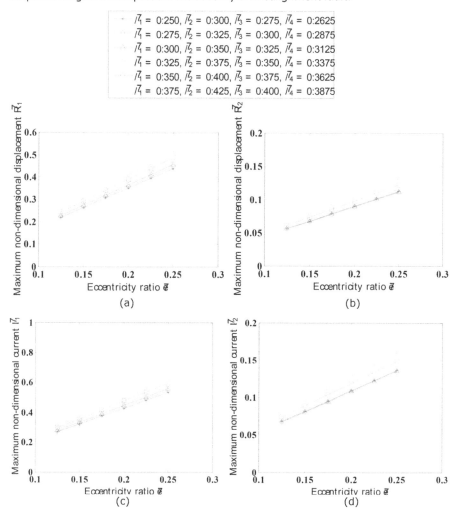

Figure 6. Peak values of dimensionless frequency response versus eccentricity ratio for each misalignment ratio (a) Rotor displacement at AMB1 (b) Rotor displacement at AMB2 (c) Controlling current at AMB1 (d) Controlling current at AMB2.

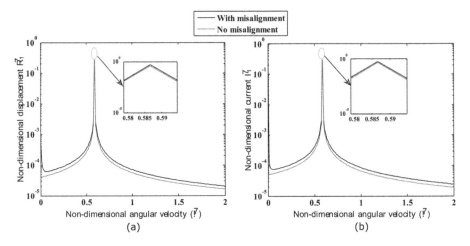

Figure 7. Influence of AMB misalignment on the rotor system (a) Frequency response of dimensionless displacement at AMB1 (b) Frequency response of dimensionless controlling current at AMB1.

4   CONCLUSIONS

This paper concludes with development of mathematical model of the four-degrees-of freedom rigid rotor system levitated by two misaligned active magnetic bearings at the end positions. An unbalanced disc is present at the offset position of the rigid shaft, which causes gyroscopic coupling at higher spin speeds. Forces due to misaligned AMBs has been obtained from linearization concept and found to consist an extra constant force term along with the terms related to force-displacement and force-current stiffnesses of AMBs. Dimensionless equations of motion of the unbalanced and misaligned rotor system have been derived as function of multiple non-dimensional system and faults parameters. These equations have been solved by developing SIMULINK™ model, which generates the dimensionless displacement and controlling current at AMBs positions. The study based on numerical simulation has been explored to investigate the dynamic nature of the rotor system under the combination of different ranges of disc eccentricity and misalignment ratios. Furthermore, the AMB misalignment effect on the rotor performance has also been studied from the frequency response plots. From the numerical investigation, it is concluded the peak values of frequency response of non-dimensional displacement and current rapidly increases with the slight increment in the dimensionless fault parameters values. Misalignment of both AMBs also causes an increase in the magnitude of the frequency responses of the rotor system. Experimental validation on the numerically obtained results of the considered rigid rotor-AMB system is quite interesting to work for the future (as rotor-AMB test rig set up is in progress). This test set up includes various mechanical and electrical components such a shaft, one disc, two number of eight pole actuators, two rolling element bearings (work as touchdown bearings), one motor, one variable frequency drive (VFD), one coupling, sensor stands, base plate, dSPACE controller, eddy current proximity sensors, eight number of amplifiers (four for each AMB for levitation purpose), etc.

REFERENCES

[1] J. K. Sinha, "Health monitoring techniques for rotating machinery," University of Wales, Swansea, 2002.

[2] S. Edwards, A. Lees, and M. Friswell, "Estimating rotor unbalance from a single run-down," in *IMECHE CONFERENCE TRANSACTIONS*, 2000, pp. 323–334.

[3] M. De Queiroz, "An active identification method of rotor unbalance parameters," *Journal of Vibration and Control*, vol. 15, pp. 1365–1374, 2009.

[4] R. Markert, R. Platz, and M. Seidler, "Model based fault identification in rotor systems by least squares fitting," *International Journal of Rotating Machinery*, vol. 7, pp. 311–321, 2001.

[5] R. Platz, R. Markert, and M. Seidler, "Validation of online diagnostics of malfunctions in rotor systems," in *IMECHE conference transactions*, 2000, pp. 581–590.

[6] G. Sudhakar and A. Sekhar, "Identification of unbalance in a rotor bearing system," *Journal of Sound and Vibration*, vol. 330, pp. 2299–2313, 2011.

[7] J. Yao, L. Liu, F. Yang, F. Scarpa, and J. Gao, "Identification and optimization of unbalance parameters in rotor-bearing systems," *Journal of Sound and Vibration*, vol. 431, pp. 54–69, 2018.

[8] A. Shrivastava and A. R. Mohanty, "Identification of unbalance in a rotor system using a joint input-state estimation technique," *Journal of Sound and Vibration*, vol. 442, pp. 414–427, 2019.

[9] D. Dewell and L. Mitchell, "Detection of a misaligned disk coupling using spectrum analysis," *Journal of Vibration, Acoustics, Stress, and Reliability in Design*, vol. 106, pp. 9–16, 1984.

[10] J. K. Sinha, A. Lees, and M. Friswell, "Estimating unbalance and misalignment of a flexible rotating machine from a single run-down," *Journal of Sound and Vibration*, vol. 272, pp. 967–989, 2004.

[11] A. Lees, J. K. Sinha, and M. I. Friswell, "Estimating rotor unbalance and misalignment from a single run-down," in *Materials Science Forum*, 2003, pp. 229–236.

[12] M. Lal and R. Tiwari, "Multi-fault identification in simple rotor-bearing-coupling systems based on forced response measurements," *Mechanism and Machine Theory*, vol. 51, pp. 87–109, 2012.

[13] S. K. Kuppa and M. Lal, "Characteristic Parameters Estimation of Active Magnetic Bearings in a Coupled Rotor System," *Journal of Verification, Validation and Uncertainty Quantification*, vol. 4, pp. 031001–031012, 2019.

[14] R. S. Srinivas, R. Tiwari, and C. Kannababu, "Model based analysis and identification of multiple fault parameters in coupled rotor systems with offset discs in the presence of angular misalignment and integrated with an active magnetic bearing," *Journal of Sound and Vibration*, vol. 450, pp. 109–140, 2019.

[15] R. S. Srinivas, R. Tiwari, and C. Kannababu, "Application of active magnetic bearings in flexible rotordynamic systems–A state-of-the-art review," *Mechanical Systems and Signal Processing*, vol. 106, pp. 537–572, 2018.

[16] F. W. da Silva Tuckmantel and K. L. Cavalca, "Vibration signatures of a rotor-coupling-bearing system under angular misalignment," *Mechanism and Machine Theory*, vol. 133, pp. 559–583, 2019.

[17] P. Kumar and R. Tiwari, "Development of a Novel Approach for Quantitative Estimation of Rotor Unbalance and Misalignment in a Rotor System Levitated by Active Magnetic Bearings," *Iranian Journal of Science and Technology-Transactions of Mechanical Engineering*, 2020.

[18] P. Kumar and R. Tiwari, "A Numerical Study on the Effect of Unbalance and Misalignment Fault Parameters in a Rigid Rotor Levitated by Active Magnetic Bearings," in *ASME 2019 Gas Turbine India Conference*, 2019.

[19] W. Zhao, C. Hua, D. Wang, and D. Dong, "Fault Diagnosis of Shaft Misalignment and Crack in Rotor System Based on MI-CNN," in *Proceedings of the 13th International Conference on Damage Assessment of Structures*, 2020, pp. 529–540.

[20] Y. Hori and R. Uematsu, "Influence of misalignment of support journal bearings on stability of a multi-rotor system," *Tribology International*, vol. 13, pp. 249–252, 1980.

[21] H. Bleuler, M. Cole, P. Keogh, R. Larsonneur, E. Maslen, Y. Okada, *et al.*, *Magnetic Bearings: Theory, Design, and Application to Rotating Machinery*. Verlag Berlin Heidelberg: Springer Science & Business Media, 2009.

[22] C. D. X. Longxiang, "Influence of Machining Error on the Performance of Active Magnetic Bearing," *Journal of Mechanical Engineering*, vol. 6, 2009.

[23] R. Tiwari, *Rotor Systems: Analysis and Identification*. Boca Raton: CRC Press, 2017.

Modal parameters evaluation of a rolling bearing rotor using operational modal analysis

G.C. Storti, N.A.H. Tsuha, K.L. Cavalca, T.H. Machado

Department of Integrated Systems, School of Mechanical Engineering, University of Campinas, Brazil

ABSTRACT

Operational Modal Analysis when applied to rotating machinery has issues that must be considered in order to have a satisfactory result. The focus of the paper is to identify modal parameters of a rotor with rolling bearings in an operational condition and compare it with a numerical model. In the experimental rotor, it is shown that only the operating condition without external excitation is not sufficient for OMA application and by applying simple techniques, without the need for input measurement and control, using both time and frequency domain, a comparison between experimental data and theoretical model showed good compliance for the identification of natural frequencies and damping factors.

1   INTRODUCTION

Investigating the modal properties of structures has become a common approach for numerical models' parameters updating, structural damage identification, or by means of structural health monitoring systems, for instance. Therefore, the demand for this type of information has stimulated a fast expansion of modal analysis procedures and created the necessity for methods capable of giving a fast and cost-effective analysis of modal parameters. Initially, modal parameters used to be identified using Experimental Modal Analysis (EMA). Although there are several experimental setups and excitation forms, EMA tests are usually carried out under impact hammer and/or shaker excitation and modal parameters are. extracted from methods of curve fitting of resonance peaks from frequency response functions (FRF).

The major drawback of these procedures is that all input forces need to be measured, which makes it difficult to analyse large structures and systems under operating conditions. To overcome the limitations of traditional EMA, a new methodology to analyse and extract modal parameters of structures under operating conditions has been proposed since the late 1990s and is commonly referred to as operational modal analysis (OMA). In OMA, systems inputs are not necessarily controlled and measured, but instead, they are represented by the assumptions of stochastic loads distributed over a broad frequency band, i.e. ideally excited by white noise, to identify natural frequencies, damping ratios and mode shapes of structures under operational conditions.

This approach was initially applied to large civil structures, with several studies conducted on bridges, dams, buildings, etc. due to difficulties over the input generation equipment necessary and the natural excitation sources often present in real cases, such as wind, traffic, and waves. With successful studies case (1–3), the method naturally expanded to new areas, with outstanding development in the testing of offshore wind generator structures (4). Nowadays, vibration analysis has become one of the primary tools in the condition monitoring of rotating machinery (5). One of the main limitations of OMA when applied to this type of system is the addition of harmonic components to the stochastic realization of the inputs. In this way, these forces can bias the modal identification or even be mis-identified as a system pole, limiting its direct application in practical cases. It is necessary, therefore, to study the influence of these components on common OMA algorithms and to propose ways to overcome its limitation to correctly iden-tify and remove, if necessary, these deterministic forces in operation conditions of rotating machinery.

The methods for modal identification of structures under operational loads are usually divided into two main groups based on the domain in which the analysis is performed (6). OMA, like the traditional EMA, has a large amount of time domain techniques. Among these techniques, there are basically three categories that stand out: methods based on the Natural Excitation Technique (NExT) (7), on Auto-Regressive Moving Average (ARMA) (8) models or problems of stochastic subspace identification (SSI) (9). When dealing with frequency domain, the methods are usually based on the analysis of power spectral matrix of response, obtained from measured data, and can be grouped into two types: those directly based on the response spectrum, non-parametric (10), as the vastly applied Fre-quency Domain Decomposition (FDD) (11), or parametric methods, based on opti-mization, as it is the case of the Enhanced Frequency Domain Decomposition (EFDD), Maximum Likelihood (12) and polynomial and transmissibility functions (13), for example.

In this context, the main objective in this paper is to investigate two commonly applied OMA techniques, EFDD and SSI, to extract modal parameters of a rotor supported by cylindrical roller bearings, and compare the results with a numerical finite element model to analyse the accuracy of the extracted natural frequencies and damping ratios. Besides, ways of identifying harmonics are introduced to make sure that the extracted modes are in fact related to the rotor and not influenced by these components.

2 OPERATIONAL MODAL ANALYSIS

The focus of this section is to give a brief overview of two recurrent methods for modal analysis: EFDD and SSI, to provide a basis for understanding the results that will be analyzed for the rotor in the following sections.

2.1 Enhanced Frequency Domain Decomposition (EFDD)

The EFDD technique was first introduced by Brincker et al. (14) and it is based on the Singular Value Decomposition of the spectral matrix. As an extension of the basic fre-quency-domain method, commonly known as Peak-Picking (10), the technique is based on the input and output power spectral densities (PSD) relationship:

$$G_{yy}(j\omega) = \overline{H}(j\omega)G_{xx}(j\omega)H(j\omega)^T \tag{1}$$

where $G_{yy}(j\omega)$ matrix is the output PSD rxr, with r outputs; $H(j\omega)$ is the frequency response function (FRF) matrix nxn, n the number of modes, and $G_{xx}(j\omega)$ is the input PSD matrix mxm, being m the number of inputs. The overbar and T superscript denote complex conjugate and transpose, respectively. The input is assumed to be a white gaussian noise. Therefore, it is possible to rewrite the expression above, decomposing the output power spectral density only as a function of modal parameters. When the system is lightly damped, and the spectrum is analyzed in a region adjacent to a resonance peak, the contribution of m modes at a given frequency tends to be limited and considerably less than the total number of n modes, and the output spectra is recomposed just in terms of modal parameters as seen in the following expression:

$$G_{yy}(j\omega) \underset{\omega \to Sub(\omega_n)}{=} \sum_{i=1}^{m} \left(\frac{d_i \phi_i \phi_i^T}{j\omega - \lambda_i} + \frac{d_i \overline{\phi_i} \phi_i^H}{j\omega - \overline{\lambda_i}} \right) \tag{2}$$

being λ the poles, ϕ_i the matrix containing the mode shapes and d a constant value. Reference (6) presents the procedure in more detail. In the Frequency Domain Decomposition, a prior step to the EFDD, the output PSD is estimated at discrete frequencies ω_i in a pre-stablished frequency range and, then, decomposed by taking the Singular Value Decomposition of Eq. (2):

$$\widehat{G}_{yy}(j\omega_i) = U_i S_i \overline{V}_i^T \tag{3}$$

In this way, an SVD of the output spectra can be interpreted as a representation of the dominant modes in a given resonance peak. Therefore, the first singular value S_i reaches a maximum at $\omega_i = \omega_n$ and the corresponding singular vector is an estimate of the correlated mode shape. In this case, modal frequencies and mode shapes are obtained directly by the analysis of the SVD plot and damping ratios could not be extracted, at this point. To overcome this limitation, an evolution of the methodology was proposed and denominated the Enhanced Frequency Domain Decomposition: data from the discrete frequencies are selected in the neighborhood of resonant peak based on a Modal Assurance Criterion threshold (11) and, by an inverse Fourier Transform, are converted back to time domain resulting in an auto-correlation function, that represents only a single mode related to the resonant peak. Finally, damping can be obtained from this final function and all modal parameters are then estimated.

2.2 Stochastic Subspace Identification (SSI)

It is common to model mechanical systems using a second-order differential equation as a function of the coordinates associated to their degrees of freedom, $q(t)$, usually expressed in the matrix form as:

$$[M]\ddot{q}(t) + [C_d]\dot{q}(t) + [K]q(t) = f(q,t) \tag{4}$$

For a system with n degrees of freedom, $[M]_{nxn}$, $[C_d]_{nxn}$, $[K]_{nxn}$ are, respectively, the general mass, damping and stiffness matrices, $f(q,t)_{nx1}$ are the excitation forces. For application in identification problems, that is, obtaining system matrices from measured data, it is necessary to propose a different approach to deal with the available information. Likewise, it is necessary to deal with the stochastic nature of the input forces. For this reason, stochastic subspace identification algorithms have become widely used for application in OMA (15), with initial studies as in (16), pointing out the potentials of the techniques presented in De Moor and Van Overschee (9) for output-only modal identification. The approach consists of organizing the data in a state-space formulation combined with an observation equation, resulting in Eq. (5).

$$\begin{aligned} x_{k+1} &= [A]x_k + w_k \\ y_k &= [C]x_k + v_k \end{aligned} \tag{5}$$

where y_k is the measurements vector of length l, x_k the state vector of length n, w_k a vector for process and modeling inaccuracies with length n and v_k vector for the measurement noise due to sensor inaccuracy with length l. Matrix $[A]_{nxn}$ describes the system dynamics and it is known as dynamical system matrix. $[C]_{lxn}$ is the output matrix, describing how the internal state is transferred to the outside world in the measurements y_k. Both vectors w_k and v_k are assumed to be stationary and zero mean random process. Assuming the pair $[A]$ and $[C]$ observable, and $[A]$ to be also controllable, the system from Eq. (5) becomes the so-called realization problem and the objective is to extract the physical information from $[A]$ knowing only the time-histories response.

The first step in the algorithm is to gather the measurements in a block Hankel matrix (9), divided into a past reference $[Y_p]$ and future $[Y_f]$ based on a time delay for the time-history. This past and future sections are determined to enable the projection of the row space of the future into the row space of the past, defining a subspace identification capable of retaining all the information in the past that can be useful to the future (17). The projections can be carried out in terms of covariances, the SS covariance driven, or directly into the time-series, the SSI data-driven, which will be further used in the results section. The projection, done in the data-driven approach by a QR-factorization, defines the observability $[O_i]$ matrix as:

$$O_i = Y_f Y_p^T \left(Y_p Y_p^T \right)^\dagger Y_p \qquad (6)$$

where $\dagger$ denotes the Moore-Penrose pseudo-inverse. Together with weighting matrices W_1 and W_2, the next step is to perform an SVD of the matrix defined above in Eq.(6). This operation is capable of extracting the extended observability matrix Γ_i, as follows:

$$\Gamma_i = \begin{bmatrix} [C] \\ [C][A] \\ [C][A]^2 \\ \vdots \\ [C][A]^{(i-1)} \end{bmatrix} = W_1^{-1} U_1 S_1^{\frac{1}{2}} \qquad (7)$$

To complete the problem, it is necessary to combine Γ_i with a Kalman State sequence, and the problem for identification system matrix is considered complete, since Γ_i contains the information necessary for estimating $[A]$ (see (9) for further details). It is possible to demonstrate that the Kalman Filter can be directly inserted into the formulation of the stochastic problem, Eq. (5), a so-called forward innovation model. $[A]$ is obtained in a least square sense and the modal parameters are, consequently, extracted (as seen in (18)). Given that the exact state-space dimension is unknown, it is typical to analyze the poles extracted from $[A]$ matrix in a stabilization diagram, with stabilization criteria previously established to differentiate physical to numerical and noise modes.

3 TEST RIG

To investigate the OMA techniques (EFDD and SSI), the experimental tests were evaluated in a rotor system supported by two radial roller bearings located at LAMAR (Laboratory of Rotating Machinery) at the University of Campinas.

In addition to the two roller bearings (model NJ 202 by NSK®), the rotor consists of a steel shaft with a 15 mm diameter, a magnetic actuator, and a disc. The disc diameter is 94.7 mm and its length, 47.8 mm. The rotor is connected to a 3 CV WEG electric motor by a flexible coupling. Figure 1 shows the experimental test rig with its main components.

The speed control of the electric motor is done by a frequency inverter WEG-CFW-08 remotely connected with a computer. The lubrication of the roller bearings uses a mineral oil, controlled by a pumping system with a filter. The lubrication oil is the Castrol ASW 32 (ISO VG 32).

The test rig allows the acquisition of several types of signals. In this study, four Bruel & Kjaer type 4384 accelerometers were used (two for each roller bearing). After passing through conditioners, the signals are filtered in two steps. The first filter removes the static gain, and the second one is a low-pass with anti-aliasing filter. The signals are acquired by the National Instruments USB-6343 board and the data are processed in LabVIEW™. The rotor passed through alignment and balancing processes, the last one using the coefficient of influence method.

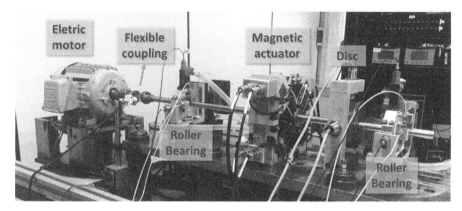

Figure 1. Experimental rotor test bench with its components.

4 NUMERICAL MODEL

In order to compare the results of EFDD and SSI, the rotor system model uses the finite element method (FEM):

$$[M]\{\ddot{q}\} + ([C] + \Omega[G])\{\dot{q}\} + [K]\{q\} = \{f\} \tag{8}$$

where $[M]$, $[K]$ and $[C]$ are respectively the mass, stiffness and damping matrices. Matrix $[G]$ is the gyroscopic matrix, $\{q\}$ is the generalized coordinates vector and $\{f\}$ is the vector of external force.

The FE model is shown in Figure 2. The rotor is divided into eleven Timoshenko beam elements. All disc and beam elements characteristics are in Table 1 and Table 2. The cylindrical roller bearings NJ 202 are at nodes 2 and 10. At nodes 3 and 11, there are pairs of proximity sensors. The disc is at node 8 and the magnetic actuator is positioned at node 5.

173

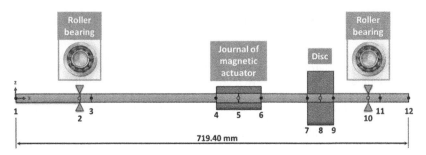

Figure 2. Rotor model by finite element method. Cylindrical roller bearings are located at nodes 2 and 10.

Table 1. Disc elements of the rotor system (FEM).

Number of nodes	Inner diameter [mm]	Outer diameter [mm]	Length of element [mm]
5	17	40.5	81
8	15	94.7	47.8

Table 2. Beam elements of the rotor system (FEM).

Number of elements	Inner diameter [mm]	Outer diameter [mm]	Length of element [mm]
1	0	15	118
2	0	15	21.55
3	0	15	229.45
4, 5	0	17	40.5
6	0	15	83.8
7, 8	0	15	23.9
9	0	15	64
10	0	15	21.55
11	0	15	52.25

To compare the experimental data with the numerical model, the equivalent stiffness and damping coefficients of the cylindrical roller bearing were inserted in as direct coefficients in finite elements matrices $[K]$ and $[C]$ using the bearing modeling proposed in (19,20). The cross stiffness and damping coefficients were considered null. Besides, when comparing experimental and numerical results, the coefficient of shaft proportional damping is $\beta = 710^{-5}$. The damping values were determined through tests and FRF adjustments of the rotor under a white noise excitation, as demonstrated in (21). The Young's modulus of steel shaft is $E = 2.110^{11} Pa$ and its density is $\rho = 7860\ kg/m^3$.

5   RESULTS

The data obtained through both the identification methodologies, namely, in frequency and time domain, were compared to analyze the accuracy of the modal parameters between OMA and finite element model of the test rig. The methodology adopted for OMA was chosen in order to present a balance between precision, speed, and simplicity, since these are fundamental characteristics for dynamic tests of real machines.

Table 3 presents the modal parameters extracted from the numerical model. The simulations considered a constant angular velocity of 30 Hz for the shaft. The evaluation of modes 2 and 3, which could have been carried out using a Campbell Diagram, for example, presents natural frequencies that have no practical interest in the analysis if compared to mode 1, which has a lower natural frequency. Therefore, modes 2 and 3 are beyond the interest of the experimental investigation that will be further conducted by OMA.

The next step is the OMA analysis, an output-only technique, requiring only the investigation of the system response, usually done under operation conditions. For this reason, it is pertinent to apply the OMA tests to the rotor operating without any type of external excitation. In this condition, there are only possible residual unbalances and

Table 3. Modal parameters obtained from rotor FEM.

Numerical Model			Natural Frequency	Damping Ratio
Mode	Shaft Angular Velocity [Hz]	Mode Direction	[Hz]	[%]
1	30	Backward	50.91	1.13
		Forward	52.26	1.13
2	30	Backward	191.27	4.19
		Forward	191.60	4.19
3	30	Backward	388.42	8.85
		Forward	390.68	8.85

misalignments inherent in test rig assembly. The acquisition time considered was based on assumptions found in Brincker and Ventura (6) leading to 60s of measurements, in a 1024 Hz sampling rate. For a preliminary step, Kurtosis and an STFT, of the acquired signal, Figure 3, are evaluated to gather useful information to guide the further identification process using both EFDD and SSI.

As previously studied by Jacobsen (22), these indicators have the potential to evaluate the rotating system to provide clear ideas on the identification of harmonics, in order to help the extraction of the modal parameters. In this case, however, due mainly to the lack of external excitation, they did not present a satisfactory result. Kurtosis should be responsible for quantifying the location of harmonics in the spectrum based on the differentiation of probability density function, while STFT has the ability, in addition to evaluating stationarity, to also help in the identification of harmonics (which appear as thin vertical lines in the frequency domain) and in the possible location of energy concentration due to the rotor's modes (thick vertical lines). However, even though it is possible to clearly distinguish the first, second and third harmonics proportional to the shaft speed, there are many other components evidenced, making it difficult to have a clear judgment about the signal content over frequency. Meanwhile, it is necessary to evaluate the OMA algorithms previously described to analyse the capability of both the frequency and time domain techniques to identify the modal parameters of the rotor in this condition. Henceforth, Figure 4 (a) shows the SVD decomposition of the spectra for the rotor without external excitation, using a resolution of 0.25 Hz and Welch's method with 66% of overlap, and Figure 4 (b) the SSI stabilization diagram.

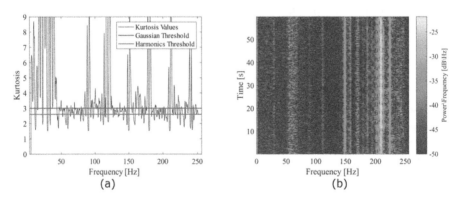

Figure 3 . Kurtosis (a) and STFT (b) for the rotor operating in 30 Hz.

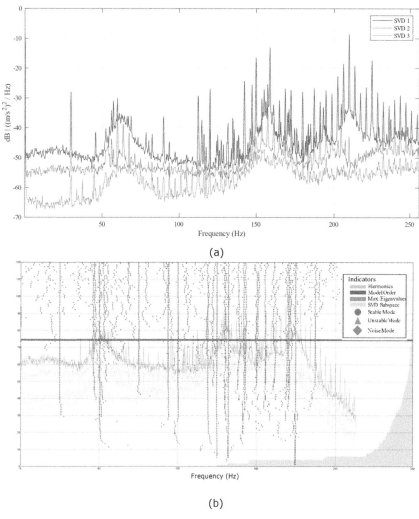

(a)

(b)

Figure 4. SVD plots (a) and SSI stabilization diagram (b) for OMA identification.

Analyzing both the spectrum and the stabilization diagram, it is not possible to distinguish possible modes. This behavior is mainly due to the low stochastic excitation present, since the rotor in the test rig does not have sufficient external forces energy that meets the necessary criteria for application of the identification methods used here, as previously indicated by Figure 3. In such cases, some techniques can be adopted to enable better identification of modal parameters. In texts such as Orlowitz and Brandt (23) and Brincker and Ventura (6), procedures called tapping, i.e. small impacts randomly distributed in time and space into the system, are often proposed as a way to insert stochastic excitations. Although previously investigated, this approach has not met yet a thorough study for rotating machinery. Thus, in a situation without ideal conditions to proceed with the extraction of modal parameters, these simple techniques still need to be explored in order to improve the operational modal analysis in this type of machinery, without implying the use of controlled inputs or any other form of more sophisticated mechanism.

Therefore, evaluating the technique into the rotor, it is necessary to check again if the indicators present any improvement, before proceeding to analyse the modal parameters. The new Kurtosis and STFT for the rotor, under the same operational conditions as tested before, but adding the tapping procedure, can be analysed in Figure 5.

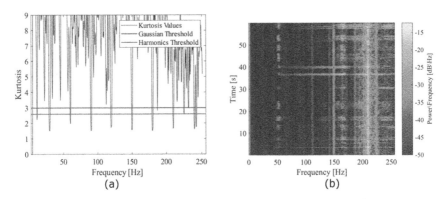

Figure 5. Kurtosis (a) and STFT (b) for tapping procedure.

In this second analysis, the added procedure allowed a considerable improvement in the indicators. Consequently, it is possible from Kurtosis values to clearly identify the harmonics related to the rotor. Besides, the STFT clearly shows that the tapping inserts energy mainly in the frequency range of the natural frequencies, becoming a highlight for the identification of the modal parameters. This relevant prior information about the operation of the rotor given by the indicators is useful to proceed with the OMA identification algorithms.

Differently from what was noted earlier, now through the SVD of the PSDs, it is possible to select a segment for use in the inverse Fourier transform, isolating the modal information related to the first mode, as expected in the EFDD procedure (Figure 6 (a)). Therefore, the desired parameters can be extracted through the frequency domain. For the identification in time domain (Figure 6 (b)), considering stabilization intervals equally established for the previous case, there was again a better identification of the modal characteristics of the first mode. Table 4 shows the values extracted from the two methods.

Comparing both Table 3 and Table 4, the results obtained from the OMA procedure showed good agreement with the modal parameters extracted from the numerical model, even when dealing with values of the damping ratios. Additionally, the indicators used, both Kurtosis and STFT were essential to provide information about the signal quality and the location of the harmonics, so that they were not incorrectly identified as modes in the proposed algorithms.

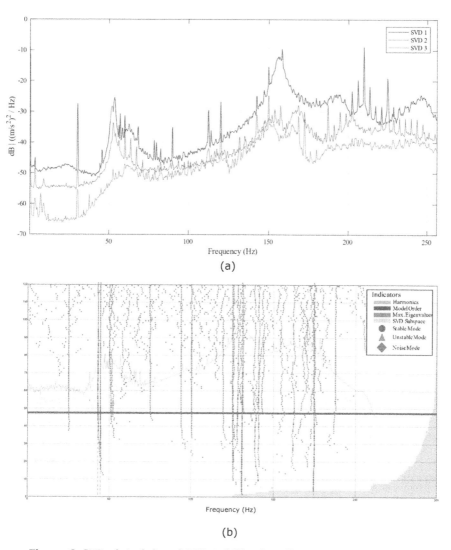

(a)

(b)

Figure 6. SVD plots (a) and SSI stabilization diagram (b) for tapping procedure.

Table 4. Modal parameters obtained from OMA tapping procedure.

OMA				Natural Frequency	Damping Ratio
Test	Shaft Angular Velocity [Hz]	Algorithm	Mode Direction	[%]	[%]
Tapping	30	EFDD	Backward	51.89	1.04
			Forward	53.54	0.95
		SSI	Backward	51.79	1.31
			Forward	53.65	1.31

6 CONCLUSIONS

The paper presents an investigation about the use of simple OMA techniques in the identification of modal parameters of a rotating system supported by roller bearings. The OMA identification when applied to rotating machinery is particularly challenging due to its inherent operating conditions, such as the presence of harmonic forces, closed-spaced modes and non-proportional damping due to the bearings. In this way, the paper explored OMA procedures in order to analyse systems conditions and ensure a good extraction of modal parameters, using two vastly known OMA algorithms in both domains, namely, EFDD and SSI.

The results show that when the OMA identification techniques are applied to a rotor in a test rig, in which there is no influence of external forces, the extraction of the modal parameters is not possible. However, through the application of small impacts, randomly distributed in time and space, over the rotor, the condition for the application of OMA becomes sufficient for obtaining the natural frequencies and damping factors, showing great agreement with the values observed through the numerical model of the same rotor. Kurtosis and STFT indicators were used to assist in the analysis of the signal in both situations and were also fundamental for the interpretation of the results.

ACKNOWLEDGEMENTS

The authors thank the funding support of CAPES, CNPq, Fapesp (#2015/20363-6) and Petrobras to this research.

REFERENCES

[1] F. Magalhães, Á. Cunha, Explaining operational modal analysis with data from an arch bridge, *Mechanical Systems and Signal Processing*. 25 (2011) 1431–1450.
[2] R. Tarinejad, M. Pourgholi, Modal identification of arch dams using balanced stochastic subspace identification, *Journal of Vibration and Control*. 24 (2018) 2030–2044.
[3] J.M.W. Brownjohn, Lateral loading and response for a tall building in the non-seismic doldrums, *Engineering Structures*. 27 (2005) 1801–1812.
[4] M. Ozbek, F. Meng, D.J. Rixen, Challenges in testing and monitoring the in-operation vibration characteristics of wind turbines, *Mechanical Systems and Signal Processing*. 41 (2013) 649–666.

[5] C. Peeters, Q. Leclère, J. Antoni, P. Lindahl, J. Donnal, S. Leeb, J. Helsen, Review and comparison of tacholess instantaneous speed estimation methods on experimental vibration data, *Mechanical Systems and Signal Processing*. 129 (2019) 407–436.

[6] R. Brincker, C.E. Ventura, Introduction to Operational Modal Analysis, John Wiley & Sons, 2015.

[7] T.G. Carne, G.H. James, The inception of OMA in the development of modal testing technology for wind turbines, *Mechanical Systems and Signal Processing*. 24 (2010) 1213–1226.

[8] L. Zhang, R. Brincker, P. Andersen, An overview of operational modal analysis: Major development and issues, *Proceedings 1st International Operational Modal Analysis Conference*. IOMAC 2005. (2005).

[9] P. Van Overschee, B. De Moor, Subspace Identification for Linear Systems, Kluwer Academic Publishers, Norwell, Massachusetts, 1996.

[10] J.S. Bendat, A.G. Piersol, Random Data, John Wiley & Sons, Hoboken, NJ, USA, 2010.

[11] R. Brincker, C.E. Ventura, P. Andersen, Damping Estimation by Frequency Domain Decomposition, *Proceedings 19th International Modal Analysis Conference*, Kissimmee, Florida., 2001: pp. 698–703.

[12] P. Guillaume, L. Hermans, H. Van der Auwerer, Maximum Likelihood Identification of Modal Parameters from Operational Data, *Proceedings 17th International Modal Analysis Conference*. (1999) 1887–1893.

[13] C. Devriendt, P. Guillaume, The use of transmissibility measurements in output-only modal analysis, *Mechanical Systems and Signal Processing*. 21 (2007) 2689–2696.

[14] R. Brincker, L. Zhang, P. Andersen, Modal Identification from Ambient Responses Using Frequency Domain Decomposition, *Proceedings 18th International Modal Analysis Conference* (2000).

[15] C. Priori, M. De Angelis, R. Betti, On the selection of user-defined parameters in data-driven stochastic subspace identification, *Mechanical Systems and Signal Processing*. 100 (2018) 501–523.

[16] L. Zhang, R. Brincker, P. Andersen, An overview of major developments and issues in modal identification, *Proceedings 22nd International Modal Analysis Conference*. (2004) 1–8.

[17] B. Peeters, G. De Roeck, Reference-based stochastic subspace identification for output-only modal analysis, *Mechanical Systems and Signal Processing*. 13 (1999) 855–878.

[18] R. Brincker, P. Andersen, Understanding Stochastic Subspace Identification, *Proceedings 24th International Modal Analysis Conference* (2006).

[19] N.A.H. Tsuha, F. Nonato, K.L. Cavalca, Stiffness and Damping Reduced Model in EHD Line Contacts, in: K. Cavalca, H. Webe (Eds.), *Proceedings 10th International Conference on Rotor Dynamics – IFToMM. IFToMM 2018. Mech. Mach. Sci.*, 2019: pp. 43–55.

[20] N.A.H. Tsuha, K.L. Cavalca, Finite line contact stiffness under elastohydrodynamic lubrication considering linear and nonlinear force models, *Tribology International*. 146 (2020).

[21] N.A.H. Tsuha, K.L. Cavalca, Stiffness and damping of elastohydrodynamic line contact applied to cylindrical roller bearing dynamic model, *J. Sound Vib.* 481 (2020).

[22] N.J. Jacobsen, P. Andersen, Operational modal analysis on structures with rotating parts, *23rd International Conference on Noise and Vibration Engineering, ISMA 2008*. 4 (2008) 2491–2505.

[23] E. Orlowitz, A. Brandt, Comparison of experimental and operational modal analysis on a laboratory test plate, *Journal of the International Measurement Confederation*. 102 (2017) 121–13.

12th International Conference on Vibrations in Rotating Machinery -
Institution of Mechanical Engineers, ISBN 978-0-367-67742-8

Rotor-angular contact ball bearing system study using EHD lubrication and comparison with experimental tests

L. Carrer, L. Bizarre, K.L. Cavalca

Department of Integrated Systems, Faculty of Mechanical Engineering, UNICAMP, Brazil

ABSTRACT

The main objective of this work is the evaluation of the performance of an EHD (elastohydrodynamic) lubrication reduced model for angular contact ball bearings. The multilevel numerical integration algorithm is applied to solve the EHD lubricated contact equations and the numerical results are optimized to adjust an equivalent force versus displacement curve of a reduced model based on equivalent nonlinear stiffness and linear damping coefficients of the bearing. The rotor-bearing finite element model is therefore built, allowing the comparison with experimental tests obtained from the literature.

1   INTRODUCTION

The aim of reducing computational time and testing costs in the design of machinery has been driving the researchers to develop more accurate and refined models to improve the prediction of system's operational conditions. In this context, this paper contributes with the application of a reduced model previously developed by Nonato and Cavalca (1) and (2) for representing the rotor behavior when supported by angular contact lubricated ball bearings, along with a comparison with experimental tests found in the literature (3).

Initially, the distributed forces on the bearing elements are estimated, taking into account, in the first step, the Hertz theory for the ball-raceway contact, considering the external load applied on each bearing, besides inertia effects due to high rotation speed and gyroscopic moment on the force equilibrium, (4) and (5). Recently, several approaches aim the complete representation of the rolling elements bearing components behavior, however, greatly increasing the computational costs (6).

Consequently, the classic formulation given in (4) has been still applied on bearing dynamics studies with high computation demand, as sensitivity analysis (7,8), fault modeling (9–11), surface finish (12–14), roughness effect (15) or models for more complex bearing configurations (16).

A common trace between earlier and recent models for angular contact ball bearing is the partial or complete lack of the stiffness and damping effects of the lubricant film thickness present at the contact interfaces, sometimes approached in a simplistic form.

The nonlinear elastohydrodynamic (EHD) contact phenomena had its earlier development in the work of Reynolds (17) and Hertz (18), further adapted by Dowson (19) and (20) and Roelands (21) to consider variations in fluid properties, namely density and viscosity, due to temperature and pressure.

However, the complete elastohydrodynamic theory leads to high computational costs, alongside the low damping and high stiffness present in heavy load conditions, which tends to increase convergence problems. Consequently, few studies have combined elastohydrodynamic contact stiffening and damping of the lubricant film with the classical bearing dynamics model.

A solution firstly proposed in (22) uses a reduced-order model for the lubricated contact in radial bearings through an implicit nonlinear formulation adjusted to a multilevel solution for the transient EHD contact point (23). Explicit formulations for the contact stiffness and damping have been studied, linearly (24) and nonlinearly (25).

In the case here investigated, the first step is the numerical solution for the EHD lubricated contact equations, obtained by Multi-Level Multi-Integration (MLMI) algorithm (see (26) and (27)).

As the EHD numerical results are highly timing costly, a reduced model is then applied by optimizing nonlinear stiffness and linear damping parameters to adjust the contact forces curves in the bearing elements and successively, in the complete bearing equivalent coefficients, as in (1) and (2).

Finally, a complete rotor-bearing system is modeled, based on the work of Aini (1) for experimental comparison purposes. The rotor model, however, comprises the shaft and disk finite element model, by the well-known Timoshenko's theory for the shaft, considering effects of inertia, elasticity and shear, developed by Nelson and McVaugh in (28) and (29) respectively.

A comparison between the simulations developed here and the experiments from the literature is carried on, being applied on a spindle supported by an angular contact ball bearing, and subject to an unbalance force at its cantilever end. The main frequencies of the complete system identified by the model are in good agreement, in amplitude and frequency, with the experimental data.

The model seems to be promising to applications involving condition monitoring on operational rotor-bearings systems, leading to extend bearings lifespan, by modeling their behavior. It also supports the development phase of bearings and rotating system by reducing computational time consuming and experimental tests costs, making it feasible to approach the physical bearing behavior by computational simulations.

2   METHODOLOGY

2.1   Contact force reduced model and bearing equilibrium

It is necessary to have a first approximation of the load distribution in the bearing spheres, which is given by the solution presented in (4) by calculating the dynamic equilibrium of the spheres with the external load. The Newton-Raphson iterative method, simultaneously applied on each sphere using the Hertzian dry contact model, determines displacements, velocities, and loads in the contact direction.

Afterward, the dimensionless parameters of Moes for load (M) and lubrication (L), described in (26), are applied as input for the contact characterization of the elastohydrodynamic (EHD) lubrication contact. A multilevel finite difference method is used to integrate a system of equation, composed by Reynolds equation, oil film thickness, viscosity-pressure, density-pressure relations, and forces equilibrium, as presented by Nonato (2).

The bearing characterization is the basis of the approach by a reduced-order model of the nonlinear restoring contact force, for internal and external raceways. The Levenberg-Marquardt optimization algorithm (30) is then applied to the static equilibrium of all spheres, using displacement ($\delta_{i,oj}$) and force ($F_{i,oj}$) data from EHD simulation. This adjustment is used to obtain the parameters of equivalent nonlinear stiffness $K_{i,o}$, displacement exponent $d_{i,o}$ and the residual force $\Delta F_{i,o}$, as shown in equation (1), according to (5).

$$F_{i,oj} = K_{i,o} \cdot \delta_{i,oj}^{d_{i,o}} + \Delta F_{i,o} \qquad (1)$$

where the subscript (i) and (o) means, respectively, inner and outer contact ball-raceway and j, the jth sphere on the bearing.

The parameters of the reduced-order model must be able to reproduce the force versus displacement curve previously obtained from the numerical simulation, as represented in Figure 1.

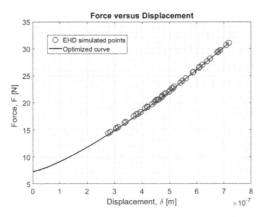

Figure 1. EHD contact on the inner raceway, five points for each sphere, and the curve representation for the reduced model of equation (1).

A spring-damper system represents the oil film in the contact area, as the scheme in Figure 2, where d_m is the bearing pitch diameter. The transient response at the EHD contacts in each sphere is used to determine the equivalent linear damping coefficient with internal and external raceways.

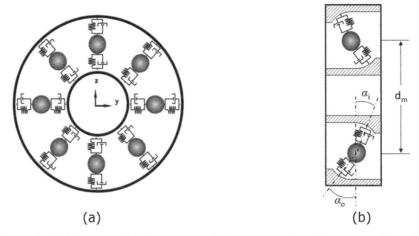

(a)   (b)

Figure 2. Ball bearing oil film representation as a spring-damper system, (a) y-z view and (b), x-z view.

The angular contact between spheres and raceways makes necessary to decompose the restoring and dissipative forces in axial and radial directions (indices A and R, respectively) to obtain the parameters of the reduced-order model in both directions. According to Equation (2), since the inner and outer contact angles, respectively α_{ij} and α_{oj}, are different when the bearing is loaded, the inner and outer reaction forces, F_{ij} and F_{oj}, are not collinear.

$$F_{i,oRj} = \sin(\alpha_{i,oj})F_{i,oj}; \quad F_{i,oAj} = \cos(\alpha_{i,oj}) \cdot F_{i,oj} \tag{2}$$

Afterward, it is possible to find the total reaction forces of the bearing in radial and axial directions, taking into account that the total displacement is the sum of the internal and external raceway displacements, as shown in Equations (3) and (4). Hence, the equivalent parameters for radial and axial bearing forces can be determined.

$$\delta_{R,A_j} = \delta_{iR,A_j} + \delta_{oR,A_j} \tag{3}$$

$$F_{Rj} = K_R \cdot \delta_{Rj}^{d_R} + \Delta F_R ; F_{Aj} = K_A \cdot \delta_{Aj}^{d_A} + \Delta F_A \tag{4}$$

The radial displacements can be represented in terms of δy_m, δz_m and ψ_j, the Azimuth angle of the jth sphere, as shown in (2). Considering P_d as the bearing diametrical clearance and directly related to the radial displacement of each sphere δ_{Rj} (Equation (5)).

$$\delta_{Rj} = \delta z_m \cos\psi_j + \delta y_m \sin\psi_j - \frac{P_d}{2} \tag{5}$$

The first time-derivative of the radial displacement of each sphere gives the velocity of the center of the spheres regarding their equilibrium position.

$$\dot{\delta}_{Rj} = -\delta z_m \dot{\psi}_j \sin\psi_j + \delta y_m \dot{\psi}_j \cos\psi_j + \dot{\delta} z_m \cos\psi_j + \dot{\delta} y_m \sin(\psi_j) \tag{6}$$

The angular velocity term in this equation,, represents the cage velocity (ω_c), given by Equation (7), where Ω is the shaft rotation, that is, the bearing inner raceway rotation, d_m, the bearing pitch diameter and D, the diameter of the spheres. Considering the outer raceway at rest.

$$\omega_c = \frac{\Omega(d_m - D\cos(\alpha))}{2d_m} \tag{7}$$

The static balance in the bearing is given by the sum of the reaction forces of the spheres in x_m, y_m and z_m directions. Since Z is the total number of spheres of each bearing, the equations (8), (9) and (10) are obtained. Considering the rotor degrees of freedom, δy_m, δz_m, and, in equation (3), (4) and (5), as y_m, z_m, and respectively.

$$F_{my} = \sum_{j=1}^{Z}\left[\left(K_R\left(z_m\cos\psi_j + y_m\sin\psi_j - \frac{P_d}{2}\right)^{d_R} + \Delta F_R\right)\cos\psi_j\right] \tag{8}$$

$$F_{mz} = \sum_{j=1}^{Z}\left[\left(K_R\left(z_m\cos\psi_j + y_m\sin\psi_j - \frac{P_d}{2}\right)^{d_R} + \Delta F_R\right)\sin\psi_j\right] \tag{9}$$

$$F_{mx} = \sum_{j=1}^{Z}\left(K_A x_m^{d_A} + \Delta F_A\right) \tag{10}$$

185

The dissipative forces of the bearing are then described by equations (11), (12) and (13), obtained with the product of equivalent bearing damping by the derivative of the position in time, given by Equations (11) and (12), for the radial direction, and directly by Equation (13) for the axial direction.

$$F_{m\dot{y}} = \sum_{j=1}^{Z} \left[(C_j(\dot{z}_m\cos\psi_j + \dot{y}_m\sin\psi_j) + \omega_c(-z_m\sin\psi_j + y_m\cos\psi_j))\cos\psi_j \right] \tag{11}$$

$$F_{m\dot{z}} = \sum_{j=1}^{Z} \left[(C_j(\dot{z}_m\cos\psi_j + \dot{y}_m\sin\psi_j) + \omega_c(-z_m\sin\psi_j + y_m\cos\psi_j))\sin\psi_j \right] \tag{12}$$

$$F_{m\dot{x}} = \sum_{j=1}^{Z} (C_j\,\dot{x}_m) \tag{13}$$

The complete equation of motion of the system is described in Equation (14). $\mathbf{M}$, $\mathbf{G}$ e $\mathbf{K}$ are, respectively, the mass, gyroscopic and stiffness matrixes. They are assembled using the element matrixes given in (31), from Timoshenko beam theory:

$$\mathbf{M} \cdot \ddot{\mathbf{u}}(t) + (\mathbf{D} + \Omega\mathbf{G}) \cdot \dot{\mathbf{u}}(t) + \mathbf{K} \cdot \mathbf{u}(t) = \mathbf{f}(\mathbf{u}, \dot{\mathbf{u}}, t) \tag{14}$$

The matrix $\mathbf{D}$ represents the shaft dissipative forces and is written as proportional to the stiffness matrix. The $\mathbf{f}(\mathbf{u}, \dot{\mathbf{u}}, t)$ vector contains all external forces and bearings non-linear reaction forces, equations 8 to 13, applied in the respective degrees of freedom, as demonstrated in equation 15.

$$\mathbf{f}(\mathbf{u}(\textbf{bearing node}), \dot{\mathbf{u}}(\textbf{bearing node}), t) = \begin{Bmatrix} F_{my} + F_{m\dot{y}} \\ F_{mz} + F_{m\dot{z}} \\ F_{mx} + F_{m\dot{x}} \end{Bmatrix} \tag{15}$$

3   RESULTS AND DISCUSSIONS

The rotor model applied in this section is shown in Figure 3, representing a spindle used in the experiments conducted by Aini (3), where nodes 2 and 4 locate the angular contact ball bearings, node 3 is the center of mass and node 1 is the tool cantilever end, subjected to rotary unbalance.

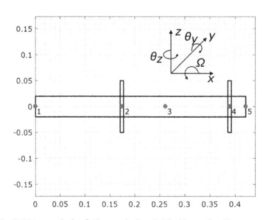

Figure 3. FEM model of the original Aini's spindle experiment.

186

The rotor's characteristics are presented in Table 1.

Table 1. Spindle Characteristics.

Shaft mass (kg)	5,5
Shaft Modulus of elasticity (Pa)	210×10^9
Shaft diameter (m)	$4,0 \times 10^{-2}$
Proporcional damping coefficient	$2,0 \times 10^{-4}$
Spindle rotational speed, Ω (RPM)	3000
Residual unbalance (kg.m)	$5,3 \times 10^{-5}$

The bearings characteristics are presented in Table 2.

Table 2. Bearing characteristics.

Number of spheres in the bearing	16
Sphere diameter (m)	7.94×10^{-3}
Bearing pitch diameter (m)	5.4×10^{-2}
Nominal angular contact (°)	15

The bearings are characterized in the operating conditions for a preload of 100N, equally applied on each one in the axial direction, and only the weight distribution in the radial direction. Then, by the optimization proposed in (5), the parameters for equivalent stiffness, K, displacement exponent, d, lift force, ΔF, damping coefficient, C, are properly determined. These results are displayed in Table 3, where the subscripts R and A represent, respectively, the radial and axial directions.

Table 3. Bearings results of Stiffness (K), nonlinear exponent (d) and offset force (ΔF).

Speed	Bearing	K_R, N/m^{dR}	d_R	ΔF_R, N	K_A, N/m^{dA}	d_A	ΔF_A, N	C, N/s²
3000	1	$1.67 \cdot 10^9$	1.37	5.86	$2.86 \cdot 10^9$	1.21	1.28	107.32
3000	2	$2.05 \cdot 10^9$	1.38	2.24	$3.21 \cdot 10^9$	1.22	1.32	107.58

Figure 4 presents the reactions forces responses from the time integration simulations of the whole system. It demonstrates the static loads in which each bearing was characterized, namely, $F_{y1} = 0$ N, $F_{z1} = 33$ N and $F_{x1} = 100$ N, and $F_{y2} = 0$ N, $F_{z2} = 22$ N and $F_{x2} = 100$ N and the time response due to unbalance. This is a simple validation of the correct loads of the bearing's characterization and the influence of unbalance on the reaction forces.

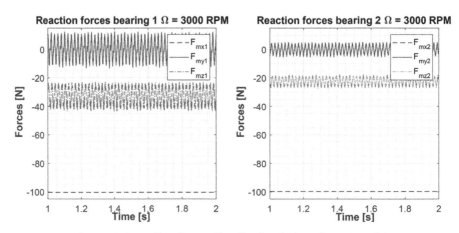

Figure 4. Reaction force distribution in bearings 1 and 2.

Figure 5 shows the DFT of the time response signal of node 1 in the z-direction. It is possible to identify the response to the rotor unbalance at the first peak at 50 Hz (rotation frequency) and its harmonics, 2X, 3X due to the radial clearance, in this case of 10mm, also identified with amplitudes close to those experimentally measured.

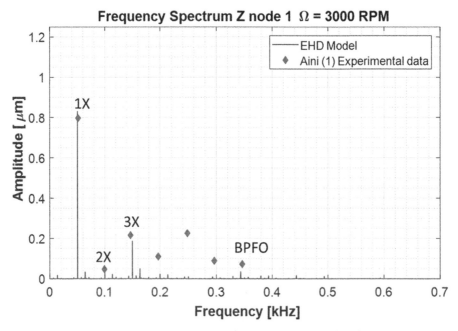

Figure 5. Frequency spectrum comparison between the simulated rotor with reduced-order bearings forces model and experimental results (3) at node 1.

The first harmonic of BPFO (Ball Pass Frequency of Outer ring), given by Equation (15), which is the frequency of passage from one sphere position to the next one, can be noted even with low amplitude, either in numerical simulations or in the experimental measurements in (3).

$$\text{BPFO} = \frac{Z \cdot \Omega}{2} \left(1 - \frac{D}{d_m} \cos(\alpha) \right) \tag{15}$$

This is an important result since the increasing amplitude in the BPFO frequency, as well as its harmonics, is a characteristic signature of rolling bearing failure.

Figure 6 shows the orbits at the finite element model nodes along with the shaft discretization, properly amplified to improve the visualization and better understanding of the rotor-bearing response. The radial clearance, P_d, of 10 mm affects the orbits, given them a characteristic flat shape.

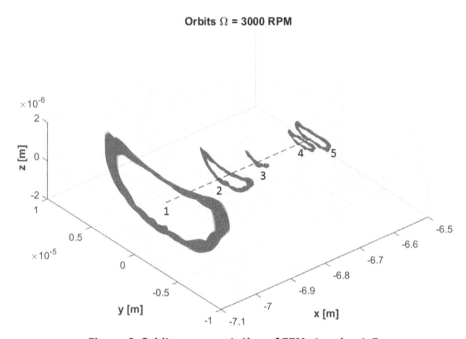

Figure 6. Orbits representation of FEM at nodes 1-5.

4   CONCLUSIONS

The comparison of the experiments conducted by Aini (3) with the theoretical results developed here demonstrates the coherence in representing the sphere-to-raceway angular contact by the reduced-order model of the EHD contact. For such representation, the insertion of contact interaction forces in the rotor system was completed by carrying the bearing coordinate system into the rotor reference system, using optimized parameters for the radial and axial directions.

The harmonics of the rotation frequency and the passing frequency of the spheres (BPFO) were identified in the numerical simulation of the rotating systems, signaling positively for the representation of these components according to the reduced-order of the non-linear forces in the sphere-raceway contacts.

As a result of bearing characterization, the reduced model provides a closed solution that can be directly introduced into the system's equations of motion, eliminating one of the most expensive numerical processes in the solution, the EHL multi-level solution. The non-linear character of the bearing forces is maintained and this process, previously developed for radial bearings (2), can be applied to angular contact bearings. The approach of applying decomposed parameters in the radial and axial direction is proved to be feasible.

The EHD modeling proposed for angular contact ball bearings, and their interaction with rotating systems, is promising for practical applications, leading to future investigations on the bearing frequencies, known to be altered in the presence of defects, which become an interesting signature for fault identification

ACKNOWLEDGMENTS

The authors thank the national council for scientific and technological development (cnpq) and petrobras for the financial support of this research.

REFERENCES

[1] F. Nonato, K.L. Cavalca, On the non-linear dynamic behavior of elastohydrodynamic lubricated point contact, J. Sound Vib. 329 (2010) 4656–4671.

[2] F. Nonato, K.L. Cavalca, An approach for including the stiffness and damping of elastohydrodynamic point contacts in deep groove ball bearing equilibrium models. Journal of Sound and Vibration, v. 333, p. 6960–6978, 2014.

[3] R. Aini, Vibration monitoring and modelling of shaft/bearing Assemblies under concentrated elastohydrodynamic condition. PhD Thesis, Kingston polytechnic, 1990.

[4] T.A. Harris, Rolling Bearing Analysis. John Wiley & Sons, New York, 1991, 1013p.

[5] L. Bizarre, F. Nonato and K.L. Cavalca, Formulation of five degrees of freedom ball bear-ing model accounting for the nonlinear stiffness and damping of elastohydrodynamic point contacts. Mechanism and Machine Theory 124, 2018: 179–196.

[6] L. Cao, F. Sadeghi, L. Stacke, A combined EFEM–DEM dynamic model of rotor-bearing-housing system, ASME. J. Tribol. (2017).

[7] I.S. Barmanov, M.N. Ortikov, Ball bearing dynamics at the interference fit on balls, Procedia Eng 176 (2017) 19–24.

[8] X. Sheng, B. Li, Z. Wu, H. Li, Calculation of ball bearing speed-varying stiffness, Mech. Mach. Theory 81 (2014) 166–180.

[9] L. Niu, H. Cao, Z. He, Y. Li, Dynamic modeling and vibration response simulation for high speed rolling ball bearings with localized surface defects in raceways, J. Manuf. Sci. Eng 136 (2014).

[10] J. Liu, Y. Shao, An improved analytical model for a lubricated roller bearing including a localized defect with different edge shapes, J. Vib. Control (2017).

[11] J. Liu, Z. Shi, Y. Shao, An analytical model to predict vibrations of a cylindrical roller bearing with a localized surface defect, Nonlinear Dynamics 89 (2017) 2085–2102.

[12] C.K. Babu, N. Tandon, R.K. Pandey, Vibration modeling of a rigid rotor supported on the lubricated angular contact ball bearings considering six degrees of

freedom and waviness on balls and races, J. Vib. Acoust. 134 (2011) pp. 011006-011006-12.

[13] L. Xu, Y. Li, Modeling of a deep-groove ball bearing with waviness defects in planar multibody system, Multibody Syst. Dyn. 33 (2015) 229–258.

[14] J. Liu, Y. Shao, Vibration modelling of nonuniform surface waviness in a lubricated roller bearing, J. Vib. Acoust. 23 (2015) 1115–1132.

[15] W. Yunlong, W. Wenzhong, Z. Shengguang, Z. Ziqiang, Effects of raceway surface roughness in an angular contact ball bearing, Mech. Mach. Theory 121 (2018) 198–212.

[16] G. Kogan, R. Klein, A. Kushnirsky, J. Bortman, Toward a 3D dynamic model of a faulty duplex ball bearing, Mech. Syst. Signal Process. 54 (2015) 243–258.

[17] O. Reynolds, On the theory of lubrication and its application to M. Beauchamps Tower's experiments, including an experimental determination of the viscosity of olive oil, Philos. Trans. R. Soc. 177 (1886) 157–234.

[18] H. Hertz, On the contact of elastic solids, in: H. Hertz (Ed.), Hertz's Miscellaneous Papers, Macmillan & Co, London, 1896, pp. 146–162.

[19] D. Dowson, A generalized Reynolds equation for fluid-film lubrication, Int. J. Mech. Sci. 4 (2) (1962) 159–170.

[20] D. Dowson, G.R. Higginson, Elasto-hydrodynamic Lubrication, SI Edition, Pergamon Press, Oxford, 1977.

[21] C.J.A. Roelands, Correlational Aspects of the Viscosity-Temperature-Pressure Relationship of Lubricating Oils Ph.D. Thesis, Technical University Delft, Delft, The Netherlands, 1966.

[22] Y.H. Wijnant, J.A. Wensing, G.C. Vannijen, The influence of lubrication on the dynamic behavior of ball bearings, J. Sound Vib. 222 (1999) 579–596.

[23] Y.H. Wijnant, Contact Dynamics in the Field of Elastohydrodynamic Lubrication Ph.D Thesis, University of Twente, 1998.

[24] G. Dong, Y. Liu, M. Jing, F. Wang, H. Liu, Effect of elastohydrodynamic lubrication on the dynamic analysis of ball bearing, in: Proceedings of the Institution of Mechanical Engineers, Part K: J. Multi-body Dyn., 2015.

[25] W.-z. Wang, L. Hu, S.-g. Zhang, Z.-q. Zhao, S. Ai, Modeling angular contact ball bearing without raceway control hypothesis, Mech. Mach. Theory 82 (2014) 154–172.

[26] C.H. Venner, A.A. Lubrecht, Multilevel Methods in Lubrication. Elsevier, Tribology Series, vol 37, 2000, 400p.

[27] C.H. Venner, G. Popovici, P. Lugt, M. Organisciak, Film thickness modulations in starved elastohydrodynamically lubricated contacts induced by time-varying lubricant supply. ASME J. Tribol. 130, 041501 (2008).

[28] Nelson, H.D. and McVaugh, J.M. (1976) The Dynamics of Rotor-Bearing Systems Using Finite Elements. ASME Journal of Engineering for Industry, 98, 593–600.

[29] Nelson, H.D. (1980) A Finite Rotating Shaft Element Using Timoshenko Beam Theory. ASME Journal of Mechanical Design, 102, 793–803.

[30] Marquardt, D. W., An Algorithm for Least-Squares Estimation of Nonlinear Parameters. Journal of the Society for Industrial and Applied Mathematics, v. 11, n. 2, p. 431–441, 1963.

[31] Tuckmantel, F. W., Integração de sistemas rotor-mancal hidrodinâmico-estrutura de suporte para resolução numérica, Master Dissertation, University of Campinas 2010 (in Portuguese).

12th International Conference on Vibrations in Rotating Machinery -
Institution of Mechanical Engineers, ISBN 978-0-367-67742-8

Effect of journal bearing preload caused by bearing-housing interference fit on nonlinear vibration of a flexible rotor supported by a journal bearing

N. Koondilogpiboon, T. Inoue

Graduate School of Engineering, Nagoya University, Japan

ABSTRACT

The interference fit between the journal bearing and the housing could reduce the bearing's vertical clearance. However, its effect on the nonlinear vibration, such as Hopf bifurcation type, has not been rigorously discussed. This paper presented the effects of the reduced bearing vertical clearance on the nonlinear vibration under various rotor parameters, such as disk mass and position. Model reduction by component mode synthesis, shooting method with parallel computing, arclength continuation, and Floquet multiplier analysis were implemented to efficiently obtain the limit cycles and their stability of the nonlinear rotor-bearing system. The experiment was performed to verify the calculation results.

1   INTRODUCTION

It is a well-known fact that subsynchronous vibration can occur in the flexible rotor supported by journal bearings (JBs) due to the nonlinearity of the JBs. This subsynchronous vibration can cause rotordynamics instability, especially at the rotational speed above the onset speed of instability (OSI) (1, 2). Even the OSI can be obtained by the eigenanalysis around the equilibrium position, which is the standard calculation in rotor-bearing system design. However, the eigenanalysis can only predict the stability in the linear analysis sense, and it cannot predict whether the rotor's whirling amplitude is bounded (supercritical bifurcation) or unbounded (subcritical bifurcation) (3) when the rotor's equilibrium position becomes unstable. The subcritical bifurcation is particularly should be avoided because of its sudden increase in the rotor's whirling amplitude, which can also occur even at speed lower than the OSI if the rotor is sufficiently perturbed. Hence, nonlinear analysis is essential to apprehend this phenomenon of the rotor system, as stated by several research papers on the nonlinear vibration of the flexible rotor supported by the journal bearing (3-5).

As noted by several prior research, the bearing profile can affect the system's stability (1, 6). However, the actual profile can be different from the designed one due to manufacturing error, as stated by Kirk (7), or interference from the bearing housing. Allaire and Flack (6) observed that the non-preloaded axial groove bearing tended to be more stable than predicted due to the preload caused by the bearing being crushed vertically by the housing. However, to the authors' knowledge, the investigation of nonlinear vibration, such as Hopf bifurcation type, caused by this kind of preload bearing has not yet been investigated.

In typical rotordynamics analysis, the rotor is modelled by finite elements (FE) with only a few nodes subjected to the nonlinear forces from bearing or seal. Hence model reduction is advantageous. Friswell and Penny (8) reviewed the accuracy of several model reduction schemes and stated that component mode synthesis (CMS) provided the most accurate orbit of the reduced model. CMS (9-11) considers the retained mode-generalized coordinates of the substructure constrained at the boundary as additional DOF. CMS was successfully implemented to analyze the stability of the periodically-excited rotor-bearing systems that had local nonlinearities (12), determine the response (13), bifurcation, and stability (14, 15). Koondilogpiboon and Inoue (16) demonstrated that using CMS, even with only the two lowest modes of the internal degrees of freedom (DOF) retained, was sufficient to predict the OSI and bifurcation type that agreed with the experimental results. Apart from the model reduction, the nonlinear calculation speed could be further improved by utilizing the shooting method with parallel computing, as done by Kim and Palazzolo (17). Using the shooting method was also advantageous as the by-product of the numerical procedure can be used to calculate the Floquet multiplier to determine the limit cycle's stability.

In this study, real mode CMS was applied to the full rotor FE model. Then, the shooting method with parallel computing was applied to the reduced rotor model to obtain the limit cycle, and Floquet multiplier analysis was used to determine the stability of the limit cycle. The selected model was a flexible rotor supported by one self-aligning rolling element bearing (REB) and one axial groove JB. Eight cases of combination of disk mass, position, and bearing preload were investigated. The experiment on the test rig was carried out to verify the calculation results.

2   MATHEMATICAL MODELS

2.1   Flexible rotor model

The rotor was modelled by 1-D finite elements (FE) derived from the Euler-Bernoulli beam into six nodes. It consisted of one large disk with selectable positions and two small disks attached. The rotor was supported by a self-aligning rolling element bearing (REB) and a journal bearing (JB), as shown in Figure 1. Additional rotor and bearing parameters are presented in Table 1.

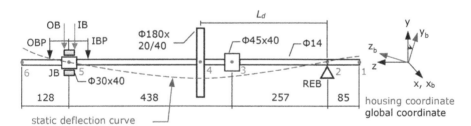

Figure 1. Rotor Model.

The equation of motion (EOM) of the balanced rotor in fixed global coordinate (x, y, z) can be written as:

$$\hat{\mathbf{M}}_r \ddot{\hat{\mathbf{q}}} + \left(\hat{\mathbf{C}}_r + \hat{\omega}\, \hat{\mathbf{G}}_\mathbf{r} \right) \dot{\hat{\mathbf{q}}} + \left(\hat{\mathbf{K}}_r + \hat{\mathbf{K}}_{reb} \right) \hat{\mathbf{q}} = \hat{\mathbf{F}}_g + \hat{\mathbf{F}}_{jb} \tag{1}$$

where $\hat{\mathbf{M}}_r, \hat{\mathbf{C}}_r, \hat{\mathbf{G}}_r, \hat{\mathbf{K}}_r$ are mass, damping, gyroscopic and stiffness matrices of the rotor, $\hat{\mathbf{K}}_{reb}$ is REB's stiffness matrix, $\hat{\mathbf{F}}_g, \hat{\mathbf{F}}_{jb}$ are gravitational and JB's forces vectors, and $\hat{\mathbf{q}} = \left\{ \hat{x}_1, \hat{y}_1, \hat{\theta}_{x1}, \hat{\theta}_{y1}, ..., \hat{x}_6, \hat{y}_6, \hat{\theta}_{x6}, \hat{\theta}_{y6} \right\}^T$ is the displacement vector. Note that "^" denotes dimensional variables. A static deflection curve of the rotor in the pin-pin support condition, shown by the dashed line in Figure 1, can be obtained by considering only $\hat{\mathbf{K}}_r$, $\hat{\mathbf{q}}$ and $\hat{\mathbf{F}}_g$ terms of Eq.(1) with $\hat{x}_2 = \hat{y}_2 = \hat{x}_5 = \hat{y}_5 = 0$. The calculated JB load, journal inclination ($\hat{\theta}_{x5st}$) and vertical deflection of the large disk ($\hat{y}_{4st}$) in each large disk mass ($\hat{m}_d$) and position ($\hat{L}_d$) is shown in Table 2. The bearing housing was tilted according to the obtained journal inclination ($\hat{\theta}_{x5st}$) to satisfy the parallel journal-bearing assumption.

Table 1. Rotor and bearing parameters.

Parameter	Variable	Value	Unit
Large disk mass	$\hat{m}_d$	4, 8	kg
Large disk position from REB	$\hat{L}_d$	0.340, 0.575	m
Shaft and disk density	$\hat{\rho}_s$, $\hat{\rho}_d$	7930	kg/m^3
Shaft's Young modulus	$\hat{E}_s$	191.4×10^9	Pa
Shaft diameter	$\hat{D}_s$	14×10^{-3}	m
REB's direct stiffness	$\hat{k}_{rebxx}$, $\hat{k}_{rebyy}$	1.34×10^7	N/m
JB length	$\hat{L}_{jb}$	12, 18×10^{-3}	m
Journal diameter	$\hat{D}_{jb}$	30×10^{-3}	m
JB's pad radial clearance	$\hat{C}_{rb}$	60×10^{-6}	m
Pad arc length	$\hat{\theta}_{arc}$	140	deg
Oil viscosity (VG32)	$\hat{\mu}$	40×10^{-3}	Pa-s

Table 2. JB load, journal inclination, and large disk's vertical deflection.

$\hat{m}_d$ (kg)	4		8	
$\hat{L}_d$ (mm)	340	575	340	575
JB load (N)	27.43	40.79	46.79	73.52
$\hat{\theta}_{x5st}$(deg)	-0.22	-0.16	-0.41	-0.29
$\hat{y}_{4st}$(mm)	-0.89	-0.30	-1.65	-0.54

2.2 Journal bearing model

In this study, the crushed bearing was assumed to have the same pad curvature as the normal one. However, the pad's center of curvature was shifted by the housing's clamping force, as shown in Figure 2(a). According to the observation of Koondilogpiboon and Inoue (16), the parallel journal-bearing assumption, which requires only calculation in half of the axial length, could correctly predict the bifurcation type in the cases that JB has length-to-diameter (L/D) ratio of 0.4. The film thickness in the bearing (housing) coordinate (xb, yb, zb) of the ith pad, shown in Figure 2(b), can be represented as:

$$\hat{H}_{pi}\left(\hat{\theta}_{zb}\right) = \hat{C}_{rp} - \hat{x}_b \cos\hat{\theta}_{zb} - \hat{y}_b \sin\hat{\theta}_{zb} - (\hat{C}_{rp} - \hat{C}_{rb})\cos(\hat{\theta}_{zb} - \hat{\theta}_{pi}) \tag{2}$$

where $\hat{C}_{rp}$, $\hat{C}_{rb}$ are pad and bearing assembly radial clearance, respectively. In this investigation, the $\hat{C}_{rb}$ values of 60 and 45 μm were selected for normal and crushed bearing, respectively. The film thickness was inserted into the Reynolds equation, defined as:

$$\frac{1}{\hat{R}_{jb}^2}\frac{\partial}{\partial\hat{\theta}_{zb}}\left(\frac{\hat{H}_{pi}^3}{12\hat{\mu}}\frac{\partial\hat{P}}{\partial\hat{\theta}_{zb}}\right) + \frac{\partial}{\partial\hat{z}_b}\left(\frac{\hat{H}_{pi}^3}{12\hat{\mu}}\frac{\partial\hat{P}}{\partial\hat{z}_b}\right) = \frac{\hat{\omega}}{2}\frac{\partial\hat{H}_{pi}}{\partial\hat{\theta}_{zb}} + \frac{\partial\hat{H}_{pi}}{\partial t} \tag{3}$$

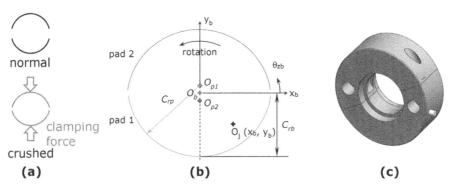

Figure 2. a) Crushed Bearing, b) Geometry, c) CAD Drawing.

A finite difference form of Eq.(3) was then derived and then solved by the successive over-relaxation (SOR) method (18, 19). The pressure at pad edges was set to zero as a fixed boundary condition, as stated by Lund and Thomsen (20). Reynolds boundary condition was used as the cavitation boundary condition due to its simplicity in implementation to the program while giving good agreement with the experimental data. It was applied by replacing negative pressure with zero before calculating the pressure at the successive node in every iteration step of SOR (18, 19). After the pressure iteration converged, bearing forces in the fixed bearing coordinate (xb, yb, zb) can be calculated and then transformed to the fixed global coordinate (x, y, z) used in the EOM. The equilibrium position of the rotor and the linear dynamic coefficients were determined to obtain the onset speed of instability (OSI), which was used as the starting point for the nonlinear analysis.

3   ROTOR MODEL REDUCTION BY CMS

In order to reduce DOF of the rotor system for numerical integration in the shooting method, Eq.(1) was rearranged as:

$$\begin{bmatrix}\hat{\mathbf{M}}_{II} & \hat{\mathbf{M}}_{IB} \\ \hat{\mathbf{M}}_{BI} & \hat{\mathbf{M}}_{BB}\end{bmatrix}\begin{Bmatrix}\ddot{\hat{\mathbf{q}}}_I \\ \ddot{\hat{\mathbf{q}}}_B\end{Bmatrix} + \left(\begin{bmatrix}\hat{\mathbf{C}}_{II} & \hat{\mathbf{C}}_{IB} \\ \hat{\mathbf{C}}_{BI} & \hat{\mathbf{C}}_{BB}\end{bmatrix} + \hat{\omega}\begin{bmatrix}\hat{\mathbf{G}}_{II} & \hat{\mathbf{G}}_{IB} \\ \hat{\mathbf{G}}_{BI} & \hat{\mathbf{G}}_{BB}\end{bmatrix}\right)\begin{Bmatrix}\dot{\hat{\mathbf{q}}}_I \\ \dot{\hat{\mathbf{q}}}_B\end{Bmatrix} + \begin{bmatrix}\hat{\mathbf{K}}_{II} & \hat{\mathbf{K}}_{IB} \\ \hat{\mathbf{K}}_{BI} & \hat{\mathbf{K}}_{BB}\end{bmatrix}\begin{Bmatrix}\hat{\mathbf{q}}_I \\ \hat{\mathbf{q}}_B\end{Bmatrix} = \begin{Bmatrix}\hat{\mathbf{F}}_I\left(\hat{t}\right) \\ \hat{\mathbf{F}}_B\left(\hat{t}\right)\end{Bmatrix}$$

$$\tag{4}$$

where $\hat{\mathbf{q}}_I$ is the internal DOF (those of node 1-4 and 6, total 20 DOF), and $\hat{\mathbf{q}}_B$ is the boundary DOF (those of node 5, total 4 DOF), which is the node that is subjected to the nonlinear forces ($\hat{\mathbf{F}}_{jb}$) vector. The constrained normal modes can be obtained by solving the free vibration problem of the internal DOF with the boundary DOF in Eq.(4) set to zero (9, 10). In general, the eigenvalues and eigenvectors of Eq. (4) are speed-dependent and complex. So, the obtained modal coordinates are complex, which requires complex variable integration. However, Kobayashi and Aoyama (11), and Sundararajan and Noah (14) found that neglecting $\hat{\mathbf{C}}_{II}$ and $\hat{\mathbf{G}}_{II}$ terms in Eq. (4) gives real eigenvalues and eigenvectors and causes only small changes in the result. If $\hat{\mathbf{C}}_{II}$ and $\hat{\mathbf{G}}_{II}$ are neglected, the internal DOF yield real eigenvectors (modes) $\varphi_1, \varphi_2, \ldots, \varphi_{20}$, which are arranged in ascending order of the corresponding real eigenvalues $\lambda_1, \ldots, \lambda_{20}$. The matrix of retained eigenvectors up to the hth retained lowest modes can be defined as:

$$\boldsymbol{\eta}_h = [\varphi_1, \varphi_2, \ldots, \varphi_h] \tag{5}$$

For the static relationship between $\hat{\mathbf{q}}_I$ and $\hat{\mathbf{q}}_B$, by assuming no external force acts on $\hat{\mathbf{q}}_I$, the static version of Eq.(4) leads to Guyan static reduction:

$$\hat{\mathbf{q}}_I = -\left[\hat{\mathbf{K}}_{II}^{-1} \hat{\mathbf{K}}_{IB}\right] \hat{\mathbf{q}}_B = \mathbf{X}\hat{\mathbf{q}}_B \tag{6}$$

By considering $\hat{\mathbf{q}}_I$ as summation of Guyan static reduction and the displacements of the retained modes, Eq.(6) could be expanded to:

$$\begin{Bmatrix} \hat{\mathbf{q}}_I \\ \hat{\mathbf{q}}_B \end{Bmatrix} = \begin{bmatrix} \boldsymbol{\chi} & \boldsymbol{\eta}_h \\ \mathbf{I} & \mathbf{0} \end{bmatrix} \begin{Bmatrix} \hat{\mathbf{q}}_B \\ \hat{\mathbf{a}}_h \end{Bmatrix} = \boldsymbol{\Psi}\hat{s} \tag{7}$$

where $\hat{\mathbf{a}}_h = \{\hat{a}_1, \ldots, \hat{a}_h\}$ and $\hat{a}_h$ is the retained h^{th} mode-generalized coordinate. The reduced EOM can be obtained by substituting Eq.(7) into Eq.(4):

$$\hat{\mathbf{M}}_{RD}\ddot{\hat{S}} + \left(\hat{\mathbf{C}}_{RD} + \hat{\omega}\,\hat{\mathbf{G}}_{RD}\right)\dot{\hat{S}} + \hat{\mathbf{K}}_{RD} + \hat{S} = \hat{\mathbf{F}}_{RD}\left(\hat{t}\right) \tag{8}$$

where

$$\hat{\mathbf{M}}_{RD} = \boldsymbol{\Psi}^T \begin{bmatrix} \hat{\mathbf{M}}_{II} & \hat{\mathbf{M}}_{IB} \\ \hat{\mathbf{M}}_{BI} & \hat{\mathbf{M}}_{BB} \end{bmatrix} \{\boldsymbol{\Psi}, \quad \hat{\mathbf{F}}_{RD}(\hat{t}) = \boldsymbol{\Psi}^T \begin{Bmatrix} \hat{\mathbf{F}}_I(\hat{t}) \\ \hat{\mathbf{F}}_B(\hat{t}) \end{Bmatrix} \tag{9}$$

4   NUMERICAL PROCEDURE

4.1   Onset speed of instability calculation
The first critical speed (FCS) and onset speed of instability (OSI) can be obtained from the eigenanalysis of the EOM that the bearing forces are linearized to bearing stiffness and damping. The OSI is the rotational speed that the real part of the positive-frequency first forward mode's eigenvalue becomes zero. For the full rotor model, free vibration of Eq.(1) becomes:

$$\hat{\mathbf{M}}_r\ddot{\hat{\mathbf{q}}} + \left(\hat{\mathbf{C}}_r + \hat{\mathbf{C}}_b + \hat{\omega}\,\hat{\mathbf{G}}_r\right)\dot{\hat{\mathbf{q}}} + \left(\hat{\mathbf{K}}_r + \hat{\mathbf{K}}_b\right)\hat{\mathbf{q}} = 0 \tag{10}$$

where $\hat{\mathbf{C}}_b$ and $\hat{\mathbf{K}}_b$ are combined damping and stiffness matrices of REB and JB. The eigenvalues of the reduced rotor model were obtained similarly by using the free vibration of Eq.(8). The obtained results are presented in Table 3. From this table, it could be seen that the reduced model yielded almost identical FCS and OSI to those obtained from the full model. For the case of $\hat{m}_d$=4 kg, reducing $\hat{C}_{rb}$ increased the OSI substantially in both $\hat{L}_d$ cases, especially when $\hat{L}_d$=575mm. However, if $\hat{m}_d$=8 kg, reducing $\hat{C}_{rb}$ did not increase the OSI noticeably when $\hat{L}_d$=340mm and even slightly lowered the OSI when $\hat{L}_d$=575mm.

Table 3. FCS and OSI of the full and reduced rotor model.

$\hat{m}_d$ (kg)	4				8			
$\hat{L}_d$ (mm)	340	340	575	575	340	340	575	575
$\hat{C}_{rb}$ (μm)	60	45	60	45	60	45	60	45
FCS (Full)	1000	1000	1630	1629	733	733	1263	1263
FCS (Reduced)	1000	1000	1636	1635	733	733	1269	1269
% Diff. (F-R)/F	0.00	0.00	-0.38	-0.38	0.00	0.00	-0.44	-0.44
OSI (Full)	2259	2509	3446	4024	1989	2003	2517	2513
OSI (Reduced)	2258	2507	3462	4045	1979	1994	2527	2522
% Diff. (F-R)/F	0.07	0.06	-0.46	-0.52	0.48	0.49	-0.38	-0.37

4.2   Shooting method, parallel computing, and arclength continuation

The shooting method was used due to its efficiency in solving periodic solutions and ability to obtain an unstable solution (21, 22). However, before the shooting method was applied, the reduced EOM had to be nondimensionalized to improve the numerical accuracy. The nondimensional version of Eq.(8) became:

$$\mathbf{s}'' + (\mathbf{C}_{RD} + \omega\mathbf{G}_{RD})\mathbf{s}' + \mathbf{K}_{RD}\mathbf{s} = \mathbf{F}_{RD}(t) \tag{11}$$

where

$$t = \hat{\omega}_{ref}\,\hat{t},\ \omega = \frac{\hat{\omega}}{\hat{\omega}_{ref}},\ ' = \frac{d}{dt},\ \mathbf{S} = \frac{\hat{\mathbf{s}}}{\hat{C}_{rb}},\ S' = \frac{\hat{\mathbf{s}}}{\hat{\omega}_{ref}\hat{C}_{rb}},\ \mathbf{S}'' = \frac{\hat{\mathbf{s}}}{\hat{\omega}_{ref}^2\hat{C}_{rb}},$$

$$\mathbf{C}_{RD} = \frac{\hat{\mathbf{M}}_{RD}^{-1}\hat{\mathbf{C}}_{RD}}{\hat{\omega}_{ref}},\ \mathbf{G}_{RD} = \frac{\hat{\mathbf{M}}_{RD}^{-1}\hat{\mathbf{G}}_{RD}}{\hat{\omega}_{ref}},\ \mathbf{K}_{RD} = \frac{\hat{\mathbf{M}}_{RD}^{-1}\hat{\mathbf{K}}_{RD}}{\hat{\omega}_{ref}},\ \mathbf{F}_{RD}\ (t) = \frac{\hat{\mathbf{M}}_{RD}^{-1}\hat{\mathbf{F}}_{RD}\left(\hat{t}\right)}{\hat{\omega}_{ref}^2\hat{C}_{rb}} \tag{12}$$

In Eq.(12), the OSI of the reduced rotor FE model of the corresponding rotor-bearing configuration, shown in Table 3, was used as $\hat{\omega}_{ref}$ in the particular rotor-bearing configuration. Since the investigated rotor-bearing system was self-excited, it was autonomous and had an unknown period. For the periodic solution, the first-order Taylor series of the periodic motion's residual vector was (21):

$$\mathbf{U}_T\left(\mathbf{U}_0^k, T_0^k\right) - \mathbf{U}_0^k + \left[\frac{\partial\mathbf{U}_T\left(\mathbf{U}_0^k, T_0^k\right)}{\partial\mathbf{U}_0}\bigg|_k - \mathbf{I}\right]\left\{\mathbf{U}_0^{k+1} - \mathbf{U}_0^k\right\}$$

$$+ \frac{\partial\mathbf{U}_T\left(\mathbf{U}_0^k, T_0^k\right)}{\partial T_0}\bigg|_k \left(T_0^{k+1} - T_0^k\right) = 0 \tag{13}$$

where $\mathbf{U} = \left\{ \mathbf{s}^T, \mathbf{s}^{T'} \right\}^T$, subscripts T and 0 represent the ending and starting point of the periodic motion, and superscript k is an index of iteration. From Eq.(13), there were 8 unknowns for $\mathbf{q}_B$ and $\mathbf{q}'_B$, 4 unknowns for $\mathbf{a}_h$ and $\mathbf{a}'_h$, and 1 unknown for T_0 with 12 equations. To make the equations solvable, the velocity constraint $x'_0 = 0$ was defined and the terms related to partial derivative respected to x'_0 in Eq.(13) were removed. As a result, $\mathbf{U}_0^{k+1}$ and T_0^{k+1} can be obtained simultaneously. The partial derivative part in Eq. (13) was solved by central difference method which required 24 integrations in each iteration step to find $\mathbf{U}_T$ of the perturbated $\mathbf{U}_0$ and T_0. However, each integration can be done parallelly to reduce the calculation time as accomplished by Kim and Palazzolo (15, 17). The guessed initial value for $\mathbf{U}_0$ and T_0 was obtained from the steady-state solution using numerical integration of Eq.(11) at the rotational speed above the OSI of the reduced rotor model obtained from the eigenanalysis in case the stable limit cycle can be obtained. However, if the steady-state solution could not be obtained, $\mathbf{U}_0$ and T_0 were selected from the point in time expansion of Eq. (11) at the rotational speed below the OSI that made the shooting method yielded a periodic solution. The contact between the journal and the bearing was checked from the journal lateral displacement at the IB and OB sides of the bearing using lateral and angular displacements of node 5, shown in Figure 1, under the rigid journal assumption. The arclength continuation (21) was applied to make the solutions able to transverse the saddle-node point (if occurred). $\mathbf{U}_0$ and T_0 were defined as functions of position s along the arc. The continuation was terminated when the contact between the journal and the bearing was detected.

4.3   Floquet multiplier analysis

After the solution for $\mathbf{U}_0$ and T_0 were obtained, the stability of the solution was verified by calculating the monodromy matrix ($\mathbf{\Phi}$). For an autonomous system, the mono-dromy matrix is defined by (22):

$$\mathbf{\Phi}(T_0) = \frac{\partial \mathbf{U}_T(\mathbf{U}_0, T_0)}{\partial \mathbf{U}_0} \tag{14}$$

which can be readily obtained from a part of the third term in Eq.(13). The periodic solution is stable if the spectral radius of the monodromy matrix is less than 1.

5   TEST RIG

The test rig, shown in Figure 3(a), that imitates the parameters used in the simulation was set up to verify the calculation results. The bearing housing was inclined to the value of $\hat{\theta}_{x5st}$ depicted in Table 2 by inserting shims under the outboard side of the bearing housing. Two JBs with $\hat{C}_{rb}$=60 and 45 μm were manufactured by EDM with ±5 μm tolerance. The interference fit between the bearing and housing was less than 10 μm, so it was reasonable to assume that the pre-load was caused by bearing profile only. Engineer's blue paste was applied to observe the contact, which also indicated the parallelism between the journal and the bearing, as shown in Figure 3(b). The bearing's inlet oil temperature was controlled so that the steady-state bearing's drain temperature at speed about 500 rpm below the calculated OSI was in the range of 30-32°C, which gives μ=42-38 mPa-s according to Vogel's equation (23). Shaft relative displacements in x and y-directions were measured at OBP and IBP positions shown in Figure 3(a) and then used to calculate the displacements at OB and IB.

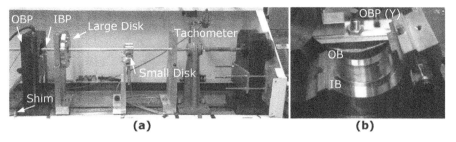

Figure 3. a) Test Rig, b) Contact Pattern.

6 RESULTS

6.1 Nonlinear calculation results

Figure 4 to Figure 7 show comparison of the maximum value of the limit cycle's eccentricity ratio at the IB side (The OB side's plots were omitted as the journal touched the bearing at IB side in all cases) when $\hat{C}_{rb}$=60 and 45 µm of each $\hat{m}_d$ and $\hat{L}_d$ case. In these figures, the solid and dashed lines represent stable and unstable solutions, respectively, whereas the square represents the OSI obtained from the eigenanalysis of the reduced model. The bifurcation point was defined by the point the limit cycle converged to the equilibrium position. It was observed that this point was identical to the OSI obtained from the eigenanalysis, which confirmed the linear analysis's validity in the infinitesimal motion range around the equilibrium position. The stability of the limit cycle connected to the bifurcation point was used to identify the Hopf bifurcation type. The Hopf bifurcation type was supercritical or subcritical when the limit cycle was stable or unstable, respectively.

The results of the cases $\hat{m}_d$=4kg, which can be considered as a lightly-loaded bearing, indicated that the reduced $\hat{C}_{rb}$ raised the OSI, especially when $\hat{L}_d$=575mm, as depicted in Figure 5. However, it changed the bifurcation type from supercritical to subcritical in this case, which means the rotor's vibration amplitude will suddenly jump when the OSI is exceeded, or the rotor is sufficiently perturbed even at speed below the OSI. These results indicated that the reduced $\hat{C}_{rb}$ could increase the OSI of the rotor, which is quite a well-known fact, but it can change the bifurcation type. When the bearing load was raised by increasing $\hat{m}_d$ to 8 kg, however, the OSI of the reduced $\hat{C}_{rb}$ cases did not substantially change from that of the normal $\hat{C}_{rb}$ cases. The reduced $\hat{C}_{rb}$ changed the bifurcation type from subcritical to supercritical in the case $\hat{L}_d$=575mm, as shown in Figure 7. Note that in the $\hat{m}_d$=8kg cases, the eccentricity ratio of the journal equilibrium in the reduced $\hat{C}_{rb}$ case was noticeably higher than that of the associated normal $\hat{C}_{rb}$ case. The higher operating eccentricity ratio contributed to a more stable operation as indicated by the wider gap between the equilibrium position and the unstable limit cycle at speed below OSI in Figure 6. In addition, it could suppress the whirling's sudden growth of subcritical bifurcation, as shown in Figure 7. Hence, the reduced $\hat{C}_{rb}$ could change the bifurcation type in specific rotor configurations, and the nonlinear analysis was essential in predicting how the vibration amplitude of the rotor grows when it becomes unstable or perturbed at speed near the OSI.

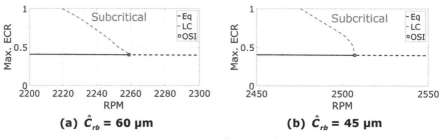

Figure 4. Limit Cycle: $\hat{m}_d$=4kg, $\hat{L}_d$=340 mm.

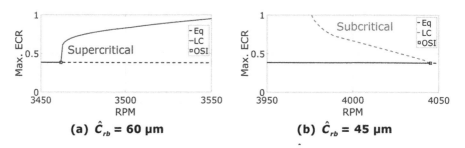

Figure 5. Limit Cycle: $\hat{m}_d$=4kg, $\hat{L}_d$=575 mm.

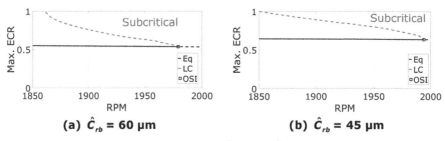

Figure 6. Limit Cycle: $\hat{m}_d$=8kg, $\hat{L}_d$=340 mm.

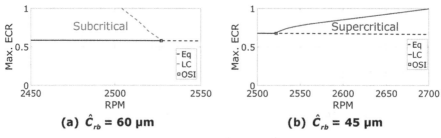

Figure 7. Limit Cycle: $\hat{L}_d$=8kg, $\hat{L}_d$=575 mm.

6.2 Experimetal Results

The vibration amplitude at the frequency around FCS of the IB side was plotted versus shaft rotation speed, as shown in Figure 8 to Figure 11. In these figures, the upper and lower triangles represent the data during run up and run down, respectively. FCS, OSI, and bifurcation type could be detected from these figures. The bifurcation type was identified by the stability of the subsynchronous vibration. If its amplitude was stable at a constant speed above the OSI, the case was identified as supercritical bifurcation and vice versa. These values were then summarized and compared with the calculated values, as shown in Table 4. It can be seen that the experimented FCS, OSI, and bifurcation type agreed well with the calculation results.

Table 4. Calculated and Experimented FCS, OSI and Bifurcation Type.

$\hat{m}_d$ (kg)	4				8			
$\hat{L}_d$ (mm)	340	340	575	575	340	340	575	575
$\hat{C}_{rb}$ (μm)	60	45	60	45	60	45	60	45
FCS (Calc, Reduced)	1000	1000	1636	1635	733	733	1269	1269
FCS (Exp)	980	1000	1611	1638	741	755	1240	1250
% Difference	2.00	0.00	1.53	-0.18	-1.09	-3.00	2.29	1.50
OSI (Reduced)	2258	2507	3462	4045	1979	1994	2527	2522
OSI (Exp)	2297	2500	3468	4060	1852	1986	2545	2520
% Difference	-1.73	0.28	-0.17	-0.37	6.42	0.40	-0.71	0.08
Bifurcation (Calc)	sub	sub	super	sub	sub	sub	sub	super
Bifurcation (Exp)	sub	sub	super	sub	sub	sub	sub	super

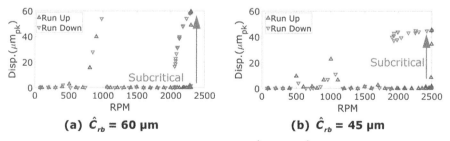

(a) $\hat{C}_{rb}$ = 60 μm

(b) $\hat{C}_{rb}$ = 45 μm

Figure 8. Vibration Amplitude: $\hat{m}_d$=4kg, $\hat{L}_d$=340 mm.

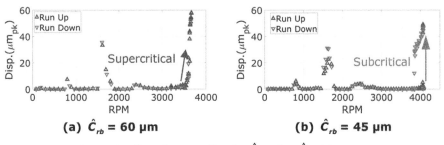

(a) $\hat{C}_{rb}$ = 60 μm

(b) $\hat{C}_{rb}$ = 45 μm

Figure 9. Vibration Amplitude: $\hat{m}_d$=4kg, $\hat{L}_d$=575 mm.

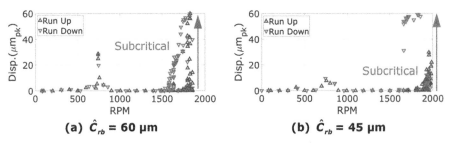

(a) $\hat{C}_{rb}$ = 60 μm

(b) $\hat{C}_{rb}$ = 45 μm

Figure 10. Vibration Amplitude: $\hat{m}_d$=8kg, $\hat{L}_d$=340 mm.

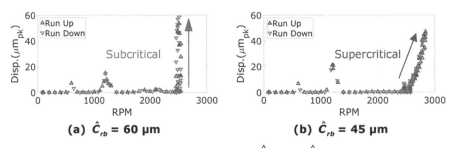

(a) $\hat{C}_{rb}$ = 60 μm

(b) $\hat{C}_{rb}$ = 45 μm

Figure 11. Vibration Amplitude: $\hat{m}_d$=8kg, $\hat{L}_d$=575 mm.

7   CONCLUSIONS

The effect of the journal bearing preload (reduction in $\hat{C}_{rb}$) caused by bearing-housing interference fit on Hopf bifurcation type of a flexible rotor supported by a journal bearing (JB) was investigated under various cases of disk mass/position. The real mode component mode synthesis (CMS) that retained only the two lowest modes of the internal degrees of freedom (DOF) of the rotor-bearing system was applied to reduce the rotor model. Shooting method with parallel computing and Floquet multiplier analysis were then applied to the reduced model to obtain the limit cycle and its stability to identify the bifurcation type. The experiment on the test rig with identical parameters to those used in the numerical calculation was performed to verify the calculation results.

The calculation and experimental results agreed well with each other in all cases. They showed that even though the reduced $\hat{C}_{rb}$ noticeably raised the OSI when $\hat{m}_d=$ 4kg and $\hat{L}_d$=575mm, it changed the bifurcation type from supercritical to subcritical. However, when $\hat{m}_d$ was increased to 8 kg, the reduced $\hat{C}_{rb}$ changed the bifurcation type from subcritical to supercritical but slightly lowered the OSI. This reversal in trend could be caused by higher operating eccentricity of the bearing with reduced $\hat{C}_{rb}$ in the case of $\hat{m}_d=$ 8kg that stabilized the rotor and suppressed the suddenly increasing unstable vibration. Hence, the reduced $\hat{C}_{rb}$ could change the bifurcation type in particular rotor configurations, and the nonlinear analysis was essential in predicting how the vibration amplitude of the rotor grows when it becomes unstable or perturbed.

REFERENCES

[1] Muszynska, A., 1988, "Stability of Whirl and Whip in Rotor/Bearing Systems," J. Sound Vib., 127(1), pp. 49–64.

[2] Childs D., 1993, "Turbomachinery Rotordynamics: Phenomena, Modeling, & Analysis," John Wiley & Sons, New York.

[3] Noah, S. T., 1995, "Significance of Considering Nonlinear Effects in Predicting the Dynamic Behavior of Rotating Machinery," J. Vib. Control, 1, pp. 431–458.

[4] Castro, H. F., Cavalca, K. L., and Nordman, R., 2008, "Rotor-Bearing Systems Instabilities Considering a Non-Linear Hydrodynamic Model," J. Sound. Vib., 317, pp. 273–293.

[5] Zheng, T., and Hasebe, N., 1999, "Nonlinear Dynamic Behaviors of a Complex Rotor-Bearing System," ASME. J. Appl. Mech., 67(3), pp. 485–495.

[6] Allaire, P. E., and Flack, R. D., 1981. "Design of Journal Bearings for Rotating Machinery," Texas A&M University Turbomachinery and Pump Symposia, pp. 25–45., doi: https://doi.org/10.21423/R1FH40.

[7] Kirk, R. G., 1978, "The Influence of Manufacturing Tolerances on Multi-Lobe Bearing Performance in Turbomachinery," In Topics in fluid film bearing and rotor bearing system design and Optimization, pp. 108–129.

[8] Friswell, M. I., and Penny, J. E. T., "Reduced Order Models for Non-linear Rotating Machines," Vibrations in Rotating Machinery, VIRM 11, Manchester, 13–15 September 2016, pp. 105–117.

[9] Craig R., and Bampton, M., 1968, "Coupling of Substructures for Dynamic Analyses," AIAA J., 6 (7), pp. 1313–1319.

[10] Genta, G., 2005, "Dynamics of Rotating Systems," Springer-Verlag, New York.

[11] Kobayashi, M., and Aoyama, S., 2011, "Transient Response Analysis of a High-Speed Rotor Using Constraint Real Mode Synthesis," JSME. J. Environment and Engineering., 6(3), pp. 527–541.

[12] Glasgow, D. A., and Nelson, H. D., 1980, "Stability Analysis of Rotor-Bearing Systems Using Component Mode Synthesis," ASME. J. Mech., 102(2), pp. 352–359.

[13] Nelson, H. D., Meacham, W. L., Fleming, D. P., and Kascak, A. F., 1983, "Nonlinear Analysis of Rotor-Bearing Systems Using Component Mode Synthesis," ASME. J. Eng. Power. 105(3), pp. 606–614.

[14] Sundararajan, P., and Noah, S. T., 1998, An algorithm for response and stability of large order non-linear systems - Application to rotor systems," J. Sound Vib. 214(4), pp. 695–723.

[15] Kim, s., and Palazzolo, A., 2018, "Shooting/continuation based bifurcation analysis of large order nonlinear rotordynamic system," The 14th International Conference on Vibration Engineering and Technology of Machinery (VETOMAC XIV).

[16] Koondilogpiboon, N., and Inoue, T., 2019, "Nonlinear Vibration Analysis Using Component Mode Synthesis of a Flexible Rotor Supported by an Axial groove Journal Bearing Considering Journal Angular Whirling Motion", Proceedings of the ASME 2019 International Design Engineering Technical Conferences and Computers and Information in Engineering Conference (IDETC/CIE 2019), August 18–21, 2019, Anaheim, CA, USA, IDETC2019–97404.

[17] Kim, S., and Palazzolo A.B., 2017, "Bifurcation Analysis of a Rotor Supported by Five-Pad Tilting Pad Journal Bearings Using Numerical Continuation," ASME. J. Tribol., 140(2).

[18] Khonsari, M. M., and Booser, E. R., 2001, "Applied Tribology," John Wiley & Sons, New York.

[19] Szeri, A. Z., 2010, "Fluid Film Lubrication: 2nd ed," Cambridge University Press, Cambridge.

[20] Lund, J. VV., and Thomsen, K. K., 1978, "A Calculation Method and Data for the Dynamic Coefficients of Oil- Lubricated Journal Bearings," Topics in Fluid Film Bearings and Rotor Bearing System Design and Optimization, ASME Book No. 100118, pp. 1–28.

[21] Nayfeh, A. H. and Balachandran, B., 1995, "Applied Nonlinear Dynamics," John Wiley & Sons, New York.

[22] Seydel, R., 1994, "From Equilibrium to Chaos: Practical Bifurcation and Stability Analysis", Springer-Verlag, New York.

[23] Stachowiak, G. and Batchelor, A., 2013, "Engineering Tribology," Elsevier, Burlington.

Validation of the stochastic response of a rotor with uncertainties in the AMBs

G. Garoli[1], R. Pilotto[2], R. Nordmann[2], H. de Castro[1]

[1]Department of Integrated Systems, School of Mechanical Engineering, University of Campinas, Brazil
[2]Fraunhofer Institute for Structural Durability and System Reliability LBF, Germany

ABSTRACT

Active Magnetic Bearings (AMB) can support and act as an actuator in a rotor system. These machines are subjected to uncertainties. To consider them, Monte Carlo methods are usually implemented. However, they may take an extensive processing time. In this work, the generalized Polynomial Chaos Expansion is used to approximate the stochastic response of a rotating system supported by AMB. It is considered uncertainties in the bearing parameters and unbalance force. The results are validated by comparison with experimental data. After the validation, a global sensitivity analysis is performed through Sobol indices.

1   INTRODUCTION

The study of rotor-bearing systems is demanded due to its extensive use in the industry, especially in power plants. The high costs of production and maintenance make the development of mathematical models and simulations necessary. Finite element method is normally used to model these systems. Nelson and Macvaugh (1) developed a finite element model of a beam using the Rayleigh model, the model included the effects of rotatory inertia, gyroscopic moments and axial load. Years later, Nelson (2) based on the previous work and the Timoshenko beam, determined the shape functions considering the internal damping effect.

An important component of a rotating system is the bearing, which connects the rotating shaft to the foundation of the machine. The ball and fluid types are commonly used, due to their price and they are suitable for a great number of rotor systems. However, they provide a passive support of the shaft and the friction between the shaft and the balls or fluid drains energy from the system. The Active Magnetic Bearings (AMB) supports the shaft through a magnetic field, therefore it does not have contact with the shaft. It allows the use of a controller, as well. Wamba (3) and Wamba and Nordmann (4) uses such components to keep the vibration of a test rig in a acceptable level while Active Balancing Devices balances the system.

These machines are subjected to uncertainties in its geometry and operating parameters. These uncertainties must be modelled and included in the mathematical model of the problem. The Monte Carlo method can be used to evaluate the stochastic response, but it may have a high computational cost. Therefore, other method to evaluate the stochastic response is necessary.

Wiener (5) first proposed the use of an orthogonal polynomial basis to propagate uncertainties. In his work, the stochastic solution considering Gaussian process was approached by a Hermite polynomials basis. Xiu and Karniadakis (6) expanded the concept developed by Wiener for non-Gaussian process, known as genelaized Polynomial Chaos Expansion (gPCE). The authors related hypergeometric orthogonal polynomials of the Askey scheme with some random variables. This correlation is made by the weighting function of the orthogonality relation of each polynomial, for example Laguerre polynomials and gamma random variables.

The use of the polynomial expansion resumes the problem into evaluate the expansions coefficients, which can be made by the Stochastic Collocation. This method uses pre-chosen nodes, evaluates the deterministic problem and the polynomials at these nodes and then uses the least square method to calculate the expansion coefficients.

After the evaluation of the expansion coefficients, a statistical analysis can be directly carried out from them. Furthermore, a variance-based sensitivity analysis of the can be performed through the expansion coefficients as proposed by Sudret (7). The Sobol indices are evaluated directly from the expansion coefficients and the effects of the inputs uncertainties on the stochastic response can be measure.

It is proposed the use of the generalized Polynomial Chaos Expansion to evaluate the stochastic response of a rotor system supported by two AMB's. Uncertainties are considered in the AMB's parameters and unbalance parameters. These uncertainties are modelled by Gausian distributions. Therefore, other methods as Kriging or a nonparametric one, such as Soize proposed (8), can evaluate the stochastic response faster than Monte Carlo methods. However, the gPCE allows the direct evaluation of the statistical moments of the response and the execution a sensitivity analysis.

After the evaluation of the expansion coefficients, the simulated response considering the uncertainties are compared with experimental data. Then, a sensitivity analysis is performed.

2   ROTOR SYSTEM

The studied rotor system is a low-pressure shaft of a helicopter engine supported by two active magnetic bearings (AMB). The hollow shaft has 1100 mm with diameter of 25 mm and wall thickness of 3,5 mm. There is a two turbine stages without the blades. The rotor system has three active balancing devices (ABD), which are not used in this work. The rotor system weights a total of 25.4 kg, taking in account the shaft, ABD's and the turbine. The AMB's act as actuators and sensors, measuring the radial displacements of the shaft inside of it in the vertical and horizontal directions. Figure 1 shows the test rig that contains the rotor system. In the experimental data and simulation results, the bearing nearer the motor drive is called bearing 1 and the one nearer the turbine is called bearing 2. The same test rig was used in the study of Wamba and Nordmann (4).

2.1   Mathematical Model
The rotor system is modelled by the finite element method. The gyroscopic, mass and stiffness equivalent matrices of the system are evaluated by the matrices presented in Nelson (2) using properties and dimensions of the rotor system. The damping is proportional to the stiffness, by a factor 0.0005. It is considered the unbalance force, AMB forces and the rotor weight as external forces. Therefore, the motion equation of the system has the form of eq. 1.

$$[M]\{\ddot{q}\} + ([C] + \omega[G])\{\dot{q}\} + [K]\{q\} = \{F_{unb}\} + \{F_{AMB}\} + \{W\} \tag{1}$$

In which, $[M]$, $[C]$, $[G]$, $[K]$ are, respectively, the mass, damping, gyroscopic and stiffness matrices, ω is the rotating speed, $\{\ddot{q}\}$, $\{\dot{q}\}$, $\{q\}$ are the acceleration, velocity and displacement of the degrees of freedom, $\{F_{unb}\}$ is the unbalance force, $\{F_{AMB}\}$ is the AMB force and $\{W\}$ is the weight.

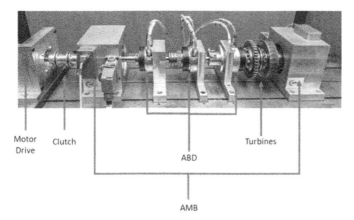

Figure 1. Studied rotor system.

2.1.1 *Unbalance force*

The unbalance is one of the most common faults and it is modelled as a residual mass displaced from the rotating axis. The residual mass and eccentricity can be combined into one variable called unbalance moment me_{unb} and the phase is θ_{unb}. It adds a force in the system, which is formulation is presented by eq. 2.

$$\{F_{unb}\} = \begin{Bmatrix} F_{unb}^{y} \\ F_{unb}^{z} \end{Bmatrix} = -me_{unb} \cdot \omega^2 \cdot \begin{Bmatrix} \sin(\omega \cdot t + \theta_{unb}) \\ \cos(\omega \cdot t + \theta_{unb}) \end{Bmatrix} \qquad (2)$$

2.1.2 *AMB force*

According to eq. 3, the AMB forces is a function of the air gap between the shaft and the stator and the current passing through the bearing.

$$\{F_{AMB}\} = [K_s] \cdot \{q\} + [K_i] \cdot \{i\} \qquad (3)$$

In which, $[K_s]$ is the diagonal matrix with the equivalent stiffness of the AMB and $[K_i]$ is the diagonal matrix with the current gain. The AMB adds a negative stiffness in the system, which make it unstable. Therefore, a control is needed to stabilize the system. The test rig already has a PID controller and its terms were used for the simulations.

3 UNCERTAINTY ANALYSIS

Uncertainties are assumed in the unbalance parameters and AMB parameters. Therefore, a stochastic method is needed. Due to the small number of uncertainties, the generalized Polynomial Chaos Expansion can be implemented. To achieve the convergence of the stochastic response, the gPCE requires less samples than a Monte Carlo method. Furthermore, from the expansion coefficients, a sensitivity analysis can be directly performed.

3.1 Generalized polynomial chaos expansion

Given a finite random space Λ, let λ be a random variable of this space, with probability density function $\rho(\lambda)$. Lets the orthogonality condition of a set of orthogonal polynomials $\{\Phi_m(\lambda)\}$ be:

$$\int_A \Phi_m(\lambda) \cdot \Phi_n(\lambda) \cdot \rho(\lambda) \, d\lambda = h_m^2 \cdot \delta_{mn} \tag{4}$$

where the normalization factor is

$$h_m^2 = \int_\Lambda \Phi_m^2(\lambda) d\lambda$$

and δ_{mn} is the Kronecker delta.

Then, ρ in Equation 4 is the integration weight, which determines the type of the orthogonal polynomial set. For Gaussian distributed random variable, which will be used in this work, its pdf defines Hermite polynomials.

To define a basis of orthogonal polynomial with maximum degree P in a N-variable random space the Equation 5 is use:

$$\Phi_M^P = \overset{\otimes}{\underset{|d| \leq P}{}} \Phi^{i,d_i}, M = \dim\left(\Phi_N^P\right) = \binom{N+P}{N} \tag{5}$$

The solution of the eq. 1 will be approach by:

$$q \cong q_N^P(x, \lambda) = \sum_{m=1}^M \hat{q}_m(x) \Phi_m(\lambda) \tag{6}$$

The expansion coefficients $\hat{q}_m(x)$ are determinate through the least square method:

$$\begin{Bmatrix} q(x,\lambda^1) \\ \vdots \\ q(x,\lambda^Q) \end{Bmatrix} = \begin{bmatrix} \Phi_0(\lambda^1) & \cdots & \Phi_M(\lambda^1) \\ \vdots & \ddots & \vdots \\ \Phi_0(\lambda^Q) & \cdots & \Phi_M(\lambda^Q) \end{bmatrix} \cdot \begin{Bmatrix} \hat{q}_1 \\ \vdots \\ \hat{q}_M \end{Bmatrix} \tag{7}$$

where $q(q, \lambda^i)$ is the solution of eq. 1 and $\Phi_m(\lambda^i)$ is the polynomial m, both calculated at the node λ^i. Different methods can be used to pre-choose the nodes λ^i, such Latin Hypercube Sampling, random samples, quadrature rules, etc. In this work, the samples were equally spaced in the variable support.

3.2   Sensitivity analysis

A global sensitivity analysis can be carried out, the effects of each random input variable on the stochastic response can be evaluated. The Sobol indices provide accurate information for most models. Sudret (7) shows how to evaluate the indices through generalized polynomial chaos expansion, using the expansion coefficients. A brief description of the author's developments will be presented next.

Let $f(\lambda_1, \ldots, \lambda_n)$ be a function with a scalar output and dependent on N independent random variables λ. The Sobol decomposition of $f(\lambda)$ into summand of increasing dimensions is:

$$f(\lambda_1, \ldots, \lambda_N) = f_0 + \sum_{i=1}^N f_i(\lambda_i) + \sum_{1 \leq i < j \leq N} f_{ij}(\lambda_i, \lambda_j) + \cdots + f_{1,\ldots,N}(\lambda_1, \ldots, \lambda_N) \tag{8}$$

with f_0 as the mean value of the function and the integral of each summand over any of its variables is zero. The Sobol indices, s_i, are defined as:

$$s_{i_1,\dots,i_s} = \frac{\sigma^2_{i_1,\dots,i_s}}{\sigma^2} \tag{9}$$

In which, $\sigma^2_{i_1,\dots,i_s}$ are the partial variances. Therefore, the sum of all Sobol indices is equal to 1.

The polynomial expansion can be reorganized as the Sobol decomposition and the sobol indices are calculated by:

$$s_{PC_{i_1,\dots,i_s}} = \sum_{\mathbf{j} = \xi_{i_1,\dots,i_s}} \frac{\hat{q}^2_j}{\sigma^{P^2}_N} \tag{10}$$

With

$$\xi_{i_1,\dots,i_s} = \left\{ \mathbf{j} : \begin{matrix} j_k > 0, \forall k = 1,\dots,N, & k \in (i_1,\dots i_s) \\ j_k = 0, \forall k = 1,\dots,N, & k \notin (i_1,\dots i_s) \end{matrix} \right. \tag{11}$$

Putting into words, $\xi_{i_1,\dots,i_s}$ corresponds to the polynomials that only depend on random variables $\lambda_1,\dots,\lambda_s$.

4 RESULTS

Experimental data were collected from the test rig. The horizontal and vertical displacements of the shaft inside the two AMB were gathered. From (4), it is known that the first critical speed is around 1700rpm. Therefore, experimental data were collected at three rotating speeds: before, near and after the critical speed. The values of the rotating speeds are 1000, 1500, 2000 rpm. After the steady state were reached, 1024 samples were gathered with $5,001 \cdot 10^{-4}$ s between each one, for a full time of 0,51 s.

It is considered two cases for the simulations: one with uncertainties in the unbalance parameters and one with uncertainties in the AMB parameters. Table 1 presents the mean and standard deviation of the truncated Gaussian distribution considered for each parameter. According to (6), the Hermite polynomials will be used to evaluate the stochastic response for each case. A maximum degree of 3 is used, resulting in 10 term of the expansion, and 100 samples equally spaced were used to evaluate the coefficients.

Table 1. Mean and standard deviation considered.

Parameters	Mean	Standard deviation
AMB Stiffness (N/m)	$871 \cdot 10^3$	$87,1 \cdot 10^3$
Current gain (N/A)	214	21,4
Unbalance moment (kg·m)	$25 \cdot 10^{-5}$	$5 \cdot 10^{-5}$
Phase (rad)	0	0,6

4.1 AMB parameters with uncertainties

Figure 2 to Figure 4 presents the stochastic time response of the shaft inside each bearing. In each figure it is shown the experimental results, the mean of the stochastic response and the 95% confidence region. In all rotating speeds, the experimental data is within the confidence bounds majority of the time. Effects of higher harmonic components are observed in the experimental data, which can be better seen at 1000 and 2000 rpm (Figure 2 and Figure 4). These effects are reduced at 1500 rpm due to the proximity to the critical speed, which excites the unbalance fault. At this speed, the experimental data is more like the stochastic response mean.

4.2 Unbalance parameters

Figure 5 to Figure 7 shows the stochastic time response of the shaft inside each bearing. Again, the 95% confidence region is presented. The confidence region is larger than the ones of the previous case. Therefore, the variance of the unbalance uncertainties has more effects in the variance of the response than the AMB parameters uncertainties. The experimental response is almost entirely inside the confidence bounds, even with the higher harmonic components' effects.

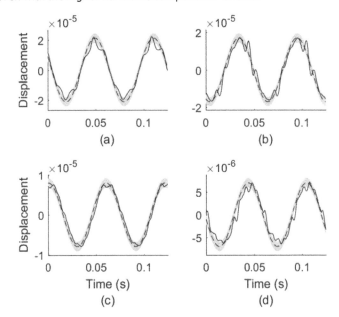

Figure 2. Displacements inside the AMB's at 1000 rpm with AMB uncertainties; horizontal (a) and vertical (b) directions of bearing 1; horizontal (c) and vertical (d) directions of bearing 2.

Even with few terms, the polynomial approach can approximate the stochastic response of the system. The considered uncertainties do not predict the exact behaviour of the rotor machine. However, the stochastic response fully covers the experimental response. Therefore, the stochastic model could be used to simulate the test rig.

4.3 Sensitivity analysis

For each case, the Sobol indices were evaluated from the expansion coefficients. Table 2 presents the indices for each case. The indices are similar in all rotating speeds. For the unbalance parameters, the unbalance moment has a higher influence in the variance of the stochastic response. For the AMB parameters, both has a similar influence in the stochastic response, with the current gain slightly higher.

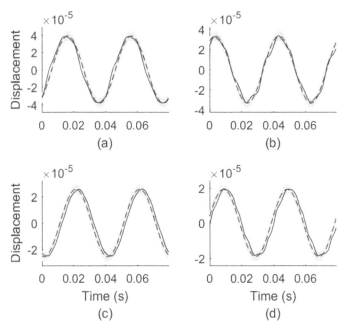

Figure 3. Displacements inside the AMB's at 1500 rpm with AMB uncertainties; horizontal (a) and vertical (b) directions of bearing 1; horizontal (c) and vertical (d) directions of bearing 2.

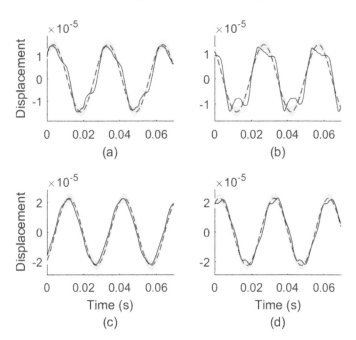

Figure 4. Displacements inside the AMB's at 2000 rpm with AMB uncertainties; horizontal (a) and vertical (b) directions of bearing 1; horizontal (c) and vertical (d) directions of bearing 2.

Table 2. Sobol indices for each case.

	1000 rpm	**1500 rpm**	**2000 rpm**
me_{unb}	0,73	0,77	0,75
θ_{unb}	0,26	0,23	0,24
Both	0,01	0,00	0,01
k_s	0,45	0,44	0,45
k_i	0,55	0,56	0,55
Both	0,0	0,0	0,0

In future analysis of this test rig configuration, uncertainties in the unbalance phase may be neglected and just the unbalance moment is considerer uncertain. The uncertainties of the AMB parameters has similar effects and both should be considered.

5   CONCLUSION

The stochastic time response of a rotor supported by AMB's was analysed. The AMB's were inserted in the model as a negative stiffness and a magnetic force. It was considered two cases, one with uncertainties in the AMB parameters and another with uncertainties in the unbalance force parameters. To evaluate the stochastic response, the generalized Polynomial Chaos Expansion was implemented, and the expansion coefficients were estimated by a least square method.

The stiffness and the current gain of the AMB model and unbalance moment and phase of the unbalance force were considered with uncertainties. They were model by truncated Gaussian distributions and the Hermite polynomials were used to build the polynomial basis. It was collected experimental data in three different rotating speeds, before, near and after the first critical speed. Then the simulated stochastic responses were compared to the experimental time responses and a sensitivity analysis is performed.

For the case with uncertainties in the AMB parameters, a good part of the experimental data was inside the 95% confidence bound in all three rotating speeds. The experimental data presented higher harmonic components, which effects were decreased when near the first critical speed. Even though, the stochastic model presented satisfactory results.

When the uncertainties were considered in the unbalance force, a larger confidence region is observed. Therefore, the unbalance force uncertainties have a higher influence in the variance of the stochastic response than the AMB uncertainties. However, this makes the experimental data almost entirely inside the confidence bound.

The global sensibility analysis through Sobol indices presents a higher influence of the unbalance moment than the phase in the stochastic time response. For the AMB parameters, the values are similar with the current gain slightly higher than the stiffness.

The results of the gPCE are satisfactory and shows that this method can be used to simulate the stochastic response of a rotor supported by AMB's. For future work, the uncertainties of the AMB and unbalance fault will be estimated by the Bayesian Inference with gPCE approximation. And a new sensitivity analysis will be performed to check if it is relevant to consider the uncertainties of the AMB in future models.

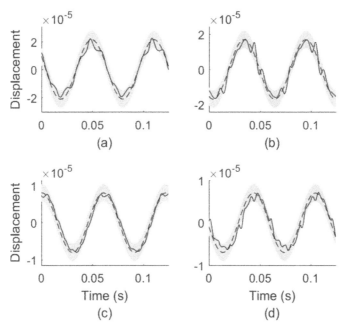

Figure 5. Displacements inside the AMB's at 1000 rpm with unbalance uncertainties; horizontal (a) and vertical (b) directions of bearing 1; horizontal (c) and vertical (d) directions of bearing 2.

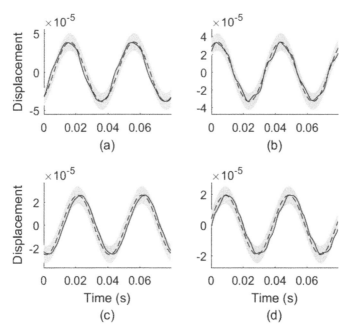

Figure 6. Displacements inside the AMB's at 1500 rpm with unbalance uncertainties; horizontal (a) and vertical (b) directions of bearing 1; horizontal (c) and vertical (d) directions of bearing 2.

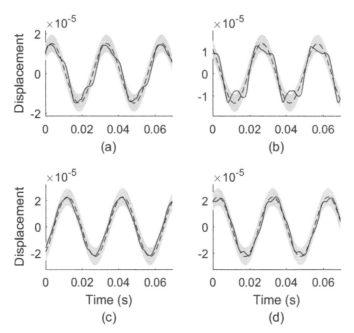

Figure 7. Displacements inside the AMB's at 2000 rpm with unbalance uncertainties; horizontal (a) and vertical (b) directions of bearing 1; horizontal (c) and vertical (d) directions of bearing 2.

ACKNOWLEDGMENTS

The authors would like to thank FAPESP grants #2015/20363-6 and #2016/13223-6 for the financial support to this research

REFERENCES

[1] Nelson, H.D., McVaugh, J.M., The Dynamics of Rotor-Bearing Systems Using Finite Elements, Journal of Engineering for Industry, Vol. 98 (2), pp.:593–600, 1976.

[2] Nelson, H.D., A Finite Rotating Shaft Element Using Timoshenko Beam Theory. Journal of Mechanical Design, Vol. 102, pp.: 793–803, 1980.

[3] Wamba, F.F., Automatische Auswuchtstrategie für einen magnetgelagerten elastischen Rotor mit Auswuchtaktoren, PhD Dissertation, Technischen Universität Darmstadt, Darmstadt, Germany, 2009.

[4] Wamba, F.F., Nordmann, R., Active Balancing of a Flexible Rotor in Active Magnetic Bearing, Proceedings of the 11th International Symposium on Magnetic Bearings, Nara, Japan, 2008.

[5] Wiener, R., The Homogeneous Chaos, American Journal of Mathematics, Vol. 60 (4), pp. 897–936, 1938.

[6] Xiu, D., Karniadakis, G., The Wiener–Askey Polynomial Chaos for Stochastic Differential Equations, Journal on Scientific Computing, Vol. 24 (2), pp.: 619–644, 2002.

[7] Sudret, B., Global sensitivity analysis using polynomial chaos expansions, Reliability Engineering and System Safety 93, pp. 964–979, 2008.

Active chatter suppression in robotic milling using H_∞ control

R. Zhang, Z. Wang, P. Keogh

Department of Mechanical Engineering, University of Bath, UK

ABSTRACT

Low-stiffness-induced vibration in industrial robots is problematic especially for robotic machining processes. In robotic milling, the spindle may experience chatter with varying characteristics under different working conditions, leading to unsatisfactory surface finishes. Active vibration control using robust controllers may therefore be an ideal solution for chatter suppression. This paper demonstrates that with two inertia actuators and one accelerometer attached near the spindle, on the robot, chatter suppression may be achieved using simple yet powerful H_∞ controllers with feedback of measured accelerations. The implementation of the controller follows a procedure, including characterisation and model identification of the robot system dynamics, selection of weighting functions, H_∞ synthesis and controller verification. Robotic milling tests of an aluminium block using a KUKA KR120R2500 PRO robot were performed to show that the designed H_∞ controller targets a low chatter frequency around 26 Hz and reduces vibration by up to 80%. The surface finishes with and without active chatter suppression are compared, confirming an improvement of over 60% in the surface roughness. The proposed active chatter suppression could benefit large volume manufacturing, where the system dynamics change during the operation.

1   INTRODUCTION

Robotic machining systems feature low costs, versatility, small footprints and large volume manufacturing capability compared with conventional machining systems. However, serial-link robots have relatively low mechanical stiffness and are therefore prone to experience machining induced vibrations. Therefore, their practical applications are hindered unless effective vibration suppression is achieved for precision machining.

Passive dynamic vibration absorbers (DVAs) are used as one common approach for vibration suppression, which transfer the energy of vibration to the absorber at its natural frequency. While classic DVAs are effective only around a narrow frequency band, modern semi-active and active DVAs offer controllable damping and/or stiffness [1, 2] and are capable of tracking and adapting to vibration frequency bands. Applications of DVAs cover a wide range such as in conventional machining systems [3, 4, 5, 6]. However, only a few DVAs are used for vibration absorption in robot machining systems. It has been shown that a magnetorheological elastomer-based semi-active DVA could be used to suppress chatter frequencies ranging from 7 to 20 Hz in robotic milling [7]; similarly, eddy current dampers (ECDs) have been used to improve damping properties and attenuate vibration in robotic polishing [8] and robotic milling [9]. DVAs and ECDs are proved to be effective when the dynamics of the system remain unchanged. However, these devices have to be designed and fabricated to target particular frequencies and dynamic modes of a system. It is difficult to operate these devices when a robot operates in different poses with variable modal characteristics.

Other approaches have been developed specifically for vibration suppression in robotic machining including optimising the system structure [10], the robot configuration and trajectory planning [11, 12], spindle speed [13], and tool path and feed rate [14, 15]. These approaches are promising, but reply heavily on prior experiments and calculations, making implementations difficult. Furthermore, they must adapt to changes in machining configurations, therefore may not be applicable under all working conditions. In contrast, active vibration control does not have such limitations and has the potential to suppress vibrations with broader attenuation frequency bands [16, 17]. In a robotic machining system with relatively low stiffness, this control method offers a solution to cope with the changing dynamics of the system.

This work furthers the development by using a pair of inertial actuators for comprehensive vibration suppression in robotic milling. H_∞ controllers are designed with acceleration feedback, which are tested in eccentric mass experiments and milling experiments to demonstrate effective vibration reduction under different working conditions.

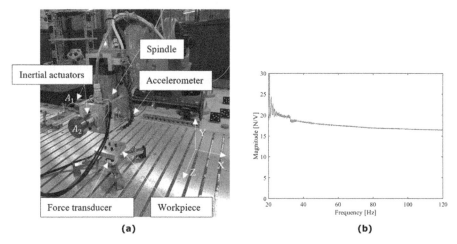

(a)   (b)

Figure 1. (a) Experimental set-up of the robot end effector and on the base plate. (b) Frequency response of the inertial actuator.

2   EXPERIMENTAL SET-UP

Figure 1(a) shows the experimental set-up. On the end effector of the robot, two inertial actuators, each consisting of one voice coil actuator (from H2W Technologies™) and one moving mass), were mounted and aligned in Directions X and Z next to the spindle. An accelerometer (from DJB™) was mounted on the back face of the end effector. On the base plate, a workpiece was mounted on the top of a force transducer (from Kistler™). All data acquisition, feedback and controller implementation were routed through a processor (from dSPACE™).

In milling, the spindle experiences forces as the rotating tool cuts the workpiece, consisting of a basic synchronous frequency, other harmonics and modal frequencies. Tests with an eccentric mass attached to the spindle may be used for inertial controller performance. In this paper, a voice coil actuator was used as the oscillator because of its broad bandwidth. Experiments were carried out to find relationships between actuation force outputs and driving voltages of inertial actuators, showing frequency responses over 20 -120 Hz, as in Figure 1(b), and a natural frequency at 14 Hz (not shown).

Table 1. Configurations of robot joints at pose 0, evaluated for models used for controller design.

Joint No.	1	2	3	4	5	6
Pose 0	-9.66°	-46.45°	114.09°	-9.95°	68.08°	3.76°

Table 2. Milling parameters.

Workpiece	Aluminium block (100 mm ×100 mm × 20 mm)
Tool	6 mm 3 flute end milling cutter
Machining type	climb milling
Coolant	None
Depth of cut	3 mm
Width of cut	0.5 mm
Spindle speed	1720 rev/min
Feed rate	1 mm/s
Path length	100 mm

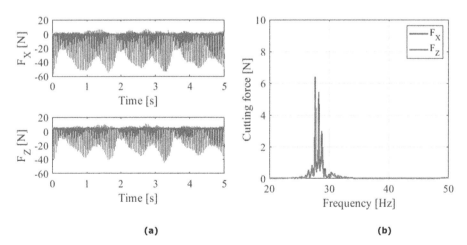

(a)   (b)

Figure 2. Measured cutting forces in directions X and Z from a milling test with chatter. (a) In the time domain, and (b) In the frequency domain.

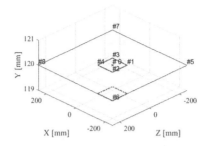

Figure 3. Configuration of poses 1-8 for fixed pose characterisation. Red dotted lines mark two regions of the workpiece in milling experiments.

3   MEASUREMENT OF MACHINING FORCES

With a particular pose of the robot (as in Table 1) and a machining trajectory in the Direction Z, it was observed that the robot system experienced mode coupling chatter in milling experiments with the parameters listed in Table 2. Figure 2(a) shows the workpiece experience larger machining forces measured in Direction X compared with those in Direction Z and a low frequency modulation in the amplitudes of the machining forces in both directions. Figure 2(b) shows multiple frequency components superimposed predominately around 27 Hz, which are associated with the spindle rotational frequency at 28 Hz and a robot natural frequency at 26 Hz. Compared with vibrations at higher frequencies, these frequency components contribute the most to the movement of the spindle, therefore they need to be suppressed.

4   SYSTEM CHARACTERISATION

System identification was undertaken for the robot system, by establishing the relationships between the driving voltages, V_1 and V_2, applied to the inertial actuators and the measured accelerations, A_X and A_Z, in Directions X and Z. While in practice a milling process may be carried out around a fixed pose of the robot system, vibration suppression may need to cover a small or large work domain depending on the configuration of the process. Therefore, it is important to identify the variation in dynamics

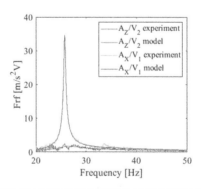

Figure 4. Evaluated frequency responses and identified models at Pose 0 in directions X and Z.

of the robot system within the specified work space. This working condition is typical for large volume manufacturing, which was taken into consideration by performing milling tests on small workpieces at two different poses in the X-Z plane, one close to and one away from the specified fixed pose. The first pose was defined as Pose 0, where the configuration of robot joints is listed in Table 1; the second pose was defined where the system dynamics of the robot system change the most within a work plane of 500 mm× 500 mm.

The robot system was characterised experimentally over 8 fixed poses as shown in Figure 3, covering a 500 mm × 500 mm work plane that is centred at the configuration of the robot as listed in Table 1. At each fixed pose, the robot system was excited using one inertial actuator at a time, with a chirp sequence over the frequency range 20-120 Hz over 60 seconds, to find the characteristic that corresponds to the axial actuation in each direction. It is assumed that the coupling effect is negligible in general. Identified system dynamics are designated by the transfer function $G^{\{m,d\}}_{pose\ \#,\ \{X,Z\}}(s)$, where the superscripts correspond with model-fitted from measurement data (m) or based on unprocessed measurement data (d). System dynamics at Pose 0 were then evaluated based on dynamics at 4 neighbouring grid points as

$$G^{\{m,d\}}_{pose\ 0,\ \{X,Z\}}(s) = \frac{1}{4}\left(G^{\{m,d\}}_{pose\ 1,\ \{X,Z\}}(s) + G^{\{m,d\}}_{pose\ 2,\ \{X,Z\}}(s) + G^{\{m,d\}}_{pose\ 3,\ \{X,Z\}}(s) + G^{\{m,d\}}_{pose\ 4,\ \{X,Z\}}(s)\right) \quad (1)$$

Total additive uncertainties, $\Delta_{pose\#,\ \{X,Z\}}$, in system identification between each pose and pose 0 were evaluated in terms of modelling error, $\Delta^m_{pose\#,\ \{X,Z\}}(s)$, and measurement error (i.e. variations in system dynamics), $\Delta^e_{pose\#,\ \{X,Z\}}(s)$, and as

$$\Delta_{pose\#,\ \{X,Z\}}(s) = \Delta^m_{pose\#,\ \{X,Z\}}(s) + \Delta^e_{pose\#,\ \{X,Z\}}(s) \quad (2)$$

where

$$\Delta^m_{pose\#,\ \{X,Z\}}(s) = \left|G^d_{pose\ 0,\ \{X,Z\}}(s) - G^m_{pose\ 0,\ \{X,Z\}}(s)\right| \quad (3)$$

$$\Delta^e_{pose\#,\ \{X,Z\}}(s) = \left|G^d_{pose\ \#,\ \{X,Z\}}(s) - G^d_{pose\ 0,\ \{X,Z\}}(s)\right| \quad (4)$$

Figure 4 Shows evaluated frequency responses at Pose 0. An 8th order and a 6th order transfer function were identified for system dynamics in Directions X and Z, correspondingly. Both frequency responses have similar curve shapes but with different scales. Resonance in both cases peaks occurs at 26 Hz. The frequency response in Direction X is over 10 times higher in magnitude compared with that in Direction Z. Given that the inertial actuators and their amplifiers are almost identical, it confirms that the robot system has significantly lower stiffness, therefore more susceptible to vibrations, in Direction X.

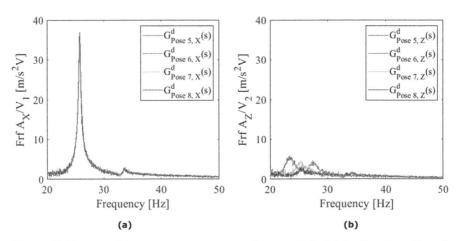

Figure 5. Measured frequency responses at poses 5-8. (a) in direction X, and (b) in direction Z.

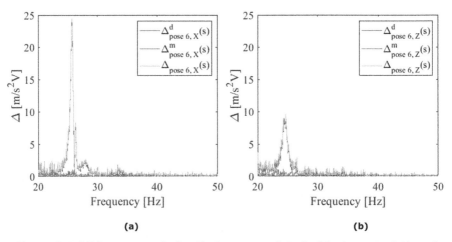

Figure 6. Additive uncertainties that are associated with characteristics of the robot system at pose 6. (a) in direction X, and (b) in direction Z.

Figure 5 shows measured characteristics of the robot system at 4 corners of the specified work plane. Variations in system dynamics are associated with a natural frequency of the robot system around 26 Hz. Figure 6 shows differences in system dynamics that is maximised at Pose 6 relative to Pose 0. Pose 6 is therefore defined as the second pose for performing milling experiments. Because fitted models are accurate, the additive uncertainties in modelling, $\Delta^m_{pose\ \#,\ \{X,Z\}}$ is small compared with that in measurement, $\Delta^e_{pose\ \#,\ \{X,Z\}}$. As for the total additive uncertainty, $\Delta_{pose\ \#,\ \{X,Z\}}$, maxima of 24.5 and 9.5 occurred at 26 Hz in Directions X and Z respectively, indicating the highest variation in system dynamics is at a resonance of the robot system.

5   CONTROLLER DESIGN

In this work, H_∞ controllers were synthesized in MATLAB based on identified system models and selected weighting functions that were designed separately for two inertial actuators, using the same equations but different coefficients as

Table 3. Configurations of H_∞ controllers. Parameter tuning excludes the calibration factor of $5g$ in acceleration measurement.

	$w_{1,\{X,Z\}}$	f_c	$f_{1,\{X,Z\}}$	$\zeta_{1,\{X,Z\}}$	$w_{2,\{X,Z\}}$
A_1	0.4	10	26	0.03	0.65
A_2	0.7	10	26	0.03	0.2

Table 4. Measured vibration reduction in eccentric mass experiment, with a spindle speed of 1,580 RPM.

Pose No.	Residual vibration in % $[X/Z]$
5	22%/12%
6	42%/14%
7	28%/22%
8	34%/9%
0	28%/11%

$$W_{1,\{X,Z\}} = w_{1,\{X,Z\}} \left(\frac{\tau s}{\tau s + 1} \right)^2 \left(\frac{\omega_{1,\{X,Z\}}^2}{s^2 + 2\zeta_{1\{X,Z\}}\omega_{1,\{X,Z\}}s + \omega_{1,\{X,Z\}}^2} \right) \tag{5}$$

$$W_{2,\{X,Z\}} = w_{2,\{X,Z\}} \tag{6}$$

where

$$\tau = \frac{1}{2\pi f_c}, \omega_{1,\{X,Z\}} = 2\pi f_{1,\{X,Z\}} \tag{7}$$

In $W_{1,\{X,Z\}}$, $w_{1,\{X,Z\}}$ is the gain of W_1 that determines vibration reduction performance of the H_∞ controller; the first term, $\left(\frac{\tau s}{\tau s + 1} \right)^2$, is a second-order high-pass filter that protect actuators from responding to low frequency noises with a cut-off frequency, f_c; the second term, $\left(\frac{\omega_{1,\{X,Z\}}^2}{s^2 + 2\zeta_{1\{X,Z\}}\omega_{1,\{X,Z\}}s + \omega_{1,\{X,Z\}}^2} \right)$, defines the attenuation frequency band with a damping coefficient, $\zeta_{1\{X,Z\}}$, and a natural frequency, $f_{1,\{X,Z\}}$, which was chosen to be a narrow band around the natural frequency of the robot system. $W_{2,\{X,Z\}}$, was designed to be a constant to include all additive uncertainties in modelling and characteristics of the robot system. The weighting function $W_{3,\{X,Z\}}$ was therefore set to zero. Table 3 shows the tuned parameters for weighting functions. It was set with expected vibration reduction of about 80% while all actuations are within stroke limits of inertial actuators.

6 ECCENTRIC MASS EXPERIMENTS

H_∞ controllers were tested in eccentric mass experiments with a spindle speed of 1,580 RPM, the natural frequency of the robot system, at Pose 0 and four corners of the work plane. Table 4 shows that H_∞ controllers suppress vibrations down to 20% in Direction X at all poses. The worst performance was found at Pose 6, where residual vibration was measured to be 42% in Direction X. This observation agrees with results in system identification since Pose 6 has the largest variation in system relative to Pose 0.

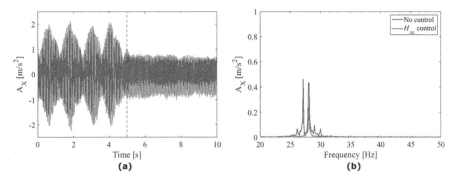

Figure 7. Acceleration measurement in direction X during milling test at pose 0. (a) In the time domain, and (b) In the frequency domain. The red dotted line marks the application of H_∞ control.

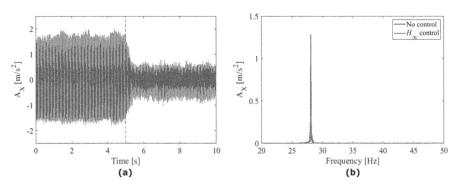

Figure 8. Acceleration measurement in direction X during milling test at pose 6. (a) In the time domain, and (b) In the frequency domain. The red dotted line marks application of H_∞ control.

7 MILLING EXPERIMENTS

Milling experiments were performed with vibration suppression using H_∞ controllers around Pose 0 and Pose 6. At Pose 0, Figure 7(a) shows measured accelerations in Direction X with H_∞ controllers applied from t = 5 s. When uncontrolled, a modulation effect is observed in accelerations with an overall amplitude of 2 m/s². With vibration control using H_∞ controllers, the modulation effect is suppressed, reducing the overall amplitude of acceleration to 0.7 m/s². Figure 7(b) shows the frequency component at

26 Hz was suppressed completely and the frequency component at 27 Hz, the sub-harmonic of cutting frequency, was reduced down to 60%. Figure 8 shows that at Pose 6, same milling exhibits a single dominant frequency component at 27 Hz. H_∞ controllers reduced the overall amplitude of acceleration by 75% within 1 second. Figure 9 shows corresponding force measurements from milling experiments at Pose 0 and Pose 6. With or without the mode coupling chatter, machining forces are halved due to vibration suppression using H_∞ controllers.

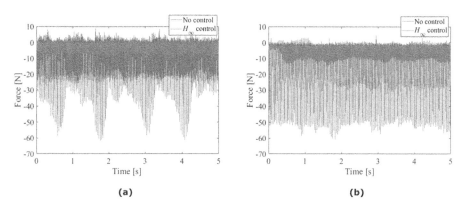

(a)   (b)

Figure 9. Force measurement in direction X during milling test. (a) At pose 0, and (b) At pose 6.

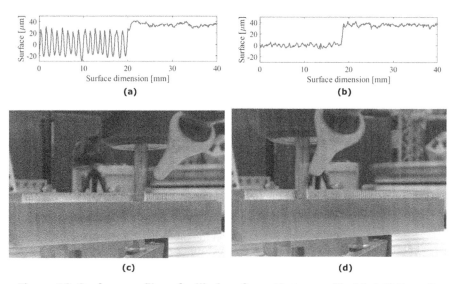

(a)   (b)

(c)   (d)

Figure 10. Surface profiles of milled surfaces that are milled (a) At Pose 0, and (b) At pose 6. Recorded photos during the milling experiment at Pose 0, (c) Uncontrolled, and (d) With H_∞ control. (milling occurred from the left side to the right side).

223

Changes in motion of the end effector also are reflected in surface finishes in these milling processes. Figures 10(a) and 10(b) show profiles of surfaces that were milled at Pose 0 and Pose 6, with vibration suppression applied on the second half of surfaces. At Pose 0, because H_∞ controllers suppress the modulation effect completely along with the resonance of the robot system, chatter marks are removed from the surface, reducing RMS surface waviness by 85% from 16.3 μm to 2.4 μm. At Pose 6, because same milling process does not cause resonance of the robot system and the modulation effect, using H_∞ controllers does not affect the surface waviness. Moreover, Figures 10(a) and 10(b) also show that vibration suppression also corrects the dimensional error of milled surfaces in both cases by reducing the offset between milled surfaces and tool paths. At Pose 0 and Pose 6, the surface is 40μm higher when using H_∞ controllers than that without any control. Figures 10(c) and 10(d) show photos taken on these surfaces, where such changes in surface waviness and surface level are visible.

8   CONCLUSIONS

This work proposes the use of H_∞ controllers to cope with the change in dynamics of a robot system for chatter suppression in robotic milling. The controller design is based on experimentally identified models of the robot system at one fixed pose and selected weighting functions that treat modelling errors and variations in dynamics of the robot system at different poses as additive uncertainties. H_∞ controllers were designed to supress vibration at the natural frequency of the robot system over a relatively large work plane; their performance characteristics were evaluated in eccentric mass experiments and milling experiments.

The robot system was found to have various stiffness within the defined work plane, is therefore more susceptible to vibration in one direction than that in another. Between different configurations of the robot system, variations in system dynamics were associated mostly with its natural frequency at 26 Hz, affecting any milling processes with sub-harmonics at similar frequencies. In eccentric mass experiments, H_∞ controllers reduced vibration by 58-80%. In milling experiments, it was found that same milling process may or may not cause mode coupling chatter, depending on the pose of the robot system. When model coupling chatter occurs, because the sub-harmonic of milling and resonance have similar frequencies, superimposition of these frequency components causes a modulation effect in motion of the end effector and produces chatter marks on workpieces. H_∞ controllers were able to suppress the resonance of the robot system completely, removing chatter marks on workpieces and reducing surface waviness of milled surfaces by over 85%. In addition, effective vibration reduction using H_∞ controllers also corrected dimensional errors by 40 μm regardless of the occurrence of mode coupling chatter because of reduced offsets between milled surfaces and tool paths.

REFERENCES

[1] S. Huyanan, N. Sims, Vibration control strategies for proof-mass actuators, Journal of Vibration and Control 13 (2007) 1785–1806.

[2] S. Sun, J. Yang, T. Yildirim, H. Du, G. Alici, S. Zhang, W. Li, Development of a nonlinear adaptive absorber based on magnetorheological elastomer, Journal of Intelligent Material Systems and Structures 29 (2018) 194–204.

[3] H. Moradi, F. Bakhtiari-Nejad, M. Movahhedy, Tuneable vibration absorber design to suppress vibrations: An application in boring manufacturing process, Journal of Sound and Vibration 318 (2008) 93–108.

[4] E. Daz-Tena, L. Lpez De Lacalle Marcaide, F. Campa Gmez, D. Chaires Bocanegra, Use of magnetorheological fluids for vibration reduction on the milling of thin floor parts, Procedia Engineering volume 63 (2013), pp. 835–842.

[5] N. Saadabad, H. Moradi, G. Vossoughi, Global optimization and design of dynamic absorbers for chatter suppression in milling process with tool wear and process damping, Procedia CIRP volume 21 (2014), pp. 360–366.

[6] J. Scheidler, M. Dapino, Stiffness tuning of FeGa structures manufactured by ultrasonic additive manufacturing, Proceedings of SPIE - The International Society for Optical Engineering volume 9059 (2014) 905907.

[7] L. Yuan, S. Sun, Z. Pan, D. Ding, O. Gienke, W. Li, Mode coupling chatter suppression for robotic machining using semi-active magnetorheological elastomers absorber, Mechanical Systems and Signal Processing 117 (2019) 221–237.

[8] F. Chen, H. Zhao, D. Li, L. Chen, C. Tan, H. Ding, Contact force control and vibration suppression in robotic polishing with a smart end effector, Robotics and Computer-Integrated Manufacturing 57 (2019) 391–403.

[9] F. Chen, H. Zhao, Design of eddy current dampers for vibration suppression in robotic milling, Advances in Mechanical Engineering 10 (2018) 11.

[10] Y. Guo, H. Dong, G. Wang, Y. Ke, Vibration analysis and suppression in robotic boring process, International Journal of Machine Tools and Manufacture 101 (2016) 102–110.

[11] L. Tunc, D. Stoddart, Tool path pattern and feed direction selection in robotic milling for increased chatter-free material removal rate, International Journal of Advanced Manufacturing Technology 89 (2017) 2907–2918.

[12] S. Mousavi, V. Gagnol, B. Bouzgarrou, P. Ray, Dynamic modeling and stability prediction in robotic machining, International Journal of Advanced Manufacturing Technology 88 (2017) 3053–3065.

[13] I. Zaghbani, V. Songmene, I. Bonev, An experimental study on the vibration response of a robotic machining system, Proceedings of the Institution of Mechanical Engineers, Part B: Journal of Engineering Manufacture 227 (2013) 866–880.

[14] I. Tyapin, K. Kaldestad, G. Hovland, Off-line path correction of robotic face milling using static tool force and robot stiffness, IEEE International Conference on Intelligent Robots and Systems volume 2015 December, pp. 5506–5511.

[15] C. Dumas, S. Caro, S. Garnier, B. Furet, Joint stiffness identification of six-revolute industrial serial robots, Robotics and Computer-Integrated Manufacturing 27 (2011) 881–888.

[16] J. Munoa, I. Mancisidor, N. Loix, L. Uriarte, R. Barcena, M. Zatarain, Chatter suppression in ram type travelling column milling machines using a biaxial inertial actuator, CIRP Annals - Manufacturing Technology 62 (2013) 407–410.

[17] A. Bilbao-Guillerna, I. Azpeitia, S. Luyckx, M. Loix, J. Munoa, Low frequency chatter suppression using an inertial actuator, International Conference on High Speed Machining Volume 2012 pp. 1–6.

225

12th International Conference on Vibrations in Rotating Machinery -
Institution of Mechanical Engineers, ISBN 978-0-367-67742-8

Fast estimation of classical flutter stability of turbine blade by reduced CFD modelling

Chandra Shekhar Prasad, Luděk Pešek

Institute of Thermomechnanics of the CAS, v.v.i., Prague, Czech Republic

Václav Sláma

Doosan Škoda power s.r.o, Pilzen, Czech Republic

ABSTRACT

The paper presents a medium fidelity reduced ordered numerical model for the calculation of aeroelastic stability diagram of 3D blade cascade of low pressure stage of steam turbine. The aeroelastic stability in steam turbine blades are calculated for the classical flutter problem. The calculation of the stability diagram for the problem of classical flutter is evaluated with assumption of running waves. Running waves will be simulated by the inter-blade phase shift approach between the blades in the cascade. Panel method based boundary element type flow solver is employed for calculation of unsteady aerodynamic forces and model the flow flied. This method is good compromise of speed and accuracy for the estimation of the stability of the blades on a classical flutter. The estimated results are compared with experimental and the high fidelity computational fluid dynamic model data.

1   INTRODUCTION

To tap more energy from the high volume steam produced in modern day's power station, larger steam turbines are now being installed. To ensure the safety and the aeroelastic stability in these modern day's large power turbine is a major design challenge, because the larger rotor blades of these machines have greater tendency towards flow induced mechanical vibration or flutter. In the large steam turbine assembly particularly in the low pressure (LP) rotor stage the classical flutter or low amplitude and low frequency flow induced vibration is one of the frequently occurring and dominating aeroelastic stability issue. Therefore, it need to be addressed carefully in the preliminary design stage of the turbomachinery system. Therefore, the study of aeroelastic stability in power turbine remains an active research topic for researchers and engineers in past and recent years [1, 2] to ensure uninterrupted and safe power production in power plants. One of the parameters which dominates the aeroelastic stability is aerodynamic damping (AD) of the rotor blade system. Especially for the steam turbine's long and thinner blades where material damping is limited, therefore, the damping from the surrounding fluid is the main damping contributor. This leads to more detail design study on aerodynamic damping. A fair amount of design iterations may be required to optimize the blade geometry and the arrangement of blade cascade for aeroelastic stability in a short period of time.

Due to higher cost and longer time involved in the experimental process, at the preliminary design stage the numerical modeling and simulations are preferred over physical experimental model, for design iteration to save both time and money. For the numerical modeling purpose, in present days there are well developed Navier-Stokes based computational fluid dynamics and structural dynamics or in short (CFD-CSD) models available. However, they are computationally very expensive, especially fluid solver part takes large amount of computational time in CFD-CSD type numerical modeling. Furthermore, it also demands highly skilled manpower and large supercomputers which dose not come very cheap either. Therefore, in this research work a medium fidelity reduced ordered aeroelastic numerical tool is developed to analyze the aerodynamic stability for low amplitude and low frequency flutter problem in steam turbine's LP stage. The main characters of this new approach will include computationally cheap and easy to adopt for the complex geometries. Therefore, in the present research paper application of computationally less expensive boundary element based flow solver is developed to model the flow in low pressure (LP) turbine region and to estimate the aeroelastic stability parameters e.g. aerodynamics damping for the blade cascade. The boundary element method e.g. panel method (PM) is one of the such method and it is first proposed by Hess and Smith [3] to model the lifting and non-lifting potential flow around slender bodies. PM methods are good compromise of speed and accuracy. Moreover, PM can be used for complex geometry unless the flow fulfill the criteria of potential flow and without separation which is the case of classical flutter or low amplitude low frequency flutter. However, modified PM can be used for the separated flow case [4]. These methods are widely adopted for aeroelastic modeling of wind turbines, helicopter rotors, and aircraft aeroelasticity [5]. The more details about theoretical and numerical implementation can be found in Katz and Plotkin [6]. However, instead of great potential of the PM to be used in LP turbine aeroelasticity very few researchers have used it in cascade aeroelastic modeling. An improved version of PM is used by [7, 8] for 2D cascade design and flow modeling, in this work only pressure distribution is estimated, whereas [9] the similar technique to estimate the aeroelastic stability parameters is used, but in this work only camber surface of the blade is modeled not the actual geometry. Therefore, in the present research work 3D surface PM is adopted which can model real geometry of the blade cascade. To estimate the aerodynamic damping, the most common method is traveling/running wave mode (TWM) method and less known aerodynamic influence coefficient (AIC) method [10, 11]. The TWM method is used by many research to estimate the aerodynamic damping for vibrating cascade. The estimated aerodynamic damping using either of the methods are theoretically identical. For the present research work TWM method is adopted because it is more representative to actual vibration pattern in the power turbine rotor. A detail explanation of both the methods can be found in Fransson [10]. The estimation of AD in TWM using panel method in 2D case is also adopted by [12].

Apart from the reduced order flow solver the structural part of the aeroelastic tool is reduced by modal model approach and calculated from full FEM model of rotating turbine bladed disk with long blades. For the fluid structure coupling a partitioned based coupling strategy is used where flow solver (PM) is loosely coupled with kinematic motion prescribed by harmonic motion of the selected mode.

2   UNSTEADY 3D SURFACE PANEL METHOD FOR FLOW MODELLING

3D Surface panel methods solve the attached flow around lifting and non-lifting surfaces using the potential flow assumptions, which are 1) flow field is inviscid (Viscosity (η) = 0) and 2) irrotational ($\times \vec{V}$ = 0). The rectangular singularity panel elements are placed on the entire solid surface and in the wake, these singularity elements are the known solutions to Laplace's equation (1).

$$\Box^2\emptyset(x,y,z) = \frac{\partial^2\emptyset}{\partial^2 x} + \frac{\partial^2\emptyset}{\partial^2 y} + \frac{\partial^2\emptyset}{\partial^2 z},$$
(1)

where Φ(x,y,z) is the scalar velocity potential and it is function of (x,y,z) such that ($\vec{V} = \nabla\Phi$). Apart from singularity distributions appropriate boundary conditions are required, to find the unique solution of Laplace's equation (1), The required BCs are, (1) enforcing the impermeability flow at solid or a fixed normal flow condition ($\nabla\Phi.\mathbf{n}$ = V.$\mathbf{n}$ = 0), (2) Far field: The disturbance created by the singularities on the body and wake must disappear at infinity ($\lim\limits_{r\to\infty} \nabla(\phi - \phi_\infty)$). (3) the Kutta condition and Kelvin's theorem, are imposed in order to obtain a complete description of the flow field and the resulting aerodynamic loads. The pressure field (p) is then obtained solving the unsteady Bernoulli's equation (2)

$$\frac{\partial \emptyset}{\partial t} + \frac{p}{\rho} + \frac{V^2}{2} = const. \tag{2}$$

In the PM to construct the flow field and find the solution of the field variables, singularity panel elements are placed over the upper and lower surface of the wing and in the wake as shown in the Figure 1.

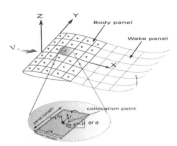

Figure 1. 3D unsteady Panel method discretization scheme using BEM.

Each wing panel includes a constant strength source (σ) and a doublet (μ) distribution while the wake panels feature only doublet panels because the wake do not carry any aerodynamic load. Boundary conditions are imposed on control points lying on the mid-point of each panel, figure 1. In figure 1 the flow field around an isolated 3D wing is modeled using unsteady surface PM. A typical surface singularity element is highlighted in the left. The collocation point is at center of the rectangular panel where tangents t_x, t_y and normal (η) are defined. The time varying wake panel only contains doublet singularity elements, because wake do not carry any load. More detailed discussion on numerical implementation and estimation of the aerodynamic loads can be found in Katz and Plotkin [6]. Also the analytical solution of induced potential for the 3D constant strength source and doublet singularity element is given in the Katz and Plotkin [6].

The above-mentioned boundary conditions are imposed in order to find the unknown perturbation potential at each collocation point on the wing and on the wake panel's corner points. Katz and Plotkin [6] show that the induced potential at any point can be estimated by applying of Green's identities on the close surface around wing and wake surface, in the present formulation internal Dirichlet boundary condition is used, therefore, the potential induced at point **P** can be written as

$$\frac{1}{4\pi} \int\limits_{S_{body}} \mu_b \boldsymbol{n} \cdot \nabla\left(\frac{1}{r}\right) ds - \frac{1}{4\pi} \int\limits_{S_{body}} \sigma\left(\frac{1}{r}\right) ds + \frac{1}{4\pi} \int\limits_{S_{wake}} \mu_w \boldsymbol{n} \cdot \nabla\left(\frac{1}{r}\right) ds = 0, \tag{3}$$

where μ_b is the strength of the body doublet distribution, σ the strength of the source distribution and μw the strength of the wake doublet distribution, while r = |r|. The Eq. 3 with three unknowns, μ_b, σ and μw. The Eq. 3 is itself integral form and cannot be solved for a general blade geometry. Therefore, the solution is to discretize the geometry, as discussed earlier, and replace the integrals by algebraic sums, and apply Eq. 3 to the control point of each panel, such that

$$A\mu_b + B\sigma + C\mu_w = 0, \tag{4}$$

where $\mu_b = [\mu b_1 \ldots \mu b_{Nb}]^T$ is the vector of the unknown doublet strengths of the body surface panels, $\sigma b = [\sigma 1 \ldots \sigma N_b]^T$ is the vector of the unknown source strengths of the body surface panels, $\mu_w = [\mu w_1 \ldots \mu w_{Nw}]^T$ is the vector of the unknown doublet strengths of the wake panels, Nb is the number of body panels, Nw is the number of wake panels and A, B, C are influence coefficient matrices. More details about solution of Eq. 4 can be found in [6]. The solution of Eq. 4 for each collocation points give the values of three unknowns, μ_b, σ and μw. Once the perturbation potential or unknown μ_b is known, the unsteady pressure coefficient can be calculated using instantaneous Bernoulli's equation and given as

$$\tilde{c}_p(x, y, t) = 1 - \frac{Q_{total}^2}{Q_\infty^2} - \frac{2}{Q_\infty^2}\frac{\partial \mu}{\partial t} \tag{5}$$

where

$$Q_{total} = V_{local-kinematic} + q_{perturbation}$$

More detail discussion on numerical implementation to estimate the aerodynamic loads can be found in [6]. Also the analytical solution for the 3D constant strength source (σ) and doublet (μ) singularity element is given in the [6].

3   MODAL MODEL FROM FULL 3D FEM BLADE MODEL

In order to calculate the classical flutter, the mode shapes and the position of the blade is extracted from the FEM modal analysis in the present approach. The flow solver is not directly coupled with any structural solver. Therefore, the amplitude, mode shape and the frequency of oscillation is extracted from the modal analysis of the 3D cascade and fed into the flow solver as input at each time step. To further reduce the computational time, the modal model reduction technique is applied, where, only few point is selected on the blade profile in the complete FEM model where the number or elements are way higher. The points are selected such a way that it should not compromise the accuracy of the real model with higher numbers of elements, and thus, the mode shapes and the displacement and frequency is mapped and fed to the flow solver for the aeroelastic computation. The method to extract mode shapes, frequency and the displacement from real blade geometry FEM model analysis is presented below. The rotor blade is used in one of the multi megawatt steam turbine produced by Doosan Skoda power s.r.o.

3.1   Methodology for easy data extraction for flow solver from structural modal model

The original geometry of the blade was specified by 100 points in the 32 cross-sections (called profiles). Profiles describe the blade along its height (z-coordinate). The edge points of the blade were determined as 50th and 1th points, respectively, of each profile. The FE model geometry of the blade with detected edge nodes (black circles) of leading and trailing edges of the blade profiles are in the Figure 2. Red crosses show next the detected edge nodes at 8 cross-sections along the height of the blade which were chosen for this study case.

By detecting the edge nodes, i.e. their node numbers of the computational model, we can associate their displacements from the modal data. Their displacements than can be used for calculation of profile displacement and its rotation (figure 3). The profile motion is associated with the leading edge displacements u_x, u_y of x_1,y_1 to x_2,y_2 coordinate system translation and rotation of the profile by angle Φ from x_2,y_2 to x_1, y_1 coordinate system with respect to the leading edge. Angle Φ was calculated by formula for scalar product of the vectors LE and L'E'.

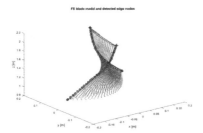

Figure 2. Transposed FE model geometry of the blade –with detected nodes (red crosses) of leading and trailing edges of the blade.

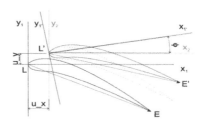

Figure 3. Definition of fundamental motions (u_x, u_y displacements and ϕ angle) of the profile.

The two dominant mode shapes, i.e. one bending and torsional, are selected for FSI simulation from the FEM modal analysis. The amplitude of motion scaled up by the scale factor five for the display purposes. The corresponding mode shapes are given in the figure 4.

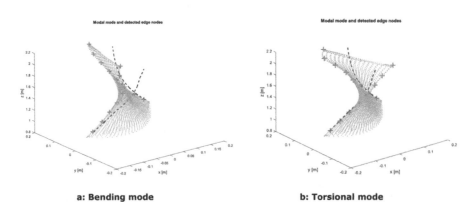

a: Bending mode   b: Torsional mode

Figure 4. Modal mode with detected edge nodes (red crosses). Dashed lines depict the edge lines of unreformed geometry of the blade. (a: bending mode) and (b: torsional mode).

4   CALCULATION OF AERODYNAMIC DAMPING IN THE CASCADE

For the calculation of aerodynamic damping similar approach is adopted as described by Vogt [14], and brief description of the method is presented here. The balance between the structure and aerodynamic forces in the system is represented by the aeroelastic Eq. 6, Vogt [14].

$$[M]\{\ddot{X}\} + [G]\{\dot{X}\} + [K]\{X\} = \{F_{ad}(t)\} = \{F_{disturbance}(t) + F_{damping}(t)\} \tag{6}$$

where [M], [G], and [K] are the modal mass, modal damping and modal stiffness matrices respectively, X represents the modal coordinate vector and finally Fad(t) is the modal unsteady aerodynamic force vector which contains two terms: $F_{disturbance(t)}$ that includes the aerodynamic disturbances upstream and downstream of the blade and $F_{damping(t)}$ which represents the aerodynamic damping resulting from the interaction between the blade and the flow. For flutter analysis, only the aerodynamic forces due to the vibration of the blade are considered resulting in $F_{disturbance(t)} = 0$ and above Eq. 6 can be simplified into below Eq. 7

$$[M]\{\ddot{X}\} + [G]\{\dot{X}\} + [K]\{X\} = \{F_{damping}(t)\} \tag{7}$$

The solution of the above equation Eq.7 allows the determination of the aerodynamic damping and thus the aeroelastic stability of the system. There are different numerical methods can be used to approximate the aerodynamic characteristics of the system. One of them is "Time Linearized Method" assumes that the unsteady perturbations in the flow are small compared with the mean flow. Then the unsteady flow can be approximated by small harmonic perturbations around a mean value. Verdon [11] shows that for the small perturbation the unsteady pressure due to harmonic motion can be represented as harmonic oscillation with respect to the blade around a time averaged steady value and can be given as

$$p(x,y,t) = \bar{p}(x,y) + \tilde{p}(x,y,t) = \bar{p}(x,y) + \hat{p}.e^{i(\omega t + \varphi_{p \to h})}, \tag{8}$$

where $\bar{p}(x,y)$ is the steady mean pressure at an arbitrary location, $\tilde{p}(x,y,t)$ the respective unsteady perturbation pressure and $\hat{p}$ the amplitude of complex pressure due to harmonic oscillation. The harmonic nature of the perturbation unsteady pressure consequently results in harmonic unsteady aerodynamic forces, therefore, can be written by Eq. 9

$$\hat{\tilde{F}} = \hat{\tilde{F}}.e^{i(\omega t + \varphi_{p \to h})}. \tag{9}$$

And thus, the work per oscillation cycle (T) can obtained by integrating the force and motion during the cycle and can be given by Eq. 10

$$W_{cycle} = \int_T \hat{\tilde{F}}.\hat{\tilde{h}}.dt = \int_T \hat{\tilde{F}}.\hat{\tilde{h}}.e^{i\omega t}dt, \tag{10}$$

where $\hat{h}$ is the complex motion of the cascade blade. In the present case only pitching motion of the cascade blade is analyzed therefore, it is corresponding to torsion motion. In the present research work potential flow based PM method is used as a linearized flow equations assuming the small unsteady perturbation. TWM method with principle of linear superimposition is used to estimate the aerodynamic damping [13]. Using unsteady PM along with principle of superimposition $\tilde{p}(x, y, t)$ is estimated on the reference blade which also includes the effect of other blades in the cascade. More details discussion about principle of superimposition can be found in [6].

Therefore, aerodynamic damping parameter is given by work done per cycle and normalized by π and amplitude of oscillation "A" and is given in Eq. 11

$$Aerodynamic - damping(\Xi) = -\frac{W_{cycle}}{\pi A^2}. \tag{11}$$

The positive value of Ξ indicates the flow acting in stabilizing manner whereas negative value can cause flutter, therefore, unstable.

5   TEST CASES FOR THE AERODYNAMIC DAMPING IN THE 3D CASCADE

In this section, description of the experimental and numerical test case is presented. A brief description of experimental measurement and validation of the numerical results are documented here. The detail discussion of the numerical results validation is also the part of this section. At the first step the newly developed BEM PM based flow solver is employed to estimate the AD in the linear cascade test case performed in KTH university [14]. In the second step, the same method is used for the estimation of the AD for the Doosan Škoda Power s.r.o blade's bundle.

1st Test case description: For the experimental test case a non-rotating 3D annular cascade section is selected. The experiment is carried in KTH university's turbomacinery test facility by Vogt [14] in 2005. The annular cascade is consisting of 7 blades and setup is presented in Figure 5 (left) and the schematic diagram of test setup is presented in Figure 5 (right). The more details about the experimental set is given in the [14].

The main flow conditions are 1) nominal inflow angle (α) = -26°, 2) Mach=0.28 and 3) k=0.1 4) Oscillation type= Torsional. The aerodynamic damping vs IBPA is measured on the instrumented blade.

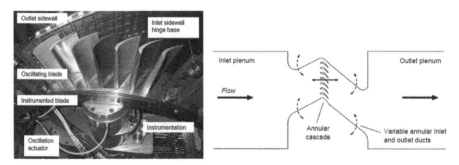

Figure 5. Experimental Setup (left) of 3D Blade cascade and schematic diagram of test setup (right) [14].

Simulation setup description: To model the flow around 3D annular cascade and estimate the aeroelastic stability parameters PM based solver is adopted. The fundamental methodology and the assumptions are similar to that of isolated wing case as describe in the Sec. 2. However, for the 3D annular cascade, method of linear superimposition is taken into account to include the effects of neighboring oscillating blades on the reference blade. Each blade in the cascade is discretized into 50 chord-wise and 10 span-wise rectangular panel elements Figure 7 (left). To simulate the wall bounded internal flow with solid wall boundaries using PM, extra source panels are placed on the tunnel boundary walls as presented in the Figure 7 (left), there are no need of doublet panels because tunnel wall have no lifting effect on the flow field. The similar technique for internal flow modeling for 3D blade cascade using panel is also described by Katz [6] for calculation of steady surface pressure. The simulation is carried out in TWM oscillation, which means all the blades, will oscillate at same frequency and amplitude but certain phase difference w.r.t reference blade. In principle, the AD results from INC and TWM methods are identical. The simulation flow conditions are similar to the chosen test case and unsteady simulation is ran for 300 time steps (enough for convergence) with time step size $\Delta t = 0.025$ sec. Only zero tip clearance case is considered in the PM model simulation. The estimated AD using PM based aeroelastic model is compared with experimental data and presented in the Figure 6 (right).

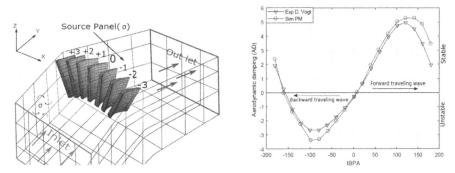

Figure 6. BEM based numerical model of the 1st test case (left) and the AD vs. IBPA comparison with experimental result (right) (Prasad et al. [15]).

Comments: In the Figure 6 the simulated AD using 3D PM flow model at different IBPA in torsional vibration in TWM agrees well with experimental result and PM based solve accurately catch the S-curve running from unstable to stable zone, (-180° to +180°). However, the PM model overestimates the magnitude of AD values between IBPA -50° to -100° in unstable zone and for IBPA +100° and +180° in stable zone. This is caused by phase lead between unsteady aerodynamic force and period of oscillation at these IBPA. Furthermore, at above mentioned IBPA for the given flow condition mild flow separation from the suction side of the blades may take place, since PM cannot simulate the separated flow, thus, overestimate the magnitude of the AD values at those IBPA, also the effect of acoustic resonance cannot be denied. Since there is no flow separation or acoustic resonance model implemented in the present version PM flow solver, therefore, it over estimates the AD magnitude values. Therefore, to deeply understand the phenomena and to find out the root cause of the overshoot of AD values, a detail CFD based (higher fidelity model). analysis will be required which is computationally very expensive. Nevertheless, the AD results are accurate enough to be used for preliminary design stage and the computation time taken is approx. 45 CPU mins which is way less than the CPU time taken, if CFD based models have been adopted for simulation.

2nd Test case description: For the 2nd test case a blade bundle of LP stage steam turbine from Doosan Škoda Power s.r.o is selected. The LP stage complete wheel assembly and the blade geometry is given in the Figure 7. The LP stage rotor has 66 blades assembly. A number of researchers have tried to understand the extent of coupling during cascade flutter [13,14,16,17]. In their study they have demonstrated the fact that influence decreases rapidly with increasing distance from the reference blade (blade number '0' in Figure 8) and convergence can be reached after blade pair ± 2. The effect of ± 3 pair is almost negligible. The contribution from ± 2 pair is generally of one order of magnitude less than the ones from blades 0 and ± 1, which leads to a characteristic S-shape. Therefore, only five bladed 3D cascade bundle out of 66 blade assembly is selected for the classical flutter analysis test case here. The blade numbering is given in the Figure 8 (left). To model the flow around 3D blade cascade using BEM PM same strategy is used as in the 1st test case. The flow conditions are provided by Skoda steady CFD analysis and BC are as, average axial flow velocity at the blade leading edge (V_x =140 m/sec), radial flow velocity (V_r = 65 m/sec) and at the revolution speed 3000 rpm. For the same BC at first the ANSYS/CFX is used with estimate the AD S-curve for the different IBPA. In the later stage the PM based flow solver is employed for steady and unsteady simulation for the torsional flutter case. The estimated AD using PM is compared with CFX calculated AD and presented in the Figure 8 (right). Furthermore, the steady state pressure along 4 span wise position is calculated using Navier-Stoke based CFD model and it is compared against chord wise steady pressure distribution estimated using BEM PM method and presented in the Figure 9.

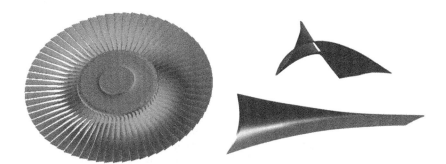

Figure 7. Geometry of Doosan Škoda Power LP turbine wheel assembly (left) and Rotor blade geometry (right).

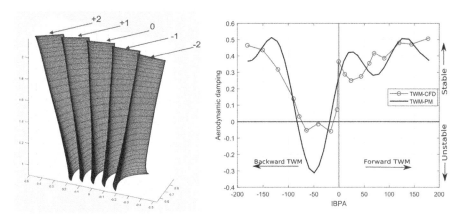

Figure 8. BEM PN based numerical discretization model of the 2ˢᵗ test case- (left) and the AD vs. IBPA comparison with CFD/CFX simulation with PM result (right).

Comments: In the Figure 8 (right) the simulated AD using 3D PM flow model and ANSYS/CFX at different IBPA in torsional vibration in TWM is compared. Results from both simulations are close; however, the PM based solver overestimated the AD magnitude by the largest margin between -100° to -25°, where the cascade is prone to get into instability. Furthermore, there are noticeable discrepancy in the both end of the curve close to peak amplitude of AD magnitude. One of the reason of for this type of discrepancy between CFX result and PM results can be answered by looking at the steady state pressure distribution curve in the Figure 9. In the Figure 9 the chord wise pressure distribution at four span wise position is compared, as we can see in the Figure 9, that at 5% span both the pressure data have good agreement and this agreement deteriorates as we move towards the tip of the blade along the span. For the given flow condition and revolution speed the flow become supersonic as we approach to the tip and compressibility effect become dominant, since the PM is based on the assumption of potential flow, therefore, compressibility effect is neglected, and thus both the results have large disagreement. Another reason for mismatch in both pressure and AD is the large flow separation, which occurs in half of the blade along the span starting from upper mid half up to the blade tip. Classical PM cannot simulate the flow separation and therefore, it leads to overestimation of aerodynamic forces. The lower accuracy of the PM method is a price for short computational time that can be, however, advantage in the preliminary design phase where many iterations required. The execution time taken by Navier-Stokes based CFD numerical model (18 CPU hrs. on 6 Intel-i7 processors) is significantly higher than the BEM PM based model to obtain the same result outputs. In addition, the PM results can be further improved by adding compressibility effect and modifying PM to include the flow separation.

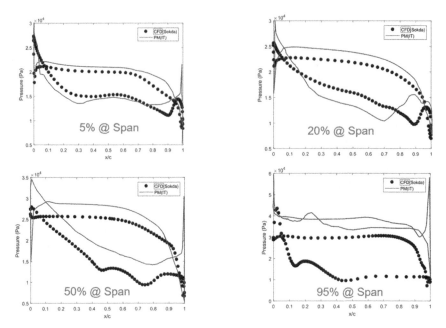

Figure 9. Comparison of chordwise steady state pressure distribution at 4 different spanwise potions for the Skoda blade test case, estimated using Navier Stoke based CFD model vs BEM PM.

6   CONCLUSION

The aeroelastic stability in terms of classical flutter phenomena is studied in the 3D blade cascade for the LP stage steam turbines. Aerodynamic damping and surface pressure distribution for steady state for low pressure turbine 3D annular cascade is estimated using medium fidelity potential flow based model and compared with experimental and the high-fidelity CFD results. For the low speed and low amplitude flutter case where the flow field around the 3D blade cascade is linear (1st test case), the PM based solver results show good agreement with measure AD data at different IBPA. However, for the more realistic flow conditions and the complex blade geometry case as in the 2nd test the method performs less accurate compared to Navier-Stokes based CFD solvers. The main cause for lower accuracy is the flow separation and the compressibility effects exist in the flow field. As mentioned earlier that the BEM PM not suitable to model the separated flow cases and moreover there is limited chance to account for compressibility effect because PM is based on assumption of inviscid flow. Since for the 2nd test case the compressibility effects and the flow separation have dominating effect on more than 40% for the blade length along the span due to higher rotational speed. At rotational speed 3000 rpm with blade span 1.25 m, the blade linear velocity along the blade span is very high and become more than 1.14 Mach at the tip. Moreover, the compressibility effect cannot be neglected as flow speed reaches close to 0.4 Mach or above, since current version of PM cannot take compressibility effect in account, thus, the PM solver overestimates the pressure and thus AD value. Nevertheless, in this research work, classical flutter is the main focus of study, which assumes low amplitude and low frequency flutter, and flow field is free from any flow separation, hence, PM performs satisfactory in classical flutter cases as shown in first test case.

Therefore, as an overall conclusion it can be said that the proposed BEM based PM based CFD solver is good compromise of speed and accuracy. The simulated results for aerodynamics damping (key stability parameter) has good agreement with experimental results. Hence, the present method can be applied to estimate the aeroelastic stability parameters for low frequency and low amplitude oscillation cases. In addition, it can a versatile design tool which can have significant impact on the computational time reduction and give researchers and engineers freedom to iterate different blade profile in very short time period at preliminary design stage of LP stage blade for optimized blade. However, the use of CFD based aeroelastic model are inevitable for the flow conditions where significant amount of non-linearity can exist within the flow field e.g. in transonic flows, separated flows or large amplitude high frequency flow in the cascade.

ACKNOWLEDGMENT

This research is supported by the research project of Technology Agency of the Czech Republic(TAČR) under the NCK framework No. TN1000007, sub-project V1.7.3: "The procedure for calculating the flutter of long blades". Also authors are also grateful to Doosan Škoda Power, s.r.o. for providing bladed wheel geometry and the FEM modal analysis data.

REFERENCES

[1] H Atassi and T Akai. Effect of blade loading and thickness on the aerodynamics of oscillating cascades. In 16th Aerospace Sciences Meeting, page 277. A1AA, 1978.

[2] J Panovsky and R. E Kilelb. A design method to prevent low pressure turbine blade flutter. Journal of engineering for gas turbines and power, 122(1):89–98, 2000.

[3] J L Hess. Calculation of potential flow about arbitrary three-dimensional lifting bodies. Final Technical Report MDC J5679-01, Naval Air Systems Command, Department of the Navy, 1972.

[4] C S Prasad and G Dimitriadis. Application of a 3d unsteady surface panel method with flow separation model to horizontal axis wind turbines. Journal of Wind Engineering and Industrial Aerodynamics, 166:74–89, 2017.

[5] C S Prasad, Q-Z Chen, O Bruls, F DAmbrosio, and G Dimitriadis. Aeroservoelastic simulations for horizontal axis wind turbines. Proceedings of the Institution of Mechanical Engineers, Part A: Journal of Power and Energy, 231(2):103–117, 2017.

[6] J Katz and A Plotkin. Low-Speed Aerodynamics. Cambridge University Press, 2nd edition, 2001.

[7] E R McFariand. Solution of plane cascade flow using improved surface singularity methods. Journal of Engineering for Power JULY, 104:669, 1982.

[8] Z Qi Lei and G Z Liang. Solution of turbine blade cascade flow using an improved panel method. International Journal of Aerospace Engineering, page Article (ID 312430), 2015.

[9] F Barbarossa, A B Parry, J S Green, and L di Mare. An aerodynamic parameter for low-pressure turbine flutter. Journal of Turbomachinery, 138(5):051001, 2016.

[10] T Fransson. Aeroelasticity in axial flow turbomachines. Von Karman Institute for Fluid Dynamics, Brussels, Belgium, 1999.

[11] J M Verdon. Review of unsteady aerodynamic methods for turbomachinery aeroelastic and aeroacoustic applications. AIAA journal, 31(2):235–250, 1993.

[12] C S Prasad and L Pesek. Analysis of classical flutter in steam turbine blades using reduced order aeroelastic model. In The 14th International Conference on

Vibration Engineering and Technology of Machinery (VETOMAC XIV), pages 150–156, Lisabon, Portugal, Sept 2018.

[13] Y Hanamura, H Tanaka, and K Yamaguchi. A simplified method to measure unsteady forces acting on the vibrating blades in cascade. Bulletin of JSME, 23 (180):880–887, 1980.

[14] D Vogt. Experimental Investigation of Three-Dimensional Mechanisms in Low-Pressure Turbine Flutter. PhD thesis, 2005, KTH University, Stockholm, Sweden.

[15] C S Prasad, and L Pesek. "Classical flutter study in turbomachinery cascade using boundary element method for incompressible flows." IFToMM World Congress on Mechanism and Machine Science. Springer, Cham, 2019.

[16] F O Carta and A O St.Hilaire. Effect of Inter-Blade Phase Angle and Incidence Angle on Cascade Pitching Stability" ASME J. of Engineering for Gas Turbines and Power, Vol. 102, 1980, pp.391–396.

[17] M Novinsky and J Panovsky. Flutter Mechanisms in Low Pressure Turbine Blades" ASME J. of Engineering for Gas Turbines and Power, Vol. 122, 2000, pp.82–88.

Suppression and control of torsional vibrations of the turbo-generator shaft-lines using rotary magneto-rheological dampers

T. Szolc[1], A. Pochanke[2], R. Konowrocki[1], D. Pisarski[1]

[1]Institute of Fundamental Technological Research of the Polish Academy of Sciences, Warsaw, Poland
[2]*Faculty of Electrical Engineering of the Warsaw University of Technology, Warsaw, Poland*

ABSTRACT

Torsional vibrations of steam turbo-generator rotor-shaft-lines coupled with bending vibrations of exhaust blades still constitute an important operational problem for this type of rotor-machines. Therefore, this work proposes a relatively simple approach for efficient suppression and control of transient and steady-state turbo-generator shaft torsional vibrations excited by short circuits in a generator or power-lines, faulty synchronization, negative sequence currents and by sub-synchronous resonances in the turbo-generator-electric network system. This target has been achieved by means of semi-actively controlled rotary dampers with the magneto-rheological fluid. Regular operation of such devices installed in a given turbo-generator rotor-shaft line enables effective minimization of amplitudes of dangerous torsional oscillations.

1   INTRODUCTION

Transient and steady-state torsional vibrations of steam turbo-generator rotor-shaft lines coupled with bending vibrations of low-pressure-rotor exhaust blades and excited by various electrical disturbances still constitute an important operational trouble for this type of rotor-machines. This problem has been already investigated for many decades by many authors using numerous theoretical and experimental methods. In the commonly cited work (1) published in 1981 the most frequent electric network and generator faults were specified and turbo-generator electromechanical models convenient to investigate the abovementioned effects were described. But from Nordmann's industrial report (2) it turned out that after another 35 years all these problems are still important, although the torsional oscillations of the turbo-generator shaft are usually more or less effectively mitigated by several anti-overload and protection systems. As it follows from the quite recent literature items (3)-(6), research in this field have been mostly focused on rotor-shaft material strength and fatigue life assessment, where excitations due to the electric disturbances were taken into consideration in the form of 'a-priori' assumed combinations of exponential and harmonic functions with parameters gained from practical observations. In (3)-(4) common spring-mass models of the turbo-generator rotor-shaft-lines are improved in order to achieve higher computational efficiency using the modified Riccati torsional transfer matrix combined with Newmark's-β method. In (5) a 3D finite element mechanical modelling was applied for more accurate stress determination. In turn, in (6) the classical 1D FEM formulation is used and several cases of synchronization faults are taken into consideration in the form of harmonic external excitations.

The presented work proposes a relatively simple approach for efficient suppression and control of transient and steady-state turbo-generator shaft torsional vibrations excited by short circuits in a generator or power-lines, synchronization faults, negative sequence currents and by sub-synchronous resonances occurring in the turbo-generator-electric network system. This target is going to be achieved by means of semi-actively controlled rotary dampers with the magneto-rheological fluid (MRF). The theoretical investigations, based on experimental measurements carried out on the real rotary MRF dampers, are performed by means of an advanced electro-mechanical model of the synchronous generator and turbo-generator shaft-line with exhaust blades.

2   MODELLING OF THE TURBOGENERATOR-ROTOR-SHAFT SYSTEM

The object of considerations is a typical 135 MW steam turbo-generator rotor-shaft system which includes the high-, intermediate- and low-pressure turbine, synchronous generator and exciter, as shown in Figure 1. This rotating system is expected to operate with the rotational speed of 3000 rpm.

Figure 1. Rotor-shaft of the 135 MW steam turbo-generator.

2.1   Hybrid structural modelling of the mechanical system

To study the electromechanical phenomena listed above, a possibly realistic and reliable mechanical model of the turbo-generator rotor-shaft line is applied. In this paper, similarly as e.g. in (7)-(8), dynamic investigations of the entire mechanical system are performed by means of a one-dimensional hybrid structural model consisting of finite continuous visco-elastic macro-elements and rigid bodies. In this model, by means of the torsionally deformable cylindrical macro-elements of continuously distributed inertial-visco-elastic properties successive cylindrical segments of the turbo-generator stepped shaft and coupling disks are substituted. In order to obtain a sufficiently accurate representation of the real object, the visco-elastic macro-elements in the hybrid model are characterized by the geometric cross-sectional polar moments of inertia responsible for their elastic and inertial properties as well as by the separate layers responsible for their inertial properties only.

Torsional vibrations of the large steam turbo-generator rotor-shaft lines are usually affected by at least two first eigenmodes of natural vibrations of typical low-pressure turbine exhaust blades. According to (9), an interaction between the exhaust blades oscillating 'in-phase' in their rims and the torsionally vibrating rotor-shaft can be taken into account in an equivalent form of the sum of movements of individual independent inertial-elastic oscillators with one degree of freedom each. These oscillators are fixed in the rotor-shaft cross-sections which correspond to the appropriate low-pressure turbine blade rims, as shown in Figure 2a. Each dynamic oscillator is characterized by the natural frequency Ω^j and modal mass m^j of the j-th eigenform of the respective exhaust blade rim.

In the hybrid model torsional motion of cross-sections of each visco-elastic macro-element is governed by the local hyperbolic partial differential equation of the wave type. Mutual connections of the successive macro-elements creating the stepped shaft as well as their interactions with the discrete oscillators are described by equations of boundary conditions. These equations contain geometrical conditions of conformity for rotational displacements of the extreme cross sections of the mutually adjacent visco-elastic macro-elements as well as equations of equilibrium for external torques, and for inertial, elastic and external damping moments. The solution for forced vibration analysis has been obtained using the analytical-computational approach described e.g. in (7)-(8). Solving the differential eigenvalue problem and an application of the Fourier solution in the form of series in orthogonal eigenmode functions lead to the set of modal equations for time coordinates $\xi_m(t)$:

$$\ddot{\xi}_m(t) + \left(\beta + \tau\omega_m^2\right)\dot{\xi}_m(t) + \omega_m^2\xi_m(t) = \frac{1}{\gamma_m^2}\left(X_m^T \cdot M_T(t) - X_m^G \cdot T_{el}(t)\right), \quad m = 1, 2, ... \qquad (1)$$

where ω_m are the successive natural frequencies of the rotor-shaft system, β denotes the coefficient of external damping assumed here as a proportional one to the modal masses γ_m^2, τ is the retardation time, $T_{el}(t)$ denotes the external retarding torque produced by the electric generator, $M_T(t)$ is the steam-turbine resultant driving torque and X_m^G, X_m^T are the modal displacement functions scaled by proper maxima and distributed respectively along the electric generator-rotor and along the successive turbine-rotors in the hybrid model. A fast convergence of the applied Fourier solution enables us to reduce the number of the modal equations to solve in order to obtain a sufficient accuracy of results in the given range of frequency. Here, it is necessary to solve only a few or at most a dozen or so modal equations (1), even in cases of very complex mechanical systems, contrary to the classical one-dimensional beam finite element formulation leading usually to large numbers of motion equations corresponding each to more than one hundred or many hundreds degrees of freedom (if the artificial and often error-prone model reduction algorithms are not applied).

2.2 Modelling of the generator

From the viewpoint of electromechanical coupling investigation properly advanced circuit model of the electric generator seems to be sufficiently accurate. In the case of a symmetrical three-phase synchronous generator electric current oscillations in its windings are described by six circuit voltage equations transformed next into the system of Park's equations in the so called 'd-q' reference system, (10):

$$\begin{bmatrix} R_l + L_d\frac{d}{dt} & -\Omega(t)L_q & -\sqrt{\frac{3}{2}}M_{lf}\frac{d}{dt} & -M_{lD}\frac{d}{dt} & \Omega(t)M_{lQ} \\ \Omega(t)L_d & R_l + L_q\frac{d}{dt} & -\Omega(t)\sqrt{\frac{3}{2}}M_{lf} & -\Omega(t)M_{lD} & -M_{lQ}\frac{d}{dt} \\ -\sqrt{\frac{3}{2}}M_{lf}\frac{d}{dt} & 0 & R_f + L_f\frac{d}{dt} & M_{fD}\frac{d}{dt} & 0 \\ -M_{lD}\frac{d}{dt} & 0 & M_{fD}\frac{d}{dt} & R_D + L_D\frac{d}{dt} & 0 \\ 0 & -M_{lQ}\frac{d}{dt} & 0 & 0 & R_Q + L_Q\frac{d}{dt} \end{bmatrix} \cdot \begin{bmatrix} i_d(t) \\ i_q(t) \\ i_f(t) \\ i_D(t) \\ i_Q(t) \end{bmatrix} = \begin{bmatrix} -u_d(t) \\ -u_q(t) \\ u_f \\ 0 \\ 0 \end{bmatrix}, \qquad (2)$$

where:
$$u_d(t) = \sqrt{\frac{2}{3}}\left\{\left[U_a + \frac{1}{4}(U_b + U_c)\right]\cos\Theta(t) \cdot \sin(\omega_e t) - \frac{3}{4}(U_b + U_c)\sin\Theta(t) \cdot \cos(\omega_e t) + \right.$$
$$\left. + \frac{\sqrt{3}}{4}(U_b - U_c)\cos(\Theta(t) + \omega_e t)\right\},$$
$$u_q(t) = -\sqrt{\frac{2}{3}}\left\{\left[U_a + \frac{1}{4}(U_b + U_c)\right]\sin\Theta(t) \cdot \sin(\omega_e t) + \frac{3}{4}(U_b + U_c)\cos\Theta(t) \cdot \cos(\omega_e t) + \right.$$
$$\left. + \frac{\sqrt{3}}{4}(U_b - U_c)\sin(\Theta(t) + \omega_e t)\right\},$$

R_l is the armature phase resistance, L_d, L_q are the armature phase self-inductances reduced to the electric field equivalent axes d and q, M_{lf} denotes the armature phase-to-field winding mutual inductance, M_{lD} is the armature phase-to-direct axis D damper-winding mutual inductance, M_{lQ} denotes the armature phase-to-quadrature axis Q damper-winding mutual inductance, M_{fD} is the field-winding-to-direct axis D damper-winding mutual inductance, L_f, L_D, L_Q and R_f, R_D, R_Q are respectively the self-inductances and resistances of: field-winding, direct axis D damper-winding and quadrature axis Q damper-winding, $i_d(t)$, $i_q(t)$ are the armature phase currents reduced to the electric field equivalent axes d and q, $i_f(t)$, $i_D(t)$, $i_Q(t)$ denote the electric currents in the field-winding, direct axis D damper-winding and quadrature axis Q damper-winding, respectively, u_f is the feeding voltage of the exciter, $\Omega(t)$ and $\Theta(t)$ denote respectively the instantaneous values of the generator rotor electrical angular velocity and rotation angle which include their average and vibratory parts, ω_e is the armature voltage circular frequency, and U_a, U_b, U_c are the amplitudes of armature phase voltages, (10). Then, the electromagnetic retarding torque produced by such synchronous generator can be expressed by the following formula:

$$T_{el}(t) = -p\left[\left(L_d\,i_d(t) - \sqrt{\tfrac{3}{2}}M_{lf}\,i_f(t) - M_{lD}\,i_D(t)\right)i_q(t) - \left(L_q\,i_q(t) - M_{lQ}\,i_Q(t)\right)i_d(t)\right], \qquad (3)$$

where p is the number of pairs of the generator magnetic poles.

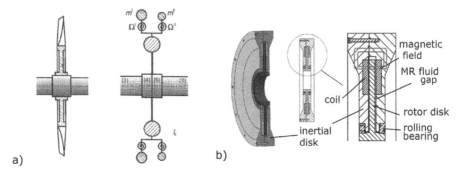

a)   b)

Figure 2. Modelling of the exhaust blade rims (a) and rotary dampers (b).

2.3   Modelling of the rotary dampers

To mitigate transient and steady-state torsional vibrations excited by electric per-turbations in the generator and electric network, rotary dampers with the magneto-rheological fluid (MRF) are applied. Such devices consist of heavy inertial disks rotata-bly mounted on selected rotor-shaft segments via MRF-film layers with controllable viscous properties, as illustrated in Figure 2b. Here, vibrations are going to be sup-pressed by means of attenuating the difference between the vibratory and the average rotor-shaft motion. The magneto-rheological fluids are functional fluids whose effect-ive viscosity depends on externally provided magnetic field. This feature makes them perfectly suitable for large rotary dampers with controllable damping characteristics. Then, besides an ability to generate large damping torques, important advantage of these devices is a low power consumption.

Here, the appropriately assumed and identified rheological model of the magneto-rheological fluid plays a fundamental role. For this aim the Bingham model of the magneto-rheological fluid is applied. Then, dynamic properties of such magneto-rheological fluid are regarded as a parallel combination of a viscous and frictional damper. In the paper the rotary dampers for the turbo-generator shaft line have been built using proper extrapolations of geometrical dimensions of the real laboratory magneto-rheological rotary damper of an analogous structure tested in (11) and with rheological properties experimentally identified in (12). By means of these identification results the following simple mathematical description of the damping torque generated by the proposed devices has been formulated. Namely, assume that there are N controllable dampers with freely rotating inertial disks, each with the independently controllable damping coefficient $c_j(i(t))$, $j=1,2,\ldots,N$. Then, the j-th rotary damper generates the following damping torque

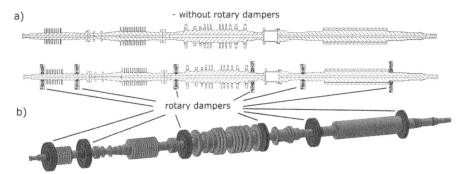

a)
- without rotary dampers

b)
rotary dampers

Figure 3. Turbo-generator rotor-shaft without- (a) and with rotary dampers (b).

$$M_j^D(i(t)) = -M_j^F(i(t)) \cdot sgn\big(\Delta\omega_j(t)\big) - c_j(i(t))\Delta\omega_j(t),$$ (4)

where $\Delta\omega_j(t) = \Omega(t) + \sum\limits_{m=1}^{\infty} \dot{\theta}_m(x_j,t) - \omega_j(t), j = 1,2,\ldots,N,$

$M_j^F(i(t))$ and $c_j(i(t))$ denote respectively the control static frictional damping torque and the viscous damping coefficient of the magneto-rheological fluid, both depending on the programmable, time-dependent control electric current $i(t)$ in device coils, $\Delta\omega_j(t)$ is the difference between angular velocities of the torsionally vibrating shaft and the rotary damper inertial disk, x_j denotes the location of the j-th damper on the rotor-shaft line, and $\omega_j(t)$ is the rotational speed of the j-th inertial disk, which obeys

$$J_j\dot{\omega}_j(t) = -c_j(i(t))\left[\omega_j(t) - \Omega(t) - \sum\limits_{m=1}^{\infty} \dot{\theta}_m(x_j,t)\right].$$ (5)

Here, in general any effective and practically usable control strategy has to be open- or closed-loop. But in the case of a synchronous generator the external excitation during electrical faults is usually characterized by two constant frequency fluctuating components: synchronous and double-synchronous. Then, electrical faults induce transient torsional vibrations of the rotor-shaft line with its fundamental natural frequencies which are also constant. According to the above, the locally optimum open-loop control functions $c_{0j}=c_j(i_0)$ can be determined for $i(t)=i_0=$const with respect to the frequency response function (FRF) of the modal equations of motion (1) augmented by the control torques (4) and the equations of motion of the inertial disks (5). Thus, in this study the open-loop control, for which the damping coefficients remain constant during the whole electrical fault process, will be applied as the most reasonable and convenient in a turbo-generator exploitation practice. Here, their values are optimum with respect to the resonant frequencies, as defined by

$$\mathbf{c_0} = \arg\ min_{\mathbf{c}}\ max_f FRF(f, \mathbf{c}) \tag{6}$$

The control damping torques $M_j^D(t)$ defined by (4) should be treated in the hybrid model as system response dependent external excitations imposed in a concentrated form on the rotor-shaft cross-sections corresponding respectively to the $j=1,2,...,N$ locations of the rotary dampers. In order to include these excitations in the modal equations of motion (1), they must be transformed into an orthogonal base of eigenfunctions of the turbo-generator rotor-shaft hybrid model. Then, using the Fourier solution one obtains generalized, response dependent external excitations in the form of series in these orthogonal eigenfunctions. As a result, the separate modal equations (1) become mutually coupled by angular speed dependent terms to form the following set of second order ordinary differential equations in modal coordinates $_m(t)$ contained in vector $\mathbf{r}(t)$:

$$\mathbf{M}\,\ddot{\mathbf{r}}(t) + \mathbf{C}(i_0)\,\dot{\mathbf{r}}(t) + \mathbf{K}\,r(t) = \mathbf{F}(t), \tag{7}$$

where $\mathbf{M}$ and $\mathbf{K}$ are the constant diagonal modal mass and stiffness matrices which contain respectively modal masses γ_m^2 and modal stiffness $\gamma_m^2\omega_m^2$ standing in Eq. (1). Symbol $\mathbf{C}(i_0)$ denotes the constant control matrix which, in addition to diagonal structural damping terms $\gamma_m^2(\beta+\tau\omega_m^2)$, is supplemented by diagonal and out-of-diagonal terms with coefficients $c_{0j}=c_j(i_0)$ of damping generated by the rotary dampers. Vector $\mathbf{F}(t)$ contains modal external excitations caused by driving torques produced by successive steam turbines and by the retarding electrical generator torque. The total number of equations (7) corresponds to the number of torsional eigenmodes taken into consideration in (1) in the frequency range of interest.

3 COMPUTATIONAL EXAMPLES

The subject of calculations is the electromechanical model of the rotor-shaft line of the above-mentioned 135 MW steam turbo-generator. This object is characterized by the total length of 23.74 m, polar mass moment of inertia of 10487.6 kgm^2 and the maximum nominal rated torque equal to 429734.4 Nm. In the mechanical model subsequent cylindrical sections of the real rotor-shaft of this turbo-generator have been substituted by 184 finite continuous macro-elements. In all computational examples numerical values of the external damping coefficient β and the retardation time τ in Equations (1), (7) correspond to the loss-factor of 0.0048 and the logarithmic decrement of the rotor-shaft free torsional vibrations equal to 0.015 which take into consideration material losses, frictional resistance in bearings and aerodynamic drags.

3.1 Determination of optimal rotary damper parameters
Before starting to perform numerical simulations of various torsional vibration scenarios, the basic parameters of the rotary magneto-rheological dampers will be determined for the system under study. Here, the most fundamental are the polar mass moment of inertia J_j of the inertial disks and optimum values of the damping coefficients c_{0j}, $j=1,2,...,N$. Considering the structure of the actual rotor-shaft line of the turbo-generator in question, $N=6$ rotary dampers were adopted, respectively on both sides of the HP-turbine, LP-turbine and the generator rotor, as shown in Figure 3.

Initially, it was assumed that all six inertial disks are identical and controlled collectively, i.e.:

$$J_j = J \quad \text{and} \quad c_{0j} = c_0, \quad j = 1, 2, ..., 6. \tag{8}$$

As can be intuitively predicted, the higher the inertial disk mass moment of inertia J, the greater ability to effectively reduce amplitudes of rotor-shaft torsional vibrations. Furthermore, from the design of the rotary damper presented in Figure 2b it follows that large mass moments of inertia are associated with larger dimensions of the inertial disk, which results in greater area of the magneto-rheological fluid gap, and therefore in higher realistic values of damping coefficient c_0 possible to achieve. However, it should be remembered that the geometrical dimensions of the inertial disk and the entire rotary damper cannot be too large and must be rationally adapted to the size of the entire turbo-generator rotor-shaft. In the case of the 135 MW turbo-generator in question, it was possible to design a rotary damper with the inertial disk mass moment of inertia $J=83$ kgm^2. For this value, by means of the open-loop control defined by equation (6), frequency response functions of the torsionally vibrating rotor-shaft system have been determined and demonstrated in Figure 4. Here, the damping coefficient c_0 is constrained to the interval 0-5·10^4 Nms/rad within excitation frequency range of 0-100 Hz, which together include FRF minima of the fundamental eigenmodes with resonant frequencies of 26.37 and 49.47 Hz. Figure 4a and 4b present FRFs of the angular velocity amplitudes at the HP-turbine and exciter rotor-shaft free ends, respectively. In Figures 4c and 4d there are illustrated FRFs of the amplitudes of dynamic torques transmitted respectively by the shaft couplings between the IP-LP turbines and between the LP-turbine-generator rotor. The all plots demonstrate significant amplitude peaks associated with the first two torsional eigenvibration modes with the abovementioned natural frequency values. Moreover, at zero excitation frequency the FRFs of the rotor-shaft angular velocities are characterized by the additional peaks which are associated with the system rigid body mode, and of constant values – independent of the damping coefficients c_0 realized by the rotary dampers. From the computational result presented in Figure 4 it follows that an effective suppression of the system dynamic response can be obtained for damping coefficients greater than ~2500 Nms/rad to reach minimum between 0.5·10^4-2·10^4 Nms/rad. But for c_0 higher than ca. 3·10^4 Nms/rad amplitudes of the first natural vibration mode noticeably increase due to a gradual sticking of the relatively light-weight inertial disks to the torsionally vibrating shaft. Analogous computations have been also performed for greater inertial disk mass moments of inertia J than the abovementioned 83 kgm^2. Then, as shown in Figure 5 (in the identical way as in Figure 4), for $J=155$ kgm^2 within the damping coefficient range under consideration the observed two resonance amplitude peaks surprisingly decay slower with the rise of c_0 from zero to reach minima at ca. 3·10^4 Nms/rad and to remain almost unchanged for greater damping coefficient values, because the greater J, the higher c_0 is required to stick a heavy inertial disk to the torsionally vibrating shaft. In the case of rotary dampers with greater mass moments of inertia of their inertial disks and larger geometrical dimensions it is easier to achieve greater damping coefficient values, but then such devices are characterized by too large outer diameters which can be troublesome for a design of the entire turbo-generator. According to the above, for further investigations the constant value of the optimal and realistic to obtain damping coefficient $c_0=2500$ Nms/rad has been chosen basing on the system frequency response functions depicted in Figure 4 and obtained for $J=83$ kgm^2 which seems to be a reasonable value from the design viewpoint of the turbo-generator being tested.

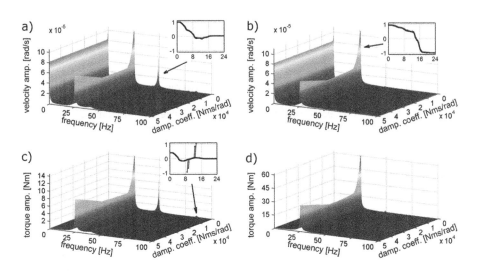

Figure 4. Frequency response functions of the turbo-generator rotor-shaft system for the rotary dampers with the inertial disk mass moment of inertia of 83 kgm^2.

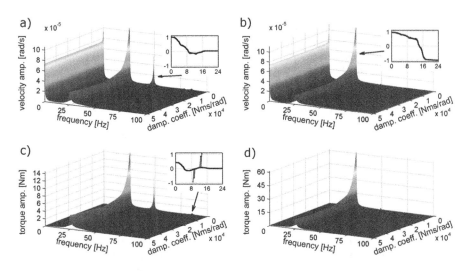

Figure 5. Frequency response functions of the turbo-generator rotor-shaft system for the rotary dampers with the inertial disk mass moment of inertia of 155 kgm^2.

3.2   Simulation of transient vibrations in time domain

In order to demonstrate suppression ability of torsional vibrations by means of the MRF rotary dampers two electrical fault scenarios will be assumed: namely, transient dynamic responses caused by a symmetrical three-phase short circuit and a two-phase short circuit in the generator. In the first case a typical automatic high-speed-reclosing after standard number of cycles is applied and the second investigated fault will be cleared after 200 cycles of its duration. For the case of the three-phase short circuit in Figure 6 there are presented time-histories of the dynamic torques transmitted by the most heavily affected turbo-generator rotor-shaft couplings, i.e. between the IP-LP turbines (Figure 6a) and between the LP-turbine-generator-rotor (Figure 6b), of the reaction torque transmitted by attachments of the longest exhaust blade rims on the LP-rotor (Figure 6c), and the angular velocity vibratory component of the shaft free end at the HP-turbine rotor side (Figure 6d). In all of these figures there are plotted dynamic responses obtained for the system without rotary dampers (using red lines) and analogous responses determined for the turbo-generator rotor-shaft equipped with the rotary dampers (by means of the green lines). Consequently, Figures 6e-h demonstrate respectively in the identical way dynamic responses of the same quantities registered during simulation of the double-phase short circuit scenario.

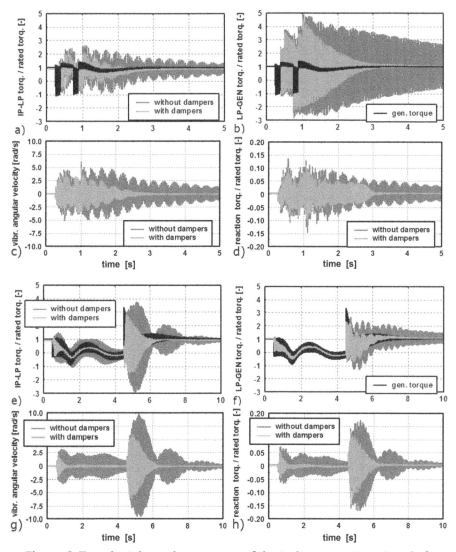

Figure 6. Transient dynamic responses of the turbo-generator rotor-shaft system for: the three-phase (a-d) and double-phase (e-h) short circuit in the generator.

Because of relatively weak structural damping, which usually characterize torsionally vibrating turbo-generator rotor-shafts, the all transient dynamic responses obtained for the system without rotary dampers, and demonstrated in Figure 6, very slowly fade over time. This can be very dangerous for a material fatigue of the most responsible elements, like shaft-couplings or exhaust blade attachments. However, when the MRF rotary dampers are used, in the case of both electric fault scenarios the amplitudes of the torques and velocity almost immediately decay over time for the same constant damping coefficient value of the magneto-rheological fluid film, which depends on the constant control current a-priori adopted for the tested object.

4 FINAL REMARKS & FURTHER WORK

In the paper there is proposed a practical method of suppression of turbo-generator rotor-shaft line torsional vibrations induced by various electrical faults. In the study presented here an attenuation of oscillation amplitudes is realized by introduction of additional damping into the mechanical system by means of magneto-rheological rotary dampers with adjustable dissipative properties. The obtained results of exemplary computations have confirmed that using the proposed approach severe transient torsional vibrations can be very effectively mitigated when constant damping coefficients of the MRF dampers are properly adopted. Optimal numerical values of them correspond to minima of the frequency response functions which take into consideration the most excitable fundamental eigenmodes of the hybrid model of the rotor-shaft line determined for the turbo-generator being tested. Considering the viscous properties of typical magneto-rheological fluids, their constant damping coefficients can be realized by constant control current values, which makes the proposed strategy relatively simple, cheap and robust in an operational practice.

The entire methodology presented in this work is based on a realistic and relatively transparent electro-mechanical model of the investigated object. The computational examples were limited here to simulations of transient vibrations caused by short circuits in the generator only, although using this model numerous scenarios of torsional vibrations induced by other electrical faults can be effectively studied. For example, steady-state oscillations caused by an unbalanced load of the generator can be spectacularly investigated for turbo-generator rotor-shaft lines sensitive to excitations with double-synchronous frequency, which unfortunately does not apply to the tested object. Moreover, in order to study the phenomenon of sub-synchronous resonances, the electromechanical model used here must be extended to include a relatively simple but reliable model of the electric transmission network. Anyway, in such cases the use of magneto-rheological rotary dampers to mitigate torsional vibrations also seems extremely justified, which will be the subject of further research in this field.

REFERENCES

[1] Berger, H., Kulig, T.S., (1981) Simulation models for calculating the torsional vibrations of large turbine-generator units after system electrical faults, *Siemens Forsch.- u. Entwickl.-Ber.*, Bd. 10, Nr. 4, pp. 237–245.
[2] Nordmann, R., (2016) Torsional and stator vibrations in turbines and generators – analysis and mitigation, *Energiforsk*, ISBN 978-91-7673-295-3.
[3] He, Q., Du, D., (2010) Modeling and calculation analysis of torsional vibration for turbine generator shafts, *Journal of Information & Computational Science 7*, 10, pp. 2174–2182.
[4] Xiang, L., Yang, S., Gan,. C., (2012) Torsional vibration of a shafting system under electrical disturbances, *Shock and Vibration*, 19, pp. 1223–1233.

[5] Gu, Y., Xu, J., (2013) Analysis on torsional stresses in turbo-generator shafts due to two-phase short-circuit fault, *Applied Mechanics and Materials*, Vols. 397-400, pp. 427–430.

[6] Bangunde, A., Kumar, T., Kumar, R., Jain, S.C., (2018) Torsional vibration analysis in turbo-generator shaft due to mal-synchronization fault, *Materials Science and Engineering*, 330, 012093, IOP Publishing.

[7] Szolc, T., (2000) On the discrete-continuous modeling of rotor systems for the analysis of coupled lateral-torsional vibrations, *International Journal of Rotating Machinery*, 6(2), pp. 135–149.

[8] Szolc, T., Tauzowski, P., Knabel, J., Stocki, R., (2009) Nonlinear and parametric coupled vibrations of the rotor-shaft system as fault identification symptom using stochastic methods, *Nonlinear Dynamics*, 57, pp. 533–557.

[9] Okabe, A., Otawara, Y., Kaneko, R., Matsushita, O., Namura, K., (1991) An equivalent reduced modelling method and its application to shaft-blade coupled torsional vibration analysis of a turbine-generator set, Proc. Instn. Mech. Engrs (IMechE), Vol. 205, pp. 173–181.

[10] IEEE Standards, (2003) *IEEE Guide for Synchronous Generator Modeling Practices and Applications in Power System Stability Analyses*, IEEE Std 1110[TM]-2002, *IEEE Power Engineer Society*, New York.

[11] Pręgowska, A., Konowrocki, R., Szolc, T., (2013) On the semi-active control method for torsional vibrations in electro-mechanical systems by means of rotary actuators with a magneto-rheological fluid, *Journal of Theoretical and Applied Mechanics*, 51, 4, pp. 979–992.

[12] Gorczyca, P., Rosół, M., (2011) A semi-active suspension system model, Proc. of the 10th Conference on Active Noise and Vibration Control Methods, Cracov, pp. 206–213.

12th International Conference on Vibrations in Rotating Machinery -
Institution of Mechanical Engineers, ISBN 978-0-367-67742-8

Coupling between axial, lateral and torsional vibration modes of a flexible shaft with flexible staggered blades

G. Tuzzi, C.W. Schwingshackl, J.S. Green

Imperial College London, Mechanical Engineering, UK
Rolls Royce Plc, UK

ABSTRACT

Rotating flexible shafts carrying multiple flexible bladed discs are the core component of gas turbines. Due to the ever more flexible nature of shafts and discs in modern applications, dynamic coupling of shafts and discs is becoming more common. Traditional analysis assumptions of rigid discs on flexible shafts, and flexible discs on rigid shafts, are no longer valid, and a combined analysis is needed. It is well known that disc/blade zero and one Nodal Diameter (ND) modes can couple with shaft axial and bending modes respectively, but a previous work by the authors has shown that a third coupling behaviour between shaft bending and disc 0ND modes exists in the presence of an asymmetric axial-radial bearing supporting structure.

This paper investigates the previous findings in a much more realistic configuration, adding a flexible bladed disc with variable stagger angle to the analysis. The results show that, in the presence of an axial-radial coupled bearing support on the shaft, both shaft axial and bending modes can combine with blade 0ND and 1ND modes into "fully coupled" modes involving all these vibrating patterns. In addition, the blade stagger angle can lead to a further coupling behaviour of these new "fully coupled" modes with shaft torsion.

1   INTRODUCTION

Rotating assemblies composed by flexible shafts and bladed discs are extensively used in industry, with applications ranging from aircraft propulsion and power gas turbines to compressor and wind turbines [1-2]. Vibration is a highly undesirable phenomenon in these systems, since it affects performance and leads to premature fatigue and wear [1, 3]. The dynamic analysis of rotors has historically been carried out by first addressing the shaft dynamics with rigid discs [1, 5], and the blades dynamics was then investigated separately [4], assuming a rigid shaft. This assumption is valid as long as the blades and shaft resonance frequencies are well separated. However, modern applications and new trends in aero-engine design introduce larger, so more compliant bladed discs on shorter, stiffer shafts [2]. In these cases, the uncoupled assumption is no longer valid, since blade frequencies can approach the shaft frequencies, and a coupled dynamic analysis is required to correctly predict the system's dynamic response [5]. An unknown coalescence of shaft bending and disc "zero Nodal Diameter (0ND)" modes has also been observed during an engine development test.

Extensive theoretical research has been carried out in the past on shaft and disc/blade dynamic interaction [8-15]. Detailed studies have shown that disc/blades 0ND and 1ND modes tend to couple with shaft axial and bending modes, respectively. When blades vibrate with a 0ND pattern, a net torque is transmitted to the shaft, exciting its torsional mode(s) [8, 13]. If the blades are staggered and vibrating with a 0ND pattern, then an axial force is also transmitted to the shaft, which can excite the shaft axial mode(s), [4, 9]. A more comprehensive model with both disc and blades has been developed in [16-18, showing that also a 1ND pattern on the bladed disk leads to a bending moment transmitted to the shaft, which in turn can excite the shaft bending modes. This behaviour is also impacted by the gyroscopic effect, which leads to whirling shaft modes that couple with disc-blades 1ND [6, 16-18]. The above investigations use simplified modelling approaches [7], which do not allow to capture 3D effects. In addition, the bearings supporting structure is often neglected and the shaft lateral, axial and torsional dynamics are considered uncoupled, with the only exception of [19], where an axial-lateral shaft coupling due to the thrust bearing is discussed.

Extending on the above, previous work from the authors [20, 21] has shown that shaft axial and lateral modes can combine to axial-lateral "Mixed Modes" in the presence of a general asymmetric bearing supporting structure [20]. These coupled shaft modes can then further couple simultaneously with flexible disk 0ND and 1ND modes, explaining the aforementioned unknown coupling, which is further impacted by the gyroscopic effect [20, 21]. This paper aims to extend the above findings to a shaft carrying a flexible disc with blades for a more realistic investigation. The behaviour of this system in the presence of axial-lateral coupling due to the bearing supports is investigated to better understand how the 0ND and 1ND blade modes interact with the shaft dynamics. The impact of the stagger angle is also investigated highlighting a further interaction with the shaft torsional dynamics.

2 MODEL DESCRIPTION

In order to capture the coupling effects described above, the simple shaft-bladed disc assembly from Figure 1a will be investigated as a full 3D finite element model. The shaft is of solid Aluminium and circular, with a length of 0.8m and a diameter of 0.02m.

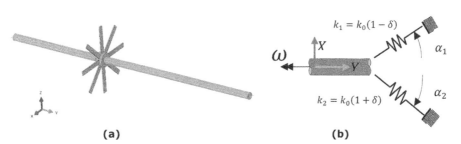

(a)   (b)

Figure 1. Assembly under study – a) solid geometry b) lumped parameters approximation of the bearing supporting structure [20].

Figure 2. Shaft with flexible disc from [20-21].

In order to assure continuity with previous works from the authors [20-21], the following choices have been made:

1. The total mass of the bladed disc (hub and blades) was set to match the one from the previous system's (Figure 2) flexible disc (0.26 kg)
2. The blades first bending frequency was set to match the frequency of the flexible disc's first 0ND modes, equal to 327 Hz

Consequently, the chosen geometry consists of a thick hub with a 4cm diameter and a 2cm thickness, which carries 8 cantilevered blades whose length is 8 cm, width is 1.2cm and thickness is 0.24 cm (Figure 1a). The blades are mounted onto the shaft with a stagger angle β (see Figure 3). In more detail, $\beta = 0$ when the blades are in plane with the disc surface. This case will be analysed firstly in section 3.1, then a parametric study for other values of β will be carried out in section 3.3.

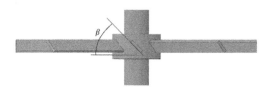

Figure 3. Blades stagger angle.

The shaft is supported by identical plain bearings at each end which were introduced in [20]. Each bearing is then supported by four rods in XY and YZ planes (see Figure 1b). In the YZ plane, the two rods are perpendicular to the shaft, whilst in the X-Y plane they are inclined towards the shaft axial direction as shown in Figure 1b, to introduce a coupling between axial and lateral shaft motion. The combined effect of bearing and rods is approximated using linear spring elements. A stiffness asymmetry is then introduced on the springs in the XY plane with the following expression $k_{1,2} = k_0(1 \pm \delta)$.

Since the shaft-blades dynamic interaction depends on forces transmitted between the shaft and the bladed disc, the stress distribution between blades and hub and hub and shaft must be computed very accurately. In the past beam and shell elements as well as analytical solutions proved unable to capture such behaviours and consequently a full 3D quadratic order tetrahedral FEM model was implemented. A combination of eigenvalue extraction and forced response analyses was carried out to obtain the presented results. Concerning the latter, the shaft is excited at its left bearing location with different forcing conditions, and radial strain on the blades and displacements on the shaft were chosen as output quantities, mimicking a typical setup in experimental rotor dynamics [1, 15], where blades are equipped with strain gauges, while shaft displacements are measured with laser sensors or accelerometers. To investigate the effect of the gyroscopic effect on the coupling behaviour, a final modal analysis including the rotation was conducted.

3   NUMERICAL RESULTS

3.1   Baseline analysis – symmetrical supports - $\delta = 0, \beta = 0$
The initial analysis is carried out with $\beta = 0$ and $\delta = 0$ (see Figure 1b and section 2). Results from the nonrotating modal analysis are summarised in Table 1 and Figure 4.

Table 1. Full assembly frequencies – asymmetrical supports with $\delta = 0$ – comparison with results with flexible disc from [20] and [21].

	Flexible disc [20]		Flexible blades (Figure 1a)	
	Frequency	Dominant mode	Frequency	Dominant mode
#1	50.47 Hz	Shaft 1st Bending – XY	50.51 Hz	Shaft 1st Bending – XY
#2	50.83 Hz	Shaft 1st Bending – YZ	50.96 Hz	Shaft 1st Bending – YZ
#3	207.62 Hz	Shaft 2nd Bending - XY	233.30 Hz	Shaft 2nd Bending - XY
#4	211.39 Hz	Shaft 2nd Bending – YZ	240.13 Hz	Shaft 2nd Bending – YZ
#5	244.74 Hz	Shaft – axial	284.89 Hz	Shaft – axial
#6	264.00 Hz	Disc 1ND – XY	318.25 Hz	Blades 1ND – XY
#7	266.91 Hz	Disc 1ND – YZ	318.80 Hz	Blades 1ND – YZ
#8	416.97 Hz	Disc 0ND	350.42 Hz	Blades 0ND
#9	429.52 Hz	Disc 2ND	318.80 Hz	Blades ND > 1
#10	219.42 Hz	Shaft torsion	394.66 Hz	Shaft torsion

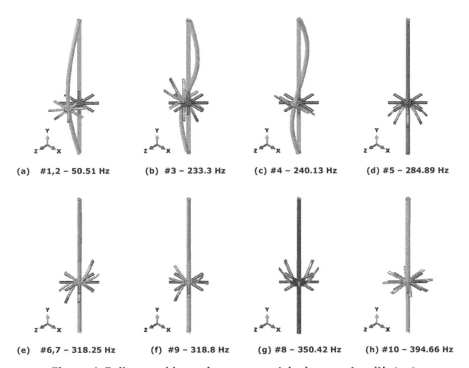

(a) #1,2 – 50.51 Hz   (b) #3 – 233.3 Hz   (c) #4 – 240.13 Hz   (d) #5 – 284.89 Hz

(e) #6,7 – 318.25 Hz   (f) #9 – 318.8 Hz   (g) #8 – 350.42 Hz   (h) #10 – 394.66 Hz

Figure 4. Full assembly modes – symmetrical supports with $\delta = 0$.

Modes #1-2 (Figure 4a) are dominated by the shaft first bending mode and the blades behave as a rigid body. Modes #3-4 are instead dominated by the shaft second bending mode and the blades oscillate on a 1ND pattern. Compared to the case with a flexible disc [20,21], the frequencies are higher. There are two reasons behind this behaviour. First, even if the mass of the blades was set to be the same of the disc in [20,21], the inertia moment is different, so is the frequency of the second bending mode. The second reason is that in a flexible disc the frequencies of 1ND and 0ND modes are well separated (264 Hz and 325 Hz) whilst they are the same when considering blades, as the first flapping mode dominates (see Figure 5). This is also the case for modes #6-7 (Figure 4e), which are dominated by the blades 1ND pattern. Modes #5 and #8 are dominated by shaft axial and blades 0ND modes, respectively. It is interesting to notice that, even if the blades frequency has been tuned to match this disc mode, there is still a high difference in frequency between the two cases. The reason behind that is the different inertia in the two cases. Figure 4f (#9) groups together all the blades modes with ND > 1, which do not couple with the shaft [9, 16-18] because they do not apply a net force/moment onto the shaft. Mode #10 is dominated by shaft torsion, and the blades slightly vibrate in the edgewise direction. The results shown here highlight, that a shaft carrying blades can have a very different dynamic behaviour from the one carrying a flexible disc, and, although they share some features, they should not be used interchangeably for approximation purposes.

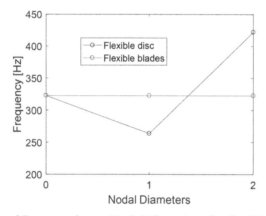

Figure 5. Natural Frequencies vs Nodal Diameters for flexible disc/blades.

3.2   Baseline analysis – asymmetrical supports – $\delta \neq 0, \beta = 0$

In this section, the results of a modal analysis with asymmetric supports in the XY plane are presented ($\delta = 0.5$). Since the YZ plane remains symmetric, only modes occurring in the XY plane are shown in Figure 5. Previous work [20] has shown that, in presence of asymmetric axial-radial supports, the shaft second bending mode and the axial rigid shaft mode combine to two axial-lateral "Mixed Modes", which show both axial and lateral vibration. These modes were shown to couple with both 0ND and 1ND disc modes due to the transmission of axial force and lateral moment, respectively.

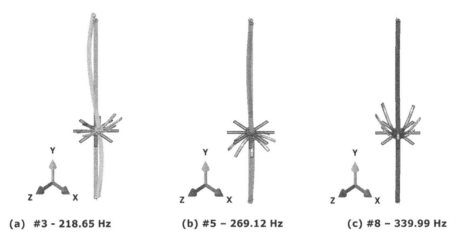

(a) #3 - 218.65 Hz   (b) #5 – 269.12 Hz   (c) #8 – 339.99 Hz

Figure 6. Full assembly modes – asymmetrical supports with $\delta \neq 0$.

Figure 6 shows the relevant modes for the shaft-blades assembly, that lead to bending-axial shaft coupling. Mode #3 (Figure 6a) is dominated by the shaft first mixed mode, which shows both axial and lateral features with the blades oscillating in a 0ND/1ND superposed pattern. Mode #5 (Figure 6b) is dominated by the shaft second mixed mode, which drives a blade 0ND pattern. Mode #8 (Figure 6c) is dominated by the blades 0ND mode with the shaft oscillating following the second mixed mode pattern. The shaft has a relatively low amplitude compared to the blades, due to the distance in frequency between shaft (255 Hz) and blades (325 Hz) modes, in addition to the lower inertia of the blades, resulting in a low amplitude of the force transmitted at the interface, which then drives the shaft mode.

3.3   Stagger angle parametric study

In this section, the impact of the blades stagger angle β (already shown in Figure 3) on the system's dynamics is investigated. In order to do so, a forced response analysis is carried out on the non-rotating system. A virtual strain gauge is applied on blade 1 in the radial direction and the shaft is excited at the left bearing with a lateral force, an axial force and a torque, respectively. Results with different stagger angles are shown in Figure 7. When blades are at $0°$ stagger angle and the shaft is driven with an axial or lateral force (7a), the shaft mixed modes are excited, which in turn drive blade flapping modes due their axial component. Instead, a torsional excitation on the shaft with $0°$ stagger angle blades transmits in-plane forces onto the blades, which then respond with low amplitude edgewise vibration (see Figure 3h) and no radial strain. The four peaks in Figure 7a correspond, respectively, to modes #3, #4, #6 and #8 from Figure 3. Since an axial or lateral excitation does not drive any torsional component, it can be concluded that blades flap and shaft torsion are not coupled in this configuration

At the other extreme, for a stagger angle of $\beta = 90°$, the opposite situation can be observed in Figure 7c): a torque on the shaft leads to a system of in-plane forces, which then strongly drive blades flap modes. The coupled mode occurs at 260 Hz. The response curves for lateral and axial excitation show very little response, which arises from an edgewise vibration of the blades in this configuration. On the other hand, the flapping motion of the blades is fully uncoupled from shaft axial and lateral.

255

When $\beta = 45°$, the system experiences a superposed situation: both torque and axial-lateral forces on the shaft lead to blade flap vibration, leading to the response in Figure 7c) where all excitations drive all modes. Given the linear nature of the system, in reverse, this behaviour means that when blade flap motion is excited (e. g. via an Engine Order excitation), both a torque and an axial force are applied to the shaft, which can then drive axial-lateral and torsional shaft modes. This results in coupled axial, lateral and torsional shaft dynamics which is a rather unique and previously unobserved dynamic combination. It is worth noticing that, when $\beta = 45°$, the natural frequency of the blades 0ND dominated mode (#8) goes down to 290.38 Hz due to the additional flexibility introduced by the shaft torsion.

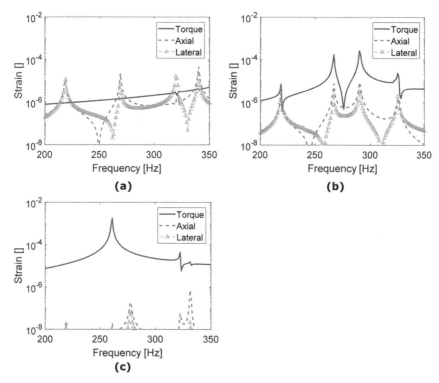

Figure 7. Blade 1 radial strain frequency response from different shaft excitation, considering a stagger angle β equal to a) 0° degrees b) 45° degrees c) 90° degrees.

In order to confirm this new coupling behaviour, an additional FRF analysis was performed on the system with $\beta = 45°$. A unitary torque was applied on the left end bearing and shaft lateral (X-direction) and axial displacements were extracted at $Y = 0.2L_{shaft}$. The results in Figure 8 clearly show that all four responses peaks are present in the shaft response for the given torque input confirming the strong coupling between axial, lateral and torsional vibration in this configuration, which is maximum in modes #5 (dominated by shaft second mixed mode and occurring at 266 Hz) and #8.

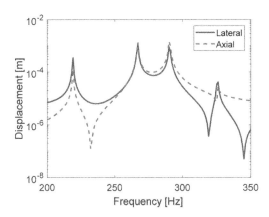

Figure 8. Shaft torque response with $\beta = 45°$.

3.4 Rotating system analysis

The final part of this study consists in a rotating analysis of the system with $\beta = 45°$ to better understand the impact of the rotation on this system's dynamics. Gyroscopic effects and centrifugal stiffening are included in this analysis and a complex eigen-values extraction is performed in the rotating frame of reference, as the assembly is not axisymmetric.

The resulting Campbell Diagram, showing the evolution of the natural frequencies over the rotational speed is provided in Figure 9. In agreement with literature [6, 16], mode #1 and #2 combine themselves into a forward and backward travelling wave due to the gyroscopic forces. As previously stated, blades behave as rigid bodies in these modes, and consequently no blades-shaft interaction is observed. Particular attention is to be paid to the next four modes. Modes #3, #5 and #8, occurring in the XY plane, combine with mode #4 in plane YZ (Figure 3c) into four travelling waves. Mode #3 evolves into a forward travelling wave given the decreasing trend of its frequency, whilst modes #4, #5 and #8 turn into backward travelling waves since their frequency increases. These modes have an additional interesting behaviour: the analysis in the XY plane from Figures 7b) and 8) has proved that axial, torsional and lateral vibration in the XY plane are fully coupled, but the gyroscopic effect then couples the bending vibration in XY and YZ planes as well, leading once more to a new family of coupling modes, with a shaft travelling wave that experiences axial and torsional vibration at the same time. Modes previously grouped under the label #9, have $ND > 1$, so they do not couple with the shaft.

Due to the low polar moment of inertia of the blades, the impact of the gyroscopic effect on the blade vibration is generally negligible. Instead, they show a moderate quadratic order increase in their natural frequencies due to centrifugal stiffening (see Figure 9b). Modes #6 and #7 are instead dominated by blades 1ND. Due to the high distance in frequency from the shaft bending modes and the low inertia, the coupling with the shaft is very low, so is the frequency splitting due to gyroscopic effect. On the other hand, they are also subjected to a moderate centrifugal stiffening, resulting in the behaviour shown in Figure 9b).

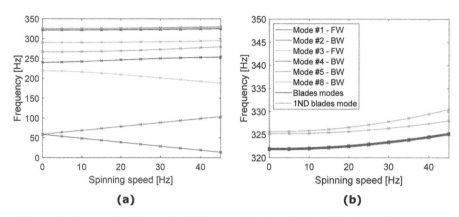

Figure 9. System full Campbell diagram in a) and detailed view of the blade modes in b).

4   DISCUSSION

Previous works from the authors [20-21] had discovered a novel coupling between axial and lateral vibration in a flexible shaft- flexible disc assembly. The obtained results showed that axial-lateral shaft modes can couple with disc 0ND and 1ND modes, leading to a new family of coupled modes. Gyroscopic effects led to Travelling Waves with axial components, which exhibit further coupling with disc 0ND and 1ND.

To better understand the impact of these new mode families on the response of a real turbo-machinery rotor, this work extended the previous findings to a shaft carrying a flexible bladed disc. A comparison with the previous results showed that while the natural frequencies of 0ND-1ND-2ND modes of the flexible disc were well separated, for the blades these NDs are dominated by the blades first flapping mode, so they occur at the same frequency (see Figure 4). Although the masses were kept the same, the mass distribution and consequently the mass moment of inertia were different. It is also worth noticing that the first edgewise bending mode of the blades has a lower frequency compared to the flexible disc first in-plane mode, leading to a light inter-action with the shaft torsional mode (Figure 3h), when the stagger angle $\beta = 0°$.

Considering the case with $\beta = 0°$ and asymmetric supports, it highlighted that the blade 0ND modes for the investigated blades can couple with shaft axial-bending modes, whilst blade 1ND modes show little coupling with shaft bending modes due to the high distance in frequency (325 to 190 Hz). It was also found that shaft torsion can couple with blade edgewise modes due to a torque transmission at the shaft disc interface. As expected in a bladed disc with no mistuned blades [10], NDs > 1 do not couple with the shaft.

With $\beta = 90°$ the behaviour was reversed: blade edgewise modes lightly coupled with axial-bending modes, whilst blade 0ND flapwise modes showed a strong coupling with shaft torsion, which was confirmed by the strong response shown in Figure 7b).

When the blades were staggered ($\beta = 45°$), the two coupling behaviours observed previ-ously become superposed: a blade 0ND mode vibration resulted in an interface axial force and torque transmission, which in turn drove simultaneously shaft axial-lateral and torsional modes, leading to an extremely complicated interaction between all modes.

When the system was rotating, all non-rotating modes involving bending deformation were affected by the gyroscopic terms and combining themselves into travelling waves; the observed modes involved shaft axial vibration components and blade 0ND and 1ND superposed patterns. 1ND modes played in this case much less of a role due to their strong frequency separation from the shaft modes. The combined effect of the axial-torsional-lateral coupling and the gyroscopic effect led to the appearance of travelling waves which also involved a torsional component.

To simplify all these complex interactions, Figure 10 summarises the observed behaviour: the gyroscopic effect provides coupling between bending modes in the two orthogonal planes, the bending supports provide coupling between axial and bending in the XY plane, whilst the stagger angle leads to coupling between blade 0ND and shaft axial and torsional modes, resulting in the full coupling described above.

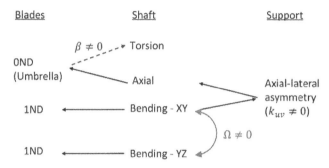

Figure 10. Full Coupling Diagram.

5 CONCLUSIONS

A flexible shaft carrying a flexible bladed disc and supported by an axial-lateral coupled bearing supporting structure has been modelled. Analogies and differences of the same system carrying respectively a flexible disc and blades have been presented, showing a strong effect of different mass distribution and dynamic properties on the coupling mechanism.

A series of interesting new coupling mechanisms were observed. Unusual Blade 0ND and shaft lateral coupling was detected, alongside a new shaft axial-lateral-torsional coupling to blade 0ND modes for staggered blades. This newly identified coupling phenomenon can lead to unexpected levels of vibration, highlighting the need for a fully coupled analysis for modern, more flexible designs, to ensure an efficient and safe operation.

ACKNOWLEDGEMENTS

The authors are grateful to Rolls-Royce plc for funding this work and granting permission for its publication.

REFERENCES

[1] Tiwari, R., A Brief History and State of the Art of Rotor Dynamics. Department of Mechanical Engineering, Indian Institute of Technology Guwahati, 781039
[2] Rolls Royce. The jet engine. John Wiley & Sons, 2005.
[3] Friswell, M. I., Dynamics of rotating machines, Cambridge University Press, 2010.

[4] Crawley, E. F., Mokadam, D. R. (1984). Stagger angle dependence of inertial and elastic coupling in bladed disks. Journal of Vibration, Acoustics, Stress, and Reliability in Design, 106(2),181–188.

[5] Crawely, E. F., Ducharme, E. H., Mokadam, D. R., Analytical and experimental investigation of the coupled bladed disk/shaft whirl of a cantilevered turbofan. Journal of Engineering for Gas Turbines and Power, 108(4),567–575.

[6] Ruffini, V., Schwingshackl, C. W., & Green, J. S. (2015). Prediction Capabilities of Coriolis and Gyroscopic Effects in Current Finite Element Software. In Proceedings of the 9th IFToMM International Conference on Rotor Dynamics (Vol. 21, pp. 1853–1862).

[7] Nissim, E. (1970). A Method for the Generation of Assumed Modes in Vibration Analysis. The Aeronautical Quarterly, 21(3),280–290.

[8] Huang, S. C., & Ho, K. B. (1996). Coupled shaft-torsion and blade-bending vibrations of a rotating shaft–disk–blade unit. Journal of Engineering for Gas Turbines and Power, 118(1),100–106.

[9] Yang, C.-H. H., & Huang, S.-C. C. (2007). Coupling vibrations in rotating shaft-disk-blades system. Journal of Vibration and Acoustics, Transactions of the ASME, 129(1),48–57.

[10] Chiu, Y.-J. J., & Huang, S.-C. C. (2007). The influence on coupling vibration of a rotor system due to a mistuned blade length. International Journal of Mechanical Sciences, 49(4),522–532.

[11] Chiu, Y. J., & Huang, S. C. (2008). The influence of a cracked blade on rotor's free vibration. Journal of Vibration and Acoustics, Transactions of the ASME, 130(5).

[12] Kim, K. T., & Lee, C. W. (2012). Dynamic analysis of asymmetric bladed-rotors supported by anisotropic stator. Journal of Sound and Vibration, 331 (24),5224–5246.

[13] Al-Bedoor, B. O. (2001). Modeling the coupled torsional and lateral vibrations of unbalanced rotors. Computer Methods in Applied Mechanics and Engineering, 190(45),5999–6008.

[14] Rządkowski, R., Drewczynski, M., Forced vibration of several bladed discs on the shaft. Proceedings of the ASME Turbo Expo, 5 PART B, 789–800.

[15] Chiu, Y.-J. J., Li, X.-Y. Y., Chen, Y.-C. C., Jian, S.-R. R., Yang, C.-H. H., & Lin, I.-H. H. (2017). Three methods for studying coupled vibration in a multi flexible disk rotor system. Journal of Mechanical Science and Technology, 31 (11),5219–5229.

[16] Ma, H., Lu, Y., Wu, Z., Tai, X., Li, H., & Wen, B. (2015). A new dynamic model of rotor–blade systems. Journal of Sound and Vibration, 357, 168–194.

[17] She, H., Li, C., Tang, Q., & Wen, B. (2018). The investigation of the coupled vibration in a flexible-disk blades system considering the influence of shaft bending vibration. Mechanical Systems and Signal Processing, 111, 545–569.

[18] Li, C., She, H., Liu, W., & Wen, B. (2017). The Influence of Shaft's bending on the Coupling Vibration of a Flexible Blade-Rotor System. Mathematical Problems in Engineering, 2017.

[19] Berger, S. S., Bonneau, O., Fre, J. et Al, Influence of Axial Thrust Bearing on the Dynamic Behavior of an Elastic Shaft: Coupling Between the Axial Dynamic Behavior and the Bending Vibrations of a Flexible Shaft. Journal of Vibration and Acoustics, 123(2),145–149.

[20] G. Tuzzi, C. Schwingshackl, J. Green, Investigation on coupling between disc umbrella mode and shaft bending modes in a rotating shaft-disc assembly, in: Proceedings of RASD - Recent Advances in Structural Dynamics, University of Southampton, 2019.

[21] G. Tuzzi, C. Schwingshackl, J. Green, Shaft bending to Zero Nodal Diameter disc coupling effects in rotating structures due to asymmetric bearing supports, Proceedings of IMAC, Society of Experimental Mechanics, 2020.

Integration of parameter sensitivity to structural optimization of helicopter rotors for minimum vibration

M.E. Bilen

Turkish Aerospace, Turkey

E. Cigeroglu, H.N. Özgüven

Department of Mechanical Engineering, Middle East Technical University, Turkey

ABSTRACT

Helicopters are notorious for their high vibration levels and the rotor systems are the main contributors to the problem. Rotor vibrations can be minimized by optimizing the rotor structure. However, the result of the optimization might require precise manufacturing of the rotor with tight tolerances leading to increased production cost and time. Moreover, due to loose manufacturing tolerances or wear from usage, the rotor might present higher vibrations during operation if the rotor vibration is sensitive to slight variations in the design variables. Hence, it is important to obtain a design point for minimum vibration that is also insensitive to such deviations. In this study, a four-bladed helicopter rotor is structurally optimized for minimum vibration and minimum blade mass. Along with these objectives, the sensitivity of each design point to variable deviations is calculated and combined into the objective function. It is aimed to obtain a design point resulting in minimum vibration and blade mass with the least variable sensitivity. Surrogate-based models are incorporated for the optimization to reduce objective function and sensitivity calculation times. For vibration minimization, vibration amplitudes along the rotor blades and at the rotor hub are considered. Furthermore, blade natural frequencies are separated from excitation frequencies to avoid any potential resonance. A comparative study is presented to provide the effect of parameter sensitivity.

NOTATION

c	Chord length	R	Rotor radius
$\mathcal{D}$	Differentiation operator	t	Lay-up thickness
e	Hinge offset distance from hub center	x	Design variable
E	Young's modulus	y_{CG}	Sectional center-of-gravity position
f	Objective function	y_{NA}	Sectional neutral axis position
F	Force	y_{SC}	Sectional shear center position
G	Shear modulus	γ	Lock number
L_b	Blade vibratory loading	ε	Sectional strain in blade span direction
L_h	Hub vibratory loading	ν	Poisson's ratio
m	Mass	ρ	Density
M	Moment	ω_i	i-th blade elastic natural frequency
p	Penalty function	Ω	Rotor rotational speed
r	Dimensional rotor station, penalty scaling factor	$/rev$	Frequency normalized to rotor rotational speed

261

1 INTRODUCTION

Helicopters are notorious for their high vibration levels. The main contribution to this problem comes from the helicopter rotors (1). During the design stage, one of the duties of helicopter rotor design engineers is to optimize the rotor to achieve minimum vibration while satisfying various static and dynamic constraints. However, the optimization will likely result in a precise design requiring a production with very tight tolerances.

For the production of helicopter rotor blades, composite materials are frequently utilized, and the manufacturing process involves a combination of computer-aided automatic processes and manual labour (2). Moreover, each of the blade components is manufactured separately and assembled at later stages. Therefore, the composite rotor blades are prone to tolerance stacking (3), which may lead to significant deviations from the optimum blade design parameters; hence, undesired rotor behaviour in terms of vibration and performance can be experienced.

Li et al. (4) proposed a methodology for the design and optimization of composite rotor blades. The methodology consists of an improved model for optimization, a parametric blade cross-section modeller, an efficient and accurate cross-sectional analysis, and an efficient optimization process. Authors constructed the objective function based on the minimization of the distances from shear and mass centers to aerodynamic center for flutter stability concerns. Moreover, the minimization of blade mass and stresses under operational loads were targeted, as well. The objective function was subjected to boundary constraints for torsion and bending stiffnesses, and blade mass. Furthermore, boundaries were introduced to limit the stress levels and structural coupling terms. As design variables, 17 parameters describing the cross-section geometry were selected. Sectional properties were obtained by using variational asymptotic beam sectional analysis (VABS) (5). The results lead to a decrease in the distance of the shear center and the mass center from the aerodynamic center by 0.49% and 2.87%, respectively. On top of that, the ratio of the maximum stress of the optimal design to the baseline design was found out to be 91%. Authors concluded that this methodology could be utilized to optimize realistic composite blades with manufacturability constraints.

Murugan and Ganguli (6) studied surrogate-based modeling for helicopter rotor blade optimization. Authors aimed to reduce helicopter vibrations via changing blade cross-sectional design parameters. For the surrogate-based modeling, polynomial function fitting was employed and for the optimization, Genetic Algorithm was used to avoid local minima. The optimization process was divided into two levels: in the first level, surrogate-based models were used to find promising regions; whereas, in the second level, the actual analysis routine were used. The authors concluded that with the use of surrogate-based modeling, the objective function can be predicted with satisfying accuracy while reducing analysis time significantly.

Bilen et al. (1) optimized a fully-articulated rotor for minimum vibration. For this purpose, two approaches were implemented namely, frequency separation approach (FSA) and direct vibration reduction approach (DVRA). Both approaches aim to minimize the blade mass while satisfying various constraints. FSA attempts to achieve minimum vibrations by separating blade natural frequencies from excitation frequencies to avoid resonances; whereas, DVRA directly targets the blade vibration amplitudes. Six variables were used for the blade cross-section shape and four variables were utilized for concentrated masses and their locations along the blades. To reduce the optimization time, surrogate-based models were implemented for blade structural property predictions. The rotor was optimized using both approaches and performances of them were compared against each other. Results show that DVRA performs better in vibration reduction while FSA provides lighter rotor blades. Moreover, the use of surrogate-based models provides significant savings in computational time while maintaining satisfactory accuracy.

Ritto et al. (7) proposed a methodology to optimize the performance of a flexible helicopter rotor-bearing system taking into account parameter uncertainties. For the minimization of rotor vibration, frequency separation approach was implemented. To integrate parameter uncertainties, probabilistic models were considered. According to the study, uncertainties obey a probabilistic distribution. The parameter uncertainty was modelled as an additional objective and both rotor and sensitivity objectives were optimized using different probability distributions. The study showed that there is not a significant difference between probability distributions, and the authors showed that the inclusion of parameter uncertainties leads to less sensitive optimal points.

In this study, a four-bladed, fully-articulated helicopter rotor is structurally optimized for minimum mass and vibration. For the optimization, three objectives are selected; namely, blade vibratory loads, hub vibratory loads, and blade mass. Moreover, various constraints are imposed to ensure optimal rotor performance. To reduce optimization times, surrogate-based models are generated for each parameter that is used to evaluate the rotor objective function. Since a deviation from the optimal point might adversely affect the optimization results, the sensitivity of the optimal point to such deviations are studied as well. A finite-difference based approach is implemented to model sensitivity. The method developed is integrated into the optimization problem as an additional optimization objective; hence, an augmented objective function involving both rotor objective and its parameter sensitivity is constructed. The contribution of rotor objectives and sensitivity is investigated by changing a weight coefficient.

2 ROTOR MODEL

2.1 Rotor topology
The rotor is fully articulated with four composite blades. In Table 1, basic rotor parameters are provided. The rotor is modeled using DYMORE; a finite-element-based multibody dynamics code (8). DYMORE can handle nonlinear flexible systems with arbitrary topologies where nonlinear elastic bodies are allowed to move arbitrarily with respect to each other. The following elements can be modelled; rigid bodies, cables, beams, shells, and joints. DYMORE employs Geometrically Exact Beam Theory (GEBT), which was introduced by Simo et al. (9) and later refined by Hodges (10). GEBT uses Generalized Timoshenko Beam Theory, which enables the beams to deform in axial, shear directions, to twist and to bend at the same time. Moreover, the theory enables the analysis of initially twisted and curved beams.

DYMORE is capable of calculating aerodynamic loads of rotor blades and/or wings internally. Simplified models based on lifting line theory and vortex wake models developed by Peters et al. (11) are used for modeling the aerodynamic loads. At each time step of the simulation, the aerodynamic loads acting on the system are computed based on the present configuration and are then used to evaluate the dynamic response. Bauchau (12) stated and showed that DYMORE can be utilized for modeling helicopter rotors. The accuracy of DYMORE in predicting the structural dynamics and the aerodynamic loads of helicopter rotors have been validated with existing test data (13)–(16). Utilizing DYMORE, the rotor topology shown in Figure 1 is modeled.

Table 1 . Basic rotor parameters.

Number of Blades	4	Airfoil Profile	NACA 0012
Rotor Speed, Ω	360 *rpm*	Pitch-Flap Coupling	11.9°, pitch up nose down
Rotor Radius, R	6000 *mm*	Chord, c	400*mm*, constant
Hinge Offset, e	5% R	Twist Rate	2°/m, linear, nose down
Root Cut-out	25% R	Tip Loss Factor	0.97

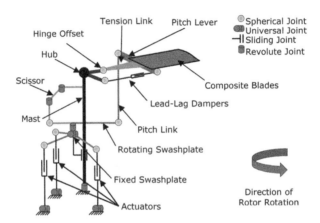

Figure 1. Topology of the Helicopter rotor.

Since the flexibility of the composite rotor blades is crucial for rotor dynamics, they are modeled using beam elements. The connection of the blades to the tension link and the pitch lever is done with a rigid interface. The tension link connects the blade to the hub via spherical joints that enable the articulation of rotor blades. The dampers are modeled using a spring and a dashpot with a spherical joint interface. The pitch link is modelled as a beam with spherical interfaces as they play an important role on the blade torsional mode frequencies. The fixed and the rotating swashplate are modeled as rigid bodies and the relative rotation is achieved using a revolute joint. The rotation of the rotor is achieved with a prescribed rotation and a scissor mechanism is used to transfer the rotor rotation to the rotating swashplate. The motions of servo actuators are prescribed so that the tilting of the swashplate, hence, the feathering of the blades can be controlled. For aerodynamic loads, a lifting line is defined from the blade root to the tip. As for the airfoil profile, NACA 0012 is selected, since it is the most frequently selected and investigated profile for rotor blades in the past (17).

Reference axis systems used for rotors are as follows; the x-axis is parallel to the far-field stream, positive downstream; the y-axis is perpendicular to the far-field stream; the z-axis is parallel to rotor mast, positive up. The hub loads are obtained with respect to this axis system. For the blade, the x-axis is coincident with the feathering axis and positive towards blade tip; the y-axis is coincident with chord line and positive towards leading edge; the z-axis is positive up. Blade sectional loads are obtained with respect to this reference frame. The reference axis systems for the hub and the blades are shown in Figure 2.

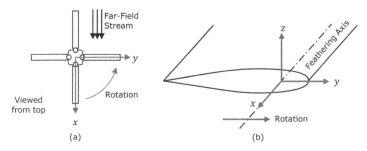

Figure 2 . Reference axis system for (a) the rotor and (b) the blades.

2.2 Blade cross-section design

This study focuses on tailoring of composite rotor blades to achieve the optimization targets. Blade cross-sections are modeled by PreVABS and are analysed using VABS. PreVABS is a pre-processing software that can effectively generate high-resolution finite element modeling data for VABS by directly using design parameters and composite laminate lay-up schema for rotor blades (18). Developed by Cesnik et. al (5), VABS can perform classical analysis for inhomogeneous, anisotropic beams with an initial twist and curvature having arbitrary reference and material properties yielding stiffness and mass matrices. Moreover, for a given loading, the three-dimensional stress and strain fields can be recovered using VABS (19). This study mainly deals with the structure of the functional region, which spans from 25%R to blade tip. The segment lying between the hinge offset and the beginning of the functional region is referred as the blade root and, in this study, the design of the blade root is taken as stiff and, during optimization, it is unchanged.

The cross-section of the functional region consists of carbon-fiber, glass-fiber, honeycomb, and titanium. Carbon-fiber lay-ups are used for skin. Glass-fiber lay-ups are used for the spar and the inner/outer wraps. Lay-up configuration for spar is [0] where the blade span axis is taken as the 0-degree lay-up angle. For the inner wrap and the outer wrap, the lay-up configurations are both [±45]. Moreover, two walls are placed in between inner and outer wraps for fortifying the cross-section. Wall lay-up configurations are [±45/0₂/∓45] glass-fiber. For avoiding crushing in the trailing edge zone, honeycomb is utilized. Titanium erosion shield is used for the protection of the blade from sand and dust abrasion. Nose block consists of 0-degree glass-fiber material. The lay-up configurations for upper and lower surfaces are identical to eliminate any coupling effects. A simple schematic is given in Figure 3. Composite material properties and their thicknesses are provided in Table 2.

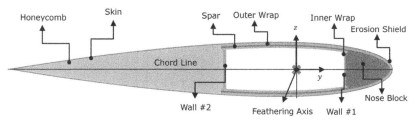

Figure 3. Cross-section design of the functional region with FEM mesh.

Table 2 . Material Properties used in the blade design.

	$E_{11}[GPa]$	$E_{22}[GPa]$	$E_{33}[(GPa]$	$G_{12}[GPa]$	$G_{13}[GPa]$	$G_{23}[GPa]$	v_{12}	v_{13}	v_{23}	$\rho[g/cm^3]$	$t[mm]$	Ref.
Glass–Fiber	45.6	16.2	16.2	5.83	5.83	5.79	0.28	0.28	0.4	1585	0.226	(20)
Carbon–Fiber	138	11	11	5.5	5.5	3.93	0.28	0.28	0.4	1750	0.133	(20)
Titanium	103	103	103	38.5	38.5	38.5	0.34	0.34	0.34	4510	0.798	(21)
Honeycomb	0.19	0.19	0.19	0.32	0.32	0.32	0.3	0.3	0.3	64.1	-	(21)

For the evaluation of the structural integrity of the blade, maximum stress and maximum strain approaches are used in the literature. Li et al. (4) used the stress approach for evaluating the structural integrity of the cross-sections; whereas, Isik and Kayran (22) used the maximum strain criterion for the optimization of the blade cross-section. In this study, maximum axial strain criterion is used for the strength evaluations. For the calculation of the strain, the centrifugal loading acting on the sectional center-of-gravity is considered. Since the functional region spans from 25% R to blade tip with identical cross-section design, the maximum centrifugal loading, F_C, occurs at 25% R and can be calculated as follows

$$F_C = m_f \Omega^2 R_f, \qquad (1)$$

where, m_f is the mass of the functional region and R_f is the spanwise center-of-gravity position of the functional region measured from the hub center.

For the calculation of Eq. (1), the rotor rotational speed is taken 125% higher to compensate for possible manufacturing defects, hot/wet conditions, material impurities, limit load considerations, and fatigue strength. Increased rotational speed also covers the forces and the moments other than centrifugal loading by creating a more conservative loading case. Moreover, the resulting centrifugal loading, F_C, is multiplied by a safety factor of 1.5 before strain calculations.

3   DESIGN VARIABLES

For the optimization of the rotor, six design variables are utilized controlling the shape of the cross-section of the blade in the functional region. The first two variables, x_1 and x_2, define the chordwise locations of Wall #1 and Wall #2, measured from the feathering axis. The third variable determines the lay-up angle of the skin such that the lay-up configuration of the upper and lower skin is [±45/±x_3/±45]. The integer variables, x_4, x_5, and x_6, are used for the number of plies in the inner wrap, the spar, and the outer wrap, respectively. All variables are illustrated in Figure 4.

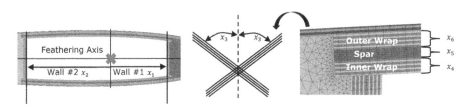

Figure 4. Illustration of design variables.

4   OPTIMIZATION OBJECTIVES AND CONSTRAINTS

For the structural optimization of the rotor, several objectives and constraints are implemented. In the following sections, each objective and constraint are detailed where x, written in bold, refers to a vector containing the design parameters i.e. $x = \{x_1, x_2, x_3, x_4, x_5, x_6\}^T$. The imposed constraints are handled using penalty function approach in which the optimization problem is penalized whenever a constraint violation occurs. Using the objective and penalty functions, rotor objective function is created as follows:

$$\overbrace{\phantom{\sum_{i=1}^{3} f_i(x)}}^{\text{objectives}} \quad \overbrace{\phantom{r\sum_{j=1}^{8} p_i(x)^2}}^{\text{penalties}}$$

$$f_{rtr}(x) = \sum_{i=1}^{3} f_i(x) + r \sum_{j=1}^{8} p_i(x)^2 , \qquad (2)$$

where $r = 100$ is the penalty scaling factor used for strict constraint enforcement. The objective and the penalty functions are normalized such that their values lie between zero and one; hence, their weight on the rotor objective function is equal. Normalization factors used are found during the surrogate-based modeling phase and detailed in the corresponding section. Subscripts *max* and *min* refer to the maximum and the minimum attainable value for a parameter within the design space, respectively.

4.1 OBJECTIVES

4.1.1 Blade mass
One of the objectives of this study is to minimize the blade mass m_b while reducing the vibration levels. The objective function is defined as follows:

$$f_1(x) = \frac{m_b(x) - m_{b,min}}{m_{b,max} - m_{b,min}} \qquad (3)$$

4.1.2 Blade Vibration
Reducing the blade vibratory loads is another target in this study. The loads acting on the blade are distributed and the reduction of blade vibration is achieved by the minimization of the maximum oscillatory load occurring along the blade span. Here, oscillatory load refers to the half-peak amplitude of the resulting forces and moments at a blade section. The oscillatory force on the x-axis is disregarded, as it is difficult to control the oscillatory force created in the centrifugal loading direction. The resulting oscillatory force and moment combination is expressed as follows:

$$L_b(x) = \sqrt{\left(F_{b,y}^{osc}(x)\right)^2 + \left(F_{b,z}^{osc}(x)\right)^2} + \sqrt{\left(M_{b,x}^{osc}(x)\right)^2 + \left(M_{b,y}^{osc}(x)\right)^2 + \left(M_{b,z}^{osc}(x)\right)^2}, \qquad (4)$$

where L_b represents the vibratory loading on the blade. The objective function for blade oscillatory load minimization is expressed as

$$f_2(x) = \frac{L_b(x) - L_{b,min}}{L_{b,max} - L_{b,min}} . \qquad (5)$$

4.1.3 Hub Vibration
Similar to the blade vibration objective, the hub vibration objective aims to minimize the oscillatory load components occurring at the center of the hub. The objective function for hub vibration minimization is as follows:

$$L_h(x) = \sqrt{\left(F_{h,x}^{osc}(x)\right)^2 + \left(F_{h,y}^{osc}(x)\right)^2 + \left(F_{h,z}^{osc}(x)\right)^2} + \sqrt{\left(M_{h,x}^{osc}(x)\right)^2 + \left(M_{h,y}^{osc}(x)\right)^2 + \left(M_{h,z}^{osc}(x)\right)^2}, \qquad (6)$$

where L_h represents the vibratory loading on the blade. The objective function for blade oscillatory load minimization is given as follows:

$$f_3(x) = \frac{L_h(x) - L_{h,min}}{L_{h,max} - L_{h,min}}. \tag{7}$$

4.2 CONSTRAINS

4.2.1 *Frequency separation*

During a steady flight, the rotor blades undergo periodic aerodynamic and inertial loading; therefore, they are exposed to excitations at frequencies corresponding to integer multiples of the rotor rotational speed. One of the major approaches for rotor vibration reduction is to separate the blade natural frequencies from the excitation frequencies to avoid resonances.

Blades are excited significantly up to frequencies tenfold of the rotor rotational speed, i.e. $10/rev$. However, the contribution of each excitation to the total vibratory load depends heavily on flight characteristics and rotor structure. In [15], Bilen et al. stated that due to resonances, significant vibration amplification can occur and the separation of frequencies is an effective measure for vibration reduction. In ideal conditions, where each natural frequency can be tailored independently, it is desirable to place each natural frequency of the blade $0.50/r$ away from the closest excitation frequency for minimum amplification. However, since the frequency of each mode is structurally coupled and there are other constraints on the blade design and production, it is unlikely to achieve this target. In this study, the frequency separation is targeted to be $\omega_{sep} = 0.20/rev$. The penalty function for frequency separation is expressed as follows;

$$p_{1,i}(x) = \begin{cases} \frac{|\omega_i(x) - \omega_{sep}|}{\omega_{sep}}, & |\omega_i(x)| < \omega_{sep}, \\ 0, & else \end{cases} \tag{8}$$

$$p_1(x) = \sum_{i=1}^{5} p_{1,i}(x), \tag{9}$$

where ω_i is the separation of the i-th blade natural frequency from the closest excitation frequency. In this study, the first five elastic blade modes are considered for separation where three of them are out-of-plane modes, one of them is an in-plane mode, and one of them is a torsional mode. The rigid modes do not contribute to vibration; hence, they are omitted. The natural frequencies are obtained under rotor operating conditions.

4.2.2 *Structural Integrity*

In this study, the maximum strain level over the cross-section in the spanwise direction, ε, is imposed as a constraint. The penalty function for the maximum strain level is taken as

$$p_2(x) - \begin{cases} 0, & \varepsilon(x) < \varepsilon_{lim} \\ \frac{\varepsilon(x) - \varepsilon_{lim}}{\varepsilon_{max} - \varepsilon_{lim}}, & \varepsilon(x) \geq \varepsilon_{lim} \end{cases}. \tag{10}$$

The maximum strain criterion is imposed on the carbon epoxy since it has the lowest ultimate tensile strength among other materials used in the blade cross-sectional modeling (23). Isik and Kayran (22) suggested that for carbon epoxy, the maximum longitudinal strain should be lower than $\varepsilon_{lim} = 5400\,\mu\varepsilon$.

4.2.3 *Lock Number*

Determined by the ratio of the aerodynamic forces to the inertial forces, many rotor performance parameters can be related to Lock number, γ (17). Higher Lock number

269

results in relatively higher aerodynamic forces on the blade causing higher flap motions of the blade (17). Calico (24) showed that the higher the lock number, the higher the tendency for the rotor blades to be unstable. Moreover, a low Lock number decreases the unnecessary flapping leading to improved fatigue lives of rotor components such as flapping bearings. In this study, Lock number below $\gamma_{lim} = 10$ is targeted. The penalty function is defined as

$$
p_3(x) = \begin{cases} 0, & \gamma(x) < \gamma_{lim} \\ \frac{\gamma(x) - \gamma_{lim}}{\gamma_{max} - \gamma_{lim}}, & \gamma(x) \geq \gamma_{lim} \end{cases}. \tag{11}
$$

4.2.4 *Torsional Deformation*

The tennis racket effect, the tension-torsion couplings of composite blades, and the aerodynamic moments generated by cambered airfoil sections are the main causes of the torsional deformation of the blades. This effect can be investigated in two parts: mean and cyclic (sinusoidal, once-per-revolution) deformations. Under torsional deformation, the blade feathering motion, dictated by the pilot to maneuver the helicopter, is degraded. The loss leads to an unnecessary increase in the pilot control to account for the deformation. Therefore, the blade needs to be torsionally tailored to compensate for this deformation. In this study, it is targeted that the mean and the cyclic deformations at 75%R do not exceed $\theta_{0,lim} = \theta_{1,lim} = 1.5°$. The penalty functions are expressed as follows:

$$
p_4(x) = \begin{cases} 0, & |\theta_0(x)| < \theta_{0,lim} \\ \frac{|\theta_0(x)| - \theta_{0,lim}}{\theta_{0,max} - \theta_{0,lim}}, & |\theta_0(x)| \geq \theta_{0,lim} \end{cases}, \tag{12}
$$

$$
p_5(x) = \begin{cases} 0, & \theta_1(x) < \theta_{1,lim} \\ \frac{\theta_1(x) - \theta_{1,lim}}{\theta_{1,max} - \theta_{1,lim}}, & \theta_1(x) \geq \theta_{1,lim} \end{cases}, \tag{13}
$$

4.2.5 *Flutter*

One of the instabilities that helicopter rotor blades can encounter is flutter, which is an aeroelastic instability involving coupled bending and torsional motion of the blades (17). This phenomenon can be avoided by bringing the sectional center-of-gravity, y_{CG}, closer to the leading edge than the sectional aerodynamic center (17). Moreover, the sectional shear center, y_{SC}, also plays an important role in flutter phenomenon, and the distance between aerodynamic center and y_{SC} should be minimum. y_{CG} and y_{SC} are measured from the feathering axis and the positions on the y-axis are considered only due to sectional symmetry about y-axis. The penalty function for y_{CG} is defined as follows:

$$
p_6(x) = \begin{cases} \frac{y_{CG}(x) - y_{CG,max}}{y_{CG,max}}, & y_{CG}(x) < 0 \\ 0, & y_{CG}(x) \geq 0 \end{cases}. \tag{14}
$$

For NACA 0012, the aerodynamic center can be assumed to be on the feathering axis; therefore, the penalty function for y_{SC} becomes

$$
p_7(x) = \begin{cases} \frac{|y_{SC}(x)| - y_{SC,max}}{y_{SC,max}}, & |y_{SC}(x)| 0 \\ 0, & |y_{SC}(x)| \geq < 0 \end{cases}, \tag{15}
$$

4.2.6 *Position of sectional neutral axis*

For cross-sections with anisotropic materials, it is likely that the neutral axis, y_{NA}, and y_{CG} do not coincide. The centrifugal loading acts on the y_{CG}; hence, the offset between y_{NA} and y_{CG} creates a steady bending moment under constant rotor rotational speed. To reduce the bending moment, it is targeted to minimize the distance between y_{CG} and y_{NA}, i.e., y_{NACG}. The penalty function is defined as

$$p_8(x) = \frac{|y_{NA}(x) - y_{CG}(x)|}{y_{CGNA,max}}. \tag{16}$$

5 INTEGRATION OF PARAMETER SENSITIVITY

In this study, helicopter rotor vibrations are minimized; also, during the optimization process, an insensitive design point is searched. To achieve this, a finite-difference based approach is implemented. The optimization process is numerical and the optimization problem cannot be expressed explicitly; therefore, the sensitivity at a design point can be estimated using numerical differentiation. In Figure 5, an example parameter sensitivity calculation for $f(x)$ at point $\tilde{x}$ is illustrated.

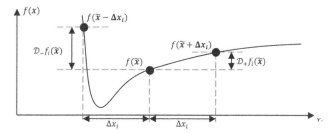

Figure 5. Illustration of sensitivity calculation.

The sensitivity for an i-th dimension at point $\tilde{x}$ is defined as the difference between the perturbed point and the original point and it can be calculated by forward and backward numerical differentiation using the following formulations

$$\mathcal{D}_+f_i(\tilde{x}) = f(\tilde{x} + x_i) - f(\tilde{x}), \tag{17}$$

$$\mathrm{D}_-f_i(\tilde{x}) = f(\tilde{x} - x_i) - f(\tilde{x}), \tag{18}$$

where $\mathcal{D}$ is the differentiation operator, $\mathcal{D}_+f_i$ is the forward differentiation, $\mathcal{D}_-f_i$ is the backward differentiation, i is the dimension of differentiation, and x_i is the perturbation vector. The perturbation vector x_i is constructed by perturbing the i-th dimension while the rest is zero.

Figure 5 shows that single forward or backward differentiation might underestimate the actual parameter sensitivity since a deviation can occur towards either way at any point within the specified perturbation interval. Hence, the differentiation is made at additional increments to minimize potential underestimation. For this study, the perturbation interval is divided into four discrete points on each side. The points, at which the differentiations are calculated, are illustrated in Figure 6 and the formulation for differentiation can be written as follows

$$\mathcal{D}_j f_i(\boldsymbol{x}) = f\left(\boldsymbol{x} + \frac{j}{4} \boldsymbol{x}_i\right) - f(\boldsymbol{x}) \; j = -4, -3, \ldots, +3, +4, \tag{19}$$

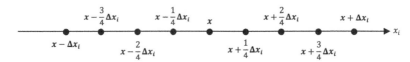

Figure 6. Illustration of differentiation points.

Additionally, at specific points, the function differentiation values might become negative; hence, to be conservative, the absolute values are considered. Therefore, in this study, the sensitivity of the objective function for the i-th dimension is calculated as

$$\mathcal{D}f_i(\boldsymbol{x}) = \max\left(\left|\mathcal{D}_j f_i(\boldsymbol{x})\right|\right) j = -4, -3, \ldots, +3, +4. \tag{20}$$

The total function sensitivity is found using

$$\mathcal{D}f(x) = \sum_{i=1}^{N} \mathcal{D}f_i(x), \tag{21}$$

where N is the problem dimension and $\mathcal{D}f(x)$ denotes the sensitivity of $f(x)$ at x. Since this study focuses on minimizing the objective function along with its sensitivity to the design parameters, Eq. (21) can be regarded as an additional objective function to be minimized. To be conservative, regardless of how large the perturbation magnitude is, each sensitivity value calculated within the perturbation range has equal weight on the calculation. In other words, a uniform probability distribution is assumed.

6   PREPARATION OF SURROGATE-BASED MODELS

In the preliminary design stages, the assessment of vibration requires high-fidelity models and solutions (25). Moreover, the achievement of minimum vibration is frequently done through optimization, which requires numerous analyses of the helicopter rotor with various design alternatives. Therefore, the process can be overwhelming in terms of computational time and effort. To reduce optimization times, surrogate-based models are utilized using a set of sample points, which are obtained by discretizing the design space in a grid manner. Each design variable has four discrete points, and through the combination of these points, 4096 points are sampled. In Table 3, the discrete points of design variables are given.

Table 3. – Discretization of design variables.

Variable	Description	Discrete Points				Lower Boundary	Upper Boundary	Unit
x_1	Wall #1 Location	20	37	54	71	20	71	mm
x_2	Wall #2 Location	-150	-100	-50	0	-150	0	mm
x_3	Skin Lay-up Angle	0	30	60	90	0	90	°
x_4	Number of Inner Wrap Plies	4	5	6	7	4	7	-
x_5	Number of Spar Plies	4	8	12	16	4	16	-
x_6	Number of Outer Wrap Plies	4	5	6	7	4	7	-

For each set of design points, the rotor undergoing a high-speed forward flight is analysed. Suitable pilot inputs are introduced to the rotor for realistic blade feathering motion. At the end of each analysis, the parameters given in Table 4 are calculated and recorded. Eventually, a set of sample points is generated. Using the recorded parameters below, the objective and the penalty functions, described previously, can be calculated.

Table 4. – Recorded parameters.

Parameter	Description	Parameter	Description
$m_b(\boldsymbol{x})$	Blade Mass	$\theta_0(\boldsymbol{x})$	Mean Torsional Deformation
$L_b(\boldsymbol{x})$	Blade Vibration	$\theta_1(\boldsymbol{x})$	Cyclic Torsional Deformation
$L_h(\boldsymbol{x})$	Hub Vibration	$y_{CG}(\boldsymbol{x})$	Sectional Center-of-Gravity Position
$\omega_{1..5}(\boldsymbol{x})$	Blade Elastic Natural Frequencies	$y_{SC}(\boldsymbol{x})$	Sectional Shear Center Position
$\varepsilon(\boldsymbol{x})$	Maximum Spanwise Sectional Strain	$y_{CGNA}(\boldsymbol{x})$	Difference between y_{CG} and y_{NA}
$\gamma(\boldsymbol{x})$	Lock Number		

Once the sample set is prepared, surrogate-based models are constructed for each recorded parameter using Gaussian Process Model (GPM). For GPM fitting, Matlab®'s Model-Based Calibration Toolbox is used (26). Bilen et al. showed that utilization of surrogate-based modeling provides accurate predictions of the recorded parameters and the prediction error is below 3% for most of the cases (27).

Since the sample data set used for surrogate-based modeling encapsulates the design space, it also gives an idea about the maximum and minimum attainable values of the recorded parameters, which are summarized in Table 5. These results are used for the normalization of objective and penalty functions mentioned in the previous section.

Table 5. – Minimum and maximum attainable values for the recorded parameters.

Minimum Value		Maximum Value		Unit	Minimum Value		Maximum Value		Unit
$m_{b,min}$	30.8	$m_{b,max}$	67.5	kg	ε_{min}	2265	ε_{max}	8278	$\mu\varepsilon$
$L_{b,min}$	3678	$L_{b,max}$	5390	–	γ_{min}	5.86	γ_{max}	16.5	–
$L_{h,min}$	11324	$L_{h,max}$	19497	–	$\theta_{0,min}$	-1.84	$\theta_{0,max}$	0.51	°
$\omega_{1,min}$	2.61	$\omega_{1,max}$	2.72	/rev	$\theta_{1,min}$	0.27	$\theta_{1,max}$	1.85	°
$\omega_{2,min}$	4.11	$\omega_{2,max}$	5.17	/rev	$y_{CG,min}$	-34.9	$y_{CG,max}$	17.1	mm
$\omega_{3,min}$	4.72	$\omega_{3,max}$	6.65	/rev	$y_{SC,min}$	-20.9	$y_{SC,max}$	34.1	mm
$\omega_{4,min}$	7.14	$\omega_{4,max}$	8.61	/rev	$y_{CGNA,min}$	-29.0	$y_{CGNA,max}$	13.5	mm
$\omega_{5,min}$	7.67	$\omega_{5,max}$	9.05	/rev					

In Figure 7, the objective function, $f_{rtr}(x)$, and corresponding sensitivity values, $\mathcal{D}f_{rtr}(x)$, are plotted for the sampled data set. The values for both $f_{rtr}(x)$ and $\mathcal{D}f_{rtr}(x)$ are in the similar order of magnitudes; therefore, there is no need to normalize $\mathcal{D}f_{rtr}(x)$. Moreover, as $f_{rtr}(x)$ decreases, the sensitivity at that point also decreases, indicating that, for the problem at hand, the minimization of the objective function tends to reduce its sensitivity. This is because the summation of objective functions can attain the theoretical value of 3 at most, i.e. $\max\left(\sum^{f_i}(x)\right) = 3$, and the rest of the values comes from the summation of penalty functions. At the higher objective function values, the contribution of the penalty functions increases quadratically as formulated in Eq. (2); hence, the objective function becomes more sensitive to the variable deviations.

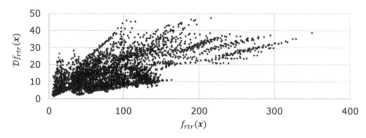

Figure 7. Objective Function and its sensitivity at sampled data points.

7 OPTIMIZATION PROBLEM

The optimization problem consists of the vibration and the blade mass minimization objectives, and several constraints to satisfy. Moreover, this study deals with the sensitivity of design parameters as an additional objective. To integrate the sensitivity, the following formulation is constructed

$$\phi(x) = wf_{rtr}(x) + (1 - w)[\mathcal{D}f_{rtr}(x)] \tag{22}$$

where $\phi(x)$ is the augmented objective function containing the rotor and the parameter sensitivity objective functions and w is the weight coefficient in the interval of $[0, 1]$. The parameter sensitivity objective function is calculated using Eq. (21) as follows

$$\mathcal{D}f_{rtr}(x) = \sum_{i=1}^{N} \mathcal{D}f_{rtr,i}(x), \tag{23}$$

where

$$\mathcal{D}f_{rtr,i}(x) = \max\left(\left|\mathcal{D}_j f_{rtr,i}(x)\right|\right) \quad j = -4, -3, \ldots, +3, +4. \tag{24}$$

By changing the weight coefficient w, the contribution of each function to the overall objective function $\phi(x)$ can be changed. For $w = 1$, the sensitivity of the optimization problem to parameter deviations is disregarded; while for $w = 0$, the least sensitive design point is searched. The augmented objective function $\phi(x)$ can be calculated by estimating the recorded parameters at requested design points using surrogate models and subsequently evaluating the objective and the penalty functions.

For wall locations x_1 and x_2, $\pm 2mm$ perturbations are used. The lay-up angle, x_3, is perturbed with $\pm 5°$. For the number of plies variables, x_4, x_5, and x_6, it is unlikely to manufacture with the wrong number of plies. However, the thickness of the plies may change slightly. To account for this, $\pm 10\%$ of a ply thickness is used for perturbations. Whenever a design variable is at its design boundary, the surrogate-based models are allowed to extrapolate outside the boundaries assuming that for small perturbations, the extrapolation does not introduce large prediction errors.

8 RESULTS AND DISCUSSION

The optimal design points, for given w, resulting in minimum augmented objective function value $\phi(x_{gb})$ are recorded, where x_{gb} refers to the global minimum. To observe the effect of the weight coefficient on the sensitivity of $f_{rtr}(x)$ and corresponding optimal design points, w is incrementally increased to various levels in the interval of $[0, 1]$. For each w, the optimization is repeated for at least 50 times to ensure the global minimum achievement. Matlab®'s Genetic Algorithm Toolbox is used as the optimization algorithm.

In Figure 8, the change of $\phi(x_{gb})$, $f_{rtr}(x_{gb})$ and $\mathcal{D}f_{rtr}(x_{gb})$ with w is provided. As w increases, the importance of the rotor objective function on the overall result increases since the optimization routine spends the available variable resources to minimize $f_{rtr}(x)$. In parallel with this, the rotor objective function becomes more sensitive to the parameter variations. For $w = 0$, where the most insensitive design is sought, the sensitivity is approximately 0.41; whereas, for $w = 1$, where sensitivity is disregarded, $\mathcal{D}f_{rtr}(x_{gb})$ becomes 2.29. This indicates an order of magnitude increase in the sensitivity of the rotor objective function. While w increases, $f_{rtr}(x_{gb})$ and $\mathcal{D}f_{rtr}(x_{gb})$

do not present a continuous trend; on the other hand, $\phi(x_{gb})$ shows a more continuous trend. When $w = 0$, the rotor optimization objective function $f_{rtr}(x_{gb})$ becomes 5.30; whereas, for $w = 1$, the value becomes 2.41. The change in the rotor objective function is much less compared to the change in the sensitivity objective function. The major jumps for $f_{rtr}(x_{gb})$ occurs for the weight coefficient between [0, 0.4]. After $w = 0.4$, the addition of sensitivity to the overall optimization problem does not affect the results significantly.

Although the optimization routine does not consider the rotor objective function for $w = 0$, the resulting $f_{rtr}(x_{gb})$ is relatively small when compared to the range of values it can attain, as shown in Figure 7. As the rotor objective function increases, it also becomes more sensitive to the perturbations. Therefore, the optimization routine unintentionally attempts to reduce the rotor objective function value along with the sensitivity objective function.

In Figure 9, the change of the objectives and the sum of penalty functions with the weight coefficient are presented. While w increases, the objective functions for the

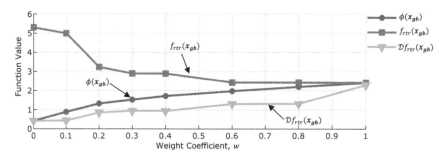

Figure 8. Change of $\phi(x_{gb})$, $f_{rtr}(x_{gb})$ and $Df_{rtr}(x_{gb})$ with weight coefficient.

blade and the hub vibrations present an increasing trend, as well. On the other hand, the blade mass objective function decreases with an increasing weight coefficient. For all weight coefficients, the penalties are almost satisfied with slight and acceptable violations. Similar to blade vibration objective function, increasing w leads to decreasing penalty function values. This is because the contribution of the penalty functions to the rotor objective function is significant due to the penalty scaling factor r.

The actual values used for the calculation of objective functions can be compared, as well. For $w = 0$, the blade mass is $49.5kg$ and for $w = 1$, it becomes $47.0kg$ indicating a slight decrease of 5%. For the blade vibration, the change is almost zero, it increases from $4{,}503N$ to $4509N$ for $w = 0$ and $w = 1$, respectively. The hub vibration increases from $12{,}677N$ to $12{,}999N$ and the corresponding percentage change is 2.5%. This shows that while the rotor objective function value increases with the addition of $Df_{rtr}(x_{gb})$ as an objective, the change in the actual physical values remain small. The rotor optimization was able to reduce the hub vibrations close to its minimum attainable value. The blade vibrations are slightly higher and the least minimization is realized with blade mass.

For the design variables, Figure 10 presents the change of normalized design variables with the weight coefficient. The largest change occurs for x_5, which corresponds to the

276

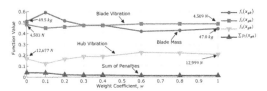

Figure 9. Change of rotor objective and sum of penalty functions with weight coefficient.

number of spar plies. For x_1, x_4, and, x_6, the change is moderate. The number of the inner and the outer wrap plies, x_4 and x_6, increased by one. In addition, the lay-up angle variable only varied within 2% range.

In Figure 11, the change of rotor objective function with variable perturbations are given. The results are presented for $w = [0, 0.4, 1]$, since these are the major values rep-

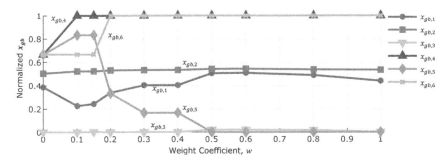

Figure 10. Change of normalized optimal Design variables with weight coefficient.

resenting the sensitivity objective function trend. The percent changes are calculated using the formulation below.

$$f_{rtr}\left(x_{gb}\right)\big|_w = \left(\frac{f_{rtr}\left(x_{gb} + x_i\right)}{f_{rtr}\left(x_{gb}\right)\big|_w} - 1\right) \times 100 \; i = 1, 2, \ldots, 6, \tag{25}$$

where $f_{rtr}\left(x_{gb}\right)\big|_w$ represents the change of rotor objective function with design variable variations, w is the weight coefficient at which the optimal values, x_{gb}, are obtained, x_i is the perturbation vector. Although affected by the change of the weight coefficient, the contribution of the design variables, except for x_3, to the sensitivity objective function is within 3%. However, for x_3, the sensitivity of the rotor objective function $f_{rtr}(x)$ is significantly large. For $w = 1$, the rotor objective function changes up to 86%; whereas, as the weight coefficient increases, the contribution of the sensitivity increases and the rotor objective function becomes less sensitive to x_3 deviations. For $w = 0$, the change is -5% at most. Although perturbation of $f_{rtr}\left(x_{gb}\right)$ may result in a decrease in the function value, the perturbed point becomes more sensitive to parameter variations. Therefore, the method presented in this study opts for the least sensitive region in both increasing and decreasing directions. Figure 11 shows that it is necessary to include the additional points within the perturbation interval. For the variable x_3 and $w = 0$, the most change occurs approximately at $-x_3/2$, signifying the importance of the method proposed in this study for the calculation of the sensitivity. The rotor objective function $f_{rtr}\left(x_{gb}\right)$ is most sensitive to skin lay-up angle variable x_3 which affects all of the recorded parameters except the blade mass, m_b, and center-of-gravity location, y_{CG}.

In Figure 12, the percentage change of the rotor objective functions and the sum of penalty functions with Δx_3 is presented. The percentage change is calculated using a similar approach used in Eq. (25). As expected, the blade mass minimization object-ive function, $f_1(x)$, is insensitive to x_3 variations. However, the blade and the hub vibra-tion objective functions $f_2(x)$ and $f_3(x)$ are affected from this variation approximately 5% and -8%, at worst, respectively. For the blade mass and vibration objective func-tions, the variations are almost independent of the weight coefficient. However, as w increases, the hub vibration objective function becomes more sensitive to variations. The most significant change occurs for the sum of penalty functions, $\sum^{p_i}(x_{gb})$. For $w = 1$, the change becomes 160% at most, indicating a remarkable increase in the vio-lations of the penalty functions. As w increases, the variation decreases due to the inclusion of the sensitivity to the overall objective function $\phi(x)$. The percentage change becomes 5% for $w = 0$, at most.

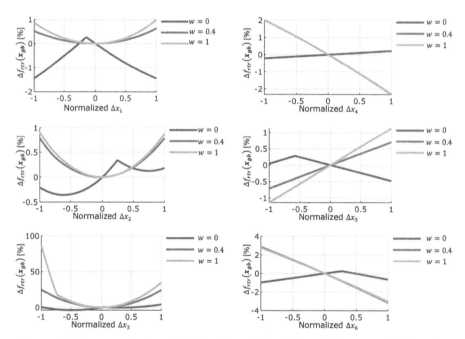

Figure 11. Change of rotor objective function with variable perturbations for different weight coefficients.

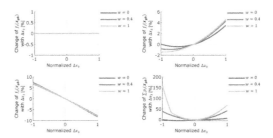

Figure 12. Change of rotor objective functions and sum of penalties with x_3 **perturbations for different weight coefficients.**

9   CONCLUSION

One of the main tasks of helicopter rotor design engineers is to minimize the helicopter rotor vibrations during the design stage. However, the optimal points determined might not be realized during the production stage. The deviation of the manufactured rotor from the designed rotor may lead to an increase in the vibration and/or the violation of the imposed constraints. In this study, the sensitivity of the rotor optimization objective function to such deviations is investigated. For this purpose, a rotor objective function is constructed. This function incorporates rotor and blade vibrations, and blade mass minimization. Moreover, several constraints are imposed to ensure satisfactory rotor performance. The integration of sensitivity is achieved first by perturbing each design variable and recording the change in the objective function. To minimize the potential underestimation of sensitivity, additional points are defined within the perturbation interval. The point resulting in the highest deviation is defined as the sensitivity of that design point. Finally, an additional objective function related to the sensitivity is derived to be minimized along with the rotor optimization objective function. To evaluate the contribution of the parameter sensitivities, a weight coefficient is used to control the relative contribution of the rotor and the sensitivity objective functions. Based on the outcomes of this study, the following conclusions can be drawn:

a) Optimal design of helicopter rotor structures may require very accurate production of rotor components which may lead to tight tolerances and increased cost.

b) Deviation from optimal design may lead to increased rotor vibration and/or constraint violations. To overcome this problem, an insensitive region can be searched along with rotor vibration minimization.

c) The sensitivity of a design point can be calculated by perturbing the rotor optimization objective function for each design variable. To capture the nature of the rotor optimization problem within the perturbation range, additional perturbation points can be used.

d) The optimization results show that it is possible to achieve insensitive regions through the addition of the sensitivity objective function. Moreover, the contribution of the sensitivity can be controlled by using a weight coefficient.

d) As the contribution of sensitivity increases, the rotor objective function value at the optimal point also increases; however, the sensitivity of the problem at that point decreases.

d) When the weight coefficient is selected such that only the least sensitive design is sought, it is possible to achieve a relatively small rotor objective function value. This is due to the fact that as the rotor objective function value increases, it also becomes more sensitive to the parameter variations.

REFERENCES

[1] M. E. Bilen, "DEVELOPMENT OF A HYBRID GLOBAL OPTIMIZATION ALGORITHM AND ITS APPLICATION TO HELICOPTER ROTOR STRUCTURAL OPTIMIZATION," Middle East Technical University, 2019.

[2] K. J. Kozaczuk, "Composite technology development based on helicopter rotor blades," *Aircr. Eng. Aerosp. Technol.*, 2018, doi: 10.1108/aeat-12-2017-0260.

[3] B. Marshall, P. Sherrill, B. K. Baskin, T. Hiros, and P. K. Oldroyd, "Rotor Blade and Method of Making Same," US8632310, 2014.

[4] L. Li, V. V. Volovoi, and D. H. Hodges, "Cross-Sectional Design of Composite Rotor Blades," *J. Am. Helicopter Soc.*, vol. 53, no. September 2016, p. 56, 2008, doi: 10.4050/JAHS.53.240.

[5] C. E. S. Cesnik and D. H. Hodges, "VABS: A New Concept for Composite Rotor Blade Cross-Sectional Modeling," *J. Am. Helicopter Soc.*, vol. 42, p. 27, 1997, doi: 10.4050/JAHS.42.27.

[6] M. S. Murugan, R. Ganguli, and D. Harursampath, "Surrogate based design optimisation of composite aerofoil cross-section for helicopter vibration reduction," *Aeronaut. J.*, vol. 116, no. 1181, pp. 709–725, 2012, doi: 10.1017/S0001924000007181.

[7] T. G. Ritto, R. H. Lopez, R. Sampaio, and J. E. Souza De Cursi, "Robust optimization of a flexible rotor-bearing system using the Campbell diagram," *Eng. Optim.*, vol. 43, no. 1, pp. 77–96, 2011, doi: 10.1080/03052151003759125.

[8] O. A. Bauchau, "DYMORE: a finite element based tool for the analysis of nonlinear flexible multibody systems," *Georg. Inst. Technol.*, pp. 1–17, 2001.

[9] J. C. Simo, "A finite strain beam formulation. The three-dimensional dynamic problem. Part I," *Comput. Methods Appl. Mech. Eng.*, vol. 49, no. 1, pp. 55–70, 1985, doi: 10.1016/0045-7825%2585%2690050-7.

[10] D. H. Hodges, *Nonlinear Composite Beam Theory*. American Institute of Aeronautics and Astronautics, 2006.

[11] D. A. Peters, S. Karunamoorthy, and W.-M. Cao, "Finite state induced flow models. I - Two-dimensional thin airfoil," *J. Aircr.*, vol. 32, no. 2, pp. 313–322, 1995, doi: 10.2514/3.46718.

[12] O. A. Bauchau, C. L. Bottasso, and Y. G. Nikishkov, "Modeling rotorcraft dynamics with finite element multibody procedures," *Math. Comput. Model.*, vol. 33, no. 10–11, pp. 1113–1137, 2001, doi: 10.1016/S0895-7177%2500%2600303-4.

[13] J.-S. Park, "Multibody analyses for performance and aeromechanics of a rotor in low-speed flight," *Aircr. Eng. Aerosp. Technol.*, vol. 86, no. 1, pp. 33–42, 2013, doi: 10.1108/AEAT-09-2012-0150.

[14] J. S. Park, S. N. Jung, Y. H. You, S. H. Park, and Y. H. Yu, "Validation of comprehensive dynamics analysis predictions for a rotor in descending flight," *Aircr. Eng. Aerosp. Technol.*, vol. 83, no. 2, pp. 75–84, 2011, doi: 10.1108/00022661111120962.

[15] J.-S. Park and Y. J. Kee, "Code-to-code comparison study on rotor aeromechanics in descending flight," *J. Mech. Sci. Technol.*, vol. 29, no. 8, pp. 3153–3163, 2015, doi: 10.1007/s12206-015-0714-9.

[16] C. B. Hoover, J. Shen, A. R. Kreshock, B. Stanford, D. J. Piatak, and J. Heeg, "Whirl Flutter Stability and Its Influence on the Design of the Distributed Electric Propeller Aircraft X-57," *17th AIAA Aviat. Technol. Integr. Oper. Conf.*, pp. 1–14, 2017, doi: 10.2514/6.2017-3785.

[17] W. Johnson, *Rotorcraft Aeromechanics*. Cambridge University Press, 2013.

[18] T. Hu, "A Validation and Comparison About VABS-IDE and VABS-GUI," 2012.

[19] W. Yu, V. Volovoi, D. H. Hodges, and X. Hong, "Validation of the variational asymptotic beam sectional analysis," *AIAA J.*, vol. 40, no. 10, pp. 2105–2112, 2002, doi: 10.2514/3.15301.

[20] P. D. Soden, M. J. Hinton, and A. S. Kaddour, "Lamina properties, lay-up configurations and loading conditions for a range of fibre reinforced composite laminates," *Fail. Criteria Fibre-Reinforced-Polymer Compos.*, vol. 58, pp. 30–51, 2004, doi: 10.1016/B978-008044475-8/50003-2.

[21] L. Li, "Structural Design of Composite Rotor Blades with Consideration of Manufacturability, Durability, and Manufacturing Uncertainties," 2008.

[22] A. A. Isik and A. Kayran, "Structural Optimization of Composite Helicopter Rotor Blades," *Am. Soc. Compos. Thirty-First Tech. Conf.*, vol. 91, pp. 399–404, 2016.

[23] D. D. Samborsky, T. J. Wilson, P. Agastra, and J. F. Mandell, "Delamination at Thick Ply Drops in Carbon and Glass Fiber Laminates Under Fatigue Loading," *J. Sol. Energy Eng.*, vol. 130, no. 3, p. 22, 2008, doi: 10.1115/1.2931496.

[24] A. Calico, W. E. Wiesel, W. Patterson, and A. Ohio, "Stabilization of Helicopter Blade Flapping," pp. 59–64.

[25] G. D. Padfield, *Helicopter Flight Dynamics: The Theory and Application of Flying Qualities and Simulation Modeling*. American Institute of Aeronautics and Astronautics, 2007.

[26] MathWorks, *Model-Based Calibration Toolbox^{TM} Reference*. 2018.

[27] M. E. Bilen, E. Cigeroglu, and H. N. Özgüven, "Evaluation of Surrogate-Based Modeling Methods for the Optimization of Helicopter Rotor Structures for Minimum Vibration," 2019.

A parametric study into the effect of variability in clearance shape and bump foil stiffness distribution in foil-air bearings

I. Ghalayini, P. Bonello

Department of Mechanical, Aerospace and Civil Engineering, University of Manchester, UK

ABSTRACT

The linearised force coefficients method (LFCM) has traditionally been used to study the stability and compute the Campbell diagrams of rotor/foil-air bearing (FAB) systems, despite reported disagreement with low-amplitude dynamics of the nonlinear system. In this paper, a new approach for extracting the Campbell diagram from the Jacobian matrix of the nonlinear state-space model that simultaneously couples the rotor, air film and foil structure domains is used for the first time in a parametric study into the effect of variability in clearance shape and bump foil stiffness distribution in foil-air bearings. The results provide guidelines for improved foil-air bearing design that raises the onset of instability speed.

1   INTRODUCTION

Experimental and numerical analysis show that the clearance and foil stiffness in foil-air bearings (FABs, also known as gas foil bearings, GFBs) play significant roles in the smooth operation of FAB-rotor systems since the system's onset of instability speed (OIS) varies significantly with varying nominal clearance and/or foil stiffness. The OIS is the speed at which the system becomes unstable with respect to small (linear) perturbations about its static equilibrium configuration, resulting in self-excited subsynchronous vibrations.

Lai et al [1] found that, for a turboexpander running on multi-decked protuberant GFBs with uniform radial clearance (i.e. circular clearance profile), a smaller radial clearance resulted in the subsynchronous vibrations being suppressed, although undersized clearance resulted in thermal runaway. Sim et al [2] similarly found that increasing the radial clearance of a circular profile FAB resulted in the reduction of the OIS, but the use of a lobed clearance profile can result in a delay in OIS for the same minimum clearance. A lobed clearance profile is defined by the mechanical preload, which is the difference between the nominal and minimum radial clearances [2] (the nominal clearance being the maximum clearance of the lobe). The lobed shape (typically with three lobes that are circumferentially equi-spaced) can be achieved either by shaping the bearing housing in this way, as done by Sim et al [2], or by the insertion of shims between the foil and the bearing housing at (typically three) equi-spaced circumferential locations [3-7]. The introduction of mechanical preload has the added benefit of enhancing the bearing load capacity [7], as is also the case when reducing radial clearance in a circular bearing [8], but this can come at the cost of increased friction torque unless the correct combination of preload and nominal clearance is used [2].

The main focus of the present study is the stability as defined above. Although it has been shown that a preload (shim thickness) of magnitude comparable to the nominal clearance (e.g. shim thickness 25 μm *vs* nominal radial clearance 32 μm) has the effect of delaying the OIS and suppressing subsynchronous in the unbalance response [5, 7], to the author's knowledge, no parametric study has been done so far that considered shims with different thicknesses and identified the optimum thickness. The present study addresses this gap and will also look into the effect on stability of introducing variability in the foil stiffness distribution.

NASA's researchers DellaCorte and Valco [9] categorised FABs into three different generations (I, II and III) according to their load capacity. These three generations are primarily distinguished by their foil structure design. Generation I bearings (1960's – 1970's) have a load capacity coefficient (defined in [9]) between 0.1 and 0.3 and their foil geometry is essentially uniform in both the axial and circumferential directions, leading to more or less uniform stiffness characteristics. Generation II bearings (1970's – 1980's) have a load capacity coefficient between 0.3 and 0.6 and the stiffness of the foil support structure varies in one direction - either axially (along the bearing length) or in the circumferential direction. Generation III bearings (1990's) have a load capacity coefficient between 0.8 and 1.0, and allow for both axial and circumferential tailoring of compliance in order to enhance bearing performance. The work in [9] focused on load capacity, but later work by Kim [10] also investigated theoretically the dynamic performance with variable stiffness distribution, considering four types of bump-foil type bearings: uniform radial clearance (circular) bearing with single foil pad of uniform stiffness; circular single-pad bearing with stiffness variation in the axial direction; lobed clearance (preloaded) bearing with three pads of uniform stiffness; preloaded three-pad bearing with pad stiffness variation the circumferential direction. Hence, for the circular single-pad bearing, stiffness variation in the circumferential direction was not considered in [10]. However, more recent theoretical work by Nielsen and Santos [11] can be regarded as involving stiffness variability in the circumferential direction since it investigated the effect of removing a section of the bump foil which supports the top foil (see Figure 1). The large sagging effect in the unsupported area creates a "shallow pocket" similar to that found in lobed bearings [11]. It was shown in [11] that proper placement of the unsupported area results in the elimination of the subsynchronous vibrations associated with large unbalance. Paradoxically however, it is also reported in [11] that this resulted in a slight *reduction* in OIS relative to the unmodified design (note that OIS is defined for the "no unbalance" condition). The best placement of the unsupported area was also the most heavily loaded region. The researchers in [11] say that this may be an issue and recommend an investigation into the stresses in the top foil.

A linearisation method provides a fast means of computing the stability and modal characteristics (including Campbell diagrams) of free perturbations of rotor systems in a parametric study involving different bearing designs. Up to now, the linearisation method used in studies on variable clearance and stiffness e.g. [10, 2, 7] has been the linear force coefficients method (LFCM), which was originally introduced by Lund [12] for rigid sleeve fluid bearings and adapted to foil bearings by Peng and Carpino [13]. However, it has been well documented e.g. [10, 14, 15] that the use of the LFCM (which works in the frequency domain) to FABs results in disagreement with low-amplitude dynamic analysis of the nonlinear system using the more time consuming approach of numerical integration in the time domain (for the orbit response). In fact, the aforementioned work by Kim [10] reported that the OIS predictions by LFCM were much lower than those determined from the orbit responses (which are the correct ones). Later works [14, 15] reported lower discrepancies, which nonetheless persist and increase with increasing foil compliance. The study in [15] provides evidence that these discrepancies are due the inaccurate consideration of the foil deflection in the linearisation process for the bearing force coefficients. Also, it is noted that the LFCM requires a spring-damper model for the foil, which means it is unsuited to more complicated foil models, which is why only time integration was used in the aforementioned study by Nielsen and Santos [11] (which used a bilinear model for the foil). The linearisation approach introduced by Bonello and Pham in [16] avoids the use of bearing force coefficients since it is based on the Jacobian matrix of the nonlinear dynamical system (i.e. the coupled rotor/air film/ foil equations). It therefore eliminates the aforementioned errors associated with the LFCM, and is thus perfectly consistent with the results of low-amplitude orbit response simulations, while being much faster [16-18]. The linearisation approach in [16] was initially used for the determination of the OIS only, but it has recently been developed by Bonello [17, 18] to provide a full modal analysis (including Campbell diagrams).

The novel contribution of this paper is therefore the use of a proven robust linearisation method for stability and modal analysis [17, 18] in a parametric study of the design of a single-pad FAB which considers:

- different preloads;
- foil stiffness variation in the circumferential direction.

2   DYNAMICAL SYSTEM MODEL

Figure 1 shows the FAB-rotor system considered, which is the same as that in [17]. The system parameters can be found in Table 1.

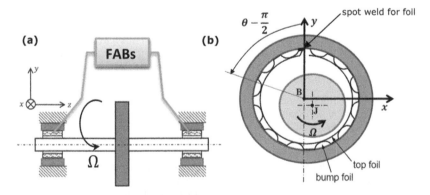

Figure 1. System considered: (a) symmetric rigid rotor-FAB system; (b) single-pad 1st generation FAB (configuration shown has a clamped leading edge/free trailing edge, as in [11]).

Table 1. FAB-rotor system parameters.

FAB-rotor system parameters	
Air viscosity μ (Ns/m^2)	1.95×10^{-5}
Air atmospheric pressure, p_a (Pa)	101,325
Damping Loss factor, η	0.25
Rotor half-mass, M (kg)	3.061
Nominal radial clearance, c (m)	32×10^{-6}
Bearing radius, R (m)	19.05×10^{-3}
Bearing length, L (m)	38.1×10^{-3}
Bump radial stiffness/Projected area, K_b (N/m^3)	4.739×10^9 (nominal value, $K_{b,nom}$)

The governing system of equations comprises the state-space (i.e. time-based first order differential) equations of the air film, foil structure and rotor and can be expressed in state-space form as [17, 18]:

$$s' = x(\tau, S) \tag{1}$$

where $()'$ denotes differentiation with respect to the time variable τ, s is the state vector which contains state variables from all three domains (air film, foil and rotor), and χ is a vector of nonlinear functions of τ and s. The specific form of eq. (1) for the symmetric rigid rotor-FAB system considered in Figure 1 is:

$$
\begin{bmatrix} \psi \\ \tilde{w} \\ \varepsilon \\ \varepsilon' \end{bmatrix}' =
\begin{bmatrix} g_{RE}(\psi, \tilde{w}, \varepsilon) \\ 2\left(P_{g,\theta}(\psi, \tilde{w}, \varepsilon)/(cK_b) - \tilde{w}\right)/\eta \\ \varepsilon' \\ \frac{4}{Mc\Omega^2}\{f_u(\tau) + f_s + f_J(\psi, \tilde{w}, \varepsilon)\} \end{bmatrix} \tag{2}
$$

where $\tau = \Omega t/2$ (Ω being the rotational speed) and the subvectors ψ, $\tilde{w}$, $(\varepsilon, \varepsilon')$ respectively contain state variables relating to the air film, the foil, and the rotor, where $\varepsilon = [x/c \ \ y/c]^T$ is the journal eccentricity vector (c being the nominal radial clearance of either FAB).

In eq. (2), f_u, f_s, f_J are 2×1 vectors containing the Cartesian components of the unbalance force, static load and air film force respectively, acting on one half of the symmetric rotor of total mass of $2M$.

The second row of eq. (2) gives the equation governing the deflection of the single-pad foil structure, according to the simple equivalent foundation model (SEFM), which only considers the stiffness and damping of the bump foil for the purpose of determining the local radial deflection w of the top foil (i.e. top foil sagging in-between bumps is neglected, as is top foil detachment from the bump foil). It is additionally assumed that w is a function only of the angular coordinate θ. $\tilde{w}$ is the $N_\theta \times 1$ vector of values $\tilde{w}_j = \tilde{w}(\theta_j)$, where $\tilde{w} = w/c$ and $\theta_j, j = 1, \ldots, N_\theta$, are discrete values of θ according to the Finite Difference (FD) grid used to discretize the Reynolds Equation (RE). K_b is the stiffness per unit area of the bump foil and the equivalent viscous damping coefficient is assumed to be $C_{damp,eq} = K_b\eta/\Omega$. $p_{g,\theta}$ is the vector of the averages of the gauge pressure p_g over the axial direction for discrete values of θ.

The first row of eq. (2) is the FD-discretised form of the RE [19]. The RE is formulated in terms of the combined state variable $\psi \equiv \tilde{p}\tilde{h}$ where $\tilde{p}$ and $\tilde{h}$ are the non-dimensional air film pressure and thickness at a position (ξ, θ) where ξ is the non-dimensional axial coordinate. The FD grid has $N_\xi \times N_\theta$ points spaced by $\Delta\xi$, $\Delta\theta$ in the axial and angular directions respectively, where $\xi = \xi_i$, $i = 1,\ldots, N_\xi$ and $\theta = \theta_j$, $j = 1,\ldots,N_\theta$. Hence, in eq. (2), ψ is the $N_\xi N_\theta \times 1$ vector of variables $\psi_{i,j} = \psi(\xi_i, \theta_j, \tau)$. The grid covers only half the axial length of the FAB due to symmetry. The atmospheric pressure constraint $\psi \equiv \tilde{h}$ is applied at the weld location (Figure 1) and therefore the air film model used is that described as "finite θ" in [19]. Additionally, the Gümbel condition [19] is applied wherein subatmospheric gauge pressures are disregarded when integrating p_g for f_j. The Gümbel condition is an assumed correction for the detachment of the top foil that is frequently used e.g. [11, 14, 15]. It should be noted however that while it has been shown to be satisfactory for single-pad FABs operating with a clamped leading edge/ free trailing edge (CLE/FTE, as shown in Figure 1b) [11], it may not be appropriate for a free leading edge/clamped trailing edge (FLE/CTE, which is the typical operating mode of single pad FABs) [19].

Eq. (2) can be analysed for the following:

a) The nonlinear response using transient nonlinear dynamic analysis (TNDA), which involves numerical integration from prescribed initial conditions $s(\tau = 0)$;
b) Free linearised vibration about the static equilibrium configuration $s = s_E$ at a given rotational speed Ω, to determine the onset of instability speed (OIS) and the Campbell diagrams.

This paper is concerned with the analysis of type b) applied to variable clearance and variable stiffness. These variations are described in section 3. The remainder of this section gives an overview of the linearisation procedure.

Following [17, 18], the static equilibrium condition $s = s_E$ can be obtained by setting $s' = 0$ for $f_u = 0$ in eq. (2) and solving the resulting set of nonlinear algebraic equations

$$\chi(0, s)|_{f_u=0} = 0 \tag{3}$$

The equation governing small (linear) free perturbations Δs about $s = s_E$ is obtained by substituting $s = s_E + \Delta s$ into eq. (1) and expanding the right hand side using the Taylor series, retaining only first order terms:

$$(\Delta s)' = J(\Delta s) \tag{4}$$

where J is the system Jacobian, defined as

$$J = \left.\frac{\partial \chi}{\partial s}\right|_{f_u=0, s=s_E} = \begin{pmatrix} \frac{\partial \chi_1}{\partial s_1} & \cdots & \frac{\partial \chi_1}{\partial s_{N_s}} \\ \vdots & \ddots & \vdots \\ \frac{\partial \chi_{N_s}}{\partial s_1} & \cdots & \frac{\partial \chi_{N_s}}{\partial s_{N_s}} \end{pmatrix}\Bigg|_{f_u=0, s=s_E} \tag{5}$$

where $s = [s_1 \ \cdots \ s_{N_s}]^T$, $\chi = [\chi_1 \ \cdots \ \chi_{N_s}]^T$. The general solution of eq. (4) is of the form

$$\Delta s = \sum_{k=1}^{N_s} C_k v_k e^{\lambda_k \tau} \tag{6}$$

where λ_k denote the eigenvalues of $\mathbf{J}$, $\mathbf{v}_k$ the associated eigenvectors and C_k arbitrary scalars. In practice, only the oscillatory part of eq. (6) is of interest and this is comprised of N_o pairs of complex conjugate eigenvalues $\lambda_{n,\mathrm{Re}} \pm \mathrm{j} \lambda_{n,\mathrm{Im}}$, $\lambda_{n,\mathrm{Im}} > 0$ ($n = 1 \ldots N_o$) and associated conjugate eigenvectors $\boldsymbol{\rho}^{(n)}$, $\boldsymbol{\rho}^{(n)*}$. The damped natural circular frequency, equivalent viscous damping ratio, and the undamped natural circular frequency of oscillatory mode no. n are given by

$$\varpi_{\mathrm{d},n} = \Omega \lambda_{n,\mathrm{Im}}/2, \zeta_n = -\lambda_{n,\mathrm{Re}} \Big/ \left(\lambda_{n,\mathrm{Re}}^2 + \varpi_{\mathrm{d},n}^2 \right)^{0.5}, \varpi_{\mathrm{u},n} = \varpi_{\mathrm{d},n} \Big/ \left(1 - \zeta_n^2 \right)^{0.5} \tag{7a, b, c}$$

The OIS is the speed for which the damping ratio ζ_n of one of the oscillatory modes becomes negative, marking the initiation of self-excited vibration (Hopf bifurcation) [18]. As shown in [17, 18], for the vast majority of the N_o oscillatory modes, the vibration of the journal is negligible relative to that of the foil pad. On the other hand, the Campbell diagram is only concerned with modes involving significant rotor vibration (whirl). Hence, for the purpose of extracting the Campbell diagram from the eigenfrequency vs speed map, a filtering technique is applied. As in [17, 18], each eigenvector is scaled so that the greatest vibration amplitude within the FAB, considering both pad and journal, is 20% of the nominal radial clearance, and the scaled mode is rejected if the amplitude of the journal vibration (average of x, y vibrations) it is not greater than 10% of this. A filtering criterion based on an upper limit for ζ_n may additionally be applied where appropriate ($\zeta_n < 0.7$ in this paper).

3 MODIFICATIONS TO CLEARANCE AND FOIL STIFFNESS PROFILES

3.1 Variable clearance shape

Based on the assumptions stated in the previous section, the non-dimensional air film thickness is given by the following expression

$$\tilde{h}(\theta) = \frac{c_s(\theta)}{c} - \varepsilon^{\mathrm{T}} \begin{bmatrix} \cos \theta \\ \sin \theta \end{bmatrix} + \tilde{w}(\theta) \tag{8}$$

where $c_s(\theta)$ is the function defining the actual clearance including the effect of the shimming. This latter function is assumed to follow the following form:

$$c_s(\theta) = c - \frac{t_s}{2} \{ 1 + \cos(N(\theta - \theta_U) + \pi) \} \tag{9}$$

where N is the number of shims (taken to be odd) and θ_U is a chosen value of θ where $c_s(\theta) = c$, the nominal radial clearance (i.e. $\theta = \theta_U$ is a location which is unaffected by the shimming). Figure 2a illustrates schematic views of the modified FAB for $N = 3$ shims and $\theta_u = \pi/2$ (i.e. the weld location). Figure 2b shows the corresponding functional variation of c_s for a nominal clearance (c) of 32 μm and shim thickness values t_s ranging from 0 to 25 μm; for the latter shim thickness it is seen that the actual radial clearance is reduced to a minimum of 7 μm at three equi-spaced circumferential locations.

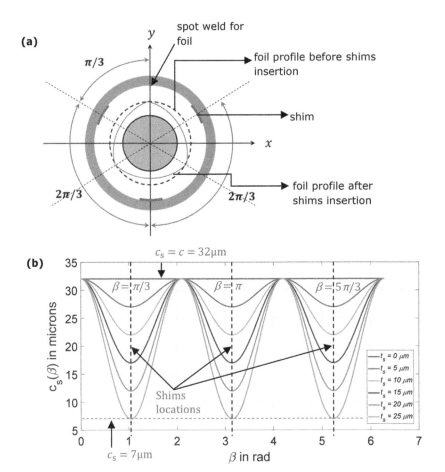

Figure 2. (a) Modified shimmed FAB (for illustration only). (b) Effect of variability in clearance shape in foil-air bearings ($\beta = \theta - \pi/2$).

3.2 Variable stiffness distribution

In this part of the analysis, it is assumed that the stiffness per unit area of the pad K_b is a function of the angular coordinate θ. Similarly to the approach taken in the previous section with respect to shimming, the assumed functional variation of K_b is taken to be such that the stiffness has maximal deviation of ΔK_b from the nominal value $K_{b,nom}$ at N circumferential locations (N assumed odd). Two cases are considered, as follows:

Case 1, $\Delta K_b \geq 0$:

$$K_b(\theta) = \begin{cases} K_{b,nom} + \Delta K_b \cos[N(\theta - \theta_{MD})], & K_{b,nom} + \Delta K_b \cos[N(\theta - \theta_{MD})] \geq K_{b,nom} \\ K_{b,nom}, & K_{b,nom} + \Delta K_b \cos[N(\theta - \theta_{MD})] < K_{b,nom} \end{cases} \quad (10)$$

Case 2, $\Delta K_{\mathrm{b}} \leq 0$:

$$K_{\mathrm{b}}(\theta) = \begin{cases} K_{\mathrm{b,nom}} + \Delta K_{\mathrm{b}}\cos[N(\theta - \theta_{\mathrm{MD}})], & K_{\mathrm{b,nom}} + \Delta K_{\mathrm{b}}\cos[N(\theta - \theta_{\mathrm{MD}})] \leq K_{\mathrm{b,nom}} \\ K_{\mathrm{b,nom}}, & K_{\mathrm{b,nom}} + \Delta K_{\mathrm{b}}\cos[N(\theta - \theta_{\mathrm{MD}})] > K_{\mathrm{b,nom}} \end{cases} \quad (11)$$

In the above equations, $\theta = \theta_{\mathrm{MD}}$ is a chosen location of maximum deviation in stiffness from the nominal i.e. where $K_{\mathrm{b}}(\theta) - K_{\mathrm{b,nom}} = \Delta K_{\mathrm{b}}$. Figures 3a, b illustrate the variation of K_{b} for cases 1, 2 respectively with $\theta_{\mathrm{MD}} = 3\pi/2$ and different values of ΔK_{b} ($K_{\mathrm{b,nom}}$ given in Table 1). The location $\theta_{\mathrm{MD}} = 3\pi/2$ is chosen since the researchers in [11] found that removing a section of the bump foil approximately in this region resulted in the suppression of sub-synchronous vibration in the TNDA response at high unbalance (as stated in the Introduction). In the present case, rather than removing a section of the bump foil, it is assumed that the pad stiffness can be reduced (or increased) in certain regions.

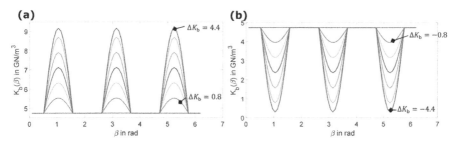

Figure 3. Bearing stiffness variation for $|\Delta K_{\mathrm{b}}| = 0, 0.8, 1.6, 2.4, 3.2, 4, 4.4\mathbf{GN/m^3}$: (a) Case 1 ($\Delta K_{\mathrm{b}} > 0$) (b) Case 2 ($\Delta K_{\mathrm{b}} < 0$). ($\beta = \theta - \pi/2$).

4   DISCUSSIONS OF SIMULATION RESULTS

In the simulations performed, the grid size (covering half axial length of the FAB) was 7×71 as per procedure explained in [19].

Figure 4 shows the loci of the static equilibrium positions (SEPs) of the journal for the speed range 5-30 krpm for two different cases: (1) when journal is supported by unshimmed FABs; (2) when journal is supported by shimmed FABs with shim thickness of 25 μm. The static equilibrium journal positions of the shimmed FABs are at a higher level relative to those of the unshimmed FABs due to the chosen positioning of the shims. The shims alter the ("undeformed") clearance profile from a circular shape to a three-lobed shape ("undeformed" refers to the condition of the foil pad prior to the application of air film pressure induced by the rotation and loading effects). It is noted that there is a deformed clearance profile corresponding to each journal SEP in Figure 4, but such profiles are not shown for clarity.

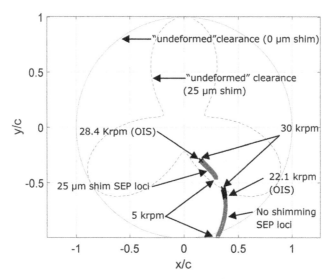

Figure 4. Loci of static equilibrium positions (SEPs) of the journal over a range of speeds for the unshimmed and shimmed FAB cases (stable positions indicated as red squares).

Free perturbations about the static equilibria in Figure 4 are now considered. Figures 5a, b respectively show the unfiltered eigenfrequency $f_{d,n}(=\varpi_{d,n}/(2\pi))$ *vs* speed map for the unshimmed and shimmed cases. Each line of dots represents all the modes at a given speed, of which very few involve significant rotor vibration, (section 2). Beyond the OIS (22.1 krpm for unshimmed FAB and 28.4 krpm for shimmed FAB) the dots change colour, as shown in Figures 4 and 5, indicating instability in one of the modes. Through the application of the filtering criteria (section 2), the "clean" Campbell diagrams for the two cases (0, 25 μm shims) are extracted and these are presented on the same axes in Figure 6a. As can be seen, for each case, there are only two modes ("Campbell modes") at each speed – mode 1 with the journal in forward whirl, mode 2 with the journal in reverse whirl for most of the speed range (transition to reverse whirl occurs beyond 8.5 krpm for unshimmed case and beyond 5 krpm for 25 μm shims). In either case, instability is due to mode no. 1, i.e. its viscous damping ratio becomes negative beyond a certain speed (see Figure 6b). The OIS values in this paper are determined as the first speed in steps of 0.1 krpm to register an instability, whereas in [17] the increment used to determine the OIS was the same as that used in the plots (0.5 krpm). It is for this reason that the OIS for the unshimmed case is presented as 22.1 krpm here and 22.5 krpm in [17]. Also, this OIS agrees with that found by Kim [10] (~24 krpm) through the long method of time integration for orbit response at different speeds, the minor difference being attributed mainly to modelling differences (e.g. the additional consideration in [10] of foil deflection variation in the axial direction).

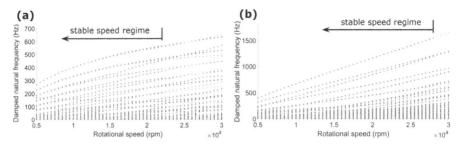

Figure 5. Unfiltered eigenfrequency $f_{d,n} \left(= \varpi_{d,n}/(2\pi) \right)$ **vs speed map for a shimmed FAB of configuration as shown in Figure 2a with shim thickness of: (a) 0 μm (b) 25 μm.**

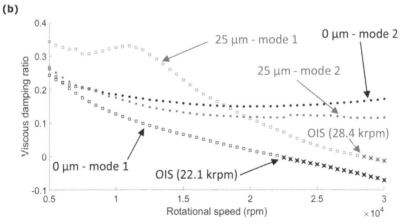

Figure 6. (a) Campbell diagrams extracted from the unfiltered eigenfrequency *vs* **speed map of Figures 5a and 5b i.e. for FABs with shim thickness of 0 μm, 25 μm respectively; (b) corresponding viscous damping ratio** *vs* **speed maps. Legend: forward whirl (□); reverse whirl (●); unstable mode points overlaid with a cross; EO (engine order).**

Figure 6a shows that shimming by 25 μm results in a substantial increase in the frequency of mode 1 for a given speed. This agrees with the observation [7] that shimming increases the rotor-bearing system's critical speed. The shimming also significantly increases the damping ratio of mode 1 (i.e. the mode becoming unstable), which is why the OIS increases substantially. Figure 7 shows Campbell modes nos. 1 and 2 for the unshimmed (a1, a2) and shimmed (b1, b2) at speeds immediately prior to the respective OIS. In either case, mode 2 is reverse whirl and significantly more damped than mode 1. It is noted that the journal/pad displacements shown in such plots comprise the actual static component plus the suitably scaled modal component with the exponential decay/growth factor in eq. (6) omitted.

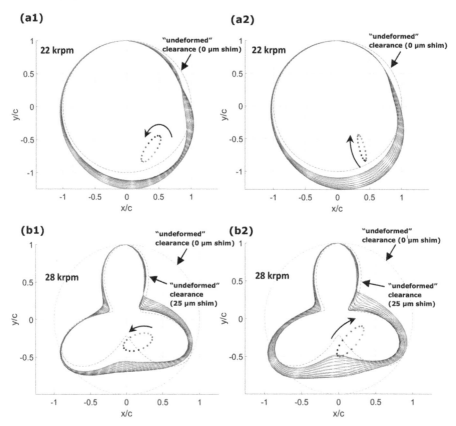

Figure 7. Modes 1 and 2 taken from Campbell diagrams in Figure 6 at speeds immediately prior to the OIS: (a1, a2) unshimmed FAB (mode 1 -
$f_{d,n} = 99.59\text{Hz}, \zeta_n = 0.0025327$, **forward whirl; mode 2 -** $f_{d,n} = 135.33\text{Hz}, \zeta_n = 0.14976$,
reverse whirl); (b1, b2) FAB with 25 μm shims (mode 1 -
$f_{d,n} = 137.33\text{Hz}, \zeta_n = 0.0039093$, **forward whirl; mode 2 -** $f_{d,n} = 165.84\text{Hz}, \zeta_n = 0.11598$,
reverse whirl) (NB: phase of vibration indicated by colour sequence black-red-green-magenta).

Table 2. Summary onset of instability speed (OIS) predictions for preload variation analysis.

Shim thickness	Onset of instability speed (krpm)
$t_s = 0$ µm	22.1
$t_s = 5$ µm	19.3
$t_s = 10$ µm	17.0
$t_s = 15$ µm	18.2
$t_s = 20$ µm	23.9
$t_s = 25$ µm	28.4

A simple modification to the FAB by inserting evenly spaced rigid shims under the bump foil layer delays the onset of instability speed (OIS) by 28.5% from 22.1 krpm to 28.4 krpm when shims of thickness of 25 µm are used. However, repeating this study for different intermediate shim thickness values shows for the first time that, considering just one shim thickness case is not enough to conclude that preload invariably improves the stability of the system. As shown in Table 2, the predictions indicate that there is a region where the presence of the shims with thickness < 20 µm has a negative effect on the system's stability with the OIS decreasing from 22.1 krpm to 17 krpm when inserting shims with a thickness of 10 µm. It should be noted that the results in Table 2 apply for the configuration in Figure 2a (i.e. relative positions of shims and foil spot weld location as shown, rotor static load along y-axis) – this is equivalent to the lobed bearing in [2] but is somewhat different from that in [7].

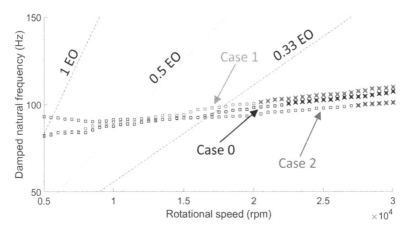

Figure 8. Frequency of Campbell mode no. 1 (only) *vs* speed for three cases: Case 0 (uniform foil stiffness, $\Delta K_b = 0$); Case 1 with $\Delta K_b = 4 \times 10^9 \text{N/m}^3$; Case 2 with $\Delta K_b = -4 \times 10^9 \text{N/m}^3$. $K_{b,\text{nom}} = 4.739 \times 10^9 \text{N/m}^3$, $N = 3$, $\theta_{\text{MD}} = 3\pi/2$; forward whirl (□); unstable mode points overlaid with a cross; EO (engine order).

Attention now turns to the stiffness variability study results. The loci of the journal SEPs (as in Figure 4) for different values of ΔK_b were calculated and the free linearised vibration about these SEPs analysed as before to extract the Campbell modes. Figure 8 shows frequency *vs* speed curves for mode no. 1 *only* with three alternative cases of ΔK_b. It is seen that $\Delta K_b > 0$ does not enhance stability since it reduces the OIS, whereas $\Delta K_b < 0$ enhances stability by delaying the OIS. This conclusion holds over a range of $|\Delta K_b|$ values, as shown in Table 3. Figure 9 shows modes nos. 1 and 2 at speeds immediately prior to the OIS for three different cases of ΔK_b in GN/m^3: 4, −4, −4.4. For the latter two cases, the three regions of foil stiffness reduction are clearly evident from the associated pockets of increased pad deflection. Two of these involve outward deflection of the pad, and the third (in the top right quadrant) involves inward deflection due to sub-atmospheric pressure in this region. These two cases result in increases of OIS of 23% and 37% respectively. For the latter case, as seen in Figure 9c1, 9c2, the static deflection in the most heavily loaded region is around twice the nominal clearance (i.e. 0.064mm), which appears reasonable. As mentioned in the introduction, researchers in [11] removed a section of the bump foil in this region but no information was provided on the predicted top foil deflection.

Table 3. Summary onset of instability speed (OIS) predictions for stiffness variation analysis.

Stiffness peak variation (GN/m^3)	Case 1 OIS (krpm)	Stiffness peak variation (GN/m^3)	Case 2 OIS (krpm)
$\Delta K_b = 0$	22.1	$\Delta K_b = 0$	22.1
$\Delta K_b = 0.8$	21.6	$\Delta K_b = -0.8$	22.8
$\Delta K_b = 1.6$	21.2	$\Delta K_b = -1.6$	23.7
$\Delta K_b = 2.4$	20.8	$\Delta K_b = -2.4$	24.6
$\Delta K_b = -3.2$	20.5	$\Delta K_b = -3.2$	25.8
$\Delta K_b = 4.0$	20.2	$\Delta K_b = -4.0$	27.2
		$\Delta K_b = -4.4$	30.3

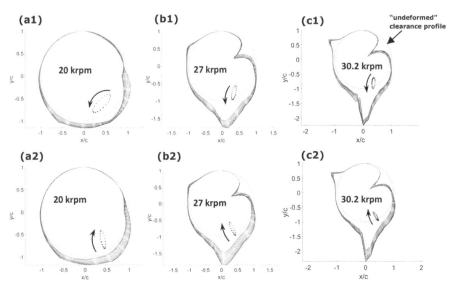

Figure 9. Campbell modes nos. 1 and 2 at speeds immediately prior to the OIS for three different cases of foil stiffness variation with
$K_{b,nom} = 4.739 \times 10^9 \text{N/m}^3$, $N = 3$, $\theta_{MD} = 3\pi/2$: **(a1, a2) Case 1 with** $\Delta K_b = 4 \times 10^9 \text{N/m}^3$ **(mode 1 -** $f_{d,n} = 100.44\text{Hz}$, $\zeta_n = 0.002185$; **mode 2 -** $f_{d,n} = 141.86\text{Hz}$, $\zeta_n = 0.17634$); **(b1, b2) Case 2 with** $\Delta K_b = -4 \times 10^9 \text{N/m}^3$ **(mode 1 -** $f_{d,n} = 99.10\text{Hz}$, $\zeta_n = 0.001654$; **mode 2 -** $f_{d,n} = 123.58\text{Hz}$, $\zeta_n = 0.10563$); **(c1, c2) Case 2 with** $\Delta K_b = -4.4 \times 10^9 \text{N/m}^3$ **(mode 1 -** $f_{d,n} = 98.20\text{Hz}$, $\zeta_n = 0.000624$; **mode 2 -** $f_{d,n} = 120.57\text{Hz}$, $\zeta_n = 0.11271$). **(NB: phase of vibration indicated by colour sequence black-red-green-magenta).**

5   CONCLUSION

This paper has, for the first time, applied a proven robust linearisation method for stability and modal analysis in a parametric study into the effect of preload and foil stiffness distribution in a rotor system with single-pad FABs. The study into the effect of preload found that an arbitrary preload did not necessarily result in an increase in the OIS, but an optimum preload could increase the OIS by a substantial amount (28.5%). The study into foil stiffness variability in the circumferential direction found that, by reducing the stiffness at specific locations, particularly the most heavily loaded region, one can achieve a substantial increase in OIS (up to 37%) while keeping the foil deflection within a reasonable level. However, research is needed into the actual foil design and the static and dynamic stresses it is required to withstand.

ACKNOWLEDGMENTS

The authors acknowledge the Engineering and Physical Sciences Research Council (EPSRC) and Dyson Technology Ltd for their support.

REFERENCES

[1] Lai, T., Chen, S., Ma, B., Zheng, Y. and Hou, Y., 2014. Effects of bearing clearance and supporting stiffness on performances of rotor-bearing system with multi-decked protuberant gas foil journal bearing. *Proceedings of the Institution of Mechanical Engineers, Part J: Journal of Engineering Tribology, 228*(7), pp.780–788.

[2] Sim, K., Lee, Y.B. and Kim, T.H., 2013. Effects of mechanical preload and bearing clearance on rotordynamic performance of lobed gas foil bearings for oil-free turbochargers. *Tribology transactions, 56*(2), pp.224–235.

[3] Feng, K., Guan, H.Q., Zhao, Z.L. and Liu, T.Y., 2018. Active bump-type foil bearing with controllable mechanical preloads. *Tribology International, 120*, pp.187–202.

[4] Hoffmann, R. and Liebich, R., 2018. Characterisation and calculation of nonlinear vibrations in gas foil bearing systems–An experimental and numerical investigation. *Journal of Sound and Vibration, 412*, pp.389–409.

[5] Hoffmann, R. and Liebich, R., 2017. Experimental and numerical analysis of the dynamic behaviour of a foil bearing structure affected by metal shims. *Tribology International, 115*, pp.378–388.

[6] Hoffmann, R., T. Pronobis, and R. Liebich. *A numerical performance analysis of a gas foil bearing including structural modifications by applying metal shims.* in *Proceedings of the 11. International Conference on Schwingungen in Rotierenden Maschinen (SIRM2015), Magdeburg, Germany.* 2015.

[7] Kim, T.H. and Andres, L.S., 2009. Effects of a mechanical preload on the dynamic force response of gas foil bearings: measurements and model predictions. *Tribology transactions, 52*(4), pp.569–580.

[8] Radil, K., Howard, S. and Dykas, B., 2002. The role of radial clearance on the performance of foil air bearings. *Tribology transactions, 45*(4), pp.485–490.

[9] DellaCorte, C. and Valco, M.J., 2000. Load capacity estimation of foil air journal bearings for oil-free turbomachinery applications. *Tribology Transactions, 43*(4), pp.795–801.

[10] Kim, D. (2007), "Parametric studies on static and dynamic performance of air foil bearings with different top foil geometries and bump stiffness distributions", *Journal of Tribology.* **129(2)**, 354–364. doi:10.1115/1.2540065.

[11] B.B. Nielsen, I.F. Santos, Transient and steady state behaviour of elasto-aerodynamic air foil bearings, considering bump foil compliance and top foil inertia and flexibility: a numerical investigation, Proc. IMechE Part J: Journal of Engineering Tribology. 231(10) (2017) doi:10.1177/1350650117689985.

[12] J.W. Lund, Calculation of stiffness and damping properties of gas bearings, Journal of Lubrication Technology (1968) 793–804.

[13] J.P. Peng, M. Carpino, Calculation of stiffness and damping coefficients for elastically supported gas foil bearings, Journal of Tribology. 115 (1993) 20–27.

[14] T. Pronobis, R. Liebich, Comparison of stability limits obtained by time integration and perturbation approach for Gas Foil Bearings, J. Sound Vib. 458 (2019) 497–509. https://doi.org/10.1016/j.jsv.2019.06.034.

[15] S. von Osmanski, J.S. Larsen, I.F. Santos, Multi-domain stability and modal analysis applied to Gas Foil Bearings: Three approaches, J. Sound Vib. 472 (2020) 115174. doi: 10.1016/j.jsv.2020.115174.

[16] Bonello, P. and Pham, H.M., 2014. The efficient computation of the nonlinear dynamic response of a foil–air bearing rotor system. *Journal of Sound and Vibration, 333*(15), pp.3459–3478. doi: 10.1016/j.jsv.2014.03.001.

[17] Bonello, P., *A new method for the calculation of the Campbell diagram of a foil-air bearing rotor model*, in *In SIRM 2019 13th International Conference on Dynamics of Rotating Machines.* 2019.

[18] Bonello, P., 2019. The extraction of Campbell diagrams from the dynamical system representation of a foil-air bearing rotor model. *Mechanical Systems and Signal Processing*, *129*, pp.502–530.

[19] Bonello, P. and Hassan, M.B., 2018. An experimental and theoretical analysis of a foil-air bearing rotor system. *Journal of Sound and Vibration*, *413*, pp.395–420.

Uncertainties in the calibration process of blade tip timing data against finite element model predictions

M.E. Mohamed[1, 2], P. Bonello[1], P. Russhard[3]

[1]Department of Mechanical Aerospace and Civil Engineering, University of
 Manchester, UK
[2]Faculty of Engineering, Cairo University, Cairo, Egypt
[3]EMTD Ltd, UK

ABSTRACT

Blade stresses are determined from blade tip timing (BTT) data by relating the meas-ured tip deflection to the stresses via Finite Element (FE) models. This process includes some uncertainties due to the following: 1) the shift in the equilibrium pos-ition of the blade tip due to steady deflection and/or movements, 2) the change in effective stiffness due to rotation-induced inertia, which affects the BTT-stress calibra-tion factors, 3) the assumption of constant speed over a single revolution that is made in most of BTT algorithms, which is not appropriate for rapid speed rates. This study shows the effect of such uncertainties on the vibration measurement and blade stress estimations.

1   INTRODUCTION

The determination of frequency and amplitude of vibration of turbomachinery blades using BTT algorithms constitutes an essential step towards a full under-standing of the dynamic behaviour. The ultimate step involves the estimation of the stress levels corresponding to such measured vibration data, in order to define the fatigue limits of the blades. This latter step requires a validated FE model [1] since BTT vibration data is limited to the blade tip deflection. Unfortu-nately, the link between FE and BTT is not as yet established or standardised due to complications arising from a number of uncertainties in BTT measurement and data processing on the one hand, and FE modelling on the other. The required calibration factor (i.e. stress-to-displacement ratio in MPa/mm) is determined from FE model predictions of tip nodes displacements and blade stress distribution, and is then applied to the measured BTT displacement to compute the corresponding stress [1]. The predicted tip displacement used in the determination of the calibration factor has to be determined at the node that is nearest to the measurement position, while the predicted stress value can be at any desired location on the blade surface (typically the location of the max-imum stress).

Table 1. BTT Uncertainties.

BTT Uncertainties		FEM Uncertainties

- Measurement position, due to:
   • blade steady movement
   • probe/blade offset
- Coherence (least squares error)
- Signal to noise ratio
- Number of probes
	- Condition number
	- Filtering
	- Zeroing
	- Averaging
	- Time resolution
		- Measurement position
		- Model validation tolerances
		- Non-homogenous material properties
		- Mistuning

The uncertainties associated with BTT measurements and analysis, in addition to those of FE modelling, have to be considered in order to achieve the best results. If all uncertainties are controlled, the current BTT capability could produce end-to-end stress measurements with uncertainty of only +/- 2.5% or better [2]. The sources of uncertainty associated with both BTT and FEM are listed in Table 1 based on previous studies [1–5]. Jousselin et al. [6] presented a method for establishing the levels of uncertainty for the blade tip displacement amplitudes measured by a BTT system. Russhard [2] studied the levels of most of the aforementioned uncertainties and showed that mostly depend on the operator, who must be highly experienced to avoid extreme uncertainty levels [2].

Uncertainty in the mean position of the probe (the sensing position) relative to the blade tip affects both the BTT measurements and the calibration of the data against FE predictions. It arises from two sources [7]: (a) positional errors (offsets) during installation of the probes and/or blades; (b) steady (non-oscillatory) deviations of the blade tip from its mean position and orientation, which are caused by the changes in the speed dependant operating conditions, such as thermal expansion and axial float of the rotor, bearing wear, and non-uniform gas loading [7]. The measurement position uncertainty considered in this paper is of type (b) since it is more problematic due to the variation of the associated positional error with speed (in contrast to (a), where the positional error is independent of speed) [7]. This shift (of the probe measurement point from its nominal position relative to the blade tip) can have a great impact on the calibration factor relating tip deflection to stress, since both the original and new points may have different amplitudes of vibration, thus introducing errors into the stress and the fatigue limit estimates.

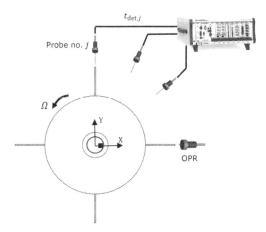

Figure 1. BTT Measurement System.

Turbomachinery blades are typically modelled using rotating cantilever beam models, which have vibration characteristics that vary significantly with speed changes [8]. One of the factors responsible for this variation is the centrifugal inertia force, which causes a significant stretch in the blades, resulting in a change in their effective bending stiffness. Some studies have considered centrifugal stiffening, and other rotation-induced inertia effects in shaft-blisk (i.e. bladed disk) assemblies, and showed their effects on the natural frequencies of the rotating systems [9,10]. However, stress stiffening does not only change the natural frequencies, but also affects the calibration factors, as will be shown in this work.

Traditional methods of BTT data processing assume that the rotational speed is constant over the course of a single revolution [11]. In reality, it is impossible for the rotational speed to keep constant [12]. Moreover, this assumption is incorrect in the case of transient rotor speeds [11] especially when the rate of change of speed ("speed rate" in rpm/rev) is high. Especially in the latter case, it is therefore expected that the constant speed assumption will result in noticeable errors in the estimated stress values as well as the vibration levels. Currently, some research studies are concerned with the development of measurement systems for applications involving significant speed rate [11–15].

This study shows for the first time the effects of three sources of uncertainty on the BTT vibration measurement and blade stress estimations: 1) Probe measurement position, 2) Centrifugal stiffening, and 3) High-rate speed change. A BTT simulator that has been presented by the authors in [16,17] is used to generate the simulated BTT data including the considered effects under controlled conditions. A new approach for the calculations of both strain and stress values simultaneously with the BTT displacement data using the simulator is presented.

2 BLADE TIP TIMING

BTT systems involve the use of a number of circumferentially distributed non-contact probes $j, j = 1, \ldots, N_{Pr}$ and a data acquisition system as shown in Figure 1 to detect and acquire the arrival times $t_{det,j}$ of all blades at the probe angular locations during every revolution [16]. The arrival time is the time at which a blade tip passes within the range of a probe. Another probe, known as the once per revolution (OPR) probe, is typically used to detect the start/end time of every revolution, from which the following information can be calculated: (1) the average rotational speed over the revolution; (2) the number of completed revolutions; (3) the expected times of arrival $t_{exp,j}$ of the blades past the probes if the assembly rotates as a rigid body. It is noted that some researchers have proposed to generate such information from the blade arrival times without need of the OPR [18,19]. The collected BTT data are then processed in order to determine the blade tip displacements as follows

$$d_j = \Omega R \Delta t_j \tag{1}$$

where d_j is the tip displacement in the plane of rotation ($X - Y$ plane) at probe no. j, Ω is the average angular speed (rad/s) over a single revolution [20], R is the radius of the blade tip measured from the centre of rotation, and Δt_j is the difference between the blade's expected and detected arrival times.

$$\Delta t_j = t_{exp,j} - t_{det,j} \tag{2}$$

The BTT displacement data are analysed in order to determine the vibration parameters (amplitude, frequency, and phase), and the equivalent stress levels are then calculated using FE-based calibration factors.

3   PROBE MEASUREMENT POSITION ERROR

Probe measurement position errors have significant effect on the estimation of stress levels in blades from BTT measurements. The errors considered are those arising from the blade steady movement, which can be resolved into three components (rotor axial shift, blade lean, and blade untwist). Figure 2 shows a blade tip that shifts axially (parallel to the axis of rotation Z). Initially (with no steady movement), the optical probe's laser beam intersects with the blade tip at point (1) which has the same axial position of the probe. Once their angular positions agree, the time is recorded as $t_{\mathrm{det},j})_1$, and the displacement is calculated from Eq (1) using the value of $t_{\mathrm{exp},j})_1$, where the difference between both times $(\Delta t_j)_1)$ is proportional to the instantaneous displacement at point (1). If the blade moves axially by an amount $\Delta u^{(\mathrm{off})}$, the measurement position will shifted to point (2) and the recorded arrival time is denoted as $t_{\mathrm{det},j})_2$. The expected arrival time for point (2) will be different from point (1) but, since the movement is unknown, the instantaneous displacement at (2) is calculated using $\Delta t_j)_2 = t_{\mathrm{exp},j})_1 - t_{\mathrm{det},j})_2$ in Eq (1). There are therefore two errors: 1) a "DC" error due to the use of $t_{\mathrm{exp},j})_1$; 2) the vibration is being measured at point (2) rather than point (1) and the vibration amplitudes at these points are typically different. The DC error can be removed by the method presented in [7], while the second error will result in wrong stress estimation unless the calibration factor is updated to refer to point (2). The method of [7] makes this possible since it allows the quantification of the steady movement and thus the location of the new sensing point (2) on the blade tip.

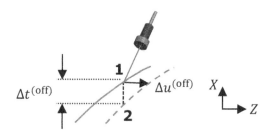

Figure 2. Probe Measurement Position Error.

4   CENTRIFUGAL STIFFENING

The centrifugal inertia tension N_{P_k} on a blisk segment with mass centre P_k $(R_{\mathrm{P}_k}, \vartheta_{\mathrm{P}_k}, Z_{\mathrm{P}_k})$ and radial extent ΔR_{P_k} can be calculated as [17]:

$$N_{\mathrm{P}_k} = \Omega^2 \left\{ \sum_{j=1}^{J} \Delta m_j R_j \right\} \bigg|_{R_1 = R_{\mathrm{P}_k}} \tag{3}$$

where the summation is applied to radial segments with mass centres located along the same angular and axial position of P_k $(\vartheta_{\mathrm{P}_k}, Z_{\mathrm{P}_k})$ and proceeding radially outward. With reference to Figure 3(a,b), the centrifugal tension N_{P_k} stiffens the segment by inducing moments in the edge-wise (UV) and flap-wise (UZ) planes ($M_{z\mathrm{P}_k}$, $M_{v\mathrm{P}_k}$ respectively) which oppose the corresponding angular deformations $\theta_{z\mathrm{P}_k}$, $\theta_{v\mathrm{P}_k}$:

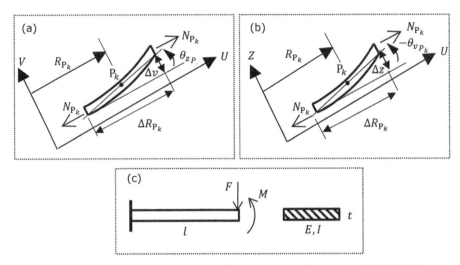

Figure 3. Centrifugal Stiffening, a) Blisk Segment (*UV*-plane), b) Blisk Segment (*UZ*-plane), c) fixed-free beam with force and moment loading.

$$M_{zPk} = -N_{Pk}\Delta v = -N_{Pk}\Delta R_{Pk}\frac{\Delta v}{\Delta R_{Pk}} = -N_{Pk}\Delta R_{Pk}\theta_{zPk}$$

$$M_{vPk} = N_{Pk}\Delta z = N_{Pk}\Delta R_{Pk}\frac{\Delta z}{\Delta R_{Pk}} = -N_{Pk}\Delta R_{Pk}\theta_{vPk}$$

(4)

where Δv, and Δz are the deflections in V and Z directions respectively. In order to understand the effect of the above moments on the calibration factors and estimated values of stresses from BTT measurements, the simple cantilever shown in Figure 3(c) is considered. The cantilever is subjected to a force (F) at the free end. The maximum deflection and maximum stress of the segment can be respectively calculated as

$$d_{\max})_F = \frac{FL^3}{3EI}, \sigma_{\max})_F = \frac{FL(t/2)}{I}$$

(5)

The calibration factor C can then be determined as

$$C_1 = \frac{\sigma_{max})_F}{d_{max})_F} = \frac{3E(t/2)}{L^2}$$

(6)

Now, considering a moment M applied at the free end of the segment, in addition to F, the maximum deflection and stress due to the total effect become

$$d_{\max})_{tot} = \frac{FL^3}{3EI} - \frac{ML^2}{2EI}, \sigma_{\max})_{tot} = \frac{FL(t/2)}{I} - \frac{M(t/2)}{I}$$

(7)

The new calibration factor will be

$$C_2 = \frac{\sigma_{\max})\text{tot}}{\sigma_{\max})\text{tot}} = \frac{\left(\frac{FL(t/2)}{I} - \frac{M(t/2)}{I}\right)}{\left(\frac{FL^3}{3EI} - \frac{ML^3}{2EI}\right)} = \frac{(FL - M)}{\left(FL - \frac{3}{2}M\right)} \times \frac{3E(t/2)}{L^2} \tag{8}$$

Substituting from Eq (6) into Eq (8)

$$C_2 = \frac{(FL - M)}{\left(FL - \frac{3}{2}M\right)} \times C_1 \tag{9}$$

From Eq (9) it should be clear that, for the actual blade whose generic segment is shown in Figure 3(a,b), the moments induced by the centrifugal effect will affect the calibration factor (reference location stress to tip deflection).

5   HIGH SPEED RATES

BTT processing algorithms assume that the rotational speed is constant over a single revolution [20], which is not valid in case of time-varying speed. In such a case, assuming linear variation in speed from Ω_i rad/s to Ω_f rad/s over a duration T_s, the rotational speed at any time t can be calculated using the following formula:

$$\Omega(t) = \Omega_i + \left(\frac{\Omega_f - \Omega_i}{T_s}\right) t \tag{10}$$

Therefore, the BTT displacement at each probe should be calculated from Eq (1) using different values of speed based on Eq (10). The general form of the blade tip displacement d_j measured at probe no. j after processing (minimization of noise and offset) is assumed to follow the following form [16]

$$d_j = a_0 + \sum_{m=1}^{M} \left(a_{1m} \sin\left(EO_m \Omega t_j\right) + a_{2m} \cos\left(EO_m \Omega t_j\right)\right) \tag{11}$$

where EO_m is the engine order of excitation no. $m, m = 1, 2, ..., M$, whose frequency $\omega_m = EO_m \Omega$, and t_j is the sampling time at probe no. j during revolution no.n according to the above assumed fit. For a constant rotational speed Ω, $t_j = t_{j,n=1} + (n-1)2\pi/\Omega$, $n = 1, 2, ...$ and so, the term $EO_m \Omega t_j = EO_m \Omega\{t_{j,n=1} + (n-1)2\pi/\Omega\}$ always represents the same angle $EO_m \Omega t_{j,n=1}$ provided EO_m is a fixed integer. Hence, Eq. (11) can then be rewritten in terms of the fixed probe angular position $\theta_j = \Omega t_{j,n=1}$:

$$d_j = a_0 + \sum_{m=1}^{M} \left(a_{1m} \sin\left(EO_m \theta_j\right) + a_{2m} \cos\left(EO_m \theta_j\right)\right) \tag{12}$$

The representation of Eq. (12) is used in some of the main BTT analysis algorithms, such as the two parameter plot [21] and sine fitting with data preparation methods [22]. As shown above, it is only valid if the rotational speed Ω is either constant or assumed to be approximately constant over one revolution, while varying gradually over time. Therefore, this assumption is not valid in case of high speed rates (rpm/rev), and may therefore result in large amount of errors in the estimated stress values. The speed rate can be calculated as

$$\text{Speed rate (rpm/rev)} = \frac{60(\Omega_f - \Omega_i)}{\left(\frac{\Omega_f + \Omega_i}{2}\right) T_s} \tag{13}$$

6 BTT SIMULATION

The following sections include descriptions of the FE model, and the BTT simulator used to generate simulated BTT displacements and blade stresses, in order to examine the effects of the abovementioned sources of uncertainty on the stress estimates.

6.1 FE model

The shaft-blisk system shown in Figure 1 was analysed before in [17,23] to study the effect of rotation-induced inertia on the natural frequencies of the rotating assembly. The system consists of a stepped circular cross-section shaft supported by two bearings (with equal stiffness in the two orthogonal directions $1.5 \times 10^7 \text{N/m}$) and carrying a disk with four blades, the material is the same for all components (density 7800kg/m^3, Young Modulus 200GPa). The FE model has been created using *ANSYS*, and the total number of degrees of freedom is 1800858. Both ends of the shaft were assumed to be constrained from axial motion and the left hand end constrained from torsional motion. Modal analysis has been carried out, and some of the resulting zero-speed natural frequencies and their corresponding mode shapes are shown in Table 2 and Figure 4 respectively.

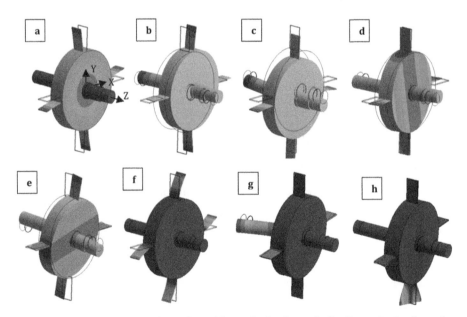

Figure 4. Mode shapes: a) mode 1, b) mode 2, c) mode 3, d) mode 4, e) mode 5, f) mode 6, g) mode 10, h) mode 15.

Table 2. Natural Frequencies of the shaft-Blisk Assembly.

Mode	1	2	3	4	5	6	10	15
Natural frequency (Hz)	121.7	133.8	133.8	266.5	266.5	364.3	1156.5	1464

6.2 BTT simulator

A simulator that is based on a realistic (FE-derived) multi-modal model of the blisk was presented by the authors in [16], and upgraded for the inclusion of rotation-induced inertia effects in [17]. The following section includes a brief description of the simulator.

If $\mathbf{u}(t)$ denotes the $3K \times 1$ vector containing the instantaneous absolute coordinates of the nodes P_k $(k = 1 \ldots K)$ on the tip of a given blade, the simulator computes $\mathbf{u}(t)$ by adding the three components that contribute to it: (A) the absolute coordinates after rigid body rotation from a reference angular position; (B) the steady shift; (C) the dynamic response. The last component is determined by calculating the dynamic excitation response $\Delta\mathbf{u}^{(\mathrm{def})}(t)$ of the non-rotating blisk in the reference angular position and then transforming it according to the rigid rotation angle prior to addition to (A) and (B). $\Delta\mathbf{u}^{(\mathrm{def})}(t)$ is calculated from a transformation to modal space using as basis functions the first H natural undamped modes of vibration of the non-rotating blisk in the reference angular position:

$$\Delta\mathbf{u}^{(\mathrm{def})}(t) = \mathbf{H}_\mathrm{P}\mathbf{q}(t) \tag{14}$$

where the $H \times 1$ vector of modal co-ordinates $\mathbf{q}(t)$ is governed by the modal equations of motion:

$$\ddot{\mathbf{q}}(t) + \mathrm{diag}([\cdots 2\zeta_r\varpi_r \cdots])\dot{\mathbf{q}}(t) + \mathrm{diag}([\cdots \varpi_r^2 \cdots])\mathbf{q}(t) = \mathbf{H}_\mathrm{f}^\mathsf{T}\mathbf{f}(t) \tag{15}$$

$\mathbf{H}_\mathrm{P}$, $\mathbf{H}_\mathrm{f}$ are modal transformation matrices whose H columns are mass-normalised eigenvectors corresponding to the natural circular frequencies ϖ_r $(r = 1, \ldots, H)$, ζ_r is the modal damping ratio, added to ensure decay of transients. $\mathbf{f}(t)$ is the vector of dynamic excitation forces applied to the blisk. In the above approach, the blisk is divided into radial/angular segments and the rotational inertia effects from the individual segments, with mass centres at selected FE nodes P_k $(k = 1 \ldots K)$, are summed over the entire system and added as additional "external" forces to the modal equation (Eq (15)).

The simulator is implemented in *Matlab/Simulink* and solves Eq. (15) using a numerical integration routine with automatic time-step control. At each time step it determines $\mathbf{u}(t)$ by combining its three components. For a given configuration of probes, $t_{\mathrm{exp},j}$ is determined by locating the passing time of the node on the blade tip that coincides with the angular and axial position of probe no. j using only the rigid rotation component of $\mathbf{u}(t)$. The actual arrival time $t_{\mathrm{det},j}$ is determined in the same way but using $\mathbf{u}(t)$. The blade tip displacements d_j are then determined as per Eqs. (1) and (2) and corrupted with Gaussian white noise of prescribed noise-to-signal ratio (NSR) to simulate AC measurement noise.

6.3 Determination of strain and stress values using the simulator

Figure 5(a) shows the stress map on a blade surface using FEM and indicates the position of maximum stress. This is usually the way the positions of the strain gauges (SGs) on the blade surface are selected. A strain rosette (a set of three SGs a, b, and c) is placed and glued to the surface at the desired location as shown in Figure 5(b) with a known orientation of each SG with respect to a specified reference. It can be assumed that the surface area including the strain rosette is flat due to the small sizes of the SGs, and thus the rosette is assumed to be in one plane $x'y'$. In the case of the simulator, the strain values can be obtained in a similar way to that used with physical SGs as follows.

-An FE node o is selected at the desired position on the blade surface.

-Three other nodes $\mathrm{a, b, c}$ are selected as close as possible to the node o as shown in Figure 5(c), so they can be considered in one plane $x'y'$ where x' and y' are orthogonal axes intersecting at the node o.

$-oa, ob,$ and oc represent the strain rosette, and the distances $\overline{oa}, \overline{ob},$ and $\overline{oc}$ can be calculated using the coordinates of the nodes before deformation with respect to the rotating local reference frame xyz as

$$\overline{oa} = \sqrt{(x_a - x_o)^2 + (y_a - y_o)^2 + (z_a - z_o)^2}$$

$$\overline{ob} = \sqrt{(x_b - x_o)^2 + (y_b - y_o)^2 + (z_b - z_o)^2} \tag{16}$$

$$\overline{oc} = \sqrt{(x_c - x_o)^2 + (y_c - y_o)^2 + (z_c - z_o)^2}$$

The angles $\theta_a, \theta_b, \theta_c$ are calculated by assuming that x' coincides with the vector $\overline{oa}$, so $\theta_a = 0$, and then the angles θ_b, and θ_c of the vectors $\overline{ob}$ and $\overline{oc}$ respectively are obtained, knowing that all the vectors are assumed to be in the same plane:

$$\cos\theta_b = \frac{\overline{ob} \cdot \overline{oa}}{|\overline{ob}| \cdot |\overline{oa}|} \tag{17}$$

Where

$$\overline{ob} \cdot \overline{oa} = (x_b - x_o)(x_a - x_o) + (y_b - y_o)(y_a - y_o) + (z_b - z_o)(z_a - z_o) \tag{18}$$

– The same distances are calculated again at a time t including deformation:

$$\overline{oa}(t) = \sqrt{(x_a(t) - x_o(t))^2 + (y_a(t) - y_o(t))^2 + (z_a(t) - z_o(t))^2}$$

$$\overline{ob}(t) = \sqrt{(x_b(t) - x_o(t))^2 + (y_b(t) - y_o(t))^2 + (z_b(t) - z_o(t))^2} \tag{19}$$

$$\overline{oc}(t) = \sqrt{(x_c(t) - x_o(t))^2 + (y_c(t) - y_o(t))^2 + (z_c(t) - z_o(t))^2}$$

where $x(t) = x + \Delta x(t)$, $y(t) = y + \Delta y(t)$, $z(t) = z + \Delta z(t)$, and $\Delta x(t), \Delta y(t), \Delta z(t)$ are the elements of $\Delta\mathbf{u}^{(\mathrm{def})}(t)$ corresponding to the nodes o, a, b, c.

Finally, the strain values can be calculated as

$$\varepsilon_{oa}(t) = \frac{\overline{oa}(t) - \overline{oa}}{\overline{oa}}, \ \varepsilon_{ob}(t) = \frac{\overline{ob}(t) - \overline{ob}}{\overline{ob}}, \ \varepsilon_{oc}(t) = \frac{\overline{oc}(t) - \overline{oc}}{\overline{oc}}$$

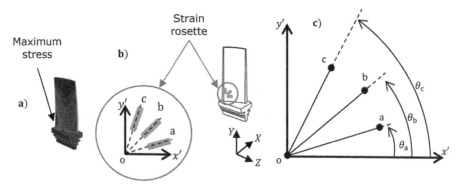

Figure 5. A) Blade stress map, b) SG rosette, c) Strain vectors.

Now, for the calculations of the stresses, the strain values are transformed into the orthogonal directions x' and y' using the following equations

$$
\begin{aligned}
\varepsilon_{oa} &= \varepsilon_{x'}\cos^2\theta_a + \varepsilon_{y'}\sin^2\theta_a + \gamma_{x'y'}\sin\theta_a\cos\theta_a \\
\varepsilon_{ob} &= \varepsilon_{x'}\cos^2\theta_b + \varepsilon_{y'}\sin^2\theta_b + \gamma_{x'y'}\sin\theta_b\cos\theta_b \\
\varepsilon_{oc} &= \varepsilon_{x'}\cos^2\theta_c + \varepsilon_{y'}\sin^2\theta_c + \gamma_{x'y'}\sin\theta_c\cos\theta_c
\end{aligned}
\tag{21}
$$

which can be formulated in a matrix form as

$$
\begin{bmatrix}
\cos^2\theta_a & \sin^2\theta_a & \sin\theta_a\cos\theta_a \\
\cos^2\theta_b & \sin^2\theta_b & \sin\theta_b\cos\theta_b \\
\cos^2\theta_c & \sin^2\theta_c & \sin\theta_c\cos\theta_c
\end{bmatrix}
\begin{bmatrix}
\varepsilon_{x'} \\
\varepsilon_{y'} \\
\gamma_{x'y'}
\end{bmatrix}
=
\begin{bmatrix}
\varepsilon_{oa} \\
\varepsilon_{ob} \\
\varepsilon_{oc}
\end{bmatrix}
\tag{22}
$$

or, $\mathbf{Ba} = \mathbf{c}$

By solving the above equation as $\mathbf{a} = \mathbf{B}^{-1}\mathbf{c}$, The values of normal and shear strain in $x'y'$ plane are obtained. Assuming plane stress conditions, and by using the stress-strain relations

$$
\varepsilon_{x'} = \frac{1}{E}\left[\sigma_{x'} - v\sigma_{y'}\right], \ \varepsilon_{y'} = \frac{1}{E}\left[\sigma_{y'} - v\sigma_{x'}\right], \ \gamma_{x'y'} = \frac{\tau_{x'y'}}{G}
\tag{23}
$$

The stress values can be determined, and then the principal plane stresses and maximum in-plane shear stress are calculated as

$$
\begin{aligned}
\sigma_{1,2} &= \frac{\sigma_{x'} + \sigma_{y'}}{2} \pm \sqrt{\left(\frac{\sigma_{x'} - \sigma_{y'}}{2}\right)^2 + \tau_{x'y'}{}^2} \\
\tau_{max} &= \sqrt{\left(\frac{\sigma_x - \sigma_x}{2}\right)^2 + \tau_{xy}{}^2}
\end{aligned}
\tag{24a, b}
$$

Finally, the equivalent (Von Mises) stress is calculated as

$$
\sigma_e = \frac{1}{\sqrt{2}}\left[(\sigma_1 - \sigma_2)^2 + (\sigma_2 - \sigma_3)^2 + (\sigma_3 - \sigma_1)^2\right]^{1/2}
\tag{25}
$$

where $\sigma_3 = 0$.

7 RESULTS AND DISCUSSIONS

7.1 Effect of probe measurement position error

Considering the torsional mode shape of blade shown in Figure 4(h), the amplitude of deformation at the blade tip is seen to vary significantly, being maximum at both ends and tending to zero at the midpoint. Assuming that a probe is located at one of the ends (initial measurement point), and calculating the modal displacements at all tip locations relative to the initial measurement point, the percentage error in BTT displacement due to the shift of the blade along the axial direction (axis of rotation Z) is as shown in Figure 6. The data in Figure 6 were computed using the formula in Eq (26), where d_{end} is the displacement at the tip's end (initial sensing location), and d_z is the displacement along Z-axis at any other location on the tip.

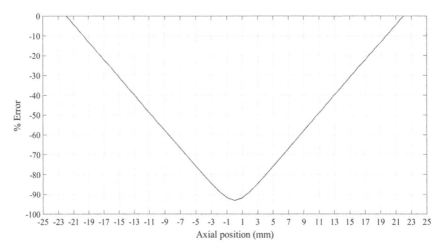

Figure 6. Percentage error of tip displacement along the axial direction relative to the displacement at the tip end.

$$\%\text{error} = \frac{d_z - d_{\text{end}}}{d_{\text{end}}} \times 100 \qquad (26)$$

Table 3. Effect of stagger angle on probe measurement position error.

Stagger angle (degrees)	0	10	20	30	40	45
% error/mm	4.37	4.42	4.64	5.03	5.69	6.16

It is clear from Figure 6 that the error in displacement value would be about 4.37% per 1 mm shift in measurement position. The rate of error was also examined at different stagger angles of blade (angle between the blade's chord and the line parallel to the axis of rotation i.e. Z-axis) and the results listed in Table 3 show that the error increases to a value of 6.16%/mm at an angle of 45°. Moreover, complex blade profiles in which the blade tip nodes have different distances from the blade root can result in similar errors in the case of blade bending modes, due to the different displacement values at the tips. These errors will then be propagated to the estimated stress values.

7.2 Effect of centrifugal stiffening

For preliminary checking, the stress/tip displacement calibration factor in the absence of centrifugal stiffening was first determined by performing harmonic analysis in *ANSYS*. A harmonic force was applied at the tip of one of the blades in the x-direction with a frequency equal to the 1st natural frequency (i.e. 121.7 Hz, see Table 2). The displacement at the tip, and the corresponding maximum stress values, were then extracted and the calibration factor calculated as 136.7 MPa/mm.

In order to study the effect of centrifugal stiffening on the calibration factor as discussed in section 4, the BTT simulator (section 6.2) was then used to generate the dynamic response values at the desired FE nodes of both the blade surface and tip, and the stress value was then calculated at the position of maximum stress using the method described in section 6.3. The simulation was implemented using the same harmonic force excitation as the above described *ANSYS* study. In the absence of centrifugal inertia effects, the resulting calibration factor was 131.24 MPa/mm, which is only 4 % less than the one determined by *ANSYS*. The simulation was then repeated by considering the centrifugal stiffening of the blisk. The Campbell diagram generated first in order to determine the new value of the critical speed with respect to the first mode, and it is found to be 434.6 rev/s (i.e. up from 121.7 rev/s at no rotation). The excitation frequency was then updated, and the responses were extracted and processed to calculate the new calibration factor, which was found to be 142.6 MPa/mm i.e. 8.6 % higher than the initial one. This agrees with the explanation presented in section 4.

7.3   Effect of high speed rate

It has been explained in section 5 that the significant variation in speed over one revolution results in an error in the calculated values of vibration amplitude, due to inaccurate displacement values at the probes (unless Eq (10) is used in Eq (1)), and, more significantly, the invalid assumption of the BTT displacement form (Eq (12)). A number of simulations were done with the same excitation conditions and the same range of speeds (0 to 250 rev/s) that includes the first critical speed (Engine Order (EO)=1), but with different speed rates. The generated BTT displacements were analysed using the multi-frequency sine fitting with data preparation method [22], and the resulting amplitude of vibration was compared to the zero-to-peak value of vibration response time history, giving the error values listed in Table 4.

Table 4. BTT displacement error vs speed rate.

Speed rate rpm/rev	24	120	1200	2400
% error	1.6	4.2	25	42

These results clearly show the significant adverse effect of high speed rate on the accuracy of the estimated values of tip displacement. Such errors would then of course be propagated to the estimated blade stresses.

8   CONCLUSIONS

The determination of blade vibration using BTT, and the estimation of the corresponding stresses, are subject to a number of uncertainties related to both measurement and FE modelling. Three sources of uncertainty were analysed with the aim of quantifying their effects, which were then illustrated for three specific cases. Probe measurement position error resulted in an error in tip displacement of about 4% per mm shift of the blade, and this error increased with the stagger angle of the blade. Centrifugal stiffening resulted in an increase of 8.6% in the stress/tip deflection calibration factor. A high speed rate of 2400 rpm/rev may result in more than 40% error in vibration amplitude. Unless such sources of uncertainty are considered in the end-to-end process for the estimation of stress levels from BTT data, their combined effect may result in a large margin of error.

ACKNOWLEDGMENT

This work is part of the EU research project Batista (Blade Tip Timing System Validator) - Clean Sky 2 - 862034, funded by the EU Commission (H2020 CS2).

REFERENCES

[1] Russhard, P., 2010, "Development of a Blade Tip Timing Based Engine Health Monitoring System," Ph.D. dissertation, University of Manchester, Manchester, United Kingdom.

[2] Russhard, P., 2016, "Blade Tip Timing (BTT) Uncertainties," AIP Conference Proceedings, 1740.

[3] Jousselin, O., 2013, "Development of Blade Tip Timing Techniques in Turbo Machinery," Ph.D. dissertation, The University of Manchester, Manchester, United Kingdom.

[4] Lawson, C., and Ivey, P., 2003, "Compressor Blade Tip Timing Using Capacitance Tip Clearance Probes," Proceedings of ASME Turbo Expo 2003, Atlanta, Georgia, USA, pp. 1–8.

[5] Knappett, D., and Garcia, J., 2008, "Blade Tip Timing and Strain Gauge Correlation on Compressor Blades," Proceedings of the Institution of Mechanical Engineers, Part G: Journal of Aerospace Engineering, 222(4), pp. 497–506.

[6] O. Jousselin, P. Russhard, and P. Bonello, 2012, "A Method for Establishing the Uncertainty Levels for Aero-Engine Blade Tip Amplitudes Extracted from Blade Tip Timing Data," Proceedings of the 10th International Conference on Vibrations in Rotating Machinery, pp. 211–220.

[7] Mohamed, M., Bonello, P., and Russhard, P., 2019, "A Novel Method for the Determination of the Change in Blade Tip Timing Probe Sensing Position due to Steady Movements," Mechanical Systems and Signal Processing, 126, pp. 686–710.

[8] Przemieniecki, J. S., 1985, Theory of Matrix Structural Analysis., McGraw-Hill, New York.

[9] Chun, S.-B., and Lee, C.-W., 1996, "Vibration Analysis of Shaft-Bladed Disk System by Using Substructure Synthesis and Assumed Modes Method," Journal of Sound and Vibration, 189(5), pp. 587–608.

[10] Chiu, Y.-J., Li, X.-Y., Chen, Y.-C., Jian, S.-R., Yang, C.-H., and Lin, I.-H., 2017, "Three Methods for Studying Coupled Vibration in a Multi Flexible Disk Rotor System," Journal of Mechanical Science and Technology, 31(11), pp. 5219–5229.

[11] Diamond, D. H., Heyns, P. S., and Oberholster, A. J., 2019, "Improved Blade Tip Timing Measurements during Transient Conditions Using a State Space Model," Mechanical Systems and Signal Processing, 122, pp. 555–579.

[12] Chen, Z., Liu, J., Zhan, C., He, J., and Wang, W., 2018, "Reconstructed Order Analysis-Based Vibration Monitoring under Variable Rotation Speed by Using Multiple Blade Tip-Timing Sensors," Sensors, 18(10), p. 3235.

[13] Fan, C., Wu, Y., Russhard, P., and Wang, A., 2020, "An Improved Blade Tip-Timing Method for Vibration Measurement of Rotating Blades During Transient Operating Conditions," Journal of Vibrational Engineering and Technologies, (0123456789).

[14] Ji-wang, Z., Lai-bin, Z., Ke-Qin, D., and Li-xiang, D., 2018, "Blade Tip-Timing Technology with Multiple Reference Phases for Online Monitoring of High-Speed Blades under Variable-Speed Operation," Measurement Science Review, 18(6), pp. 243–250.

[15] Bouchain, A., Picheral, J., Lahalle, E., Chardon, G., Vercoutter, A., and Talon, A., 2019, "Blade Vibration Study by Spectral Analysis of Tip-Timing Signals with OMP Algorithm," Mechanical Systems and Signal Processing, 130, pp. 108–121.

[16] Mohamed, M. E., Bonello, P., and Russhard, P., 2020, "An Experimentally Validated Modal Model Simulator for the Assessment of Different Blade Tip Timing Algorithms," Mechanical Systems and Signal Processing, 136, p. 106484.

[17] Mohamed, M., 2019, "Towards Reliable and Efficient Calibration of Blade Tip Timing Measurements against Finite Element Predictions," Ph.D. dissertation, University of Manchester.

[18] Gallego-Garrido, J., Dimitriadis, G., Carrington, I., and Wright, J., 2007, "A Class of Methods for the Analysis of Blade Tip Timing Data from Bladed Assemblies Undergoing Simultaneous Resonances—Part II: Experimental Validation," International Journal of Rotating Machinery, pp. 1–10.

[19] Carrington, I., Wright, J., Cooper, J., and Dimitriadis, G., 2001, "A Comparison of Blade Tip Timing Data Analysis Methods," Proceedings of the Institution of Mechanical Engineers, Part G: Journal of Aerospace Engineering, 215(5), pp. 301–312.

[20] Russhard, P., 2014, "Derived Once per Rev Signal Generation for Blade Tip Timing Systems," Instrumentation Symposium 2014, IET & ISA 60th ..., (1), pp. 1–5.

[21] Heath, S., 2000, "A New Technique For Identifying Synchronous Resonances Using Tip-Timing," Journal of Engineering for Gas Turbines and Power-Transactions of the ASME, 122(2), pp. 219–225.

[22] Russhard, P., 2010, "Development of a Blade Tip Timing Based Engine Health Monitoring System," Eng.D. dissertation, University of Manchester, Manchester, United Kingdom.

[23] Ma, H., Lu, Y., Wu, Z., Tai, X., Li, H., and Wen, B., 2015, "A New Dynamic Model of Rotor–blade Systems," Journal of Sound and Vibration, 357, pp. 168–194.

Parametric coupled instabilities of an on-board rotor subject to yaw and pitch with arbitrary frequencies

Y. Briend[1], M. Dakel[1], E. Chatelet[1], M-A. Andrianoely[1], R. Dufour[1], S. Baudin[2]

[1]University Lyon, INSA-Lyon, CNRS UMR, LaMCoS, France
[2]AVNIR Engineering, Paris, France

ABSTRACT

Parametric excitations can be encountered in a large variety of dynamic mechanical systems. Such excitations must be taken into account at the design stage in order to avoid the frequency operating ranges in which the mechanical systems may exhibit instabilities sources of failures. Consequently, systems with parametric excitations have been widely studied over the past decades and many relevant works have been published in this research field. Among others, on-board rotors can undergo parametric instabilities as soon as they are subject to a rotational motion of their base. Nevertheless, all the cases of study available in the literature are only limited to mono-axis rotation of the base or at least multi-axes rotations with the same parametric frequency. In practice, the excitations are not generally restricted to such cases. In this context, the present paper aims to investigate the dynamics of an on-board rotor subject to multi-axes sinusoidal rotations of its base, with arbitrary frequencies. This is achieved by analyzing the finite element model of an on-board rotor, which is composed of a slender shaft with two disks, supported by two hydrodynamic bearings having finite length. The nonlinear forces related to these bearings are linearized in the vicinity of the rotor static equilibrium position for a constant speed of rotation so as to obtain a linear system. Floquet theory is finally applied in order to perform stability analysis and the influence of the amplitudes, frequencies and phases of the excitation is assessed.

This rotordynamics prediction is essential for the design of many on-board rotating machineries such as those implemented in vessels subject to roll or pitch motions induced by the waves to name just one example.

Keywords: On-board rotor, parametric instability, multi-axes excitation, base rotations, Floquet theory

1   INTRODUCTION

Rotating machinery can be found in quite a large variety of mechanical systems such as motors, turbines, turbochargers, fans. The design stage has to ensure that these rotor-bearing systems will not experience any failure and will preserve their integrity during the product lifetime. Beside the classical mass unbalance forces and the risk of resonance phenomena due to the critical speeds, rotor systems can be subject also to the external excitations induced by the motions of its base, making their rotordynamics prediction more complicated and triggering a quite recent interest in the literature. In particular, one of the first study has been achieved by Lin and Meng (1) which assessed the influence of constant or sinusoidal translational motions of the support on the time history responses of a Jeffcott rotor. Later, Duchemin *et al* (2) demonstrated both analytically and experimentally the stability of a rotor excited by a periodic rotational motion of the rigid support. Other experimental validation was carried out by Driot *et al* (3) in terms of orbits comparisons.

On-board rotors are also well-known to be associated with parametric instabilities, due to the presence of linear time-varying terms in the damping or stiffness system characteristics in case of rotational motion of the base. In this context, Dakel *et al* (4) analyzed an academic rotor composed of a shaft and one disk discretized by beam finite elements (FE) with four degrees of freedom (DOFs) per node, supported by hydrodynamic journal bearing whose forces are linearized in the vicinity of the static equilibrium position. They found, by means of the Floquet theory, the presence of large instability regions for mono-axis sinusoidal rotation of the base. Han *et al* (5) and Yi *et al* (6) also investigated stability of a similar rotor, however for multi-axis sinusoidal rotations of the base with the same frequencies and in the presence of shaft asymmetry. Lately, Soni *et al* (7) analyzed the stability of rotor supported by active magnetic bearings and excited sequentially by the three different axes of rotation of the base. Thus, all of these previous researches were limited either to sequential mono-axis motions or to multi-axes motions with only one frequency. Furthermore, they considered only shaft motion in bending.

In this context, the purpose of the present paper is to analyze an academic rotor able to be designed and constructed in order to investigate the possible coupled instabilities due to the harmonic yaw and/or pitch of its base. Such multi-axes sinusoidal rotations with arbitrary frequencies reflect a more general case and likely more representative of the real operating conditions experienced by a rotor system. Moreover, all the likely dynamic effects of the shaft are taken into account, that is to say, bending, torsion and axial effects. This is permitted by resorting to a FE mesh of the rotor shaft composed of six DOFs per node, four of them related to bending motion according to the Timoshenko beam theory and the other two for torsion and axial motions. The rotor case under study is composed of a slender shaft with two perfectly rigid disks, supported by two hydrodynamic journal bearings mounted on rigid moving. First, the construction of FE model of the on-board rotor and the kinematics of the base motion are described. Then a brief recall of the Floquet theory and its extension to multi-frequency are made. Finally, numerical simulations are performed in order to determine the stability of the system for yaw and pitch base excitation, assessed separately or simultaneously.

2 MODELING

2.1 Kinematics of the base

The rotor under study is shown in Figure 1. It is composed of a slender shaft, two identical disks and two hydrodynamic journal bearings. The latter are mounted on a rigid base able to move according to six types of deterministic motions (three translations and three rotations). To define these motions, two different frames of reference are used. The first one, so-called R_0 is considered as the inertial frame. It has an origin O_0 and three orthogonal axes $(\vec{X}, \vec{Y}, \vec{Z})$. The second one, so-called R, is fixed with the rotor base. Its origin O is coinciding with the left end of the rotor shaft at rest (i.e. without applied load) and it has three axes $(\vec{x}, \vec{y}, \vec{z})$. In this way, the translational motion of the base can be defined thanks to the vector $\overrightarrow{O_0O}$ expressed in the inertial frame according to:

$$\overrightarrow{O_0O} = \left\{ \begin{array}{c} X \\ Y \\ Z \end{array} \right\}_{R_0} \tag{1}$$

with X, Y and Z the coordinates of O in R_0. These coordinates depend on time but this is kept implicit in the sake to simplify the notation.

Figure 1. Sketch of the on-board two-disk-rotor bearing system defined with the references frames R_0 and R.

The orientation of the rotor base with respect to the frame R_0 is defined by means of the Euler angles $(\alpha_1, \alpha_2, \alpha_3)$ α_1 is the yaw rotation around the axis $\vec{Z}$, and creates a first intermediary frame R_1 with the origin O_0 and axes $(\vec{x}_1, \vec{y}_1, \vec{z}_1)$ where $\vec{z}_1 = \vec{Z}$. α_2 is the pitch rotation around the axis $\vec{x}_1$ of R_1, and creates a second intermediary frame R_2 with the origin O_0 and axes $(\vec{x}_2, \vec{y}_2, \vec{z}_2)$ where $\vec{x}_2 = \vec{x}_1$. α_3 is the roll rotation around the axis $\vec{y}_2$ of R_2, so that it creates the new frame R with $\vec{y} = \vec{y}_2$. The mobility of the rotor base with respect to the inertial frame can be seen as the three translational variables (X, Y, Z) and the three rotational variables $(\alpha_1, \alpha_2, \alpha_3)$ which all depend implicitly on time. Besides, any vector can be expressed from one frame to another by using the following transformation:

$$
\left\{ \begin{array}{c} \vec{X} \\ \vec{Y} \\ \vec{Z} \end{array} \right\} = [P_{R_0 \to R}] \left\{ \begin{array}{c} \vec{x} \\ \vec{y} \\ \vec{z} \end{array} \right\}
\tag{2}
$$

with

$$
[P_{R_0 \to R}] = \begin{bmatrix} \cos\alpha_1 \cos\alpha_3 - \sin\alpha_1 \sin\alpha_2 \sin\alpha_3 & -\sin\alpha_1 \cos\alpha_2 & \cos\alpha_1 \sin\alpha_3 + \sin\alpha_1 \sin\alpha_2 \cos\alpha_3 \\ \sin\alpha_1 \cos\alpha_3 + \cos\alpha_1 \sin\alpha_2 \sin\alpha_3 & \cos\alpha_1 \cos\alpha_2 & \sin\alpha_1 \sin\alpha_3 - \cos\alpha_1 \sin\alpha_2 \cos\alpha_3 \\ -\cos\alpha_2 \sin\alpha_3 & \sin\alpha_2 & \cos\alpha_2 \cos\alpha_3 \end{bmatrix}
$$

the orthogonal transformation matrix from the frame R_0 to the frame R. The instantaneous rotation vector $\vec{\Omega}_R^{R_0}$ of R with respect to R_0 turns is expressed as $\vec{\Omega}_R^{R_0} = \dot{\alpha}_1 \vec{z} + \dot{\alpha}_2 \vec{x}_1 + \dot{\alpha}_3 \vec{y}$. Thanks to the transformation matrix of Equation (2), this vector can be rewritten in R such that:

$$
\vec{\Omega}_R^{R_0} = \left\{ \begin{array}{c} \omega_x^0 \\ \omega_y^0 \\ \omega_z^0 \end{array} \right\}_R = \left\{ \begin{array}{c} -\dot{\alpha}_1 \cos\alpha_2 \sin\alpha_3 + \dot{\alpha}_2 \cos\alpha_3 \\ \dot{\alpha}_1 \sin\alpha_2 + \dot{\alpha}_3 \\ -\dot{\alpha}_1 \cos\alpha_2 \cos\alpha_3 + \dot{\alpha}_2 \sin\alpha_3 \end{array} \right\}_R
\tag{3}
$$

Likewise, the origin O of R can be rewritten in R instead of R_0 in the following way:

$$
\overrightarrow{O_0 O} = \left\{ \begin{array}{c} x_0 \\ y_0 \\ z_0 \end{array} \right\}_R = [P_{R_0 \to R}]^T \left\{ \begin{array}{c} X \\ Y \\ Z \end{array} \right\}_{R_0}
\tag{4}
$$

where $[.]^T$ is the matrix transpose.

Once the motions of the rigid base have been defined, the system of equations governing the motion of the on-board rotor is easily obtained by using a FE approximation of the system, and by applying the Lagrange's equations.

2.2   FE model of the on-board rotor
In order to apply the Lagrange's equations, it is first necessary to obtain the expression of the kinetic energy, the strain energy, the gravitational potential and the virtual work of the whole system, and then resorting to FE discretization to obtain the final global matrices of the system of equations. Since kinetic energy and gravitational potential are the only features affected by base motion, special care is given below to describe their derivation. Nevertheless, only the part corresponding to the disks is computed since the one of the shaft can be easily determined as an extension of the disk case. More details may be found in (8). The expressions of the strain energy and virtual work are not detailed here since they can be obtained by classical approaches.

Hence, the example of the disk represented in Figure 2 is addressed. It has a radius R_d, a thickness h, a density ρ_d and a volume V_d. The disk has its center of inertia C initially (before shaft deformation) located at the coordinates $(0, y_d, 0)$ in the frame R. In its initial configuration, the disk can be represented by any point P of coordinates (x, y, z) in R with $y \in [-h/2; h/2]$ and $x^2 + z^2 \leq R_d^2$. After deformation, the disk orientation can be described with a new frame R' fixed to the disk, with an origin C' and axes $(\vec{x}', \vec{y}', \vec{z}')$. In the same logic as for the orientation of R with respect to R_0, three new Euler angles $(\psi_d, \theta_d, \phi_d)$ can be used to denote the three rotations to obtain R' from R, characterized by the same transformation matrix as in Equation (2) where $(\alpha_1, \alpha_2, \alpha_3)$ are substituted for $(\psi_d, \theta_d, \phi_d)$. Under small rotation hypothesis, it can be assumed then that ψ_d is the rotation around $\vec{z}$ and θ_d is the rotation around $\vec{x}$. The third angle ϕ_d can be decomposed as $\phi_d = \phi^* + \beta_d$ where ϕ^* is the nominal rotation (shaft rotation angle without deformation) and β_d is the torsion angle. Regarding, the translational motion of the disk, it can be defined by the coordinates (u_d, v_d, w_d) in R of its center of inertia C that moves to C' such that $\overrightarrow{CC'} = u_d\vec{x} + v_d\vec{y} + w_d\vec{z}$. By using this kinematics, $(u_d, v_d, w_d, \psi_d, \theta_d, \beta_d)$ become the six DOFs of the disk, being respectively the deflection in $\vec{x}$, the axial motion in $\vec{y}$, the deflection in $\vec{z}$, the bending rotation around $\vec{z}$, the bending rotation around $\vec{x}$ and the torsion around $\vec{y}$. Since the disk is considered to be perfectly rigid, the point P moves to P' after the nominal rotation of the shaft such that $\overrightarrow{C'P'} = x\vec{x}' + y\vec{y}' + z\vec{z}'$. Then, P'can be located with respect to the inertial frame R_0 according to:

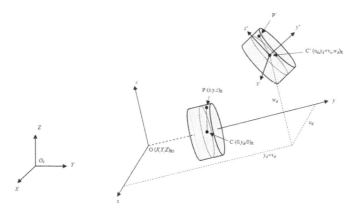

Figure 2. Typical motion of the disk due to external excitation.

315

$$\overrightarrow{O_0P'} = \overrightarrow{O_0O} + \overrightarrow{OC} + \overrightarrow{CC'} + \overrightarrow{C'P'} = \begin{Bmatrix} x_0 \\ y_0 \\ z_0 \end{Bmatrix}_R + \begin{Bmatrix} 0 \\ y_d \\ 0 \end{Bmatrix}_R + \begin{Bmatrix} u_d \\ v_d \\ w_d \end{Bmatrix}_R + \begin{Bmatrix} x \\ y \\ z \end{Bmatrix}_{R'} \tag{5}$$

The kinetic energy of the disk with respect to the inertial frame R_0 is then by definition:

$$T_d^0 = \frac{1}{2} \int_{V_d} \rho_d \left(\vec{V}_d^{R_0}(P') \right)^2 dV \tag{6}$$

where $\vec{V}_d^{R_0}(P')$ is the velocity any point P' belonging to the disk after deformation. Then, starting from Equation (5) and after some mathematical developments, it is easy to find the following expression of the kinetic energy:

$$T_d^0 = \frac{1}{2} m_d \left(v_{d,1}^2 + v_{d,2}^2 + v_{d,3}^2 \right) + \frac{1}{2} \left(I_{d,x}(\omega_{x'}^0)^2 + I_{d,y}\left(\omega_{y'}^0\right)^2 + I_{d,z}(\omega_{z'}^0)^2 \right) \tag{7}$$

with m_d the disk mass, $(I_{d,x}, I_{d,y}, I_{d,z})$ respectively its moments of inertia around the axes $(\vec{x}', \vec{y}', \vec{z}')$,

- $v_{d,1} = \dot{x}_0 + \dot{u}_d + \omega_y^0(z_0 + w_d) - \omega_z^0(y_0 + y_d + v_d)$,
- $v_{d,2} = \dot{y}_0 + \dot{v}_d + \omega_z^0(x_0 + u_d) - \omega_x^0(z_0 + w_d)$,
- $v_{d,3} = \dot{z}_0 + \dot{w}_d + \omega_x^0(y_0 + y_d + v_d) - \omega_y^0(x_0 + u_d)$

and

$$\begin{Bmatrix} \omega_{x'}^0 \\ \omega_{y'}^0 \\ \omega_{z'}^0 \end{Bmatrix} = \begin{Bmatrix} -\dot{\psi}\cos\theta\sin\phi + \dot{\theta}\cos\phi \\ \dot{\psi}\sin\theta + \dot{\phi} \\ -\dot{\psi}\cos\theta\cos\phi + \dot{\theta}\sin\phi \end{Bmatrix} + [P_{R' \to R}] \begin{Bmatrix} \omega_x^0 \\ \omega_y^0 \\ \omega_z^0 \end{Bmatrix} \tag{8}$$

Following the same kinematics, the gravitational potential $U_{g,d}^0$ of the disk is found to be expressed as:

$$U_{g,d}^0 = m_d g(-\cos\alpha_2 \sin\alpha_3(x_0 + u_d) + \sin\alpha_2(y_0 + y_d + v_d) + \cos\alpha_2\cos\alpha_3(z_0 + w_d)) \tag{9}$$

where g is the gravitational acceleration directed along $-\vec{Z}$.

Regarding the shaft energies, as mentioned above, they can be easily obtained as an extension of the disk case accounting for the y-dependency of the 6 DOFs. The equations of motion can further be derived by first FE discretizing of the whole system. Timoshenko beam theory is employed to represent the shaft, accounting for both rotation inertia of the cross sections and for shear effects. For the bending motion, the FE used has four DOFs per node, associated to cubic interpolation of the shape functions (see IIE element of Reddy (9)). The classical linear Lagrange shape functions is applied for the axial and torsion DOFs, adding two more DOFs per node. The final system of equations of motion is finally obtained by applying the Lagrange's equation with the different energies involved, as in Equations (7) and (9). Only the global form is given below with a focus on the matrices and forces vectors of the disk related to the base motion, which will already represent a good support to explain the phenomena illustrated in the simulation section. Thus the final global system of motion is:

$$[M]\{\ddot{\delta}\} + ([C] + [G] + [C_{jb}] + [C_b(t)])\{\dot{\delta}\} + ([K] + [K_{jb}] + [K_b(t)])\{\delta\} = \{F_g(t)\} + \{F_b(t)\} \tag{10}$$

with $[M]$, $[C]$, $[G]$, and $[K]$ respectively the classical mass, damping, gyroscopic (speed of rotation dependent) and stiffness matrices related to a rotor with a fixed base and $\{\delta\}$ the global vector of unknowns of size (Nx1) expressed with respect to the static equilibrium position. $[C_{jb}]$ and $[K_{jb}]$ which depend on the speed of rotation are respectively the damping and stiffness matrices coming from the linearization of the nonlinear forces of the hydrodynamic journal bearing around the static equilibrium position. $[C_b(t)]$, $[K_b(t)]$ and $\{F_b(t)\}$ are respectively extra damping matrix, stiffness matrices and force vector due to the base motion. $\{F_g(t)\}$ is the force vector related to gravitational potential that depends on time also due to base motion. These time-dependent matrices and vectors are all composed of the participation of the shaft and the two disks. The contribution of one disk to the global matrices and vectors gives the following matrices expressed in the 6-DOFs ($u_d, v_d, w_d, \psi_d, \theta_d, \beta_d$):

$$[C_{b,d}(t)] = \begin{bmatrix} 0 & -2m_d\omega_z^0 & 2m_d\omega_y^0 & 0 & 0 & 0 \\ 2m_d\omega_z^0 & 0 & -2m_d\omega_x^0 & 0 & 0 & 0 \\ -2m_d\omega_y^0 & 2m_d\omega_x^0 & 0 & 0 & 0 & 0 \\ 0 & 0 & 0 & 0 & (I_{d,y}-2I_{d,x})\omega_y^0 & I_{d,y}\omega_x^0 \\ 0 & 0 & 0 & -(I_{d,y}-2I_{d,x})\omega_y^0 & 0 & -I_{d,y}\omega_z^0 \\ 0 & 0 & 0 & -I_{d,y}\omega_x^0 & I_{d,y}\omega_z^0 & 0 \end{bmatrix} \quad (11.\text{a})$$

$$[K_{b,d}(t)] = [K_{b,d,t}(t)] + [K_{b,d,r}(t)] \quad (11.\text{b})$$

$$[K_{b,d,t}(t)] = m_d \begin{bmatrix} -\left(\omega_y^0\right)^2 - \left(\omega_z^0\right)^2 & -\dot{\omega}_z^0 + \omega_x^0\omega_y^0 & \dot{\omega}_y^0 + \omega_x^0\omega_z^0 & 0 & 0 & 0 \\ \dot{\omega}_z^0 + \omega_x^0\omega_y^0 & -\left(\omega_x^0\right)^2 - \left(\omega_z^0\right)^2 & -\dot{\omega}_x^0 + \omega_y^0\omega_z^0 & 0 & 0 & 0 \\ -\dot{\omega}_y^0 + \omega_x^0\omega_z^0 & \dot{\omega}_x^0 + \omega_y^0\omega_z^0 & -\left(\omega_x^0\right)^2 - \left(\omega_y^0\right)^2 & 0 & 0 & 0 \\ 0 & 0 & 0 & 0 & 0 & 0 \\ 0 & 0 & 0 & 0 & 0 & 0 \\ 0 & 0 & 0 & 0 & 0 & 0 \end{bmatrix} \quad (11.\text{c})$$

$$[K_{b,d,t}(t)] = \begin{bmatrix} 0 & 0 & 0 & 0 & 0 & 0 \\ 0 & 0 & 0 & 0 & 0 & 0 \\ 0 & 0 & 0 & 0 & 0 & 0 \\ 0 & 0 & 0 & (I_{d,y}-I_{d,x})\left(\left(\omega_y^0\right)^2 - \left(\omega_x^0\right)^2\right) + I_{d,y}\omega_y^0\dot{\phi}^* & (I_{d,y}-I_{d,x})\left(\dot{\omega}_y^0 + \omega_x^0\omega_z^0\right) & 0 \\ 0 & 0 & 0 & I_{d,x}\dot{\omega}_y^0 + (I_{d,y}-I_{d,x})\omega_x^0\omega_z^0 & (I_{d,y}-I_{d,x})\left(\left(\omega_y^0\right)^2 - \left(\omega_z^0\right)^2\right) + I_{d,y}\omega_y^0\dot{\phi}^* & 0 \\ 0 & 0 & 0 & -I_{d,y}\dot{\omega}_x^0 & I_{d,y}\dot{\omega}_z^0 & 0 \end{bmatrix}$$

$$(11.\text{d})$$

$$\{F_{b,d}(t)\} = - \begin{Bmatrix} m_d\left[\ddot{x}_0 - 2\left(\omega_z^0\dot{y}_0 - \omega_y^0\dot{z}_0\right) - x_0\left(\left(\omega_y^0\right)^2 + \left(\omega_z^0\right)^2\right) - (y_0-y_d)\left(\dot{\omega}_z^0 - \omega_x^0\omega_y^0\right) + y_0\left(\dot{\omega}_y^0 - \omega_x^0\omega_z^0\right)\right] \\ m_d\left[\ddot{y}_0 + 2\left(\omega_z^0\dot{x}_0 - \omega_x^0\dot{z}_0\right) + x_0\left(\dot{\omega}_z^0 - \omega_x^0\omega_y^0\right) - (y_0+y_d)\left(\left(\omega_x^0\right)^2 + \left(\omega_z^0\right)^2\right) - z_0\left(\dot{\omega}_x^0 - \omega_y^0\omega_z^0\right)\right] \\ m_d\left[\ddot{z}_0 - 2\left(\omega_y^0\dot{x}_0 - \omega_z^0\dot{y}_0\right) - x_0\left(\dot{\omega}_y^0 - \omega_x^0\omega_z^0\right) + (y_0+y_d)\left(\dot{\omega}_x^0 - \omega_y^0\omega_z^0\right) - z_0\left(\left(\omega_x^0\right)^2 + \left(\omega_y^0\right)^2\right)\right] \\ I_{d,x}\dot{\omega}_z^0 + (I_{d,y}-I_{d,x})\omega_x^0\omega_y^0 + I_{d,y}\omega_x^0\dot{\phi}^* \\ I_{d,x}\dot{\omega}_x^0 + (I_{d,y}-I_{d,x})\omega_y^0\omega_z^0 + I_{d,y}\omega_z^0\dot{\phi}^* \\ I_{d,y}\dot{\omega}_y^0 \end{Bmatrix} \quad (11.\text{e})$$

$$\{F_{g,d}(t)\} = -m_d g \begin{Bmatrix} -\cos\alpha_2 \sin\alpha_3 \\ \sin\alpha_2 \\ \cos\alpha_2 \cos\alpha_3 - 1 \\ 0 \\ 0 \\ 0 \end{Bmatrix} \tag{11.f}$$

It is to be highlighted that no mass unbalance force is taken into account since only stability is the only interest of the present paper. Likewise, all external force of Equation (10) will be removed so that only the following homogenous system:

$$[M]\{\ddot{\delta}\} + ([C] + [G] + [C_{jb}] + [C_b(t)])\{\dot{\delta}\} + ([K] + [K_{jb}] + [K_b(t)])\{\delta\} = \{0\} \tag{12}$$

is addressed.

2 PARAMETRIC STABILITY BY FLOQUET THEORY

Time-dependent matrices can be noted in the homogenous system of equations (12). This particular type of system, called parametric system or Mathieu equation in the case of parametric harmonic excitation, is well-known to be likely unstable for periodic motions with certain conditions of amplitudes and frequencies. This can be predicted by applying the Floquet Theory (10) or (14). Its principle consists in determining the monodromy matrix $[A]$ of the system and to compute its eigenvalues. Instability occurs when the module of one of these eigenvalues, also called as Floquet character-istic multipliers, is higher than unity. The monodromy matrix can be easily found by computing the solutions of Equation (12) with unitary initial conditions for all the DOFs involved, over one period of the parametric excitation. Namely, the following equation,

$$[M]\begin{bmatrix}\ddot{\Phi}\end{bmatrix} + ([C] + [G] + [C_{jb}] + [C_b(t)])\begin{bmatrix}\dot{\Phi}\end{bmatrix} + ([K] + [K_{jb}] + [K_b(t)])[\Phi] = [0] \tag{13}$$

is to be solved, where $[\Phi]$ is an $(N \times 2N)$ matrix such that $[\Phi(0)] = [[I] \ [0]]$ and $[\dot{\Phi}(0)] = [[0] \ [I]]$ with $[I]$ and $[0]$ respectively the $(N \times N)$ identity and nil matrices. Then monodromy matrix is then found as

$$[A] = \begin{bmatrix} [\Phi(T)] \\ [\dot{\Phi}(T)] \end{bmatrix} \tag{14}$$

This can be achieved with numerical integration, for instance with the classical New-mark implicit scheme (12). The stability can be investigated for several values of two parameters, for instance the amplitude and the frequency of a mono-axis sinusoidal excitation. This leads to the construction of Strutt diagrams, or stability diagrams. In case of multi-axis sinusoidal base excitation, several frequencies may be contained in the signal. Consequently, it may be of great interest to assess the stability behavior with respect to these frequencies. Let the example of a two-axes rotation of the base be considered, defined respectively by the yaw and pitch harmonic rotations $\alpha_1 = A_1 \cos(2\pi f_1 t + \varphi_1)$ and $\alpha_2 = A_2 \cos(2\pi f_2 t + \varphi_2)$ where (A_1, A_2), (f_1, f_2) and (φ_1, φ_2) are respectively the amplitudes, frequencies and phases of the rotations around the axes $\vec{Z}$ and $\vec{x}_1$. If the set of frequencies (f_1, f_2) is arbitrary, the parametric excitation may not be periodic. It will remain periodic provided the ratio f_1/f_2 is a rational number. In this case, the overall period is $T = n_1/f_1 = n_2/f_2$ where (f_1, f_2) is the set of integers such as n_1/n_2 is the irreducible ratio of f_1/f_2. By noting this, it is possible to construct stability

318

diagrams with (f_1, f_2) as the two parameters, numerically integrating Equation (13) over $t \in [0, T]$, where T must be updated with the values of these parameters. Nevertheless, the time step Δt used must be small enough to account for the linear combination of the harmonic of (f_1, f_2) that may have high values (see expressions of ω_x^0, ω_y^0 and ω_z^0 in Equation (3)).

3 NUMERICAL SIMULATIONS

The objective is to design an academic rotor, sketched in Figure 3a, able to be tested in the future on a 6-DOF shaker for investigating the possible coupled instabilities due to the combination of yaw and pitch rotation of its base. Its FE mesh is shown in Figure 3b. The rotor shaft has a length of 851 mm and 27 beam FEs, yielding $N = 168$ DOFs. The radius of the first three elements (left extremity) and last three elements (right extremity) is 12 mm while it is 12.7 mm for all others elements. The shaft and disk densities are respectively 7776.7 kg/m^3 and 7778 kg/m^3. The disks radius and thickness are 63.055 mm and 15.6 mm. The Young modulus and Poisson coefficient of the shaft are 210 GPa and 0.3. The journal bearing radius and length are 6.39 mm and 6.2 mm. The clearance with the shaft is 42.5 μm and the oil viscosity is 0.0206 Pa.s. It must be emphasized that the journal bearings geometry does not respect the short journal bearing approximation. Hence, finite-length approximation is used which implies a numerical resolution of the Reynolds equation with a finite difference scheme. Reynolds boundary conditions are employed with the help of a complementary problem as described by Xiao (13). The shaft speed of rotation will be held to 1700 rpm for all the analysis. The characteristic of the two hydrodynamic bearings for this value of speed of rotation are given in Table 1 (H.B.1 for the left hydrodynamic bearing which is the closest to the motor (left) and H.B.2 for the right one). The coupling characteristics are given in Table 2. The corresponding first natural frequencies, dimensionless damping ratio and associated motion (axial, torsion and bending in Forward Whirl or Backward Whirl) at 1700 rpm are the ones displayed in Table 3.

In order to interpret with more simplicity the stability diagram for a two-axe rotation, it is expedient first to focus on the two mono-axis rotations separately. Therefore, two different stability diagrams are constructed, each one corresponding to a yaw harmonic rotation defined such as $\alpha_1 = A_1 \cos(2\pi f_1 t + \varphi_1)$ or a pitch rotation $\alpha_2 = A_2 \cos(2\pi f_2 t + \varphi_2)$, with (A_1, f_1) and (A_2, f_2) respectively the two parameters of the first and second diagram. For both cases, the amplitudes are varying respectively within [0;0.5] rad divided with 30 equidistant iterations, the frequencies within [0;200] Hz divided with 200 equidistant iterations. The phases are both defined such as $\varphi_1 = \varphi_2 = 0°$ since for mono-axis rotation they do not have any influence. Regarding the numerical integration, 512 points per period of excitation $T = 1/f_1$ or $T = 1/f_2$ are used in the Floquet Theory.

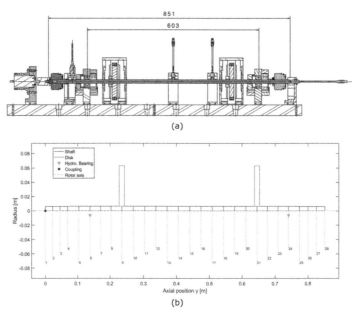

(a)

(b)

Figure 3. Academic rotor. (a) drawing (in mm), (b) FE mesh: On-board rotor FE mesh containing 27 two-node beams, 2 one-node disks, 1 one-node coupling, and two-node bearings.

Table 1 . Hydrodynamic bearings characteristics.

Stiffness coef.	H.B. 1	H.B. 2	Damping coef.	H.B. 1	H.B. 2
k_{uu}	4.31×10^6 N/m	4.19×10^6 N/m	c_{uu}	2.34×10^4 N.s/m	2.28×10^4 N.s/m
k_{uw}	-2.36×10^6 N/m	-2.31×10^6 N/m	c_{uw}	-5.62×10^3 N.s/m	-5.51×10^3 N.s/m
k_{wu}	-3.94×10^5 N/m	-3.82×10^5 N/m	c_{wu}	-5.62×10^3 N.s/m	-5.50×10^3 N.s/m
k_{ww}	8.55×10^5 N/m	8.41×10^5 N/m	c_{ww}	3.08×10^3 N.s/m	3.03×10^3 N.s/m

Table 2 . Coupling characteristics.

Stiffness coef.	Coupling	Damping coef.	Coupling
k_{uu}	0 N/m	c_{uu}	12 N.s/m
k_{vv}	2700 N/m	c_{vv}	10 N.s/m
k_{ww}	0 N/m	c_{ww}	0 N.s/m
$k_{\psi\psi}$	0 N.m/rad	$c_{\psi\psi}$	0 N.m.s/rad
$k_{\theta\theta}$	0 N.m/rad	$c_{\theta\theta}$	0 N.m.s/rad
$k_{\beta\beta}$	100 N.m/rad	$c_{\beta\beta}$	2.4×10^{-3} N.m.s/rad

Table 3 . First modes of the rotor bearing system running at 1700 rpm.

Mode number	1	2	3	4	5	6	7
Motion	Axial	Torsion	Bending (BW)	Bending (FW)	Bending (FW)	Bending (BW)	Torsion
Frequency [Hz]	4.22	18.90	35.67	37.96	90.40	92.50	94.00
Damping ratio	0.049	1.2×10^{-3}	0.041	0.019	0.084	0.013	2.91×10^{-4}

The resulting two stability diagrams are presented in Figure 4.a and 4.b. For each one, the color map denotes the value of the highest Floquet exponent of the monodromy matrix, namely the intensity of the instability. For both cases, a large instability zone can be noticed, as the one evidenced by Dakel *et al.* (4). Nevertheless, narrow instability zones also appear, almost at the same frequencies of the base excitation for the two cases, highlighting a certain symmetry level. In fact, these new regions can be easily explained by resorting to the famous formula of Hsu (14), according to which instability may occur when $F = |f_i \pm f_j|/k$ where F is the parametric excitation frequency, (f_i, f_j) are two natural modes frequencies of the system and k is the order of the instability. In this way, the instability region around 35.18 Hz of Figure 4.a is found to be a combination of the third mode of Table 3 with itself. Indeed, in the case when only the yaw is acting, the matrix $[K_{b,d}(t)]$ of Equation (11.b) simplifies as

$$[K_{b,d}(t)] = \begin{bmatrix} -m_d(\omega_z^0)^2 & -m_d\dot{\omega}_z^0 & 0 & 0 & 0 & 0 \\ m_d\dot{\omega}_z^0 & -m_d(\omega_z^0)^2 & 0 & 0 & 0 & 0 \\ 0 & 0 & 0 & 0 & 0 & 0 \\ 0 & 0 & 0 & 0 & 0 & 0 \\ 0 & 0 & 0 & 0 & -(I_{d,y}-I_{d,x})(\omega_z^0)^2 & 0 \\ 0 & 0 & 0 & 0 & I_{d,y}\dot{\omega}_z^0 & 0 \end{bmatrix} \quad (15)$$

with $\omega_z^0 = \dot{\alpha}_1 = -A_1 2\pi f_1 \sin(2\pi f_1 t + \varphi_1)$ and $\dot{\omega}_z^0 = \ddot{\alpha}_1 = -A_1(2\pi f_1)^2 \cos(2\pi f_1 t + \varphi_1)$. Thus, owing to the square on ω_z^0 present in the diagonal of this matrix, the parametric frequency is $2f_1$ for the bending or for the axial motions without coupling. Then, it occurs that $2f_1 = 70.36 \approx |35.67 + 35.67|/1$, so that it is a primary instability of bending motion only. Proceeding with the same logic, the instability region around 42.21 Hz of Figure 4.a is a combination of the first and fourth modes of Table 3. Indeed, the parametric frequency is in this case f_1 due to the extra-diagonal term $\dot{\omega}_z^0$ which couples bending and axial motion, and $f_1 = 42.21 \approx |4.22 + 37.96|/1$. For pitch excitation with α_2, the matrix $[K_{b,d}(t)]$ turns

$$[K_{b,d}(t)] = \begin{bmatrix} 0 & 0 & 0 & 0 & 0 & 0 \\ 0 & -m_d(\omega_x^0)^2 & -m_d\dot{\omega}_x^0 & 0 & 0 & 0 \\ 0 & m_d\dot{\omega}_x^0 & -m_d(\omega_x^0)^2 & 0 & 0 & 0 \\ 0 & 0 & 0 & -(I_{d,y}-I_{d,x})(\omega_x^0)^2 & 0 & 0 \\ 0 & 0 & 0 & 0 & 0 & 0 \\ 0 & 0 & 0 & -I_{d,y}\dot{\omega}_x^0 & 0 & 0 \end{bmatrix} \quad (16)$$

with $\omega_x^0 = \dot{\alpha}_2 = -A_2 2\pi f_2 \sin(2\pi f_2 t + \varphi_2)$ and $\dot{\omega}_x^0 = \ddot{\alpha}_2 = -A_2(2\pi f_2)^2 \cos(2\pi f_2 t + \varphi_2)$. In the same way, a pure bending instability appears near 34.17 Hz as exhibited in Figure 4.b, once again related to the third mode of Table 3. Regarding the instability region near 39.2 Hz, it comes from a combination of the axial mode with the third mode of Table 3 with $f_2 = 39.2 \approx |4.22 + 35.67|/1$. This shows that pitch excitation of the base favors more the participation of the backward whirl while yaw excitation favors more the forward whirl, when couple with axial motion. Likewise, the others narrow regions can be seen as other natural modes combinations. Furthermore, all of these observations were confirmed by proceeding with a numerical integration transient test with a free response to small initial condition in the base excitation configurations mentioned. This ensures first to validate the divergence occurring over time and secondly to identify the modes involved by performing Fourier analysis for instance.

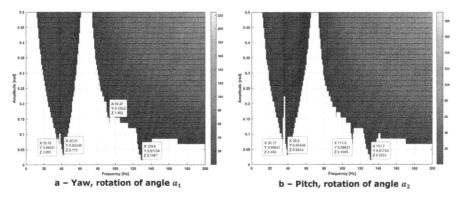

a – Yaw, rotation of angle α_1 b – Pitch, rotation of angle α_2

Figure 4. Stability diagrams due to the yaw or to pitch of the base.

Once the two rotations have been assessed separately, the two-axe rotation can be addressed. In this case, four parameters are now necessary to build complete stability diagrams, namely (A_1, f_1, A_2, f_2). Actually, there are even six parameters with the phases (φ_1, φ_2) that may now have an influence on the stability. Nevertheless, for the sake of clarity, it is expedient to restrict the diagrams to only two parameters, which will be (f_1, f_2) for the reasons stated in the introduction section. Hence, the other four parameters have to be fixed. By looking at Figure 4, interesting amplitudes of rotation can be deduced as the ones for which instability exist. For instance, with $A_1 = A_2 = 0.05$ rad, two narrow regions are present in the frequency range of [1;200] Hz. Thus these values of amplitude are kept for the multi-axes case. Regarding the two phases, arbitrary values are chosen such as $\varphi_1 = 57°$ and $\varphi_2 = 91°$. The Floquet theory is then applied, varying the two parameters (f_1, f_2) with 200 equidistant iterations for each one (so that $\Delta f_1 = \Delta f_2 = 1$ Hz). Numerical integration is performed with a time step $\Delta t = 1 \times 10^{-4}$ s.

The corresponding stability diagram is presented in Figure 5. In contrast with a system for which the two directions of excitation would not be coupled, the diagram presents horizontal and vertical narrow instability regions that are no longer constant according to the values of (f_1, f_2). In particular, variations can most be noticed at the intersections of these regions and in the highest frequencies. Furthermore, a new instability region appear for instance at $(f_2 = 33$ Hz$, f_1 = 183$ Hz$)$ and $(f_2 = 178$ Hz$, f_1 = 37$ Hz$)$, whereas there was no such instability in the respective mono-axis cases for the parameters $(A_1 = 0.05$ rad$, f_1 = 37$ Hz$)$ and $(A_2 = 0.05$ rad$, f_1 = 33$ Hz$)$. In the same way, the large instability region also appears and its boundary is diagonal. In addition, a diagonal new tiny instability region shows up at $(f_2 = 165$ Hz$, f_1 = 165$ Hz$)$, which seems to be directed by a perfect linear relation between the two parameters, such as $f_1 = 1f_2$. This can be explained by the existence of the term $\dot{\omega}_y^0$ which is no longer nil in case of a two-axes rotation, although there are only yaw and pitch rotations. Indeed, the expression of ω_y^0 in Equation (3) in this case is $\omega_y^0 = \dot{\alpha}_1 \sin \alpha_2$ (since $\alpha_3 = 0$), so that the parametric excitation frequency of the Hsu formula is actually of the type $|f_1 \pm (2q+1)f_2|$ where q is any integer (in this case $q=0$). This shows that multi-axes parametric excitation can highly enrich the stability behavior of the system and thus how essential it has to be taken into account rather than the mono-axis ones assessed separately.

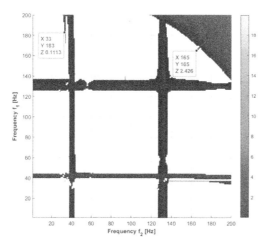

Figure 5. Stability diagrams due to the combined harmonic yaw and pitch excitations.

It is legitimate now to wonder about the influence of the others four parameters available $(A_1, A_2, \varphi_1, \varphi_2)$ in the stability results. To this end, it is proposed to build to new stability diagrams, with the same parameters as previously for Figure 5, however one with increased amplitudes such as $A_1 = A_2 = 0.1$ rad and another one with different phase such as $\varphi_1 = 57°$ and $\varphi_2 = 221°$. The latter values of phases have been chosen arbitrarily. The corresponding stability diagrams are respectively exposed in Figure 6.a and 6.b. In the former, it can be seen that the large instability region has drastically grown and represents now the most part of the graphic. It was expected since for this value of amplitude, Figure 4.a and 4.b already exhibited this behavior. Nevertheless, more variations are also noticed at the intersection of the horizontal and vertical narrow instability regions, demonstrating a higher level of coupling between the two directions of excitations. Furthermore, new laws of instability of the type $|f_1 \pm (2q+1)f_2|$ seems to occur near $f_1 = f_2 = 100$ Hz, for which one has $f_1 = 1/2f_2 + b$ or $f_1 = 2f_2 + b$ where b is the vertical intercept. Regarding Figure 6.b, no significant difference with Figure 5 can be stated, emphasizing the low impact of the phase shift. Only small local effects can be reported such as the appearance of a new diagonal narrow region near ($f_2 = 97$ Hz, $f_1 = 193$ Hz). The too low value of amplitude imposed may be one reason to explain this lack of influence. The important level of symmetry between the two excited directions may be another reason.

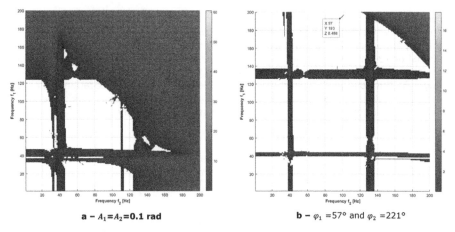

a – $A_1 = A_2 = 0.1$ rad   b – $\varphi_1 = 57°$ and $\varphi_2 = 221°$

Figure 6. Influence of the other parameters on the stability diagram.

4 CONCLUSION & FURTHER WORK

The objective of this investigation is to design an academic on-board rotor for the experimental validation of predicted coupled instabilities due to its base rotations. Therefore, in this paper was proposed the analysis of an academic on-board rotor system parametrically excited, with harmonic yaw and pitch with arbitrary frequencies that may not necessarily be multiples. The stability diagrams were obtained by the Floquet theory and by varying the two frequencies of the sinusoidal rotation around the transverse axes of the rotor. Outstanding observations were highlighted. In particular, couplings between the bending, axial and torsion were shown mathematically and their consequences were evidenced first on mono-axis parametric excitations. Secondly, variations of the corresponding narrow instability zones were seen, mostly at their orthogonal intersections and at high frequencies. Furthermore, diagonal narrow instability regions were evidenced, which confirmed the importance of accounting for multi-axes motions rather than mono-axis motion on the study of parametric instability. Finally, amplitude of the respective sinusoidal rotations has revealed a significant impact on the stability behavior, while the phase shift only bring small and local effects.

Despite stability has been investigated for periodic motions, an apparent continuity in frequency can be noticed in the multi-axes stability diagrams. Therefore, it is legitimate to think that instability may also occur in case of two non-commensurable frequencies, namely when the two frequencies (f_1, f_2) are not such as their ratio f_1/f_2 is a rational number. In this context, one can wonder if the instability still hold for a random process motion of the base, whose power spectral density would be defined in a frequency range contained within a narrow instability region. Besides, it remains to validate these results with suitable experimental validations. Nevertheless, the main drawback of practical experiments is that it is difficult to get rid of external forces that necessarily act in case of base motion and would likely mask the instability behavior.

ACKNOWLEDGMENTS

The authors gratefully acknowledge the financial support of the French National Research Agency (ANR) in the framework of the LaBCom-SME AdViTAM, ANR-16-LCV1 -0006, a joint laboratory of LaMCoS and AVNIR Engineering.

REFERENCES

[1] Lin, F., Meng, G., "Study on the dynamics of a rotor in a maneuvering aircraft", *ASME J. Vibr. Acoust*. 125 (3) (2003) 324–327.

[2] Duchemin, M., Berlioz, A., Ferraris, G., "Dynamic behavior and stability of a rotor under base excitation", *ASME J. Vibr. Acoust*. 128 (5) (2006) 576–585.

[3] Driot, N., Lamarque, C.H., Berlioz, A., "Theoretical and experimental analysis of a base-excited rotor", *ASME J. Comput. Nonlinear Dyn*. 1 (3) (2006) 257–263.

[4] Dakel, M., Baguet, S., Dufour, R., "Nonlinear dynamics of a support-excited flexible rotor with hydrodynamic journal bearings", *J. Sound Vibr*. 333 (10) (2014) 2774–2799.

[5] Han, Q., Chu, F., "Parametric instability of flexible rotor-bearing system under time-periodic base angular motions", *Appl. Math. Model*. 39 (15) (2015) 4511–4522.

[6] Yi, Y., Qiu, Z., Han, Q., "The effect of time-periodic base angular motions upon dynamic response of asymmetric rotor systems", *Adv. Mech. Eng*. 10 (3) (2018) 1–12.

[7] Soni, T., Dutt, J.K., Das, A.S., "Parametric stability analysis of active magnetic bearing-supported rotor system with a novel control law subject to periodic base motion", *IEEE Trans. Ind. Electron*. 67 (2) (2020) 1160–1170.

[8] Briend, Y., Dakel, M., Chatelet, E., Andrianoely, M. A., Dufour, R., & Baudin, S. (2020). "Effect of multi-frequency parametric excitations on the dynamics of on-board rotor-bearing systems", *Mechanism and Machine Theory*, 145, 103660.

[9] Reddy, J.N., "An Introduction to the Finite Element Method", *McGraw-Hill Education*, (2005).

[10] Nayfeh, A., and Mook, D., "Nonlinear Oscillations", *Wiley, Hoboken, NJ*, (1995).

[11] Dufour R., Berlioz A., "Parametric instability of a beam due to axial excitations and to boundary conditions", ASME – J. Vib. and Acoust., 120(2), (1998), 461–467.

[12] Bathe, K.J., "Finite Element Procedures", *Prentice Hall*, (2006).

[13] Xiao, Z., Wang, L., Zheng, T., "An efficient algorithm for fluid force and its Jacobian matrix in journal bearing", *Journal of Tribology*, 128 (2) (2006) 291–295.

[14] Hsu, C.S., "On the parametric excitation of a dynamic system having multiple degrees of freedom", *ASME J. Appl. Mech*. 30 (3) (1963) 367–372.

[15] Briend, Y., Dakel, M., Chatelet, E., Andrianoely, M.-A., Dufour, R., Baudin, S., "Extended modal reduction for on-board rotor with multifrequency parametric excitation", *ASME J. Vibr. Acoust*. 141 (6) (2019) 061009-1–061009-12.

12th International Conference on Vibrations in Rotating Machinery -
Institution of Mechanical Engineers, ISBN 978-0-367-67742-8

On the foundation dynamics and the active control of flexible rotors via active magnetic bearings

T.T. Paulsen, I.F. Santos

Department of Mechanical Engineering, Technical University of Denmark, Denmark

ABSTRACT

A high-speed blower consisting of a flexible shaft supported by two radial active magnetic bearings and an axial passive magnetic bearing is theoretically and experimentally analysed. The blower is mounted on a t-slot plate and the interactions between the shaft, the housing, and the t-slot plate are under consideration. It is investigated if the dynamical contribution from the foundation dynamics (t-slot plate plus housing) to the frequency response function of the high-speed blower can be reduced by including some of the most influential dynamics (mode shapes) of the foundation in the control object using a model-based control design.

MATHEMATICAL PARAMETERS

α – Proportional damping for mass

β – Proportional damping for stiffness

μ_0 – Permeability for vacuum

$\omega_{0,1}$ – Cross frequency for notch filter 1

$\omega_{0,2}$ – Cross frequency for notch filter 2

A – Surface area for pole legs

i_0 – Operating current in coils

N – Number of turns for coils

Q – Stop band factor for notch filter

R – Ohmic resistance

x_0 – Operating air gap in bearings

$\mathbf{x}$ – Vector x (bold small letter)

$\mathbf{X}$ – Matrix x (bold capital letter)

$[\ ]_{Axial}$ – Axial magnetic bearing related

$[\ ]_b$ – Boundary nodes

$[\ ]_{CB}$ – Craig-Bampton related

$[\ ]_{ext}$ – External related

$[\ ]_f$ – Foundation related

$[\ ]_i$ – Internal nodes

$[\ ]_{int}$ – Integral states augmented

$[\ ]_l$ – Left side transformation

$[\ ]_{MR}$ – Modal reduction related

$[\ ]_n$ – n row and m column matrix

$[\ ]_{obj}$ – Control object

$[\ ]_r$ – Rotor related

$[\ ]_{ref}$ – Reformulated matrix

$[\ ]_{ri}$ – Right side transformation

$[\ ]_{ss}$ – State space related

1   INTRODUCTION

The number of industrial applications exploiting magnetic bearing technology is increasing. Some of the great advantages in favour of the technology are low friction, low maintenance, low to zero wear during start-up and run-down, and customisable frequency response. Some setbacks for the technology are the complexity and cost,(1). For specific applications, the advantages are overwhelming. These applications have helped develop the magnetic bearing technology into how it is used today. The load carrying capabilities of magnetic bearings have been increasing due to this development. This has led to an increasing number of applications adopting magnetic bearing technology. In (2), the capabilities of the active magnetic bearing technology taking form at that time is discussed. An example of researchers trying to push the limits on carrying capacities of the technology using novel configurations is presented in (3).

For rotating machinery, some sort of support used to fixate the structure to the ground is necessary. If the support structure is flexible within the speed range of the machine it is fixating, process disturbances or residual unbalance could excite the flexible support structure. Interactions between beams or rotors and support structures or foundations are a familiar problem as presented in (4-7) among others. As the carrying capabilities for the magnetic bearing technology is increasing, so is the weight of the machines it is used for. Therefore, a greater support structure is needed. In general, the change of the mass will be more dominant than the change in stiffness for a structure as it is scaled, if the structure is not wisely designed. Therefore, if the foundation of a machine is increased in size, the flexible modes for that foundation will be lowered. Combined with the higher reaction forces needed for the magnetic bearings to support the rotor used in the machine, the flexible modes for the foundation are more likely to be excited through disturbances in the rotor within the operational speed range of the machine.

For other applications, a reduction to the total weight of the machine is desired, though the inertia of a spinning element will still be considerable. In this case, the support structure will be flexible if it is not carefully designed. Furthermore, for some applications a soft interaction between the machine housing and the machine carrier is considered an opportunity (8). This can result in amplifications of the dynamical response due to external disturbances.

If a machine were to comply with ISO 10349-1, a threshold for the mechanical vibration amplifications within the speed range of the machine is specified. If the specified threshold cannot be satisfied, the speed range will have to be compromised, else a reduction in the amplifications should be considered.

In this article, the dynamical impact from the foundation of a high-speed blower without an impeller mounted, thereby neglecting the fluid forces, is considered. The high-speed blower is investigated both mathematically and experimentally.

The experimental test setup, described in (9), consists of a rotor, an axial passive magnetic bearing, two radial active magnetic bearings, and an induction motor. The bearings and the motor are situated in a housing that is mounted to a t-slot plate. The housing and the t-slot plate is referred to as the foundation. To establish the mathematical model used to simulate the high-speed rotor interacting with the foundation, two finite element method models have been formulated, i.e. one for the rotor and one for the foundation. The interactions between the rotor and the foundation are described through a linearized mathematical model of the bearing dynamics.

In this work, a Proportional and Integral (PI) control law is used for the amplifiers to track a current reference defined by a position control law. Initially, the position control law is a Proportional, Integral, and Derivative (PID) control law using two notch filters applied to the position sensor signal. From this setup, an identification procedure for the estimated interaction coefficients can be carried out.

Using the identified bearing coefficients, model-based control laws can be used to generate the current reference signal. This will allow the control law to utilize the mathematical model conducted to help the controller place meaningful gains in the frequency domain to counteract undesired resonating peaks from the rotor-bearing-foundation interaction. Here, a Linear Quadratic Gaussian (LQG) control law using full state feedback with a Kalman filter as observer has been chosen. The PID control law will be compared against the LQG control law approach to investigate if this method is good at reducing the undesired amplifications in the frequency range.

Figure 1. Presentation of high-speed blower test facility components with included features. The shaft has a copper sheet, ①, a pair of rotor laminate stacks, ② and ④, aluminium surfaces for sensors, ③ and ⑤, and a disc with passive magnets, ⑥. The housing consists of an induction motor, ⑦, a pair of radial AMB stator geometries, ⑧ and ⑩, eddy current sensor fitting modules, ⑨ and ⑪, and axial PMB, ⑫.

Table 1. Values for AMB design.

Parameter	N	R	A	i_0	x_0
Value	$78[-]$	$0.8[\Omega]$	$754[\mathrm{mm}^2]$	$4[\mathrm{A}]$	$0.45[\mathrm{mm}]$

2   EXPERIMENTAL TEST FACILITY

The experimental test facility representing a high-speed blower, presented in figure,1, consists of a rotor and a housing. The rotor and housing are equipped with hardware components used to facilitate the use of Active Magnetic Bearings (AMBs) for a high-speed blower. An induction motor is used to rotate the shaft; two AMB geometries are used to control the shaft position with the help of four eddy current sensors. An axial Passive Magnetic Bearing (PMB) is used to keep the shaft approximately centred in the axial direction during test runs. The design parameters for the AMBs implemented in the high-speed blower are provided in Table 1.

2.1   Experimental investigations
The conducted experiments and theoretical analysis are based on an initial study for the high-speed blower controlled using a PID controller implemented with two notch filters on the position sensor signals for stabilizing the first flexible mode of the shaft. The linearized coefficients for the AMBs have been optimized through a system identification procedure similar to the one described in (10). The frequency response obtained experimentally is compared to that expected from the control object. The experimentally obtained frequency response using the PID control design is then compared to the frequency response generated from the LQG control design considering the most influential foundation dynamical response in the control objective. The comparison is carried out with the rotor of the high-speed blower levitated without the

rotor spinning. A Campbell diagram and unbalance response are simulated for both cases and the results are compared.

3   MATHEMATICAL MODEL

The global mathematical model is described from a combination of mathematical models for the components included in the high-speed blower test facility. The mathematical models include a Finite Element Model (FEM) of the shaft, a FEM of the foundation, linearized expressions for the dynamics within the AMBs and the PMB, filters applied to sensor signals, and control schemes implemented to the control objective. In Figure 2, an overview of the connections established between the different mathematical models is available. Also, the implemented model-based control design is presented in the figure. The construction of the mathematical model is also presented in (9). The rotor model is formulated using Timoshenko beam elements described in (11) with the mathematical formulation of the dynamic response described by equation (1). The equation for the dynamical response is formulated into a system of first order differential equations in equation (3) with the matrices presented in equation (2).

$$\mathbf{M}_r \ddot{\mathbf{X}}_r + (\mathbf{D}_r - \omega \mathbf{G}_r) \dot{\mathbf{X}}_r + \mathbf{K}_r \mathbf{X}_r = \mathbf{f}_{r,ext} \tag{1}$$

$$\mathbf{A}_{r,ss} = \begin{bmatrix} 0 & \mathbf{I} \\ -\mathbf{M}_r^{-1}\mathbf{K}_r & -\mathbf{M}_r^{-1}(\mathbf{D}_r - \omega \mathbf{G}_r) \end{bmatrix}, \mathbf{B}_{r,ss} = \begin{bmatrix} 0 \\ \mathbf{M}_r^{-1} \end{bmatrix}, \mathbf{C}_{r,ss} = \begin{bmatrix} \mathbf{I} & 0 \\ 0 & \mathbf{I} \end{bmatrix}, \mathbf{D}_{r,ss} = \begin{bmatrix} 0 \\ 0 \end{bmatrix} \tag{2}$$

$$\mathbf{G}_{r,ss} = \{ \begin{matrix} \dot{\mathbf{x}}_{r,ss} = \mathbf{A}_{r,ss}\mathbf{x}_{r,ss} + \mathbf{B}_{r,ss}\mathbf{u}_{r,ss} \\ \mathbf{y}_{r,ss} = \mathbf{C}_{r,ss}\mathbf{x}_{r,ss} + \mathbf{D}_{r,ss}\mathbf{u}_{r,ss} \end{matrix}, \ \mathbf{x}_{r,ss} = \left\{ \begin{matrix} \mathbf{x}_r \\ \dot{\mathbf{x}}_e \end{matrix} \right\}, \ \mathbf{u}_{r,ss} = \{\mathbf{f}_{r,ext}\} \tag{3}$$

It is assumed that the damping present in the rotor can be described from proportional damping as formulated in equation (4).

$$\mathbf{D}_r = \alpha_r \mathbf{M}_r + \beta_r \mathbf{K}_r \tag{4}$$

The foundation is described using ANSYS Solid186 elements. The mathematical formulation for the foundation extracted from ANSYS consists of mass, $\mathbf{M}_f$, and stiffness, $\mathbf{K}_f$, matrices. These matrices are reduced using a Craig-Bampton reduction scheme. In equation (5), a reformulation, $\mathbf{M}_{f,red}$ and $\mathbf{K}_{f,red}$, of the extracted matrices is presented. The reformulation describes the mass and stiffness matrices by a set of internal nodes, i, and a set of boundary nodes, b.

$$\mathbf{M}_{f,ref} \left\{ \begin{matrix} \ddot{\mathbf{u}}_i \\ \ddot{\mathbf{u}}_b \end{matrix} \right\} + \mathbf{K}_{f,ref} \left\{ \begin{matrix} \mathbf{u}_i \\ \mathbf{u}_b \end{matrix} \right\} = \mathbf{f}_{f,ref}, \begin{bmatrix} \mathbf{M}_{f,i\times i} & \mathbf{M}_{f,i\times b} \\ \mathbf{M}_{f,b\times i} & \mathbf{M}_{f,b\times b} \end{bmatrix} \left\{ \begin{matrix} \ddot{\mathbf{u}}_i \\ \ddot{\mathbf{u}}_b \end{matrix} \right\} + \begin{bmatrix} \mathbf{K}_{f,i\times i} & \mathbf{K}_{f,i\times i} \\ \mathbf{K}_{f,i\times i} & \mathbf{K}_{f,i\times i} \end{bmatrix} \left\{ \begin{matrix} \mathbf{u}_i \\ \mathbf{u}_b \end{matrix} \right\}$$
$$= \left\{ \begin{matrix} \mathbf{f}_{f,ext,i} \\ \mathbf{f}_{f,ext,b} \end{matrix} \right\} \tag{5}$$

The reformulation is used to establish the reduction matrix, $\mathbf{R}$, presented in equation, (6). The reduced eigenvector matrix, $\mathbf{V}_{i\times x}$, is calculated from solving the eigenvalue problem presented in equation (7) and choosing a number, x, of modes needing to be included in the reduced model.

$$\mathbf{R} = \begin{bmatrix} \mathbf{\Psi}_{i\times b} & \mathbf{\Phi}_{i\times x} \\ \mathbf{I}_{b\times b} & \mathbf{0}_{b\times x} \end{bmatrix}, \quad \mathbf{\Psi}_{i\times b} = -\mathbf{K}_{f,i\times i}^{-1}\mathbf{K}_{f,i\times b}, \quad \mathbf{\Phi}_{i\times x} = \mathbf{V}_{i\times x} \tag{6}$$

$$\mathbf{\Lambda}_{i\times i} = \mathbf{V}_{i\times i}^{-1}\mathbf{M}_{f,i\times i}^{-1}\mathbf{K}_{f,i\times i}\mathbf{V}_{i\times i} \tag{7}$$

For the Craig-Bampton reduction of the foundation, the number of modes is chosen as $x_{f,CB} = [1 : 100]$. The method for calculating the reduced mass, stiffness and damping matrices is presented in equation (8), assuming proportional damping. The system dynamics is then assumed to be described from equation (9). This dynamical relation can be formulated in state space as a first order system of differential equations in equation (11) with the matrices presented in equation (10).

$$\mathbf{M}_{f,red} = \mathbf{R}^T \mathbf{M}_{f,ref} \mathbf{R}, \mathbf{K}_{f,red} = \mathbf{R}^T \mathbf{K}_{f,ref} \mathbf{R}, \mathbf{D}_{f,red} = \alpha_f \mathbf{M}_{f,red} + \beta_f \mathbf{K}_{f,red} \tag{8}$$

$$\mathbf{M}_{f,red}\ddot{\mathbf{x}}_f + \mathbf{D}_{f,red}\dot{\mathbf{x}} + \mathbf{K}_{f,red}\mathbf{x}_f = \mathbf{f}_{f,ext} \tag{9}$$

$$\mathbf{A}_{f,ss} = \begin{bmatrix} \mathbf{0} & \mathbf{I} \\ -\mathbf{M}_{f,red}^{-1}\mathbf{K}_{f,red} & -\mathbf{M}_{f,red}^{-1}\mathbf{D}_{f,red} \end{bmatrix}, \mathbf{B}_{f,ss} = \begin{bmatrix} \mathbf{0} \\ \mathbf{M}_{f,red}^{-1} \end{bmatrix}, \mathbf{C}_{f,ss} = \begin{bmatrix} \mathbf{I} & \mathbf{0} \\ \mathbf{0} & \mathbf{I} \end{bmatrix}, \mathbf{D}_{f,ss} = \begin{bmatrix} \mathbf{0} \\ \mathbf{0} \end{bmatrix} \tag{10}$$

$$\mathbf{G}_{f,ss} = \begin{cases} \dot{\mathbf{x}}_{f,ss} = \mathbf{A}_{f,ss}\mathbf{x}_{f,ss} + \mathbf{B}_{f,ss}\mathbf{u}_{f,ss} \\ \mathbf{y}_{f,ss} = \mathbf{C}_{f,ss}\mathbf{x}_{f,ss} + \mathbf{D}_{f,ss}\mathbf{u}_{f,ss} \end{cases}, \mathbf{x}_{f,ss} = \begin{Bmatrix} \mathbf{x}_f \\ \dot{\mathbf{x}}_f \end{Bmatrix}, \mathbf{u}_{r,ss} = \{\mathbf{f}_{f,ext}\} \tag{11}$$

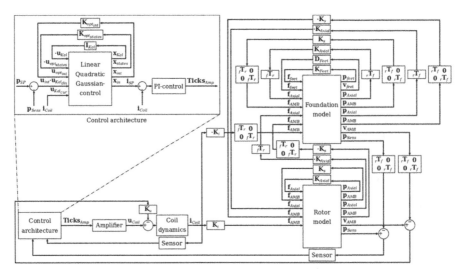

Figure 2. Schematic of the connection for the state-space models used to describe the system dynamics with implemented model-based controller.

The FEMs for the rotor and foundation are then both reduced using a modal reduction method. Two transformation matrices are deduced based on the eigenvalue problem for the model needing to be reduced, as presented in equation (12). A number, x, from a total number, n, of modes is chosen to establish the transformation. The reduced state space system is presented in equation (14) using the transformed matrices presented in equation (13).

$$\mathbf{\Lambda} = \mathbf{V}^{-1}\mathbf{A}_{ss}\mathbf{V}, \quad \mathbf{T}_l = \left(\mathbf{V}^{-1}\right)_{x \times n}, \quad \mathbf{T}_{ri} = \mathbf{V}_{n \times x} \tag{12}$$

$$\overline{\mathbf{A}}_{ss} = \mathbf{T}_l\mathbf{A}_{ss}\mathbf{T}_{ri}, \quad \overline{\mathbf{B}}_{ss} = \mathbf{T}_l\mathbf{B}_{ss}, \quad \overline{\mathbf{C}}_{ss} = \mathbf{C}_{ss}\mathbf{T}_{ri} \tag{13}$$

$$G_{-ss} = \begin{cases} \dot{\mathbf{x}}_{ss} = \mathbf{A}_{-ss}\mathbf{x}_{-ss} + \mathbf{B}_{-ss}\mathbf{u}_{ss} \\ \mathbf{y}_{ss} = \mathbf{C}_{-ss}\mathbf{x}_{-ss} + \mathbf{D}_{ss}\mathbf{u}_{ss} \end{cases} \tag{14}$$

In this specific investigation, the number of modes for the shaft is chosen as $x_r = [12345678]$, which is describing the rigid body movements and the first two flexible modes in perpendicular directions. For the foundation, the modes $x_{f,MR} = [21\,22\,25\,26]$ are chosen to be included in the mathematical model. These are some flexible modes belonging to the foundation, which are found to influence the dynamical behaviour of the combined rotor-bearing-foundation dynamics for the high-speed blower system. To describe the coil dynamics, the state space representation in equation (16) is used. The representation is established with the matrices from equation (15) and is valid for one coil in the electromagnet.

$$A_{c,ss} = -\frac{R}{L}, \mathbf{B}_{c,ss} = \begin{bmatrix} \frac{1}{L} & -\frac{k_u}{L} \end{bmatrix}, C_{c,ss} = 1, \mathbf{D}_{c,ss} = \begin{bmatrix} 0 & 0 \end{bmatrix} \tag{15}$$

$$G_{c,ss} = \begin{cases} \dot{x}_{c,ss} = A_{c,ss}x_{c,ss} + \mathbf{B}_{c,ss}\mathbf{u}_{c,ss} \\ \mathbf{y}_{c,ss} = C_{c,ss}x_{c,ss} + \mathbf{D}_{c,ss}\mathbf{u}_{c,ss} \end{cases} \tag{16}$$

The coefficients, L, k_u, k_i, and k_s are linearized for the expected position of the shaft during operation. A value for k_{Axial} has been obtained from a study using COMSOL for the passive magnet rings moving in a radial direction. In the study, it is assumed that the reaction is linearly dependent on the relative displacement between the magnet rings. In equation (17), the estimations for the coefficients are presented.

$$L = \frac{\mu_0 N^2 A}{2x_0}, k_u = \frac{\mu_0 N^2 A i_0}{2x_0^2}, k_i = \frac{\mu_0 N^2 A i_0}{x_0^2}, k_s = \frac{\mu_0 N^2 A i_0^2}{x_0^3}, k_{Axial} = 19 \left[\frac{\text{kN}}{\text{m}} \right] \tag{17}$$

For the sensors, a first or second order Butterworth filter is used depending on the implemented control scheme. The control object is obtained from combining the components as presented in Figure 2. The resulting state space formulation is described in equation (18). The matrices $\mathbf{C}_{ss,obj}$ and $\mathbf{D}_{ss,obj}$ can be designed such that one obtains the desired output states from the control objective.

$$G_{ss,obj} = \begin{cases} \mathbf{x}_{ss,obj} = \mathbf{A}_{ss,obj}\mathbf{x}_{ss,obj} + \mathbf{B}_{ss,obj}\mathbf{u}_{ss,obj} \\ \mathbf{y}_{ss,obj} = \mathbf{C}_{ss,obj}\mathbf{x}_{ss,obj} + \mathbf{D}_{ss,obj}\mathbf{u}_{ss,obj} \end{cases} \tag{18}$$

The PID control scheme, presented in equation (19), is implemented utilizing a second order Butterworth filter for the position and current sensors in the global mathematical model.

$$G_{PID-scheme}(s) = K_p \left(1 + \frac{1}{\tau_i s} + \frac{\tau_d s}{\tau s + 1} \right) \left(\frac{s^2 + \omega_{0,1}^2}{s^2 + \frac{\omega_{0,1}}{Q}s + \omega_{0,1}^2} \right) \left(\frac{s^2 + \omega_{0,2}^2}{s^2 + \frac{\omega_{0,2}}{Q}s + \omega_{0,2}^2} \right) \tag{19}$$

Table 2. Values used in PID control scheme and current tracking controller.

Parameter	K_p	τ_i	τ_d	$\omega_{0,1}$	$\omega_{0,2}$	Q	K_{cur}	τ_{cur}
Value	13 [A]	0.3 [s]	2 [ms]	466 [Hz]	470 [Hz]	2 [−]	500 [$\frac{\text{Ticks}}{\text{A}}$]	1 [s]

The output of the PID control scheme is the current reference which is tracked by a PI control design presented in equation (20).

$$G_{PI-scheme}(s) = K_{cur}\left(1 + \frac{1}{\tau_{cur}s}\right) \tag{20}$$

In Table 2, the values used in the PID control scheme and current tracking control scheme are presented. The model-based control design implemented is based on the control object having first order Butterworth filters implemented on the sensor signals. For this control object, an observer in the form of a Kalman filter is used. Using an estimation of the process noise intensities, V_1, and measurement noise intensities, V_2, the Kalman filter gains can be computed from equation (21).

$$0 = A_{ss,obj}Q + QA_{ss,obj}^T + B_{ss,obj}V_1B_{ss,obj}^T - QC_{ss,obj}^TV_2^{-1}C_{ss,obj}Q, L_{Kal} = QC_{ss,obj}^TV_2^{-1} \tag{21}$$

The model-based control scheme should include integral action on the position sensor signals. Therefore, the control object is augmented with integral states in equation (22). The controller gains are calculated in equation (23) based on weighting matrices for the states, R_1, and the input, R_2.

$$A_{ss,obj,int} = \begin{bmatrix} A_{ss,obj} & 0 \\ -C_{ss,obj} & 0 \end{bmatrix}, B_{ss,obj,int} = \begin{bmatrix} B_{ss,obj} \\ 0 \end{bmatrix} \tag{22}$$

$$0 = A_{ss,obj,int}P + PA_{ss,obj,int}^T + R_1 - PB_{ss,obj,int}R_2B_{ss,obj,int}^TP, K_{Opt} = R_2^{-1}B_{ss,obj,int}^TP \tag{23}$$

Using the system augmented with integral states, the controller dynamics are built in line with the control architecture visualized in Figure 2. Before implementation, the LQG control design is discretized, balanced, and reduced to fit the space requirements for the hardware installed in the test-facility. This reduction is based on keeping the most important singular values for the control design.

4   RESULTS

The mathematical model has been optimized using a least squares fit for the expected relative displacements between the rotor and foundation against the measured relative displacements as well as the expected current running in the coils against the measured current. The expected values are provided as outputs from the global mathematical model. The measured values are obtained with the PID control design implemented on the high-speed blower system. A comparison between the estimated frequency response from the control objective and the measured frequency response after the optimization is presented in Figure 3. It is noticed that the control object does contain some regions where the estimation of the real plant is not completely in line with the experimentally obtained response, i.e. around 150 [Hz] and between 400 [Hz] to 500 [Hz]. However, the control plant does seem to include the most prominent responses and does seem to be able to sustain a model-based control design. In Figure 4, the LQG control design is implemented to the experimental test facility. The experimentally obtained frequency response is compared to the estimated response from the control object applied with the LQG control design. The experimentally obtained frequency response for the system with the PID control scheme is plotted again for easier comparison. It is noted that the experimentally

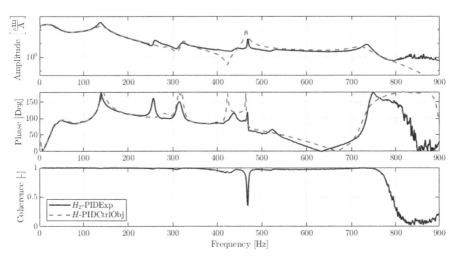

Figure 3. Experimentally obtained frequency response versus the estimated frequency response for the control object implemented with a PID control design.

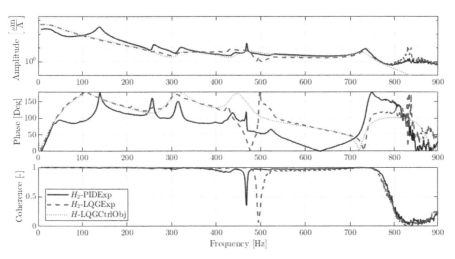

Figure 4. Experimentally obtained frequency response for PID control design versus the experimentally obtained frequency response for LQG control design and estimated frequency response for the control object.

obtained frequency response is well described by the expected frequency response in most of the frequency range. However, as it is for the control object applied with the PID control scheme as well, the amplification and phase shift are not well described in the frequency band from 400 [Hz] to 500 [Hz]. Comparing the experimental frequency response using the PID control scheme against the experimental frequency response using the LQG control design, it is noticed that while the response in the range from 0 [Hz] to 100 [Hz] has increased for the system with the LQG control design implemented,

333

the response has either decreased or remained at the same level for the rest of the frequency band. Therefore, the method of including the foundation dynamics in the control design is deemed successful.

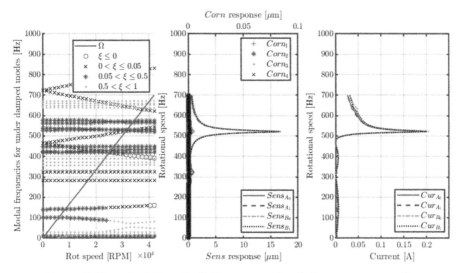

Figure 5. Expected Campbell diagram and unbalance response for high-speed blower using PID control scheme (Centre bottom: Rotor responses; Centre top: Foundation corner responses; Right bottom: Current in coils).

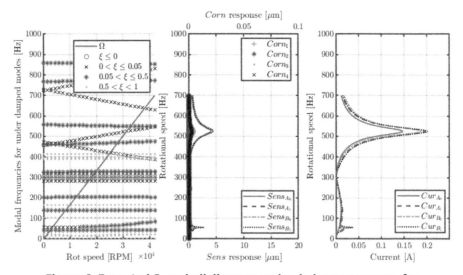

Figure 6. Expected Campbell diagram and unbalance response for high-speed blower using LQG control scheme (Centre bottom: Rotor responses; Centre top: Foundation corner responses; Right bottom: Current in coils).

In Figures 5 and 6, two unbalance responses are calculated for the system having the two different control schemes implemented. The residual unbalance mass is assumed to be in the rotor laminates for the active magnetic bearing. Figure 5 shows a very high response peak at around 520 [Hz], which is coincident with the crossing of the forward whirl for the first flexible eigenmode for the shaft. It is noticed that no corners of the foundation are expected to generate a considerable response. However, it is also noticed that the mathematical model predicts the system to be unstable at around 31500 [RPM]. In Figure 6, it is evident that the resonance peak caused by the crossing of the first flexible mode is reduced drastically but at the expense of some additional undesired resonance peaks around 50 [Hz] and a slightly higher response in the corners of the t-slot plate. Analysing the expected amplification of the current in the coils, it is noticed that the limitations of the bearing design are complied with and that the amplification of the current around 520 [Hz] is distributed over a broader frequency range. Also, the control design is expected to be stable from the assessment of the damping factors of the eigenvalues. The increased response in the corners of the t-slot plate indicates that the LQG control algorithm is dissipating some of the energy from the first flexible mode for the shaft to the foundation, thereby exciting the foundation structure as well.

5   CONCLUSIONS

In this article, it has been illustrated that foundation dynamics can be included in a model-based control design such as an LQG control scheme. The method has been compared against a PID control scheme utilizing notch filters to stabilize the shaft.

It is found that some complications regarding undesired amplifications of the response occur when the foundation response is considered in the control design. However, the overall performance of the model-based control design appears to be superior to that of a decentralized PID control scheme.

REFERENCES

[1] Schweitzer, G., Maslen, E. H., et al. (2009) "Magnetic Bearing: Theory, Design, and Application to Rotating Machinery". *Springer*, Berlin, Germany.

[2] Schweitzer, G. (2002) "Active magnetic bearings-chances and limitations". *6th International Conference on Rotor Dynamics.*, 1–14.

[3] Filatov, A. & Hawkins, L. (2015) "Comparative study of axial/radial magnetic bearing arrangements for turbocompressor applications". *Proceedings of the Institution of Mechanical Engineers. Part I: Journal of Systems and Control Engineering*, **230**(4), 300–310.

[4] Tondl, A. (1960) "The stability of motion of a rotor with unsymmetrical shaft on an elastically supported mass foundation". *Ingenieur-Archiv*, **29**(6), 410–418.

[5] Gasch, R. (1976) "Vibration of large turbo-rotors in fluid-film bearings on an elastic foundation". *Journal of Sound and Vibration*, **47**(1), 53–73.

[6] Mourelatos, Z. P. & Parsons, M. G. (1987) "A finite element analysis of beams on elastic foundation including shear and axial effects". *Computers and Structures*, **27**(3), 323–331.

[7] Ruijgrok, M., Tondl, A. & Verhulst, F. (1993) "Resonance in a rigid rotor". *Zamm*, **73**(10), 255–263.

[8] Dagnaes-Hansen, N. A. & Santos, I. F. (2019) "Magnetically suspended flywheel in gimbal mount - test bench design and experimental validation". *Journal of Sound and Vibration*, **448**, 197–210.

[9] Paulsen, T. T. (2018) "Modelling, Design, Implementation and Experimental Testing of Flexible Rotors on Active Magnetic Bearings Considering Flexible Foundations". MS Thesis, Technical University of Denmark, Kgs. Lyngby, Denmark, June.

[10] Lauridsen, J. S. & Santos, I. F. (2018) "On-site identification of dynamic annular seal forces in turbo machinery using active magnetic bearings: An experimental investigation". *Journal of Engineering for Gas Turbines and Power*, **140**(8), 1–9.

[11] Nelson, H. D. (1980) "A finite rotating shaft element using Timoshenko beam theory". *Journal of Mechanical Design*, **102**(4), 793–803.

Vibration behaviour of a 11 kW two-pole induction motor mounted on elastic steel frame foundation with actuator system

R. Wachter

Nuremberg Tech, ELSYS Institute for Power Electronic Systems, Germany

U. Werner

Nuremberg Tech, Faculty EFI, Germany

H-G Herzog

Technical University of Munich (TUM), Institute of Energy Conversion Technology, Germany

C. Bauer

Siemens AG, Process Industries and Drives Division, Large Drives, Simulation, Germany

ABSTRACT

This paper describes the vibration behaviour of a small (11 kW) two pole induction motor, which is mounted on an elastic steel frame foundation with a special actuator system. This paper presents natural frequencies as a result of experimental model analysis of the whole system – motor, actuator system and steel frame foundation – and compares them to the numerical modal analyses using a finite element model. Additionally, forced vibrations are measured using the actuators as a shaker to de rive the resonance frequencies of the system. The measured natural frequencies by modal analysis, the measured resonance frequencies and the simulated natural fre quencies are compared to each other. Additionally, the necessary electrical power of the electro-dynamic actuators is analysed.

1   INTRODUCTION

In industrial drive applications, vibrations are undesirable and may lead to damages or may reduce maintenance intervals. Large induction motors are designed for a mas sive foundation, referring to the international motor standards e.g. IEC 6003414 [1]. In practice however, induction motors are often mounted together with the load ma chine on an elastic steel frame foundation, which influences the natural frequencies of the system – motor with foundation. Nowadays, due to the efficiencydriven develop ment, more and more large induction motors are driven by converters and no longer by a rigid grid. The issue is, that vibration problems increase, because the dynamic behaviour of a motor on a soft foundation is strongly influenced by the foundation ([2] and [3]), and that the whole operating speed range may not be used any more, because of resonance problems. The consequence is that offlimit speed ranges have to be defined where steadystate operation is not allowed, which may reduce the effi ciency of the plant. To face this challenge, a special actuator system was developed, which consists of spring elements, electrodynamic actuators, and additional passive damping elements. The actuator system is mounted between an elastic steel frame foundation and a 11 kW twopole induction motor in the test lab (Figure 1). Of course, the system characteristics of such small test bench can not be compared with a large induction motor (> 1 MW) in industrial application. Therefore, a steel frame founda tion was developed in a previous study [4], to create similar conditions so that three natural frequencies occur in the

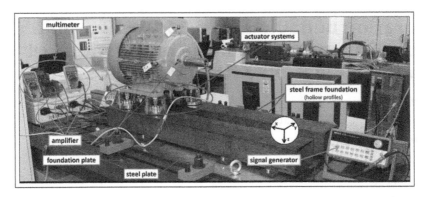

Figure 1. Test bench with foundation plate, steel plate, elastic steel frame foundation, actuator systems, motor and electrical equipment for external excitation.

operating speed range of the motor. Two square hol low profiles are used for the steel frame foundation. Additionally, small steel blocks between the motor feet and the steel frame foundation were placed to reduce the contact zone between the motor feet and foundation and to create therefore a soft foundation. The frequencies and the move- ments in the various modes are similar to large motor applications with an elastic steel frame foundation. For further develop ments and testings on the test bench another motor or load machine can be mounted and used as a brake. The motor is connected in star-connection and is driven by a converter. The possible speed range is from 0 rpm to 4 800 rpm. The main data of the motor is shown in Table 1.

Table 1. Main data of the two-pole induction motor.

Description	Symbol	Value	
Rated power	P_N	11	kW
Rated voltage in star connection	U_N in Y	400	V
Rated current in star connection	I_N in Y	21.2	A
Rated speed	n_N	2915	1/min
Rated frequency	f_N	50	Hz
Limiting frequency (mechanical)	f_{max}	80	Hz
Mass of the motor	m_{ges}	62	kg
Distance between feet	2b	254	mm
Height of the centre of gravity	h	158	mm

2   DESCRIPTION OF ACTUATOR SYSTEM

This section gives an overview about the main components of the actuator system, the electrodynamic actuator, and the mechanical frame which carries the motor.

2.1   Carrier system

The carrier system of each actuator system connects the steel frame foundation with each motor foot by a spring element, which is shown in Figure 2. The optimization of the carrier system was carried out with regard to a high stiffness in horizontal and axial direction. Four main parts are screwed together to transmit the force: The steel frame foundation with the lower plate, the lower plate with the lower part of the spring, the upper part of the spring with the upper plate and finally the upper plate with the motor foot. The entire weight of the motor is carried by the spring in the actuator system, so the electrodynamic actuator only has to control dynamic forces.

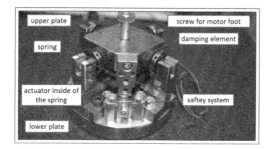

Figure 2. Carrier system with actuator inside.

The electrodynamic actuator itself has no guidance, so that the carrier system may only allow minor movements in horizontal and axial direction. In vertical direction the upper plate of the actuator system has to be movable with little resistance in the carrier system. Since the actuator only has a clearance of 0.5 mm in radial direction, the horizontal position of each upper plate which holds the core of the actuator is adjustable. The free horizontal and axial adjustability also allows an adjustment of the preload force by the damping elements within certain limits. A novelty of this system is that the weight of the motor causes no compression to the damping elements due to the design of the carrier system. Overloads on test stand are unlikely, but cannot be excluded. Therefore a mechanically safety system was included to prevent excessive movement in all directions.

2.2   Details of actuators

The actuators that are used in this application are called "voice coil actuators" consisting in general of a moveable coil holder with an electrical coil and a soft iron stator with a magnetic core. Figure 3 a) and b) show one of the actuators which is used in this application. The principle of this electrodynamic actuator type is based on the use of Lorentz force. The system is designed so that the equilibrium position of the actuator is at mid stroke (5 mm). In this position range, the highest continuous force is possible (Figure 3 c)). In a worst case scenario with an effective oscillation velocity of 7.1 mm/s, which is the limit according to DINISO108163 [5]. Based on this as sumption a maximum deflection in the first mode (12 Hz) on motor feet is lower than 0.27 mm, from peak-to-peak, if assumed that the motor feet oscillates with the same effective vibration velocity as the motor shield. Because of this small displacements a linear behaviour can be assumed, see Figure 3 c). At vibration values > 7.1 mm/s RMS, a safety mechanism switches off the motor (converter drives to 0 rpm). Another important data, especially for a pulsed control like PWM (Pulse Width Modulation), is the time constant of the actuator. It has to be much lower than the highest expected reciprocal vibration frequency. Further key characteristics are shown in Table 2. The continuous force is low with about 7 N, but sufficient, which is shown in Section 4.

Figure 3. Electrodynamic actuator assembled, b) disassembled and c) force plotted over stroke.

Table 2. Main data of the actuator (extraction).

Description	Value	
Peak force	42.09	N
Stroke	10	mm
Continuous force @ 100 °C	7.32	N
Coil resistance @ 25 °C	10.35	Ω
Inductance @ 1 kHz (inside fully)	3.47	mH
Continuous stall current @ 100 °C	0.65	A
Voltage maximum	48	V
Force constant @ Mid stroke	11.21	N/A
Time constant (electrical)	335	µs

3   NATURAL FREQUENCIES

This section describes how the natural frequencies of the system were determined. A similar way to determine the frequencies of the system is an experimental modal analysis. To make the movements of the modes in the system visible a finite element analysis was performed. However, this paper only focuses on the first six natural frequencies, which are all within the rotational frequency range of the motor.

3.1   Numerical modal analysis

For the first simulation and visualisation of the system mode shapes and natural frequencies a finite element model was created as shown in Figure 4 b). The figure shows the real system and the simplified model. The motor is replaced by a point mass in center of gravity of the motor. A model of the motor is also available as CAD geometry, so that the center of mass and the moments of inertia are included in the point mass. Since the low damping factor of the damping elements has only a minor influence on natural frequencies, the damping of the damping elements was neglected. On the other hand, the stiffness of the damping elements in vertical z-axis and horizontal y-axis was considered. Since the stiffness of the spring elements is very low in relation to the stiffness of the foundation plate only the square hollow profiles and the steel plate were taken into consideration. The foundation plate was not considered because it has a very large mass and is very stiff in comparison. The connection between the point mass and the upper plates of the actuator systems is much stiffer than other stiffnesses in the system. Therefore, in the simulation these connections were assumed to be rigid.

To avoid a complex model, with elaborate contact definitions, the stiffness of the spring and damping element were determined outside of the simulation and springs were used (Figure 4 b)). The contact areas in the simulation were all configured as"bonded" so that the adjacent nodes in the finite element model are fixed to each other. For large parallel surfaces, this may lead to an unrealistic stiffening and deviations between the simulation results and reality. For this reason, the surfaces between steel plate and the steel frame foundation were separated from each other by washers – in simulation and on the real test bench. The same was done for the connections between the steel frame foundation and the actuator systems. The installation of washers between the steel frame foundation and the actuator systems allows a comparison with the system without actuator systems. Due to these simplifications, it was expected that only the first natural frequencies would fit to the experimental modal analysis.

With regard to [4], which describes the behaviour of the system without the actuator systems, two new mode shapes in speed range occur. Figure 5 show the first six modes as a result of this simulation. Only the deflection in one direction is shown.

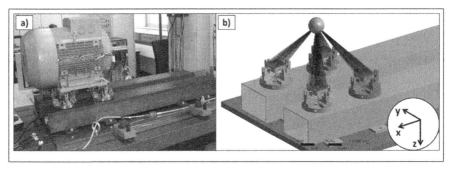

Figure 4. A) real system b) finite element model with motor as a point mass in center of gravity.

In mode 1, the point mass makes a rotation movement at the horizontal y-axis combined with a translation in axial x-direction, both in phase. The feet mainly perform a vertical movement with relatively low tipping and translational movement in axial direction. This movement leads to an elastic deformation in upper part of the steel frame foundation. Mode 2 is very similar to mode 1 with the difference in direction of movement. In this mode the point mass makes a rotation movement at the axial x-axis combined with a translation in horizontal y-direction, both in phase to each other. The feet mainly perform a vertical movement with relatively low tipping and translational movement in horizontal y-direction. In mode 3, the point mass and the feet together, perform only a vertical movement in phase with no rotation or translation in horizontal or axial direction.

Mode 4 is a special mode, because the whole motor (point mass and feet in phase) makes a rotation movement at the vertical z-axis. No vertical translational movement in z-direction occurs in this mode. The consequences of this particular mode is that the actuators have no influence to it. This fact and what can be done in this case will be explained in Section 5. In mode 5, it can be assumed that the point mass does only very small translational movement in axial x-direction because the motor does a movement like a bell. The point mass does a rotation at the horizontal y-axis. Mode 6 is very similar to mode 5, also nearly acting like a bell but the main direction of movement is in horizontal y-direction. In this mode, a translational movement of the feet in the horizontal y-direction occur, coupled with a small vertical movement. The point performs only small movements in the horizontal y-direction again.

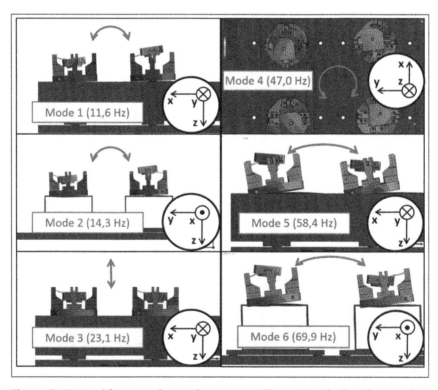

Figure 5. Natural frequencies and corresponding natural vibration modes.

3.2   Experimental modal analysis

With the visual information of the numerical modal analysis about the mode shapes, a constructive experimental modal analysis is possible.

To perform the force pulse excitation a modal hammer with an exchangeable soft tip and an additional mass is used. With the combination of a soft tip with a large mass, the lower spectrum of this structure can be excited. Figure 6 a) to d) shows the execution and the results of the experimental modal analyses. Therefore, the natural frequencies can be determined. Mode 5 and mode 6 were examined in two additional experiments. A total of six measurements were evaluated. An explanation of movement is only given, if the behaviour is different to the explanation in the previous section.

Figure 6 a) shows the measurement of an impact on the terminal box in axial x-direction. Two dominant resonances occur here – mode 1 at a frequency of 12.1 Hz and mode 5 at a frequency of 52.9 Hz. The magnitude of mode 5 is lower than for mode 1. This mode was better excited by one of the additional experiments with a greater gain about −6 dB. The second measurement of the impact is shown in Figure 6 b). This impact has excited three modes. Only mode 2 and mode 6 are ex plained here, as mode 4 is described below. The magnitude of mode 2 at a frequency of 14.6 Hz has a sharp peak. The peak of mode 6 has a low magnitude (about −18 dB). Therefore, a second additional experiment with an impact on the lower part of bearing shield in axial x-direction was done, which resulted in a sharp and tight peak (−7 dB) at a frequency at 59.6 Hz. The aim of the impact at measurement point 3 (Figure 6 c)) was to determine the natural frequency of mode 3. Since an impact in the center of mass was not possible, two more modes were excited. The nonideal impact in this experiment presumably is the reason for the low magnitude of −15 dB. However, the peak of this mode is sharp enough to read the frequency at 25.0 Hz. As the result of the fourth impact (Figure 6 d)), mode 4 was excited very good with a sharp and high peak (4 dB) at a frequency of 41.2 Hz.

343

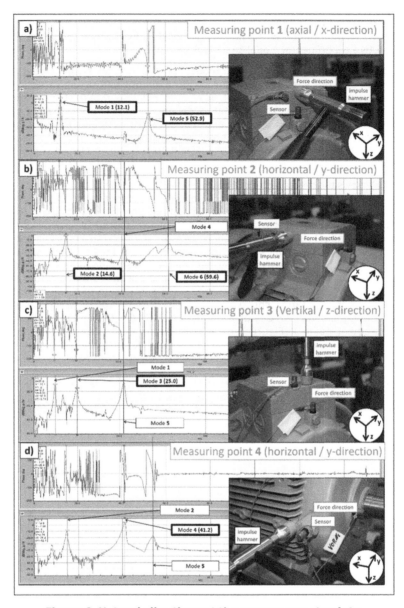

Figure 6. Natural vibrations at the measurement points.

Table 3. Summary experimental modal analysis – modes with corresponding natural frequencies.

Description	Value
Mode 1	12.1 Hz
Mode 2	14.6 Hz
Mode 3	25.0 Hz
Mode 4	41.2 Hz
Mode 5	52.9 Hz
Mode 6	59.6 Hz

As preliminary summary, Table 3 shows the natural frequencies of the experimental modal analysis. The first four rigid body modes could be successfully excited in exper iments 14 and the natural frequencies of these could be determined. To determine the frequencies of the fifth and sixth rigid body modes it was necessary to extend the impulse test by two excitations at different points below the motor center of gravity. By extension of the test these two modes could be assigned clearly to the frequencies.

4   RESONANCE FREQUENCIES AND POWER CONSUMPTION

In order to determine the resonance frequencies, the actuators are used as shakers. Based on the simulation results in Figure 5, the four actuator systems were intercon nected. By different interconnections of the actuators and the crossing of the low frequency band (up to 80 Hz) it was tried to excite all rigid body modes known from numerical simulation and to determine their resonance frequencies. The damping degree of the damping elements are usually in a range between 0.05 to 0.1. The damping of steel springs is much lower (0.005 to 0.01). Thus, it is expected that the resonance frequency is almost similar to the natural frequency. To determine the resonance frequencies a simple way was chosen, shown in Figure 7 a). Therefore, a signal generator gives a harmonic sinusoidal excitation to both of the amplifiers which amplify the signals. Each amplifier drives two actuators. In the test, the excitation frequency was increased continuously by 1 Hz and near a resonance, the resonance was passed through in steps of 0.1 Hz. Three test configurations resulted in most of the desired modes and frequencies. However, not all of the desired six modes could be excited.

In the first test configuration, the two actuator systems 1 and 2 on drive side and the two actuator systems 3 and 4 on non drive side were operated in pairs and in antiphase (180°) to each other, see Figure 7 b). Due to this configuration, mode 1 could be excited at a resonance frequency of 12.0 Hz and mode 5 at 52.2 Hz. In the second test configuration, the two lateral actuators were operated in pairs and in antiphase (180°) to each other (actuator system 4 and 1 work in antiphase to system 2 and 3). Here, mode 2 could be excited at a resonance frequency of 14.6 Hz. In third test configuration, where all of the actuators work in phase mode 3 could be effectively excited at a resonance frequency of 24.8 Hz.

345

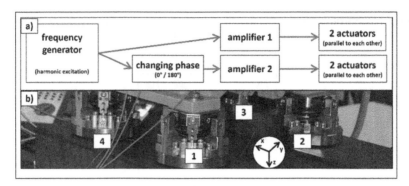

Figure 7. A) schematic of open loop control for the actuators b) numbering of actuators.

All modes, except modes 4 and 6, could be excited in this test configuration as shown in Table 4. Mode 4, which can be excited only by an excitation at the horizontal y-axis for example by a couple unbalance of the rotor, cannot be excited with the actuator systems. Furthermore, no resonance frequency could be determined for mode 6. An increase in amplitude can be observed but due to the damping effect of the damping elements the amplitude is very broad that a frequency cannot be specified for this mode.

Table 4. Summary shaker excitation – modes with the corresponding resonance frequencies.

Description	Value
Mode 1 (1st test)	12.0 Hz
Mode 2 (2nd test)	14.6 Hz
Mode 3 (3rd test)	24.8 Hz
Mode 4	- Hz
Mode 5 (1st test)	52.2 Hz
Mode 6	- Hz

Since the voltage and current were recorded in all test configurations, a reverse conclusion can be made about the electrical requirements of the system with harmonic system behaviour. As in the measurements of the resonances the four actuator systems were interconnected in the same three test configurations and the frequency band was crossed. The current was held constant by adjusting a harmonic voltage excitation at each measuring point, with actuators running in shaker mode. In each of the resonance frequencies the important values were recorded, which are shown in Table 5. The velocity values were measured at the bearing shield of drive side, in referring to [5]. The measurements show that in each resonance frequency with 200 mA current (per actuator) the velocity of the characteristic movement direction was greater than the permitted value of 7.1 mm/s. For example, Test 1 at a resonance frequency of 12.0 Hz only an electrical active power of 0.43 W is necessary for each actuator.

Table 5. Measured values for electrical requirements.

	Test 1	Test 2	Test 3	
	(DE antiphase to NDE)	(side by side)	(all same phase)	
Frequency	12.0	14.6	24.8	
I_{eff} (per actuator)	200	200	200	mA
U_{eff}	2.14	2.10	2.20	Volt
v_{eff} horizontal	–	10.2	–	mm/s
v_{eff} axial	9.1	–	–	mm/s
v_{eff} vertical	5.5	–	9.3	mm/s

5   COMPARISON OF MEASUREMENTS AND SIMULATION

To conclude the modal analyses, Table 6 shows the determined modes with their frequencies. During shaker excitation, modes 4 and 6 could not be detected due to the system and the wide band deflection. In the lower frequency band, the finite element model fits well with the results of the pulse excitation and the shaker test. Due to the shortcomings caused by the reduction of the complex motor model and the neglecting of the damping, the results of the finite element model from mode 4 are higher than the measured natural frequencies of the system. A main reason for the differences between numerical and experimental modal analysis is the additional stiffness associated with the simplification of the point mass. The frequencies of the shaker test and the impulse excitation are very close together. A comparison to the previous study [4], where the motor was directly mounted on the elastic steel frame foundation, shows that the actuator systems lowers the natural frequencies significantly. Mode 4 did not appear in the lower frequency range of the previous study and therefore, it was not considered. The frequency of mode 6 was also so high before the conversions that it did not fall within the speed range of the motor and was therefore not considered by the study.

Table 6. Comparison of the modal analyses from experiments and analyses from a previous study [4].

Mode	Finite element model	Experimental model analysis	Shaker excitation	Experimental model analysis from previous study without actuators [4]
1	11.6	12.1	12.0	30.5 Hz
2	14.3	14.6	14.6	35.3 Hz
3	23.1	25.0	24.8	52.8 Hz
4	47.0	41.2	–	– Hz
5	58.4	52.9	52.2	109.0 Hz
6	69.9	59.6	–	– Hz

6 CONCLUSION

The aim of this paper is to show the influence of an actuator system, which is mounted between an elastic steel frame foundation and a motor, on the natural frequencies and the mode shapes of a test bench. It could be shown, that already a simplified finite element model is sufficient to provide good results in relation to natural frequencies in the lower band (up to 40 Hz). A comparison between numerical modal analysis and experimental modal analysis confirms this. Above this frequency the results from nu merical modal analysis are systematically too high. In comparison to the preliminary study [4], the natural frequencies of the system are much lower than before due to the actuator systems. The order of modes changes and new modes occur. Now, six instead of four modes occur in the desired frequency range due to the use of the actu ator systems. Mode 4 is special, because this mode is the only one that has horizontal and axial, but no vertical deflections. This mode 4 has to be damped passively. In all other modes vertical movements occur, which can be influenced by the actuator systems. Furthermore, with a reverse investigation the paper has shown that the actually mounted actuators with a continuous force of 7 N are sufficient enough and only low power is necessary to influence the vibration behaviour of the whole system. Until now, the actuators have only been controlled in open loop by the signal genera tor and amplifier. As research continues, the system will be extended by a controller and a power unit to control the actuators in closed loop. Therefore, every actuator system can be controlled separately. The various control strategies can be tested in practice on a scaled but similar system in reference to [6]. For a better understanding and further measurements, additional sensors will be mounted on the motor on non drive side and within the steel frame foundation in center of the actuator systems.

REFERENCES

[1] IEC-60034-14. (2007) *Rotating electrical machines – part 14: Mechanical vibra tion of certain machines with shaft heights 56 mm and higher – measurement, evaluation and limits of vibration severity.*

[2] Genta, G. (2005) *Dynamics of rotating systems.* Springer Science and Business Media.

[3] Gasch, R. et al. (2002) *Rotordynamik.* Springer.

[4] Mathes, S., Werner, U., and Bauer, C. (2018) *Numerical and experimental vibra tion analysis of a twopole induction motor mounted on an elastic machine testbed.* ISMA conference – Noise and Vibration Engineering Conference.

[5] DIN-ISO-10816-3 (2018) *Mechanical vibration - evaluation of machine vibration by measurements on non-rotating parts - part 3: Industrial machines with nomi nal power above 15 kw and nominal speeds between 120 r/min and 15000 r/min when measured in situ* Beuth Verlag GmbH.

[6] Werner, U. (2018) *Vibration control of large induction motors using actuators between motor feet and steel frame foundation.* MSSP – Mechanical System and Signal Processing, Vol. 12.

Hybrid crankshaft control: Reduction of torsional vibrations and rotational irregularities under non-stationary operation

G. Paillot, D. Rémond, S. Chesné

University Lyon, INSA-Lyon, CNRS UMR5259, LaMCoS, Villeurbanne, France

ABSTRACT

Due to an oscillating input torque created by the cylinders, a crankshaft is submitted to a detrimental phenomenon, the rotational irregularities. Moreover, these oscillations are known to excite the crankshaft torsional modes, able to break the part if no damping is provided. In this paper, the authors introduce a hybrid control concept that consists in an electromagnetic coupling of a passive rotating tuned mass damper and a permanent magnet machine, which aims to redistribute the energy in the system depending on the operative conditions. Results are presented on the performance of such a concept in numerical simulations and a guideline for the typical behaviours that characterize the damper due to non-linear effects and inherent conditions is outlined.

1   INTRODUCTION

The working principle of a 4-stroke internal combustion engine (ICE) is well known. After the admission of fuel and air in the cylinder, the mix is compressed and ignited, creating an explosion that moves the piston, and creates a rotating motion on the propelling shaft thanks to the crank. The gas is then exhausted and the cycle starts again. However, that cycle takes place every two turns of the crankshaft, which means the torque created by the explosion is a 4n-periodic function. Moreover, the lever arm created by the crank also depends on the angular position, and so does the pressure in the cylinders that creates the torque. After calculation, it appears that the torque exerted by a cylinder on the crankshaft is a multi-harmonics periodic function of the angular position of the shaft itself. Combining the influence of the N_{cyl} cylinders, although letting aside the – non-negligible - incidence of the rotating inertias, we obtain a well-known result [1]: the input torque is a periodic angular function which spectrum displays all harmonics multiple of $N_{cyl}/2$, with an average value that is nothing else than the propelling torque.

However, the magnitude of those additional harmonics, which is naturally dependent on the pressure in the cylinders and so on the engine regime, can be high [2], especially for the lower harmonics. Consequently, the input torque is not a constant value but an oscillating function, and the influence of this oscillation appears in the dynamics of the shaft. This means in particular that the output speed of the crankshaft, and then the input speed of the rest of the driveline, is also an oscillating function, with the same harmonics as the input torque. This phenomenon, often referred to as "rotational irregularities", is a problem, as it leads to noise and early fatigue in the driveline, and especially in the gearbox. The traditional way to decrease the magnitude of the oscillations is to add inertia to the back end of the crankshaft, with a flywheel or its further developments (Double or Triple Mass Flywheel (DMF [3], TMF [4]), Centrifugal Pendulum Vibration Absorbers (CPVA [5]), variable inertia [6], etc). This solution enables a consequent mitigation of the oscillations, but is not enough and needs constant improvement to reach the present standards, which paves the way to the emerging active and semi-active alternatives [7-8-9].

The second problem created by the oscillating torque is the excitation of the critical frequencies of the crankshaft by the harmonics of the input torque. These are spatial harmonics in so far as they depend on the angular position of the shaft and not on the time. Consequently, this excitation depends on the instantaneous angular speed that we have already depicted as a harmonic function, which average is the engine regime. Among the various modes, the most dangerous one is the first torsional mode [10], which can break the shaft very quickly if no additional damping is provided. To tackle this issue, the traditional solution is a viscous damper located on the front end. It is an efficient solution, but at the cost of a vibration energy dissipated as heat. In heavy-duty vehicles, a power up to 1kW can be wasted this way.

To that extent, the oscillating torque applied on the crankshaft is a twofold problem. However, the cure is different for the two: the vibration damper is an energy sink while the newest flywheel concepts need power supply. There is then a need for an energy management that enables the vibration damper to feed a semi-active device to mitigate the rotational irregularities. The present paper offers a concept for such a type of damper, as well as some of its important behaviours observed in numerical simulations.

2   HYBRID CONTROL CONCEPT

A crankshaft is often modelled as a discrete set of inertias, torsional springs and torsional dashpots, with as many inertias as cylinders and wheels. However, as there is only one mode considered (the first torsional mode), we suppose for a matter of universality and simplicity that we have only two inertias in the crankshaft (I_1 and I_2), linked by one torsional spring (k_{Teq}) and one torsional dashpot (c_{Teq}). To that extent, only two modes are considered, namely the first torsional mode and the rigid body mode. This little detour by the modelling paves the way for the description of our concept.

This one can be summed up by the following sentence: a combination of energy harvesting and controlled damping in a coupled multi-physics and rotating device. It is indeed a multi-physics design as the easiest and most common way to harvest energy is via an electrical circuit.

2.1   Vibration damper

The first necessary adjunction to the crankshaft is a damper. The traditional viscous damper is not suited as it would be complicated to convert the heat into electricity. We introduce instead a rotational tuned mass damper (rTMD), which consists in the creation of an additional torsional degree of freedom on the lighter of the two inertias (the front one), which canalizes a part of the vibration energy and eventually decreases the magnitude of vibration at resonance. The literature about its working principle is extensive when translational motions are involved, but is quite rare about its performance in rotation. The base principle is nevertheless the same. Its inertia, stiffness and damping coefficient are determined using Den Hartog's principles for tuning [11], replacing the usual masses by inertias, and considering the equivalent modal inertia for the tuning.

This passive rTMD is designed so to enable a safe and secure operation of the crankshaft in case of failure of the whole system. Moreover, it creates a mechanical degree of freedom, which energy is easy to transform into electricity with either a piezo or an electromagnetic converter. As the magnitude in rotation of the rTMD is quite high, the electromagnetic converter suits better [12], and makes our rTMD a device close to an active mass damper (AMD). We can see this converter as a magnet set on the rTMD that moves within a coil, which itself is clamped on the front degree of freedom. The induction phenomenon creates a current in the coil that has two consequences:

- A current is indeed harvested
- There is a torque created between the coil and the magnet due to the electromagnetic interaction, that slows down the rTMD (detrimental effect) but also exerts an additional effort on the front degree of freedom

On the other hand, the reverse (AMD-like) behaviour is also possible, in which a current in the coil, provided by another source, amplifies the motion of the rTMD thanks to the torque created. In our model, this source is nothing else than the rotational irregularities damper.

2.2 Rotational irregularities damper

The design of the rotational irregularities damper part was first inspired by present academic or industrial research [8-9-13-14]. It consists in a permanent magnet synchronous machine (PMSM) with its coils located on the flywheel in a salient-pole-like fashion, and surrounded by a set of permanent magnets clamped to the frame. A sketch of the whole concept and a possible layout for the two parts is given in Figure 1, and provides a better hint of the working principle and design. This sketch is limited to a proof of concept, and does not aim to describe the actual implementation in a real system.

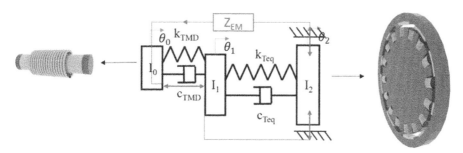

Figure 1. 3-DoF sketch of the damper concept.

This PMSM also relies on the electromagnetic interaction between the coil and the magnets, and has two possible behaviours as well. Either the current harvested by the rTMD creates a torque thanks to the induction phenomenon, which enables a mitigation of the output rotational speed oscillations depending on the current and the shape of the magnets layout. Or, the rotation of the coils in the magnetic field creates an electromotive force (EMF) that itself leads to the circulation of charges and slows down the rotation of the flywheel, including its oscillations. However, these are the two "ideal" behaviours of each part, but the coupling between the two actually leads to an "intermediate" behaviour, that must be tailored to be beneficial.

2.3   Coupled circuit

Figure 1 also shows with red lines the electrical coupling, and with red arrows the electromagnetic interactions. In the coupled circuit, we ensure the possibility to add another impedance, Z_{EM}, which purpose is to adapt the values of resistance and inductance in the circuit. Indeed, those values contribute to the phase of the current, which control is essential in order to decrease the oscillations, for we need at least a negative contribution of the electromagnetic torques created on the front end and on the back end. However, the phase of the EMF created on the flywheel is relative to an arbitrary initial position between the coils and the magnets. We consider the initial angular position, p, as the angle between the direction of an arbitrary coil and the vertical upwards direction (see Figure 2a).

3   MODELLING THE HYBRID DAMPER

3.1   Space derivative of the magnetic flux

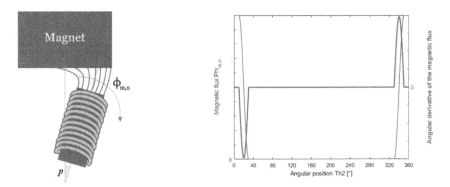

**Figure 2. a. Interaction between a coil (rotor) and a magnet (stator)
b. Magnetic flux and derivative as a spatial function.**

In order to understand the EMF created by the rotational irregularities damper, it is necessary to recall the base equation of the induction, which is Faraday's law. If we note ϕ the overall magnetic flux between the sets of coils and magnets, we have:

$$e_{FW} = -\frac{d\phi}{dt} \tag{1}$$

Due to the rotation, we know that the EMF is an angular periodic function. If we introduce the angle, θ_2 according to the sketch in Figure 1, we get:

$$e_{FW} = -\frac{d\phi}{d\theta_2}\dot{\theta}_2 \tag{2}$$

The EMF is then defined as the product of two functions, the instantaneous angular speed (IAS) $\dot{\theta}_2$ introduced in the next subsection, and the spatial derivative of the overall magnetic flux to θ_2. This latter is unfortunately not very well known. Most of the time its precise description on a given machine is either measured experimentally or simulated using a FE analysis. We can however approximate its shape using simple equations. Indeed, as we can see in Figure 2a, the local magnetic flux is maximal when a coil n is facing a magnet m, then decreases to 0 when the coil turns away, and then increases again to its maximal value after a complete rotation to start a new cycle. This description is close to the construction of the permeance model by Ostovic [15]. To that extent, we assume an analogous description of the local magnetic flux, which is given in blue in Figure 2b, with a cosine to bridge the gap between the maximal value and 0. The angular derivative, depicted in red, lets then only one positive peak and one negative peak, described with a sine function. The summation yields:

$$\frac{d\phi}{d\theta_2} = \sum_m \sum_n \frac{d\phi_{m,n}}{d\theta_2} \tag{3}$$

Of course, the exact description depends on the layout and number of magnets and coils. It is the same for the total electromagnetic torque $C_{em} = R.F_{em}$ created by the damper, R being the radius of the flywheel. The Virtual Work method as introduced in [16] and [17] recalls that the electromagnetic torque is the derivative of the magnetic co-energy with respect to the angular position. This provides:

$$C_{em,th} = \frac{\partial W'_{mag}}{\partial\theta_2} = -\frac{\partial W'_{mago}(\theta_2)}{\partial\theta_2} + \frac{\partial\phi(\theta_2)}{\partial\theta_2}i(\theta_2) + \frac{1}{2}\frac{\partial L(\theta_2)}{\partial\theta_2}i(\theta_2)^2 \tag{4}$$

Assuming a constant inductance and no initial magnetic co-energy this expression can be restricted to its main component:

$$C_{em} = \frac{d\phi}{d\theta_2}i(\theta_2) \tag{5}$$

This is also consistent with the common assumption that the coupling coefficients in the two descriptive equations of electromagnetic transducers are equal. In that case, $\frac{d\phi}{d\theta_2}$ is seen as a coupling function, denoted T_f in the remaining of the paper.

3.2   IAS approach for the description in non-stationary operation

In equation (2), the second term is the IAS $\dot{\theta}_2$, that we know as a multi-harmonic oscillating function. An example of this function is given in Figure 3, for a 6-cylinders crankshaft. The oscillating behaviour with a predominance of the third harmonic is noticeable.

This result is dependent on the excitation, and so on the engine regime. In order to describe properly the oscillation whatever the regime, it seems difficult to rely on the time variable, as it would be necessary to make assumptions about the oscillation shape and harmonics. Moreover, both the input torque and the spatial derivative of the magnetic flux are angular functions. To that extent, we adopt in our model an angular description of the dynamics instead of a time description.

In order to change the variable, we use an IAS approach [18] applied on a state-space model, as a time derivation is equivalent to an angular derivation times the IAS, as (6) puts it. This description is mainly used to detect defaults in bearings, but is also perfectly suitable here. Still, our two angular functions are not parameterized by the same angle. We choose θ_1 as the determinant variable, as the input torque applies on the first degree of freedom.

$$\frac{dX(t)}{dt} = A(t,\theta_1).X(t) + B(t,\theta_1).U(t,\theta_1) \rightarrow \frac{dX(\theta_1)}{d\theta_1}\frac{d\theta_1}{dt} = A(t,\theta_1).X(\theta_1) + B(t,\theta_1).U(t,\theta_1) \tag{6}$$

The consequence in the state-space model is then a division by $\dot{\theta}_1$, with the inherent assumption that this speed is strictly positive. To that extent, we must ensure that it remains under control, especially because of the dissipation caused by internal damping.

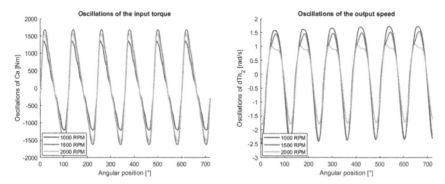

Figure 3. Torque C_a and IAS $\dot{\theta}_2$ without damping, for three regimes.

3.3 Equations of the 3-DoF model

The last point to describe is the electromagnetic interaction at the rTMD. It behaves as an electromagnetic converter, and is to that extent described in the same way, with a constant coupling coefficient T_2 linking the current to the effort between the rTMD and the front wheel, as well as the deformation speed $\dot{\theta}_1 - \dot{\theta}_0$ to the EMF created in the coil. With the three mechanical degrees of freedom, we can draw three coupled electro-mechanical equations with two excitations: the input torque C_a on the front wheel and a resistive constant torque C_r on the back wheel, to balance the average acceleration. Our coupled circuit can be described as a series circuit with a resulting inductance L_c, a resulting resistance R_c, and two voltage generators depicting the EMFs, thus yielding a fourth electrical equation:

$$
\begin{cases}
I_o\ddot{\theta}_0 + c_{TMD} \cdot \left(\dot{\theta}_0 - \dot{\theta}_1\right) + k_{TMD} \cdot (\theta_0 - \theta_1) + T_2 i = 0 \\
I_1\ddot{\theta}_1 + c_{TMD} \cdot \left(\dot{\theta}_1 - \dot{\theta}_0\right) + c_{Teq} \cdot \left(\dot{\theta}_1 - \dot{\theta}_2\right) + k_{TMD} \cdot (\theta_1 - \theta_0) + k_{Teq} \cdot (\theta_1 - \theta_2) - T_2 i = C_a(\theta_1) \\
I_2\ddot{\theta}_2 + c_{Teq} \cdot \left(\dot{\theta}_2 - \dot{\theta}_1\right) + k_{Teq} \cdot (\theta_2 - \theta_1) - T_f(\theta_2) \cdot i = -C_r \\
L_c\frac{di}{dt} + R_c i + T_f(\theta_2) \cdot \dot{\theta}_2 + T_2 \cdot \left(\dot{\theta}_1 - \dot{\theta}_0\right) = 0
\end{cases}
$$

These equations describe the general behaviour of the coupled system shaft-hybrid damper. They enable the constitution of the state-space vector $[\dot{\theta}_2, \dot{\theta}_1, \dot{\theta}_0, \theta_2, \theta_1, \theta_0, i]$ and the non-linear state-space matrix. The system is then solved for various engine regimes using MATLAB Simulink, with an output vector divided at each angular step by $\dot{\theta}_1$.

4 IMPORTANT BEHAVIOURS

4.1 Multi-harmonic excitation and harmonics superposition

As the input torque is multi-harmonic, it is interesting to understand whether the correction of one harmonic also has an influence on the other harmonics. The various input harmonics being independent, we can suppose that they are also independent in the response. However, as T_f is a harmonic function, we actually have two products of functions, one in the electric equation and one describing the electromagnetic torque created by the PMSM at the flywheel. Two consequences: if T_f does not involve the same harmonics as C_a, other harmonics are created in the response but no correction is provided for the actual excitation. If it does, there is a correction, but this product of functions yields a constant value (at frequency 0) and a doubled harmonic. Yet the harmonics of the input torque are all multiple of $N_{cyl}/2$, so a doubled harmonic may have a consequence in the correction.

Figure 4 shows the difference for three scenarios, respectively a H9 or H18 mono-harmonic excitation, one H9+H18 multi-harmonics with the same magnitudes as before and one H6+H9+H18 multi-harmonics with the same magnitudes as before. For each harmonic, the initial angular position p is adapted to achieve the minimal oscillation in torsion under a mono-harmonic excitation.

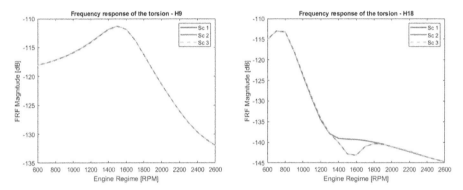

Figure 4. – Mono-harmonic and multi-harmonic excitations.

Obviously, there is no particular influence on the 9th harmonic when other harmonics are added. On the contrary, the introduction of a H9 sub-harmonic when a H18 is involved leads to a valley around 1600 RPM for the tested model. In the third scenario, the influence of the 6th harmonic does not lead to main changes, although the spectrum also displays a bunch of new harmonics that did not exist in the input torque. This discloses that there is an actual influence between the different harmonics, and that it is not possible to consider the several harmonics separately.

4.2   Behaviour in acceleration

The vocation of a crankshaft is not to rotate at a given constant regime. It undergoes phases of accelerations, decelerations and stability. To that extent, it is necessary to ensure that the damper offers efficient damping over the complete range of rotation, from idle to maximum speed (the case of starting the engine is let aside). We already know that the initial angular position p has an importance, as it modifies the phase shift between the efforts created by T_f and the other efforts.

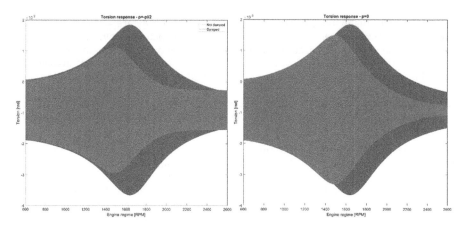

Figure 5. – Effect of the hybrid damper over a constant acceleration.

Figure 5 illustrates this statement for a simple case of constant acceleration from idle (600 RPM) to maximum speed (2600 RPM), with a H9 mono-harmonic excitation. This is merely another representation of what a Bode diagram would display. However, it paves the way for more complex acceleration schemes as well as for multi-harmonic excitations. It shows the efficiency of the damper for a magnitude of 0.2 for T_f. In the case $p=0$ for that simulation, we achieve a very good damping at high speeds (-10dB), but it is average at resonance (-3dB) and weak at low speeds. For the case $p=-n/2$, there is an amplification at high speeds (+0.5dB), but an acceptable damping at resonance (-4dB). This indicates the need for a dynamic adaptation of p over the complete speed range, in order to maximize the damping effect over the whole range. It can be also noticed that the maximum value is reached for lower speeds, which is a main feature of our damper concept.

4.3   Incidence on rotational irregularities

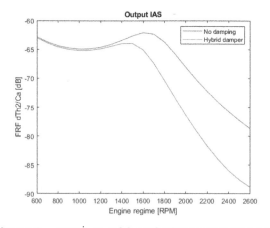

Figure 6. – FRF $\dot{\theta}_2/C_a$ with and without the damper.

We have mainly seen the case of vibration mitigation, but it is also necessary to evaluate the performance in rotational irregularities damping, which means the magnitude of the oscillations in $\dot{\theta}_2$. Again, this performance depends on the considered engine regime and on the parameters of the hybrid damper. Figure 6 illustrates the effect for the same magnitude as previously and the most favourable p. A greater magnitude would simply amplify the phenomenon. It appears that the effect is dramatic at high speeds, with up to 10dB less with the damper than without. However, the performance at low speeds, where the oscillations and so the effect of the rotational irregularities are the highest, is weak, which can be accounted for a lower current harvested at the PMSM.

5 DISCUSSION AND FUTURE WORKS

The present work shows that our damper concept successfully manages the energy to mitigate at the same time the torsional vibrations and the rotational irregularities. However, the influence on the rotational irregularities is very limited in the speed range where these irregularities are the most detrimental. A possibility to overcome this limitation could be a combination with the newest concepts on passive and active dampers, especially the DMF or the e-DMF [7]. Moreover, the present electromagnetic description for T_f and T_2 is simple and does not involve loss models or saturation that would affect the dynamics and the shape of the harvested current. The system being already non-linear, such additional non-linearity could lead to a different behaviour.

Still without this additional description, the introduced matter of phases and of multi-harmonics combination drastically complicates the prediction of an optimal design. The most suitable solution seems to be a dynamic adaptation of the initial angle p using an additional impedance Z_{EM} in the circuit, with the sole limitation that no capacitance is considered. Otherwise, the electric equation changes to a 2nd degree equation in the charge, leading to a resonant behaviour that would considerably alter the dynamics. Another improvement at low rotation speeds could be an adaptive switch on/switch off in the electrical circuit, which could activate the coupled circuit only when it is more efficient than the passive damping system brought by the rTMD, hence the importance of an effective fail-safe tuning. However, both require the addition of a control unit within the rotating system, as well as its power supply.

Future works will consist in an investigation of such improvements in the model and in the test of other acceleration-deceleration patterns to prevent any detrimental behaviours under non-stationary conditions. Additionally, the design and testing of a 3-DoF prototype is necessary in order to corroborate the numerical simulations and set up an experimental proof of concept. As the electromagnetic coupling at the rTMD tends to detune it if no external current is applied, future works could also feature an investigation of the potential benefits of passive or hybrid untuned mass dampers. It could indeed be on interest for the harmonics that do not excite the resonance frequency, even though the fail-safe property is lost.

6 CONCLUSION

In this paper, we have seen a new concept for a crankshaft damper that aims to reduce at once the torsional vibrations as well as the rotational irregularities through an electrical coupling, which enable to dispatch the energy where it is needed. This concept is pRoved efficient on numerical simulations, even though the rotational irregularities mitigation at low speeds is weak or inexistent. The several nonlinearities that characterize its dynamic behaviour limit the possibility to determine an optimal design for the damper, and the various phase shifts in operation suggest the need for an adaptive behaviour. However, the sharp damping effect at high speeds makes this design concept relevant.

REFERENCES

[1] Ligier, JL, Baron, E. "Acyclisme et vibrations: applications aux moteurs thermiques et aux transmissions". Technip, 2002.

[2] Sun, L. et al. "Research on torsional vibration reduction of crankshaft in off-road Diesel engine by simulation and experiment". JVE, 2018.

[3] Wramner, L. et al. "Vibration dynamics in non-linear dual mass flywheels for heavy-duty trucks". In Proceedings of ISMA2018, 1935 47, 2018.

[4] Suryanarayana, A. "Engine dynamics and torsion vibration reduction". Chalmers, 2015 (M.Sc. Thesis).

[5] Newland, D.E. "Developments in the design of Centrifugal Pendulum Vibration Absorbers", 24th ICSV, London, 2017.

[6] Dong, X. et al. "Magneto-rheological variable inertia flywheel". Smart Materials and Structures 27, 115015 (July 2018).

[7] Pfleghaar, J., et al. "The electrical Dual Mass Flywheel -an efficient active damping system". IFAC Proceedings Volumes, 7th IFAC Symposium on Advances in Automotive Control, 46, (September 2013): 483 88.

[8] Masberg, U. et al., "System for actively damping rotational non-uniformity of the crankshaft of a combustion engine or a shaft coupled thereto". European Patent EP 0847485 B1, filed August 31, 1996, and issued June 17, 1998.

[9] Kalev, C.M. et al., "Electromechanical Flywheels". European Patent EP 2761729B1, filed November 13, 2012 and issued August 6, 2014.

[10] Mendes, AS. et al. "Analysis of torsional vibration in internal combustion engines: modelling and experimental validation". Proceedings of the Institution of Mechanical Engineers, Part K: Journal of Multi-Body Dynamics 222, no 2 (June 2008)

[11] Den Hartog, JP. Mechanical vibrations. 4th ed. NY: McGraw-Hill, 1956.

[12] Austruy, J. "Rotor hub vibration and blade loads reduction, and energy harvesting via embedded radial oscillator", 2011 (Ph.D Thesis).

[13] Nakajima, Y. et al. "A study on the reduction of crankshaft rotational vibration velocity by using a motor-generator". JSAE Review 21, (July 2000): 335 41.

[14] Davis, R.I. et al. "Engine torque ripple cancellation with an integrated starter alternator in a hybrid electric vehicle: implementation and control". IEEE Transactions on Industry Applications 39, 6 (November 2003): 1765 74."

[15] Ostovic, V. "Dynamics of saturated electric machines", Springer, 1989

[16] Strahan R.J. "Energy conversion by permanent magnet machines and novel development of the single phase synchronous permanent magnet motor". University of Canterbury, Christchurch, New Zealand. 1998. (PhD Thesis).

[17] Pyrhonen, J. et al. "Design of rotating electrical machines", Wiley, 2008.

[18] Bourdon, A., et al. "Introducing angularly periodic disturbances in dynamic models of rotating systems under non-stationary conditions". Mechanical Systems and Signal Processing, Elsevier, 44 (1–2), 60–71. 2014.

Improved reduction methodology for rotor-dynamic systems using Modified SEREP

Ankush Kapoor

Siemens Limited, Gurugram, India

Jayanta Kumar Dutt

Department of Mechanical Engineering, Indian Institute of Technology Delhi, New Delhi, India

A.S. Das

Department of Mechanical Engineering, Jadavpur University, Kolkata, India

ABSTRACT

Reduction of model size reduces cost of calculation, processing and simulation. This paper attempts reducing size of equations of motion for a general rotor-dynamic system. A rotor-dynamic system comprises the shaft, disc or discs and support or supports. The former is a continuum with distributed inertia, dissipation and restoring properties. Discs, bearings, and supports, add locally, inertia, gyroscopic effects, restoring and stationary and rotary dissipative effects, unlike the shaft. Therefore, the shaft requires many more coordinates than the discs, bearings and supports to express its effects in the equations of motion, causing inflation of model size. Literature reports that the technique 'System Equivalent Reduction Expansion Process' (SEREP) and its modification to reduce the rotor-dynamic systems as a whole at each speed of rotation. The system matrix is generally non-symmetric, varies with spin speed, and a change in local effect, e.g. its location, properties, therefore, calls for a fresh reduction. In contrast, this paper reports reducing dynamic behaviour of only the shaft continuum and then adds the local effects subsequently, avoiding a fresh reduction. Thus, this paper proposes appropriate transformation matrices, which act upon the original system matrix to achieve the model of reduced size, that may be suitably augmented with local effects, e.g. inertia, restoring, gyroscopic effects, and dissipation properties, wherever desired, and very importantly the information of spin speed, without inflating the reduced size. A close match exists between the modal matrices of proposed reduced model and earlier published model based on Pseudo-Orthogonality Check (POC). A close match between relevant eigenvalues and response, between those predicted by Finite element method simulation and those predicted by the present methodology. This proves correctness of the reduction proposed.

Industrial Applications

Adjusting the rotor-shaft system characteristics at the design stage will be extremely simple and easy to use by following this process. Change of system properties may be easily tracked by revaluating the system. Applying control measures on a rotor-shaft system may be applied better with the help of a short model.

1 INTRODUCTION

Reduction of model size has gained importance after the successful development and use of finite element methods by academia and industry. The finite element formulation generally provides a large set of equations, handling and solution of which becomes very costly. The model size reduction in physical coordinates was introduced by Guyan (1) as a method to decrease the size of dynamic equations for making the computation easier and faster. Thereafter, many researchers worked on improvement of reduction techniques for vibratory systems. A concise description of various reduction methodologies is given by Qu in (2). The reduction techniques are generally divided in two subgroups namely physical coordinate reduction and generalized coordinate reduction. The Guyan reduction (1) and its various modifications form the basis of physical coordinate reduction. The pioneering work on generalized coordinate reduction was done by O'Callahan et al. (3) which resulted in a reduction method called SEREP standing for System Equivalent Reduction Expansion Process. All the major reduction techniques found good application in structural vibration problems whereas the application of the reduction, especially the modal coordinate reduction, on rotor-dynamic systems, has been far less in comparison. General distribution of damping, gyroscopic effects, rotating damping (originating from material damping in the rotor shaft, from fluid films in fluid film bearings, Alford's force), in rotor-dynamic systems call for representation in state-space, where system matrix becomes asymmetric in general. So, asymmetry of system matrices necessitates consideration of both right and left eigenvectors of the system matrix for the purpose of reduction. Friswell et al. (4) applied various reduction techniques including SEREP on rotating system with gyroscopic effects but the transformation matrices were formed without taking into account left and right eigenvectors thereby compromising on the bi-orthogonality conditions. Furthermore, (4) did not include the rotating damping in the system and finally concluded that the available reduction techniques were not suitable to reduce the systems with gyroscopic effects and damping. Das and Dutt (5) presented a modified version of SEREP, which formed the transformation matrix pairs as used in SEREP, however by taking both right and left eigenvectors of the non-symmetric system matrix into account. The authors presented application of modified SEREP in controlling the rotor vibrations in (6). Shrivastava and Mohanty (7) used modified SEREP technique for estimating single plane unbalance properties by employing Kalman filtering based estimation technique thereby widening the scope of the application of modified SEREP reduction technique.

Most of the reduction methodologies of rotor-dynamic systems in state space do not permit any addition or alteration of local effects, (for example, adding, removing or shifting the location of discs, bearing or supports or changing values of parameters), after the reduction is done, and the corresponding transformation matrices are obtained. Therefore, if needed, transformation matrices for the modified system (i.e. after the addition or alteration of local effects are done), may be obtained only after following the reduction methodology afresh. A fresh reduction is also called for, to obtain a reduction of model size for a different spin-speed, as the system matrix is generally dependent on spin speed due to the presence of gyroscopic effect, rotary form of damping, etc. A fresh reduction every-time thus causes unnecessary wastage of computational time and resources and is unfriendly at least in the design stage. Han et al (8) used Krylov subspace based model reduction method to make the transformation matrices derived from Krylov-basis vectors by considering non rotating undamped rotor bearing system and using them on the damped rotating system. Friswell (4) applied the similar method of using undamped stationary system to form transformation matrices and apply them on the damped rotating systems using multiple techniques including SEREP but the results obtained were not encouraging. The SEREP and modified SEREP techniques are spin speed dependent which makes their practical applications difficult. Therefore, a transformation matrix pair for SEREP reduction, independent of the influence of additions or alterations of local effects and spin speed is very much beneficial, as then the designer is able to directly find the reduced model size to see the effect of changes (addition and/or alteration, speed) in

360

design. This inspired the authors to work for a reduction methodology devoid of the influences of local effects and spin speed. Subbaiah et al.(9), Genta and Delprete (10) did add the bearings after reduction but their reduction was carried out directly using component mode synthesis (9) or directly in second order equations without taking gyroscopic effects and damping into account in reduction (10). Friswell et al (11) described a method to reduce system with local non linearity by considering it as a forcing function and using the transformation matrix of the rest of the system to reduce the forcing function matrix which includes the local non linearity. The method was inspiring and the present work extends this philosophy in quest of a reduction methodology devoid of the influences of local effects and spin speed as proposed above. The validation of reduction techniques not only involve transient and steady state analysis but also includes matching of eigenvectors. Thereby mode correlation techniques play an important role in validating a reduction methodology. Zuo et al (12) used Modal Assuarance Criteria for validation of their proposed reduction technique.

In the current work, a novel reduction methodology based on modified SEREP is presented, wherein the need of repeating the reduction process with the change in local effects, gyroscopic effects or rotating damping are eliminated. The same transformation matrices obtained once can be utilized for reduction of system having different local effects, speeds and rotating damping coefficients. A set of numerical examples are provided in support of the above. Finally, the Pseudo-orthogonality checks (POC) (13) are used to compare the eigenvectors obtained 1) by following the reduction methodology of the full model with modified SEREP (5) and 2) by following the proposed reduction methodology, to validate the current reduction model. A close match between the eigenvectors of the above methodologies proves the correctness of the proposed methodology.

2 ANALYSIS

2.1 The philosophy of reduction

A general dynamic system generates inertial, gyroscopic, restoring and both stationary and rotary dissipative effects when subject to generic dynamic forcing. These effects may be continuously distributed over the entire span or remain concentrated locally and accordingly the effects may fall under continuum or local. Therefore, the continua need many (ideally infinite, and pragmatically large but finite) degrees of freedom for representing them within an acceptable level of accuracy in comparison with the number of coordinates required for modelling the local effects. From the angle of view as above, in a rotor-dynamic system, primarily the effects generated by the shaft fall under continuum and in comparison, those generated by discs, bearings, gears and supports fall under local. An attempt of reduction of size of the continuum effects forms the basic philosophy as continua inflate the size of the model primarily. The current reduction technique like the modified SEREP (5) focuses on reducing the complete FE model of the system with both continuum and local effects, whereas it is clear from above discussion that reduction of model-size of the continuum should be the basic focus and; subsequently add the local effects appropriately. The same is proposed and attempted here. Towards this end, this paper proposes a proper set of transformation matrices (a transformation matrix pair) for appropriate addition or proper positioning of the local effects in the reduced continuum system. This technique makes the reduction process independent of local effects and allows modification (addition, alteration in position as well as properties) of local effects without repeating the reduction afresh. Finally, since the continuum of any material will produce approximately similar normalized right and left eigenvectors if the geometry remains the same, the transformation and inverse transformation matrices (transformation matrix pair) constituted from the normalized right and left eigen-vectors may be considered independent of material properties.

Based on the above discussion, it transpires that reduction methodology should include gyroscopic and rotating damping effects from the continuum in reduction process to obtain transformation matrix pair. However, a correct reduction process like modified SEREP (5), which preserves the mode shapes, allows to keep the rotating damping out of the reduction process. The effect may be added directly in the reduced system of equations because rotating damping forces and gyroscopic effects do not influence the mode shapes of the system heavily, even though these effects are spread throughout the continuum. All these make the methodology fast, spin-speed independent, flexible and help to save time.

2.2 Governing equation and reduction mechanism
The equations of motion of any generic linear rotor-bearing system discretized using finite element formulation is given by

$$[M]_{n \times n} \{\ddot{q}\}_{n \times 1} + [D]_{n \times n} \{\dot{q}\}_{n \times 1} + [K]_{n \times n} \{q\}_{n \times 1} = \{f\}_{n \times 1} \tag{1}$$

$$[M] = [M]_{translatory} + [M]_{rotary}; \quad [K] = [K]_{symmetric} + \eta \omega [K]_{circulatory} + [K]_{support}$$

$$[D] = [D]_{support} + \eta [K]_{symmetric} + \omega [G]$$

In the above $[M]_{n \times n}$ is the inertia matrix, constituting translatory and rotary inertia having n degrees of freedom. The damping matrix $[D]_{n \times n}$ has contributions from support damping $[D]_{support}$, viscous material damping with viscosity coefficient η and gyroscopic effects with $[G]$ as gyroscopic matrix and ω as the angular velocity. Similarly, the stiffness matrix $[K]_{n \times n}$ has a symmetric stiffness part $[K]_{symmetric}$, a circulatory stiffness part $[K]_{circulatory}$ (this is skew symmetric) to model the shaft element and has a contribution of stiffness from supports $[K]_{support}$. The vector $\{f\}_{n \times 1}$ represents the forcing function and the vector $\{q\}_{n \times 1}$ represents the nodal displacement. Many of these effects, which form the stiffness and damping matrices, cause them to become non-symmetric in nature. The next subsections explain the reduction methodology, which uses the modified SEREP technique (5) as the basis and so, for the sake of completeness modified SEREP (5) needs some explanation.

2.2.1 Modified SEREP- From Literature
Modified SEREP (5), uses the state space equations of motion, given in Eq. (2) and Eq. (3), corresponding to Eq. (1) as described by Meirovitch (14).

$$\{\dot{z}\}_{2n \times 1} = [A]_{2n \times 2n} \{z\}_{2n \times 1} + [B]_{2n \times r} \{u\}_{r \times 1} \tag{2}$$

$$\{y\}_{p \times 1} = [C]_{p \times 2n} \{z\}_{2n \times 1} + [D_0]_{p \times r} \{u\}_{r \times 1} \tag{3}$$

$$\{z\} = \begin{Bmatrix} q \\ \dot{q} \end{Bmatrix}_{2n \times 1} \text{ and } [A] = \begin{bmatrix} [0]_{n \times n} & [I]_{n \times n} \\ -[M]_{n \times n}^{-1}[K]_{n \times n} & -[M]_{n \times n}^{-1}[D]_{n \times n} \end{bmatrix}$$

In Eq. (2) and Eq. (3), $[B]$, $[C]$ and $[D_0]$ are input, output and direct transmission matrices respectively whereas $\{u\}_{r \times 1}$ is the forcing function vector.

To solve for free vibration case in state space, Das (5) found the eigenvalues and eigenvectors of the system matrix $[A]$. Since the system matrix $[A]$ is asymmetric and real, it has different Left and Right eigenvectors, for the same eigenvalues.

During the process of reduction, the respective modal matrices corresponding to the right and left eigenvectors $[U]_{2n \times 2n}$ and $[V]_{2n \times 2n}$ respectively, are first reduced by decreasing the number of modes in the system to "m" and thereby the first set of reduced modal matrices $[\bar{U}]_{2n}$ and $[\bar{V}]_{2n}$ are produced. Reducing the number of states to "a" causes further reduction in the matrices, which become $[\bar{U}_1]_a$ and $[\bar{V}_1]_a$ respectively. The case where a is of importance, since it is a more practical situation for reduction. The reduction process given in detail by [5], thus obtains the following reduced equations of motion

$$\{\dot{\bar{z}}\}_{a\times 1} = [\bar{A}]_{a\times a}\{\bar{z}\}_{a\times 1} + [\bar{B}]_{a\times r}\{u\}_{r\times 1} \tag{4}$$

$$\{y\}_{p\times 1} = [\bar{C}]_{p\times a}\{\bar{z}\}_{a\times 1} + [D_0]_{p\times r}\{u\}_{r\times 1} \tag{5}$$

$$[\bar{A}]_{a\times a} = [T]_{a\times 2n}[A]_{2n\times 2n}[\bar{T}]_{2n\times a}; \; [\bar{B}]_{a\times r} = [T]_{a\times 2n}[B]_{2n\times r}; \; [\bar{C}]_{p\times a} = [C]_{p\times 2n}[\bar{T}]_{2n\times a}$$

In the reduced equation of motion, the transformation matrix $[T]$ and inverse transformation matrix $[\bar{T}]$ (Transformation matrix pair) play the pivotal role. These matrices map the solution vector from full FE system subspace to reduced subspace and vice versa as shown in Eq. (6) and Eq. (7) respectively.

$$\{\bar{z}\}_{a\times 1} = [T]_{a\times 2n}\{z\}_{2n\times 1} \tag{6}$$

$$\{z\}_{2n\times 1} = [\bar{T}]_{2n\times a}\{\bar{z}\}_{a\times 1} \tag{7}$$

The transformation and inverse transformation matrices $[T]_{a\times 2n}$ $[\bar{T}]_{2n\times a}$ are given by Eq. (8) and Eq. (9) where, superscript (+) stands for pseudo-inverse.

$$[T]_{a\times 2n} = [\bar{U}_1]_a, [\bar{V}]^T_{m\times 2n} \tag{8}$$

$$[\bar{T}]_{2n\times a} = [U]_{2n\times m}[\bar{V}_1]^T_{m\times a}\left[[\bar{V}_1]^T\right]^+_{a\times m}[\bar{U}_1]^+_{m\times a} \tag{9}$$

2.2.2 *Proposed Methodology – Continuum Reduction without rotating damping and gyroscopic effects*

In the proposed methodology, the general equation of motion of any rotor-bearing system, Eq. (1) is divided into two parts; the first part contains the mass, stiffness and damping properties of the continuum and permanent local effects, which will define the system, therefore cannot be changed, and will decide the transformation and inverse transformation matrices. The second part contains the mass, stiffness and damping properties corresponding to the temporary local effects, rotating damping and gyroscopic effects; these effects do not take part in forming the transformation and inverse transformation matrix pair. Rewriting Eq. (1) in the form as described, results in Eq. (10).

$$[[M_C]_{n\times n} + [M_{PL}]_{n\times n}]\{\ddot{q}\}_{n\times 1} + [[D_{PL}]_{n\times n}]\{\dot{q}\}_{n\times 1} + [[K_C]_{n\times n} + [K_{PL}]_{n\times n}]\{q\}_{n\times 1} = \{f\}_{n\times 1} -$$
$$[M_{TL}]_{n\times n}\{\ddot{q}\}_{n\times 1} - [\eta[K_C]_{n\times n} + \omega[G_C]_{n\times n} + \omega[G_{PL}]_{n\times n} + \omega[G_{TL}]_{n\times n} + \omega[D_{TL}]_{n\times n}]\{\dot{q}\}_{n\times 1} - \tag{10}$$
$$[\eta\omega[K_{CCL}]_{n\times n} + [K_{TL}]_{n\times n}]\{q\}_{n\times 1}$$

In the above, on the left, subscripts "C" and "PL" represent the effect from basic continuum and the permanent local effects. All other effects like temporary local effects with subscript "TL", gyroscopic effects (continuum and local), effects due to dissipation and circulatory terms and non-potential forcing functions appear on the right hand side of the equation. Thus in Eq. (10), $[M_C]$, $[K_C]$, $[K_{CCL}]$ and $[G_C]$ represent the Continuum parts of Mass matrix, symmetric stiffness matrix, skew symmetric circulatory stiffness matrix, gyroscopic matrix respectively and $[M_{PL}]$, $[K_{PL}]$, $[D_{PL}]$ and $[G_{PL}]$ represent mass, stiffness, damping and gyroscopic matrices corresponding to permanent local effects. $[M_{TL}]$, $[K_{TL}]$, $[D_{TL}]$ and $[G_{TL}]$ represent mass, stiffness, damping and gyroscopic matrices of temporary local effects. All permanent local effects are supposed to remain fixed during the reduction, based on which the transformation matrix pair will be formed, whereas all other local effects may be changed after reduction and their effects on the characteristic and response may be found by applying the same pair of transformation and inverse transformation matrices.

Converting Eq. (10) to state space form results in Eq. (11).

$$\{\dot{z}\}_{2n\times1} = [A_C]_{2n\times2n}\{z\}_{2n\times1} + [N_L]_{2n\times2n}\{\dot{z}\}_{2n\times1} + [A_L]_{2n\times2n}\{z\}_{2n\times1} + [B]_{2n\times n}\{u\}_{n\times1} \quad (11)$$

$$[A_C] = \begin{bmatrix} 0_{n\times n} & I_{n\times n} \\ -[M_C + M_{PL}]_{n\times n}^{-1}[K_C + K_{PL}]_{n\times n} & -[M_C + M_{PL}]_{n\times n}^{-1}[D_{PL}]_{n\times n} \end{bmatrix}$$

$$[A_L] = \begin{bmatrix} 0_{n\times n} & 0_{n\times n} \\ -[M_C + M_{PL}]_{n\times n}^{-1}[\eta\omega K_{CCL} + K_{TL}]_{n\times n} & -[M_C + M_{PL}]_{n\times n}^{-1}[\eta K_c + \omega G_c + \omega G_{PL} + \omega G_{TL} + D_{TL}]_{n\times n} \end{bmatrix}$$

$$[N_L] = \begin{bmatrix} 0_{n\times n} & 0_{n\times n} \\ 0_{n\times n} & -[M_C + M_{PL}]_{n\times n}^{-1}[M_{TL}]_{n\times n} \end{bmatrix}; [B_{2n\times n}] = \begin{bmatrix} 0_{n\times n} \\ [M_C + M_{PL}]_{n\times n}^{-1} \end{bmatrix}; \{u\}_{n\times1} = \{f\}_{n\times1}$$

Equation (11), in state space has two components, the first component consists of the effects of basic continuum with permanent local effects and the second component consists of temporary local effects rotating damping and gyroscopic effects of the complete system. Only the first component of the equation given by Eq. (12) will be reduced, i.e. the transformation matrix pair will be found based on the following equation.

$$\{\dot{z}\}_{2n\times1} = [A_c]_{2n\times2n}\{z\}_{2n\times1} \quad (12)$$

The modified SEREP methodology as described in Eq. (2-9), is applied to reduce Eq. (12). If all the local effects are deemed temporary in nature and only the continuum part, which has free-free boundary condition, undergoes reduction, then the first few modes, depending upon the FEM discretization, will be rigid body modes having zero natural frequency.

The reduction methodology helps to get the transformation matrix pair, $[T_C]$ and $[\bar{T}_C]$ respectively from Eq. (8) and Eq. (9). It must be noted that although the transformation matrix pair is obtained by reducing Eq. (12) which contains only the continuum and permanent local part of the system, these matrices are still utilized to redistribute the temporary local effects appropriately into reduced system. Thus, these transformation matrices are used to reduce the full Eq. (11). Therefore, using the transformation matrices $[T_C]$ and $[\bar{T}_C]$ on Eq. (11), the reduced equation in state-space is obtained and written as in Eq. (13).

$$\{\dot{\bar{z}}\}_{a\times1} = [\bar{A}C]_{a\times a}\{\bar{z}\}_{a\times1} + [\bar{N}L]_{a\times a}\{\dot{\bar{z}}\}_{a\times1} + [\bar{A}L]_{a\times a}\{\bar{z}\}_{a\times1} + [\bar{B}]_{a\times n}\{u\}_{n\times1} \quad (13)$$

$$[\bar{A}C]_{a\times a} = [T_C]_{a\times2n}[A_C]_{2n\times2n}[\bar{T}_C]_{2n\times a}; [\bar{N}L]_{a\times a} = [T_C]_{a\times2n}[N_L]_{2n\times2n}[\bar{T}_C]_{2n\times a}$$

$$[\bar{A}L]_{a\times a} = [T_C]_{a\times2n}[A_L]_{2n\times2n}[\bar{T}_C]_{2n\times a}; [\bar{B}]_{a\times n} = [T_C]_{a\times2n}[B]_{2n\times n}$$

Rearranging Eq. (13) by clubbing the terms belonging to $\{\dot{\bar{z}}\}_{a\times1}$ and $\{\bar{z}\}_{a\times1}$ separately, results in equation (14).

$$[\bar{N}T]_{a\times a}\{\dot{\bar{z}}\}_{a\times1} = [\bar{A}T]_{a\times1}\{\bar{z}\}_{a\times1} + [B]_{a\times n}\{u\}_{n\times1} \quad (14)$$

In equation (14),

$$[\bar{N}_T]_{a\times a} = I_{a\times a} - [\bar{N}_L]_{a\times a} \tag{15}$$

$$[\bar{A}_T]_{a\times a} = [\bar{A}_C]_{a\times a} + [\bar{A}_L]_{a\times a} \tag{16}$$

Equation (14) is rewritten as in Eq. (17), which is the final reduced form obtained from Eq. (11) and will be used for further numerical simulations.

$$\{\bar{z}\}_{a\times1} = [\bar{N}T]_{a\times a}^{-1}[\bar{A}T]_{a\times a}\{\bar{z}\}_{a\times1} + [\bar{N}T]_{a\times a}^{-1}[\bar{B}]_{a\times n}\{u\}_{n\times1} \tag{17}$$

The reduction methodology shows that any change in temporary local effects or properties like speed and rotating damping coefficient, η does not call for a fresh reduction process or deducing matrices $[T_c]$ and $[\bar{T}_c]$. This enables adding the above effects as desired after deriving the matrices. Including more local effects as permanent effects, results in a closer reduced model, which predicts closer characteristics of the reduced system to the unreduced system. Closeness of a reduced model to the unreduced model increases by increasing the modes of the continuum preserved in reduction, however, change of a property of continuum, like the shaft diameter, will require a fresh reduction. The transformation matrix pair helps to add the temporary local effects directly to the reduced system. **The most significant contribution of this reduction mechanism is the fact the continuum based transformation matrices can be used to reduce rotordynamic systems with local effects, gyroscopic effects, structural damping and other material properties like modulus and density that do not impact mode shapes.** Figure 1 Shows a flow chart to illustrate the algorithm. It explains the complete methodology to perform reduction using the method presented in this work.

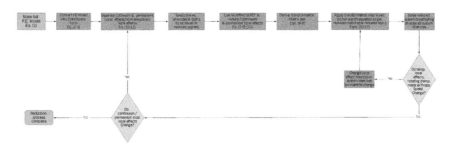

Figure 1. Flowchart illustration of the reduction process.

2.3   Validation of the reduction methodology by using Mode Correlation

Validation of the proposed reduction methodology is primarily achieved by comparison of transient and frequency response of reduced model with the full model. This is presented in the following section. In addition to this, it is of essence that the eigenvalues and eigenvectors are also checked to support the proposed technique. This may prove to be difficult since the eigenvectors of full model and reduced model are of different dimensions. Thus, it is proposed to compare the eigenvectors of the proposed reduction technique with modified SEREP, which is already published in literature as they are of same dimension. Modal assurance criteria – MAC (15,16) has been a preferred choice for this task but it is not useful in the case of bi-orthogonal systems (13).

Pseudo Orthogonality Check (POC) (13) given by eq.(18) has been utilized for quantifying the degree of correlation between modes. The POC check described in (13) is mass normalized. However, the eigenvectors used in the current work are bi-orthonormal with respect to each other. Therefore, there is no mass normalization in Eq. (18).

$$POC = [Y_{ms}]^T_{m \times a}[X_r]_a \qquad (18)$$

In Eq. (18), $[Y_{ms}]^T$ is the transpose of the reduced left modal matrix from the modified SEREP technique and $[X_r]$ is the reduced right modal matrix corresponding to the proposed reduction methodology of this paper. Reference (16) explains that good correlation and bi-orthogonality exist between a pair of experimental real vectors (the elements of which are real numbers), if upon scaling the diagonal terms are 1 and in comparison the off-diagonal terms are ≤ 0.1. Since the eigenvectors and correspondingly the elements of POC matrix are complex, so, absolute values of the elements of the POC matrix are scaled to the check the correlation and orthogonality as above.

3   RESULTS AND DISCUSSION

To illustrate the flexibility of the proposed reduction technique with respect to the choice of permanent local effects in addition to the continuum in a step by step manner, Table 1 gives a set of five test cases; each reduced by using the proposed reduction methodology. For all the test cases, the Lalanne's rotor bearing system (17) or its modification are used. Figure 2 shows the basic Lalanne's rotor bearing system, which has a uniform shaft, Table 2 and Table 3 give the properties. Difference of properties from those mentioned in Table 2 and Table 3 for specific test cases is mentioned appropriately. For the FE analysis of the full model, the Lalanne's rotor shaft model is built by using thirteen similar Timoshenko beam elements with 4-DoF's per node and its equations of motion are derived with the help of the formulation provided by Nelson (18). The viscous form of internal material damping given in (19) is used to model the rotating damping with $\eta_v = 0.0002s$

Table 1. List of Analysis test cases.

Case No.	Case Description	Continuum Effect	Permanent Local Effect	Temporary Local Effect
I	Rotor bearings with 3 discs (Lalanne's rotor)	All shaft elements	Nil	All bearings and discs
II	Overhung version of Lalanne's rotor bearing system with 3 discs	All Shaft elements	Nil	All bearings and discs
III	Rotor bearings with 3 discs (Lalanne's rotor)	All Shaft elements	All bearings	All discs
IV	Rotor bearings with 3 discs (Lalanne's rotor)	All Shaft elements	All discs	All bearings
V	Rotor bearings with 3 discs (Lalanne's rotor)	All Shaft elements	All bearings & all discs	Nil

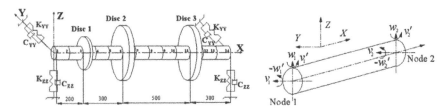

Figure 2. Finite element model of Lalanne's rotor-bearing system (14).

Table 2. Properties of Lalanne's Rotor-Bearing system (14).

Parameter	Value	Parameter	Value
Density, ρ; Modulus, E	7800 Kg/m3; 200 GPa	Rotor Length, L; Shaft Dia., D	1.3 m; 0.1 m
Kyy; Kzz	7 x 10^7 N/m; 5 x 10^7 N/m	Element Length; Element No.	0.1 m; 13
Cyy; Czz	500 Ns/m; 700 Ns/m	Viscous damping Coef., ηv; Poisson's Ratio	0.0002s; 0.3

Table 3. Disc details for standard Lalanne's rotor bearing system.

Disc	1	2	3
Outside Dia. (m)	0.24	0.4	0.4
Thickness (mm)	5	5	6
Density (Kg/m3)	78000	7800	7800
Mass Unbalance (Kg m)	0	0.2 x 10-3	0
Position from rotor Left End (m)	0.2	0.5	1.0

3.1 Case I- Lalanne's rotor bearing system with no permanent local effects

The case I is the standard Lalanne's rotor bearing system; Table 2 gives its properties. For the reduction, all the local effects (due to all the three discs and two bearings at the ends) are taken as temporary in nature. Therefore, the transformation matrix pair is derived from the system matrix corresponding to the equation of motion of only the free-free shaft continuum. The reduced system matrix is augmented subsequently by adding the local effects due to inertia of the discs, restoring effects of the bearings, all gyroscopic effects and both stationary and rotary dissipation effects by using the transformation matrix pair generated. The force due to unbalance is also suitably modified by the transformation matrix pair. Figure 3, Figure 4 and Figure 5 show the Unbalanced response amplitude (UBR), Campbell Diagram and Decay rate plot for Case-I respectively. The UBR plot clearly shows the match between the solution obtained for full FE model using 56 equations and that predicted by the reduced model having six and eight modes by using only 16 reduced equations. The shaded area shows the region where the system is unstable, and is unnecessary to plot, due to negative cross-coupled stiffness arising from the rotor material damping. This plot may satisfy reader's interest to check the accuracy of this reduction methodology at higher spin speeds. The Campbell diagram (Figure 5) shows that the first two eigen-frequencies of reduced model match the eigen-frequencies of the full (FE) model closely while preserving both six modes and eight modes. However, the next two eigen-frequencies match only by preserving eight modes during reduction. This is because while reducing the free-free system, the first four eigenvectors correspond to the rigid body modes. So saving m modes during reduction actually means saving $(m-4)$ flexural modes. Again considering the four rigid body modes redundant is not practical as they also contain the characteristics of the system. The decay rate plot shown in Figure 5, also matches closely preserving six or eight modes during reduction; as a result, the reduced model predicts the stability limit speed very close to that predicted by the Finite-element model.

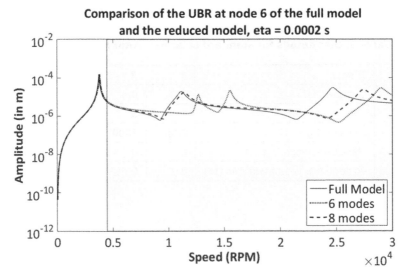

Figure 3. Comparison of UBR between full and Reduced model at node 6 for case I. 8 DoF's saved.

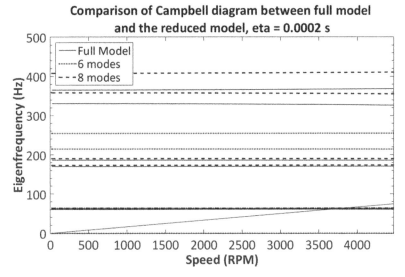

Figure 4. Campbell Diagram comparison between full and reduced model for case I. 8 DoF's saved.

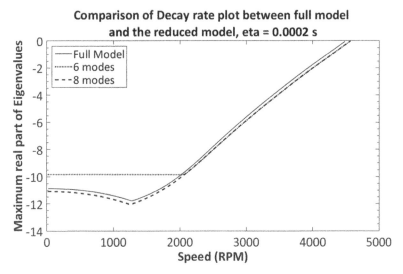

Figure 5. Decay rate plot comparison between full and reduced model for case I. 8 DoF's saved.

3.2 Case II- Overhung version of Lalanne's rotor bearing system with no permanent local effects

As proposed earlier, the present reduction methodology is almost free from any influence due to a change of local effect. This example shows that change of position of a local effect is marginal. To this end, an overhung version of Lalanne's rotor (in Figure 2) created by shifting the right end bearing from node 14 to node 9 as shown in Figure 6 serves the purpose. In this case, an unbalance of $0.2 \times 10^{-3} Kgm$ is assumed on disc 3 instead of disc 2. The same transformation matrix pair as used in Case-I, are utilized to construct the reduced model as the transformation matrix pair is derived by reducing the free-free shaft continuum as before. All the local effects like discs and bearings are considered later to augment the system matrix, the forcing function is modified as above to obtain the reduced equation of motion of the whole system. Figure 7 shows a comparison of UBR at node 11 predicted by the reduced model and the Finite-element model; a close match proves that considering local effects subsequent to the reduction does not impair the prediction. This shows that the proposed methodology is applicable

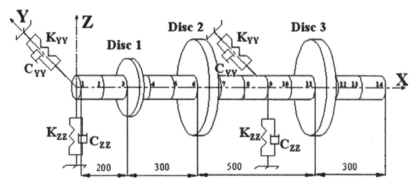

Figure 6. Schematic Diagram for case II.

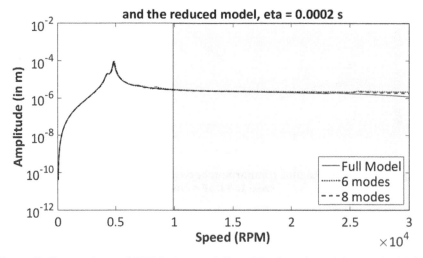

Figure 7. Comparison of UBR between full and Reduced model at node 11 for case II. 8 DoF's saved.

in different cases of changing local effects, which do not need different transformation matrix pairs for reduction. This is the greatest advantage of this methodology.

3.3 Case III-V Lalanne's rotor bearing system with bearing effects and disc effects considered permanent on an individual basis as well as together

Transformation matrix pair obtained by reducing only the free-free shaft continuum helps to get a reduced model, which closely predicts the UBR, as shown in Case I. However, the transformation matrix pair derived by considering additionally the local restoring (bearing Stiffness) or inertia effects (Disc Mass & inertia) as permanent, yield even closer results to actual UBR than Case-I. To this end, in case III to V, the two bearings and three discs form permanent local effects, first separately (Case III and IV) and then together (Case V), and are included in formation of transformation matrix pair. The system matrices corresponding to these cases are different from each other and those used for Case-I and Case-II. These matrix pairs are used to reduce the system matrices and augment them with the other effects like the inertial effect of discs (case III), stiffness and damping effects from bearings (Case IV), all gyroscopic effects and both stationary and rotary dissipation effects, not considered earlier, to deduce the reduced model of the total system, (the system matrix) and the modified forcing function. Figures 8, 9 and 10 shows the UBR plots for the Case III, Case IV and Case V respectively. Comparison of Figure 3, Figure 8 and Figure 9 shows that the Case III predicts much closer match with the FE model than either of Case I (Figure 3) or Case IV (Figure 9). The results from Case IV match much more closely to the only continuum reduction Case I. This leads us to conclusion that fixing position and even the approximate properties of the bearings during the design of rotor-dynamic system lead to closer reduced models; the influence of position of inertia altering local effects is marginal. . However, in all the cases, it is clear that increasing the number of modes in the reduction process always leads to more accurate reduced model.

Moving on to the Case V and its response shown by Figure 10, this Case investigates the influence of gyroscopic effects and rotary dissipation effects from all the sources on accuracy of reduced model. In this case, in addition to the inertia and restoring effect of the continuum, those from all the local effects due to three discs and two end bearings appear in the system matrix; gyroscopic effect and rotary dissipation effects do not appear in the system matrix. Transformation matrix pair is deduced from this system matrix to reduce the whole system. UBR plotted in Figure 10 shows that the influence of rotating damping and gyroscopic effects is somewhat noticeable only in the case when six modes are saved during reduction and that too at speeds near 30000 RPM, deep in the unstable region. Even this effect seems almost disappearing as the number of modes saved in the reduced model increases. This leads to the conclusion that leaving out gyroscopic and all dissipation effects during forming the system matrix has a very marginal impact on accuracy of reduction and this degradation almost vanishes by saving more modes during reduction. This helps to form a very important conclusion that using the transformation matrix pair which is spin speed independent, as both gyroscopic and rotary damping effects depend on rotor-spin-speed are excluded in making the system matrix, has marginal or negligible influence on the reduction accuracy.

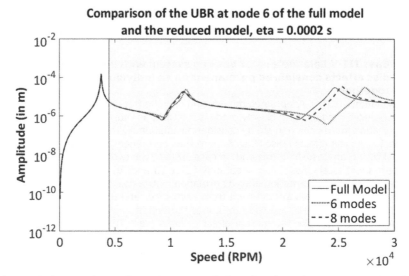

Figure 8. Comparison of UBR between full and reduced model at node 6 for case III. 8 DoF's saved.

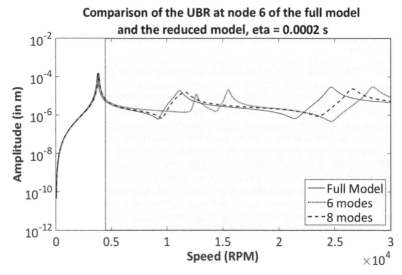

Figure 9. Comparison of UBR between full and reduced model at node 6 for case IV. 8 DoF's saved.

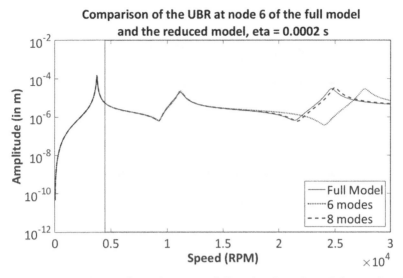

Figure 10. Comparison of UBR between full and reduced model at node 6 for case V. 8 DoF's saved.

3.4 Mode correlation results for reduced system

The Campbell diagram and decay rate plot only for Case I are given to keep the paper short. These allow correlating the eigenvalues obtained from the proposed reduction methodology with the finite element model. However, a comparison of the eigenvectors is also essential to ensure that the proposed eigenvectors also match the eigenvectors of the actual system. Various mode correlation techniques discussed in literature are MSF(Modal Scale Factor)(16), MAC (Modal Assurance Criteria) (15) and few others but for the current bi-orthogonal system, Pseudo Orthogonality check (POC) (13) as given in Eq. (18) is one of the better ways to go for mode correlation. Comparison of eigenvectors obtained from proposed reduction methodology with those from finite element method is not possible since there is a mismatch in degrees of freedoms. Therefore, this paper attempts correlation of modes obtained by the present reduction methodology with those obtained through modified SEREP (5), which report a near exact match between the UBR of reduced and unreduced (FE) models as shown in Figure 12. This work attempts correlation of eigenvectors by using the Pseudo Orthogonality Check (POC), suitable for bi-orthogonal systems, as given in Eq. (18).

Table 4 and Table 5 provide the modulus of the POC matrices as defined by Eq. (18) at 4000 RPM after preserving six modes and eight modes respectively. For ensuring bi-orthogonality between eigenvectors, the scaled POC matrix as in Eq. (18) should have diagonal elements as 1 and off diagonal elements should be ≤ 0.1. Table 4 shows that only first two modes are biorthogonal with respect to each other. Correspondingly, UBR plot for Case-I in Figure-3 shows that there is no match between the reduced and full model after the first two natural frequencies when preserving six modes. Table 4 also shows that the third mode has very bad bi-orthogonal relationship with the first two modes. Table 5 describes the POC check when 8 modes are preserved during reduction. The table shows that the first four modes are very much bi-orthogonal with respect to each other and then the higher modes become gradually less correlated as compared to the case when preserving only six modes during reduction. So mode correlation through POC proves good correlation between the eigenvectors of the proposed reduction methodology and modified SEREP; the correlation improves with preserving more modes during reduction.

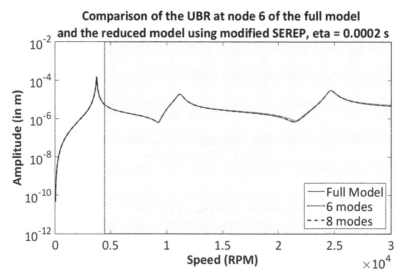

Figure 11. Comparison of UBR between full and reduced model at node 6 for case II using modified SEREP[5]. 8 DoF's saved.

Table 4. Absolute POC matrix for case I when 6 modes and 8 DoF's saved.

Modes	1	2	3	4	5	6
1	1.000	0.011	0.006	0.004	0.006	0.004
2	0.018	1.000	0.005	0.003	0.008	0.004
3	1.218	1.424	1.000	0.061	0.722	0.742
4	1.264	0.885	0.366	1.000	0.951	0.274
5	0.983	1.522	0.029	0.024	1.000	0.270
6	1.543	0.996	0.021	0.024	0.252	1.000

Table 5. Absolute POC matrix for case I when 8 modes and 8 DoF's saved.

Modes	1	2	3	4	5	6	7	8
1	1.000	0.013	0.020	0.015	0.152	0.107	0.002	0.053
2	0.015	1.000	0.032	0.024	0.117	0.253	0.069	0.041
3	0.006	0.006	1.000	0.033	0.270	0.194	0.203	0.092
4	0.005	0.007	0.027	1.000	0.128	0.246	0.135	0.213
5	0.011	0.002	0.085	0.070	1.000	0.243	0.183	0.170
6	0.005	0.006	0.012	0.084	0.157	1.000	0.098	0.177
7	0.013	0.030	0.211	0.116	0.749	0.801	1.000	0.100
8	0.027	0.011	0.160	0.268	0.846	0.887	0.249	1.000

4   CONCLUSIONS

The discussions in previous sections lead to the following conclusions

a) The reduction of shaft continuum is primary in nature. Transformation matrix pairs are derived by considering only distributed inertial and stiffness effects. The same matrix pair helps to place all secondary effects like local or point inertias, bearing stiffness, their locations and parametric values, all local and distributed gyroscopic effects, stationary or rotary dissipation effects, to augment the system matrix appropriately. The same matrix pair helps to obtain the reduced external generalized force vector. Thus reduced system matrix and force vector form the total reduced model in reduced state space.

b) The proposed reduction methodology is a novel, fast, simple, cost effective and flexible tool in the design stage of rotor-shaft systems. Transformation matrix pair is independent of the shaft material, all secondary effects, and rotor-spin-speed. Thus user may freely choose the parametric values of secondary effects, their locations and rotor-spin-speed to examine the modal and response behaviours of the reduced system with the matrix pair deduced once. The proposed reduction methodology is a novel, fast, simple, cost effective and flexible tool for dynamic design of rotor-shaft systems.

c) Retaining additionally the locations and at least tentative values of local stiffness, rather than local inertias in the system matrix containing inertia and stiffness properties of shaft continuum, leads to more accurate transformation matrix pair leading to closer reduced model than the situation if the effects are not considered. Therefore, the location of bearings and tentative parametric values are beneficial for getting closer reduced models.

d) The reduction methodology is independent of the position of the active or master degrees of freedom of the system as long as the degrees of freedom are more than the modes saved during reduction. Furthermore, the accuracy of prediction of the reduced model increases with the number of modes retained.

REFERENCES

[1] GUYAN, R. J., "Reduction of Stiffness and Mass Matrices," *AIAA ournal.*, 3(2), 1965, pp. 380–380.
[2] Qu, Z., *Model Order Reduction Techniques: With Applications in Finite Element Analysis*, Springer, 2004.

[3] O'Callahan, J., Avitabile, P., and Riemer, R., "System Equivalent Reduction Expansion Process (SEREP)," *Proceedings of 7th International Modal Analysis Conference (IMAC VII)*, Las Vegas, 1989, pp. 29–37.

[4] Friswell, M., Penny, J. E. T., and Garvey, S. D., "Model Reduction for Structures with Damping and Gyroscopic Effects," *Proceedings of the International Seminar on Modal Analysis*, pp. 1151–1158.

[5] Das, A. S., and Dutt, J. K., 2008, "Reduced Model of a Rotor-Shaft System Using Modified SEREP," *Mechanics Research Communication*, 35(6), 2001, pp. 398–407.

[6] Das, A. S., and Dutt, J. K., "A Reduced Rotor Model Using Modified SEREP Approach for Vibration Control of Rotors," *Mechanical Sysems and Signal Processing*, 26(1), 2012, pp. 167–180.

[7] Shrivastava, A., and Mohanty, A. R., "Estimation of Single Plane Unbalance Parameters of a Rotor-Bearing System Using Kalman Filtering Based Force Estimation Technique," *Journal of Sound and Vibration*, 418, 2018, pp. 184–199.

[8] Han, J. S., "Krylov Subspace-Based Model Order Reduction for Campbell Diagram Analysis of Large-Scale Rotordynamic Systems," *Structural Engineering and Mechanics*, 50(1), 2014, pp. 19–36.

[9] Subbiah, R., Bhat, R. B., and Sankar, T. S., "Dynamic Response of Rotors Using Modal Reduction Techniques," Journal of Vibration, Acoustics, Stress and Reliability, 111, 1989.

[10] Genta, G., and Delprete, C., "Acceleration through Critical Speeds of an Anisotropic, Non-Linear, Torsionally Stiff Rotor with Many Degrees of Freedom," *Journal of Sound and Vibration*, 180(3), 1995, pp. 369–386.

[11] Friswell, M. I., Penny, J. E. T., and Garvey, S. D., "Using Linear Model Reduction to Investigate the Dynamics of Structures with Local Non-Linearities," *Mechanical Sysems and Signal Processing*, **9**(3), 1995, pp. 317–328.

[12] Zuo, Y., Wang, J., Ma, W., Zhai, X., and Yao, X., "Method for Selecting Master Degrees of Freedom for Rotating Substructure," *Proceedings of the ASME Turbo Expo*, 2014.

[13] Lal, H. P., Jith, J., Gupta, S., and Sarkar, S., "Reduced Order Modelling in Stochastically Parametered Acousto-Elastic System Using Arbitrary PCE Based SEREP," *Probabilistic Engineering Mechanics*, **52**(December 2017), 2018, pp. 1–14.

[14] Meirovitch, L., "Fundamentals of Vibrations by Leonard Meirovitch," *McGraw-Hill*, 2001, pp. 345–355.

[15] Pastor, M., Binda, M., and Harčarik, T., "Modal Assurance Criterion," *Procedia Engineering*, **48**, 2012, pp. 543–548.

[16] Choi, S. H., Mukasa, S. B., Kwon, S. T., and Andronikov, A. V., "Sr, Nd, Pb and Hf Isotopic Compositions of Late Cenozoic Alkali Basalts in South Korea: Evidence for Mixing between the Two Dominant Asthenospheric Mantle Domains beneath East Asia," *Chemical Geology*, **232**(3–4), 2006, pp. 134–151.

[17] Parkinson, A. G., "Rotordynamics Prediction in Engineering," *Journal of Sound and Vibration*, John Wiley & Sons, 1991, pp. 547–548.

[18] Nelson, H. D., "A Finite Rotating Shaft Element Using Timoshenko Beam Theory," *Journal of Mechanical Design*, 102(4), 1980, pp. 793.

[19] Nelson, H. D., "Finite Element Simulation of Rotor · BearingSystems With Internal Damping," *Journal of Engineering for Power*, (76), 1976, pp. 1–6.

[20] Ali, A., and Rajakumar, C., *The Boundary Element Method: Applications in Sound and Vibration*, A.A. Balkema Publishers, 2004.

12th International Conference on Vibrations in Rotating Machinery -
Institution of Mechanical Engineers, ISBN 978-0-367-67742-8

Effect of L/D ratio and clearance of 3-lobe taper land bearing on stability of flexible rotor system

S. Braut, A. Skoblar, G. Štimac Rončević, R. Žigulić

Faculty of Engineering, University of Rijeka, Rijeka, Croatia

ABSTRACT

The present study analyzes the effects of L/D ratio and bearing radial clearance of 3-lobe taper land bearing on stability of flexible rotor system. The pressure distribution in a fluid film journal bearing was calculated using incompressible Reynolds equation, implemented via finite difference method (FDM). Stiffness and damping coefficients were calculated with respect to Sommerfeld number. As an example of flexible rotor system a high speed steam turbine is used. The rotor was modeled with 1d finite elements based on Euler-Bernoulli beam theory. Stability of the flexible rotor system was estimated using logarithmic decrement. Six cases were analyzed and recommendations for optimal bearing configuration were proposed.

1 INTRODUCTION

The current trends in turbomachinery leads to high speed operation in order to make them compact as well as to reduce their mass. The ordinary journal bearing solutions in form of circular bearings, which are most common type of the bearings, are found to be unstable at high speeds. Different kind of bearing designs such as multi-lobe bearings (lemon, multi-lobe) (1) or multi-lobe pressure dam bearings (2), could offer better rotor stability for light rotors rotating at higher speeds.

Multi-lobe taper bearings also could offer higher stability, especially for high-speed rotors. Hargraeves (3) deals with performance characteristics of tri-taper journal bearings operating in high-speed step up gearbox which increases rotating speed to about 13000 [rpm] from reaction turbine to a three-stage gas compressor. He has studied influence of manufacturing tolerances, misalignment and turbulence on the steady state rotation of shaft supported in tri-taper journal bearings, providing the designers to predict characteristics of similar bearings. The main reason in increasing of oil flow rate and decreasing of load-carrying characteristics he found in over-sized journal ramps. Rao et all (4), theoretically investigated the whirl instability of an rotor in tri-taper journal bearings, operating under steady, periodic and variable rotating load, using the nonlinear - transient approach. In the case of unidirectional constant load and variable rotating load, they found that journal locus attained a stable position at the faster rate, with an increase in ramp size, but an increase of L/D ratio resulted in larger excursions of journal center. For periodic load, the journal locus ended in a limit cycle and was suspectible to instability. The same authors obtained that for the given ramp size, an increase of L/D ratios leads to the increasing of region of stability, which is also true for reverse case i.e. an increase in ramp size for the given L/D ratio. Stability characteristics of such as bearings was also investigated by Pai et all (5) which variated ramp sizes and L/D ratios and drew the conclusion that tri-taper bearings with higher ramp size and higher L/D ratio results in an increased mass parameter and improved stability margin of the bearing.

In this paper a stability of rotor supported with two equal 3-lobe taper land bearing (3-LTL) was studied. The focus of analysis was to determine the effects of L/D ratio and bearing radial clearance of 3-LTL bearing on stability of flexible rotor system. Stiffness and damping coefficients were calculated with respect to Sommerfeld number. As an example of flexible rotor system a high speed steam turbine was used. The rotor was modeled with 1d finite elements. Stability of the flexible rotor system was estimated by using logarithmic decrement. According to results of first three analyzed cases it is established that increasing of L/D ratio has beneficial effect on rotor system stability. Also decreasing of bearing radial clearance has another favorable influence on flexible rotor stability.

2 3-LOBE TAPER LAND BEARING

2.1 Bearing geometry

3 lobe taper land bearing analyzed in this paper is shown in Figure 1a). This bearing geometry is somewhat different from similar bearings found in literature (3-5). As opposed to the classic symmetric configuration 3 x 120° (with the same tapered part of segments 3 x 60°), Figure 1a) shows bearing configuration which can be roughly described as (90° + 90° + 180°) with tapered angles (54°, 60°, 114°). Bearing was made up of two shells. First, covers first two taper land and the second covers third

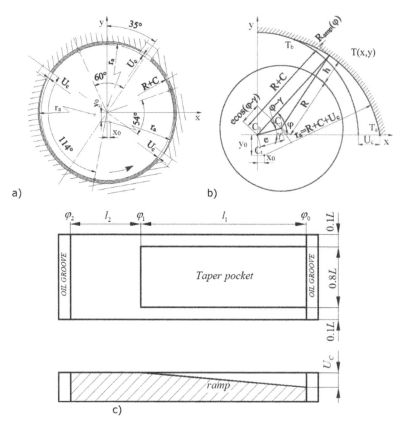

a)   b)

c)

Figure 1. 3-lobe taper land bearing, a) geometry, b) oil film thickness definition. Figure 1 3-lobe taper land bearing, c) schematic view and cross section of i-th segment.

taper land area. After assembling the bearing shells are rotated to an angle of 35 degrees in the opposite direction of rotation of the rotor. The width of taper arc segment is 0.8 L, where L is bearing width.

From geometry of each lobe (Figure 1b), the following relationships between taper's Descartes coordinates x and y and it's belonging angles φ as well as Descartes coordinates of taper arc center x_0 and y_0, could be established:

$$x(\varphi) = \frac{y(\varphi)}{\tan \varphi} = \frac{x_0 + y_0 \, \tan \varphi \pm \sqrt{\left(r_a^2 - y_0^2\right) + 2x_0 y_0 \tan \varphi + (r_a^2 - x_0^2) tan^2 \varphi}}{1 + \tan^2 \varphi} \tag{1}$$

where, from the Figure 1, taper arc radius r_a depends on shaft radius R, radial clearance C as well as undercut Uc. Having in mind all previous quantities the depth of tri - taper pocket at the referent angle φ, could be defined

$$R_{amp}(\varphi) = \sqrt{x^2 + y^2} - R - C \tag{2}$$

Taper depth, R_{amp}, for a point $T(x,y)$ at taper edge is defined from the fact that the point T is at the intersection of straight line from the bearing center $C_b(0,0)$ with angle φ and circle with radius $r_a = R+C+U_C$ with center $C_t(x_0,y_0)$, where undercut U_C is equal to maximum value of R_{amp}. Center $C_t(x_0,y_0)$ is positioned on distance r_a from the beginning and end point of taper (points T_a and T_b, Figure 1) and point T_b (taper end) is also at distance $R+C$ from bearing center. From the distance between point T and the bearing center, shaft radius R and bearing clearance C, taper depth R_{amp} can be calculated with expression (2).

2.2   Oil film thickness and pressure distribution
Pressure distribution in hydrodynamic journal bearing in the case of incompressible, non-viscous laminar flow is defined by means of Reynolds equation:

$$\frac{\partial}{\partial x}\left(h^3 \frac{\partial \bar{p}}{\partial x}\right) + \frac{\partial}{\partial z}\left(h^3 \frac{\partial p}{\partial z}\right) = 6\eta U \frac{\partial h}{\partial x} + 12\eta \frac{\partial h}{\partial t} \tag{3}$$

where h and p are oil film thickness and pressure, x and z are Descartes coordinates of points on bearing surface, η is oil dynamic viscosity, U is sum of tangential velocities on the rotor's and stator's surface and t is time. Dimensionless form of equation (3) is:

$$\frac{\partial}{\partial \varphi}\left(\bar{h}^3 \frac{\partial \bar{p}}{\partial \varphi}\right) + \frac{1}{4}\left(\frac{D}{L}\right)^2 \frac{\partial}{\partial \bar{z}}\left(\bar{h}^3 \frac{\partial \bar{p}}{\partial \bar{z}}\right) = 6 \frac{\partial \bar{h}}{\partial \varphi} + 12\Omega \frac{\partial \bar{h}}{\partial \tau} \tag{4}$$

where τ and Ω are dimensionless time and dimensionless rotating speed. Other dimensionless variables in equation (4) as well as further in paper, are marked with straight bar placed above variable. Oil film thickness in each segment, depending on eccentricity e, angle φ^* measured from +Ox axis and assumed value of attitude angle ϒ, is defined as follows:

$$h = C + e \cos(\varphi^* - \gamma) + R_{amp}(\varphi^*) \tag{5}$$

Dimensionless oil film thickness is equal to:

$$\bar{h} = 1 + \varepsilon \cos(\phi^* - \gamma) + \bar{R}_{amp}(\phi^*) \tag{6}$$

where ε is eccentricity ratio and $\overline{R}_{amp}$ is dimensionless depth of tri-taper pocket.

2.3 Pressure boundary conditions

Pressure boundary conditions are connected with atmospheric pressure along bearing edges at any angle φ:

$$\bar{p}_0(\bar{z}=0) = \bar{p}_0(\bar{z}=1) = 0 \tag{7}$$

Second boundary condition through whole domain refers to cavitation which is allowed to occur at ambient pressure, taken into account in calculations through setting negative pressures to zero as iterative solution scheme based on conditions:

$$\bar{p}_0 = 0, \ \frac{\partial \bar{p}}{\partial \varphi} = 0 \tag{8}$$

Dimensionless supply pressure from oil grooves is set to $\bar{p}_s = 0.1362$.

2.4 Reynolds equation under steady state conditions

The first part of calculation is connected with obtaining the attitude angle $\Phi0$ which should be equal to assumed value of attitude angle Υ connected with angle ϕ measured from the +Ox axis. In that case the force components along (W_r) and perpendicular (W_φ) to the line of centres C_bC_j, as well as its dimensionless versions, are equal to:

$$\overline{W}_r = \frac{W_r C^2}{\eta U R^2 L} = -\int_0^1 \int_0^{2\pi} \bar{p}_0 \, \cos(\varphi^* - \gamma) d\varphi d\bar{z} \tag{9}$$

and

$$\overline{W}_\varphi = \frac{W_\varphi C^2}{\eta U R^2 L} = \int_0^1 \int_0^{2\pi} \bar{p}_0 \, \sin(\varphi^* - \gamma) d\varphi d\bar{z} \tag{10}$$

From equations (9) and (10), the dimensionless load carrying capacity $\overline{W}$ (11) as well as real attitude angle $\Phi0$ (12) could be calculated

$$\overline{W} = \sqrt{\overline{W} + \overline{W}} \tag{11}$$

$$\Phi_0 = \arctan\frac{\overline{W}_\varphi}{\overline{W}_r} \tag{12}$$

Calculated attitude angle $\Phi0$ is after each calculation of force components along and perpendicular to the line of centres C_bC_j, using the steady state version of Reynolds equation

$$\frac{\partial}{\partial \varphi}\left(\bar{h}_0^3 \frac{\partial \bar{p}_0}{\partial \varphi}\right) + \frac{1}{4}\left(\frac{D}{L}\right)^2 \frac{\partial}{\partial \bar{z}}\left(\bar{h}_0^3 \frac{\partial \bar{p}_0}{\partial \bar{z}}\right) = 6\frac{\partial \bar{h}_0}{\partial \varphi} \tag{13}$$

compared with assumed value of attitude angle Υ and procedure is iterative with small incremental increase of Υ.

2.5 Reynolds equation under dynamic conditions

Reynolds equation under dynamic conditions uses small perturbations of rotor steady state position defined with dimensionless eccentricity $\varepsilon0$ as well as attitude angle $\Phi0$:

$$\varepsilon = \varepsilon_0 + \varepsilon_1 e^{it}, \quad \Phi = \Phi_0 + \Phi_1 e^{it} \tag{14}$$

Keeping in mind that $\varepsilon 1 \ll \varepsilon 0$ and $\Phi 1 \ll \Phi 0$, the dimensionless pressure and film thickness can be written as:

$$\bar{p} = \bar{p}_0 + p_1 \varepsilon_1 e^{i\tau} + p_2 \varepsilon_0 \Phi_1 e^{i\tau}, \quad \bar{h} = h_0 + \varepsilon_1 e^{i\tau} \cos\varphi + \varepsilon_0 \Phi_1 e^{i\tau} \sin\varphi \tag{15}$$

Returning a perturbed dimensionless pressure and film thickness from (14) in (4), a following three equations are obtained:

$$\bar{h}_0^3 \frac{\partial^2 \bar{p}_0}{\partial \varphi^2} + 3\bar{h}_0^2 \frac{\partial \bar{h}}{\partial \varphi} \frac{\partial \bar{p}}{\partial \varphi} + \frac{1}{4}\left(\frac{D}{L}\right) \bar{h}_0^3 \frac{\partial^2 \bar{p}_0}{\partial \bar{z}^2} - 6\frac{\partial \bar{h}}{\partial \varphi} = 0 \tag{16}$$

$$\bar{h}_0^3 \frac{\partial^2 \bar{p}_1}{\partial \varphi^2} + 3\bar{h}_0^2 \cos\varphi \frac{\partial^2 \bar{p}_0}{\partial \varphi^2} - 3\bar{h}_0^2 \sin\varphi \frac{\partial \bar{p}_0}{\partial \varphi} + 6\bar{h}_0 \cos\varphi \frac{\partial \bar{h}_0}{\partial \varphi} \frac{\partial \bar{p}_0}{\partial \varphi} + 3\bar{h}_0^2 \frac{\partial \bar{h}_0}{\partial \varphi} \frac{\partial \bar{p}_1}{\partial \varphi}$$
$$\frac{1}{4}\left(\frac{D}{L}\right)^2 \bar{h}_0^3 \frac{\partial^2 \bar{p}_1}{\partial \bar{z}^2} + \frac{3}{4}\left(\frac{D}{L}\right)^2 \bar{h}_0^2 \cos\varphi \frac{\partial^2 \bar{p}_0}{\partial \varphi \bar{z}^2} + 6\sin\varphi - 12i\Omega\cos\varphi = 0 \tag{17}$$

$$\bar{h}_0^3 \frac{\partial^2 \bar{p}_2}{\partial \varphi^2} + 3\bar{h}_0^2 \sin\varphi \frac{\partial^2 \bar{p}_0}{\partial \varphi^2} - 3\bar{h}_0^2 \cos\varphi \frac{\partial \bar{p}_0}{\partial \varphi} + 6\bar{h}_0 \sin\varphi \frac{\partial \bar{h}_0}{\partial \varphi} \frac{\partial \bar{p}_0}{\partial \varphi} + 3\bar{h}_0^2 \frac{\partial \bar{h}_0}{\partial \varphi} \frac{\partial \bar{p}_2}{\partial \varphi}$$
$$+\frac{1}{4}\left(\frac{D}{L}\right)^2 \bar{h}_0^3 \frac{\partial^2 \bar{p}_2}{\partial \bar{z}^2} + \frac{3}{4}\left(\frac{D}{L}\right)^2 \bar{h}_0^2 \sin\varphi \frac{\partial^2 \bar{p}_0}{\partial \varphi \bar{z}^2} + 6\cos\theta - 12i\Omega\sin\varphi = 0 \tag{18}$$

2.6   Oil film stiffness and damping coefficients

Fields of perturbed dimensionless pressures $\bar{p}_1$ and $\bar{p}_2$, could be numerically calculated by solving equations (17) and (18) by means of finite differences mothed. They will be further used for calculation of dimensionless oil film stiffness and damping coefficients in the polar coordinate system:

$$\bar{K}_{rr} = -\text{Re}\left(\int_0^1 \int_0^{2\pi} \bar{p}_1 \cos\phi \, d\phi \, d\bar{z}\right), \quad \bar{K}_{\varphi r} = -\text{Re}\left(\int_0^1 \int_0^{2\pi} \bar{p}_1 \sin\phi \, d\phi \, d\bar{z}\right),$$

$$\bar{K}_{r\varphi} = -\text{Re}\left(\int_0^1 \int_0^{2\pi} \bar{p}_2 \cos\phi \, d\phi \, d\bar{z}\right), \quad \bar{K}_{\varphi\varphi} = -\text{Re}\left(\int_0^1 \int_0^{2\pi} \bar{p}_2 \sin\phi \, d\phi \, d\bar{z}\right),$$

$$\bar{C}_{rr} = -\text{Im}\frac{\left(\int_0^1 \int_0^{2\pi} \bar{p}_1 \cos\phi \, d\phi \, d\bar{z}\right)}{\Omega}, \quad \bar{C}_{\varphi r} = -\text{Im}\frac{\left(\int_0^1 \int_0^{2\pi} \bar{p}_1 \sin\phi \, d\phi \, d\bar{z}\right)}{\Omega},$$

$$\bar{C}_{r\varphi} = -\text{Im}\frac{\left(\int_0^1 \int_0^{2\pi} \bar{p}_2 \cos\phi \, d\phi \, d\bar{z}\right)}{\Omega}, \quad \bar{C}_{\varphi\varphi} = -\text{Im}\frac{\left(\int_0^1 \int_0^{2\pi} \bar{p}_2 \sin\phi \, d\phi \, d\bar{z}\right)}{\Omega} \tag{19}$$

The relations between dimensionless stiffnesses and dampings in polar and Descartes coordinates can be established by following matrix transformations:

$$\begin{bmatrix} \bar{K}_{xx} & \bar{K}_{xy} \\ \bar{K}_{yx} & \bar{K}_{yy} \end{bmatrix} = \begin{bmatrix} \cos\Phi_0 & -\sin\Phi_0 \\ \sin\Phi_0 & \cos\Phi_0 \end{bmatrix} \begin{bmatrix} \bar{K}_{rr} & \bar{K}_{r\varphi} \\ \bar{K}_{\varphi r} & \bar{K}_{\varphi\varphi} \end{bmatrix} \begin{bmatrix} \cos\Phi_0 & \sin\Phi_0 \\ -\sin\Phi_0 & \cos\Phi_0 \end{bmatrix} \tag{20}$$

and

$$\begin{bmatrix} \bar{C}_{xx} & \bar{C}_{xy} \\ \bar{C}_{yx} & \bar{C}_{yy} \end{bmatrix} = \begin{bmatrix} \cos\Phi_0 & -\sin\Phi_0 \\ \sin\Phi_0 & \cos\Phi_0 \end{bmatrix} \begin{bmatrix} \bar{C}_{rr} & \bar{C}_{r\varphi} \\ \bar{C}_{\varphi r} & \bar{C}_{\varphi\varphi} \end{bmatrix} \begin{bmatrix} \cos\Phi_0 & \sin\Phi_0 \\ -\sin\Phi_0 & \cos\Phi_0 \end{bmatrix} \tag{21}$$

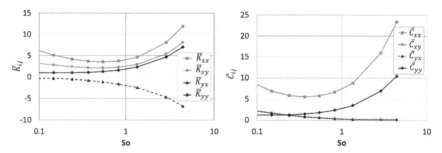

Figure 2. Dimensionless stiffness and damping coefficients of 3-lobe taper land bearing.

Stiffness and damping coefficients (K_{ij}, C_{ij}) can be obtained from calculated dimensionless stiffness and damping coefficients (equations 20 and 21) as follows:

$$K_{ij} = \frac{LDp_s}{C}\overline{K}_{ij}, \ C_{ij} = \frac{LDp_s}{C\omega}\overline{C}_{ij}, \ i = \{x,y\}, j = \{x,y\} \tag{22}$$

Commonly stiffness and damping coefficients as well as its dimensionless versions, are shown in diagrams (Figure 2), depending on Sommerfeld number defined by:

$$So = \frac{\eta NLD}{\psi^2 W} \tag{23}$$

where N, s^{-1} is rotational speed, W, N is external load and ψ is clearance ratio.

3   TURBINE ROTOR DESCRIPTION

The subject of the analysis is a 1.35 MW steam turbine with a nominal speed of 12500 rpm and with distance between radial bearings of 1.5 m, Figure 1. The calculation of flexural vibrations is based on the 1D finite element method and takes into account the physical characteristics (masses, moments of inertia, stiffnesses and dampings) of shaft, disks (with blades), and bearings.

Equation of motion of rotor modeled with finite elements is given by

$$M\ddot{q} + H\dot{q} + Kq = f. \tag{24}$$

where: **q**, $\dot{q} \ and \ \ddot{q}$ are displacement, speed and acceleration vectors of the turbine rotor,

$M = M_T + M_R$ translational and rotational inertia matrices,

$H = C + \Omega G$ the sum of the damping matrix and the gyroscopic matrix,

K   stiffness matrix,

f   vector of generalized forces.

The analyzed rotor model was divided into 41 beam finite elements i.e. had 42 nodes or 168 degrees of freedom. The material of rotor was steel 21CrMoV5-7 with density ρ = 7800 kg/m^3.

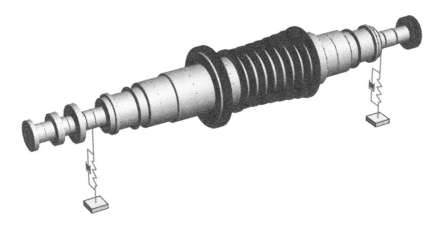

Figure 3. Steam turbine FE model.

4   CRITICAL SPEEDS AND STABILITY

Within this analysis, the eigenvalue problem defined by the associated homogeneous form of the equation (24) will be discussed i.e. by the equation,

$$M\ddot{q} + H\dot{q} + Kq = 0 \qquad (25)$$

In cases where vibrations are not caused by operation at a critical rotor speed but by instabilities caused by one of causes of self-excited vibrations, it is a common strategy to try to add greater external damping to the system This can be done by proper selection of hydrodynamic journal bearing. In other words, a stability calculation performs verification of proper selection of journal bearings. The eigenvalues for the damped system are obtained in the form $\lambda = \alpha + \beta i$, i.e. in the form of a complex number, wherein the general solution of the vibration of the rotor for a particular form of vibration is (6):

$$x = Ve^{\lambda t} = Ve^{\alpha t}(cos\,\beta t + i\,sin\,\beta t) \qquad (26)$$

where **V** is the eigenvector of the amplitudes of the individual nodes of the finite element model, the member $e^{\lambda t}$ represents the change in amplitudes in time, and β the natural frequency. When, for the considered vibration form $a > 0$ unstable vibrations occur i.e. amplitudes grow with time. In this paper, logarithmic decrement was used to express the stability of the system $\delta = -2\pi\alpha/\beta$ and natural frequency, $f = \beta/2\pi$ whereby an unstable situation will be indicated by $\delta < 0$.

Figure 4 shows first seven normal modes of rotor supported by two equal 3-lobe taper land bearings, running at 12500 rpm. First mode has natural frequency equal to zero at whole speed range of interest, therefore it is called mode 0. For some other bearing configuration like lemon bearing, analysis showed that natural frequency of this mode has value zero at lower speeds but at higher speeds including nominal rotor speed, this mode has real values. Real first mode f1 represent bearing oil film or so called "circular" rigid body mode Figure 4 b). At the nominal rotor speed this first mode starts to behave as elastic shaft mode.

The second mode f2, Figure 4 c), is actually 1st elastic shaft mode and behave similarly over the entire considered speed range. The third mode (f3, Figure 4d) is again bearing oil film rigid body "conical" mode. From the fourth mode and on (including fifth and sixth), all modes are elastic shaft modes.

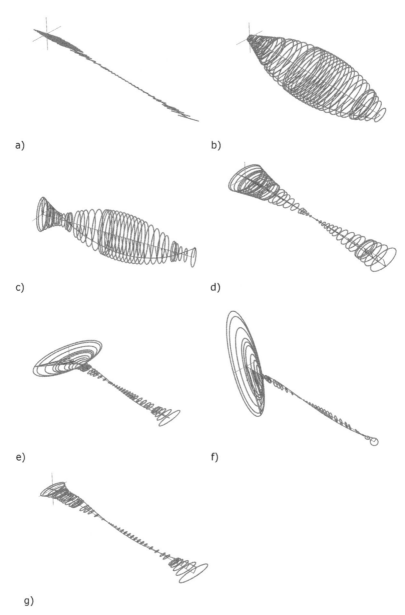

a)

b)

c)

d)

e)

f)

g)

Figure 4. Rotor normal modes at nominal speed n = 12500 rpm, a) f_0 =0.0 Hz, b) f_1 =97.6 Hz, c) f_2 =133.0 Hz, d) f_3 =160.8 Hz, e) f_4 =345.3 Hz, f) f_5 =382.4Hz. Figure 4 Rotor normal modes at nominal speed, g) f_6 =460.11 Hz.

5   RESULTS AND DISCUSSION

Stiffness and damping coefficients for 6 different 3-lobe taper land bearing (3-LTL) configurations are calculated in order to simulate rotor vibration behavior over the considered speed range (0 – 13750 rpm). The performance of rotor system was

Table 1. 3 lobe taper land bearing configurations.

No.	D, mm	L, mm	C, mm	Uc, mm
1	80	40	0.068	0.095
2	80	45	0.068	0.095
3	80	50	0.068	0.095
4	80	48	0.056	0.095
5	80	48	0.064	0.095
6	80	48	0.072	0.095

assessed according to two conditions, avoiding resonance with 1x harmonic at nominal speed and its stability at considered speed range, with special focus on nominal speed (12500 rpm) and overspeed event (13750 rpm). In all simulation it is taken characteristics of ISO VG 46 oil with constant dynamic viscosity at a temperature of 65 degrees Celsius at a value of 0,017 Pa.s.

Simulations showed that only 1 mode (f1) shape was critical at higher rotational speeds. Therefore, log decrement diagrams are presented only for 1st rotor mode shape and for different bearing configurations. Table 1 shows analyzed 3 lobe taper land bearing configurations. First three version analyze L/D ratio, while the other three versions consider the influence of bearing clearance. As the analyzes showed that no configuration was in danger of excitation with 1st harmonic, only one Campbell diagram for version No. 6, is shown in Figure 5. Dashed line represent 1x excitation harmonic. Figure 6 shows log decrement of a rotor system as a measure of stability for first three cases mentioned in Table 1. Results of log decrement presented in Figure 6 show influence of L/D ratio of bearing. It can be concluded that, for the considered parameters, increasing the bearing width has a favorable effect on stability. So, in this first part of relative comparison, the biggest stability has case No. 3 with the greatest width. In the second part of comparison, radial clearance of bearing were changed while all other parameters were kept constant. The analysis results showed that

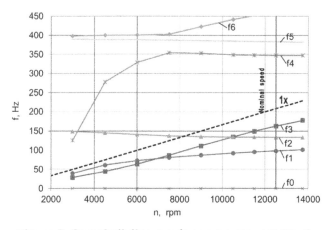

Figure 5. Campbell diagram for parameter set No. 6.

385

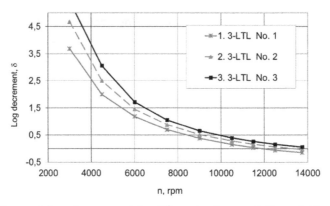

Figure 6. Log decrement for 1st mode shape – L/D variation.

decreasing of radial clearance has another beneficial effect, so case No. 4 had the biggest log decrement throughout the range of speeds analyzed.

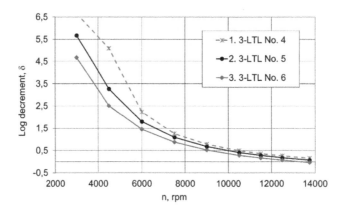

Figure 7. Log decrement for 1st mode shape – clearance variation.

6   CONCLUSIONS

In this paper a stability of flexible rotor supported with two equal 3-lobe taper land bearing (3-LTL) was studied. At the beginning, the whole procedure for calculation of stiffness and damping coefficients were described in detail.

The focus of analyses was to determine the effects of L/D ratio and bearing radial clearance of 3-LTL bearing on stability of rotor system. Beside stability a critical speeds in Campbell diagram were investigated in order to discover potential resonances at nominal rotor speed. Analyses showed that no configuration was in danger of excitation with 1st harmonic.

As an example of flexible rotor system a high speed steam turbine was used. The rotor was modeled with 1d finite elements. Stability of the flexible rotor system was assessed using logarithmic decrement.

It can be concluded that, for the considered parameters, increasing the bearing width has a favorable effect on stability of rotor 1st mode. In the second part of comparison, radial clearance of bearing were changed while all other parameters were kept constant. The analysis results has shown that decreasing of radial clearance has also beneficial effect on stability.

ACKNOWLEDGMENT

This work has been fully supported by the University of Rijeka under the project number uniri-tehnic-18-225.

REFERENCES

[1] Butković, M., Žigulić, R., Braut, S., "Contribution to the Numerical Solving of the Reynolds Equation by Finite Difference Method Using the Highly Convergent Iterative Methods", Proceedings of Tenth World Congress on the Theory of Machines and Mechanisms, Vol 7, 1999., pp. 2783–2788.

[2] Bhushan, G., Rattan, S.S, Mehta, N.P., "Effect of L/D Ratio on the Performance of a Four-Lobe Pressure Dam Bearing", International Journal of Mechanical and Mechatronics Engineering, 1, 2007, pp. 401–405.

[3] Hargreaves, D.J., "Predicted performance of a tri-taper journal bearing including turbulence and misalignment effects", Proceedings of the Institution of Mechanical Engineers, Part J: Journal of Engineering Tribology, 209, 1995, pp. 85–97.

[4] Rao, D.S., Shenoy, B.S., Pai, R.S., Pai. R., "Stability of tri-taper journal bearings under dynamic load using an on-linear transient method", Tribology International, 43, 2010, pp. 1584–1591.

[5] Pai, R., Rao, D.S., Shenoy, B.S., Pai, R.P., "Stability Characteristics of a Tri-taper Journal Bearing: a Linearized Perturbation Approach", Journal of Materials Research and Technology, 1, 2012, pp. 84–90.

[6] Someya, T., 'Journal-Bearing Databook', Springer-Verlag Berlin Heidelberg GmbH, 1989.

12th International Conference on Vibrations in Rotating Machinery -
Institution of Mechanical Engineers, ISBN 978-0-367-67742-8

Data combination for a consolidated diagnosis of rotor and bearing faults

K.C. Luwei, A. Yunusa-Kaltungo

Dynamics laboratory, School of Mechanical, Aerospace and Civil Engineering,
The University of Manchester, UK

ABSTRACT

Vibration-based condition monitoring is useful for fault detection in rotating machines. The time and frequency domain feature unification with signals from multiple sensors for fault diagnosis to achieve a single analysis have been well established in earlier studies. An initial study developed data fusion of acceleration and velocity features (dFAVF) with a useful machine condition diagnosis. However, these studies focused on rotor faults only. This study tends to incorporate bearing data in the developed dFAVF model. The aim is to diagnose an extensive range of machines' faults in a single analysis. Signals observed were from a test rig operating with multiple speeds below and above its first critical speed. The preliminary result showed useful identification of the machines' faults. The usefulness of this approach in industrial-scale fault diagnosis may improve the simplicity of understanding the machine's overall behaviour with regularly measured vibration data.

1   INTRODUCTION

Vibration-based condition monitoring (VCM) techniques have been useful in detecting faults, especially in rotating machines critical parts (1). Current successes in fault diagnosis, especially on rotating machinery, stems from earlier research. Some of which various rotor faults were analysed individually (2,4-18.), same for bearing defect diagnosis (19-27). Even in studies with a combined rotor and bearing fault diagnosis (1,3), there is also individual analysis for the specific condition. However, this paper explores data combinations of VCM time and frequency domain parameters to diagnose a wide range of rotating machines faults (Rotor and bearing) in a single analysis.

This study stems from some earlier studies (9-17) that focused on data fusion for a single analysis in identifying machine conditions. Jyoti and Elbhbah (10) proposed the fusion of signals from multiple sensors with frequency domain features from higher-order spectra (HOS), i.e., bispectrum, developed from the composite spectrum (CS) gave for useful analysis. This method showed promising practical application. However, diagnostic features considered only the bispectrum amplitude as the phase got lost in CS computation. Yunusa-Kaltungo et al. (11) further improved this approach using the cross-powers spectrum density in computing a proposed poly coherent composite spectrum (pCCS). The approach retains both amplitude and phase from all measured signals in its computation. Observation showed an excellent representation of machine behaviour in comparison to the CS. Yunusa-Kaltungo et al. (12) further proposed a method whereby data combination of poly coherent composite higher-order (*pCCHO*S), i.e., bispectrum and trispectrum, showed useful analysis. The study observed a similar rotating machine installed with varying foundation flexibilities or rotating speeds resulting in proper diagnosis that enhances the machine's reliability. Nembhard and Sinha (13), proposed the unified multispeed approach (UMA), where extracted time and frequency domain features from a rigid test rig classified various machine conditions. To test the transferability of this approach, further investigation (14) employed signals from a modified relatively flexible flange-based test rig, obtaining an improved analysis. Nembhard et al. (15) improved the UMA to include multiple foundations. Observation showed the approach presented encouraging analysis. However, these studies only covered acceleration signals in the diagnosis when the machine ran below its first critical speed. In a further study, Luwei et al. (16) proposed optimisation of UMA (13), to improve machine faults diagnosis, where a combined vibration acceleration and velocity features diagnosis, yielded useful analysis. However, the study (16) focused on signals obtained below the machine's first critical speed.

Machines may run at multiple speeds, such as is observed in large turbochargers and aero-engines, operating most times with speeds either below or above their natural frequencies (3). At such performances, vibration may increase, especially as the machine runs through the critical speeds. Thus, faults begin to develop in the machine over time with a continuous operation (3). The faults show up around critical components such as the rotor or bearings. Further study led to the investigation of the machine operating below and above the machine's first critical speed using the optimized approach. Thus, Luwei et al. (17) proposed the data fusion of acceleration and velocity features (dFAVF) model in which pattern recognition using principal component analysis (PCA) gave a useful diagnosis of various rotor-related faults. However, this and other background studies considered only rotor faults.

Vibration signals in velocity, or acceleration are useful in observation machine behaviour. According to Bruel & Kjaer (21), the acceleration tends toward high-frequency components while the velocity signal covers a frequency range of 10 Hz to 1000 Hz, most suitable for vibration severity indication. At such, acceleration covers bearings and gear faults, which occurs at high frequencies, and velocity covers a frequency range at which rotor-related fault occurs (22). Thus, a combination of features from acceleration and velocity should significantly distinguish a consolidated rotor-related and bearing faults in a single analysis. In this study, the velocity spectra computed was with the omega arithmetic method. This correlative method converts spectra density in acceleration to velocity, considering the frequency (Hz). In rotodynamic, omega (ω) represents frequency measured in radian/second. Therefore, by dividing the acceleration spectra density with omega (ω), the velocity spectra density is computed (28). The acceleration and velocity parameter can also reflect faults' signal behaviour, which may be in the low or high-frequency range. The 'low' frequency contains information from rotor related faults, and the 'high' frequency has information from bearing faults.

This current study tends to investigate the inclusion of features from bearing analysis into the dFAVF model. The aim is to use data combinations to detect rotor and bearing faults in a single analysis. In order to extract relevant parameter from bearing diagnosis, the envelope analysis was applied and observed in both the time and frequency domain. The machine faults considered are misalignment, rub, crack shaft, and bearing cage defect. Observation from this preliminary analysis gave a useful clustering of both rotor and bearing defects. The intent is that this approach would contribute to solving the diagnosis problem in the industry. At such, an extensive range of faults in rotating machines can be easily and quickly detected from observation of the changes in the overall behaviour of the machine using regularly measured vibration signals.

2   ROTATING MACHINES' CRITICAL COMPONENT FAULTS

Rotor-related faults show up as unbalance, misalignment, looseness, shaft bow, crack shaft (3), and so on, while bearing faults can be detected in either its inner race, outer race, cage or its balls. Bearing fundamental frequency helps to determine the frequency of defect for each case, based on its geometry. They include ball pass frequency inner-race (BPFI), ball pass frequency outer-race (BPFO), fundamental train frequency (FTF), and ball spin frequency (BSF) (3), respectively. According to Nakhaeinejad and Ganeriwala (4), 70% of machine faults are rotor related. Whereas, most bearings undergo premature failure due to different reasons (19). About 80% of bearing failure stems from improper lubrication; others include improper installation, contaminations, and production error during the manufacturing of connecting parts, handling unskilled personnel (20). Some VCM techniques for rotor-related fault diagnosis include power spectral density, wavelet analysis, orbit plot, and among others. Some VCM approaches for bearing defect diagnosis include crest factor analysis, shock pulse monitoring, kurtosis spectrum analysis, demodulated resonance analysis, and envelope power spectral density analysis. Since there is the incorporation of bearing fault in this study, a brief explanation of the rolling element ball bearing is discussed in subsection 2.1.

2.1 Rolling element bearing

Figure 1 shows a diagram representing a typical ball bearing's internal structure, the inner race, outer race, balls, and cage represented clearly. The fundamental frequencies can be calculated based on the geometry of the bearing (3).

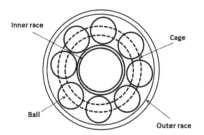

Figure 1. Typical structure of a ball bearing.

During operation, the bearing housing's natural frequency is excited due to impact loading per rotation (3). It ranges from 1 to 5 kHz. However, the bearing housing natural frequency in this study is around 2.4 kHz. In bearing fault diagnosis, power spectrum density (PSD) analysis may not show the related fundamental frequency, especially when the defect is small. The envelope analysis helps to mitigate such a situation by extracting impacts with low energy (3). In this study, the envelope analysis on bearing signals helps to examine the behaviour of a damaged bearing cage. Also, since the study considers bearing cage defect, the FTF is calculated mathematically using (3);

$$FTF = \frac{f_r}{2}\left(1 - \frac{d_b}{d_p} \cos\beta\right) Hz \qquad (1)$$

Here, ball diameter and pitch circle diameter represent d_b and d_p, respectively. The f_r represents relative speed between the inner race and outer race i.e., shaft speed, and β is the ball contact angle (3).

3   TEST RIG AND EXPERIMENTS

Earlier sections presented the context for this study. The experimental approach also presents with a description of the test rig and experiments conducted for rotor-related and bearing fault in subsections 3.1 and 3.2, respectively.

3.1   Test rig

Figure 2 shows a picture of the laboratory test rig used in this study, located in the Dynamics Laboratory at the University of Manchester. The test rig is a spring-based pedestal (SBP), modified from an earlier flange-based pedestal (FBP) test rig. This modification helps to manipulate the systems' critical speeds so that investigation of machine's behaviour from signals obtained below and above its critical speeds can be achieved. The SBP test rig is made up of two shafts 1m and 0.5m long, both having 0.02m diameter and joined by a rigid coupler. The long shaft is joined to a motor by a flexible coupler. The motor operates by a PC based speed controller (NEWTON TESLA CL750 and FR configurator SW3) with which the user determines the operating speed of the shaft. There are three balance discs on the shafts with two on the long shaft and one on the short shaft. Four spring-based bearing pedestal (SBP) installed on a lathe bed with dampers, are connected to the shafts at different locations. One accelerometer on each of the bearing pedestal placed at 45° have been used to strengthen earlier study (10), reducing amount of sensor per bearing. Modal testing carried out experimentally gave a couple of natural frequencies i.e., 11.52 Hz, 18.62Hz, 30.75 Hz, 49.13 Hz, and 85.83 Hz. Details of the test rig and dynamic characterization are in Luwei et al. (17) work. Selected for this study are three machine running speeds, based on avoidance of running on the natural frequencies, they are, 450 rpm (7.5 Hz) which is below the first critical speed, 900 rpm (15 Hz) and 1350 rpm (22.5 Hz) which are above the machines first critical speed.

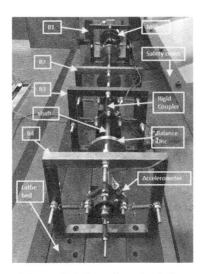

Figure 2. Laboratory test rig.

3.2 Test conducted

Figure 3 shows the schematics of the test rig and the various faulty conditions simulated in this study. The SPB 1 to 4 represents B1 to B4, and others are clearly labelled. The misalignment was done first with two shims of 0.0008m thick, place under B1 pedestal, creating a parallel misalignment, as shown in Figure 3. Next is the rub fault using an apparatus that holds firmly two Perspex-sheet above and below the shaft, thus causing blade rub when the machine runs. The rub fault was simulated at 0.26m from B1. The next was the crack shaft condition simulated using the electro-discharge machining (SDM) wire erosion process by cutting a 0.00034m deep and 0.004m wide notch. A 0.00033m shim glued in the notch helps to create a breathing crack. The shaft crack is 0.16m from B1. Finally, the bearing fault simulated employed a Dremel engraver, which created some notches on the bearing cage. The notches were on the bearing in B2 pedestal.

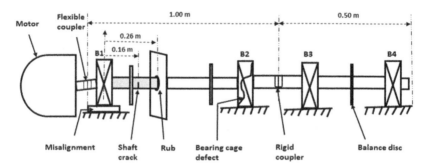

Figure 3. Schematics of test rig with the simulated faulty conditions and locations.

4 SIGNAL PROCESSING

The signals collected are the baseline and simulated faults, obtained at all three speeds. Each with 20 sets of data, collected for 2 minutes at a sampling frequency of 10 kHz. Note that each condition has been simulated independently of the other. The computational parameters selected present an adequate comparison for rotor and bearing conditions. Thus, an equal length of data collected at a similar time with the same sampling frequency helped manage any concerns. Other computational parameters included the number of data points for Fourier transform (N), which was 16,384; frequency resolution (df) was 0.6104 Hz; the number of averages used was 287, for both rotor and bearing conditions respectively.

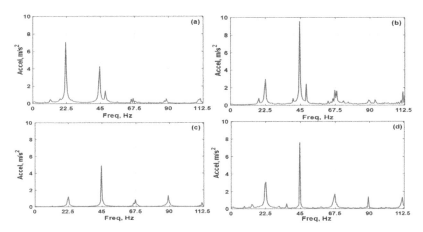

Figure 4. Typical spectrum plots at 1350 rpm for (a) baseline-RMRU (b) Misalignment (c) Crack (d) Rub.

Figure 4 shows typical plots from B2 pedestal with the baseline-residual misalignment and residual unbalance (RMRU) and various rotor faults, i.e., misalignment, crack and rub faults when the machine ran at 1350 rpm (22.5 Hz), above its first critical speed. In Figure 4(a) observation shows the presence of 1x and its harmonics. The presence of a high amplitude at 2x may be due to residual misalignment, and residual unbalance (RMRU), thus representing baseline condition (12). Small peaks observed before 1x (22.5 Hz) may be due to the natural frequency at 11.52 Hz and 18.62 Hz, respectively. The peak after the 1x is due to the natural frequency at 30.75 Hz. Also, the peak after the second harmonic is due to the natural frequency at 49.13 Hz. Other small peaks observed around further harmonics may similarly be due to the closeness of the natural frequency. The appearance of these peaks in most of the other spectrum for the faulty conditions, as in Figure 4(b) - (d), may also result from the same natural frequency observed in the baseline-RMRU spectrum. The spectrum of the faulty rotor conditions shown in Figure 4(b) – (d) all have 1x at various amplitudes with their harmonics appearing either with higher or lower amplitude. A similar observation may be at other bearing locations, depending on their mode shape around that location.

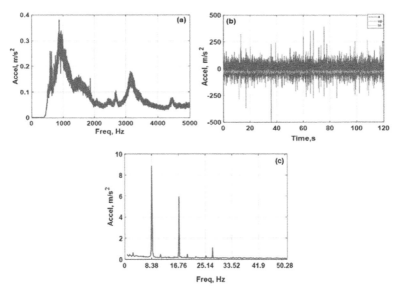

Figure 5. Bearing defect at 1350 rpm for (a) Spectrum without envelope (b) Time domain with envelope (c) Envelope spectrum.

Similarly, Figure 5 shows typical plots of bearing analysis. Figure 5(a) gives a typical spectrum plot of bearing defect after a high filter at 500 Hz. The filtering helps to remove the low frequencies, which may be rotor related or noise. After that, the envelope analysis on the time domain gives useful information, as presented in Figure 5(b). Here, **a** represents the original signal while **up** and **lo** represents the upper and lower envelope signal, respectively. Figure 5(c) shows the spectrum plot for the envelope analysis of the signal. Worthy of note is that calculations of the fundamental train frequencies (FTF) gave 2.79 Hz, 5.58 Hz, and 8.38 Hz with the machine running speeds at 7.5 Hz, 15 Hz, and 22.5 Hz respectively. All the typical plots represented were from signals at B2 pedestal when the machine ran at 1350 rpm (22.5 Hz). Plots from other speeds showed relatively similar observation at the single bearing. Thus, giving an overall representation of the machine behaviour both in time and frequency domain. Also, the velocity-based computation showed similarity except for its unique characteristics (17).

In this study, the linear scale has been used in spectra analysis because it represents the actual vibration amplitude. In comparison, the log scale would be useful to observe the fundamental frequencies, especially in bearing analysis where low amplitude frequencies are amplified, and high frequencies compressed during a visual investigation. The log scale may not represent the correct frequency amplitude. In VCM fault diagnosis, typical time and frequency domain analysis may give some useful insight into the machine's behaviour (3). However, as changes occur over a long period of machine operation with changing speeds, data is continuously recorded. The effect is that too much VCM data becomes available (14). The analysis may become cumbersome for the vibration analyst, thus, creating poor human judgment and inefficient diagnosis. So, to improve fault diagnosis (FD), data fusion-based approach has been considered in various studies (2,12-18) with valuable outcomes. Therefore, this study tries to improve FD by developing a data combination model of acceleration-based time domain and velocity-based frequency domain features from the rotor and bearing parameters for a single analysis. It is expected that the model represents a simplified approach which effectively detects an extensive range of machine faults.

5   FEATURES SELECTION AND PCA-BASED MODEL

Selected features for this study included the time domain root mean square (RMS), crest factor (CF) and kurtosis (Ku), and the frequency domain spectrum energy (SE) and 1x-5x amplitude (17). The features are used to build a data matrix loaded into the principal component analysis (PCA) pattern recognition-based fault diagnosis model. PCA is a multivariate statistical tool that reduces large interrelated datasets to a small number of variables while retaining the variability in the original data. Principal components (PCs) are the outcome of such computation so that the first few PCs retain the variability of the original data (7, 13-17). PCA reveals the existing variance present in an original data, identified as observations (e.g., the quantity of measured vibration signal) and variables (e.g., CF, Ku, 1x-5x). In this study, PCA has been used to investigate the relationship of an extensive range of experimentally simulated rotating machine faults in a single analysis. Given a data matrix K, features $kaT1$, $kaT2$, $kaT3$, ... $kaTn$ represents acceleration-based time domain parameter while $kvF1$, $kvF2$, $kvF3$, ... $kvFn$ represents velocity-based frequency domain parameter. The recorded vibration data constituted the observation Dm, with machine conditions Ck, and rotation speed Sp, where m is 1, 2, 3 ... m, k is 1, 2, 3 ... k, and, p is 1, 2, 3 ... p, respectively. Equations (2) and (3) represent the data matrix computed to develop the model;

$$K_{C_k S_p} = \begin{bmatrix} k_{aT1_{D_1}} & \cdots & k_{aTn_{D_1}} & \cdots & k_{vF1_{D_1}} & \cdots & k_{VFn_{D_1}} \\ \vdots & \ddots & \vdots & \ddots & \vdots & \ddots & \vdots \\ k_{aT1_{Dm}} & \cdots & k_{aTn_{Dm}} & \cdots & k_{vF1_{Dm}} & \cdots & k_{VFn_{Dm}} \end{bmatrix}_{C_1 S_1} \tag{2}$$

$$K = \begin{bmatrix} K_{C_1 S_1} & \cdots & K_{C_1 S_p} \\ \vdots & \ddots & \vdots \\ K_{C_1 S_p} & \cdots & K_{C_k S_p} \end{bmatrix} \tag{3}$$

Equation (2) shows the initial build-up of the model. The computation is such that $C_1 S_1$ represents condition one at speed one, say baseline-RMRU at speed 450 rpm (7.5 Hz). This computation extends to all conditions C_k at all speeds S_p, as shown in equation (3). Acceleration-based time domain and velocity-based frequency domain parameters made up 9 features per bearing, and with 4 bearings in the test rig where the measurement was done at 3 running speeds a total of 9 x 4 x 3 = 108 features computed. Similarly, 20 sets of data per machine condition formed the observation. Since the simulated machine conditions were 5, all cases' total observation becomes 20 x 5 = 100. Thus, data fusion of acceleration and velocity features (dFAVF) at all speeds and all experimentally tested conditions gave a 108 x 100 data matrix. Data normalization is achieved here by converting each element in **K** to zero mean and unit variance. The normalization helps create a common scale to avoid misrepresentation of the data (13). Thereafter, the PCA-based pattern recognition of machine condition was employed to analyse the computed data matrix (17).

6   OBSERVATION AND DISCUSSION

Figure 6 shows a representation of the PCA-based pattern recognition using the data fusion model from section 5. PC1 and PC2 contain a larger variance in the data matrix (17). In plotting principal components (PCs), one should note that each PC has a single direction with a midpoint at zero. A positive or negative PC gives the direction of the variable in that PC regarding a single dimension vector; thus, PCs may be positive or negative (7,13). The simulated conditions represented in Figure 6 (a) are the baseline residual misalignment, and residual unbalance (baseline-**RMRU**), misalignment (**M**), crack close to B1 (**C1**), shaft rub (**R**) and bearing cage defect at B2 (**Bc2**). Observation showed that baseline-**RMRU, M, C1, R,** and **Bc2** had useful clustering and separation. The baseline-**RMRU** had a separate cluster from other conditions, however, it is closer to the rotor conditions than the bearing. Also, while the **M, C1,** and **R**, which are rotor conditions, stayed close to each other, **Bc2** a bearing defect is separated from other conditions while showing some spread in its cluster. This separation of the bearing from other conditions may be due to differences in the frequency range (21), i.e., 'low' frequency faults cluster around the same region, further away from the 'high' frequency faults. On the other hand, the spread seen in **Bc2** may be due to variation in the impact load during machine operation. The rotor faults seem to

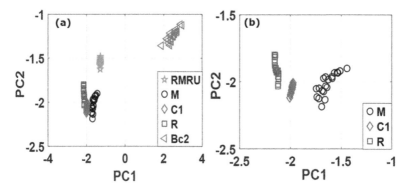

Figure 6. Single classification of rotor-related and bearing fault in a lab rotating rig (a) combined view (b) zoomed view.

overlap each other as observed in Figure 6 (a). However, a zoomed view showed clear separation of each condition as seen in figure (6) b.

In this study, the consolidated fault identification analysis of rotor-related and bearing fault using the PCA-based pattern recognition model was achieved. Therefore, this model's application may provide useful diagnostics information and a better understanding of individual faults behaviour at the various frequency ranges. This information may be helpful to vibration analyst for early decision making in the plant.

7   CONCLUSION

In this study, the data fusion of acceleration and velocity features (dFAVF) model further incorporated bearing features in its analysis. The aim was to develop a consolidated fault detection model for an extensive range of machine faults (rotor and bearing) in a single analysis. The acceleration-based time domain and velocity-based frequency domain features from vibration signals were selected to populate the model. The signals obtained from multiple speed operation i.e., below and above the machine's first critical speed gave good analysis. The investigation from this preliminary study showed useful clustering of individual conditions. Further, observation showed significant separation between the rotor-related and bearing faults, which may represent the 'low' frequency and 'high' frequency ranges of faults. The result is the outcome from a single analysis. This development presents a robust background for further studies in which more faulty conditions can be captured and thus extended to industrial-scale fault detection.

ACKNOWLEDGMENTS

The authors recognise Professor Jyoti K. Sinha for the original ideas and intellectual properties (IP) in this study. Also, the authors deeply appreciate him for his supervisory role offered during this study.

REFERENCES

[1] Saimurugan, M., & Nithesh, R. (2016). Intelligent fault diagnosis model for rotating machinery based on fusion of sound signals. *International Journal of Prognostics and Health Management*, ISSN 2153-2648, 2016 018. https://pdfs.semanticscholar.org/0a03/f30012526e37c23734ee33798d7a1e203ce5.pdf

[2] Luwei, K. C., and Yunusa-kaltungo, A., Sha'aban Y. A. (2018). Integrated fault detection framework for classifying rotating machines fault using frequency domain data fusion and artificial neural network. *MDPI - Machines* 6(4), 59. https://doi.org/10.3390/machines6040059

[3] Sinha, J. K. (2015). "Vibration analysis, Instruments, and signal processing. Taylor and Francis Group, Boca Raton, FL.

[4] Nakhaeinejad, M., Ganeriwala, S. (2009). Observation of Dynamic Responses of Misalignment. TechNote, SpectraQuest Inc (Sept). Available at; https://issuu.com/osanoothu/docs/observations_of_dynamic_responses_o

[5] Felten, D. (2003). Understanding bearing vibration frequencies. *Mechanical Field Service Department L&S Electric Inc*, Schofield, Wisconsin. Available at; https://dokumen.tips/documents/understanding-bearing-vibration-frequencies.html

[6] Shaeffler, (no date). Bearing failure: causes and cures. Available at; https://www.schaeffler.com/remotemedien/media/_shared_media/08_media_library/01_publications/barden/brochure_2/downloads_24/barden_bearing_failures_us_en.pdf

[7] Bruel & Kjaer (1982). Measuring vibration. Available at; https://www.bksv.com/media/doc/br0094.pdf

[8] Marcal, R. F. M., Negreiros, M. et al. (2000). Detecting faults in rotating machines. *IEEE Instrumentation & Measurement Magazine*, pp 24–26. Available at; https://www.lume.ufrgs.br/bitstream/handle/10183/27565/000295687.pdf?sequence=1&locale-attribute=es

[9] Luwei, K. C., Sinha, J. K. and Yunusa-kaltungo, A et al. (2018). Data fusion of acceleration and velocity features (dFAVF) approach for diagnosis in rotating machines. *14th International conference on Vibration Engineering and Technology of Machinery*, Lisbon, Portugal, pp 1–6. https://doi.org/10.1051/matecconf/201821121005

[10] Zhu, Y., Jiang, W., and Kong, X. et al. (2014). An accurate integral method for vibration signal based on feature information extraction. Shock and vibration. pp 13 https://doi.org/10.1155/2015/962793

[11] Adewusi, S.A. and Al-bedoorm B.O. (2000). Wavelet analysis of vibration signals of an overhang rotor with a propagating transverse crack. *Journal of sound and vibration*, 246(5),777–793. DOI: 10.1006/jsvi.2000.3611

[12] Song, G. F., Yang, Z. J. et al. (2013). Theoretical-experimental study on rotor with a residual shaft bow. *Mechanism and Machine Theory*, 63 (2013), pp. 50–58. DOI: 10.1016/j.mechmachtheory.2013.01.002

[13] Luwei, K., Yunusa-Kaltungo, A. and Sinha, J. (2016). A simplified rotor-related faults detection approach based on a combination of time and frequency domain features. *Journal of Maintenance Engineering*. 1 ed. Aylesbury, Buckinghamshire: Shieldcrest Publishing, Vol. 1, (10) (2016), pp. 138–147.

[14] Sinha, J.K. and Elbhbah, K. (2012). A future possibility of vibration based condition monitoring of rotating machines. Mechanical System and Signal Processing. 34 (2015), pp 231–240. http://dx.doi.org/10.1016/j.ymssp.2012.07.001

[15] Nembhard, A. D. and Sinha, J.K. (2015). Unified Multi-speed analysis (UMA) for the condition monitoring of aero-engines. Mechanical Systems and Signal Processing. 64-65 (2015), pp. 84–99. https://doi.org/10.1016/j.ymssp.2015.04.027

[16] Nembhard, A. D., Sinha, J.K. and Yunusa-Kaltungo, A. (2014). Development of a generic rotating machine fault diagnosis approach insensitive to machine speed and support type. Journal of Sound and Vibration. 337 (2014), 321–341. http://dx.doi.org/10.1016/j.jsv.2014.10.033

[17] A.D. Nembhard, J.K. Sinha, Vibration-based condition monitoring for rotating machinery with different flexible supports, Proceedings of the 10th International Conference on Vibration Engineering of Machinery (VETOMACX), Manchester, UK, 2014.

[18] Malhi, A. and Goa, R. X. (2004). PCA-based feature selection scheme for machine defect classification. IEEE Transactions on Instrumentation and Measurement, Vol 53(6), pp. 1517–1525. DOI: 10.1109/TIM.2004.834070

[19] Yunusa-kaltungo, A., Sinha, J. K. and Elbhbah, K. (2014). An improved data fusion technique for fault diagnosis in rotating machines. Measurement. 58 (2014),27–32. http://dx.doi.org/10.1016/j.measurement.2014.08.017

[20] Yunusa-Kaltungo, A., Sinha, J.K., and Nembhard, A.D. (2015). A novel fault diagnosis technique for enhancing maintenance and reliability of rotating machine. *Structural Health Monitoring*, 14(6) 604–621. https://doi.org/10.1177/1475921715604388

[21] Luwei, K. C., Sinha, J. K. and Yunusa-kaltungo, A. (2017). Optimisation of different acceleration and velocity features for fault diagnosis in rotating machines. Proceeding of 2nd *International Conference of Maintenance Engineering, Manchester (InCoME II)*, 2nd ed. Aylesbury, Buckinghamshire: Shieldcrest Publishing

[22] Luwei, K. C., Sinha, J. K. and Yunusa-kaltungo, A. (2018). Poly-coherent composite bispectrum analysis for fault diagnosis in rotating machines. Proceeding of 3rd *International Conference of Maintenance Engineering (InCoME III)*, 3rd ed. Aylesbury, Buckinghamshire: Shieldcrest Publishing

[23] Zhang, Y., and Randall, R.B. (2009). Rolling element bearing fault diagnosis based on the combination of genetic algorithm and fast kurtogram. *Mechanical System and Signal Processing* 23. 1506–1517. DOI:10.1016/j.ymssp.2009.02.003

[24] Randall, R.B., Antoni, J., and Chobsaard, S. (2001). The Relationship between spectral correlation and envelope analysis in the Diagnosis of bearing faults and other cyclostationary machine signal. *Mechanical Systems and Signal Processing* (2001)15(5) 945–962. DOI: 10.1006/mssp.2001.1415

[25] Randall, R.B. Antoni, J. and Chobsaard, S. (2000) A Comparison of cyclostationary and envelope analysis in the diagnosis of rolling element bearing pp 3882–3885. DOI: 10.1109/ICASSP.2000.860251

[26] Attoui, I., Oudjani, B., Boutasseta, N. et al. (2020). Novel predictive features using a wrapper model for rolling bearing fault diagnosis based on vibration signal analysis. *Internationa Journal of advanced Manufacturing Technology*, *106*, 3409-3435. https://doi.org/10.1007/s00170-019-04729-4

[27] Misra, R., Shinghal, K., Saxena, A., Agarwal, A. (2020). Industrial motor bearing fault detection using vibration analysis. In: Singh Tomar G., Chaudhari N., Barbosa J., Aghwariya M. (eds) *International Conference on Intelligent Computing and Smart Communication 2019*. Algorithms for Intelligent Systems. Springer, Singapore

[28] Mercer, C. (2006). Acceleration, velocity and displacement spectra – Omega Arithmetic. PROSIG Signal Processig Tutorials. Available at: http://prosig.com/wp-content/uploads/pdf/blogArticles/OmegaArithmetic.pdf

12th International Conference on Vibrations in Rotating Machinery -
Institution of Mechanical Engineers, ISBN 978-0-367-67742-8

Asynchronous rotor excitation system (ARES) – A new rotor dynamic test facility at Imperial College London

C.W. Schwingshackl[1], L. Muscutt[1], M. Szydlowski[1], A. Haslam[1], G. Tuzzi[1], V. Ruffini[2], M. Price[3], A. Rix[3], J. Green[3]

[1]Imperial College London, UK
[2]University of Bristol, University Walk, Clifton, UK
[3]Rolls-Royce Plc., London, UK

ABSTRACT

Long flexible shafts with bladed discs are a fundamental component of any gas turbine, and as such have been investigated in detail for many decades. With the emergence of the next generation of gas turbines, with shorter, stiffer shafts, and larger and more three-dimensional blades, current understanding of the underlying physical phenomena will need to be re-evaluated and where necessary extended. To support the ongoing research into blade-shaft coupling and cross-shaft coupling, a new rotor dynamic test facility has been developed over the last few years at Imperial College London, providing the ability to study the dynamic interaction of individual rotor components during synchronous and asynchronous excitation in a well-controlled environment.

The design and features of the test facility are discussed in detail, focusing on the novel features of the rig. Initial results for a bladed disc with staggered blades show the ability of the rig to operate under a wide range of conditions and highlight the good quality data that can be obtained.

1   INTRODUCTION

Shafts and flexible bladed discs form the backbone of many rotor systems, with applications ranging from aircraft propulsion and power gas turbines to vacuum cleaners. New requirements such as weight reduction, better performance and extended life, lead to a more and more demanding dynamic environment in which these systems operate. Shafts and discs have historically been studied independently of each other due to their large frequency separation, analysing the flexible shaft with rigid discs, and the flexible bladed disc with a rigid shaft.. This uncoupled analysis is considered valid as long as the disc resonances are well above the shaft ones and the shaft interaction is kept to a minimum. However, modern design trends, particularly in the aero engine sector, lead to more flexible discs and blades, shorter and more rigid shafts, and much more flexible casings, bringing the natural frequencies of the different components much closer together, and consequently leading to much more dynamic interaction between these components.

To study the underlying coupling phenomena and allow a high-quality validation of existing and emerging analysis tools, a unique test facility has been designed and commissioned at Imperial College London, allowing the detailed study of the partici-pating coupling phenomena in a well-controlled environment.

2   ARES TEST RIG

The principle idea behind the Asynchronous Rotor Excitation Rig (ARES) is to provide for the first time a well-controlled rotor dynamic test environment that allows the study of blade-shaft and cross shaft coupling behaviour at the same time.

2.1   Rig requirements and basic design

During the initial phases of the design, a series of requirements were defined for the test facility. The main requirements thereby were to provide a (i) long and flexible (ii) single or dual shaft system with (iii) synchronous or asynchronous rotor excitation [1], [2], that could be supported by (iv) rigid or flexible, (v) symmetric or asymmetric bearing supports, to study advanced blade shaft coupling phenomena [3]. The rig needed to be (vi) easily re-configurable to test a multitude of setups and (vii) allow accurate measurements of the shaft response. In addition, the rig needed to house (viii) a flexible bladed disc on the shaft, to study blade shaft interaction, without the influence of (ix) aero-elastic effects.

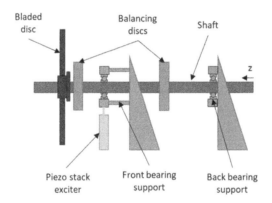

Figure 1. Axial locations of different rig components.

Based on these initial requirements, the concept rig design in Figure 1 was developed. It consists of a single flexible shaft, which is mounted on two bear-ings. The back bearing thereby Is a radially rigid bearing that floats in the axial direction, while the front bearing can be mounted in a rigid or flexible configur-ation. Two balancing discs are added to the shaft to allow dynamic balancing of the rig and synchronous excitation via an out of balance mass. A flexible bladed disc is also mounted on the shaft via a lock and release mechanism. In order to simplify the rig design, it was decided to simulate the asynchronous excitation coming from a second shaft via a well-controlled excitation of the flexible bear-ing housing. The shaft is driven from the back (z=0 mm) via a flexible coupling, and the entire system is enclosed in a low-pressure environment to reduce the aeroelastic effects on the bladed disc, and provide containment in case of a critical event.

Based on the requirements and the initial design concept the detailed design in Figure 2 was developed, addressing all the described features in a single test facility. The rig design was heavily influenced by the experience obtained during the design and operation of the AFRODITE (Advanced Flexible Rotating Disc TEst) facility [4], which allows the testing of a flexible bladed disc on a rigid shaft. In the following a detailed discussion of the main components of the ARES facility will be provided.

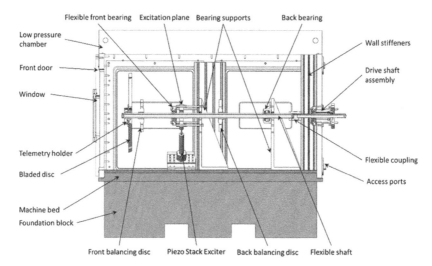

Figure 2. The fully assembled ARES test facility.

2.2 ARES Components

The rig is located on a large, rigid inertial concrete block, which is isolated from the ground via air springs (see Figure 3). The inertial block has a weight of approx. 2 tons and ensures minimal vibration transmission. It is fitted with a machine bed with T-slots to provide a flat reference surface and allow easy attachment of the test rig components and its supporting instrumentation. Directly attached to the inertia block is a large, rectangular aluminium chamber (L 1725 x W 980 x H 945 mm), which is sealed via O-rings to provide a low-pressure environment for testing without aeroelastic interaction. The chamber in Figure 3 has been designed to withstand a pressure difference of up to 1 bar (0.95 bar achieved during commissioning). In order to achieve this pressure requirement, the chamber is strengthened with aluminium ribs and an internal Aluminium profile frame, leading to less than 1mm deflection in the fully evacuated state. The chamber has five large doors with fitted polycarbonate windows, giving access to the test rig from all sides. The back of the chamber is permanently mounted and is fitted with a series of access ports and a centrally fitted sealed shaft bearing assembly to transmit the externally fitted motor motion to the shaft.

a)   b)

Figure 3. The Foundation block and low-pressure chamber a) in closed state and b) with open doors.

Two large and rigid steel-L shapes are mounted to the test bed (see Figure 2), which act as the bearing supports for the back and front bearing. The rear bearing support has a permanently mounted floating self-aligning bearing (SKF 2208-EKTN9) attached at a height of 400 mm, leading to a maximum bladed disc diameter of 700 mm. The front bearing assembly is of a much more complicated design, since it needs to be reconfigurable in order to allow a transition from rigid to very flexible in multiple stages and needs to incorporate the ability to excite the shaft with an asynchronous excitation. A dedicated bearing housing has been designed for the self-aligning SKF (SKF 2208-EKTN9) bearing, which can be attached via four flexible rods of variable dimensions and shapes to the L-shapes (see Figure 4a). This arrangement allows for a wide selection of bearing configurations, where round rods of variable length provide symmetrical bearing housing stiffness, while the bars with machined flat sides in Figure 4b) introduce asymmetric bearing housing stiffness. The two bearings are attached to the shaft via a keyless bushing, that provides a strong connection between inner race and shaft and allows to control the clearance in the bearing at the same time.

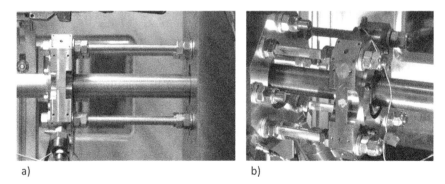

a)   b)

Figure 4. Flexible front bearing assembly a) symmetric and b) asymmetric.

The 1300mm long hollow shaft is made of stainless steel. It has an outer diameter of 35 mm and an inner diameter of 22.2 mm. The inner hole thereby allows to route cables from the back end of the shaft to the front, where a flexible disc and a telemetry system can be mounted. Two identical balancing discs (see Figure 5a) are mounted to the shaft via a Trantorque GT keyless bushing which allows an easy repositioning of the discs along the shaft. The balancing discs have an outer diameter of 240 mm, and a mass of 7.2 kg. A series of circular patterns of threaded holes allow a large range of different balancing weights to be attached to the balancing discs during dynamic balancing to ensure a smooth running of the system and provide synchronous shaft excitation when needed.

A flexible disc can be mounted to the shaft via a telemetry/disc holder that is attached to the shaft via an additional Trantorque GT keyless bushing (see Figure 5b). The telemetry holder houses up to three telemetry systems and acts as the interface for any excitation power that may be required in the rotating frame. It also doubles as the flange to which a flexible bladed disc of variable design can be mounted.

By choosing the friction based mounting systems for the bearings and all the discs a shaft of uniform diameter can be used. This enables a very flexible rig configuration, allowing to change bearing and disc locations depending on the needs of the test setup.

a) b)

Figure 5. Detail of a) balancing disc with Trantorque GT mounting and b) four blade disc configuration with telemetry holder.

2.3 Excitation

In order to study the dynamic response of the ARES rotor, a well-controlled excitation and an accurate measurement of the response is required. Three types of excitation are available on the rig: (i) synchronous excitation via an out of balance mass on one of the two balancing discs, (ii) asynchronous excitation of the front flexible bearing housing via two piezo stack exciters, and (iii) blade excitation via piezo patches on the rotating blades.

The synchronous excitation is thereby the easiest to provide, since a small mass, added to one of the balancing discs normally suffices to generate a force, synchronous to the rotor speed. The location of the force input can thereby be varied, since the balancing discs can be moved around the shaft within the available space.

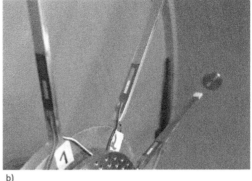

a)   b)

Figure 6. Excitation of a) bearing housing via piezo stack exciters and b) bladed disc via piezo patch excitation.

A much more challenging excitation is the asynchronous system which must allow excitation of the shaft via the flexible front bearing at any chosen excitation frequency and any orbit to simulate forcing coming from a secondary shaft system. It needs to be able to transmit large forces into the bearing, without impacting the flexibility of the bearing support. Two piezo stack exciters (Piezosystem Jena PSt 1000/35/200 VS45) attached to two amplifiers (Piezosystem Jena RCV1000/7) were chosen for this purpose. They are arranged at a 45° angle (see Figure 6a) between the machine bed and the bearing housing and are attached to the shaft assembly via strong push rods. The piezo stack exciters were chosen as exciters due to their large forces (>40 kN blocking force) and their ability to operate in a low pressure environment. Their high stiffness required a careful design of the front bear housing support, since attaching the piezos directly to the bearing housing negated the flexible bearing support, grounding the bearing once more. To address this issue an additional excitation plane between the L-Shape and the bearing housing was introduced (see Figure 6a) to which the piezos are attached. The excitation plane is thereby attached via flexible rods to the L-shape, allowing it to be actuated in all radial directions via the two attached piezo stack exciters. The bearing housing is then connected via flexible rods to the excitation plane, which when excited inputs base excitation into the front bearing housing, which in turn transmits the forces via the bearing into the shaft. Although this setup may sound somewhat complicated, it proved to be highly effective, eliminating the impact of the piezos on the stiffness of the front bearing housing, without compromising too much on the achievable excitation of the system.

The third excitation system available in the system is a blade excitation system shown in Figure 6b), similar to the one presented in [4]. It consists of a purpose built amplifier box that can drive up to nine piezo patches simultaneously. The piezo power is transmitted into the rotating frame via a liquid slipring (Jordll Technic - Rotrans) and distributed via the telemetry holder to the individual blades on the disc. This allows excitation of up to nine individual blades at the same time, enabling harmonic standing and traveling wave excitation, different Nodal Diameter excitation, and random and impact excitation of the blades.

2.4   Instrumentation

The dynamic response of the system to the previously described excitation mechanisms, can include rotor speed, rotor vibration orbits, and, if present, bladed disc vibration.

a)   b)

Figure 7. Measurement setup a) LDTV shaft measurements and b) Strain gauge telemetry system.

Currently four highly accurate laser displacement probes (Keyance LK-H022) form the backbone of the measurement system. They allow a contactless measurement of the shaft orbits at two different locations along the shaft. Two probes are thereby mounted orthogonal to each other to measure one location, as shown in Figure 7a). Due to the high reflectivity of the stainless-steel shaft, excellent measurement readings can thereby be obtained at all running speeds. The motion of the bearing housings is being monitored via individual accelerometers (PCB 353B03) that are aligned with the shaft centre line in the vertical and horizontal direction. The rotor speed is measured via a once per revolution tachometer, and a 4096 pin encoder (British Encoder – Model 776), to ensure an accurate knowledge of the instantaneous rotor speed.

The measurement of the blade vibration is based on a set of strain gauges on the blades. Up to six strain gauges can thereby be directly attached to the liquid slip ring (no on board amplification required due to excellent signal to noise ratio), if no piezo blade excitation is required, or a low cost digital telemetry system shown in Figure 7b) from Transmission Dynamics can be used to wirelessly transmit up to 12 strain measurements from the rotating to the static frame.

Finally the forces introduced by the piezo stack exciters to the excitation plane are being monitored via two dynamic force gauges (PCB 208C05).

2.5   Drive and control system

The ARES test facility is powered by a 7.5 kW Motor (Siemens Simotics S) which provides up to 13Nm of torque and enables a maximum operational speed of 6000rpm. The motor is attached to the short, sealed drive shaft via a V-belt. A very flexible spring coupling (R+W BKH 60) is being used to connect the rigid drive shaft to the flexible test shaft inside the chamber, in order to minimise any impact on the dynamic response of the rotor.

The entire facility is being controlled via a system based on a network of National Instrument cRios [5]. The advanced implementation of the control system, in the FPGA layer of the hardware directly and in its real time operating system, allows a very quick and reliable operation of the rig, at reasonable cost. A total of 16 output channels control the motor speed, the asynchronous piezo stack exciters and the piezo patch blade excitation, while 32 input channels keep track of up to four LDTV's, two rotational input signals, two force gauges, four accelerometers, and up to 12 strain gauge channels. All input and output channels are thereby on the same clock, providing excellent phasing information between all the obtained data.

3   INITIAL ARES TESTS

The rig configuration used for the initial rotating tests of the ARES facility is shown in Figure 8. The setup was selected so that the first critical speed would be within the running range of the rig, and the blade modes would couple with the shaft response.

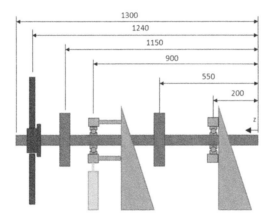

Figure 8. Axial locations of different rig components.

The first step during rig operation is its dynamic balancing. Due to the flexible arrangement possibilities of the rotor components, this process must be streamlined and reasonably quick, in order to balance the system after each modification. A dedicated Labview code, based on the influence-coefficient method, is available for this purpose, that allows a reasonably quick dynamic balancing of the system.

3.1   Unbalance excitation

Once balancing was completed successfully, its balanced state was considered the nominal state. Unbalance masses were then added to the front disc, and the rotor speed was then increased from zero in 0.1 Hz increments through the first critical speed to measure the rotor response.

Figure 9a) shows the shaft response in the location of the front balancing disc in x and y-direction for the balanced case, and two levels of increasing unbalance. The results highlight that a reasonably good level of balance was achieved initially, and clearly identifies the critical speed at 33Hz. A significant response below 20Hz could also be observed which gradually reduces to zero, which could be attributed to a runout of the shaft.

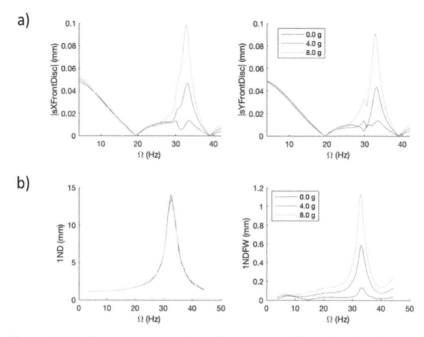

Figure 9. Unbalance response a) shaft response at front disc location and b) the bladed disc.

The bladed disc first Nodal Diameter (1ND) stationary response and 1ND forward (1ND FW) travelling response are shown in Figure 9b). It can be observed that both the stationary and FW travelling modes respond at the first critical speed, but only the FW traveling response depends on the unbalance level. This is because the 1ND stationary pattern is excited by gravity, whereas the 1ND FW travelling pattern is due to unbalance.

3.2 Asynchronous piezo-electric excitation

One key capability which is enabled by exciting the rotor asynchronously with the piezo stack exciters is that it is possible to experimentally obtain a Campbell diagram. The piezo stack exciters were used to excite the rotor over a range of frequencies, at a fixed rotor speed, which was then repeated for multiple rotor speeds. The frequency response was then extracted and plotted on a 2D plane with rotor speed and piezo stack exciters frequencies as axes.

3.2.1 *Forward traveling wave (FW) excitation*

A forward rotating force was first applied with the piezo stack exciters which leads to the response of the FW modes. The response to FW excitation is shown in Figure 10. Two FW modes can be identified which have been highlighted by the white lines in the plots; one which is shaft dominated (Figure 10a) and b), and one which is bladed disc dominated (Figure 10c). The response frequency of the bladed disk mode is much lower than the shaft frequency at low speeds, but it rises rapidly due to centrifugal stiffening so that the two modes start to interact around $\Omega = 10$ Hz, where significant curve veering can be observed.

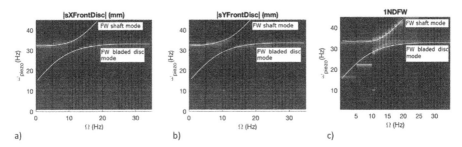

a)   b)   c)

Figure 10. Response to FW excitation a) shaft at front disc in x and b) y direction and c) the 1ND FW bladed disc response.

3.2.2 *Backward traveling wave (BW) excitation*

The phasing between the piezo stack exciters was then reversed, to apply a BW rotating force, which excited the BW modes. The response is plotted in Figure 11 below. The BW shaft-dominated mode can be clearly observed in the disc response in both directions (Figure 11a) and b). It reduces slightly with speed due to gyroscopic effects. The same shaft mode can be observed in the bladed disc response, in addition to the blade-dominated BW mode which decreases rapidly with increasing rotor speed. This reaches 0Hz at a rotor speed around 30Hz, which explains the large 1ND stationary response due to gravity in Figure 9. The two BW modes remain well-separated and therefore do not display any interaction.

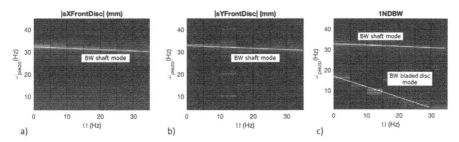

a)   b)   c)

Figure 11. - Response to BW excitation from PEAs from PEAs a) shaft at front disc in x and b) y direction and c) the 1ND BW bladed disc response.

4   CONCLUSIONS

A unique test rig has been designed by the Dynamics Group at Imperial College London dedicated to the study of the interaction of long flexible blades with shorter, stiffer shafts, which can lead to blade-shaft coupling or cross-shaft coupling. The test facility also provides the capability to study flexible symmetric and asymmetric bearing supports and has the ability to excite the rotating components synchronously via out of balance masses, asynchronously via piezo stack exciters, and with a variety of ND patterns via piezo patch exciters on the blades.

First experimental testing of the ARES facility has demonstrated the ability to excite synchronous and asynchronous rotor response, highlighting a strongly coupled dynamic behaviour between a simulated shaft excitation, the main shaft and its mounted bladed disc.

ACKNOWLEDGEMENT

Thanks to Rolls-Royce plc and the EPSRC for the support under the Prosperity Partnership Grant "Cornerstone: Mechanical Engineering Science to Enable Aero Propulsion Futures", Grant Ref: EP/R004951/1 for supporting this work and allowing to publish its outcomes.

REFERENCES

[1] Haslam A.H., Schwingshackl, C.W., Rix, A.I.J., "Influence of a rolling-element bearing on the non-linear rotor response using generalised harmonic balance".Non-linear Dynamics, Online Mar 2020, https://doi.org/10.1007/s11071-020-05470-4.

[2] Haslam A.H., Schwingshackl C.W., Rix A.I.J., "Analysis of the Dynamic Response of Coupled Coaxial Rotors", Proceedings of the International Modal Analysis Conference XXXVI, Orlando, FL, Jan. 2018.

[3] Tuzzi, G. Schwingshackl, C.W, Green, J.S., "Shaft bending to Zero Nodal Diameter disc coupling effects in rotating structures due to asymmetric bearing supports", Proceedings of the International Modal Analysis Conference XXXVIII, Houston, TX, Feb. 2020.

[4] Ruffini, V., Schwingshackl, C.W., Green, J.S., "Modal analysis of rotating structures under MIMO random excitation", Proceedings of VIRM 2016, Manchester, Sept. 2016.

[5] Szydlowski, M.J., Schwingshackl, C.W., Rix, A., "Distributed acquisition and processing network for experimental vibration testing of aero-engine structures", Proceedings of the International Modal Analysis Conference XXXVIII, Houston, TX, Feb. 2020.

Experimental investigation of non-linear stiffness behaviour of a rolling-element bearing

A. Haslam[1], C.W. Schwingshackl[1], L. Muscutt[1], A. Rix[2], M. Price[2]

[1]Imperial College London, UK
[2]Rolls-Royce Plc., UK

ABSTRACT

Rolling-element bearings are used extensively to support shafts in rotating machines due to their low friction and high load capacities. They are known to be inherently non-linear in nature due to clearance and stiffening nonlinearities, which depend on many parameters which are difficult to quantify, such as the radial clearance. As a result, the nonlinear behaviour of the bearings is challenging to predict and consequently the bearings are regularly modelled as linear springs in rotordynamic analyses. With ever more flexible rotor designs emerging, a need for more physical bearing models arises to enable advanced non-linear rotordynamic simulations, and hence an improved understanding of the non-linear bearing behaviour is required. For this purpose, a new static test rig known as the BEaring LOading System (BELOS) has been developed at Imperial College London which allows some of these unknown bearing parameters to be identified.

The design of the rig allows accurate quantification of the nonlinear deflection curves, including zero-stiffness effects during bearing clearance and increasing stiffening behaviour under higher loads. It was demonstrated that an experimental setup can measure the bearing stiffness to a good level of accuracy. Some key bearing parameters were identified by tuning the numerical model to match the experimental results, ready for use in a nonlinear rotordynamic analysis.

1   INTRODUCTION

Rolling-element bearings are used extensively to support shafts in rotating machines due to their low friction and high load capacities [1]. They are a key source of both flexibility and damping in rotor systems, and therefore have a large influence on the response [2], [3]. They are known to be inherently nonlinear in nature, primarily due to clearance [4], [5] and the non-linear Hertzian contacts [6], [7]. These depend on parameters which are highly uncertain, such as the radial clearance, making it difficult to predict the nonlinear behaviour of the overall rotor-bearing system. The current approach is often to simply model the bearings as linear springs [2], [3], but with the emergence of ever lighter and more flexible rotor designs, this fails to capture certain phenomena observed in real turbo-machinery such as jump phenomena [8] and chaotic responses [9], [10]. There is therefore a need to better understand the influence of these bearing non-linearities on the dynamics of rotors, and to develop improved high-fidelity bearing models.

A project has been underway to specifically investigate the influence of a rolling-element bearing on the response of a simplified rotor, using the Asynchronous Rotor Excitation System (ARES) test rig at Imperial College London [11]. It consists of a single shaft with two balancing discs, supported by 2 self-aligning bearings. The rotor can be excited synchronously by an out-of-balance mass on one disc, or asynchronously by piezo-electric actuators connected to the front support.

411

However, since the ARES test rig is complex with many components, it was decided that it was necessary to first validate the bearing model in a simplified setting. To this end, a separate static bearing test rig was developed, which is known as the BEaring LOading System (BELOS). This allows the non-linear bearing stiffness to be accurately measured, and some unknown bearing parameters to be identified.

2   EXPERIMENTAL SETUP

The BELOS test rig consists of a SKF 2208-EKTN9 Self-Aligning bearing, which is fitted into a rigid steel housing block using a J7 ISO tolerance, as shown in Figure 1. The bearing is mounted in a horizontal orientation, so that the bearing is not axially loaded. The steel housing block is bolted to an interface plate using $8 \times M12$ bolts, which is in turn bolted down firmly to a test bed using T-slots and $12 \times M10$ bolts.

A short 35 mm OD shaft runs through the bearing and is fixed with an SKF H308E adaptor sleeve. This device allows the internal clearance within the bearing to be varied, depending on how many turns it is tightened. The adaptor sleeve was tightened up using a torque wrench to a constant 34 Nm, to ensure the bearing clearance was consistent across the tests. Since the bearing was self-aligning, the shaft was free to tilt around the x and y-axes (shown in Figure 1). The shaft was also not constrained against rotation around the z-axis, but the small amount of friction in the bearing was found to be adequate to prevent this.

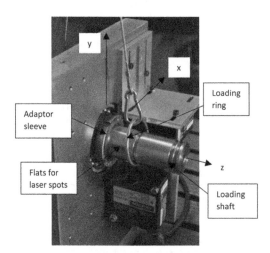

Figure 1. BELOS bearing housing.

Loads are applied to the bearing using known masses loaded onto a cable-pulley system via a short shaft running through the bearing. Loads can be applied in either the vertical or horizontal directions, and the shaft displacements are measured with laser displacement sensors. The acquired data allow force-displacement curves to be plotted, and the non-linear bearing stiffness to be measured. These systems will be discussed in more detail in the following sections.

2.1   Displacement sensors
The shaft displacement is measured by horizontally and vertically aligned probes on each side of the bearing, so that the displacement in both orthogonal directions was measured simultaneously. Keyence LK-H022 laser displacement probes were used with a 0.01 μm repeatability. These were connected to a Keyence LK-G5001 controller, which interfaced with the PC over USB. Since the laser displacement sensors are designed to work on flat surfaces, the laser spots reflect off smooth flats machined into the shaft, which maximises the accuracy of the measurements.

Since the displacements were measured on both sides of the bearing, it was possible to resolve the shaft translation in the horizontal (x) and vertical (y) directions removing the effect of any rotation around these axes. For example, in the horizontal (xz) plane, the following transformation was applied:

$$\delta_x = \frac{\delta_x^{front} + \delta_x^{rear}}{2} \tag{1}$$

$$\theta_y = \frac{\delta_x^{front} - \delta_x^{rear}}{2l_{laser}} \tag{2}$$

where δ_x is the horizontal shaft deflection, θ_y is the rotation around the y-axis. A similar transformation was applied for the data in the vertical (yz) plane.

2.2   Loading system
The bearing is loaded via steel rings encircling the shaft, which sit within machined grooves. This ensured that there was a single contact on the shaft and no torque could be applied, minimising any shaft rotation. Pre-calibrated weights applied loads to these rings via a cable-pulley system. The loading system was designed to distribute loads evenly between each side of the bearing, to minimise any tilting motion.

Table 1. BELOS rig parameters.

Parameter	Symbol	Value
Laser separation	l_{laser}	40 mm
Shaft mass	m_{shaft}	1.57 kg

2.3   Shaft compliance compensation
Although the shaft displacement was measured very close to the bearing using the laser probes (see Table 1), the compliance of the shaft could not be neglected. The bearing stiffness was predicted to be of the order $\sim 10^8$ N/m, whereas the shaft compliance was of the order $\sim 10^9$ N/m. This means there is a $\sim 10\%$ discrepancy between the deflection measured by the laser and the true bearing deflection, as depicted in Figure 2. This effect was compensated for in post-processing.

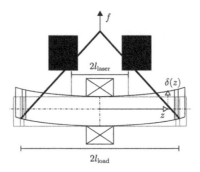

Figure 2. Shaft loading schematic.

The shaft stiffness k_{shaft} seen at the laser was derived from beam theory, by considering the equivalent cantilevered beam model shown in Figure 3.

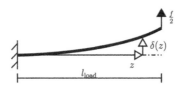

Figure 3. Equivalent shaft model using beam theory.

The lateral deflection along the beam at a given axial location is given by [12]:

$$\delta(z) = \frac{f}{12EI}z^2(3l_{load} - z)$$

where f is the total applied load, which is assumed to be split evenly between each side of the bearing. The deflection at the laser can be found by substituting in $z = l_{laser}$:

$$\delta(l_{laser}) = \frac{f}{12EI}l_{laser}^2(3l_{load} - l_{laser})$$

which can then be rearranged to yield the shaft stiffness:

$$k_{shaft} = \frac{f}{\delta(l_{laser})} = \frac{12EI}{l_{laser}^2(3l_{load} - l_{laser})} \tag{3}$$

The shaft compliance can then be compensated for by introducing a simple load-dependent correction, to give the bearing deflection $\delta_{bearing}$:

$$\delta_{bearing} = \delta(l_{laser}) - \frac{f}{k_{shaft}} \tag{4}$$

where $\delta(l_{laser})$ is the shaft deflection from the laser displacement probes, after removing the effect of any shaft rotation using (1). This correction was applied to both the vertical and horizontal deflections.

2.4 Vertical loading configuration

The bearing was initially loaded vertically, since this was the simpler test setup, as shown in Figure 4. The rig was only able to apply upward loads to the bearing in this configuration, so that only the elements at the top of the bearing were in contact and loaded, while the elements at the bottom of the bearing remained unloaded. Therefore, the bearing never crossed the dead-band due to its internal clearance, and it was not possible to measure this parameter from these tests. However, it was possible to apply higher loads, thereby allowing any change in bearing stiffness with load to be measured accurately.

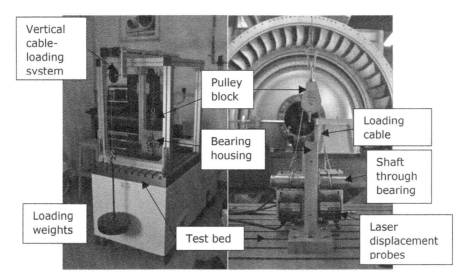

Figure 4. Vertical loading configuration.

Instead of directly connecting both sides of the bearing to the main loading cable, an intermediary pulley block was used. This ensures that an equal load was applied to each side. The pulley was kept well-greased to minimise the influence of friction. Before each round of tests, a large load was initially applied and removed to allow the rig to settle into its natural equilibrium position. The bearing was loaded from an initial 10 N, which slightly exceeded the weight of the pulley and shaft, up to a maximum load of 1300 N in 50 N increments. One complete cycle of loading and unloading was completed in each test, so that any hysteresis could be measured. This test was repeated many times to be able to quantify the precision of the measurements.

The load applied to the bearing in the vertical direction is not in fact the same as the load applied to the loading cable. This is because a certain mass is required to offset the weight of the pulley and shaft. This can easily be handled by adding an offset to the applied load:

$$f_y^b = f_{pulley} - \left(m_{shaft} + m_{pulley}\right)g \tag{5}$$

where the shaft mass is shown in Table 1, and the pulley mass in Table 2. The value of the effective shaft stiffness from (3) was pre-computed and is shown in Table 2.

Table 2. Vertical loading parameters.

Parameter	Symbol	Value
Cable loading position	l_{lshaft}	110 mm
Pulley mass	m_{pulley}	0.367 kg
Shaft stiffness	k_{shaft}	884 MN/m

2.5   Horizontal loading configuration

The rig setup was then adjusted to load the bearing in the horizontal direction. Since cables were connected to both sides of the bearing, it was possible to apply a net load to the bearing in either direction, unlike in the vertical tests. As a consequence of this, it is possible to directly observe any internal clearance in the bearing. However, there is a constant vertical load due to the shaft weight in these tests, which complicates the analysis.

In this configuration, beams were used to distribute the load between the front and rear of the bearing, as shown in Figure 5. Beams were used instead of a pulley block, in order to reduce the weight of the loading system, which would otherwise cause the loading cable to sag. The relevant parameters are shown in Table 3.

A constant 10 N load was applied to each side of the bearing to maintain tension in the cables. Additional masses were then hung from one side of the bearing from 0 N up to 1300 N in increments of 50 N, and then gradually unloaded down to 0 N. The masses were then applied in the same way to the other side of the bearing, so that the direction of loading was reversed, thus completing the loading cycle. This whole process was repeated many times as with the vertical loading tests.

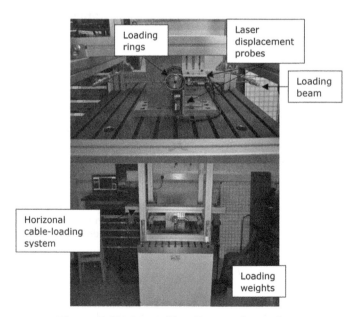

Figure 5. Horizontal loading configuration.

In this configuration, it was not necessary to offset the loads applied by the weights to compute the bearing loads, since the weight of the loading beams has little influence on the tension in the cables. However, the weight of the shaft was always present leading to a constant vertical load of $f_y = m_{shaft}g$, which must be included in the numerical model in addition to the horizontal load f_x applied by the cable system in order to accurately replicate the test conditions.

Table 3. Horizontal loading parameters.

Parameter	Symbol	Value
Cable loading position	l_{lshaft}	200 mm
Shaft stiffness	k_{shaft}	425 MN/m

As with the vertical loading case, the value of the effective shaft stiffness from (3) was pre-computed and is shown in Table 3. Note that the value has changed since the horizontal loads were applied further along the shaft, so that l_{shaft} had a different value.

3   BEARING MODEL

The experimental results were used to parameterise a numerical model implemented in MATLAB. This model was based upon the widely-used quasi-static approach introduced by Jones [13]. For the sake of brevity, this will not be derived in this paper, and the reader is referred to Lim et al. [6]. The baseline model was generated with the geometric parameters in Table 4 which were obtained from the SKF website [14]. It should be noted that since this bearing has many elements, the bearing stiffness is insensitive to rotation angle, so the so-called 'variable compliance' effect was negligible.

Table 4. SKF 2208-EKTN9 known bearing parameters [14].

Parameter	Symbol	Value
Number of balls	Z	36
Bearing width	B	23 mm
Ball diameter	D	8.73 mm
Distance between ball centres	d_m	60.1 mm
Contact angle	α_0	10.8°
Axial separation between rows	z_0	5.75 mm

However, some other key bearing parameters are unknown. For example, the contact loads Q are related to the contact deflections δ by the following non-linear Hertzian contact law [1], [6]:

$$Q = K\delta^n$$

where the constant K is the combined Hertzian contact stiffness coefficient of the inner and outer race contacts, and the exponent $n = 3/2$ for the circular contacts in a ball bearing. The stiffness coefficient K is not typically supplied by the manufacturer, although an initial estimate can be computed by assuming generic material properties of steel.

The bearing clearance also has a large influence on the bearing stiffness. This is typically expressed in terms of a radial clearance value c_r. Although manufacturers sometimes supply typical values for the clearance, it will also depend on many other variables such as the bearing housing tolerance and how much the adapter sleeve is tightened. Therefore, to get an accurate value, it needs to be measured in situ.

In order to fully parameterise the bearing model, both the Hertzian contact stiffness coefficient K and the radial bearing clearance c_r were extracted from the experimental data. This process will be discussed in the following section.

4   RESULTS

4.1   Vertical loading

4.1.1 *Stiffness characteristic*
The results from the vertical loading setup are shown in Figure 6. The raw laser displacements were converted into shaft displacements using (1) and (2), and shaft-compliance was compensated for using (4). The load from the weights were offset using (5) to compute the loads applied to the bearing.

The displacements from each of the 10 individual tests were initially offset slightly from each other, since the zero point would drift between tests, so the shaft would never start from *exactly* the same position. Each test was therefore offset in post-processing so that they would align at high loads where the behaviour was more consistent. There is a strong stiffening characteristic with increasing load, demonstrating that the bearing is non-linear. The bearing is also very stiff, since the maximum observed deflection was only 10 µm.

There are many points where the load is very low, which is where the bearing is *just* in contact. However, since the bearing was not loaded downwards, these points do not necessarily cover the whole dead-band region due to clearance. Unfortunately, this meant that it was not possible to extract the radial clearance from these points.

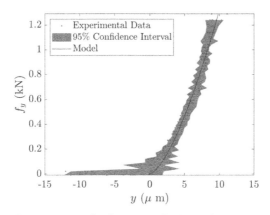

Figure 6. Results from vertical loading tests.

4.1.2 *Experimental uncertainty*

The experimental uncertainty was estimated by computing the standard deviation of the bearing deflection at each load level, and assuming a normal distribution. This is shown as the shaded region in Figure 6. The uncertainty is quite low at high loads, but very large at the lowest loads, where the bearing just starts to become loaded. This was found to be due to the shaft titling and rotating slightly during loading, leading to different initial starting conditions for each test. For a different bearing such as a deep-groove bearing which would provide some constraint against titling, it would be expected that this uncertainty would be reduced.

4.1.3 *Comparison to model*

The unknown Hertzian contact stiffness coefficient K in the bearing model was tuned by minimising the least-squares error with the experimental data, which yielded a value of $K = 5.0E9$ Nm$^{-3/2}$. This line is also plotted on Figure 6.

The model and experimental data in Figure 6 agree well at higher loads, which is also the region where the experimental uncertainty was lowest. However, there is larger scatter at lower loads, where the model appears to stiffen more rapidly than the experimental data, indicating a softer transition. It was found that varying the Hertzian contact stiffness coefficient did not improve the agreement any further in this region, indicating that the model was not able to capture all the features of the bearing loading. One explanation could be that imperfections in the bearing races caused some elements to become loaded sooner than the idealised case assumed in the model. Also, the model neglects friction, and this may not be a valid assumption at such low loads. Lastly, the model is also quasi-static, so it is assumed that elements reach their equilibrium position, but this assumption is likely to be less valid in a static non-rotating test.

4.2 Horizontal loading

4.2.1 *Stiffness characteristic*

The load-displacement relation for this loading configuration, after post-processing the deflections using (1), (2) and (3), is plotted in Figure 7. Since the deflections were so small, the zero offsets on the laser probes would drift between tests, since the laser spot would sit at a slightly different point on the shaft. To counteract this, the displacements from each test were offset by the average displacement for $|f_x| < 50$ N, so that they would be centred around $x = 0$ μm.

The radial bearing clearance is immediately obvious, leading to a dead-band around $x = 0$ μm with a width of around ~8 μm where the stiffness is very low. Outside of these regions, there is clear stiffening characteristic which is approximately symmetric as the bearing is loaded in either direction. The horizontal stiffness is also very similar to the vertical stiffness, since the bearing displaces from ~4 μm at the edge of clearance to ~14 μm at a load of 1300 N, which is similar to the ~10 μm deflection observed in the vertical loading case. This is as expected, since the bearing should be approximately axisymmetric.

4.2.2 *Experimental uncertainty*
The experimental uncertainty was estimated in the same way as for the vertical loading. It can be observed that the experimental uncertainty is much higher in the case of horizontal loading, even for the same applied loads. Unlike in the vertical loading case, the bearing was unloaded twice within one test, which allows the shaft to tilt very slightly which was found to reduce the accuracy of the results, and explains why the uncertainty is particularly high around $f_x = 0$ N. This means the bearing clearance can only be identified to a precision of around ~1 μm at best. This highlights that the vertical loading configuration is a more accurate setup.

4.2.3 *Comparison to model*
The results from the updated numerical model are overlaid on the measured data in Figure 7, after tuning the radial clearance to $c_r = 4.1$ μm to minimise the error with the experimental data in a least-squares sense. The Hertzian contact stiffness coefficient identified from the vertical loading configuration in Section 4.1 was retained. Good agreement with experimental data was obtained, with the model predicting the same clear dead-band region where the bearing is in clearance. It can be observed that stiffness in the clearance region is non-zero in both the experimental data and from the model. This is due to the presence of a small vertical load from the shaft weight. Outside of this region, the bearing stiffness rises rapidly as the shaft moves away from the central position.

It is interesting to note the experimental data shows a much sharper change in horizontal stiffness (in either direction) as the bearing starts to become loaded than in the vertical loading case (as plotted in Figure 6). The agreement was found to be poor in this region previously, but since the agreement is now much improved in the case of horizontal loading, there is evidence that the very soft transition observed in the case of vertical loading may just be due to experimental uncertainty.

On the other hand, the experimental results from the horizontal loading case in Figure 7 show some slight asymmetry, with a slightly higher stiffness for loading in the positive direction than in the negative direction. This could be due to the bearing being slightly anisotropic due to manufacturing tolerance, or that the balls have failed to settle into their quasi-static equilibrium position. However, the asymmetry could also stem from the fact that separate loading systems were used to load the bearing in each direction.

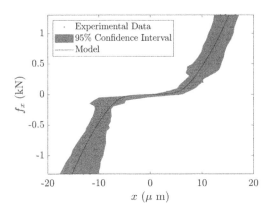

Figure 7. Results from horizontal loading tests.

4.3   Summary

The BELOS test rig has allowed two key unknown bearing parameters to be identified, which have been summarised in Table 5 below. The resulting bearing model provides a good agreement with the experimental results. However, the clearance could only be identified to a lower precision, due to experimental uncertainty.

Table 5. Identified bearing parameters.

Parameter	Symbol	Value
Radial clearance	c_r	4.1 μm
Hertzian contact stiffness coefficient	K	5.0E9 Nm$^{-3/2}$

5   CONCLUSIONS

The stiffness of a bearing was successfully measured using a relatively simple, static test rig. It was shown that the bearing stiffness is heavily non-linear, with a dead-band at low loads due clearance, and a strong stiffening characteristic outside this region, due to the Hertzian contacts.

Although the experimental uncertainty was quite high at lower loading levels, it was found that this was partly due to the low tilting stiffness of the self-aligning bearing. This would likely be less problematic for other types of bearing. It was found that the vertical loading test had a lower uncertainty at higher loads and was useful to measuring the stiffening characteristic. However, the horizontal loading was more useful for identifying the clearance in the bearing, since the loading was bi-directional. The rig could therefore be improved by allowing down-ward vertical loads, which is hoped would allow the clearance to be more accurately identified.

The obtained results were used to parameterise a quasi-static bearing model. It was possible to extract the Hertzian contact stiffness coefficient and radial clearance by tuning the model to match the experimental results, and the agreement of the updated model with the data was very good. The sensitivity to the adaptor sleeve torque will now be investigated, which should allow the radial clearance in the bearing to be varied.

REFERENCES

[1] T. A. Harris and M. N. Kotzalas, *Rolling Bearing Analysis: Essential Concepts of Bearing Technology*, 5th ed. Boca Raton, FL: CRC Press, 2006.

[2] M. I. Friswell, J. E. T. Penny, S. D. Garvey, and A. W. Lees, *Dynamics of Rotating Machines*, 1st ed. Cambridge University Press, 2010.

[3] A. Muszynska, *Rotordynamics*, 1st ed. Boca Raton, FL: CRC Press, 2005.

[4] D. W. Childs, "Fractional-Frequency Rotor Motion Due to Nonsymmetric Clearance Effects," *J. Eng. Power*, vol. 104, no. 3, p. 533, 1982.

[5] Y. B. Kim and S. T. Noah, "Bifurcation analysis for a modified Jeffcott rotor with bearing clearances," *Nonlinear Dyn.*, vol. 1, no. 3, pp. 221–241, May 1990.

[6] T. C. Lim and R. Singh, "Vibration Transmission Through Rolling Element Bearings. Part I: Bearing Stiffness Formulation," *J. Sound Vib.*, vol. 139, no. 2, pp. 179–199, 1990.

[7] T. A. Harris and M. N. Kotzalas, *Advanced concepts of bearing technology: Rolling bearing analysis*, 5th ed. Boca Raton, FL: CRC Press, 2006.

[8] S. Saito, "Calculation of Nonlinear Unbalance Response of Horizontal Jeffcott Rotors Supported by Ball Bearings With Radial Clearances," *J. Vib. Acoust. Stress Reliab. Des.*, vol. 107, no. 4, pp. 416–420, 1985.

[9] A. Kahraman and R. Singh, "Non-linear Dynamics of a Geared Rotor-bearing System with Multiple Clearances," *J. Sound Vib.*, vol. 144, no. 3, pp. 469–506, 1991.

[10] M. Tiwari, K. Gupta, and O. Prakash, "Effect of Radial Internal Clearance of a Ball Bearing on the Dynamics of a Balanced Horizontal Rotor," *J. Sound Vib.*, vol. 238, no. 5, pp. 723–756, 2000.

[11] C. W. Schwingshackl, et al.., "Asynchronous Rotor Excitation System (ARES) – A new rotor dynamic test facility at Imperial College London," in *Vibrations in Rotating Machinary*, 2020.

[12] J. Gere, *Mechanics of Materials*, 6th ed. Cengage Learning, 2003.

[13] A. B. Jones, "A General Theory for Elastically Constrained Ball and Radial Roller Bearings Under Arbitrary Load and Speed Conditions," *J. Basic Eng.*, vol. 82, no. 2, p. 309, 1960.

[14] AB SKF, "2208 EKTN9 Self-Aligning Ball Bearing." [Online]. Available: https://www.skf.com/uk/products/rolling-bearings/ball-bearings/self-aligning-ball-bearings/productid-2208 EKTN9. [Accessed: 22-Aug-2019].

An optimal frequency band selection for bearing fault diagnosis based on squared envelope analysis

L. Xu, S. Chatterton, P. Pennacchi

Department of Mechanical Engineering, Politecnico di Milano, Italy

ABSTRACT

The squared envelope spectrum (SES) is widely used to rolling element bearing (REB) fault diagnosis since it is simple and effective. Generally, the performance of SES depends on the bearing fault signal that separated from the raw signal. Bandpass filtering is a commonly used approach to separate different signal components into different frequency bands. Both the centre frequency and bandwidth affect the performance of the separation. In order to separate the bearing fault signal into a frequency band well, several bandwidths are defined in advance and the distribution of the centre frequency promises no blind span throughout the entire frequency range. The kurtosis of the squared envelope (SE) and SES are usually used to select the filtered signal that contains most of the fault information. Most of the present approaches assess the kurtosis of the entire SE or SES which are easy to be interfered with by a single or a few impulses in the SE or SES. A method based on the kurtosis of the SES is proposed, which only evaluates the kurtosis of the part of SES that is closely connected to the bearing fault. Therefore, the interference of a single or several undesired impulses in the SES can be avoided. The performance of the method is tested via simulation signals and real vibration data.

1   INTRODUCTION

It is widely organized that rolling element bearing is a critical component of a machine and prone to damage. Many machines break down are caused by the bearing damage. Since fatigue is the main reason of bearing deterioration, therefore, usually, the defect development of the bearing is a very complex process and last a relatively long period. If it is viable to detect and identify the bearing defect at an earlier stage, there would be enough time to prepare maintenance activities and avoid serious accidents.

Vibration-signal based bearing fault diagnosis is the main branch of bearing fault diagnosis and attracted the attention of many researchers (1,2). Since the strength of the component in the vibration signal caused by the incipient defect on the bearing is very weak, the work of separating this weak component from the original vibration signal determines the performance of the method to a large degree. Commonly used signal separation methods include bandpass filtering, wavelet packet transform, empirical mode decomposition, Wiener filtering. For bandpass-filtering based methods, two key parameters, centre frequency and bandwidth, control the performance of them. Some innovative approaches have been proposed to obtain the optical centre frequency and bandwidth, for instance, binary tree approach (3), fixed bandwidth of three times of the fault frequency approach (4), several predefined frequency bands approach (5). Even though these approaches can obtain a relative optimal centre frequency and bandwidth in some cases, they still have some drawbacks. For example, the binary tree approach, its performance will be affected if the optimal centre frequency is close to the defined boundaries. While the fixed bandwidth of three times of the fault frequency approach is easy to contain too much noise in some cases if the fault frequency is high. About the predefined frequency bands approch, it needs to select and determine the frequency bands for every case in advance based on the frequency spectral. Another common shortcoming of the present bandpass filtering approaches occurs in the process of selecting the optimal frequency band according to the spectral kurtosis of the filtered signal. The entire spectral is used to calculate the kurtosis which is easy to be interfered with by some undesired components in the signal.

To overcome these drawbacks, a bandpass frequency method is proposed in this paper. In this method, several frequency bandwidths are determined by the fault frequency and the centre frequency is correspondingly determined by the bandwidth. The determined frequency bandwidths vary from 1.5 times to 3.0 times the fault frequency. The corresponding centre frequencies have no blind zone. Meanwhile, only the part of SES of the filtered signal that is close to the first two harmonics of the fault frequency is used to calculate the kurtosis. It can release the interference of the undesired impulse in the SES.

2   THE ALGORITHM OF THE METHOD

The algorithm flowchart of the method is shown in Figure 1. There are four key steps of this method, including discrete the deterministic and random components, bandpass filtering, spectral kurtosis calculation, and PMFSgram display.

Step 1: Discrete the deterministic and random components. Since there is a slight slip between the roller and the races, the fault frequencies of the bearing are a bit different from the theoretical value. This introduces some random characteristic to the signal component caused by the local bearing defect. Therefore, the signal component caused by the local bearing defect is different from the deterministic signal component caused by gear meshing or shaft rotation. The goal of this step is to separate the two kinds of components according to their different characteristics. Widely used deterministic and random signal separating methods including time synchronous averaging (TSA), linear prediction, self-adaptive noise cancellation, discrete/random separation, and cepstral method are compared in (6). In this paper, TSA is adopted because its good ability and easy to carry out.

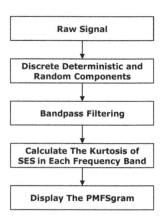

Figure 1. The Flowchart of the Method.

Step 2: Bandpass filtering. This step and the next step are the core of the method and determine the performance of this method. This step is to construct the frame of setting the frequency band and the schematic diagram is shown in Figure 2. The FIR filter is adopted because of its stability and easy implementation. Since it is difficult to detect the higher harmonics of the fault frequency in the SES when the SNR is low, in this method the maximum frequency bandwidth is three times the fault frequency. The bandwidth is an arithmetic progression and the step size is 0.3 times the fault frequency $\mathbf{f}a$, as follows:

$$\mathbf{bw}_i = (\mathbf{1.2} + i \cdot \mathbf{0.3})\mathbf{f}_a \tag{1}$$

where i = 1, 2, 3, …, 6, is the serial number of the six frequency bandwidths

The number of frequency bands $\mathbf{n}_c$ for each frequency bandwidth is the same. Then, the step size $\Delta \mathbf{fc}_i$ for each frequency bandwidth can be obtained as follows:

$$\Delta \mathbf{fc}_i = (\mathbf{F}_s/\mathbf{2} - \mathbf{bw}_i)/\mathbf{nc} \tag{2}$$

where $\mathbf{F}_s$ is the sampling frequency. The selected $\mathbf{nc}$ should make sure $\Delta \mathbf{fc}_1$ is smaller than bw_1.

Therefore, the centre frequency for each frequency band can be obtained as follows:

$$\mathbf{fc}_{ij} = (j - \mathbf{1})\ \mathbf{bw}_i + \mathbf{0.5bw}_i \tag{3}$$

where j = 1, 2, 3, …, $\mathbf{nc}$, is the serial number of the frequency band for each frequency bandwidth

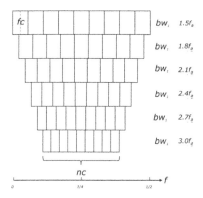

Figure 2. The schematic diagram of the bandpass filtering.

Step 3: Calculate the kurtosis of SES. Unlike most of the presented methods, in this paper, only part of the SES but not the entire SES is adopted to evaluate the kurtosis. Because it is difficult to identify the third or higher harmonic of the fault frequency and it is reasonable to make a conclusion of existing defect on the bearing if there are high amplitudes at the first two harmonics of the fault frequency in the SES. Therefore, only two segments of SES, shown in Figure 3, that is close to the first two harmonics of the fault frequency is adopted to obtain the kurtosis of SES. The two segments are centred at the corresponding fault frequency harmonic with the width of three times of the shaft rotation frequency $\mathbf{f}_r$. The SES of an arbitrary filtered signal using the frequency band shown in Figure 2 can be obtained as follows (7):

$$\mathbf{K}(\mathbf{f}_c,\ \mathbf{b}_w) = \ <(|\mathbf{SES}_{f_a}[\mathbf{n}]| - \ < |\mathbf{SES}_{f_a}| >)^4 > / <(|\mathbf{SES}_{f_a}[\mathbf{n}]| - \ < |\mathbf{SES}_{f_a}| >)^2 >^2 \tag{4}$$

where $\mathbf{f}_c$ is the centre frequency of one frequency band shown in Figure 2, $\mathbf{SES}_{fa}$ is the selected squared envelope spectrum by the way shown in Figure 3.

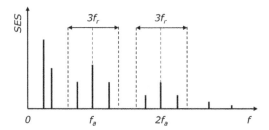

Figure 3. The selected squared envelope spectrum.

While the SES can be obtained by the Fourier transform of the squared envelope of the signal and the SE of the signal can be obtained through the analytical form of the signal by the Hilbert transform, as follows (8):

$$\mathbf{SE[n]} = |\mathbf{x[n]} + \mathbf{j} \cdot \mathbf{Hilbert}(\mathbf{x[n]})|^2 \tag{5}$$

$$\mathbf{SES} = |\mathbf{FFT(SE)}| \tag{6}$$

where **j** is the imaginary unit.

3   ACTUAL VIBRATION SIGNAL TEST

In order to test the performance of the method, real vibration signal of bearings with artificial defect obtained from a full-scale test rig of a high-speed train traction system is used. The overview of the test rig and the schematic diagram of the motor and gearbox are shown in Figure 4. The tooth number of the gear on the input shaft and output shaft are 26 and 85 respectively. More details of the test rig can be obtained in the reference (9). The damaged bearing is a tapered roller bearing BG3-GP5 on the output shaft with an artificial defect on the outer race, as shown in Figure 5. The parameters of the tested bearing are shown in Table 1.

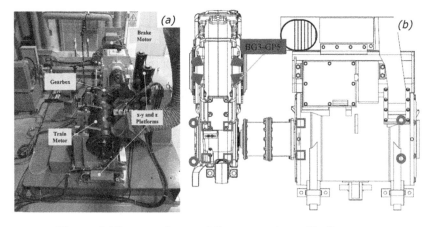

Figure 4. The overview and the core schematic diagram.

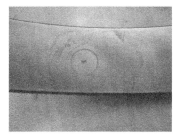

Figure 5. The outer ring of the bearing BG3-GP5.

Table 1. Parameters of the tested bearing BG3-GP5.

Bearing type	Damage position	Ball pass frequency of outer race (BPFO) in the order domain/NX
FAG-804989	Outer Race	18.4

The input speed of the motor is about 83 Hz and the corresponding torque is about 405 Nm. Both vibration acceleration signal and tacho signal are collected with the same sampling frequency of 20 kHz. The length of each signal is 5 seconds with 10^5 samples. The raw vibration signal is shown in Figure 6. Since order tracking can release the effect of the shaft rotation speed variation, the raw signal has been order-tracked using the tacho signal.

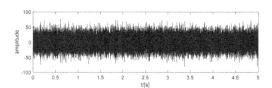

Figure 6. The raw vibration signal.

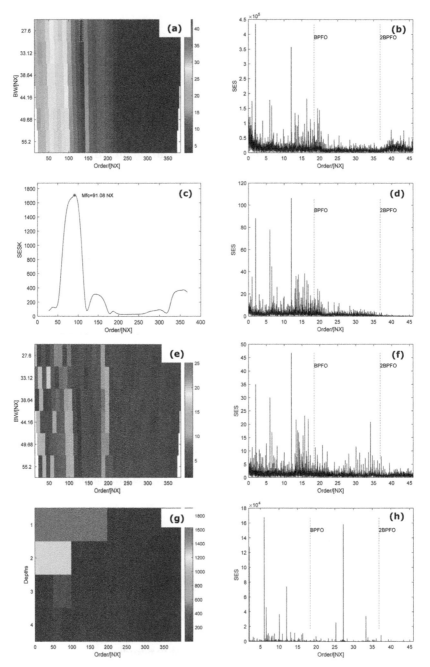

Figure 7. – Analysis results: (a) (b) PMFSgram and the SES of its optimal frequency band; (c) (d) protugram and the SES of its optimal frequency band; (e) (f) PMgram and the SES of its optimal frequency band; (g) (h) The enhanced kurtogram and the power spectrum of its optimal frequency band.

The total number of frequency bands of each bandwidth is 35 for the PMFSgram in this case study and it also adopted by the following PMgram. It is shown in Figure 7(a) that the optical frequency band (green rectangle) selected by the PMFSgram is [118.86, 146.46] NX. The corresponding SES of the filtered signal is shown in Figure 7(b) which has an obvious peak value at BPFO. It indicates the existence of the defect on the outer race. However, the optimal centre frequency selected by the protugram is 91.08 NX, shown in Figure 7(c). The corresponding SES of the filtered signal is shown in Figure 7(d) and there is no obvious peak value at the first two harmonics of BPFO. It means failed to detect the defect on the outer race. The SES of the filtered signal using the optimal frequency band ([32.42, 60.02] NX) selected by the PMgram (see Figure 7(e)) is shown in Figure 7(f), which also has no clear peak value at the first two harmonics of BPFO. It means the defect identification is unsuccessful as well. The SES of the sub-signal at the 4th level (red rectangle) obtained by the enhanced kurtogram (see Figure 7(g)) (10) is shown in Figure 7(h), which is also failed to detect the defect on the outer race.

4   DISCUSSION & FURTHER WORK

The work in this paper provides a useful method for frequency band selection when detecting the bearing defect using SES. The paper mainly focuses on two aspects that are frequency band construction and SES evaluation. The strengths of the method are also shown in the two aspects. In the frequency band construction process, the proposed frequency band building approach can obtain a relatively good frequency band and avoid the drawbacks of most of the existing methods. In the SES kurtosis evaluation process, only part of the SES that is closely connected to the fault frequency is used to calculate the kurtosis, which can avoid the interference of some impulses that are far away from the harmonic of the fault frequency.

A remaining problem of this method is that it needs to know the fault type in advance, otherwise, it needs to test one time for all four fault types. In future work, if it is possible to synthesize the optimal frequency band for each fault type in one PMFSgram would make this method more convenient for bearing fault diagnosis. This method also has the potential to filter useful signals for bearing condition monitoring.

5   CONCLUSIONS

The frequency band selection method in this paper takes advantage of the new frequency band construction approach and the new SES kurtosis assess approach. The frequency construction approach has no centre frequency blind zone and the SES kurtosis approach can avoid the interference of the impulses that far away from the fault frequency harmonics. It can be used to detect the incipient fault on the bearing. The performance and superiority of the method have been validated by actual vibration data from a test rig.

REFERENCES

[1] Rai, A., and S.H. Upadhyay, "A review on signal processing techniques utilized in the fault diagnosis of rolling element bearings", Tribology International, 2016.
[2] Greaves, M., F. Elasha, J. Worskett, D. Mba, H. Rashid, and R. Keong, "Vibration Health or Alternative Monitoring Technologies for Helicopters", 2012, pp. 187.

[3] Antoni, J., "Fast computation of the kurtogram for the detection of transient faults", Mechanical Systems and Signal Processing 21(1), 2007, pp. 108–124.

[4] Barszcz, T., and A. Jabłoński, "A novel method for the optimal band selection for vibration signal demodulation and comparison with the Kurtogram", Mechanical Systems and Signal Processing 25(1), 2011, pp. 431–451.

[5] Borghesani, P., S. Chatterton, P. Pennacchi, and A. Vania, "A novel threshold for the diagnostics of rolling element bearings", Proceedings of the ASME Design Engineering Technical Conference 8, 2014, pp. 1–8.

[6] Randall, R.B., N. Sawalhi, and M.D. Coats, "A comparison of methods for separation of deterministic and random signals", The International Journal of Condition Monitoring 1(1), 2011, pp. 11–19.

[7] Xu, L., S. Chatterton, and P. Pennacchi, "A novel method of frequency band selection for squared envelope analysis for fault diagnosing of rolling element bearings in a locomotive powertrain", Sensors (Switzerland), 2018.

[8] Xu, L., P. Pennacchi, and S. Chatterton, "A new method for the estimation of bearing health state and remaining useful life based on the moving average cross-correlation of power spectral density", Mechanical Systems and Signal Processing, 2020.

[9] Pennacchi, P., S. Bruni, S. Chatterton, et al., "A Test Rig for the Condition-Based Maintenance Application on the Traction Chain of Very High Speed Trains", World Congress on Railway Research, (2011).

[10] Wang, D., P.W. Tse, and K.L. Tsui, "An enhanced Kurtogram method for fault diagnosis of rolling element bearings", Mechanical Systems and Signal Processing, 2013.

12th International Conference on Vibrations in Rotating Machinery -
Institution of Mechanical Engineers, ISBN 978-0-367-67742-8

Improving the thrust bearing performance of turbocharger rotors using optimization methods and virtual prototypes

P. Novotný[1], J. Hrabovský[1], J. Klíma[2], V. Hort[2]

[1]Brno University of Technology, Czech Republic
[2]PBS Turbo s.r.o., Czech Republic

ABSTRACT

The aim of research activities is to develop a strategy for a thrust bearing design leading to significant benefits in real operating conditions while maintaining the existing bearing production technology. The genetic algorithms and efficient hydrodynamic solver are key elements throughout the strategy to find the design parameters of the thrust bearing. The new design of the thrust bearing and the whole chain of computational tools are verified by technical experiments in real turbocharger operating states. The applicability of the strategy is demonstrated by the new bearing design leading to savings in friction losses of approximately 20%.

1 INTRODUCTION

Mechanical efficiency is an important parameter for achieving a high overall turbocharger efficiency and hence the efficiency of the internal combustion engine (ICE). Turbocharged ICEs are often operated in steady-state or transient-state operating regimes and they are specifically designed for these regimes. There are applications where very transient operation with fast ICE response to varying power requirements is required. In this case, it is appropriate to design the ICE for transient operating regimes and this also requires increased emphasis on the mechanical efficiency of the turbocharger (TC).

The TC comprises a rotor, which is basically a shaft with turbine and compressor wheels mounted most often in a thrust bearing and a pair of journal bearings. The journal and thrust bearings significantly affect the mechanical efficiency and thus the overall efficiency of the TC. TC rotor bearings generally operate in a wide variety of transient states. It is known that journal bearings fundamentally affect rotor movement with a great impact on the rotor stability. Many studies have focused on the determination of mechanical losses in TCs and in detail on hydrodynamic bearings. The hydrodynamic bearings have been studied, for example, by Deligant et al. [1] and the thrust bearing have been found to have a greater impact on friction losses than the journal bearings. Hoepke [2] also analysed the friction losses of the TC of the passenger car ICE and determined the friction loss ratio of the thrust bearing at approximately 38%.

A thrust bearing, which mainly affects the axial movement of the rotor, can have a great influence on the lateral vibrations of the rotor. Some researchers [3][4] present a significant influence of the thrust bearing on the rotor dynamics. Their finding is that the thrust bearing can affect critical speeds and can have a positive effect on rotor stability. Vetter [5] shows that the thrust bearing mainly affects the conical shapes of the rotor vibration. The thrust bearing also has a significant effect on the flow of lubricant through the TC. Novotný et al. [6] present a study of the dynamics of the TC rotor of the heavy-duty ICE. The results show that the lubricant flow rate through the thrust bearing can be several times greater than the lubricant flow rate through the journal bearings.

A traditional design of components consists in changing only a few parameters and subsequently in verification by computational modelling tools or technical experiments. However, these approaches are beginning to conflict with the limitations imposed by the designer's ability to understand the impact of different parameters on different types of results. Parametric studies represent a certain qualitative step forward and may be a partial improvement, but still do not allow the assessment of the impact of multiple parameters.

The solution may be the use of optimization methods enabling multi-parametric optimization of components or machines. These approaches look very promising. But in the case of practical tasks due to the consideration of many parameters in the analysis, the length of the computational time can be a problem. This approach must be complemented by an efficient computational model that maintains sufficient physical depth while being acceptable in terms of computational complexity.

2   AIM OF THE WORK

An optimal design of the thrust bearing with significantly better performance is the aim of the work. This optimal design of the thrust bearing is conditioned by the ability of the manufacturer to produce it with existing production technologies.

Thrust bearing performance can be described by several integral characteristics and their required change compared to the series version of the bearing as follows:

- The friction loss of the thrust bearing must be decreased by more than 20%.
- The load capacity of the thrust bearing must not decrease.
- The lubricant flow rate must not be significantly higher.
- The bearing lubricant temperature at outlet must not be significantly higher.

The values of the above integral characteristics are influenced by many design and operating parameters. A practical problem is the choice of some strategy to analyse the effect of individual parameters over a reasonably long period of time with predefined boundary conditions.

3   ASSUMPTIONS TO ACHIEVE THE AIM

3.1   Solution strategy

Finding the optimal thrust bearing geometry could theoretically be done in a very general way. In practice, however, there are several partial restrictions. For example, if optimum bearing design parameters are found, it may not be guaranteed that the bearing design can be effectively produced by available manufacturing technologies. It is also necessary to consider influences, such as how sensitive the design will be to normal manufacturing inaccuracies or how it will respond to exceptional operating conditions.

The proposed solution strategy assumes an application of a sequence of the computational and experimental methods, the scope of which may partially overlap. These overlaps increase the verification rate of the results. This solution strategy also brings some loss of generality and undoubtedly requires a certain level of practical experience with the problem. The proposed strategy graphically depicted in Figure 1 contains the following steps:

a) A definition of the criteria for evaluating the problem.
b) Evaluations of the typical TC operating conditions and analysis of the serial version of the thrust bearing including the TC rotor dynamics.
c) Optimization of the thrust bearing parameters utilizing genetic algorithms and an efficient computational numerical model.
d) Verifications of the new design of the thrust bearing using computational approaches for steady and transition states and an experimental verification of the thrust bearing prototype.

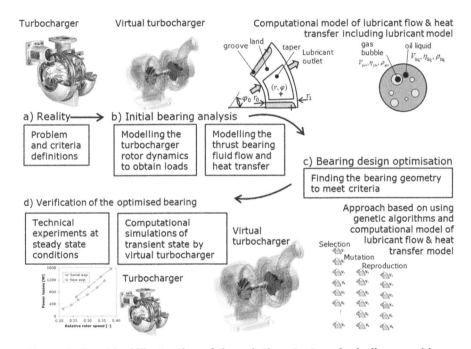

Figure 1. Graphical illustration of the solution strategy including consideration of reality, selection of criteria, initial analysis of bearing conditions, optimization process and verification of the new design of the thrust bearing.

3.2 Selection of thrust bearing concept and operating conditions

The thrust bearing of the TC (Figure 2a) is divided into thrust and counter-thrust sides (Figure 2b), each side is divided into several segments (Figure 2c). Every segment transmits forces through a suitably shaped working surface. The axial load on the bearing in most operating regimes points from the turbine to the compressor and loads the thrust side. The counter-thrust side, although mostly almost unloaded, is a source of considerable mechanical losses and significant lubricant flow rate.

The thrust bearing under consideration is a component of the TC operating in weakly transient operating modes. However, for simplification, there is chosen a defined number of steady-state regimes (n_{wg}) represented by TC rotor speeds (n_i), axial forces $(F_{ax,i})$, inlet oil pressures $(p_{in,i})$ and inlet oil temperatures $(T_{in,i})$, where $i = 1, 2, ..n_{wg}$.

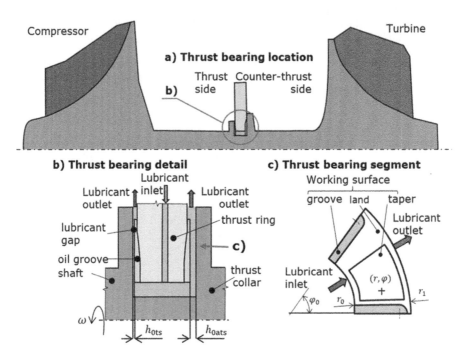

Figure 2. Thrust bearing location in the TC (a), detail of thrust bearing in section (b) and arrangement of the work surface on one bearing segment (c).

Searching for a bearing design using optimization methods needs some degree of simplification. In general, optimization methods are always limited in some way by boundary conditions. Therefore, it is necessary to choose a bearing concept. The bearing concept requires an arrangement definition of the working surface on segments both the thrust and counter-thrust sides. The selected concept of the working surface is presented in Figure 3 and includes two angle parameters and four length parameters for the bearing segment. These parameters thus clearly define geometrically the working surface.

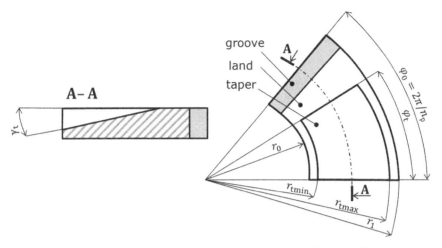

Figure 3. Schema presenting the selected concept of the working surface arrangement on one bearing segment.

3.3 Effective computational model of the thrust bearing

A computational model for bearing performance evaluation calculating integral characteristics is a key part of the proposed strategy. The model is selected according to the ability to

- describe the liquid-gas mixture flow in the thrust bearing, including the influence of cavitation,
- solve quickly the steady state regimes, lasting a maximum of seconds,
- include the effect of temperature and shear rate on fluid properties and
- include geometric features such as bore outlets, grooves, chamfers, etc.

The chosen computational model meeting the proposed abilities uses the principles presented by Novotný and Hrabovský [7]. The model is based on the generally known theory of thin lubricating layers with correction of the influence of lubricant temperature, inertia forces and turbulence. This theory assumes that the hydrodynamic pressure in the thin lubricating layer is invariable through the thickness of the lubricating layer and that Newtonian fluid adheres to the surfaces. The fluid flow is laminar (with subsequent turbulence correction) with negligible inertial forces compared to viscous forces. It is also defined that fluid density and viscosity are constant throughout the lubricant volume. Most of these prerequisites are fulfilled for typical hydrodynamic bearings, but some are not, especially the condition of constant lubricant properties, which are generally spatially dependent variables. However, the literature, for example [6][7][8], shows that, despite these contradictions, the theory of thin lubricating layers can be used for most bearing lubrication regimes.

The solution of the lubricant flow in the thrust bearing is characterised by the distribution of hydrodynamic pressure in the lubricating layer (p_h) and the average temperature rise of the lubricant through the bearing (ΔT_{oil}). Integral characteristics that need to be calculated based on the hydrodynamic pressure are the bearing capacity (F_u), the outlet mass flow rate ($\dot{m}_r$), the friction moment (M_f). Equations for calculations of the integral characteristics can be found in [7].

The computation for one bearing operating state requires a solution of the force equilibrium of external axial loads and the reaction forces on both sides of the thrust

435

bearing. All forces are added in nonlinear force (f_{NL}) in discrete time step k (or also steady state load step) generally as a function of bearing relative eccentricity (ε) as

$$f_{NL,k} = F_{ax} - F_{uts,k}(\varepsilon) + F_{uats,k}(\varepsilon) \tag{1}$$

F_{ax} is a total axial force on TC rotor considering gas pressures on compressor and turbine wheels for a defined steady-state operating regime, F_{uts} is the load capacity of the bearing thrust side and F_{uats} is a load capacity of the bearing counter-thrust side. Relative eccentricity is defined as follows

$$\varepsilon = 1 - \frac{h_{0ts}}{c_{ax}} \tag{2}$$

Symbol $c_{ax} = h_{0ts} + h_{0ats}$ is a thrust bearing axial clearance as one of main bearing design parameters, h_{0ts} is the minimal thickness of lubricating gap on thrust side and h_{0ats} is the minimal thickness of lubricating gap on counter-thrust side. A conventional method using Newton's second law [9] is applied to compute axial forces based on the measured pressures upstream and downstream the impellers.

3.4 Application of genetic algorithms for optimization of thrust bearing parameters

The issue in bearing design is to find the optimal values of many design and operating parameters, i.e. from the mathematical point of view, to find the global maximum of the objective function. This problem is solved by an optimization algorithm that uses an efficient thrust bearing model [7] to express integral characteristics (load capacity, lubricant flow rate, friction moment, lubricant temperature rise) and finds optimal values for selected parameters. The optimization algorithm requirements are as follows:

- Efficiency in multidimensional space search for global maximum.
- Ability of parallel execution of computations on multiple processor cores.

In the case of the proposed strategy, a genetic algorithm (GA) is chosen for optimization. GA is a heuristic method that by applying the principles of evolutionary biology finds solutions to problems for which there is no applicable exact algorithm. GA belongs to a group of so-called population-based metaheuristic methods that works with more than one potential solution. The principle of evolution is directly related to a certain degree of stochasticity, which plays an important role in the algorithm and serves to escape from local extremes. The advantages of the genetic algorithm include simplicity of its implementation, autonomy of the algorithm, high probability of finding a solution lying near the global extreme and high independence on the type of problem solved [10] [11][12]. The disadvantages include a relatively high number of fitness function evaluation, i.e. hydrodynamic solutions, and an exponential increase in the size of the searched space depending on the number of optimized parameters.

GA uses fixed terminology by default. In the case of hydrodynamic thrust bearing lubrication, each solution (individual) is represented by an ordered set of design parameters (genotype) that characterize the solution. Integral characteristics (fitness) are calculated for each solution in the transition to the new generation of bearing designs (population). The algorithm looks for global extreme of fitness function. The value of the fitness then expresses the quality of the individuals. According to this fitness values, individuals who are modified by genetic operators - selection, crossover and mutation - are stochastically chosen to create a new population.

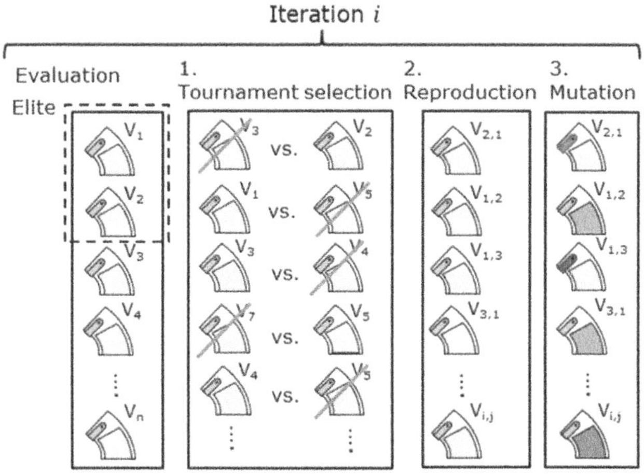

Figure 4. Diagram showing the realization of operators within one iteration and leading to finding a new generation (new group of hydrodynamic solutions).

In order to optimize the geometry of the thrust bearing working surface, the genetic algorithm model using elitism [13] is created and used. In this model, during each iteration, a new population of individuals is formed while maintaining a predetermined proportion of the best individuals of the previous generation (elite individuals). The use of elitism does not potentially lose a given percentage of the best solutions, thus ensuring a monotonically increasing sequence of maximum fitness values in each population.

Tournament selection [14] is applied as a selection operator, which does not create such a high selection pressure that could lead to a loss of diversity in the population and to the possibility of being trapped at the local extreme.

The crossover operator, which is applied after the selection operator is applied to the selected individuals and is responsible for combining the genotypes of these individuals to create new individuals and gradually the whole new population. Combining two individuals with a high fitness function is expected to give new and different individuals containing quality genotypes inherited from their parents.

The third operator is a mutation in which the genotype is randomly modified. Because too frequent mutation can lead to the loss of quality genes, this operator is only performed with a relatively low probability. In the optimization presented in this paper, the probability is inversely proportional to the magnitude of the searched space, i.e. the number of genotypes [15].

The search for optimal parameter values begins with a random selection of the parameter values and operator applications. Once these operators are done, the iteration is completed, and a new generation of individuals is created. A schematic of one iteration is shown in Figure 4. This process is iteratively repeated, thus increasing the quality of individuals in the population. The algorithm usually stops when sufficient (defined level of solution quality) solution quality is achieved, or after a predetermined number of iterations.

3.5   Virtual prototype of the turbocharger

The thrust bearing parameters are optimized for steady state regimes of TC. However, the TC often works even in transient states with time-variable external loads, so it is necessary to verify the bearing performance even in these transient operating regimes. The computational model allowing the solution of transient states, i.e. generally time dependent nonlinear model, requires a somewhat different approach. For the purposes of this strategy, a higher-level computational model, the so-called virtual turbocharger, is used.

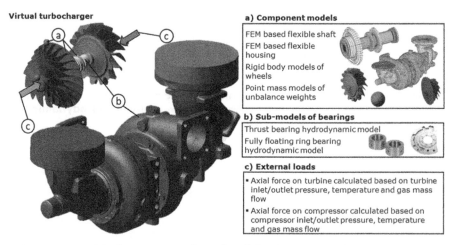

Virtual turbocharger

a) Component models
- FEM based flexible shaft
- FEM based flexible housing
- Rigid body models of wheels
- Point mass models of unbalance weights

b) Sub-models of bearings
- Thrust bearing hydrodynamic model
- Fully floating ring bearing hydrodynamic model

c) External loads
- Axial force on turbine calculated based on turbine inlet/outlet pressure, temperature and gas mass flow
- Axial force on compressor calculated based on compressor inlet/outlet pressure, temperature and gas mass flow

Figure 5. Graphical representation of multibody computational model. The model includes component models (a), sub-model of bearings (b), constrains and external loads (c).

The virtual turbocharger is a multi-physical turbocharger computational model built in a multibody dynamics software environment. This model is generally a three-dimensional (3-d) formulated in a time domain and includes sub-models of bodies of different levels. The body models are a mass point model used for unbalance, a rigid body model used for non-deformable bodies, or a 3-d model and TC housings. A rotor shaft requires to include elasticity and therefore 3-d flexible models are used. The 3-d flexible models of components are based on reduced finite element (FE) models, using component mode synthesis (CMS) according to Craig [16]. The interaction between the rotor and housing is solved using specialized sub-models of bearings. The radial bearing computational model used for transient simulations is described by Novotný et al [6] and a similar thrust bearing computational model can be found in Novotný et al [17]. The axial load of the rotor is calculated based on the measured pressures and temperatures upstream and downstream of the turbine and the compressor according to the procedure described by Nguyen-Schäfer [9]. The graphical representation of the virtual turbocharger is shown in Figure 5.

The virtual turbocharger is being used for simulations of TC rotor transient operating states. For this case, the rotor speed is increased up to the maximum speed of the TC over defined time period. The hydrodynamic bearing sub-models include a database of reaction forces in the lubricating layer. This database is created by separate hydrodynamic bearing lubrication solutions according to the methodology presented for radial bearings in [6] and for thrust bearings in [17]. The sub-model also includes the calculation of the average temperature rise of the lubricating layer, which updates the lubricant properties (dynamic viscosity, density and specific heat capacity) based on actual operating integrals (lubricant friction losses and mass flow rate).

4   RESULTS AND DISCUSSION

The proposed solution strategy assumes verification of the new bearing performance against the serial bearing version in transient operating states of the TC. To evaluate the entire bearing, it is also necessary to consider the thrust and counter-thrust sides.

Rotor dynamics simulation using the virtual turbocharger assumes a transient revolution sweep starting from the minimum to the maximum rotor speed ($n_{\max}$) of the TC and in 10 seconds. The axial load of the rotor is entered directly as a function of time. Lubricant properties used for the solution are presented in Tab. 1. Bearing performance described by integral characteristics is evaluated relative to the serial variant.

Table 1. Lubricant properties used for calculations.

Oil pressure at inlet, p_{in} [bar]	2.7
Oil pressure at outlet, p_{out} [bar]	1.0
Oil temperature at inlet, T_{in} [°C]	80
Lubricant specification [-]	SAE 10W40

The total friction torque is fundamentally influenced by the thrust bearing, especially the thrust side of the bearing. This is mainly due to the prevailing axial force in the direction from the turbine to the compressor. The new bearing design due to the optimized working surface shows significantly lower friction losses. The lower friction losses are also achieved by the reduced outer diameter of the bearing on the counter-thrust side. In principle the counter-thrust side assumes the negative bearing load capacity and an outer diameter reduction is a positive change. On the other hand, the reduced working surface diameter on the counter-thrust side leads to a certain increase in the lubricant flow rate, especially at lower operating speeds, in which the inlet pressure influence prevails. Comparisons of the relative friction torques and lubricant flow rates determined by the virtual turbocharger over the entire TC speed range is presented in Figure 6. The relative friction torque ($\bar{M}_{\text{f}}$) is related to the friction torque of the serial version (M_{fs}) by equation $\bar{M}_{\text{f}} = M_{\text{f}}/M_{\text{fs}}$. Similarly, the relative lubricant flow rate of the TC ($\dot{m}_{\text{r}}$) is related to the flow rate of the serial TC ($\dot{m}_{\text{rs}}$) by equation $\dot{m}_{\text{r}} = \dot{m}_{\text{f}}/\dot{m}_{\text{fs}}$. Relative rotor speed is related to the maximal rotor speed of the TC by relation $\bar{n} = n/n_{\max}$, where n is a rotor speed.

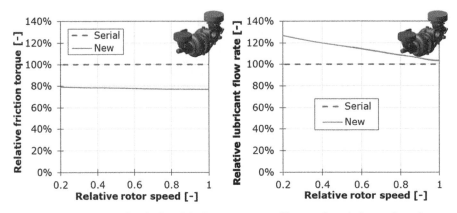

Figure 6. Computed relative friction torques of journal and thrust bearings and relative lubricant flow rates of the TC for the serial version of the thrust bearing (Serial) and for the new version of the thrust bearing (New).

The results of the computational model of both bearing versions show approximately 20% savings in friction losses. However, computational methods assume certain simplifying assumptions when describing a given physical problem. Therefore, in technical practice, the factor that then determines the correctness of the design is a technical experiment.

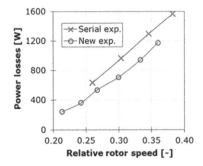

Figure 7. Overall power losses of the TC driven by compressed air determined by technical experiments on the serial version of the thrust bearing (Serial exp.) and the new version of the thrust bearing (New exp.).

An experimental verification on the TC in so-called cold operation regime is chosen to verify the new bearing design. The turbine is driven by externally supplied compressed air and thus the turbine side of the TC is not heated according to standard operating regimes. In this way, the power losses of the lubricating system can be determined with a certain error by measuring the temperature of the lubricant at the inlet and outlet. Compared to the normal operating regimes, the heating component is missing due to heat transfer from the surrounding bearing walls. A certain disadvantage is the limitation of the maximum speed of the TC, as often a sufficiently powerful source of compressed air is not necessary to turn the TC to maximum speed. Overall power losses of the TC are presented in Figure 7. A comparison of experimentally and computationally determined relative power savings is shown in Figure 8. Finally, relative savings in friction losses of the TC over 20% are archives.

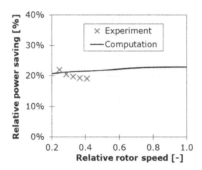

Figure 8. Relative savings in power losses determined by the virtual turbocharger computations and technical experiments.

Experimental results have proved the significantly better performances of the new thrust bearing compared to serial one. The friction losses are reduced more than chosen criteria. Probably slight differences in results can be expected for standard operating regimes, but overall tendencies should be maintained.

5   CONCLUSIONS

The developed solution strategy is designed to allow the assessment of the impact of many design and operating parameters on thrust bearing performance. The optimal parameter values are found by means of GA incorporating the efficient computational model of the thrust bearing. The results of the computations and technical experiments proved the applicability of this solution strategy not only for the selected operating states, but also for the whole range of operating regimes of the TC.

The solution strategy also shows some limitations. This is mainly a limitation in the selection of the thrust bearing concept of the working surface arrangement and its parameterization. Genetic algorithm searches only given parameterized concept, no other variant is considered during optimization. As a result, it is necessary to carry out a study of various conceptual solutions by CFD tool for example and then optimize the selected concept.

The optimization algorithm computes the integral characteristics using an efficient thrust bearing model. However, this effective model also has its limitations, as stated by Novotný and Hrabovský [17]. As a result, the selected description is only suitable for thin lubrication gaps and exhibits high inaccuracies in the case of thick thicknesses described e.g. by high Reynolds numbers. These limitations follow the chosen lubrication theory not to correctly describe 3-d nature of unsteady lubricant flow and heat transfer in the thrust bearing.

ACKNOWLEDGEMENT

The authors gratefully acknowledge funding from the Technology Agency of the Czech Republic, project Thrust bearing load capability increasing, reg. No. TH03020426 and from the Specific research on Brno University of Technology, reg. No. FSI-S-20-6267.

REFERENCES

[1] Deligant, M., Podevin, P. & Descombes, G. (2012) Experimental identification of turbocharger mechanical friction losses. *Energy*, 39(1).

[2] Hoepke, B., Uhlmann, T., Pischinger, S., Lueddecke, B. & Filsinger, D. (2015) Analysis of thrust bearing impact on friction losses in automotive turbochargers. *J Eng Gas Turbines Power*. 137(8).

[3] Mittwollen, N., Hegel, T. & Glienicke, J. (1991) Effect of Hydrodynamic Thrust Bearings on Lateral Shaft Vibrations. *ASME. J. Tribol.*, 113(4).

[4] Berger S., Bonneau O., Frene J. (2000) Influence of Axial Thrust Bearing on the Dynamic Behavior of an Elastic Shaft: Coupling Between the Axial Dynamic Behavior and the Bending Vibrations of a Flexible Shaft. *ASME. J. Vib. Acoust.*, 123(2).

[5] Vetter, D., Hagemann, T. & Schwarze, H. (2014) Predictions for run-up procedures of automotive turbochargers with full-floating ring bearings including thermal effects and different bearing setups. *11th International Conference on Turbochargers and Turbocharging*.

[6] Novotný, P., Škara, P. & Hliník, J. (2018) The effective computational model of the hydrodynamics journal floating ring bearing for simulations of long transient regimes of turbocharger rotor dynamics. *International Journal of Mechanical Sciences*. 148.

[7] Novotný, P. & Hrabovský J. (2020) Efficient computational modelling of low loaded bearings of turbocharger rotors. *International Journal of Mechanical Sciences*, 174.

[8] Hori, J. (2006) *Hydrodynamic Lubrication*. Tokyo: Springer Verlag, ISBN 978-4-431-27898-2.

[9] Nguyen-Schäfer, H. (2015) *Rotordynamics of Automotive Turbochargers*. 2nd edition. Heidelberg: Springer. ISBN 978-3-319-17643-7.

[10] Tang, K. S., Man, K. F., Kwong, S. & HE, Q. (1996) Genetic Algorithms and their Applications. *IEEE Signal Processing Magazine*. 13(6).

[11] Stodola, P. & Stodola, J. (2020) Model of Predictive Maintenance of Machines and Equipment. *MDPI Applied Sciences*, 10(1).

[12] FURCH, Jan (2016). A model for predicting motor vehicle life cycle cost and its verification. Transactions of FAMENA.40(1), ISSN 1333-1124.

[13] Baluja, S. & Caruana R. (1995) Removing the Genetics from the Standard Genetic Algorithm. *Machine Learning Proceedings*. Morgan Kaufmann. ISBN 9781558603776.

[14] David, E. G. & Deb, K. (1991) *A Comparative Analysis of Selection Schemes Used in Genetic Algorithms*. Morgan Kaufmann Publishers, Inc. ISBN 1-55860-170-8.

[15] Mühlenbein, H., & Schlierkamp-Voosen, D. (1993) Predictive Models for the Breeder Genetic Algorithm I. Continuous Parameter Optimization. *Evolutionary Computation*, 1(1).

[16] Craig, R. R. (1981) *Structural Dynamics*. First edition. John Wiley & Sons.

[17] Novotný P., Hrabovský J., Juračka J., Klíma J. & Hort, V. (2019) Effective thrust bearing model for simulations of transient rotor dynamics. *International Journal of Mechanical Sciences*, 148.

Characteristics of a high speed thin film fluid lubricated bearing

N.Y. Bailey

Department of Mechanical Engineering, University of Bath, UK

ABSTRACT

A fluid lubricated bearing model is derived for operation under extreme operating conditions, including velocity slip boundary conditions appropriate for very small bearing face separation and retention of centrifugal inertia effects. Both compressible and incompressible Reynolds equations are formulated to model the fluid film and the fluid flow characteristics are examined for the steady state case.

Coupling the fluid flow to the bearing structure, where the rotor and stator are modelled as spring-mass-damper systems, allows the dynamics to be examined when the bearing is subject to an external harmonic force. This replicates forces the bearing may be subject to when situated within a larger complex dynamical system.

1   INTRODUCTION

Fluid lubricated bearing and seal technology comprises two structural components; namely a rotor and stator, which are separated by a thin fluid film that experiences relative rotational motion. The set up is of a thrust/axial bearing with radial flow and this type of technology is also described as non-contacting, gas-lubricated or film-riding and those containing an air film termed air-riding seal. Next generation bearing and seal technology aims to provide a considerable improvement in efficiency for applications characterised by higher rotational speeds and smaller operating clearances, possibly down to the order of several microns.

To completely capture the dynamics of a fluid-lubricated bearing, the fluid flow and bearing structure need to be appropriately coupled together. If an external axial force is imposed on the bearing, a hydrodynamic force is typically generated by the normal motion of the faces, enhancing the local fluid film pressure, causing the fluid film to be maintained. Etison modelled a fluid lubricated device, identifying the hydrodynamic and hydrostatic components of the air film pressure and showed that the squeeze film behaviour (incorporated in the hydrostatic component) was potentially able to maintain the air film between the rotor and stator [1]. For highly vibrating operations, Salbu examined the effect of significant axial disturbances by describing the rotor-stator clearance with oscillatory motion and confirmed squeeze films have a load carrying capacity [2]. The dynamics of a coupled high speed air lubricated bearing were examined by Garratt et al., where the effects of centrifugal inertia were retained and one bearing face underwent small prescribed amplitudes, and the other responded to the induced film dynamics [3].

The effect of non-parallel faces, which can arise through design (to enhance performance) or operation (due to over pressurisation) has been investigated. Stability studies have identified optimal coning angles for a range of practical bearing configurations [1] and the effect of the bearing geometry on the ability to maintain a fluid film thickness has been investigated [4-7].

The effects of compressibility on the bearing dynamics has been examined by Parkins et al. through comparison of experimental results and a coupled model for an oil squeeze film bearing, showing limitations in the model due to neglecting compressibility effects [8]. Conditions for compressibility effects to be neglected for bearing flow have previously been derived by Bailey et al. showing compressible flow in the bearing behaves as if it were incompressible flow for sufficiently small radial and azimuthal speeds [5].

The importance of very thin fluid film analysis for bearing/seal dynamics has been highlighted by Sayma et al. [9], where steady-state solutions to the system of Navier-Stokes equations with no-slip boundary conditions for gaps of the order of 10 microns or less are not possible to be found. This gap is typical of some hydrodynamic seals, resulting in the flow having a high Knudsen number, $K_n = \tilde{\lambda}/\hat{h}_0$, with $\tilde{\lambda}$ as the mean free molecular path and $\hat{h}_0$ the characteristic fluid thickness. Therefore, due to surface related phenomena becomes increasingly dominant, a slip flow condition could emerge as a consequence of an insufficient number of molecules in the sampling region [10]. In this slip flow regime, $10^{-3} \geq Kn \leq 10^{-1}$ a continuum model with a slip boundary condition is usually employed, opposed to typical consideration of a continuum flow with no-slip boundary conditions for $Kn < 10^{-3}$.

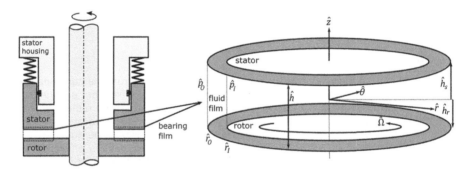

Figure 1. Schematic of a parallel axisymmetric fluid lubricated bearing in a cylindrical polar coordinate system $(\hat{r}, \hat{\theta}, \hat{z})$, with pressures imposed at the inner and outer radii; $\hat{p}_I$ at $\hat{r} = \hat{r}_I$ and $\hat{p}_o$ at $\hat{r} = \hat{r}_o$, respectively. The rotor has relative angular rotation $\hat{\Omega}$ with axial height $\hat{h}_r$, whereas the stator height is $\hat{h}_{ss}$ giving the rotor-stator height as $\hat{h}$.

Navier proposed a first order slip model which was based on a linear relationship between the tangential shear rate and the fluid-wall velocity difference. The jump velocity at the wall is linearly proportional to the first order derivatives of the fluid velocity with the proportionality constant being the slip length [11]. Park et al. considered a non-axisymmetric bearing with slip flow, using a classical Reynolds equation for the gas flow with both slip and no-slip conditions examined [12]. Coupling to the bearing structure, small axial amplitude rotor displacements were examined showing a slip condition is associated with a smaller load carrying capacity and decrease in fluid stiffness and damping coefficients. Higher order slip models have also been used when studying a coupled gas journal bearing, showing similar trends [13, 14].

In Section 2 the model geometry is presented, the formulation of the fluid flow governing equations for compressible and incompressible flow is presented in Section 3 and the fluid flow characteristics are examined in Section 4. The structural model is given in Section 5, where the axial stator and rotor displacement equations are modelled as spring-mass-damper systems, and an external harmonic force can be imposed on the rotor. Solving the coupled model of a compressible flow bearing requires the modified Reynolds equation to be solved simultaneously with the structural dynamic equations, with the different numerical schemes presented in Section 6. Dynamics results are presented in Section 7, showing the differences between a compressible and incompressible flow approximation.

2   MODEL GEOMETRY

Due to industrial bearing and seal designs being complex, a simplified mathematical model is formulated through retaining key features of the fluid lubricated bearing geometry, which is an axial/thrust bearing with radial flow. A parallel rotor-stator configuration is shown in Figure 1 in a cylindrical polar coordinate system $(\hat{r},\hat{\theta},\hat{z})$ where the rotor and stator are modelled as a pair of coaxial axisymmetric annuli separated by a thin fluid film. The stator and rotor are mounted flexibly to a stationary housing and high speed shaft, respectively, allowing them to have axial motion in response to the film dynamics; the rotor has a relative angular rotation $\hat{\Omega}$. The stator height is denoted by $\hat{h}_s$ and the rotor by $\hat{h}_r$, giving the rotor-stator height as $\hat{h} = \hat{h}_s - \hat{h}_r$. A pressure gradient can be imposed across the annuli through setting a pressure at the inner and outer radii; $\hat{p}_I$ at $\hat{r} = \hat{r}_I$ and $\hat{p}_o$ at $\hat{r} = \hat{r}_o$, respectively. This replicates the bearing being subject to the conditions of the wider system that it is placed within.

A radial taper on the rotor is incorporated into the model, denoted by $\hat{\beta}$, which may arise due to over-pressurisation of the bearing. A schematic of this geometry is shown in Figure 2, together with an external axisymmetric harmonic axial force $N(t)$ imposed on the rotor, which may arise due to the dynamics in bearing surroundings. The rotor and stator will move axially in response to the external force/induced film dynamics with displacements modelled using spring-mass-damper systems. The rotor height will also be a function of radial taper angle and radius with definition

$$\hat{h}_r\left(\hat{r},\hat{\beta},\hat{t}\right) = \hat{h}_r(\hat{t}) - (\hat{r} - \hat{r}_1)\hat{\beta} \text{ for } \hat{\beta} > 0, \text{ or } \hat{h}_r(\hat{r},\hat{\beta},\hat{t}) = \hat{h}_r(\hat{t}) - (\hat{r} - \hat{r}_o)\beta \text{ for } \hat{\beta} < 0 \quad (1)$$

The positive and negative radial taper angles are separated due to it being assumed they arise due to over-pressurisation of the bearing. Therefore, internal pressurisation $(\hat{p}_I > \hat{p}_o)$ will result in a positively radial tapered bearing (PTB) and external pressurisation $(\hat{p}_I < \hat{p}_o)$ a negatively radial tapered bearing (NTB). Thus, the pressure gradient corresponds to a diverging channel.

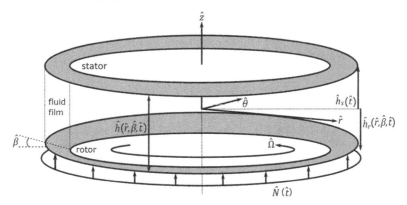

Figure 2. Schematic of an axisymmetric radial tapered fluid lubricated bearing experiencing an axisymmetric axial force $\hat{N}(\hat{t})$ in cylindrical polar coordinate system $(\hat{r},\hat{\theta},\hat{z})$. The radial taper angle of the rotor is represented by $\hat{\beta}$.

445

3   FLUID FLOW MODEL

Both compressible and incompressible flow models have been used to model the fluid film depending on the applications and operation of a bearing or seal. Both flow models are derived to examine the differences in fluid flow characteristics and full coupled dynamic behaviour. Conditions under which compressibility effects can be neglected have been derived [5], showing that when the conservation of mass reduces to the statement of solenoidal velocity field, i.e. for sufficiently small radial and azimuthal speeds, compressible flow behaves as if it were incompressible flow in a fluid lubricated bearing.

The fluid flow models are derived from the Navier-Stokes momentum equations and conservation of mass equation in cylindrical polar coordinates; it is assumed to be isothermal flow. The governing system of Navier-Stokes equations for the fluid flow are expressed in dimensionless variables in terms of rotor velocity $\widehat{\Omega r}$, outer bearing radius $\widehat{h}_0$, typical pressure $\widehat{p}$, time scale $\widehat{T}$ and unperturbed air density $\widehat{p}_0$. The equilibrium rotor-stator height $\widehat{h}_0$ is taken to be the equilibrium position when no external forces are acting, but the rotor has angular rotation, and has a value of 5×10^{-5} m; several orders of magnitude larger than the mean free path of air, $\widehat{l} = 6.8 \times 10^{-8}$ m. However, if experiencing an external force with a significantly large amplitude, the thickness of the fluid film may become comparable to the mean free path of a fluid molecule, thus reaching the limitations of the continuum limit model. Therefore, slip effects are included on the fluid-solid interface through the velocity boundary conditions, giving the fluid flow in the slip regime as characterised by $10^{-3} \leq K_n = \widetilde{\lambda}/\widehat{h}_0 \leq 10^{-1}$.

Dimensionless velocities are given by $\widehat{u}/\widehat{U}, \widehat{v}/(\widehat{\Omega r}_0)$ and $\widehat{w}/(\widehat{h}_0 \widehat{T}^{-1})$ and the dimensionless pressure, height and radius are taken to be $p = \widehat{p}/\widehat{p}, z = \widehat{z}/\widehat{h}_0$ and $r = \widehat{r}/\widehat{r}_0$, respectively. The dimensionless slip length is given by $l_s = \widehat{l}/\widehat{h}_0$ and the density by $\rho = \widehat{p}/\widehat{p}_0$; note in the case of incompressible flow $\widehat{\rho} = \widehat{\rho}_0$. The imposed pressures are scaled by $p_I = \widehat{p}_I/\widehat{P}$ and $p_I = \widehat{p}_I/\widehat{P}$ and the inner and outer radii as $a = \widehat{r}_1/\widehat{r}_0$ and 1, respectively. The dimensionless stator and rotor heights are denoted by $h_s = \widehat{h}_s/\widehat{h}$ and $h_r = \widehat{h}_r/\widehat{h}_0$ and the rotor-stator height by $h = h_s/h_r$.

The radial taper angle is assumed small, such that $\sin\widehat{\beta} = o(\delta_0)$ and $\cos\widehat{\beta} = 1 + o(\delta_0)$ resulting in the scaling $\widehat{\beta} = \beta\delta_0$ with $\widehat{\beta} = o(1)$ ensuring that the lubrication condition still holds. This produces a dimensionless rotor height of

$$h_r(r,\beta,t) = h_r(t) - (r-a)\beta \text{ for } \beta > 0, \text{ or } h_r(r,\beta,t) = h_r(t) - (r-1)\beta \text{ for } \beta < 0, \quad (2)$$

with $\partial h_r/\partial r = -\beta$.

The associated radial Reynolds number and Reynolds number ratio Re* are given, together with the aspect ratio δ_0 and Froude number Fr, respectively, as

$$Re_U = \frac{\widehat{U}\widehat{h}_0}{v}, \ Re^* = \frac{\widehat{\Omega r}_0}{\widehat{U}}\delta_0^{-1}, \ \delta_0 = \frac{\widehat{h}_0}{\widehat{r}_0} \text{ and } Fr = \frac{\widehat{U}}{\sqrt{\widetilde{g}\widehat{h}_0}}, \quad (3)$$

where $\widetilde{g}$ denotes the acceleration due to gravity and $v = \mu/\widehat{\rho}_0$ the kinematic viscosity. For thin fluid film bearings $\delta_0 << 1$, and a lubrication approximation is used. The importance of the gravitational effects relative to the radial inertia are described by the Froude number Fr, however, gravity can be neglected if $Re_U\delta_0 Fr^{-2} << 1$, which is consistent with the lubrication theory provided the Froude number is $O(1)$. Due to the reduced Reynolds number $Re_U\delta_0 << 1$, classical lubrication theory neglects inertia. However, the effects of viscosity are retained at leading order as high speed operation is considered which means the ratio of the Reynolds numbers $(Re^*)^2$ is not always negligible and must be considered. The pressure is scaled as $\hat{P} = \mu \hat{r}_0 \hat{U}/\hat{h}_0{}^2$, to ensure the effects of viscosity are retained at leading order.

Applying a lubrication condition, and the stated assumptions, to the compressible Navier-Stokes momentum equations together with the conservation of mass equation and equation of state, results in the leading order equations

$$-\lambda \rho \frac{v^2}{r} = -\frac{\partial p}{\partial r} + \frac{\partial^2 u}{\partial z^2}, \quad 0 = \frac{\partial^2 v}{\partial z^2}, \quad 0 = \frac{\partial p}{\partial z}, \quad \frac{\partial \rho}{\partial t} + \frac{1}{\sigma r}\frac{\partial}{\partial r}(r\rho u) + \frac{\partial}{\partial z}(\rho w) = 0 \quad P = K_s \rho \qquad (4)$$

The dimensionless ideal gas constant is defined by $K_s = RT_0 \hat{h}_0{}^2/(v \hat{r}_0 \hat{U} M)$ and connects the pressure and density field; R, M and T_0 are the ideal gas constant, molar mass and fluid temperature, respectively. The speed parameter is defined by $\lambda = Re_U \delta_0 (Re^*)^2 = r_0 \hat{h}_0{}^2 \hat{\Omega}^2/(v \hat{U})$ and the classical lubrication equations for compressible flow in axisymmetric cylindrical coordinates are retained when $\lambda = 0$. Letting $\sigma = \hat{h}_0{}^2/(v \hat{T} Re_U \delta_0)$ be of $O(1)$, implies $(\hat{h}_0{}^2 v)/\hat{T}$ in the derivation has to be of $o(\delta_0)$. Thus, the time scale of the flow field $\hat{T}$ must be much slower than the time scale for the vorticity to diffuse over the film thickness $\hat{h}_0{}^2$; $\tau = \hat{h}_0{}^2/v$.

Similarly, the incompressible Navier-Stokes momentum equations together with the conservation of mass equation are derived and given by

$$-\lambda \frac{v^2}{r} = -\frac{\partial p}{\partial r} + \frac{\partial^2 u}{\partial z^2}, \quad 0 = \frac{\partial^2 U}{\partial z^2}, \quad 0 = \frac{\partial p}{\partial z}, \quad \frac{1}{\sigma r}\frac{\partial}{\partial r}(ru) + \frac{\partial w}{\partial z} = 0. \qquad (5)$$

A first-order Navier slip model is implemented, in which the velocity boundary conditions comprise of tangential components where continuity of the velocity across the fluid-solid boundary is modified by a slip condition induced by the wall shear, combined with a normal component of a no flux condition. To leading order the dimensionless velocity boundary conditions are

$$\begin{array}{llll} u = -l_s \dfrac{\partial u}{\partial z}, & v = -l_s \dfrac{\partial v}{\partial z}, & w = \dfrac{dh_s}{dt} & \text{at } z = h_s, \\[2mm] u = l_s \dfrac{\partial u}{\partial z}, & v = r + l_s \dfrac{\partial v}{\partial z}, & w = \dfrac{\partial h_r}{\partial t} - \dfrac{Re_U \delta_0}{\kappa} u\beta & \text{at } z = h_r, \end{array} \qquad (6)$$

In equation (6), l_s is a dimensionless slip length and proportionality constant between the fluid velocity components tangential to the wall and the wall shear stress [6]. The limiting cases of $l_s \to \infty$ represents a total slip model denoting a zero tangential wall fluid shear rate and $l_s = 0$ represents no-slip conditions.

The dimensionless pressure boundary conditions are defined as

$$p = pI \text{ at } r = a, \quad \text{and } p = po \text{ at } r = 1. \qquad (7)$$

The fluid flow velocities derived from a compressible flow model are readily found from the leading order Navier-Stokes equations (4), with the radial, azimuthal and axial velocities given by

$$u = \frac{1}{2}\frac{\partial p}{\partial r}(z^2 - (h_s + h_r)z + h_s h_r - l_s h)$$

$$- \frac{\lambda pr}{12K_s(h + 2l_s)^2}((z - h_r)(z - h_s)(z^2 + (h_r - 3h_s)z + 3h_s^2 - 3h_s h_r + h_r^2)$$

$$+ l_s((z - h_s)(-4(z - h_s)^2 + 6h^2) - h^3) + l_s^2(6(z - h_s)(z - h_r) - 6h) - 6hl_s^3,$$

$$v = -\frac{r}{(h + 2l_s)}(z - h_s - l_s),$$

$$w = -\frac{1}{p}\frac{\partial}{\partial t}\left((z - h_r)p\right) - \frac{1}{12\sigma pr}\frac{\partial}{\partial r}\left(pr\frac{\partial p}{\partial r}(2(z - h_r)^3 - 3(z - h_r)^2 h - 6(z - h_r)hl_s)\right) \qquad (8)$$

$$+ \frac{\lambda}{120\sigma K_s pr}\frac{\partial}{\partial r}(\frac{p^2 r^2}{(h + 2l_s)^2}(2(z - h_r)^5 - 10(z - h_r)^4 h + 20(z - h_r)^3 h^2 - 15(z - h_r)^2 h^3)$$

$$+ 10l_s(-(z - h_r)^4 + 4(z - h_r)^3 h - 3(z - h_r)^2 - 3(z - h_r)h^3)$$

$$+ 10l_s^2(2(z - h_r)^3 - 3(z - h_r)^2 h - 6(z - h_r)h^2) + 10l_s^3(-6z - h_r)h)\Big)$$

where dependence on the radial taper angle is implicit through expressions the rotor height h_r and rotor-stator height h. Removing the terms in blue results (pressure terms and ideal gas constant) in the velocity equations (8) gives the corresponding equations for incompressible flow.

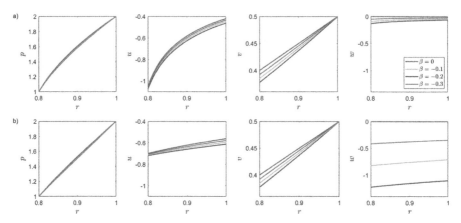

Figure 3. Pressure and velocities at z = 0.5 for a) compressible and b) incompressible flow for increasing magnitude of radial tapered bearing $0 \le \beta \le -0.3$; $a = 0.8$, $\lambda = 0.005$, $\sigma = 1$, $a = 1.5$, $l_s = 0.01$, $K_s = 1$, $K_z = 12$, $D_a = 1.5$, $p_I = 1$ and $p_O = 2$.

Integrating the conservation of mass equation in equation (4) between the stator and rotor, applying the Leibniz integral rule and slip velocity boundary conditions (6), results in the modified Reynolds equation for compressible flow as

$$\frac{\partial}{\partial t}(ph) - \frac{1}{12\sigma r}\frac{\partial}{\partial r}\left(pr\frac{\partial p}{\partial r}h^2(h + 6l_s)\right) + \frac{\lambda}{12K_s\sigma r}\frac{\partial}{\partial r}\left(p^2 r^2\frac{h^2\left(\frac{3}{10}h^3 + 3h^2 l_s + 7hl_s^2 + 6l_s^3\right)}{(h + 2l_s)^2}\right) = 0. \qquad (9)$$

Similarly, dependence on the radial taper angle is given implicitly in the rotor-stator height h. The corresponding Reynolds equation for incompressible flow is obtained by removing the terms in blue. For the traditional no-slip velocity condition, the slip length is equal to zero $l_s = 0$ and in the case of zero speed parameter $\lambda = 0$, the centrifugal effects are neglected, however the rotor and stator still experience relative rotational motion due to the velocity boundary conditions.

In the case of incompressible flow, an analytical solution to the Reynolds equation subject to the pressure boundary condition (7) can be obtained, see [6] for full details. However, in contrast for the case of compressible flow, the solution is not easily available and it is usually necessary to utilise to numerical techniques.

4   FLUID FLOW BEHAVIOUR

The fluid flow behaviour of the thin film can be investigated when the bearing is in its steady state; the stator and rotor are axially fixed in their equilibrium position, but the rotor has constant azimuthal velocity. Results are presented for an externally pressurised bearing, with the stator fixed at $h_s = 1$ and rotor being parallel or a NTB, $h_r = -(r - 1)\beta$ due to it being assumed that the radial taper arises due to over-pressurisation.

The pressure across the bearing radius and the velocity values approximately half-way between the rotor and stator, at $z = 0.5$, are shown in Figure 3 for both compressible and incompressible flow in the case of increasing magnitude of radial tapered bearing, with a small speed parameter and slip length. The azimuthal velocities are identical for both compressible and incompressible flow which increases from the inner to outer radius. The radial taper causes a decrease in azimuthal velocity, which is greatest nearer the inner radius. The pressure field is similar for both flow types, increasing monotonically from the inner to outer radius, with the radial taper having little effect. However, the pressure gradient is significantly different; causing considerable differences in the radial and axial velocities for compressible and incompressible flow. The radial velocity increases monotonically from the inner radius across the bearing for both flow types with an increasing radial taper causing the flow magnitude to increase. However, compressible flow has an increase of 0.65, whereas incompressible flow only increases by 0.1, with a value lying within the range of compressible flow. The axial velocity for incompressible flow has a significantly larger magnitude than compressible flow; both cases have zero value for a parallel bearing.

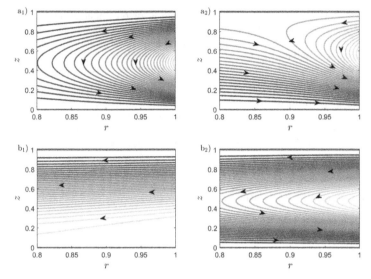

Figure 4. Streamlines for a) compressible and b) incompressible flow for 1)
$\lambda_{cc} = 12.841$ and 2) $\lambda_{ci} = 18.5264$; $\beta = 0$, $a = 0.8$, $\sigma = 1$, $a = 1.5$, $l_s = 0.01$,
$K_s = 1$, $K_z = 12$, $D_a = 1.5$, $p_I = 1$ and $p_O = 2$.

The mass flux of fluid through the bearing is given by

$$Q(r, \lambda, l_s) = \int_0^{2\pi} \int_{h_r}^{h_s} ru \, dzd\theta$$

$$= 2\pi \left(\frac{\lambda pr^2}{12K_s(h + 2l_s)^2} \left(\frac{3}{10}h^3 + 3h^2 l_s + 7hl_s^2 + 6l_s^3 \right) - \frac{r}{2} \frac{\partial p}{\partial r} h^2(h + 6l_s) \right) \qquad (10)$$

The radial path of the fluid flow through the bearing is given by the streamfunction ψ, defined by,

$$u = \frac{1}{K_s pr} \frac{\partial \psi}{\partial z}, \quad w = -\frac{1}{\sigma K_s pr} \frac{\partial \psi}{\partial r} \qquad (11)$$

resulting in

$$\begin{aligned}
\Psi(r, \lambda, l_s, z) = &\frac{rp}{12K_s} \frac{\partial p}{\partial r} (2(z - h_r)^3 - 3(z - h_r)^2 h - 6(z - h_r)hl_s) \\
&- \frac{\lambda r^2 p^2}{120K_s^2(h + 2l_s)^2} (2(z - h_r)^5 - 10(z - h_r)^4 h + 20(z - h_r)^3 h^2 - 15(z - h_r)^2 h^3) \\
&+ 10l_s(-(z - h_r)^4 + 4(z - h_r)^3 h - 3(z - h_r)^2 h^2 - 3(z - h_r)h^3) \\
&+ 10l_s^2(2(z - h_r)^3 - 3(z - h_r)^2 h - 6(z - h_r)h^2) + 10l_s^3(-6(z - h_r)h)),
\end{aligned} \qquad (12)$$

using the radial and axial velocity, given in (8).

Using the mass flux equation (10), a critical speed parameter can be calculated which gives a net mass flux of zero at the outer radius of the bearing. This arises due to the rotation of the rotor forcing fluid to the outer radius, but the external pressurisation forcing fluid to the inner radius. For compressible and incompressible flow, these are found to be λ_{cc} = 12.8415 and λ_{ci} = 18.5264, relatively. Figure 5 shows the streamlines for a parallel bearing with compressible and incompressible flow for the two critical speed parameters. In the case of compressible flow with λ_{cc} = 12.8415, there is not net mass flux at r = 1, where fluid flows into the bearing near the stator and recirculates, before flowing out near the rotor. For the same speed parameter but an incompressible flow assumption, fluid flows from the outer radius to inner radius only, with a mass flux of −0.719. When the speed parameter λ_{ci} = 18.5264 is used with incompressible flow, a zero net mass flux is achieved with a similar trend to the compressible flow with λ_{cc} but with greater amounts of fluid passing through the bearing. Examining the compressible flow with λ_{ci} gives a small amount of flow entering the bearing at r = 1 near the stator, which the exits the bearing at the outer radius after recirculating. A larger amount of fluid passes through the bearing from the inner to outer radius giving a net mass flux at the outer radius of 0.582.

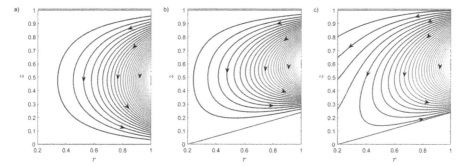

Figure 5. Streamlines for compressible flow with a) parallel faces β = 0 and no-slip condition l_s = 0, b) radial tapered faces β = −0.3 and no-slip condition l_s = 0 and c) radial tapered faces β = −0.3 and slip condition l_s = 0.2; a = 0.2, λ_{cc} = 4.81569, σ = 1, a = 1.5, K_s = 1, K_z = 12, D_a = 1.5, p_I = 1 and p_O = 2.

The effect of a radial tapered rotor and slip condition on the fluid flow behaviour is investigated, with streamlines shown in Figure 5 for a wide annulus bearing with compressible flow; parallel faces β = 0 and no-slip condition l_s = 0; radial tapered faces β = −0.3 and no-slip condition l_s = 0; radial tapered faces β = −0.3 and slip condition l_s = 0.2. The speed parameter is chosen so that there is zero net mass flux at the outer radius for case of parallel faces and no-slip condition; note that this plot differs from Figure 5a), due to the different annulus width. Including a radial taper on the rotor causes the fluid to flow further into the bearing before recirculating and exiting near the rotor, but maintains an effective zero net mass flux at the outer radius. Additionally, incorporating a slip condition causes some of the fluid to flow completely through the bearing near the stator from outer radius to inner, giving a net mass flux at r = 1 of −0.0145.

5 STRUCTURAL MODEL

To examine the dynamics of a fluid lubricated bearing with an imposed external force, the axial displacement of the stator and rotor is coupled to the fluid flow through the axial force imposed from the fluid on the bearing faces. The axial height of the bearing faces is modelled using standard spring-mass-damper models incorporating the axial bearing pressure variation from the fluid film; it is assumed the rotor and stator have identical damping, effective restoring force and force coupling properties. The external periodic force imposed on the rotor is defined as $N(t) = \in \sin t$, where $\in$ is a measure of the amplitude of the forcing. The dimensionless tator displacement equation and force from the is fluid film $F(t)$ are given by

$$\frac{d^2 h_s}{dt^2} + D_a \frac{dh_s}{dt} + K_z(h_s - 1) = \alpha F(t), \quad \text{with } F(t) = 2\pi \int_a^1 (p - p_a) r\, dr, \tag{13}$$

and the rotor displacement equation by

$$\frac{d^2 h_r}{dt^2} + D_a \frac{dh_r}{dt} + K_z h_r = -\alpha(F(t) - N(t)) \qquad \text{for } \beta \geq 0,$$

$$or \tag{14}$$

$$\frac{d^2 h_r}{dt^2} + D_a \frac{dh_r}{dt} + K_z(h_r - (1 - \alpha)\beta) = -\alpha(F(t) - N(t)) \qquad \text{for } \beta < 0.$$

The dimensionless parameters are defined such that $p_a = \hat{p}_a/\hat{P}$ is the reference pressure, $\alpha = v^3 \hat{U} Re U^2 \hat{T}^2 / \hat{m} \hat{h}_0^4 \delta_0$ the force coupling parameter and is the mass of a face. The dimensionless effective restoring force parameter and damping are given by, $K_z = \hat{K}_z Re_U^2 \delta_0^2 v^2 \hat{T}^2 / \hat{h}_0^4$ and $D_a = \hat{D}_a Re_U \delta_0 v \hat{T} / \hat{h}_0^2$ respectively.

6 NUMERICAL METHODS

The modified Reynolds equation (9) is solved simultaneously with the structural equations (13)-(14) via numerical techniques for compressible flow. In contrast, for the case of incompressible flow, an explicit analytical solution of the corresponding Reynolds equation for the pressure can be formulated and used in the structural equations (13)-(14). Details of the two different numerical schemes are given.

It is mathematically convenient to de ne the time dependent minimum face clearance (MFC) by

$$g(t) = h_s(t) - h_r(t) = h_s - \varepsilon \sin t. \tag{15}$$

This gives a parallel bearing having the MFC equal to the rotor-stator height *(g(t) = h(t))*, and a radially tapered bearing having the MFC at the inner radius for a PTB *(h(r,β,t) = g(t)+(r −a)β)* and outer radius for a NTB *(h(r,β,t) = g(t) + (r − 1)β)*.

Rewriting the compressible governing equations (9) and (13)-(14) in terms of the MFC and discretising the Reynolds equation in the spatial variable using a second order central finite-difference approximation, allows a stroboscopic map solver to be implemented as periodic solutions are sought due to the rotor forcing being periodic. The system of coupled first-order ordinary differential equations for a NTB is

$$\frac{dg}{dt} = z,$$

$$\frac{dh_r}{dt} = y,$$

$$\frac{dZ}{dt} = -D_{as}z - K_{zs}(g-1) - (D_{as} - D_{ar})Y - (K_{zs} - K_{zr})h_r + K_{zr}(1-a)\beta + (a_s + a_r)F_f - a_rN(t),$$

$$\frac{dY}{dt} = -D_{ar}Y - K_{zr}(h_r - (1-a)\beta) - a_rF_f + a_rN(t),$$

$$\frac{dp_i}{dt} = -\frac{p_iZ}{(g+(r_i-1)\beta)} + \frac{(g+(r_i-1)\beta)(g+(r_i-1)\beta+6l_s)}{12\sigma}$$
$$\left(\frac{p_{i+1} - p_{i-1}}{4r_i\delta_r} + \frac{p_{i+1}^2 - 2p_i^2 + p_{i-1}^2}{2\delta r^2}\right) + \frac{p_ir_i(p_{i+1} - p_{i-1})}{24\sigma\delta r}(3\beta(g+(r_i-1)\beta+4l_s))$$
$$-\frac{\lambda}{12\sigma K_s}\frac{(g+(r_i-1)\beta)}{(g+(r_i-1)\beta+2l_s)^2}\left(\frac{p_{i+1}^2r_{i+1}^2 - 2p_i^2r_i^2 + p_{i-1}^2r_{i-1}^2}{2r_i\delta r^2}\right)$$
$$\left(\frac{3}{10}(g+(r_i-1)\beta)^3 + 3(g+(r_i-1)\beta)^2l_s + 7(g+(r_i-1)\beta)l_s^2 + 6l_s^3\right)$$
$$-\frac{\lambda}{12\sigma K_s}\frac{p_i^2r_i^2\beta}{(g+(r_i-1)\beta+2l_s)^3}$$
$$\left(\frac{9}{10}(g+(r_i-1)\beta)^4 + 9(g+(r_i-1)\beta)^3l_s + 31(g+(r_i-1)\beta)^2l_s^2 + 42(g+(r_i-1)\beta)l_s^3 + 24l_s^4\right)$$

$$(16)$$

In equation (16) $i = 2: (M-1)$ are the finite difference collocation points in the Reynolds equation (9) and the total number of discretization points are denoted by M; the pressure boundary conditions impose $p_1 = p_I$ and $p_M = p_O$. The fluid force on the faces $F(t)$ is approximated by numerical quadrature, with weighting function w_i.

Solutions to the system of equations in (16) are denoted by the vector $\boldsymbol{g}(g(t), h_r(t), Z(t), Y(t), p(t))$. The stroboscopic map advances an initial condition $\boldsymbol{g_0}$ at initial time t_0 by a time T, through integrating the system of equations (16) forward by one period of the external forcing; an ordinary differential solver based on a variable-step, variable-order solver based on the numerical differentiation formulas is used. The periodic solutions are then identified by the fixed points of the map $\boldsymbol{g}(t) = \boldsymbol{g}(t+T)$, which are found iteratively through an iterative Newton's method [7], until a prescribed tolerance is reached, $| \boldsymbol{g}(T) - \boldsymbol{g_0}(t_0) | \leq tol$ and thus a periodic solution is achieved.

For the case of incompressible flow, the solution technique is significantly less computationally costly. This is because an explicit analytical solution for the modified Reynolds equation in terms of the pressure can be found through integration of the modified Reynolds equation and application of the pressure boundary conditions (7). Substituting the pressure equation into the structural equations (13)-(14) leads to a system of nonlinear second-order ordinary differential equations for the face clearance and rotor height

$$\frac{d^2g}{dt^2} + (D_a - 2\alpha B(g, ls))\frac{dg}{dt} + K_z(g-1) - 2\alpha A(g, \lambda, ls) + \alpha N(t) = 0,$$
$$\frac{d^2h_r}{dt^2} + D_a\frac{dh_r}{dt} + K_zh_r + \alpha\left(A(g, \lambda, ls) + B(g, ls)\frac{dg}{dt}\right) - \alpha N(t) = 0,$$

$$(17)$$

with

$$A(g, \lambda, l_s) = \pi \left((1 - a^2)(p_I - p_a) + (p_O - p_I) \left(\frac{(1 - a^2)}{2 \ln a} + 1 \right) \right.$$
$$\left. - \lambda \frac{g^3 + 10g^2 l_s + \frac{70}{3} g l_s^2 + 20 l_s^3}{4(g + 6 l_s)(g + 2 l_s)^2} \left(1 - a^4 + \frac{(1 - a^2)^2}{\ln a} \right) \right), \tag{18}$$

$$B(g, l_s) = -\frac{\pi \sigma}{8 g^2 (g + 6 l_s)} \left(1 - a^4 + \frac{(1 - a^2)^2}{\ln a} \right).$$

Equation (18) identifies that a steady bearing with negligible inertial effects, $\lambda = 0$, has a pressure field and force on the stator which is independent of the slip length.

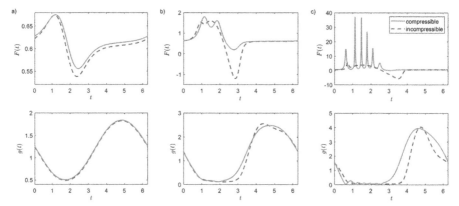

Figure 6. Fluid force and MFC for compressible and incompressible flow for increasing amplitude of forcing a) $\in$ = 5, b) $\in$ = 10 and c) $\in$ = 15; a=0.8, $\lambda = 0.005$, $\sigma = 1$, $\alpha = 1.5$, $l_s = 0.01$, $K_s = 1$, $K_z = 12$, $D_a = 1.5$, $p_I = 1$ and $p_O = 2$.

7   DYNAMIC BEHAVIOUR

The dynamic fluid force and MFC for a bearing with compressible and incompressible flow are given in Figure 6, for increasing magnitude of the amplitude of forcing. The smaller amplitude of forcing $\varepsilon = 5$ corresponds to a case near the limiting condition of a compressible flow being dynamically represented as incompressible flow, giving both flows having similar dynamics. The fluid force on the faces has a small magnitude with a maximum when the rotor is forced upwards and a minimum when the rotor moves downwards, away from the stator. The MFC follows the path of a negative sine curve, with minimum values of 0.507 for compressible flow and 0.492 for incompressible flow.

Increasing the amplitude of the external force to $\varepsilon = 10$ results in a small difference between compressible and incompressible flow. When the MFC is greater than the equilibrium value, the fluid force has a small magnitude and is effectively constant. Else, the fluid force fluctuates with a peak of greater magnitude when the rotor is pushed towards the stator, and a minimum when the rotor moves away from the stator, which is much larger for incompressible flow. This behaviour results in a thin fluid film being maintained; the case of compressible flow has a very small MFC between $t = 1.1$ and $t = 2.1$ with a minimum value of 0.117 and incompressible flow between $t = 1.1$ and $t = 2.7$ with minimum value 0.125.

For a larger amplitude of external force $\varepsilon = 15$, the compressible and incompressible flow bearings now have significantly different dynamics. For compressible flow, when the rotor and stator come into close proximity there is a sharp increase in the fluid force around $t = 0.45$ ensuring a face clearance is maintained; for incompressible flow this is around $t = 0.6$. In the case of compressible flow, the faces the move slightly apart before becoming close together again, but because the external force is still forcing the rotor towards the stator, there is another sharp increase in the fluid force to maintain a face separation. This trend continues for a total of six instances of the rotor and stator coming very close together, thus inducing a flapping motion of the faces when they are in very close proximity. Considering incompressible flow conditions gives the stator following the path of the forced rotor, without any flapping motion, maintaining an effectively constant film thickness until the rotor pulls away from the stator around $t = 3.5$. Comparing the smallest value of the MFC gives a compressible flow bearing having a value of 0.00967and incompressible flow of 0.0523.

To verify the flapping motion in the MFC for compressible flow when the fluid film is very thin, an alternative numerical technique was implemented. This was based on the methodology in Garratt et al. [3], namely a time stepping algorithm (initial boundary value solver) which is less robust and more computationally expensive than the stroboscopic map used in this work. However, the transient solver also predicts the flapping

motion when the rotor is in close proximity with the stator. This result was not identified in the work by Garratt et al. as the rotor and stator did not become close enough. Previous experimental or numerical studies were not found that examined the limit of very small fluid film thickness, almost contact condition, to provide direct validation of the predicted flapping motion.

8   CONCLUSIONS

A fluid lubricated bearing model is formulated, appropriate for a very small bearing gap (velocity slip boundary conditions are included) and high speed operation (retains the leading order effects of centrifugal inertia relevant for high speed rotational flows). Modified compressible and incompressible Reynolds equations are derived to model the fluid film, together with explicit expressions for the fluid velocity. The fluid flow behaviour of the steady state case, where the stator and rotor are fixed axially, but the rotor has rotational velocity, is examined. Despite the pressure field being similar for both compressible and incompressible flow, the pressure gradient is significantly different, causing large differences in the radial and axial fluid velocities.

The fluid flow can be coupled to the bearing faces (modelled as spring-mass-damper systems), through the axial force imposed on the bearing faces by the fluid, enabling the dynamics of the bearing to be studied. An external harmonic force with an amplitude larger than the equilibrium face clearance is imposed on the rotor, used to simulate possible destabilising excitations. Periodic solutions are found through solving the modified Reynolds equation and structural equations simultaneously using a stroboscopic map solver; different solvers have been developed for compressible and incompressible flow, due to an analytic solution for the Reynolds equation being readily available for the latter. Results are given for extreme operating conditions, where for a small amplitude of rotor forcing prediction for the bearing dynamics are similar for compressible and incompressible flow, however, for a large amplitude of the rotor forcing predictions differ significantly. The compressible flow bearing has the faces undergoing a flapping motion when the external force on the rotor is sufficiently large, whereas as an incompressible flow bearing has the faces remaining effectively a constant distance apart with a very thin fluid film.

REFERENCES

[1] Etison I. Squeeze effects in radial face seals. Journal of Lubrication Technology, 102(2): 145 151, 1980.

[2] Salbu, E.O.J. Compressible squeeze films and squeeze bearings. Journal of Basic Engineering, 86(2): 355 364, 1964.

[3] Garratt J.E., Cliffe K.A., Hibberd S. and Power H. Centrifugal inertia effects in high speed hydrostatic air thrust bearings. Journal of Engineering Mathematics, 76(1): 59 80, 2012.

[4] Bailey N.Y., Cliffe K.A., Hibberd S. and Power H. On the dynamics of a high speed coned fluid lubricated bearing. IMA Journal of Applied Mathematics, 79(3): 535 561, 2013.

[5] Bailey N.Y., Cliffe K.A., Hibberd S. and Power H. Dynamics of a parallel, high-speed, lubricated thrust bearing with Navier slip boundary conditions. IMA Journal of Applied Mathematics, 80: 1409 1430, 2014.

[6] Bailey N.Y., Cliffe K.A., Hibberd S. and Power H. Dynamics of a high speed coned thrust bearing with a Navier slip boundary condition. Journal of Engineering Mathematics, 97:1 24, 2016.

[7] Bailey N.Y., Hibberd S. and Power H. Dynamics of a small gap gas lubricated bearing with Navier slip boundary conditions. Journal of fluid Mechanics, 818:68 99, 2017.

[8] Parkins D.W., and Stanley W.T. Characteristics of an oil squeeze film. Journal of Lubrication Technology, 104:497 502, 1982.

[9] Sayma A.I., Breard C., Vahdati M. and M. Imregun, Aeroelasticity analysis of air-riding seals for aeroengine applications. Journal of Tribology, 124(3):607 616, 2002.

[10] Gad-el-Hak, M., MEMS: Introduction and Fundamentals, 2nd edition. Taylor and Francis, Boca Raton, 2006.

[11] Navier C.L.M.H., M'emoire sur les lois du mouvement des fluids. Mem. Acad. Sci. Inst. Fr., 6:389 416, 1829.

[12] Park D.J., Kim C.H., Jang G.H. and Lee Y.B., Theoretical considerations of static and dynamic characteristics of air foil thrust bearing with tilt and slip flow. Tribology International, 41:282 295, 2008.

[13] Huang H., Investigation of slip effect on the performance of micro gas bearings and stability of micro rotor-bearing systems. Sensors, 7:1399 1414, 2007.

[14] Zhang, W.M., Meng, G., Huang, H., Zhou, J.B., Chen J.Y. and Chen D. Characteristics analysis and dynamic responses of micro-gas-lubricated journal bearings with a new slip model. Journal of Physics D: Applied Physics, 41:155305 (16pp), 2008.

Simulation model to investigate effect of support stiffness on dynamic behaviour of a large rotor

E. Kurvinen[1], R. Viitala[2], T. Choudhury[1], J. Sopanen[1]

[1]Department of Mechanical Engineering, LUT University, Finland
[2]Department of Mechanical Engineering, Aalto University, Finland

ABSTRACT

In this study, the effect of support stiffness on the dynamic behaviour of a large industrial rotor is studied using simulation models and experimental test setup having a mechanical apparatus for changing the horizontal support stiffness. Two different simulation models with different levels of complexity were created to simulate the behaviour of the system. Such verified simulation models can be used to predict the critical speeds and minimum vibration operating speed areas. The achievements of this paper could be utilized in the industries using large rotors when operating conditions can be taken into consideration already in rotor system development.

1   INTRODUCTION

In rotating machinery, supports have a critical role as they affect the critical speeds of the system, e.g. in turbomachinery industry (1, 2) and power industry (3). Similarly, in the paper and steel industries, the supports have a significant effect on the dynamic behaviour of relatively large rotors, as subcritical frequencies are on the operational speed range and there are several sources of excitations due to the manufacturing inaccuracies, e.g. non-circular bearing geometries. To avoid these critical frequencies, one approach is to adjust the support stiffness and minimize the operational responses in the paper and steel industry applications.

The large rotors for paper and steel industries are typically made by bulk manufacturing process with very limited customization possibilities. However, rotors manufactured in the same production line can have a significant variation in their dynamic behaviour. This is related to the inaccuracies of manufacturing process that cause e.g. the thickness variation in a rotor body. In large rotating machines, the critical speeds are naturally low which increase the requirements of supporting structure; the supports of large machines need to be massive and highly stiff to preserve high enough critical speeds. Otherwise, the critical speed or its subcritical frequencies decrease, and resonances are possible.

To gain a deeper understanding of the root causes of the dynamic behaviour of a manufactured rotor a physics based white-box model together with measurements can be utilized (4). Nevertheless, the functionality of a rotor is not guaranteed even if the rotor is perfectly manufactured; the operational conditions of a rotor has a remarkable effect on rotor behaviour. Therefore, dynamic rotor behaviour could be only estimated, since the eventual effect of the operating conditions of the rotating system are unknown.

The support structure has a great impact on the dynamic behaviour of a rotating structure (5), which have been studied using physical (6, 7) as well as modal coordinates

(8). Depending on the case, there might be significant cross-coupling between the supports. A few traditional techniques for identifying the foundation parameters are experimental modal analysis (6), modal model-based methods which used response from bearing pedestals with least squares (9) or Extended Kalman Filter (10) to compare bearing forces with measured response, or comparison based on shaft displacements (11). In the case where the cross-coupling between supports is minor a concentrated parameter approach can be utilized by simplifying the support in to horizontal and vertical spring-mass-damper elements (12, 13). In addition, for cases with cross-coupling of supports, the transfer matrices approach can be combined with measured data or even simulation depending on availability of accurate geometry. The frequency response function can be used to obtain modal parameters of the structure by describing the transfer function in various directions at support locations (12) or simple curve fitting methods (14).

The idea of manufacturing for the operating conditions is discussed earlier in several publications where the rotors are optimized to correspond to the demands of the final installation location. For example, Kuosmanen (15) and Widmaier (16) have proven that the dynamic run-out of a rotor could be decreased when the operating conditions were taken into account. In (15) and (16), the rotor feedback measurement was conducted in operating conditions and then with 3D grinding, the dynamic run-out caused by the thickness variation and thermal run-out were compensated. A simulation model for a large rotor was created by Sopanen et al. (17) using flexible multibody formulation along with nonlinear bearing model. Heikkinen et al. (18) proposed a model based on Timoshenko beam elements including the effect of nonlinear bearings with asymmetric rotor properties. They studied the waviness of bearing inner ring and showed that it has significant effect on the dynamic run-out. Viitala et al. (19) developed a predictive method that minimize the excitation from bearing inner ring waviness. In their research, the roundness error of bearing inner ring were reduced by the compensative grinding of the bearing surface.

In this paper, a novel simulation application is presented that focuses on the simulation of the behaviour of a rotor under different speeds and support stiffnesses. The objective of this paper is to study the operational speed map (OSP). The OSP defines the operation speed range and support stiffness values and response amplitudes. These are studied by using two simulation models having different levels of complexity. The first one is a simplified three degree of freedom model that includes the rotor, the supports and the bearings with waviness excitation. Second model is a high-fidelity model proposed by Heikkinen et al. (18). In this paper, the model proposed in (18) is further developed for computational efficiency and parallel computing is utilized to calculate transient behaviour under different speeds and support stiffnesses. The model based on Timoshenko beam elements and nonlinear spherical roller bearings proposed by (20) is computationally efficient, which allows to study the operational speed map for given bearing excitations at various support stiffness and speed conditions in matter of hours. In the actual experimental setup, even if an adjustable stiffness device is available, the measurement process takes a week due to time consumed in changing the support stiffness and speed and capturing the steady state response. Operational speed maps obtained using simulation models are compared with the measured OSP. With the developed simulation method, the effects of support stiffness on the rotor behaviour can be predicted and by using the yielded information, the rotor and support design can be optimized.

The simulation model can be used as a virtual sensor to determine the support stiffness and thus additional stiffness measurement in the current system can be avoided. This can be accomplished as the response peaks that are visible at subharmonic frequencies, e.g. from the acceleration measurements, can be compared to the operational speed map. The virtual sensor can be used to identify the system on the field, which in turn would help in optimizing the dynamic behaviour. In addition, the simulation model can aid in designing the rotor, as in early design phase, the operational speed map can be studied and thus, optimal stiffness values and operational speeds can be selected. As one additional possibility for already manufactured large rotor systems, the support stiffness properties can be identified, and then either stiffened or loosened by adding or removing material from the supports for optimal dynamic behaviour for the application.

2   ROTORDYNAMICS, SIMULATION MODELS AND CASE STUDY

2.1   Rotordynamics
The test rotor used in this study has been considered as a flexible rotor. This means that the test rotor is assumed to have flexural modes. The determination affects specially to the balancing of a rotor, since the flexible rotors have practically infinite number of bending modes with their own critical speeds. However, a significant impact is typically only in the first modes. A rotor can be modelled and balanced also as a rigid rotor, if flexural modes are not remarkable. ISO 21940-12 (21) gives guidance for the determination of a correct rotor type in relation to an operation speed. In there a rotor can be considered rigid if the first flexural resonance speed exceeds the maximum service speed by at least 50 %.

Subharmonic frequencies are integer fractions of the critical speeds. Those occur at 1/2, 1/3, 1/4, ..., etc. times a critical speed. Subharmonic frequencies are harmful for large rotating machines that usually uses the frequencies under the critical speed. At these frequencies, harmonic excitation causes resonance in which the excitation frequency coincidence with the critical speed. The harmonic excitation produces the excitation as a function of rotating speed. Typical source for the harmonic excitation can be e.g. the thickness variation of a rotor body that produces excitation two times per revolution. This kind of excitation causes the resonance at speed of 1/2 times the critical speed and, thus, this speed is also called as half critical speed. Similar mechanism excites also lower subharmonic frequencies.

2.2   Simulation model descriptions
Two simulation models with different levels of complexity are utilized in this study. The first one (simple) is a three degree of freedom model that can describe the first flexible mode of a rotor-bearing-support system. The second model (detailed) is a beam element model of the rotor including rotor's asymmetry due the shell thickness variation as well as nonlinear spherical roller bearing model. Both simulation models include description of bearing waviness excitation. The simplified model is solved in frequency domain while the detailed model is solved with time transient numerical integration.

2.2.1 *Three degree of freedom model (simple)*
The three degree of freedom model is constructed in such a way that it can represent the first flexible mode of the roll-support system. Since the first flexible mode is symmetric, the model represents only one half of the actual system. Supported roll and free-free roll behaviour is depicted in Figure 1.

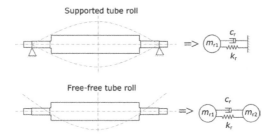

Figure 1. Simplified 2DOF rotor models for supported and free-free rotor.

The supported rotor's critical frequency (f_{supp}) and free-free frequency (f_{free}) can calculated as

$$f_{supp} = \frac{1}{2\pi}\sqrt{\frac{k_r}{m_{r1}}} \tag{1}$$

$$f_{free} = \frac{1}{2\pi}\sqrt{\frac{k_r(m_{r1}+m_{r2})}{m_{r1}m_{r2}}}, \tag{2}$$

where k_r is the support stiffness, m_{r1} represents the middle part of the rotor and m_{r2}, represents the end part of the rotor.

Figure 2 shows the schematic of the model. In the model, one half of the roll is further divided into two parts (m_{r1} and m_{r2}) that are connected with spring (k_r) and damper (c_r). The roll is connected to support (m_s) with bearings having spring coefficient (k_b) and damping coefficient (c_b), while the support mass (m_s) is connected with spring (k_s) and damper (c_s) to the ground.

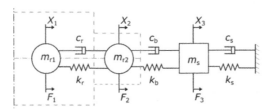

Figure 2. Schematic of simplified model of 3 DOF representing half of the rotor and one support.

To capture the first natural frequency of the roll properly with the simplified model, the following relations can be created:

1.) One-half of total mass of the roll is $m/2 = m_r = m_{r1} + m_{r2}$
2.) Supported roll frequency, $f_{supp} = 30$ Hz
3.) 1st free-free frequency of the roll, $f_{free} = 75$ Hz

Using the above relations and Eqs. (1) and (2), the parameters for the simplified roll model can be obtained. The damping coefficient for the roll part (c_r) can be obtained by assuming 1% modal damping ratio. Support mass, m_s, is 190 kg. The bearing stiffness in horizontal direction is $2.5 \cdot 10^8$ N/m and the bearing damping (c_b) is assumed to be 0.25% of the bearing stiffness. The support stiffness is $2 \cdot 10^8$ N/m in vertical direction and in horizontal direction, it is varied from 2.14 MN/m to 18.34 MN/m. For support damping, c_s, 2% damping ratio to the critical damping coefficient is used.

Bearing inner ring waviness is assumed to cause harmonic excitation forces that are included in F_2 and F_3. Bearing waviness excitation mechanism is fairly complicated, but in this study, it is assumed to cause forced displacement excitation between the roll and support. In general, bearing waviness generates several harmonic excitations and the kth harmonic waviness excitation can be written as

$$\delta_k = A_k \cos(k\omega t + \phi_k), \tag{3}$$

where A_k the amplitude and ϕ_k is the phase of the kth is the order waviness, ω is the rotor rotating frequency. Waviness can be considered base motion, in which the excitation forces are transferred through the bearing stiffness and damping. Therefore, harmonic spring and damping forces for the kth order waviness excitation can be written as follows

$$F_2(t) = \underbrace{(-k_b A_k \sin \phi_k - c_b A_k k\omega \cos \phi_k)}_{F_s} \sin(k\omega t) + \underbrace{(k_b A_k \cos \phi_k - c_b A_k k\omega \sin \phi_k)}_{F_c} \cos(k\omega t) \tag{4}$$

Corresponding counteraction force is applied to the support (F_3). Considering multiple waviness excitations, the equation of motion of system can be written as

$$\mathbf{M}\ddot{\mathbf{x}}(t) + \mathbf{C}\dot{\mathbf{x}}(t) + \mathbf{K}\mathbf{x}(t) = \sum_k \left(\mathbf{F}_s^k \sin k\omega t + \mathbf{F}_c^k \cos k\omega t\right) \tag{5}$$

Superposition principle can be used to obtain the final steady state solution as follows

$$\mathbf{x}(t) = \sum_k \left(\mathbf{a}^k \sin k\omega t + \mathbf{b}^k \cos k\omega t\right) \tag{6}$$

where the coefficient vectors $\mathbf{a}^k$ and $\mathbf{b}^k$ can be solved for each harmonic excitation k as follows

$$\begin{bmatrix} \mathbf{a}^k \\ \mathbf{b}^k \end{bmatrix} = \begin{bmatrix} \mathbf{K} - (k\omega)^2\mathbf{M} & -k\omega\mathbf{C} \\ k\omega\mathbf{C} & \mathbf{K} - (k\omega)^2\mathbf{M} \end{bmatrix}^{-1} \begin{bmatrix} \mathbf{F}_s^k \\ \mathbf{F}_c^k \end{bmatrix}. \tag{7}$$

2.2.2 High-fidelity simulation model (detailed)
In this section, the model developed by Heikkinen et al. (18) is implemented to the test case. The simulation model included asymmetric 3D beam elements based on Timoshenko beam theory. The asymmetry is induced by varying the thickness of the tube section based on ultrasonic measurement similarly as in (22). The thickness variation is implemented in the model by defining the thickness profiles of the cross sections along the length of the tube section, thus affecting to the area moment of inertia.

The rotor is supported with two SKF 23124 CCK/W33 spherical roller bearings. The spherical roller bearings are modelled with nonlinearity included, similarly as in Ghalamchi et al. (20). In the simulations the nominal bearing clearance of 60 μm is used, and measured waviness profiles of measured sections are shown in Table 1. Waviness components from twice to six times per revolution were measured from the bearing inner ring for the service end and drive end of the machine.

Table 1. Measured bearing inner ring roller path waviness amplitudes.

Service end	First roller element path			Second roller element path		
	k	Amplitude [μm]	Phase [rad]	k	Amplitude [μm]	Phase [rad]
	2	0.4647	7.1075	2	0.5757	5.5588
	3	0.1501	1.6304	3	0.1304	7.5205
	4	0.2520	2.8345	4	0.2024	2.8113
	5	0.3197	4.0324	5	0.4019	3.8472
	6	0.1475	1.8118	6	0.0479	3.7840
Drive end						
	2	0.4361	2.9981	2	0.4802	7.6331
	3	0.4186	3.6507	3	0.4092	3.6553
	4	0.3121	7.7142	4	0.1651	7.4677
	5	0.3346	2.5530	5	0.1176	1.6024
	6	0.1569	5.1329	6	0.1283	4.2674

The measured roll section thickness variations are included in the simulation model. The supports are modelled as mass-spring-damper elements, individually in horizontal and vertical directions. Damping for support structures has 2% damping ratio for horizontal and 3% for vertical direction. Modal damping ratios 1.5% (1st), 2% (2nd), 2.5% (3rd) and 3.0% (4th to 6th) where the value in parenthesis corresponds to the flexible mode number. In the transient analysis, model reduction is applied and the number of retained modes is 16. Simulation runs are conducted for 9 seconds with a sampling rate of 2000 Hz (time-step of 0.0005s). The resulting response are captured at the bearing locations. The computational time is approximately 300 seconds per single simulation of 9 seconds. However, the computation time is dependent on the system parameters, such as parameters of nonlinear bearing model. Calculations are conducted with a Triton cluster (Node Dell PowerEdge C6420 with 2x20 core Xeon Gold 6148 at 2.4 GHz and 192 GB DDR4-2667 memory), having two nodes, both with 40 CPU's parallel computing. With that setup, the computation time is 15 hours.

2.3 System and measurement setup

The test setup has a possibility to vary the horizontal stiffness and thus, similar data set could be measured as generated in the simulation. The adjustable stiffness in the test setup is implemented by using a similar structure as in balancing machines where a rotor is supported by plate springs. Subsequently, the stiffness for the support is provided through an external beam. The structure is implemented at both rotor ends. The mechanism for adjustable stiffness and force sensor locations are depicted in Figure 3.

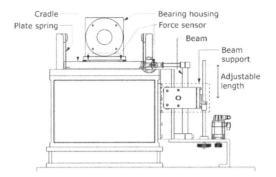

Figure 3. Mechanism for adjusting the support stiffness in horizontal direction.

The studied rotor is a paper machine guiding roll (tube roll). The mass of the rotor is 720 kg with a total length of 5 m and a 4 m long tube section. Figure 4 shows the significant dimensions of the rotor. Other additional masses that affect to the horizontal movement of the shaft ends are the bearing housing masses and the masses of the parts that constitute the adjustable stiffness device (m_s).

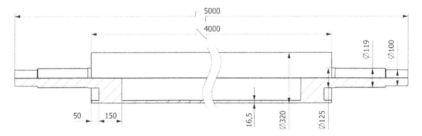

Figure 4. Studied rotor cross-section with dimensions (in mm).

The verification measurement is conducted by using the radial bearing force sensors at both bearings. The measured rotation speed range is 4 – 18 Hz in which the rotor is accelerated with 0.05 Hz increments. The acceleration ramp is repeated with 31 different stiffnesses. The acceleration is limited to the maximum speed below the critical speed. The measurement procedure is conducted as follows:

1. The stiffness is adjusted at the lowest point
2. The rotor is accelerated at 4 Hz
3. 100 rounds are measured
4. The rotor speed is increased with the increment of 0.05 Hz
5. Steps 4 and 5 are repeated until the last rotating speed is reached
6. The stiffness is increased
7. Steps 2 to 6 are repeated until the last stiffness level is reached.

The measured data is post processed using time synchronous averaging (TSA) that is originally developed by (23, 24) The method enables the averaging of a phase locked signal and representing it as a single round. From this averaged signal the maximum values of each direction (horizontal and vertical) are captured. The details of the force measurement setup can be found in Viitala's et al. (25) research in which same measurement method was used.

3   RESULTS

3.1   Simulation model and validation

The simulation models are validated with free-free frequencies and supported frequencies measured from actual machine, i.e. 75 Hz for free-free rotor, and supported frequencies of 20.9 Hz in horizontal direction and 30 Hz in vertical direction. The first and second bending modes (horizontal) of the rotor are shown in Figure 5. In the vertical direction, the first bending mode occurs at 30 Hz (30.5 Hz with infinitely stiff support).

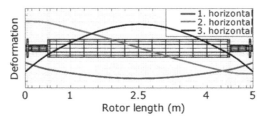

Figure 5. First and second horizontal bending modes at 20.9 Hz and 34.3 Hz with maximum support stiffness.

3.1.1   *Operational speed map with simple simulation model*

Figure 6 depicts the simple simulation response results. In the model, the waviness amplitudes of the service end at the first roller path are used. Due to the computational simplicity of the model, the results are calculated with 100 different stiffness values and speed range from 0 to 18 Hz. The numbering e.g. "1-2" refers to 1st bending mode and 2nd fraction of critical speed etc. As can be seen in the figure the simple model can only include the first bending mode.

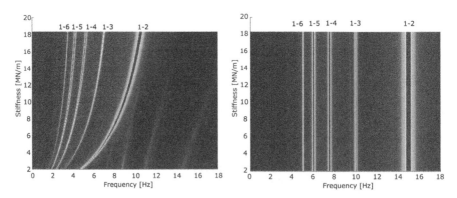

Figure 6. Simplified model responses in horizontal (left) and vertical (right) directions.

3.1.2   *Operational speed map with detailed simulation model*

In the analysis, 281 points from 4 Hz to 18 Hz (increment of 0.05 Hz) are analysed along with 31 different horizontal support stiffness values, yielding to a network of 8711 calculations. From these simulations, the maximum amplitude is captured and plotted with the speed and stiffness values used in the analysis. Figure 7 depicts the operation speed map for horizontal and vertical directions.

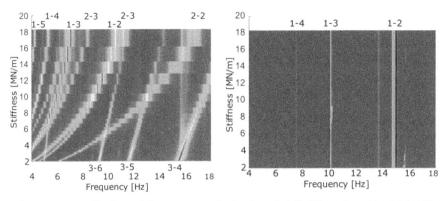

Figure 7. Detailed model responses in horizontal (left) and vertical (right) directions.

3.1.3 *Operational speed map with measurement*

In the measurement, with the highest stiffness value set in the horizontal direction, the full range of 4 Hz to 18 Hz with 0.05 Hz increments is measured. As the support stiffness is decreased the maximum speed is decreased to avoid resonance with the 1x component in the measurements.

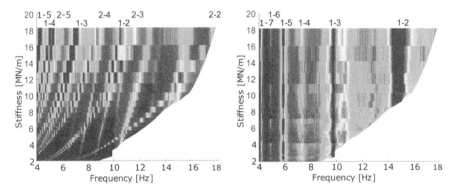

Figure 8. Measured responses in horizontal (left) and vertical (right) directions.

3.2 Results comparison

Table 2 depicts the results comparison at the highest stiffness 18.32 MN/m and at the lowest stiffness 2.04 MN/m. In the table the frequency peaks shown in Figure 6-Figure 8 are collected and compared.

465

Table 2. Comparison of simulated and measured horizontal subharmonic frequencies.

Support stiffness (MN/m)	Mode # - harmonic	Simulated (Simple) (Hz)	Simulated (Detailed) (Hz)	Measured (Hz)	Difference measured to simple [%]	Difference measured to detailed [%]
18.32	1-2	10.60	10.60	10.58	0.19	0.19
	1-3	7.05	7.05	7.06	-0.14	-0.14
	1-4	5.30	5.35	5.32	-0.38	0.56
	1-5	4.25	4.25	4.27	-0.47	-0.47
	2-2	-	16.80	17.84	-	-6.19
	2-3	-	11.20	11.63	-	-3.84
	2-4	-	8.45	8.75	-	-3.55
	3-5	-	14.40	12.82	-	10.97
	3-6	-	11.60	10.83	-	6.64
	3-7	-	-	9.15	-	-
	3-8	-	-	8.05	-	-
2.04	1-2	4.80	4.65	4.97	-3.44	-3.44
	2-2	-	6.25	6.90	-	-10.46
	2-3	-	4.15	4.62	-	-11.25
	3-4	-	15.55	-	-	-
	3-5	-	11.65	-	-	-
	3-6	-	9.25	8.49	-	8.22
	3-7	-	-	7.20	-	-
	3-8	-	-	6.31	-	-

4   DISCUSSION

A comparison between the simplified and detailed simulation models shows that the accuracy is almost same for both models when considering the 1st bending mode. Both agree well with the measurements and have less than 1% error compared to measured results in case of the highest support stiffness and, correspondingly, 3.5% error in case of the lowest support stiffness. The simplified simulation model only consists of the first bending mode and thus is not able to show higher modes.

With the 2nd bending mode, the detailed simulation model gives approximately 4-6% lower frequencies compared to measured ones at the highest support stiffness and 11% lower frequencies in case of the lowest support stiffness. For the 3rd bending mode, the detailed simulation model estimates 6-10% higher frequencies than measured at the highest support stiffness and 8% higher frequency in case of the lowest support stiffness. With lowest support stiffness the support behaviour becomes more dominant when compared to the highest stiffness where the rotor flexibility is more dominating. The higher order waviness components are visible well in both simulation models.

As can be interpreted from the results, variation of the horizontal support stiffness affects differently to different bending critical frequencies of the modes. In simulation results of both models, the relative frequency change is remarkably similar as in the verification measurements. In the measurements, the support stiffness was lowered 88.9% from its highest value and this yielded 53.6%, 60.7% and 21.1% change in the first, second and third measured bending mode frequencies, respectively. This result reveals the sensitivity of each bending critical speed to the variation of support stiffness. The corresponding changes in the detailed simulation were 56.1%, 62.8%, 19.1%. Thus, the largest relative error 2.5% was in the first mode. The simple simulation model yielded 54.4% change in the first mode frequency that is even closer than the relative change in detailed simulation model.

Finally, it should be noted that the 7th and 8th harmonic response due to bearing waviness can be seen in the measurements but those are not included in the simulations. However, from practical point of view the higher order waviness components does not have significant role.

5   CONCLUSIONS

The two simulation models were created and their capacity of producing operational speed map in a case study were reported. The study shows that the simple model can reproduce the operational speed map within 0.5% accuracy when compared to measured operational speed map. However, this model can describe only the first flexible bending mode. Furthermore, when the simple and complex simulation cases were compared, there was no major difference in the results to the first bending mode, i.e. the asymmetry and nonlinearity is not causing major effect to the results. The addition that the detailed model gives is the inclusion of second and third bending mode and their effect is visible in the operational speed map. To summarize, the simple model can generate operational speed map with a high accuracy and from computationally point of view, the calculation can be made within seconds, even with very fine computational grid. To design the large rotors for specific application, such maps helps to optimize the application specific rotation speed to designed support stiffness value.

ACKNOWLEDGEMENT

The research was funded by the Academy of Finland (project no. 313675 and 313676). We acknowledge the computational resources provided by the Aalto Science-IT project.

REFERENCES

[1] J. Hong, K. Shaposhnikov, D. Zhang, and Y. Ma, "Theoretical modeling for a rotor-bearing-foundation system and its dynamic characteristics analysis," in *Proceedings of the 9th IFToMM International Conference on Rotor Dynamics*. Springer, 2015, pp. 2199–2214.

[2] W. Yan, K. Shaposhnikov, P. Yu, Y. Ma, and J. Hong, "Experimental investigation and numerical analysis on influence of foundation excitation on the dynamics of the rotor system," in *ASME Turbo Expo 2015: Turbine Technical Conference and Exposition*. American Society of Mechanical Engineers Digital Collection, 2015.

[3] P. Paturu, I. Vinoth Kanna, and G. Mallela, "A detailed analysis of free vibration on 70 MW hydro power turbine rotor," *International Journal of Ambient Energy*, pp. 1–8, 2019.

[4] M. Cocconcelli, L. Capelli, J. Cavalaglio Camargo Molano, and D. Borghi, "Development of a methodology for condition-based maintenance in a large-scale application field," *Machines*, vol. 6, no. 2, p. 17, 2018.

[5] A. Lees, J. Sinha, and M. Friswell, "Model-based identification of rotating machines," *Mechanical Systems and Signal Processing*, vol. 23, no. 6, pp. 1884–1893, 2009.

[6] J. K. Sinha, A. Lees, M. Friswell, and R. Sinha, "The estimation of foundation models of flexible machines," in *Proceedings of the Third International Conference Identification in Engineering* Systems, *Swansea, UK*, 2002, pp. 300–310.

[7] J. K. Sinha, M. Friswell, A. Lees, and R. Sinha, "An alternate method for reliable modelling of flexible rotating machines," *Proceedings of VETOMAC*, vol. 2, 2002.

[8] M. Smart, M. Friswell, A. Lees, and U. Prells, "Estimating turbogenerator foundation parameters," *Proceedings of the Institution of Mechanical Engineers, Part C: Journal of Mechanical Engineering Science*, vol. 212, no. 8, pp. 653–665, 1998.

[9] A. Lees, "The least squares method applied to identify rotor/foundation parameters," in *IMechE., Conference," Vibrations in rotating machinery*, 1988.

[10] R. Provasi, G. A. Zanetta, and A. Vania, "The extended kalman filter in the frequency domain for the identification of mechanical structures excited by sinusoidal multiple inputs," *Mechanical Systems and Signal Processing*, vol. 14, no. 3, pp. 327–341, 2000.

[11] N. Feng and E. Hahn, "Including foundation effects on the vibration behaviour of rotating machinery," *Mechanical Systems and Signal Processing*, vol. 9, no. 3, pp. 243–256, 1995.

[12] K. Cavalca, P. Cavalcante, and E. Okabe, "An investigation on the influence of the supporting structure on the dynamics of the rotor system," *Mechanical Systems and Signal Processing*, vol. 19, no. 1, pp. 157–174, 2005.

[13] P. F. Cavalcante and K. Cavalca, "A method to analyse the interaction between rotor-foundation systems," in *SPIE proceedings series*, 1998, pp. 775–781.

[14] R. Stephenson and K. Rouch, "Generating matrices of the foundation structure of a rotor system from test data," *Journal of sound and vibration*, vol. 154, no. 3, pp. 467–484, 1992.

[15] P. Kuosmanen, *Predictive 3D roll grinding method for reducing paper quality variations in coating machines*. Helsinki University of Technology, 2004.

[16] T. Widmaier, *Optimisation of the roll geometry for production conditions*. Aalto University Publication Series Doctoral Dissertations 156/2012. 184 p, 2012.

[17] J. Sopanen, J. Heikkinen, and A. Mikkola, "Experimental verification of a dynamic model of a tube roll in terms of subcritical superharmonic vibrations," *Mechanism and Machine Theory*, vol. 64, pp. 53–66, 2013.

[18] J. E. Heikkinen, B. Ghalamchi, R. Viitala, J. Sopanen, J. Juhanko, A. Mikkola, and P. Kuosmanen, "Vibration analysis of paper machine's asymmetric tube roll supported by spherical roller bearings," *Mechanical Systems and Signal Processing*, vol. 104, pp. 688–704, 2018.

[19] R. Viitala, T. Widmaier, and P. Kuosmanen, "Subcritical vibrations of a large flexible rotor efficiently reduced by modifying the bearing inner ring roundness profile," *Mechanical Systems and Signal Processing*, vol. 110, pp. 42–58, 2018.

[20] B. Ghalamchi, J. Sopanen, and A. Mikkola, "Simple and versatile dynamic model of spherical roller bearing," *International Journal of Rotating Machinery*, vol. 2013, 2013.

[21] "ISO ISO21940-12 Mechanical vibration. Rotor balancing. Part 12: Procedures and tolerances for rotors with flexible behaviour," International Organization for Standardization, Geneva, CH, Standard, 2016.

[22] J. Juhanko, E. Porkka, T. Widmaier, and P. Kuosmanen, "Dynamic geometry of a rotating cylinder with shell thickness variation," *Estonian Journal of Engineering*, vol. 16, no. 4, p. 285, 2010.

[23] P. McFadden, "A revised model for the extraction of periodic waveforms by time domain averaging," *Mechanical Systems and Signal Processing*, vol. 1, no. 1, pp. 83–95, 1987.

[24] P. McFadden and M. Toozhy, "Application of synchronous averaging to vibration monitoring of rolling element bearings," *Mechanical Systems and Signal Processing*, vol. 14, no. 6, pp. 891–906, 2000.

[25] R. Viitala, R. Viitala, and P. Kuosmanen, "Method and device for large rotor bearing force measurement," in *2019 IEEE International Instrumentation and Measurement Technology Conference (I2MTC)*, May 2019, pp. 1–6.

Identification of frame dynamics of vertically oriented high-speed steam generator using model update procedure for reduced-order model

E. Sikanen[1], J. Heikkinen[1], T. Sillanpää[1], E. Scherman[2], J. Sopanen[1]

[1]Lappeenranta-Lahti University of Technology, Lappeenranta, Finland
[2]Saimaa University of Applied Sciences, Lappeenranta, Finland

ABSTRACT

Active Magnetic Bearing (AMB) systems are common in high-speed rotating machinery. Bearing control system input is the relative displacement data measured between the sensor and the measurement surface. In case of frame resonance, controller may create a feedback loop which may amplify the frame vibration modes. The frame vibration dynamics of 1 MW steam generator with AMB-support is studied. Vibration mode identification using FE-based model are discussed. Model-order reduction techniques are utilized to solve a large-scale numerical problem. The paper shows an update procedure for a computationally efficient reduced-order model that can be used for identifying measured unclear vibration modes.

1   INTRODUCTION

Active Magnetic Bearing (AMB) systems are often considered as the most suitable bearing solution for rotating high-speed applications. As the AMB system utilize active feedback control, the quality of the position signal is essential for the stability of the bearing control system. In AMB systems, rotor position signal is measured between the stator frame and the rotor lamination yielding a relative displacement. In cases where frame dynamics significantly contributes to system dynamics, the relative displacement of AMB sensor should not be considered as absolute reading since in worst case it can cause amplified vibration due to a possible feedback loop in the conventional AMB control system [1]. Often, as the power rating of the high-speed machines increase, the size and mass of the machine structure is increased, and thus, the need of analysing the frame dynamics becomes more important. Electrical machine frame vibration dynamics with magnetically suspended rotor seems to be a topic not often discussed in literature.

The finite element method (FEM) is often used for solving numerical vibration problems. The use of three-dimensional (3D) solid element models is nowadays a standard practice. Often, due to fine discretization of the structure, the problem size may become very large, thus causing very long computation times in solution. Therefore, model-order reduction becomes necessary. Vibration analysis in modal domain has its own requirements compared to time domain analysis in which contacts and constraints may be expressed simply as time-dependent forces at boundaries. In modal domain, individual components are required have physical boundary degrees-of-freedom for body-to-body coupling to be utilized. The concept of superelement-based techniques is well known, and perhaps the most well-known method utilized is the Craig-Bampton method [2]. When using substructuring [3] and modelling individual subdomains as single superelements, a very large problem can be reduced significantly without losing much accuracy [4]. Although, there are multiple model-order reduction methods [5] available, such as System Equivalent Reduction Expansion

Process (SEREP) [6] and Improved Reduced System (IRS) [7], the Craig-Bampton method has advantages as the superelement can have variable boundary constraints, and thus, assembling of multiple superelements is enabled. Generating and assembling superelements is possible in commercial FE-software, but the model update options for reduced-order model seems to be very limited [8].

As a case example, the frame dynamics of vertically oriented steam generator is investigated. In the case example, a harmful frame resonance was detected at operating speed range during the AMB commissioning. The problematic frame resonance frequency was identified with both axial and radial AMB sensors, but the mode of this frequency was still unknown. FEM was utilized for modelling the complete generator assembly. In order capture the machine dynamics properly the mesh have to be dense. As a result, the size of finite element (FE) problem became too large to be solved within reasonable time. Thus, model-order reduction techniques were utilized. Substructuring was utilized in order to generate submodels for the Craig-Bampton transformation. The generated superelements were assembled and eigenvectors were solved, and then expanded into global degrees-of-freedom in order to visualize the vibration modes. The problematic frame vibration mode was then identified.

In addition, the suspected axial-radial coupling of the vibration mode was studied using harmonic response analysis. For this analysis, a reduced-order model update procedure is proposed. The model update referred here is modal-based update for dynamic problems in modal domain focusing on eigenmode frequency and corresponding modal damping ratio updates. A customizable 3D solid FEM-based code implemented in MATLAB environment is utilized for the modelling and solving of eigenproblem with multiple superelements [9]. The main reason for using in-house made FEM code arises from the need of fully customizable solver programming in order to make all steps required, generation of superelements, assembly and model update, in the same modelling environment. As result, the proof of existence of the suspected axial-radial vibration coupling in the case study was revealed based on the model update procedure proposed in this work.

2   SUBSTRUCTURING AND REDUCED-ORDER MODEL UPDATE

Basic idea of substructuring is to divide the large model into several submodels. Every submodel is transformed into an arbitrary shaped element, generally called as superelement, having a number of selected boundary degrees-of-freedom. Common method for generating a superelement is the Craig-Bampton transformation. Craig-Bampton method produces physical boundary degrees-of-freedom, thus, allowing superelement to be connected with other elements. Craig-Bampton transformation allows full expansion of the reduced-order solution. Thus, it is ideal method for studying large complex-shaped structures in frequency domain. Still, updating dynamic properties of individual submodel is not directly possible due to the characteristics of Craig-Bampton method. Assembled superelement-based reduced-order model can be updated after modal decoupling.

2.1   Craig-Bampton method
Due to the large size of the vibration problem discussed in this work, substructuring of a full FE-model is utilized. Craig-Bampton method is used for generating superelements. The Craig-Bampton transformation matrix for generating a superelement can be written as follows [2]:

$$\mathbf{T}_{\mathrm{CB}} = \begin{bmatrix} \boldsymbol{\Phi}_{ik} & \boldsymbol{\Psi}_{ib} \\ \mathbf{0} & \mathbf{I}_{bb} \end{bmatrix} \tag{1}$$

where $\boldsymbol{\Psi}_{ib}$ is the coupling matrix between the internal and boundary degrees of freedom, which can be written as follows:

$$\boldsymbol{\Psi}_{ib} = -\mathbf{K}_{ii}^{-1}\mathbf{K}_{ib} \tag{2}$$

and $\boldsymbol{\Phi}_{ik}$ is the matrix containing a subset of eigenvectors as follows:

$$\boldsymbol{\Phi}_{ik} = [\boldsymbol{\Phi}_{i1}\,\boldsymbol{\Phi}_{i2}\ldots\boldsymbol{\Phi}_{ik}] \tag{3}$$

The eigenvectors used for making Craig-Bampton transformation can be solved as follows:

$$(\mathbf{K}_{ii} - \lambda_n\mathbf{M}_{ii})\boldsymbol{\Phi}_n = 0 \tag{4}$$

Using an elastic stiffness matrix as an example, the structural matrices shall be re-arranged for the Craig-Bampton transformation as follows:

$$\mathbf{T}_R^{\mathrm{T}}\mathbf{K}^e\mathbf{T}_R = \mathbf{K}_R^e = \begin{bmatrix} \mathbf{K}_{ii} & \mathbf{K}_{ib} \\ \mathbf{K}_{bi} & \mathbf{K}_{bb} \end{bmatrix} \tag{5}$$

where $\mathbf{T}_R$ is the transformation matrix for matrix reordering, $\mathbf{K}^e$ is the original elastic stiffness matrix and $\mathbf{K}_R^e$ is the re-arranged elastic stiffness matrix derived for purpose of generating a superelement.

Using an elastic stiffness matrix as an example, the Craig-Bampton transformation is completed and deriving a superelement can be written as follows:

$$\mathbf{K}_{\mathrm{CB}}^e = \mathbf{T}_{\mathrm{CB}}^{\mathrm{T}}\mathbf{K}_R^e\mathbf{T}_{\mathrm{CB}} \tag{6}$$

2.2 Expanding eigenvalue solution

The reduced-order eigenvalue problem is solved based on assembled FE-model made of individual superelements. Using the Craig-Bampton transformation, the reduced-order solution can be expanded into original size as follows:

$$\boldsymbol{\Phi}_{n,G} = \mathbf{T}_R\left(\mathbf{T}_{\mathrm{CB},k}\boldsymbol{\Phi}_{n,k}\right) \tag{7}$$

where the subscripts k and G indicates the index of superelement and the expanded original global order of the degrees-of-freedom, respectively.

2.3 Assembly of superelements

Assuming a typical rotating machine construction consisting three main parts: support, frame and rotor, the assembled equations of motion can be written as follows:

$$
\begin{bmatrix} \mathbf{M}_{CB,S} & \mathbf{0} & \mathbf{0} \\ \mathbf{0} & \mathbf{M}_{CB,F} & \mathbf{0} \\ \mathbf{0} & \mathbf{0} & \mathbf{M}_{CB,R} \end{bmatrix} \begin{bmatrix} \ddot{\mathbf{x}}_S \\ \ddot{\mathbf{x}}_F \\ \ddot{\mathbf{x}}_R \end{bmatrix} + \begin{bmatrix} \mathbf{C}_{CB,S} + \mathbf{C}_S + \mathbf{C}_c & -\mathbf{C}_c & \mathbf{0} \\ -\mathbf{C}_c & \mathbf{C}_{CB,F} + \mathbf{C}_c + \mathbf{C}_b & -\mathbf{C}_b \\ \mathbf{0} & -\mathbf{C}_b & \mathbf{C}_{CB,R} + \Omega\mathbf{G}_{CB,R} + \mathbf{C}_b \end{bmatrix}
$$

$$
\begin{bmatrix} \dot{\mathbf{x}}_S \\ \dot{\mathbf{x}}_F \\ \dot{\mathbf{x}}_R \end{bmatrix} + \begin{bmatrix} \mathbf{K}_{CB,S}^e + \mathbf{K}_S + \mathbf{K}_c & -\mathbf{K}_c & \mathbf{0} \\ -\mathbf{K}_c & \mathbf{K}_{CB,S}^e + \mathbf{K}_S + \mathbf{K}_c & -\mathbf{K}_b \\ \mathbf{0} & -\mathbf{K}_b & \mathbf{K}_{CB,R}^e \end{bmatrix} \begin{bmatrix} \mathbf{x}_S \\ \mathbf{x}_F \\ \mathbf{x}_R \end{bmatrix} = \boldsymbol{F}
$$

(8)

where matrices $\mathbf{M}$, $\mathbf{K}$, $\mathbf{C}$ and $\mathbf{G}$, in general, contribute to mass, stiffness, damping and gyroscopic effects, respectively, the subscripts CB, S, F and R contribute to Craig-Bampton transformation, support, frame assembly and rotor, respectively, Ω is the rotor angular velocity, x is the vector of nodal coordinates and $\mathbf{F}$ is the vector of external forces. The coupling stiffness and damping matrices $\mathbf{K}_S$ and $\mathbf{C}_S$, $\mathbf{K}_c$ and $\mathbf{C}_c$ and $\mathbf{K}_b$ and $\mathbf{C}_b$ contribute to foundation-to-support, support-to-frame and frame-to-rotor couplings.

2.4 Assembled model update

FE-model update based on measured dynamic properties becomes straightforward. The eigenfrequency and modal damping ratio can be updated based on measured data to match the corresponding eigenmode. The assembled superelement expression is used as basis. Equation (8) can be rewritten as follows:

$$
\mathbf{M}_A\ddot{x}_A + \mathbf{C}_A\dot{x}_A + \mathbf{K}_A x_A = \boldsymbol{F}_A \tag{9}
$$

where the subscript A indicates assembled matrices. After solving a subset of lowest undamped eigenvectors of Equation (9) and mass normalizing the eigenvectors, the structural modal coordinate matrices can be written as follows:

$$
\begin{aligned}
\widetilde{\mathbf{M}} &= \boldsymbol{\Phi}_A^{\mathrm{T}}\mathbf{M}_A\boldsymbol{\Phi}_A \\
\widetilde{\mathbf{K}} &= \boldsymbol{\Phi}_A^{\mathrm{T}}\mathbf{K}_A\boldsymbol{\Phi}_A \\
\widetilde{\mathbf{C}} &= \boldsymbol{\Phi}_A^{\mathrm{T}}\mathbf{C}_A\boldsymbol{\Phi}_A
\end{aligned} \tag{10}
$$

where $\widetilde{\mathbf{M}}$, $\widetilde{\mathbf{K}}$ and $\widetilde{\mathbf{C}}$ are the modal mass, stiffness and damping matrices, respectively. Updated modal matrices for modal harmonic response analysis can be written as follows:

$$
\begin{aligned}
\widetilde{\mathbf{M}} &= \mathbf{I} \\
\widetilde{\mathbf{K}} &= \boldsymbol{\omega}^{\mathrm{T}}\boldsymbol{\omega} \\
\widetilde{\mathbf{C}} &= 2\boldsymbol{\omega}\boldsymbol{\xi}
\end{aligned} \tag{11}
$$

where ω and ξ are diagonal matrices containing the modal angular frequencies and modal damping ratios, respectively.

The update procedure presented is valid only for updating the eigenfrequency and modal damping ratio of the particular eigenpair, not for updating the eigenmode vector itself. Updating individual eigenvector has certain limitations as the initial selection of the boundary degrees-of-freedom will have influence on the outcome, and thus, the Craig-Bampton transformation must be redone.

2.5 Harmonic response

The equation for harmonic problem using modal coordinate expression can be written as follows [10]:

$$\begin{bmatrix} \widetilde{\mathbf{K}} - \omega^2 \widetilde{\mathbf{M}} & -\omega \widetilde{\mathbf{C}} \\ \omega \widetilde{\mathbf{C}} & \widetilde{\mathbf{K}} - \omega^2 \widetilde{\mathbf{M}} \end{bmatrix} \begin{bmatrix} \widetilde{\boldsymbol{p}}_s \\ \widetilde{\boldsymbol{p}}_c \end{bmatrix} = \begin{bmatrix} \boldsymbol{\Phi}_A^{\mathsf{T}} F_{A,s} \\ \boldsymbol{\Phi}_A^{\mathsf{T}} F_{A,c} \end{bmatrix} \tag{12}$$

The physical displacement vector of the assembly $\mathbf{x}_A$ can be written as follows:

$$\boldsymbol{x}_A = \boldsymbol{\Phi}_A (\tilde{\boldsymbol{p}}_s \, sin\theta + \tilde{\boldsymbol{p}}_c \, cos\theta) \tag{13}$$

3   CASE EXAMPLE: HIGH-SPEED STEAM TURBINE

The studied 1 MW 12,500 rpm high-speed turbine-generator construction under investigation is illustrated in Figure 1. Vertical design was selected in order to have less bearing load in radial direction, thus, minimizing the length of radial AMB laminations on the rotor. In addition, rotor has inner water-cooling channels, which were more functional in case of vertically oriented rotor. In the lower end of the frame construction is located the turbine housing, and on the shaft the double-sided four-phase radial steam turbine impeller. The center of mass of the machine is relatively high due to the need for connecting the steam outlet tubing that would be fastened at the bottom of turbine housing.

In order to visualize the problematic frame vibration mode, a complete generator assembly including the full support leg construction was modelled using FEM. Linear time invariant system is assumed. In order to reduce the problem size, undamped eigenvalues are solved. Regardless of the simplifications to the model of the generator assembly, the FE-problem size was too large to be solved without model-order reduction in sufficient time. Therefore, substructuring of full assembly was utilized. The assembly was split into three submodels, and superelements were generated based on every submodel using Craig-Bampton transformation. The details of the connectivity of superelements are discussed. Using reduced-order assembled superelement-based model, the solution of eigenvalue problem was obtained, and the problematic frame vibration was managed to be visualized using reduced-order solution expansion based on the Craig-Bampton transformation. Also, harmonic response analysis was performed in order to study further the possible axial-radial coupling of this frame vibration mode in question. For this response analysis, the reduced-order model dynamical properties were updated using the measured vibration data so that mode frequencies and damping ratios were obtained from the measurement data and updated into the numerical model.

Figure 1. 1 MW hermetic high-speed steam generator. On the left is the detailed view of the CAD model, and on the right is the fully instrumented generator in laboratory.

3.1   Model substructuring and assembly of superelements

The complete generator assembly illustrated in Figure 1 was divided into three substructures: support legs, frame assembly and rotor. These submodels are visualized in Figure 2. The number of degrees-of-freedom of different submodels, and the original and final reduced-order models are given in Table 1. The discretization of individual submodels is made using quadratic tetrahedron elements. The connections between components of a single subassembly are linearized.

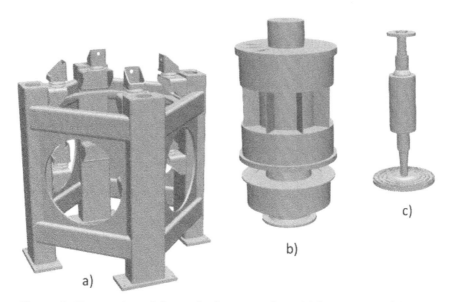

Figure 2. Three submodels used: a) support legs, b) frame assembly and c) rotor.

Table 1. Original and reduced-order model size.

Submodel	Original size (dofs)	Superelement size (dofs)
Support legs	805,767	744
Frame assembly	765,036	684
Rotor	1,154,202	204
Total	2,725,005	1,632

The size of individual superelements is based on two factors: the number of flexible modes used and the number of the boundary dofs included into the element. The number of the lowest flexible modes per submodel was set to 24, and the number of nodes per individual connection was set to be approximately 50% of all boundary nodes. The use of reduced number of boundary nodes for contact reduced the size of the transformation matrix effectively without yielding too soft boundary connection. The number of connections is different, in this case, for every submodel as follows:

- Support legs
  - 8 x body-to-ground connections
  - 8 x body-to-body connections between legs and frame

- Frame assembly
  - 8 x body-to-body connections between legs and frame
  - 3 x body-to-body connections between frame and rotor

- Rotor
  - 3 x body-to-body connections between frame and rotor

The leg support body-to-ground connections are assumed to be stiff, thus, the foundation flexibility is neglected in this case. It is reasonable since the body-to-body connections between the support legs and frame are very soft due to installed vibration isolators. Based on identified axial rigid body mode frequency of the frame and the total mass of the frame assembly, the spring stiffness of a single degree-of-freedom mass-point system was calculated, and the equivalent stiffness of a single support-to-frame contact describing single vibration isolator was calculated to be $0.668 \cdot 10^6$ N/m. The body-to-body connections between the frame and rotor are the linearized AMB stiffnesses. In order to obtain accurate values, both radial and axial stiffnesses of the AMB were identified by tuning the model to match the measurements. In radial direction, the model was tuned to match the measured first lateral bending mode of rotor and, as a result, a radial AMB stiffness of $4 \cdot 10^6$ N/m was obtained. Similarly, the axial rigid mode of the rotor was identified based on measurement data, and from the frequency and the mass of the rotor the axial AMB stiffness of $5.16 \cdot 10^6$ N/m was obtained.

3.2 Model update for harmonic response analysis

Once the problematic frame vibration mode is identified, the modal reduced FE-model is updated using the method presented in Section 2.4 to match the measured eigenmode frequencies, and the corresponding modal damping ratios are calculated from the measured data. Then, modal harmonic problem is solved, and the coupled axial-lateral mode responses are compared against the measured axial and lateral vibration frequency responses. The solution of the harmonic response is along the inertial coordinate system used, although, the results will be processed so that only relative displacements between the sensor housings and measurement surfaces are presented.

4 RESULTS

The results of the problematic frame vibration mode among the other measured resonances are studied and discussed. The mode of the problematic frame vibration is identified by using both experimental data and the numerical simulation results.

4.1 Experimental data

Experimental frequency responses were measured in both axial and radial directions. The data is based on the AMB sensor data, and thus, the nature of the position data recorded is relative. The excitation force is generated using AMBs for system identification purposes for non-rotating machinery. The axial sensor was located in the N-end near the axial bearing. Two radial sensor per lateral axis were used at both radial bearing locations. Logically, the N-end, or ND-end, is the upper end, and the D-end, or drive-end, is the turbine end. The measured data is converted into frequency domain, and the resolution in the axial channel is 0.5 Hz, while in radial channels 1/3 Hz. The frequency responses of one axial and two radial channels are presented and relevant resonance peaks are marked. The modes of these identified frequencies are presented in Table 2. From axial frequency response in Figure 3, it can be seen that in N-end there is a strong resonance at frequency of 117 Hz. This frequency corresponds to the problematic frame vibration. Identification of the other frequencies in the axial frequency response requires the use of FEM-based vibration model.

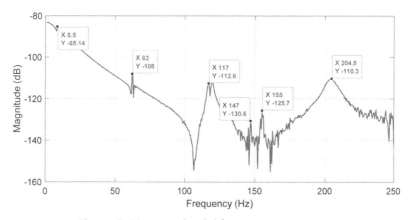

Figure 3. Measured axial frequency response.

Previously discussed axial-radial vibration mode coupling is indeed visible in the radial frequency response. In Figure 4, considering the resolution of the plot, the frequency of the coupled axial-radial mode is approximately at 117...118 Hz. Similarly, in the case of axial frequency response, the identification of other modes requires the use of FEM-based vibration model. The modes of these identified frequencies are presented in Table 2.

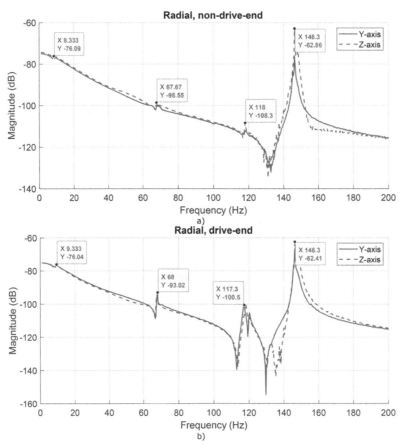

**Figure 4. Measured radial frequency responses: a) non-drive-end, and
b) drive-end.**

4.2 Identification of eigenmodes

Previously identified resonance at the frequency of 117 Hz seemed to be happening at the both ends of the generator frame in both axial and radial directions. Thus, the solved undamped eigenmodes of the reduced-order model were investigated and the identified mode based on the description derived based on the measured data is visualized in Figure 5. As seen in the figure, both bearing end plates are deforming in the axial direction. In addition, the upper bearing end plate seems to be bending non-symmetrically as illustrated with the red dashed line in the Figure 5. This non-symmetric bending provides proof for the axial-radial coupled vibration seen in the measured sensor data.

Reason for the non-symmetric upper bearing end plate bending seems to be due to the power cable inlet located on one side of the upper bearing end plate, as seen in Figure 2. Further investigations revealed that the total deformation of the axial frame mode at 117 Hz is not centralized only to the bearing end plates, but, in fact, the whole frame tube was compressing and expanding in axial direction.

478

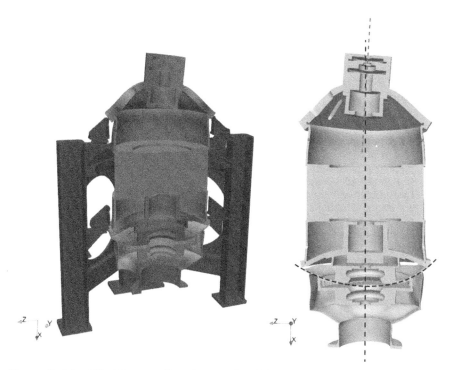

Figure 5. Identified frame vibration mode of the resonance at 117 Hz. On the left is the mode visualization of complete assembly using great deformation factor, and on the right is the visualization of the skewed local bending of the upper bearing end plate.

The relevant resonance peaks were identified from the measured sensor frequency response with the numerical model. Even though, the accuracy of the solved eigenfrequencies was not perfect due to many simplifications in the geometry, the measured eigenfrequencies were able to be matched to the corresponding simulated mode shapes. These relevant mode shapes were selected for harmonic analysis. The corresponding eigenfrequencies were updated to the modal coordinate model based on the measured sensor frequency responses. Further reduced modal coordinate expression was utilized for purposes of updating the exact modal damping ratios based on the measured data. Calculated damping ratios are presented in Table 2. The original model having 2,725,005 dofs is reduced to 1,632 dofs in superelement-based model and then further reduced to modal coordinate model having only 7 dofs. This drastic reduction could be performed without noticeable loss of accuracy compared to the full-sized problem. The accuracy of Craig-Bampton method based superelement expression, in general, is dependent on the number of selected modes for the transformation, while a good choice of the boundary degrees-of-freedom is also important. In addition, the reduced 7 dof modal coordinate model can be easily expanded back into the original full model with over 2.7 Mdofs, as presented in Section 2.3.

In Figure 6 are plotted the relative axial and radial frequency responses at the axial and D-end radial bearing sensor locations simulating the real measuring conditions. As theorized, the problematic frame vibration mode at 117 Hz is, indeed, visible on both axial and radial relative frequency responses. The other identified modes are described in Table 2. By comparing the measured and simulated response amplitudes, the following can be concluded:

- Frame rigid axial mode is barely visible in the simulated response due to large modal damping, the frequency of resonance peak of this mode seems to be a bit lower than in measured case as seen in Table 2.
- Steam outlet pipe axial mode is also in both measured and simulated radial responses. Although, the radial steam outlet pipe response is not visible in the simulated radial response.
- The problematic frame vibration at frequency of 117 Hz is visible in both axial and radial directions in the simulated response as well, and thus, proofing the existence of global coupled axial-radial vibration mode. Naturally, this mode is mainly axial as seen by comparing the magnitudes in Figure 6.
- Highly damped second axial frame mode at 205 Hz is lower in magnitude than in the measured response.

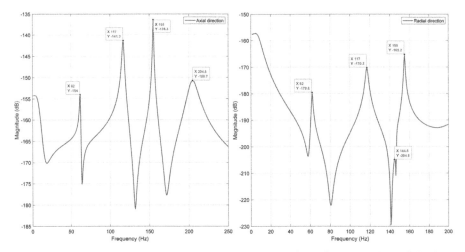

Figure 6. Simulated relative responses at sensor locations: on the left is the upper axial sensor frequency response, and on the right is the D-end radial sensor frequency response.

480

Table 2. The identified eigenmodes based on the numerical model.

Measured frequency	Calculated damping ratio	Direction	Location	Mode shape
8.5 Hz	58.82%	Axial	Frame global	Rigid axial
62 Hz	0.97%	Axial	Steam outlet pipe	Umbrella
68 Hz	0.66%	Radial	Steam outlet pipe	Bending
117 Hz	1.07%	Axial/ Radial	Frame global	1. axial mode
146 Hz	0.24%	Radial	Rotor	1. lateral bending
155 Hz	0.34%	Axial	Rotor	Rigid axial
205 Hz	3.18%	Axial	Frame global	2. axial mode

5   CONCLUSIONS

Model-order reduction, solution expansion and Craig-Bampton transformation-based reduced-order model update for large structural dynamic problems is discussed. As a case example, 1MW steam generator frame dynamics is studied. The generator model is divided into three submodels, and superelements are generated from every submodel. Measured frequency responses from different AMB sensor locations indicated large vibration amplitude at 117 Hz in different directions between the sensor and rotor. Using superelement-based reduced-order model, the unknown frame vibration mode was identified by means of expanding the reduced-order solution in order to visualize the vibration mode using full-sized model. Once the problematic and other relevant vibration modes were identified by expanding the solved eigenvectors of Craig-Bampton transformation-based assembled superelement model, the reduced-order model was updated based on measured frequency responses. The measured mode frequencies and modal damping ratios were updated into the numerical model, and simulated model-based harmonic responses were calculated. These simulated frequency responses indicated the proof for the existence of axial-radial coupled vibration mode of the generator frame.

REFERENCES

[1] Schweitzer, G., Maslen, E.H., "Magnetic Bearings", *Springer*, 2009.
[2] Blades, E.L., Craig, Jr., R.R., "A Craig-Bampton Test-Analysis Model", *Proceedings of SPIE - The International Society for Optical Engineering*, 1997, pp. 1386–1391.
[3] Seshu, P., "Substructuring and Component Mode Synthesis", Journal of Shock and Vibration, 4 (3), 1997, pp. 199–210.
[4] Boo, S.-H., Kim, J.-H., Lee, P.-S., "Towards improving the enhanced Craig-Bampton method", *An International Journal of Computers & Structures*, 196, 2018, pp. 63–75.

[5] Qu, Z.-Q., "Model Order Reduction Techniques with Applications in Finite Element Analysis", *Springer*, 2004.

[6] Wijker, J.J., "Mechanical Vibrations in Spacecraft Design", *Springer*, 2004.

[7] Friswell, M.I., Garvey, S.D., Penny, J.E.T., "Model reduction using dynamic and iterated IRS techniques", *Journal of Sound and Vibration*, 186(2), 1995, pp. 311–323.

[8] ANSYS, Inc. ANSYS Mechanical APDL Theory Reference. *ANSYS, Inc.*, 2013.

[9] Sikanen, E., "Dynamic analysis of rotating systems including contact and thermal-induced effects," Doctoral Dissertation, *LUT University*, 2018.

[10] Lalanne, M., Ferraris, G., "Rotordynamics Prediction in Engineering", *John Wiley & Sons*, 1998.

Comparison of different time integration schemes and application to a rotor system with magnetic bearings in MATLAB

M. Kreutz, J. Maierhofer, Th. Thümmel, D.J. Rixen

Chair of Applied Mechanics, Technical University of Munich, Germany

ABSTRACT

Evaluating the dynamic properties of rotor systems in an early design stage requires simulation. At the Chair of Applied Mechanics at the Technical University of Munich, the rotor simulation program AMrotor was developed using Timoshenko-beam elements to discretize the rotor. Components with non-linear reaction forces can be added on the right-hand side of the equation to the external loads. This requires considering the differential equations in the time domain. In order to get an approximate solution, direct time integration is used, as it is applicable for systems with general excitation.

In this work, different time integration schemes are applied to a rotor system and validated with experimental results from an academic test rig.

Examples of Runge-Kutta, backward differentiation formula and Newmark methods are implemented in the rotor simulation program. The methods are compared using Fourier transforms of resulting displacements and validated using experimental frequency response functions.

Experiments are performed on a rotor test rig, which consists of a simple rotor levitating in active magnetic bearings. A discrete PID controller, which is included in the simulation, controls the electric current in the electromagnets to stabilize the rotor.

This discussion helps the user of a finite element rotor simulation code to decide for a suitable time integration method. Finding a fast and reliable integration method can drastically improve the simulation speed, making the code suitable for real-time simulation. In a next step, the code may be used for real-time hybrid sub-structuring (Hardware-in-the-Loop).

1   INTRODUCTION

The theory of time integration for mechanical systems is well developed. However, choosing a suitable integration method is still hard for users as the simulation model's properties have to be considered. Example simulations from literature can help the engineer in choosing a time integration for his problem.

Some examples of time integration for flexible rotor systems are given. In [1], a study of Runge-Kutta methods for lumped-mass rotor models is presented. In [2] and [3], Runge-Kutta methods are used for integrating a reduced set of equations of motion for rotor systems. In [4], a Newmark method for state-space equations of motion for a rotor system consisting of finite beam elements is developed. The equations of motion are not reduced. The Newmark method is preferred, because of its unconditional stability.

MAIERHOFER [5] mentions some dedicated rotor simulation software. Here, a summary of their time integration methods is given. DYNROT, MADYN 2000, Dyrobes and XLRotor are examples of commercial programs. DYNROT and MADYN 2000 are toolboxes developed for MATLAB. DYNROT is capable of performing time integration using central difference scheme or Runge-Kutta methods, cf. [6]. MADYN 2000 uses Runge-Kutta methods, more specifically the MATLAB functions ode45 and ode23, cf. [7], [8]. Dyrobes has a wide variety of available time integration methods, i.e. Gear's, Runge-Kutta, Newmark-beta, Wilson-theta and Newmark-modified methods, cf. [9]. XLRotor is a commercial package for Microsoft Excel. For transient analysis, it uses implicit methods. Wilson-Theta, Newmark with or without numerical damping and general-ised-alpha-method are used with fixed step sizes. Rosenbrock's method is imple-mented with an adaptive step size, cf. [10, p. 126f].

The open-source program VibronRotor for Octave is not capable of time integration, cf. [5]. The well-developed open-source simulation program ROSS [11] for python uses the time integration code *lsim* from the scipy signal package [12] for linear system time integration. However, ROSS does not include active components in the simulation, which reduces the demands on the time integration method.

The open-source simulation toolbox AMrotor [5] for MATLAB is used in this contribu-tion. Here different time integration methods can be chosen. This contribution com-pares different time integration schemes and provides a practical example of simulation of a rotor in active magnetic bearings with discrete PID-controllers. This can serve as a guide for engineers on the pros and cons of different methods for their special rotordynamic applications.

2   DIRECT TIME INTEGRATION

Engineering problems involving vibrations are often tackled using the finite element method (FEM). Arising equations are second-order ordinary differential equations (ODEs) in time. They are formulated as initial value problems (IVP):

$$M\ddot{x} + D\dot{x} + Kx = F$$
$$x(t = 0) = x_0, \dot{x}(t = 0) = \dot{x}_0 \tag{1}$$

involving mass, damping and stiffness matrix on the left-hand side of the equation and general forces on the right-hand side.

The system of equations can be solved by direct time integration. Several time integra-tion schemes exist for first order ODEs, because it is the most general case. Here only time stepping schemes are considered. Important methods are the Runge-Kutta-methods or backward differentiation formulas. These are also used in the time integration procedures of MATLAB-functions ode45 and ode15s, respectively. SHAMPINE [13] gives an overview of the methods used in the MATLAB-ODE-suite. In structural dynamics, methods for second-order differential equation are often preferred, as they can be dir-ectly applied to systems in the form of eq. (1). Most widely used is the Newmark-method.

2.1   First order ODE
The most common time integration schemes have been developed for first order systems:

$$\dot{y} = f(t, y). \tag{2}$$

In order to use these methods for mechanical systems, eq. (1) must be transformed into a first order system, e.g. by

$$\dot{y} = \begin{bmatrix} 0 & I \\ -M^{-1}K & -M^{-1}D \end{bmatrix} y + \begin{bmatrix} 0 & 0 \\ 0 & -M^{-1} \end{bmatrix} \begin{bmatrix} 0 \\ F \end{bmatrix}, \text{with } y = \begin{bmatrix} x \\ \dot{x} \end{bmatrix}. \tag{3}$$

For this investigation, the equation is implemented as in eq. (3). Integration of eq. (2) from t_k to t_{k+1} yields

$$y(t_{k+1}) - y(t_k) = \int_{t_k}^{t_{k+1}} f(t, y(t)) dt \tag{4}$$

The right-hand side can be approximated by a quadrature formula. Because of the approximation, the left-hand side will be replaced by $y_{k+1} - y_k$. Note that $y(t_{k+1})$ denotes the exact solution, while y_{k+1} denotes the approximate solution. Depending on the quadrature formula used, different integration schemes are deduced.

2.1.1 *Runge-Kutta*
The Runge-Kutta method is a generalization of one-step methods for first order ODEs. It provides a way of building quadrature formulas for the integral in (4).

The slope of function $f(t, y)$ is approximated by k_j at m points in the integration interval. The value y_{k+1} is then obtained by weighting the slopes.

$$k_j = f\left(t_k + c_j h, y_k \cdot [a_{j1}k_1 + a_{j2}k_2 + \ldots + a_{jm}k_m]\right), \text{for } j = 1 \ldots m. \tag{5}$$

$$y_{k+1} = y_k + h[b_1 k_1 + b_2 k_2 + \ldots + b_m k_m] \tag{6}$$

The classical Runge-Kutta method is an explicit scheme of fourth order with parameters found for example in [14, p. 225f].

For a method with an adaptive step size, an estimation of the local truncation error is needed. This is achieved by forming pairs of RK-formulas of orders p and $p-1$ and comparing the results for an error estimation, cf. [13, 14]. The local truncation error e_{k+1} is then estimated by the difference of results obtained by the two methods.

$$y_{k+1}^* = y_k + h\left[b_1^* k_1 + b_2^* k_2 + \ldots + b_{m-1}^* k_{m-1}\right] \tag{7}$$

$$e_{k+1} = y_{k+1} - y_{k+1}^* \tag{8}$$

Using the same slopes k_j for both formulas makes the error evaluation cheap, as they only must be calculated once. The resulting formulas are then called Runge-Kutta pairs. The MATLAB function ode45 uses the Dormand-Prince (4,5) pair, cf. [13, 15].

2.1.2 *Backward differentiation formula*
A class of implicit time integration schemes are backward differentiation formulas (BDF). They approximate the solution $y(t)$ by a polynomial $P(t)$. The polynomial of order m interpolates $y_{k+1}, y_k, y_{k-1}, \ldots, y_{k+1-m}$. We require that the polynomial satisfies the differential equation eq. (2):

$$\dot{P}(t_{k+1}) = f(t_{k+1}, y_{k+1}) \tag{9}$$

This yields the backward differentiation formula of order m: BDFm. BDFs are consistent with a consistency order of m, cf. [14, p. 237]. BDF of orders $m \leq 2$ are A-stable. BDFs of higher orders $3 \leq 6$ are $A(\alpha)$-stable. BDFs of order 7 or higher are not stable, cf. [16, p. 68]. Also, all the convergent BDFs are $L(\alpha)$-stable, which means the numerical integration has a strong damping behaviour for large negative eigenvalues of the equations of motion.

As BDFs are implicit methods, computing the solution requires solving a system of equations.

BDFs are predominantly used for stiff ODEs. A stiff problem has multiple decaying solution components which differ greatly in their decaying speed. The efficiency of solving stiff problems is limited by stability but not by accuracy.

The MATLAB function ode15s, that uses a BDF with variable order, is suitable for stiff problems, because of its high stability. It is also possible to use the MATLAB function ode45, which is an explicit Runge-Kutta method, for moderately stiff problems. The error control forces a small time step, as the error increases strongly for an unstable integration. A small step size can keep the method stable, but the method is inefficient because it.

The recommended procedure for solving IVPs with MATLAB is to first try ode45. If ode45 converges slowly or does not converge at all, it can be suspected that the problem is stiff. Then ode15s should be used, see also MATLAB's documentation. The MATLAB-function ode15s uses BDF schemes with variable orders ranging from 1 to 5, cf. [16, p. 74].

2.2 Second order ODE – Newmark

We have discussed some explicit and implicit methods for first order ODEs. In this section we want to introduce a time integration method, that is directly applicable to second order ODEs like eq. (1): The Newmark method. Its derivation and properties are shown in [17, p. 522ff].

Applying the Newmark method to the linear mechanical system, eq. (1), yields an equation system for the acceleration at time t_{k+1}

$$
\underbrace{(M + \gamma h D + \beta h^2 K)}_{S} \ddot{x}_{k+1} =
$$
$$
= F_{k+1} - D[\dot{x}_k + (1 - \gamma)h\ddot{x}_k] - K[x_k + h\dot{x}_k + (1/2 - \beta)h^2\beta\ddot{x}_k] \tag{10}
$$

The matrix S is called time stepping matrix. In order to solve the equation system (10), S must be factorized. In case that S is symmetric positive definite, it is possible to perform a Cholesky factorization. In this contribution, the time integration methods must be able to include gyroscopic effects in rotor systems. When $D = (C + G)$ includes gyroscopic effects the matrix S will lose its symmetry because of the skew-symmetric gyroscopic matrix G. So, in this contribution a LU factorization is used. The LU factorization is computationally more expensive than the Cholesky factorisation, but it does not require S to be symmetric.

The analysis of the Newmark scheme in terms of stability and accuracy is presented in [17, p. 525ff] for mechanical systems.

The average constant acceleration method ($\gamma = 0.5, \beta = 0.25$) is the best unconditionally stable method. It is well-suited for stiff problems. It provides no numerical damping and has a periodicity error proportional to $\omega^2 h^2 / 12$.

More sophisticated methods to include numerical damping are e.g. the Hilber-Hughes-Taylor-α-method or the Generalized-α-method, cf. [17].

486

3 NUMERICAL COMPARISON

Throughout this contribution, the time integration schemes are used as described above. The goal of the comparison is to assess the behaviour of these time integration schemes for non-linear, rotordynamic problems, that include gyroscopic effects.

The classic Runge-Kutta formula of fourth order (RK4), the BDF2-scheme and Newmark's average constant acceleration method are implemented as described above. For simplicity the unconditional stable average constant acceleration scheme will be referred to as Newmark's method. The Jacobian for BDF2 is obtained by finite differences. For simplicity the number of Newton iterations to solve the in general non-linear equations of motion, in each time step is fixed to 4. Here, we use the implementation of BDF2 for non-linear problems, although here in chapter 3 the problem is linear.

This section provides examples, that help in understanding the properties of the different time integration schemes. The test problem is the homogenous two-mass-oscillator, which is shown in Figure 1. Its equations of motion are given as:

$$\begin{pmatrix} m_1 & 0 \\ 0 & m_2 \end{pmatrix} \begin{pmatrix} \ddot{x}_1 \\ \ddot{x}_2 \end{pmatrix} + \begin{pmatrix} d_1 + d_2 & -d_2 \\ -d_2 & d_2 \end{pmatrix} \begin{pmatrix} \dot{x}_1 \\ \dot{x}_2 \end{pmatrix} + \begin{pmatrix} k_1 + k_2 & -k_2 \\ -k_2 & k_2 \end{pmatrix} \begin{pmatrix} x_1 \\ x_2 \end{pmatrix} = \begin{pmatrix} 0 \\ 0 \end{pmatrix} \quad (11)$$

with the parameter values $m_1 = 2$kg, $m_2 = 1$kg, $k_1 = 400$N/m, $k_2 = 100$N/m and the initial conditions $\dot{x} = 0$, $x = (1, 1)^T$. The values of the damping coefficients will determine if the system is stiff. The results of fixed step integration formulas will be compared, as well as the results of using built-in MATLAB functions ode15s and ode45, using standard tolerances. An analytical solution can be obtained by modal superposition for this simple system.

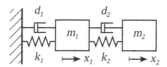

Figure 1. Two-mass-oscillator.

A large step size of $h = 0.1$ s is chosen for all integration schemes, so the effects of the integration formulas are easily observable. The solutions are computed for a time span of 100 seconds. To see how the physical behaviour of the system is changed in terms of amplitude and frequency a discrete Fourier transform (DFT) of the results without windowing is computed.

3.1 Non-stiff system

Here, a damping matrix $D = 10^{-4}K$ is used for eq. (11), which describes a non-stiff, weakly damped system with damping ratios 0.083% and 0.042% for the modes. The DFT for the system is shown in the upper part of Figure 2.

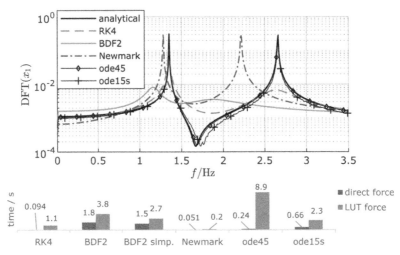

Figure 2. Comparison of results for non-stiff test system using fixed step size.

The Runge-Kutta-method shows no periodicity error. It shows good results for lower frequencies, but a strong damping behaviour for higher eigenfrequency. BDF2 has large numerical damping and shows the biggest error in periodicity. Higher order BDFs are not stable for the chosen time step size. Newmark's method shows no numerical damping, i.e. no amplitude error. It can be clearly seen that its periodicity error increases with frequency.

The simulation times are also given in Figure 2. The first bar shows the simulation time, if the force is given directly as zero. The second bar shows the simulation times if the force is evaluated in every function call by a look-up table (LUT), that only contains zeros. This shows in which way a costly ODE function evaluation influences the simulation time. The simulation time for BDF2, with a simplified Newton iteration is shown. There, the Jacobian is recomputed for every tenth time-step.

First the simulation times for the direct force approach are compared. The Runge-Kutta method and Newmark's method are the fastest. Newmark's method uses a Cholesky factorization of its iteration matrix, which, for linear systems, only has to be computed once and can then be reused. Because of this, its simulation time is very low, although it is an implicit method. The BDF2 method takes the longest, because it performs a LU factorization in every time step, which is computationally expensive.

These results are compared with the simulation times in case the ODE function evaluation is more expensive, which is here represented by force evaluation by look-up table. The biggest influence can be seen in RK4, which needs the most function evaluations for each time step. Newmark's method takes only about 10 times as long and still is the fastest method. The effect on BDF2 is the lowest. Here, time can be saved by using the simplified Newton iteration, which uses less factorizations.

The built-in MATLAB functions ode15s and ode45 can also be used. The results from ode45 agree very well with the analytical results. Results from ode15s slightly underestimate the amplitudes outside of the resonances. A comparison of the simulation times of ode45 and ode15s is also given in Figure 2. It shows a similar effect as before when comparing RK4 and BDF2, respectively. For the direct force evaluation ode45 is much faster than ode15s, because it is an explicit method. The influence of an expensive function evaluation is very strong on ode45 as it uses a lot of evaluations, while the influence on ode15s is smaller, because of less evaluations. The effect is so strong, that the implicit ode15s is even faster than the explicit ode45 for costly look-up table force evaluation. So, it can be advantageous to use the implicit method if the function evaluation of the ODE is expensive.

3.2 Stiff system

To simulate a stiff problem, different damping coefficients $d_1 = 1$ Ns/m, $d_2 = 10^4$ Ns/m are used in eq. (11). Here, only the cheaper direct force evaluation is used. The ratio of the largest to the smallest magnitude of the real part of the system's eigenvalues is called S. According to [14, p. 249f] a problem is called stiff, if its stiffness ratio $S > 10^3$. The chosen parameter values yield $S = 1.5 \cdot 10^6$.

The Runge-Kutta method is unstable for $h = 0.1$ s. BDFs of orders 3 and higher are also unstable for this problem. The DFTs of the results from BDF2 and Newmark's method are shown in Figure 3. Only one peak is visible, because one eigenfrequency is over-critically damped and does not show.

The BDF2 method remains stable, but because of the extremely large time step it represents the analytical solution only poorly. The Newmark method yields sensible results, even for the chosen step size. The evaluation of Newmark's method takes just about 0.023 s, while BDF2 takes about 1.7 s, for the same reasons as in the previous section. So, the Newmark method yields the best results, while being much faster than BDF2.

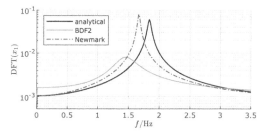

Figure 3. Comparison of results for stiff test system using fixed step size.

The stiff problem is also used to check the performance of the adaptive MATLAB-methods ode45 and ode15s. They yield good results, with no visible difference to the exact solution. However, the evaluation of ode45 takes 12.5 s, because it is forced to choose small step sizes to remain stable. The implicit ode15s algorithm only takes 0.28 s. It can choose large step sizes, because it is not restricted by a finite stability region. This clearly shows the advantage of implicit formulas for stiff problems.

4 PRACTICAL EXAMPLE

After illustrating the time integration method's behaviours for simple examples, they shall be applied to a more sophisticated realistic problem. They are tested in simulation of the rotor system in the magnetic bearing test rig. First, the test rig is presented, and its modelling is explained. The integration methods are compared to one another for a numerical example. Lastly, the simulation model is validated by comparison with experimental results.

4.1 Test rig

The magnetic bearing test rig is shown in Figure 4. The active magnetic bearings (AMBs) use a bias current of $2,5A$. The electrical control current in the AMBs is provided by a PID controller, using the rotor's position. The positions are measured by eddy current sensors, that are integrated in each AMB. An additional laser triangulation sensor measures the position of the mass disc in the middle of the rotor.

The test rig control is designed in MATLAB/Simulink and transferred to the real-time test-rig management system dSPACE, which also is used for data acquisition.

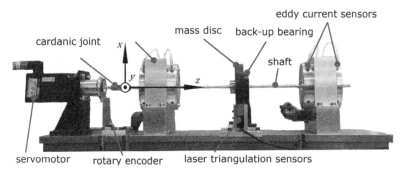

Figure 4. Magnetic bearing test rig.

4.2 Simulation program and modelling

The simulation of the rotor system uses a finite element formulation. The rotor is discretized using Timoshenko-beam elements, that also include gyroscopic effects, cf. [18]. The resulting equations from the discretization take the form of eq. (1).

Modelling of the rotor's geometry including additional mass discs was validated with results from an experimental modal analysis, cf. [19, 20]. There, the active magnetic bearings were modelled as linear spring-damper elements. Here the magnetic bearings are modelled in more detail, including the discrete controller for the bearing's electrical current in the real system. The bearing forces are written on the right-hand side of the equations of motion, which yields

$$M\ddot{q} + D\dot{q} + Kq = F_{extern} + F_{AMB} \tag{12}$$

The evaluation of the ODE function, especially force evaluation is relatively expensive. Magnetic bearing forces F_{AMB} are defined for lateral degrees of freedom x, y of corresponding nodes. The magnetic bearing forces for the lateral degrees of freedom are given by

$$F_{AMB,j} = -k_x x_j + k_i i_j \tag{13}$$

With the displacement x_j at the corresponding dof and the control current i_j. The displacement factor $k_x = -10^5 \mathrm{N/m}$ and the current factor $k_i = 50 \mathrm{\,N/A}$ are determined from static measurements of the magnetic bearings, cf. [21].

The displacement factor k_x can be included in the stiffness matrix K on the left-hand side. Because of its negative value it has a destabilizing effect on the system. The control current is computed with a fixed controller frequency f_{cntr} using a PID-control law, here $f_{cntr} = 1000$ Hz, so $1/f_{cntr} = 1$ ms.

$$i_j = Pe_{disp} + Ie_{cumu} + De_{velo}, \; j \in N \tag{14}$$

The variables $e_{disp}, e_{cumu}, e_{velo}$ denote the displacement, cumulative displacement and velocity error, respectively. The values P, I, D are the controller coefficients ($P = 5000$ A/m, $I = 1500$ A/(ms), $D = 5$ As/m).

The implementation of the controller depends on the integration method. When using a MATLAB solver, a time vector $(0 : 1/f_{cntr} : T)$ can be specified for the integration. At the timepoints in the vector, a function computes the controller current.

When using a fixed step size, the time step size must be $h = 1/(n \cdot f_{cntr})$ with $n \in \mathbb{N}$. If the time $t_{j \cdot n} = j/f_{cntr}$, a new controller current i_j is computed with $y_{j \cdot n}$, which holds until $t_{(j+1) \cdot n}$. The value $y_{j \cdot n}$ is then recomputed once, to account for the new controller current i_j ($t_{j \cdot n}, y_{j \cdot n} \to i_j \to y_{j \cdot n} \to y_{j \cdot n+1} \to y_{j \cdot n+2} \to ...$, until $t_{(j+1) \cdot n}$).

4.3   Numerical investigations

The performance of different time integration schemes is compared by simulation of the test rig. Therefore, a constant rotational speed of 800 rpm is assumed and an unbalance of $m_u r_u = 2.75 \cdot 10^{-4}$kg m is applied. The initial conditions are stationary solutions of the unbalanced rotor system. Between 500 ms and 510 ms a constant force of 10N is applied in x-direction at the left AMB's position. This pulse causes excitation in a broad frequency band. The considered time is 7.5 s.

The problem is stiff, as an evaluation using ode45 is very slow and the classical Runge-Kutta method fails even for small integration step sizes (e.g. $h = 1\text{ms}/100$). Thus, the results of these methods are not shown here. Only the results for ode15s and the unconditionally stable methods BDF2 and Newmark's method are compared. Here the simplified Newton iteration is used for BDF2. Suitable step sizes are chosen based on simulation speed and on accuracy of the results. BDF2's simulation time is relatively long, so a large step size was chosen. Newmark's method is fast, so a smaller step size could be chosen. The evaluation times for the different methods and the chosen time steps are collected in the lower part of Figure 5. For ode15s relaxed tolerances ($RelTol = AbsTol = 10^{-5}$) are chosen to improve the computation time.

Newmark's method provides the best computation time. BDF2 takes more than 12 times as long. The variable step solver ode15s is in between.

A DFT of the disc's displacement in x-direction without windowing is computed for $t = 0.5 \dots 6.5 \, s$ which corresponds to 80 revolutions of the shaft since the force pulse starts. The results are shown in the upper part of Figure 5.

At $800/60$ Hz ≈ 13.33 Hz is the peak from the unbalance. It is the same for all methods, the stationary deflection is thus correctly represented by all methods. The peak at approx. 17Hz corresponds to the first bending eigenfrequency of the rotor system. It is well resolved by all three methods.

The resonances appear at approx. 174 Hz and 182 Hz with the Newmark and ode15s method. The split of the eigenfrequencies results from gyroscopic effects. These eigenfrequencies are not adequately captured by BDF2. For BDF2 the peaks cannot be distinguished and appear as a single peak at a lower frequency. This is an effect of the large numerical damping of the method. Eigenfrequencies between 17 Hz and 170 Hz do not produce significant peaks, as they are not sufficiently excited and their corresponding eigenmodes show vibration nodes in the vicinity of or exactly at the disc's position. Newmark's method shows a very small peak at about 113 *Hz*. This peak cannot be seen for ode15s. When zooming in, the entire curve for ode15s looks less smooth than the other curves, because of switching the order of the underlying BDF method and the variable step sizes of the method inside of the integration algorithm.

In summary Newmark's method shows the best results and it has the shortest computation time. It also represents resonances of the system well.

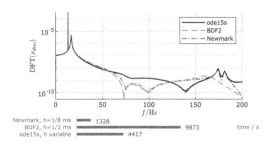

Figure 5. Comparison of results for test rig simulation.

Table 1. Computation times for test rig simulation with Newmark's method.

h	$1\ ms/8$	$1\ ms/16$	$1\ ms/32$	$1\ ms/64$	$1\ ms/128$
CPU time	1328 s	2566 s	5016 s	9983 s	19892 s

Because Newmark's method shows very good results for a relatively short computation time, its behaviour for decreasing the step size is investigated. The step size is reduced by $h = 1\ ms/2^n$, with $n \in \mathbb{N}$.

Table 1 shows the CPU times for smaller step sizes. The simulation time increases approximately linear with the number of time steps $\sim 2^n$. Newmark's results converge for decreasing the step size.

4.4   Experimental validation

For validation of the simulation a comparison to experimental results from the magnetic bearing test rig is performed. The results will be compared by evaluating an exemplary frequency response function (FRF) for the test rig. Figure 6 and Figure 7 show amplitudes of an FRF for excitation using the left magnetic bearing and response measurement of the leftmost eddy current sensor that is integrated in the housing.

The setup is analogue to the previous section (800 *rpm*, unbalance). As excitation signal a chirp signal from 0 to 200 Hz is applied, instead of the pulse in the previous section. The signal is $1.5 \cdot \text{s}$ long and is repeated 4 times. Similar to section 4.3, MATLAB's ode15s, simplified BDF2 with $h = 1/2$ ms and Newmark's method with $h = 1/8$ ms are used. The FRF is computed by using the open-source MATLAB experimental-modal-analysis-toolbox Abravibe [22], that accompanies BRANDT's text book [23]. For simulation a rectangular window is used.

In the experiment a burst random excitation with band-limited white noise (limited to 200 Hz) is used. Excitation for 0.5s is followed by 1.5s pause. The excitation signal is repeated 100 times. The force is not measured at the test rig, but it is estimated from a static model using the electric current of the AMB and the displacement of the shaft inside of the AMB: $F = f(x, I)$. A half-sine window and 67% overlap is applied to the experimental signals to reduce leakage. The resulting FRF is also computed using Abravibe. The results are compared in Figure 6 and Figure 7.

Similar to section 4.3, the ode15s integration and Newmark's integration yield similar results. The BDF2 integration cannot properly reproduce higher eigenfrequencies with the chosen step size.

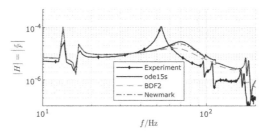

Figure 6. Frequency response function for the test rig in double log representation.

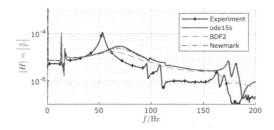

Figure 7. Frequency response function for the test rig.

Table 2. Comparison of eigenfrequencies with experimental results.

Experiment	17 Hz	53 Hz	95 Hz	109 Hz	163 Hz	170 Hz
Simulation	17 Hz	70 Hz	99 Hz	113 Hz	173 Hz	181 Hz

Table 2 compares the eigenfrequencies of the system, that are read from the peak's positions of Newmark's curve in Figure 7.

The first peak in the FRF results from the unbalance force. It does not correspond to an eigenfrequency of the system. The first eigenfrequency at approx. 17Hz is well

493

represented by the simulation. The second eigenfrequency in Table 2 appears to have much more damping in the simulation. The third and fourth eigenfrequency only show very small peaks in the simulated FRF. All simulation eigenfrequencies are higher than in the experimental case. This corresponds to the results of [19], where the underlying rotor model was built and compared to experimental results performing modal analysis. For a better agreement, the rotor model has to be updated to better represent the experimental system.

For frequencies over approx. 70 Hz the simulation FRF has a much higher amplitude, than in the experimental case. This may be due to errors in the positioning of the displacement sensors. Positioning has large influence, since the measurement position is close to a node of vibration modes above 70 Hz, induced by the bearings. Also, the forces for the experimental FRF are statically estimated from the desired electrical current. The real current in the coils is not measured here. It is expected, that for rising frequency the amplifier cannot support the desired current, so $|i_{real}/i_{desired}|$ decreases for rising frequency, cf. [24, p. 129]. This leads to an overestimation of the forces in this experiment, i.e. an underestimation of the frequency response function amplitude $|H| = |x/F|$ in the experiment.

The FRF-amplitude for 0 Hz agree reasonably well between experiment and simulation. It can be concluded, that the AMB's coefficients k_i, k_x, see eq. (13) are sufficiently identified. However, the simulation of the AMBs using a discrete PID control law results in bearings that provide more damping than in the experimental case.

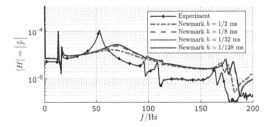

Figure 8. Frequency response function for the test rig with Newmark's method.

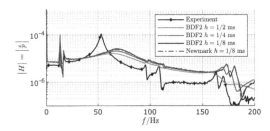

Figure 9. Frequency response function for the test rig with BDF2-method.

Figure 10. Comparison of integration times for decreasing step size.

Decreasing the time step size, does not improve the agreement between experimental and simulated results. In fact, for Newmark's method, a decrease in the step size does not significantly change the simulation results, as already implied in section 4.3. So, a step size of $h = 1/8$ ms for Newmark's method is sufficient. A comparison of resulting FRFs is depicted in Figure 8.

The bad results from the BDF2 method in Figure 7 can be improved by reducing the time step. The results are shown in Figure 9. Newmark's method is given as a reference. However, for BDF2 the simulation times are too large compared to Newmark's method and ode15s in order to be reasonable for practical use.

A comparison of simulation times is given in Figure 10. Newmark's method shows the best performance here. It uses little computation time and the results converge to the results of the other methods. However, there is a distinct difference between the overall simulation results for all integration methods and the experimental results.

The result of the validation is that the general dynamic behaviour of the system can be recognized in the measurements, but further investigations are needed to detect the strong differences in the FRF amplitudes.

5   CONCLUSION

In this contribution typical time integration schemes for dynamic simulation of mechanical systems were compared. The fixed-step methods Runge-Kutta method, backward differentiation formulas, Newmark's constant average acceleration method and MATLAB specific adaptive time stepping algorithms ode45, ode15s were compared.

Using a simple two-mass-oscillator as example showed that the explicit Runge-Kutta method and ode45 provide very good results, if the problem is not stiff and if the ODE function evaluation is not too expensive. However, if the problem is stiff or if the ODE function evaluation is expensive, implicit methods (Newmark, BDF, ode15s) are a better choice.

The realistic test problem of the rotor in magnetic bearings with a discrete PID-controller is a stiff problem, so only implicit methods are reasonable for this case. Newmark's method proved to be the fastest method, which also provides good results. The simulations could be validated by comparison with experimental frequency response functions. However, there are some uncertainties in the experimental setup and in parameter values for its simulative representation.

Further studies should focus on improving the experimental setup, i.e. identifying suitable parameter values.

The simulation program may be extended by applying adaptive time stepping for Newmark's methods, for example according to [25, p. 285ff].

However, for the application of time integration in real-time hybrid sub-structuring only fixed-time-step methods are applicable, because their integration time should be known in advance. A possible application could be to simulate a rotor shaft with discs and couple it with an experimental setup involving the AMBs, because the AMBs' behaviour is hard to model, while the shaft should be easier to simulate.

REFERENCES

[1] R. G. Kirk and E. J. Gunter, "Transient Response of Rotor-Bearing System," in *Design Engineering Technical Conference*, Cincinnati, Ohio, USA, 1973. DOI: 10.1115/1.3438383.

[2] K.-C. Lee, D.-K. Hong, Y.-H. Jeong, C.-Y. Kim and M.-C. Lee, "Dynamic Simulation of Radial Active Magnetic Bearing System for High Speed Rotor using ADAMS and MATLAB Co-simulation," in *8th IEEE International Conference on Automation Science and Engineering*, Seoul, Korea, 2012. DOI: 10.1109/CoASE.2012.6386492.

[3] O. Halminen, A. Kärkkäinen, J. Sopanen and A. Mikkola, "Active magnetic bearing-supported rotor with misaligned cageless backup bearings: A dropdown event simulation model," *Mechanical Systems and Signal Processing*, vol. 50, pp. 692–705, 2015. DOI: 10.1016/j.ymssp.2014.06.001.

[4] A. S. Lee, B. O. Kim and Y.-C. Kim, "A finite element transient response analysis method of a rotor-bearing system to base shock excitations using the state-space Newmark scheme and comparisons with experiments," *Journal of Sound and Vibration*, vol. 297, pp. 595–615, 2006. DOI: 10.1016/j.jsv.2006.04.028.

[5] J. Maierhofer, M. Kreutz, T. Mulser, T. Thümmel and D. J. Rixen, "AMrotor - A MATLAB Toolbox for the Simulation of Rotating Machinery," in *Vibrations in Rotating Machinery*, Liverpool, 2020.

[6] G. Genta, C. Delprete and D. Bassani, "DYNROT A finite element code for rotordynamic analysis based on complex co-ordinates," *International Journal for Computer-Aided Engineering and Software*, vol. 13, no. 6, pp. 86–109, 1996. DOI: 10.1108/02644409610128427.

[7] DELTA JS AG, "MADYN 2000 - Release notes versions 3.1 to 3.3," 16 07 2009. [Online]. Available: https://www.delta-js.ch/en/software/release-notes/. [Accessed 01 07 2020].

[8] J. Schmied and A. Fuchs, "Nonlinear analyses in rotordynamic engineering," in *International Conference on Rotor Dynamics*, Cham, 2018. DOI: 10.1007/978-3-319-99270-9_31.

[9] R. G. Kirk, E. J. Gunter and W. J. Chen, "Rotor drop transient analysis of AMB machinery," in *International Design Engineering Technical Conferences and Computers and Information in Engineering Conference*, Long Beach, California, USA, 2005. DOI: 10.1016/j.ymssp.2014.06.001.

[10] Rotating Machinery Analysis, Inc., XLRotor - Reference Guide Version 5.7, Brevard, NC, USA, 2020.

[11] R. Timbó, R. Martins, G. Bachmann, F. Rangel, J. Mota, J. Valério and T. G. Ritto, "ROSS - Rotordynamic Open Source Software," *Journal of Open Source Software*, vol. 5, no. 48, 2020. DOI: 10.21105/joss.02120.

[12] P. Virtanen, R. Gommers, T. E. Oliphant, M. Haberland, T. Reddy, D. Cournapeau, E. Burovski, P. Peterson, W. Weckesser, J. Bright, S. J. van der Walt, M. Brett, J. Wilson, K. J. Millman and M, "SciPy 1.0 - Fundamental Algorithms for Scientific Computing in Python," *Nature Methods*, vol. 17, pp. 261–272, 2020. DOI: 10.1038/s41592-019-0686-2.

[13] L. F. Shampine and M. W. Reichelt, "The MATLAB ODE Suite," *SIAM Journal on Scientific Computing*, vol. 18, no. 1, pp. 1–22, 1 1997. DOI: 10.1137/S1064827594276424.

[14] G. Bärwolff, Numerik für Ingenieure, Physiker und Informatiker, 2016. DOI: 10.1007/978-3-662-48016-8

[15] J. Dormand and P. Prince, "A reconsideration of some embedded Runge—Kutta formulae," *Journal of Computational and Applied Mathematics*, vol. 15, no. 2, pp. 203–211, 6 1986. DOI: 10.1016/0377-0427(86)90027-0.

[16] L. F. Shampine, I. Gladwell and S. Thompson, Solving ODEs with MATLAB, Cambridge University Press, 2003. DOI: 10.1017/CBO9780511615542.

[17] M. Géradin and D. J. Rixen, Mechanical Vibrations - Theory and Application to Structural Dynamics, 3 ed., Wiley & Sons, Ltd, 2015.

[18] R. Tiwari, Rotor Systems, Boca Raton: CRC Press, 2017. DOI: 10.1201/9781315230962.

[19] M. Kreutz, J. Maierhofer, T. Thümmel and D. J. Rixen, "Modaler Modellabgleich eines Rotors in Magnetlagern," in *Sechste IFToMM D-A-CH Konferenz*, Lienz, Österreich, 2020. DOI: 10.17185/duepublico/71192.

[20] J. Maierhofer, M. Gille, T. Thümmel and D. J. Rixen, "Using the Dynamics of Active Magnetic Bearings to perform an experimental Modal Analysis of a Rotor System," in *13th International Conference on Dynamics of Rotating Machinery*, Copenhagen, Denmark, 2019. DOI: 10.13140/RG.2.2.26743.16809.

[21] J. Maierhofer, C. Wagner, T. Thümmel and D. J. Rixen, "Progress in Calibrating Active Magnetic Bearings with Numerical and Experimental Approaches," in *Proceedings of the 10th International Conference on Rotor Dynamics – IFToMM*, 2018. DOI: 10.1007/978-3-319-99272-3_18.

[22] A. Brandt, "ABRAVIBE - A MATLAB toolbox for noise and vibration analysis," 2019. [Online]. Available: http://www.abravibe.com.

[23] A. Brandt, Noise and Vibration Analysis: Signal Analysis and Experimental Procedures, Wiley, 2011. DOI: 10.1002/9780470978160

[24] G. Schweitzer and E. H. Maslen, Magnetic bearings, Berlin, Heidelberg: Springer, 2009. DOI: 10.1007/978-3-642-00497-1.

[25] M. Geradin and A. Cardona, Flexible Multibody Dynamics A Finite Element Approach, John Wiley & Sons Ltd., 2001.

12th International Conference on Vibrations in Rotating Machinery -
Institution of Mechanical Engineers, ISBN 978-0-367-67742-8

Vibration monitoring of a large rotor utilizing internet of things based on-shaft MEMS accelerometer with inverse encoder

I. Koene, R. Viitala, P. Kuosmanen

Department of Mechanical Engineering, Aalto University, Finland

ABSTRACT

Typically, accelerometer-based vibration measurements of rotating machinery are conducted with sensors mounted to a static part of the machine. Now, with increasing accuracy of compact and low powered microelectromechanical systems (MEMS) accelerometers, on-shaft vibration measurements have become an interesting research topic. MEMS sensors are optimal for internet of things (IoT) applications and wireless measurements, which makes on-shaft measurements more convenient. However, typically in wireless applications, the sample clock is time-based, and thus the data is not bound to the phase of the rotating rotor. In this research, a novel wireless sensor unit with an inverse encoder is mounted to the end of a large rotor to investigate the dynamic behavior of the rotor. In addition, a method to separate the vertical and horizontal vibration from the sensor data is studied.

1   INTRODUCTION

Vibration measurement is a crucial part of the condition monitoring of a rotating machine and accelerometers are commonly used to measure the vibrations. Typically, these sensors are wired piezoelectric accelerometers, and they are mounted to a static part of the machine, such as a bearing housing or frame (1)–(5). However, in recent years cost-effective, compact and low powered microelectromechanical systems (MEMS) accelerometers have become a popular option to replace the piezoelectric sensors. Several different types of MEMS accelerometer sensing schemes have been developed, such as piezoresistive (6) and capacitive (7), (8), (9), (10) based sensing elements. Capacitive sensing elements are commonly used due to their good noise performance, high sensitivity and low temperature sensitivity (7). Piezoelectric and MEMS accelerometers have been studied and compared to determine whether the MEMS sensor can measure the same phenomenon as a piezoelectric sensor (1), (2), (4), (5), (11). MEMS accelerometers have shown promising results, and in many situations, the piezoelectric accelerometer could be replaced with a MEMS sensor. However, the noise level has been higher compared to the piezoelectric accelerometers, as Koene et al. (4) pointed out in their research.

The low power consumption of the MEMS sensors enables using a battery as a power source. With a battery as a power source, wireless data transfer is beneficial as the cables can be eliminated. The wireless measurement gives a new opportunity to measure rotor behavior by mounting the sensors directly to the rotating part of the machine. Elnady et al. (12), (13), Jiménez et al. (14) and Feng et al. (15) have studied condition monitoring of a rotating machine by mounting a MEMS accelerometer to the rotor. Elnady et al. (12), (13) mounted a two-axis accelerometer to the surface of the rotor, as well as two one-axis accelerometers to the bearing housing of the same rotor, to study lateral vibrations. Because the sensor was not aligned with the central axis of the rotor, the accelerometer measured radial and tangential accelerations. They conducted a sweep measurement and noticed that the vibration peaks did not occur at the same frequency in the bearing housing accelerometer data and on-shaft accelerometer data. In the on-shaft accelerometer data, the vibration peaks were observed at two different frequencies, which had a mean frequency coinciding with the corresponding peak frequency from the bearing housing (12). Jimémez et al. (14) observed the same phenomenon when they did impact tests with their test rotor.

Jimémez et al. (14) aligned a two-axis MEMS accelerometer with the rotor central axis. They analyzed the measurement data to solve the radius of the rotor orbit and the velocity of the rotor. They also conducted impact measurements to observe the natural frequency of the system. Feng et al. (15) studied compressor condition monitoring by mounting a MEMS accelerometer to the flywheel of the compressor. They studied a method on how to eliminate the gravitational acceleration from the rotating accelerometer data. Their results indicate that the on-shaft sensor can be used to detect the common fault types of their test compressor. In addition to rotor condition monitoring, wind turbine condition monitoring has adopted piezoelectric and MEMS accelerometers (16)–(19). With wind turbines, the accelerometers can be mounted to the wind turbine frame (18), (19), or to the rotating blades itself (16)–(18). Typically, in these cases, several accelerometers are used simultaneously, which requires designing and studying sensor networks (18), (19). It is also possible to adopt these sensor networks to another rotating machine condition monitoring application, where monitoring several rotating parts is required.

This paper presents a novel on-shaft MEMS accelerometer unit, which has an inverse encoder, and a novel method to separate the vertical and horizontal vibration from the accelerometer data. The inverse encoder signifies an encoder, which rotates with the rotor, and the encoder shaft is mounted to a static part of the machine. The code wheel of the encoder is mounted to the encoder shaft; hence the code wheel is fixed in the Earth's coordinate system and does not rotate as with typical encoders. In the inverse encoder, the light sensors measuring the code wheel rotates with the rotor, which eliminates the need for slip rings when using an encoder as a sample clock with an on-shaft sensor. The internal timer of the sensor unit can be used as a sample clock as well, which enables the possibility to study two different sampling clock methods.

2   METHODS

2.1   Proposed sensor unit and reference sensor

The proposed sensor unit consists of ESP32 internet of things (IoT) platform (Espressif Systems), AD7682 analog to digital converter (ADC) (Analog Devices), ADXL354 3-axis MEMS accelerometer (Analog Devices) and AMT102 encoder (CUI devices) with 1024 pulses per revolution and reference pulse, which occurs once per revolution. Figure 1 presents the sensor unit. The specification of the parts can be found from Tables 1, 2 and 3. The measurement range of ADXL354 in the measurements was ±8 g, which signifies the sensitivity of 100 mV/g. The sensor unit can also be defined to use a 10 kHz time-based sample clock (TBSC) for the measurement instead of the encoder. The TBSC was used to study the phase tracking ability of the sensor unit when the phase information was purely deducted using the MEMS data, which shows the direction of gravity. In these comparisons, the phase accuracy was compared to the inverse encoder data.

Figure 1. A) Sensor unit and b) components of the sensor unit.

To validate the measurements, a similar IoT sensor, which is verified by Koene et al. (4), was used. It had ADXL355 MEMS accelerometer, which has the same sensing elements as in ADXL354. Specifications of ADXL355 are presented in Table 3, and the same measurement range was used as with the ADXL354.

Table 1. Specifications of ESP32 IoT platform.

Specifications	ESP32
Microcontroller Unit	Tensilica Xtensa 32-bit LX6
Cores	2
Clock frequency (MHz)	240
SRAM (KiB)	520
802.11 b/g/n WI-FI	HT40
GPIO	36
SPI/I2C interfaces	4/2

Table 2. Specifications of AD7682 ADC.

Specifications	AD7682
Channels	4
Resolution	16 bits
Throughput	250 kS/s
Interface	SPI

Table 3. Specifications of ADXL354 and ADXL355 accelerometers. Used values are bolded.

Specifications	ADXL354	ADXL355
Measurement range	±2/**8** g	±2/4/**8** g
Axis	3	3
Bandwidth	1500 Hz	1000 Hz
Sensitivity	400/**100** mV/g	3.9/7.8/**15.6** µg/LSB

2.2   Test setup

Tests were conducted with full-size paper machine roll, which weighs around 700 kg and is 5 meters long. The sensor unit was mounted to the rotor end, and the reference sensor was mounted to the bearing housing. Figure 2 a) presents the test rotor, and Figure 2 b) presents the sensor setup.

Figure 2. A) Test rotor, which weighs around 700 kg and is 5 meters long. b) Sensor setup with the sensor unit and reference sensor. The coordinate system with u- and v-axis indicate the orientation of two of the axes of the MEMS accelerometer inside the sensor unit. One axis of the MEMS accelerometer was coaxial with the axis of the rotor, but it was not used during the present study.

The measurements were done with a rotor velocity of 500 rpm (8.33 Hz), and at least 100 rounds were measured. Two measurements were made with the sensor unit: one with the encoder-based sample clock (EBSC) and one with the time-based sample clock (TBSC). During both measurements, the reference sensor was measuring simultaneously as well.t

2.3   Data analyses

In the case of TBSC measurement, the rotor phase was calculated from the gravitational component of the signal from both u- and v-axis. The calculation process is presented below and in Figure 3 as a block diagram.

1. Lowpass filter the frequencies above the rotation speed from the data
2. Determine the phase of a measurement sample with the following equations:

$$a(\alpha) = a_u(n) \cdot \cos(\alpha) + a_v(n) \cdot \sin(\alpha) \tag{1}$$

$$\alpha(n) = \arg\max_{\alpha}(a(\alpha)) \tag{2}$$

Where a_u is the u-axis acceleration value, a_v is the v-axis acceleration value, a is the angle and n is the sample number. The equation (2) finds the a value, which has the highest acceleration based on equation (1) and that a is the phase of the sample. The phase is calculated with 0.1 degree accuracy.

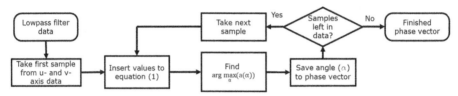

Figure 3. Method how to calculate a phase with the gravitation component of the sensor data.

Typically, lateral vibrations are measured in horizontal and vertical directions; however, in a case where the sensor rotates with the rotor, the sensor axis direction does not stay constant in the Earth's coordinate system. Hence, here is presented a method to separate the horizontal and vertical vibration from rotating sensor data with a rotation matrix:

$$\begin{bmatrix} x(n) \\ y(n) \end{bmatrix} = \begin{bmatrix} \cos(\alpha(n)) & -\sin(\alpha(n)) \\ \sin(\alpha(n)) & \cos(\alpha(n)) \end{bmatrix} \cdot \begin{bmatrix} a_u(n) \\ a_v(n) \end{bmatrix} \tag{3}$$

Where x is horizontal vibration, y is vertical vibration, a is the phase, a_u and a_v are the sensor data, and n is the sample number. The sensor axis u and v are marked in Figure 2 b). The vibration frequencies of the vertical and horizontal data were calculated with fast Fourier transform (FFT) (20).

2.4 Calibration
The sensor unit was calibrated after mounting it to the rotor by utilizing the gravitation. The calibration revealed the relation between the ADC output and the acceleration. Calibration measurement was made by rotating the test rotor 100 rounds with 10 rpm velocity. The encoder was used as a sample clock. Because the sensor was aligned with the rotor central axis and the rotor was rotated with a very low velocity, the only significant acceleration component affecting the sensor was the gravitation. The sensor was calibrated by analyzing the measurement data, as presented in Figure 4 and below:

1. Calculate one round average from the ADC output
2. Find the maximum, minimum and mean values for the one round average
   a. The maximum value is equal to 1 g
   b. The minimum value is equal to -1 g
   c. The mean value is equal to 0 g

3. Calculate the conversion coefficient to convert the ADC output to m/s² with the following equation:

$$k = \frac{9.81 \frac{m}{s^2} \times 2}{max - min} \quad (4)$$

Where max is the axis maximum value and min is the axis minimum value. The conversion coefficient is calculated to both u- and v-axis. The final equation to convert the ADC output to m/s² is for the u-axis:

$$a_u(n) = (ADC_u(n) - mean_u) \times k_u = (ADC_u(n) - 32702) \times 0.0028 \quad (5)$$

Where ADC_u is the u-axis ADC output, $mean_u$ is the u-axis mean value, k_u is the u-axis conversion coefficient and n is the sample number. For v-axis, the equation is as follows:

$$a_v(n) = (ADC_v(n) - mean_v) \times k_v = (ADC_v(n) - 32737) \times 0.0027 \quad (6)$$

Where ADC_v is the v-axis ADC output, $mean_v$ is the v-axis mean value, k_v is the v-axis conversion coefficient and n is the sample number. The $mean_u$ and $mean_v$, and k_u and k_v values are calculated from the calibration measurement data and those same values were used to convert the measurement data from the ADC output to m/s². The phase offset of the encoder is determined by using the maximum value of the one round average, which indicates a 90-degree phase.

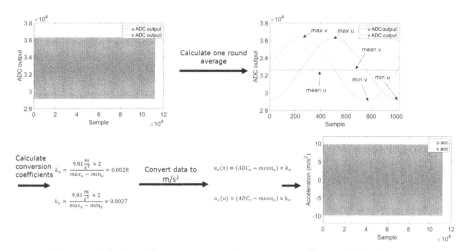

Figure 4. Calibration process and conversion from ADC values to acceleration.

3   RESULTS

Figures from 5 to 11 presents the test results from the 500 rpm measurements. The sensor unit measurement results are presented in three different ways, which are listed below:

1. measurements with the encoder-based sample clock (EBSC) and encoder-based phase (Figures 5 a), 6 a), 7 a), 8 a), 9 a) and 11 a)),
2. measurements with the time-based sample clock (TBSC) and gravity-based phase (Figures 8 b) and 9 b)),
3. and measurements with the encoder-based sample clock (EBSC) and gravity-based phase (Figure 11 b)).

The last category is used to compare how two differently determined phases effect to the coordinate transformation of the same measurement data.

3.1 Encoder-based sample clock measurement

Figures 5 and 6 present the sensor unit data with the EBSC and encoder-based phase after the coordinate transformation, as well as the reference sensor data. In Figure 5, the results are similar in the horizontal direction (x-axis). The same vibration peaks can be observed from both the proposed and reference sensor data, and the horizontal natural frequency is visible as well, being approximately 20.9 Hz. However, the amplitudes differ from each other.

When comparing the vertical vibrations (Figure 6), there are more differences. The amplitudes have more deviation between the sensors: the reference sensor has approximately 20 times lower highest vibration peak than the sensor unit data. The vertical natural frequency is approximately 29.2 Hz, and it is the highest peak in the reference sensor measurements. However, in the sensor unit measurement, the natural frequency peak is visible, but amplitude is low compared to the other peaks.

Figure 7 b) presents the raw u-axis acceleration data of the same EBSC measurement and Figure 7 a) presents the coordinate transformed data shown in the Figure 5 a). In the raw acceleration data, the vibration frequencies are presented as two peaks (sidebands), which are approximately 16.6 Hz (two times the rotation frequency, 8.3 Hz) apart from each other. The average of the peaks is the frequency measured from the bearing housing or observed after the coordinate transformation. The phenomenon is caused by an amplitude modulation. Because the sensor is rotating with the rotor, the direction of the sensor axes changes in the Earth's coordinate system. However, the vibration direction stays constant in the Earth's coordinate system, which causes the vibration amplitude to modulate in the sensor coordinate system depending of the sensor angle in the Earth's coordinate system. The amplitude change occurs at the rotation frequency, which causes the sidebands to appear at both sides of the vibration frequency with an offset of rotation frequency.

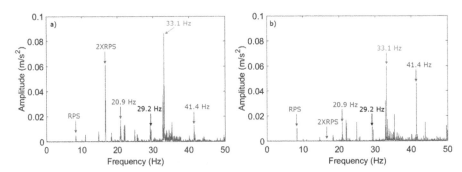

Figure 5. Horizontal (x-axis) vibration at 500 rpm (8.3 Hz) in the frequency domain. a) Sensor unit measurement with the encoder-based sample clock and b) reference sensor measurement from the bearing housing. RPS means rounds per second, which is equal to the rotation frequency.

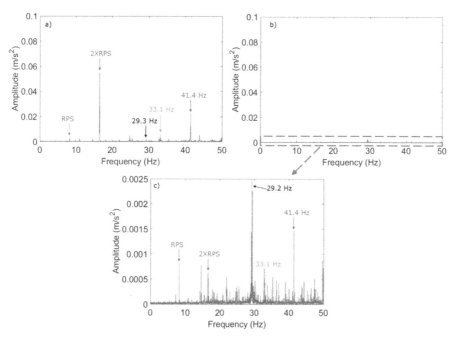

Figure 6. Vertical (y-axis) vibration at 500 rpm (8.3 Hz) in the frequency domain. a) Sensor unit measurement with the encoder-based sample clock, b) reference sensor measurement from the bearing housing with the same y-axis scale as in a) and c) scaled plot from the reference sensor measurement. RPS means rounds per second, which is equal to the rotation frequency.

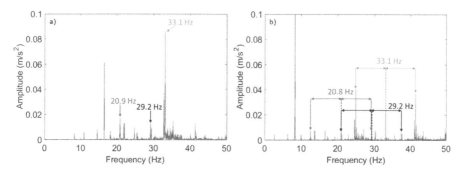

Figure 7. Comparison between horizontal coordinate transformed and raw u-axis sensor unit data in the frequency domain from the encoder-based sample clock measurement. a) horizontal coordinate transformed data and b) raw u-axis sensor unit data. The averages of the marked frequency peaks in b), are the same values as in the coordinate transformed data. The values presented in b), are the average of the frequency peaks where the same colored arrows point. This is caused by an amplitude modulation of the measured signal due to the sensor rotating.

3.2 Time-based sample clock measurement

Figures 8 and 9 present the same data from the EBSC measurement as in Figures 5 and 6, and the measurement data from the TBSC measurement, with the gravity-based phase, after the coordinate transformation. Some peaks can be observed in both figures in the vertical and horizontal directions. Most significant differences occur at the rotation frequency and two times rotation frequency. In the horizontal vibrations, the amplitude of the two times rotation frequency is significantly lower in the EBSC measurement than in the TBSC. In the vertical vibrations, the magnitudes of the amplitude differences were not as high.

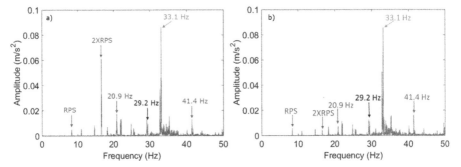

Figure 8. Horizontal (x-axis) vibration at 500 rpm (8.3 Hz) in the frequency domain. a) Sensor unit measurement with the encoder-based sample clock and b) sensor unit measurement with the time-based sample clock. RPS means rounds per second, which is equal to the rotation frequency.

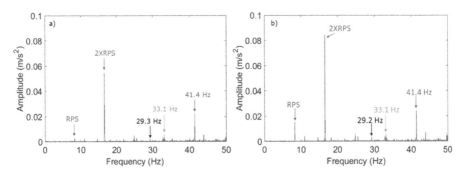

Figure 9. Vertical (y-axis) vibration at 500 rpm (8.3 Hz) in the frequency domain. a) Sensor unit measurement with the encoder-based sample clock and b) sensor unit measurement with the time-based sample clock. RPS means rounds per second, which is equal to the rotation frequency.

3.3 Comparison of phase calculating methods

Figure 10 presents the phase difference between the two different methods for calculating the phase: one that utilized the encoder and one that utilized the gravitational components of the MEMS sensor data. Both methods were applied to the 500 rpm EBSC measurement. The phase was constantly shifted approximately 1.9 degrees, and there was an oscillation in the form of a sine wave. It was assumed that the encoder based phase determination was correct; this suggests that gravity-based phase determination lags ca 1.9 degrees and has oscillation in the range of 1-2.5 degrees.

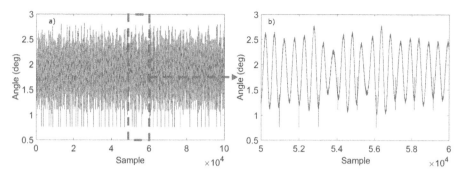

Figure 10. Difference of encoder-based sample clock sensor unit measurement with both phase calculation methods. a) is from 100000 samples and b) is a closeup of 50000 to 60000 samples. The x-axis is the sample count of the measurement.

Figure 11 presents the horizontal vibration of the data of the EBSC measurement with both phase-calculating methods. Figure 11 a) presents the results with the encoder-based phase and Figure 11 b) presents the results with the gravity-based phase. Similar results can observe as in Figure 8: the most significant difference appears in the amplitude of the two times rotation frequency (2XRPS, 16.6 Hz) peak. However, there is an amplitude difference in the horizontal natural frequency as well. Nevertheless, both methods present the same frequency peaks, even though there was a small difference in the phases depending on the calculating method.

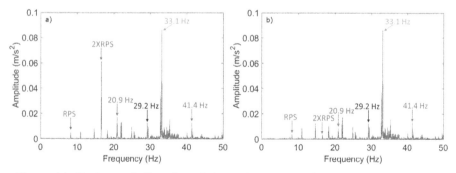

Figure 11. Horizontal vibration of the 500 rpm encoder-based sample clock measurement in the frequency domain. a) presents the results of a coordinate transformation made with the encoder-based phase and b) presents the results of coordinate transformation made with the gravity-based phase.

4   DISCUSSION

The results showed that it is possible to separate the horizontal acceleration data from the rotating sensor data; however, there were some difficulties in separating the vertical data. The reason for this may be that the amplitudes of the signal in vertical direction were significantly lower. The lower amplitudes can be observed from the reference measurement from the bearing housing. The highest amplitude in the vertical direction was approximately 0.0025 m/s^2 when in the horizontal direction, it was approximately 0.06 m/s^2. The amplitude differences are caused by the different stiffnesses in the vertical and horizontal directions of the test bench foundation. In the vertical direction, there were some of the same frequency peaks, which were observed in the horizontal direction. The higher magnitude vibration of the horizontal direction may have caused the vibrations to be visible in the vertical direction as well. However, the horizontal natural frequency is not visible in the vertical data, which indicates that the separation of the vertical data from the horizontal one may be possible. Nevertheless, the horizontal results were promising. Both on-shaft and reference sensors observed the same frequency peaks.

The comparison between the TBSC and EBSC sensor unit measurements showed that no significant differences occur depending on the sampling clock method. The most significant differences appear in the amplitudes of the rotation frequency and at twice the rotation frequency. The reason for that may be the different methods for calculating the phase. The horizontal natural frequency (20.9 Hz) can be observed from both figures. The vertical natural frequency can be observed from both measurements as well, but the amplitudes are quite low. However, if the vertical natural frequency peak amplitudes of the time and encoder-based sample clock sensor unit data is compared to the amplitude of the reference measurement, they are in the same range. The EBSC measurement, TBSC measurement and the reference measurement have a vertical natural frequency amplitude of 0.0020 m/s^2, 0.0033 m/s^2, and 0.0025 m/s^2, respectively. However, the reference sensor and proposed sensor unit were mounted to different parts of the machine, which may affect the amplitudes.

The results showed a promising start for separating the horizontal and vertical vibration of a rotating two-axis accelerometer. The horizontal vibration was separated well from the raw measurement data; however, the vertical vibration was not as clear as the horizontal. Hence, separating the vertical and horizontal vibration needs to be studied more, and one way to do it is to use the sensor unit in a test rotor bench were the magnitude of the vertical and horizontal vibrations are more similar to each other.

Still in many applications, the accelerometer mounted to the static part of the machine is better way of measuring the machine vibrations. However, in some machines it is not possible to measure the vibrations effectively from a static part of the machine. In these applications, on-shaft accelerometers could be used. On-shaft accelerometers also opens new possibilities to study machine condition monitoring and to measure other aspects of machines such as, lateral vibrations or central point movement. Lateral vibrations could be measured from the rotor, if the accelerometer is mounted with a small offset from the central axis. This would be a cost-effective method to measure the lateral vibrations with minimal measurement setup process. The central point movement is typically not measured in rotating machines because it is expensive, or the measurement instruments cannot fit close enough to the rotating part. The sensor unit could be mounted inside of a rotor or even several places in one rotor. This way, the central point movement could be determined from different positions of the rotor and find the modal shapes affecting it.

5   CONCLUSION

This study presents a novel on-shaft sensor unit with a MEMS accelerometer, inverse encoder and wireless data transfer. The encoder rotates with the rotor, and the encoder shaft i.e. code wheel is mounted to the static part of the machine. The light sensors measuring the code wheel are rotating with the rotor, hence enabling to use of the encoder pulse as a sample clock without slip rings. The sampling can also be clocked with time, which enables testing of gravity-based phase calculation and comparing the result to the encoder-based phase calculations. There were some differences in the phases with the different calculating methods. However, when the coordinate transformation was made with the different phases, the results did not differ significantly.

The separation of the horizontal and vertical axis from the data of the rotating sensor showed promising results. The horizontal vibration was well visible from the data and presented the same vibration peaks, as did the reference measurement. However, the vertical vibration did not occur as clearly as it did in the reference measurement; hence the presented method needs more research.

ACKNOWLEDGEMENTS

This research was supported by Academy of Finland (Digital Twin of Rotor System, under Grant 313675) and Business Finland (Reboot IoT Factory, under Grant 4356/31/2019).

REFERENCES

[1] A. Albarbar, S. Mekid, A. Starr, and R. Pietruszkiewicz, "Suitability of MEMS Accelerometers for Condition Monitoring: An experimental study," *Sensors*, vol. 8, pp. 784–799, 2008.

[2] A. Albarbar, A. Badri, J. K. Sinha, and A. Starr, "Performance evaluation of MEMS accelerometers," *Measurement: Journal of the International Measurement Confederation*, vol. 42, no. 5, pp. 790–795, 2009.

[3] J. D. Son, B. H. Ahn, J. M. Ha, and B. K. Choi, "An availability of MEMS-based accelerometers and current sensors in machinery fault diagnosis," *Measurement: Journal of the International Measurement Confederation*, vol. 94, pp. 680–691, 2016.

[4] I. Koene, R. Viitala, and P. Kuosmanen, "Internet of Things Based Monitoring of Large Rotor Vibration With a Microelectromechanical Systems Accelerometer," *IEEE Access*, vol. 7, pp. 92210–92219, 2019.

[5] I. Koene, R. Viitala, and P. Kuosmanen, "Vibration analysis of a large rotor over industrial internet," in *59th Ilmenau Scientific Colloquium*, pp. 1–9, 2017.

[6] A. Vogl, D. T. Wang, P. Storås, T. Bakke, M. M. V. Taklo, A. Thomson, L. Balgård, "Design, process and characterisation of a high-performance vibration sensor for wireless condition monitoring", *Sensors and Actuators A: Physical*, vol. 153, pp. 155–161, 2009.

[7] C. Acar, A. Shkel, "Experimental evaluation and comparative analysis of commercial variable-capacitance MEMS accelerometers", *Journal of Micromechanics and Microengineering*, vol. 13, pp. 634–645, 2003.

[8] A. Aydemir, Y. Terzioglu, T. Akin, "A new design and a fabrication approach to realize a high performance three axes capacitive MEMS accelerometer", *Sensors and Actuators, A: Physical*, vol. 244, pp. 324–333, 2016.

[9] B. Tang, K. Sato, S. Xi, G. Xie, D. Zhang, Y. Cheng, "Process development of an all-silicon capacitive accelerometer with a highly symmetrical spring-mass structure etched in TMAH + Triton-X-100", *Sensors and Actuators A: Physical*, vol. 217, pp. 105–110, 2014.

[10] Z. Mohammed, G. Dushaq, A. Chatterjee, M. Rasras, "An optimization technique for performance improvement of gap-changeable MEMS accelerometers", *Mechatronics*, vol. 54, pp. 203–216, 2018.

[11] Y. J. Chan and J.-W. Huang, "Multiple-point vibration testing with micro-electromechanical accelerometers and micro-controller unit," *Mechatronics*, vol. 44, pp. 84–93, 2017.

[12] M. E. Elnady, J. K. Sinha, and S. O. Oyadiji, "Identification of Critical Speeds of Rotating Machines Using On-Shaft Wireless Vibration Measurement," *Journal of Physics: Conference Series*, vol. 364, pp. 1–10, 2012.

[13] M. E. Elnady, A. Abdelbary, J. K. Sinha, and S. O. Oyadiji, "FE and Experimental Modeling of On-shaft Vibration Measurement," in *Proceedings of the 15th International Conference on Aerospace Sciences & Aviation Technology*, no. May, pp. 1–18, 2013.

[14] S. Jiménez, M. O. T. Cole, and P. S. Keogh, "Vibration sensing in smart machine rotors using internal MEMS accelerometers," *Journal of Sound and Vibration*, vol. 377, pp. 58–75, Sep. 2016.

[15] G. Feng, N. Hu, Z. Mones, F. Gu, and A. D. Ball, "An investigation of the orthogonal outputs from an on-rotor MEMS accelerometer for reciprocating compressor condition monitoring," *Mechanical Systems and Signal Processing*, vol. 76–77, pp. 228–241, Aug. 2016.

[16] O. O. Esu, J. a Flint, and S. J. Watson, "Condition Monitoring of Wind Turbine Blades Using MEMS Accelerometers," *Renewable Energy World Europe*, pp. 1–12, 2013.

[17] M. Mollineaux, K. Balafas, K. Branner, P. Nielsen, A. Tesauro, et al., "Damage Detection Methods on Wind Turbine Blade Testing with Wired and Wireless Accelerometer Sensors," EWSHM - 7th European Workshop on Structural Health Monitoring, IFFSTTAR, Inria, Université de Nantes, Nantes, France, 2014.

[18] R. Simon Carbajo, E. Simon Carbajo, B. Basu, and C. Mc Goldrick, "Routing in wireless sensor networks for wind turbine monitoring," *Pervasive and Mobile Computing*, vol. 39, pp. 1–35, Aug. 2017.

[19] G. Kilic and M. S. Unluturk, "Testing of wind turbine towers using wireless sensor network and accelerometer," *Renewable Energy*, vol. 75, pp. 318–325, 2015.

[20] J. W. Cooley and J. W. Tukey, "An Algorithm for the Machine Calculation of Complex Fourier Series," *Mathematics of Computation*, vol. 19, no. 90, p. 297, 1965.

Mechanical design of rotor-bearing system in a high-speed 20 kW range extender for battery electric vehicles

H. Kim[1], J. Nerg[2], A. Jaatinen-Värri[3], J. Pyrhönen[2], J. Sopanen[1]

[1]Mechanical Engineering, School of Energy Systems, LUT University, Finland
[2]Electrical Engineering, School of Energy Systems, LUT University, Finland
[3]Energy Technology, School of Energy Systems, LUT University, Finland

ABSTRACT

Electrification of transportation is one of the most effective means to reduce emissions. However, widespread adoption of electric vehicles faces challenges because of the limited availability of raw materials of batteries. Because of these concerns, hybrid electric technology is seen as an alternative. This paper presents the mechanical design for a rotor-bearing system of a 20 kW micro-gas-turbine-based range extender (RE). The RE has an integrated permanent magnet high-speed electric generator. Rotordynamic analysis is conducted for a rotor supported by ball bearings and squeeze film dampers (SFD). From the results, requirements for the SFD design are defined.

1   INTRODUCTION

Transition from internal combustion engine (ICE) vehicles to battery electric vehicles (EV) is the main target currently in the automobile industry. However, when evaluating this transition so far, we see that the transition has not been successful because of the limitations of the battery electric vehicle. The battery electric vehicle has several drawbacks to tackle; the prolonged battery charging time, lack of charging facilities, limited driving range due to the low specific energy of Lithium-Ion (Li-Ion) batteries [1]. Furthermore, limitations in the amount of raw materials for batteries is a potential drawback limiting the pace of increasing the share of battery electric vehicles. Therefore, to overcome this problem, range extender (RE) EV is being studied by many researchers as a promising solution. New technology should, however, replace the present-day massive ICE-based hybrid technology.

A range extender is an onboard genset that can be used to recharge the car battery to extend its driving range. The traction power is generated by an electric motor only whereas the genset is purely used for recharging the battery. Hence, when using this solution, the driving range can be secured with low battery capacity. Typically, 10 kWh net energy in the battery is enough to perform daily commute and whenever the battery energy is not high enough the RE starts providing the necessary average power. A normal family car does not need more average power than 20 kW.

Variable types of range extenders are being studied to find the best solution. Available types are fuel cell, rotary engines, two- and single cylinder engines, Wankel engine and gas turbine engine. Presently, reciprocating engines have higher efficiency than gas turbines, especially at the lower ends of the power spectrum.

Consequently, gas turbines have challenges in obtaining feasible efficiency at an acceptable cost [2]. The electric efficiencies of single-shaft gas turbines around 100 kW power is around 30 percent [3], [4], and the efficiencies at lower power are even lower. However, despite lower efficiency, small gas turbines have some advantages when compared to reciprocating engines. The partial load efficiency is higher and emissions lower [5], due to continuous combustion. They also offer a wider fuel variety (e.g. solar-power-based synthetic fuels) combined with lower emissions, and they have smaller maintenance costs.

Due to benefits in other aspects than efficiency, gas turbines in some cases can offer an alternative to reciprocating engines. If the efficiency of small-scale and micro turbines could be improved, they would be even more competitive. The efforts to improve the efficiency advances on several frontiers. For example, there have been studies how component efficiency affects the cycle performance [6], studies on how ceramic materials can be used to increase the cycle peak temperature and therefore the efficiency [7]. Furthermore, there have been suggestions that in small scale turbines, the expansion and compression, which typically are divided into stages also in small machines, could be arranged on separate shafts [8], [9]. The division of compression and expansion on separate shafts, which can be controlled independent of each other, allow some leeway in the turbomachinery design, component placement, etc. In addition, the partial load efficiency is better. Two-spool micro-gas-turbine technology is seen as a viable high-efficiency and material efficient option for the on-board power production in vehicles.

In this study, a 20 kW micro gas turbine range extender is considered. Mechanical design for this range extender is conducted based on the requirement of the system. Within limitations set by the required rotation speed and electric machine design, ball bearing cartridge with squeeze film damper (SFD) is selected for supporting the rotor system. Finally, initial design of the rotor bearing system is evaluated by conducting unbalance response rotordynamic simulation. Specifically, in this simulation, the singular value decomposition (SVD) method [10] is used for obtaining the worst-case scenario. From the results, detailed requirements for the design of the SFD are defined.

2   DESIGN REQUIREMENTS

2.1   Micro turbine range extender under study
Considering system's requirements, such as the desired efficiency and required rated power, the range extender is designed based on two-spool, high-speed micro gas turbine technology. The selected concept is presented in Figure 1. This system consists of two rotating shafts. One is a high-pressure (HP) shaft which acts as a gas generator and the other is low-pressure (LP) shaft which has the electric generator. Having the generator only on the low-pressure shaft keeps the high-pressure shaft simpler, allowing the possibility to push the turbine inlet temperature of the high-pressure turbine higher.

Initially, the process values (pressure ratios, etc.) are optimized based on the desired electrical power output, and these are the preliminary values for the turbomachinery design. The design is based on the specific speed N_s depending on the rotational speed ω, volumetric flow q_{v1} and enthalpy difference Δh_s

$$N_s = \frac{\omega \sqrt{q_{v1}}}{\Delta h_s^{0.75}} \tag{1}$$

and the specific diameter

$$D_s = \frac{D\Delta h_s^{0.25}}{\sqrt{q_{v1}}} \tag{2}$$

where D is the actual diameter. In turbomachinery design, it is commonly known that each type of turbomachine has some specific speed value within which we can obtain the peak efficiency (e.g. see Rodgers [11]), and there is an optimum corresponding specific diameter for each specific speed (Cordier line, e.g. see Casey et al. [12]). In this instance, the desired power output is relatively low (20 kW), and consequently, the design inlet volume flow q_{v1} is relatively small. In addition, as the compression and turbine work are divided into two stages, the design pressure ratio and the design isentropic enthalpy change Δh_s are moderate. The relatively small inlet volume flow and the moderate pressure ratio follows that the desired rotational speed will be high with respect to typical designs and that the physical size of the impellers is small compared to typical designs.

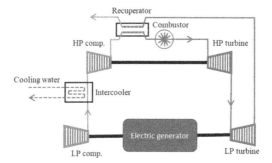

Figure 1. Schematic diagram of gas turbine range extender under study.

2.2 Requirements for the rotor-bearing system of LP gas turbine with electric generator

According to the design requirement for the whole range extender system, the permanent magnet (PM) synchronous machine is selected as the type of the electric generator. Its main dimensions and rotation speed are decided with the design of the turbine and compressor. Here, a cylindrical permanent magnet with 30 mm diameter is used in the electric generator excitation. Turbine and compressor are designed as radial type for high efficiency. In conclusion, the fundamental design requirements for the LP gas turbine with electric generator is defined as in Table 1. Specifically, the allowable vibration amplitude is defined as 30 μm (0-pk). This criterion will be used to evaluate the designed rotor system.

From these requirements, firstly, the magnet retaining sleeve must be designed by using stress analysis. This design problem aims to minimize the thickness of the sleeve within the constraints that a) the designed sleeve must endure the stress generated during high-speed operation, b) radial stress inside the magnet should not exceed tensile strength during high-speed operation, c) shrink-fit should remain at compression in all conditions. As a result, its material (Inconel 718) and thickness (4 mm) was selected as shown in Table 2. The details of this stress analysis are not presented in this paper. Secondly, the rotor-bearing system is designed with selection of a suitable bearing type.

Table 1. Main parameters for RE system design.

Parameters (unit)	Values
For whole RE system design	
Electricity generation efficiency target, %	Min. 40
Rated output electric generated power, kW	22
For rotor-bearing system design	
Rated speed, rpm	101000
Rotor PM material	PM having 1.2 T remanence at 80 °C
PM size, mm	ø 30×100 (Cylindrical shape)
Turbine mass, g	74
Turbine outer diameter, mm	71
Turbine moment of inertia, kg-m^2	$2.56 \cdot 10^{-5}$
Compressor mass, g	54
Compressor outer diameter, mm	66
Compressor moment of inertia, kg-m^2	$1.88 \cdot 10^{-5}$
Allowable vibration amplitude (0-pk), µm	30

3   BEARING SELECTION AND ROTORDYNAMIC CASE STUDY

In this section, firstly, the ball bearing with SFD is selected to satisfy the required specification of the machine, and then initial rotor design is suggested. However, in this design stage, it is difficult to estimate the dynamic characteristics of the SFD accurately because it is dependent on detailed design of the SFD. Hence, in this study, the objective is to find the required stiffness and damping coefficients of the SFD by conducting unbalance response analysis. Singular value decomposition (SVD) theorem is utilized in finding the worst-case unbalance configuration.

3.1   Bearing selection

To select the bearing type, two factors in addition to low cost are considered. The first is to minimize the bearing loss. The second is to reserve the stability of the high-speed rotating system. Within the required speed range, available options are gas foil bearing, active magnetic bearing and ball bearing. The oil-suspended ring bearings used widely in turbochargers consume too much friction power for the range-extender application. A gas foil bearing is a promising option for a high-speed operation because of low bearing friction loss, and the bearing has a simple structure. Meanwhile, the damping coefficient of the gas foil bearing is low, therefore, a rotor system supported by gas foil bearings can experience instability as a result of external excitations. Active magnetic bearing is also a good option for high-speed operation because it is operated in a noncontact condition and has controllable coefficients in bearing stiffness and damping. However, it has a complex structure and causes high cost. Meanwhile, ball bearing is also quite useful for high-speed operation. Furthermore, this type of a bearing is widely used in many applications and its stability is guaranteed. Ball bearing has a high stiffness coefficient, but its damping coefficient is low. To overcome this shortage, a squeeze film damper (SFD) can be integrated with the ball bearing. This is studied by several researchers [13], [14]. Because of this advantage, the ball bearing with a squeeze film damper became a promising option for automobile turbochargers.

In conclusion, the ball bearings with SFD are selected for the machine studied. When comparing it with gas foil bearing, this option can be better in rotordynamic stability and better in economic feasibility than active magnetic bearing. The concept of the ball bearing cartridge with SFD used in automobile turbocharger is presented in Figure 2. In this example, decoupling ring is applied as both seal and centering spring. In this design concept, SFD's dynamic characteristics can be adjusted by changing the clearance and decoupling design.

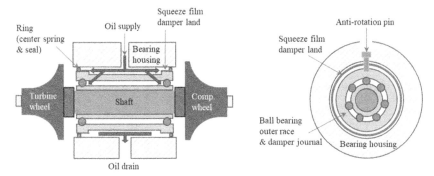

Figure 2. Schematic diagram of ball bearing cartridge with SFD for vehicle turbocharger.

3.2 Designed rotor model

Based on the designed impellers, electric generator, rotor sleeve and the selected bearing type, the rotor-bearing system is designed. Its main parameters are presented in Table 2 and the finite element model for rotordynamic analysis is shown in Figure 3. In this rotor system, the electric generator is located at the center of the machine and the two impellers are assembled at both ends. Two ball bearing cartridges with SFD support the rotor system. The bore diameter of the ball bearings is 10 mm. This cartridge has several advantages; it enables the electric generator to be located far from the high-temperature turbine. Secondly, it can make the bending mode frequencies to be far from the operation speed by decreasing the first two bending mode frequencies. In this study, for simple calculation, this bearing cartridge is modeled with several assumptions.

1) Ball bearing has constant stiffness and damping coefficients and has no cross-coupling term. Its coefficients are estimated using the simple method proposed by Gargiulo [15] and Krämer [16].
2) SFD has constant centering spring stiffness and it is assumed to be within the range of 10^5 to 10^7 N/m.
3) SFD has constant damping and it is assumed to be within the range of 0 to 1,000 Ns/m. It was roughly estimated using analytical calculation for the expected SFD design [17], [18].
4) SFD journal is rigid and it has two translational degrees of freedom. Its rotation is constrained.

Based on these assumptions, the rotor system supported by the ball bearing cartridges with SFD is modeled as in Figure 3. The shaft and the permanent magnet rotor are modeled with beam finite elements. The model has four degrees of freedom per node and it is assumed that there is no displacement in the axial direction and no rotation around the rotor axis. The ball bearing cartridge with SFD is modeled with the concept as in Figure 4. The shaft and SFD journal are connected by the constant spring and damper have ball bearing's stiffness and damping coefficient. According to the above assumption, the SFD journal is modeled as one mass point having the mass moment of inertia. Finally, the SFD journal is connected to the rigid base by a constant spring and damping of the SFD. To add the effect of impellers, turbine and compressor impellers are modeled as mass points including the mass moment of inertia. The locations of these mass points were adjusted from the end of shaft considering the center of mass. The parameters used for the modeling of the rotor system are presented in Table 2.

Table 2. Main parameters for designed rotor system.

Parameters (unit)	Values
Permanent magnet material	NdFeB
Sleeve & shaft material	Inconel 718
Sleeve thickness, mm	4
Inner diameter of ball bearing, mm	10
Length of ball bearing cartridge, mm	55.2
Ball bearing stiffness coefficient, N/m	10^6
Ball bearing damping coefficient, N-s/m	25
SFD stiffness coefficient, N/m	$10^5 \sim 10^7$
SFD damping coefficient, N-s/m	$0 \sim 1,000$

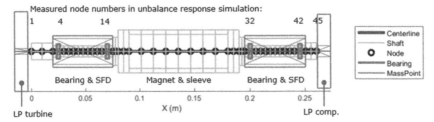

Figure 3. Beam finite element model of a rotor-bearing system with two impellers and permanent magnet.

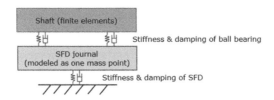

Figure 4. Modeling of bearing cartridge with SFD.

3.3 Unbalance response simulation using SVD

As presented in the design requirement for the rotor system, allowable vibration amplitude is 30 μm. To evaluate the rotor system based on this requirement, unbalance response simulation is conducted. Generally, unbalance response can be obtained by directly solving the motion equation including the unbalance force term [17]. When the rotor is modeled with finite elements, several unbalance masses are added to certain nodes for simulating the condition of distributed unbalance mass. The phase angles of unbalance masses must also be defined because that affect the result. Meanwhile, when an engineer designs rotor system, the worst-case situation needs to be considered with a conservative point of view, which can be generated within allowable manufacturing condition, e.g., unbalance mass distribution. However, according to the existing simulation method, because the result is very dependent on the definition of the unbalance mass, furthermore, it is not clear to know what condition of unbalance mass can make the worst unbalance response. Therefore, many cases must be conducted to study the worst-case result.

In this situation, the SVD is a good solution to simulate the unbalance response at the worst condition without multiple simulation for many cases. This tool was proposed by Cloud et al. [10] for the application to practical rotordynamics problems. Therefore, in this study, using the same process studied previously, the unbalance response is studied. The details about this method can be found in [10]. Furthermore, unbalance response using existing method is also simulated and the result is compared with that using SVD. For the simulation, total unbalance mass (0.254 g-mm) is defined based on a G2.5 balance grade. Then, in existing method, for maximizing the response at 2nd bend mode frequency, total unbalanced mass is distributed to two nodes near bearings with opposite phase, which is defined based on 2nd bend mode shape shown in Figure 7.

3.4 Rotordynamic simulation result

Using the method explained in the previous section, unbalance responses versus rotor speed up to 140000 rpm are obtained for the selected SFD stiffness and damping cases. From these results, it is expected to evaluate the rotordynamics for the designed system and, further, to find the suitable stiffness and damping level of the SFD to minimize the rotor vibration, which is used in the detailed design of the SFD.

Figure 5 shows the maximum displacements at the rated speed and three natural frequencies (2nd rigid, 1st and 2nd bending modes). These results are obtained as a function of stiffness and damping of the SFD to find requirements for that minimizes the displacements. From the results, it can be found that when SFD is designed so that its stiffness varies from 10^5 to 10^6 N/m and its damping varies from 250 to 500 N-s/m, the vibration conditions can be the best.

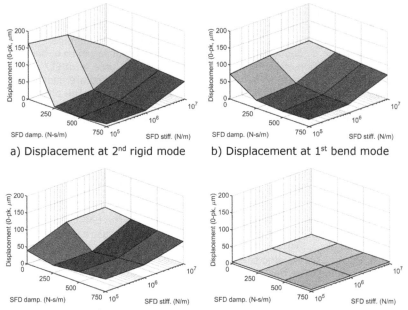

a) Displacement at 2nd rigid mode   b) Displacement at 1st bend mode

c) Displacement at 2nd bend mode   d) Displacement at rated speed (101000 rpm)

Figure 5. Max. displacements as a function of stiffness and damping of the SFD at three natural frequencies and rated speed.

Figure 6 shows the unbalance response results versus rotor speed. It is shown that the rotor has to pass two rigid and two bending-mode frequencies before reaching the rated speed. The mode shapes at the natural frequencies are shown in Figure 7. Comparing the results at two different conditions (without SFD and with SFD), it is found that the vibration peaks at the natural frequencies are dampened well by SFD. Therefore, it can be concluded that the designed rotor system can be stable when the SFD with 10^6 N/m stiffness and 500 Ns/m damping coefficients is designed. When comparing the results by using two different methods, it is shown that the response amplitude using SVD is about four times higher than that using existing method. Furthermore, the responses with backward whirling mode also appear significantly in the result using SVD.

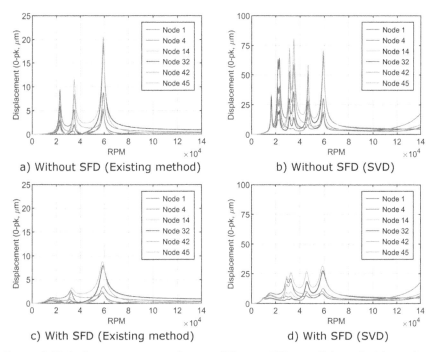

a) Without SFD (Existing method) b) Without SFD (SVD)

c) With SFD (Existing method) d) With SFD (SVD)

Figure 6. Unbalance responses in two different SFD conditions (without SFD, with SFD which has 500 N-m/s damping and 10^6 N/m stiffness coefficients) using two different simulation methods.

4   CONCLUSIONS

This paper presented the mechanical design study of the rotor-bearing system for an electric vehicle range extender. Design requirements were extracted from the selected concept of gas-turbine range extender with permanent magnet electric generator. Then, based on these requirements, the mechanical design for the rotor-bearing system was studied. Firstly, initial design of the rotor-bearing system was suggested with selection of bearing type. Then, the designed rotor-bearing system was evaluated through rotordynamic simulation method and furthermore, requirements for the SFD design were found. Summary of findings from this study are as follows.

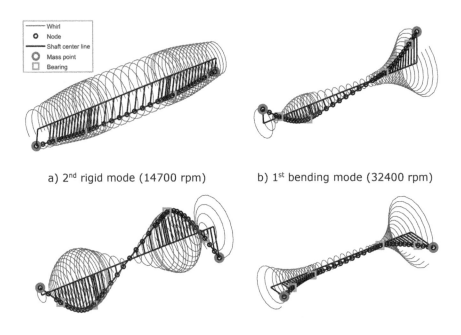

a) 2nd rigid mode (14700 rpm)   b) 1st bending mode (32400 rpm)

c) 2nd bending mode (58900 rpm)   d) At rated speed (101000 rpm)

Figure 7. Deformed shapes at second rigid mode, first bending mode, second bending mode and rated speed.

(1) Ball bearing cartridge with SFD is selected to support the rotor system. When considering rotordynamic stability at high-speed operation, low bearing loss and economic efficiency, this type of bearing can be a suitable option for this 20 kW range extender system.

(2) Studying unbalance response and deformed mode shape at critical speeds through the rotordynamic simulation, the designed rotor system was evaluated. From the results, it was shown that the designed rotor's critical speeds are far from the rated speed, but this rotor must go through two rigid and two bending mode frequencies to reach the rated speed operation. However, vibration at the critical speeds are dampened well by using SFD. Conducting rotordynamic study for expected range of stiffness and damping coefficients of the SFD, desired SFD stiffness and damping values for efficient damping effect are found. This result can be used for detailed design of SFD in the future.

REFERENCES

[1] F. X. Tan, M. S. Chiong, S. Rajoo, A. Romagnoli, T. Palenschat, and R. F. Martinez-Botas, "Analytical and Experimental Study of Micro Gas Turbine as Range Extender for Electric Vehicles in Asian Cities," *Energy Procedia*, vol. 143, 2017, pp. 53–60.

[2] G. B. de Campos, C. Bringhenti, A. Traverso, and J. T. Tomita, "Thermoeconomic optimization of organic Rankine bottoming cycles for micro gas turbines," *Appl. Therm. Eng.*, vol. 164, 2020, p. 114477.

[3] M. Montero Carrero, W. De Paepe, S. Bram, A. Parente, and F. Contino, "Does humidification improve the micro Gas Turbine cycle? Thermodynamic

assessment based on Sankey and Grassmann diagrams," *Appl. Energy*, vol. 204, 2017, pp. 1163–1171.

[4] F. Caresana, G. Comodi, L. Pelagalli, and S. Vagni, "Micro Combined Plant With Gas Turbine and Organic Cycle," *In Proceedings of ASME Turbo Expo*, GT2008-51103, June 9-13, Berlin, Germany, 2008

[5] M. L. Ferrari, A. Traverso, and A. F. Massardo, "Smart polygeneration grids: Experimental performance curves of different prime movers," *Appl. Energy*, vol. 162, 2016, pp. 622–630.

[6] L. Galanti and A. F. Massardo, "Micro gas turbine thermodynamic and economic analysis up to 500 kWe size," *Appl. Energy*, vol. 88, no. 12, 2011, pp. 4795–4802.

[7] F. Sadegh Moghanlou, M. Vajdi, A. Motallebzadeh, J. Sha, M. Shokouhimehr, and M. Shahedi Asl, "Numerical analyses of heat transfer and thermal stress in a ZrB2 gas turbine stator blade," *Ceram. Int.*, vol. 45, no. 14, 2019, pp. 17742–17750.

[8] M. Malkamäki, A. Jaatinen-Värri, J. Honkatukia, J. Backman, and J. Larjola, "A high efficiency microturbine concept," *11th Eur. Conf. Turbomach. Fluid Dyn. Thermodyn. ETC* 2015, pp. 1–12.

[9] A. Jaatinen-Värri et al., "Design of a 400 kW Gas Turbine Prototype." *In proceedings of ASME Turbo Expo*, GT2016-56444, June 13-17, Seoul, South Korea, 2016.

[10] C. H. Cloud, G. Li, E. H. Maslen, L. E. Barrett, and W. C. Foiles, "Practical applications of singular value decomposition in rotordynamics," *Aust. J. Mech. Eng.*, vol. 2, no. 1, 2005, pp. 21–32.

[11] C. Rodgers, "Specific speed and efficiency of centrifugal impellers. In Performance prediction of centrifugal pumps and compressors," *The 25th Annual international gas turbine conference and exhibit and the 22nd annual fluids engineering conference*, March 9-13, New Orleans, Louisiana, USA, 1980.

[12] M. Casey, C. Zwyssig, and C. Robinson, "The Cordier Line for Mixed Flow Compressors," *In proceedings of ASME Turbo Expo*, June 14-18, Glasgow, UK, 2010

[13] F. Y. Zeidan, L. San Andres, and J. M. Vance, "Design and Application of Squeeze Film Dampers in Rotating Machinery," *Proc. twenty-fifth Turbomach. Symp.*, 1996, p. 20.

[14] K. Kim and K. Ryu, "Rotordynamic Analysis of Automotive Turbochargers Supported on Ball Bearings and Squeeze Film Dampers in Series : Effect of Squeeze Film Damper Design Parameters and Rotor Imbalances," *Tribol. Lubr.*, vol. 34, no. 1, 2018, pp. 9–15.

[15] E. P. J. Gargiulo, "A simple way to estimate bearing stiffness," *in Machine Design*, 1980, pp. 107–110.

[16] E. Krämer, "Dynamics of Rotors and Foundations," *Springer-Verlag Berlin Heidelberg GmbH*, 1993.

[17] W. J. Chen and E. J. Gunter, "Introduction to Dynamics of Rotor-Bearing Systems," Vol. 175. *Victoria: Trafford*, 2007.

[18] J. Vance, "Rotordynamics of turbomachinery," *Jonh Wiley & Sons*, 1988.

Computational rotordynamics considering shrink fits

N. Wagner[1], H. Ecker[2]

[1]INTES GmbH, Stuttgart, Germany
[2]Technical University of Vienna, Vienna, Austria

ABSTRACT

Pre-stressed structures can be found in many technical applications, such as rotor and bladed disk assemblies. Press fits in particular are commonly used in high-speed rotating machinery. However, this design leads to questions about contact modelling, contact behavior and contact friction/damping, which have to be addressed during the design process.

In a first step, press fits are evaluated in a contact analysis based on a finite element approach by using solid elements. This method allows a realistic description of the contact in comparison with simplified beam models. Due to temperature and centrifugal loads, the contact gaps are generally not completely closed for increasing speeds of the rotor. These effects must be taken into account in a subsequent dynamic analysis. For this purpose, the active contacts are replaced by multipoint constraints. The following static analysis with linearized contacts serves to provide the internal stress state for the calculation of the geometric stiffness matrix. The solution of the generalized eigenvalue problem considers both the geometric and the convective stiffness matrix.

The last step is to solve the complex eigenvalue problem using a state-space representation of the reduced second-order system under consideration of the gyroscopic matrix for the speed range of interest. To lower the computing time in industrial applications, the system matrices are usually transformed into modal space using the real eigenmodes. A special procedure performs mode-tracking of the changing eigenfrequencies with increasing rotational speed.

This contribution is based on recently published works and presents further improvements. An implementation is demonstrated based on the FEA package PERMAS. Computations and examples are all carried out by using this commercially available software to emphasize the relevance of the findings for industrial applications.

1   INTRODUCTION

Finite element technique has become a popular tool in rotor dynamic analysis. Dynamic studies of rotating machines are generally performed using, on the one hand, beam element models representing the position of the rotating shaft and, on the other hand, three-dimensional solid rotor-stator models. A specific advantage of solid models is the inclusion of stress stiffening, spin softening, contacts and temperature effects in the rotor dynamics analysis.

Nowadays, CAD models of rotors are becoming more and more detailed. The tedious and time-consuming task of building equivalent beam models is omitted by using solid models. Lateral vibrations, and in particular bending vibrations of the shaft and attached components, are in general the most important vibrations in high-speed machinery. Understanding and controlling these lateral vibrations is important because excessive vibrations may lead to all kind of problems, and ultimately to a failure of the machinery. In extreme cases, lateral vibrations also can cause the rotating parts of a machine to come into contact with stationary parts, with potentially disastrous consequences.

An interference fit, also known as friction fit or press fit, is a fastening between two parts, which is achieved by friction forces after the parts are pushed together. This type of connection has been analyzed in numerous publications. Chen et. al. (1,2,3) used a finite element model to study the effect of a hot-fit on the dynamic behavior of a rotor-shaft assembly by performing a pre-stressed modal analysis. Hao and Hao and Wang (5) conducted a 3D finite element contact model to study the failure of an interference-fit planet carrier and shaft assembly. Other types of pre-stressed structures in rotor dynamics are laminated rotors (7). The laminations of the stack are axially pre-stressed either on a central shaft, by peripheral tie rods (4,15) or a combination of both of these technologies. Bolted rotors with curved couplings (16,17) are widely used in the aero-engine industry (10) and heavy-duty gas turbines.

2   GOVERNING EQUATIONS

Since the numerical solution will be computed by a Finite Element code, the model must be defined accordingly. Therefore, the rotating structure has to be meshed and for that, a grid has to be established. Since structured grids are superior over unstructured grids considering numerical and computational efficiency this type of mesh is employed.

Another decision has to be made at the beginning of the modelling procedure concerning the reference system. Rotordynamic systems can be described with respect to the inertial reference frame or in a rotating reference system. Efficiency would call for a model description in a co-rotating reference frame, since several advantages of the model and its formulation can be taken. However, restrictions apply to this kind of model, and one of them is that the rotational speed of the rotor needs to be constant. Consequently, transient behavior cannot be considered, but this is not needed in this study. Another one is that anisotropic bearings lead to position-dependent and, in combination with a constant rotor speed, to time-periodic bearing forces. If the focus of investigation lies on the bearing characteristics or on the support structure, and if the rotor is axisymmetric, then using the inertial reference frame is advantageous. Fortunately, the choice of a reference frame is not a one-way decision, since a transformation matrix exists and allows switching between the two descriptions, see (6).

For convenience, the rotor system is described in a fixed (non-rotating) reference frame. However, for the following example the alternative representation would also be possible. Moreover, only the linearized system will be considered, and steady state conditions are assumed throughout this study. Internal and external damping is included, as well as gyroscopic effects. The final rotordynamic model is a large FE-model, derived as described in (6) and the references therein. The final governing equations of motion used for this axisymmetric rotor system supported by isotropic bearings in a fixed (inertial) reference frame are given by

$$\mathbf{M}\ddot{\mathbf{u}} + (\mathbf{D} + \mathbf{D_b} + \mathbf{G})\dot{\mathbf{u}} + (\mathbf{K} + \mathbf{K_b} + \mathbf{K_g} + \mathbf{K_c})\mathbf{u} = \mathbf{f}(t, \Omega^2) \qquad (1)$$

where $\mathbf{M}$ denotes the symmetric mass matrix, $\mathbf{D}$ a viscous damping matrix, $\mathbf{D_b}$ the bearing viscous damping matrix, $\mathbf{G}(\Omega)$ the gyroscopic matrix, $\mathbf{K}$ structural stiffness matrix, $\mathbf{K_b}$ bearing stiffness matrix and $\mathbf{f}(t, \Omega^2)$ external forces including possible unbalance forces and centrifugal loads. The matrices $\mathbf{K_c}(\Omega)$ (convective stiffness) and $\mathbf{K_g}$ (geometric stiffness) are specific for a 3D Finite-Element model and are not present in the familiar models with lumped masses and FE-beam-elements stretching along the rotor axis, see (6) for details.

The first computational step is a pure static analysis considering the contact between shaft and attachment for the basic model to determine the contact status and stress distribution under centrifugal loads. The second step is again a static analysis, where active contacts are replaced by general multipoint constraints. That is basically a linearization step to conduct a subsequent eigenvalue analysis. It is also a prerequisite for the calculation of the geometric stiffness matrix $\mathbf{K_g}$. The next step is the calculation of the real eigenmodes to set up the modal matrix $\mathbf{Y}$

$$\mathbf{Y} = [\mathbf{y}_1, \ldots, \mathbf{y}_r] \tag{2}$$

including geometric and convective stiffness matrices

$$\mathbf{M\,Y} = \left(\mathbf{K} + \mathbf{K_b} + \mathbf{K_g} + \mathbf{K_c}\right)\mathbf{Y}\Lambda^{-1} \text{with } \Lambda = \operatorname{diag}\,(\lambda_i),\ \lambda_i = \omega_i^2 \tag{3}$$

The equations of motion (1) are transformed into modal space by means of the coordinate transformation

$$\mathbf{u} = \mathbf{Y}\,\eta \tag{4}$$

resulting in

$$\tilde{\mathbf{M}}\,\eta + \left(\tilde{\mathbf{D}} + \tilde{\mathbf{D}}_b + \tilde{\mathbf{G}}\right)\dot{\eta} + \left(\tilde{\mathbf{K}} + \tilde{\mathbf{K}}_b + \tilde{\mathbf{K}}_g + \tilde{\mathbf{K}}_c\right)\eta = \tilde{\mathbf{f}}(\mathbf{t}) \tag{5}$$

The transformation into modal space enables to reduce the number of equations of motion by deleting unnecessary high-frequency components (modes) and thus allows for the efficient solution of the complex eigenvalue problem. By introducing $\zeta = \dot{\eta}$, the second-order form (5) is transformed into a state-space representation:

$$\begin{bmatrix} \tilde{\mathbf{M}} & \mathbf{0} \\ \mathbf{0} & \mathbf{I} \end{bmatrix}\begin{bmatrix} \dot{\zeta} \\ \dot{\eta} \end{bmatrix} + \begin{bmatrix} (\tilde{\mathbf{D}} + \tilde{\mathbf{D}}_b + \tilde{\mathbf{G}}) & (\tilde{\mathbf{K}} + \tilde{\mathbf{K}}_b + \tilde{\mathbf{K}}_g + \tilde{\mathbf{K}}_c) \\ -\mathbf{I} & \mathbf{0} \end{bmatrix}\begin{bmatrix} \zeta \\ \eta \end{bmatrix} = \begin{bmatrix} \tilde{\mathbf{f}}(\mathbf{t}) \\ 0 \end{bmatrix} \tag{6}$$

The state-space form is used to solve the complex eigenvalue problem for different rotational speeds to obtain the Campbell diagram. Additional static mode shapes may be added to enrich the modal space. Of course, if the free response is considered, $\tilde{\mathbf{f}}(t) = \mathbf{0}$ holds.

A typical feature of the equations of motion (6) of a rotor system is the skew-symmetric pseudo-damping matrix $\tilde{\mathbf{G}}$, which introduces gyroscopic effects and forces. The particular form of the matrix results in two kinds of complex eigenmodes. One group is termed *forward modes*, having increasing eigenfrequencies with increasing rotational speed. The second group are the so-called *backward modes* having the opposite behavior, i.e. decreasing eigenfrequencies with increasing speed. It can be shown that the unbalance vector cannot feed energy into the so-called backward whirl, if direct stiffness symmetry is present in the system. However, in certain configurations of stiffness asymmetry this may not be the case, see (18).

Since eigenfrequencies depend in a non-linear manner on the rotor speed, cross-over of different eigenfrequency functions in the frequency domain occur regularly. Therefore, keeping track of a specific eigenfrequency and the associated mode is very important. The software (19,20) used for this study has a mode tracking algorithm implemented to sort and arrange the complex eigenvalues in a useful manner.

Figure 1. FE-model of the rotor shaft and the shrink-fitted impeller.

3 EXAMPLE

This numerical example is taken from the literature, see Refs.(11,12,13). Other than the unstructured mesh (based on tetrahedral elements) used there, a *structured mesh* is employed here for efficiency reasons. The finite element model consists of 73520 hexahedral and 936 pentahedron elements, see Figure 1. The impeller is incompatibly meshed and it is connected to the elastic shaft by a frictional contact. Two isotropic bearings (not shown in Figure 1) are idealized by spring-damper elements which are connected to the shaft by multipoint constraints. The material parameters of the shaft/impeller assembly are given in Table 1. Bearing properties are given in Table 2. The profile of the impeller ensures that the centrifugal forces are not transmitted uniformly, see the distribution of the centrifugal forces in Figure 2.

Table 1 . Material parameters.

Part/Material	Density [kg/m^3]	Youngs modulus [MPa]	Poisson ratio
Shaft/Steel Alloy	7850	200000	0.3
Impeller/Aluminum alloy	2810	71700	0.33

Table 2 . Bearing parameters.

Bearing	Stiffness k_{yy} [N/mm]	Stiffness k_{zz} [N/mm]	Viscous damping [Ns/mm]	Viscous damping [Ns/mm]
1	10^5	10^5	1	1
2	10^5	10^5	1	1

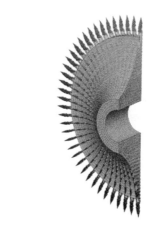

Figure 2. Equivalent nodal point forces due to a centrifugal load.

The non-uniform stress distribution (along the rotor axis) within the shrink fit area ensures partial opening of the contact between shaft and impeller as the rotor speed increases. The occurrence of a gap reduces the number of active contacts, which in turn leads to a reduction of the algebraic constraint equations. A different treatment of normal and tangential directions of frictional contacts is available during the linearization process in order to mimic a realistic behaviour. A further kind of fuzziness can be introduced by using the current contact pressure, see (14). Once the actual contact pressure exceeds a certain threshold a multipoint constraint is introduced. It should be noted that the undeformed structure is used to linearize the active contacts. This causes the natural frequencies to change because the entire system becomes softer due to the reduced length of the contact interface. This also changes the Campbell diagram and the complex eigenmodes. To study these effects in detail a design space exploration is performed in which the centrifugal load is varied.

3.1 Design space exploration

The investigation starts with the contact analysis. A surface to node contact is established between the shaft and the impeller, respectively. The geometric overlap ($\delta = -0.2$mm) between shaft and impeller due to a press fit is assumed to be constant. Likewise, a spatially varying initial contact gap would be possible, which can be described by a function. Coulomb's friction law is used and the friction coefficient μ is assumed to be constant as well. Now the rotational speed of the rotor is increased in discrete steps and the influence on the contact is analysed. Several variables can be output, such as the contact gap width, contact pressure and contact state to name a few. For the contact state we distinguish between passive and active contacts. When a contact is active, the underlying contact node can stick or slide. Here we first focus on the contact gap. The contact is completely closed over the entire contact length (80mm ≤ x ≤ 150mm) for rotational speeds at and below $\Omega=3000$ rad/s. By increasing the angular velocity, the contact opens more and more starting at the right end of the contact zone in Figure 3. At the maximum speed of $\Omega=10,000$ rad/s, only a contact length of ~5mm remains. This clearly will change the dynamic behaviour of the rotor system significantly.

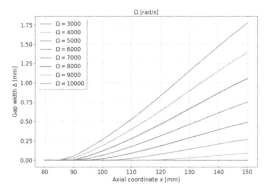

Figure 3. Gap width as a function of the rotor angular speed.

The strain energy distribution as shown in Figure 4 is quite useful to detect the nature of a vibrational mode. Each column holds for a certain eigenfrequency of the rotating system. Bending modes appear pairwise due to the symmetry of the rotor-bearing system. The fifth and the eight mode represent torsional modes. The impeller contributes to the strain energy at higher modes, while the lower modes are dominated by the shaft and the bearings. Figures 5 and 6 depict mode shapes of some of the most significant modes. These mode shapes are computed for the non-rotating system with no gyroscopic forces applied.

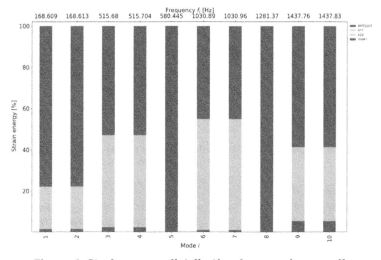

Figure 4. Strain energy distribution (no angular speed).

Figure 5. Mode shape #6 - "Impeller mode" (no angular speed).

Figure 6. Mode shapes: (top left) Mode #1, (top right) Mode #3,(bottom left) Mode #5, (bottom right) Mode #9.

The rates of change of the natural frequencies, see Table 3, due to gap development at increasing speed is depicted in Figure 7. The lower modes are mainly characterized by bending of the shaft. Clearly, the higher order modes and mode shapes are more sensitive to the changing contact interface at the press fit. This is due to the fact, that the number of general multipoint constraints introduced by contact linearization decreases. Note the onset of eigenfrequency decreasing starting at $\Omega \simeq 3000$ rad/s and compare with Figure 3.

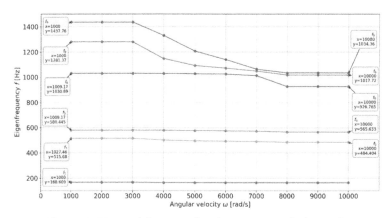

Figure 7. Natural frequencies (no gyroscopic forces).

Table 3 . Natural frequencies (no angular speed).

Natural Frequency [Hz]	1, 2	3, 4	5	6, 7	8	9, 10
No. Type	1st bend. mode	2nd bend. mode	1st tors. mode	3rd bend. mode	2nd tors. mode	4th bend. mode
f$_i$	168.6	515.7	580.4	1030.9	1281.4	1437.8

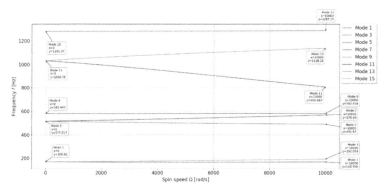

Figure 8. Campbell diagram, only valid for Ω<3000 rad/s.

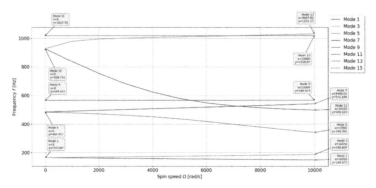

Figure 9. Campbell diagram, only valid @ Ω=10.000 rad/s.

So far the analysis was based on the eigenvalue problem as defined by Eq.(3). The contact problem has been solved and the dependency of the real eigenvalues on the decreasing contact area, when the angular speed increases, has been captured. The final step in the numerical analysis is to solve the complex eigenvalue problem of the linearized but full model as established in Eqs.(5),(6).

The results of such numerically expensive calculations is presented as Campbell diagrams. Figure 8 shows the eigenfrequencies of the rotor system as a function of the rotational speed Ω. Note the well known splitting of frequencies due to gyroscopic forces, which are included in the calculation now. However, this diagram is valid only within the speed range $0 \leq \Omega \leq 3000$ since the press fit is in contact over the full length for this speed range. Therefore, the eigenvalue functions beyond only hold if the press fit is not treated as shown before.

Figure 9 is similar to Figure 8, but here the contact conditions are employed for the maximum speed of $\Omega_{max} = 10000$ rad/s. Consequently, valid eigenfrequencies for the full model including the effect of the shrink fit are found in the vicinity of Ω_{max} only. The eigenfrequency functions hold for a system with a speed-independent contact length as computed for Ω_{max}. Nevertheless, the fictional part of the diagram helps to better understand the dynamics of that system.

The importance of the final results of this study is seen especially if the various results of the eigenfrequency (EF) of the 3rd bending mode,"impeller mode" #6 (see Figure 5 and Table 3) is studied. For low rotational speeds this EF is 1030.9 Hz. Including only the influence of the shrink fit, this EF drops down to 926,8 Hz (Figure 6), whereas considering only gyroscopic effects results in an increase for the forward whirl to 1138 Hz. For the full model with all effects considered, this EF decreases to 1031 Hz (Figure 9), which is, only by chance, the starting point of this comparison.

4   CONCLUSIONS

Preload effects are widespread in rotor systems and need to be considered in a thorough dynamic analysis. This study, although limited in several respects, enhances the understanding of how shrink fits of rotor assemblies have an effect on rotordynamics.

It is shown that the static behavior including the active contact zone is directly affected by the applied centrifugal load. The higher the centrifugal load, the lower the coupling between shaft and impeller. This in turn has a direct effect on the dynamic response behavior of the rotor.

By efficiently computed parameter variations, a multitude of analyses can be performed in a relatively short time. Thus, uncertainties in the form of a bandwith for system parameters can be considered. Moreover, sensitivity analyses are possible to assess the influence of different parameters on the dynamic behavior of the rotor.

ACKNOWLEDGMENT

The authors would like to thank Dr. Eerik Sikanen for providing an unstructured mesh of the rotor model used in this study.

REFERENCES

[1] Chen, S.-Y., Kung, C., Liao, T.-T., Chen, Y.-H., "Dynamic effects of the interference fit of motor rotor on the stiffness of a high speed rotating shaft", *Transactions of the Canadian Society for Mechanical Engineering*, 34 (2), 2010, pp. 243–261.

[2] Chen, S.-Y., Kung, C. Hsu, J.-C., "Dynamic analysis of a rotary hollow shaft with hot-fit part using contact elements with friction", *Transactions of the Canadian Society for Mechanical Engineering*, 35 (3), 2011, pp. 461–474.

[3] Chen, S.-Y., "An equivalent direct modelling of a rotary shaft with hot-fix components using contact element modal analysis results", *Computers and Mathematics with Applications*, 64, 2012, pp. 1093–1099.

[4] Gao, J., Yuan, Q., Li, P., Feng, Z., Zhang, H., Lv, Z., "Effects of bending moments and pretightening forces on the flexural stiffness of contact interfaces in rod-fastened rotors", *Journal of Engineering for Gas Turbines and Power*, 134, 2012.

[5] Hao, D., Wang, D., "Finite-element modelling of the failure of interference-fit planet carrier and shaft assembly", *Engineering Failure Analysis*, 33, 2013, pp. 184–196.

[6] Kirchgäßner, B., *"Finite Elements in Rotordynamics"*, Procedia Engineering 144, 2016, pp. 736–750.

[7] Mogenier, G., Baranger, T., Dufour, R., Besso, F., Durantay, L., "Nonlinear centrifugal effects on a prestressed laminated rotor", *Mechanism and Machine Theory*, 46, 2011, pp. 1466–1491.

[8] Özel, A., Temiz, S., Aydin, M. D., Sen, S., "Stress analysis of shrink-fitted joints for various fit forms via finite element method", *Materials and Design*, 26, 2005, pp. 281–289.

[9] Qin, Z., Han, Q., Chu, F., "Bolt loosening at rotating joint interface and its influence on rotor dynamics", *Engineering Failure Analysis*, 59, 2016, pp. 456–466.

[10] Shuguo, L., Yanhong, M., Dayi, Z., Jie, H., "Studies on dynamic characteristics of the joint in the aero-engine rotor system", *Mechanical Systems and Signal Processing*, 29, 2012, pp. 120–136.

[11] Sikanen, E., Heikkinen, J.E., Sopanen, J., "Shrink-fitted joint behaviour using three-dimensional solid finite elements in rotor dynamics with inclusion of stress-stiffening effect", *Advances in Mechanical Engineering*, 10 (6), 2018, pp. 1–13.

[12] Sikanen, E., Heikkinen, J.E., Sopanen, J., "Effect of operational temperature on contact dynamics of shrink-fitted compressor impeller joint", Proceedings of the

15th IFToMM World Congress on Mechanism and Machine, Advances in Mechanism and Machine Design, 2019, pp. 3341–3351.

[13] Sikanen, E., "Dynamic analysis of rotating systems including contact and thermal-induced effects", PhD thesis, Lappeenranta University of Technology, 2018.

[14] Wagner, N, Helfrich, R., "How to cope with uncertainties in boundary conditions and couplings of substructures", NAFEMS World Congress, San Diego, 2015.

[15] Wagner, N., Helfrich, R., "Static and dynamic analysis of a rod-fastened rotor", NAFEMS World Congress, Stockholm, 2017.

[16] Yuan, S., Zhang, Y., Zhang, Y., Jiang, X., "Stress distribution and contact status analysis of a bolted rotor with curvic couplings", *Proc. IMechE Part C J. Mechanical Engineering Science*, 224, 2009, pp. 1815–1829.

[17] Yuan, S., Zhang, Y., Fan, Y., Zhang, Y., "A method to achieve uniform clamp force in a bolted rotor with curvic couplings", *Proc. IMechE Part E J. Process Mechanical Engineering*, 230, 2016, pp. 335–344.

[18] Greenhill, L.M., Cornejo, G.A., "Critical speeds resulting from unbalance excitation of backward whirl modes", DE-Vol. 84-2, 1995 DETC Volume 3-Part B, ASME 1995.

[19] PERMAS Version 18 User's Reference Manual, INTES Publication No. 450, 2020.

[20] VisPER Version 18, User's Reference Manual, INTES Publication No. 470, 2020.

12th International Conference on Vibrations in Rotating Machinery -
Institution of Mechanical Engineers, ISBN 978-0-367-67742-8

Analysis of the cavitation characteristics of elastic ring squeeze film damper

Z.F. Han, Th. Thümmel, Q. Ding, Dong li

Chair of Applied Mechanics, Technical University of Munich, Germany
Department of mechanics, Tianjin University, China

ABSTRACT

The elastic ring squeeze film damper (ERSFD) shows superior performance in vibration control than traditional squeeze film damper (SFD), it has been used in aero-engine and has potential application on gas turbine and centrifuge. Based on the Mixture theory, an unsteady gas-liquid two phase flow model of inner and outer oil films of ERSFD is established in ANSYS Fluent to investigate the influence of cavitation. The dynamic mesh technique is used to calculate the oil film pressure and vapor volume fraction under different precession radius and frequencies. The results reveal that there is cavitation phenomenon in the low-pressure zone of inner and outer oil film. Because tangential squeezing effect of the journal precession is weaked by the division of the boss of the elastic ring, the negative pressure area and the vapor volume fraction calculated by the cavitation model is bigger than that of full-film model. Vapor volume fraction of an open end damper increase with the increase of precession radius and frequency.

1   INTRODUCTION

1.1   The structure of ERSFD

Elastic ring squeeze film damper(ERSFD) is an essential damping com ponent in aero engine, which has better ability to suppress non-linear response than traditional squeeze oil film damper (SFD)[1-3]. The obvious difference between ERSFD and SFD is that the oil chamber is divided into two inner and outer oil films by an elastic ring with uniformly distributed bosses [4]. Dynamical characteristics such as force coefficients, load capacity, and stability of rotor- SFD system have been proven to be affected by fluid cavitation, whether gaseous or vaporous[5]. ERSFD, like SFD, relies on squeezed oil film to provide damping, so analyzing the cavitation characteristics of ERSFD is very meaningful for engineering design.

1.2   Theoretical method

Many theoretical models, such as the ϖ-Sommerfeld assumption, were proposed to analyze the effects of cavitation on the oil pressure, without directly simulating the cavitation process. Two-phase homogenous models in which the lubricant density and viscosity are weighted functions of the dissolved gas mass (or volume) fraction were proposed in Ref [6-8] . A considerable reduction in the load capacity of SFDs was found by experiment when the gas concentration went up in Refs [9, 10]. Zeidan [11, 12] experimentally investigated the characteristics of gas cavitation and vapor cavitation in the damper and conditions when will they occurs. By coupled the full Navier-Stokes equation and homogeneous cavitation model, C Xing et [13, 14] et developed a three-dimensional numerical model of a two-phase SFD. The pressure and direct damping coefficient calculated by this model is more accurate than that calculated by ϖ -film assumption while gas volume fraction is low. Fan[15] analyzed the effect of lubricant inertia on the cavitation for high-speed partially sealed SFD by simulation and experiment. Both Elrod algorithm and Gumbel's cavitation boundary condition were used to simulate the lubricant cavitation and the results reveal that the fluid inertia effects notably extend the area of cavitation region and change the cavitation onset. Gehannin[16] et evaluate the relative importance of the terms in the Rayleigh-Plesset equation by compare the numerical and experimental results.

1.3   Numerical method

For traditional SFD, the condition when will the transition to turbulence occurs are discussed by researchers. Tichy[17] proposed the appropriated Reynolds number of the SFD and analyzed the effect of turbulence. The turbulent transition occurs at much higher Reynolds numbers in SFD than usual lubrication turbulence models. Andres [18] analyzed the effect of fluid inertia on the force coefficients for SFD under the assumption of turbulence or laminar according to Reynolds number. Some researchers have incorporated the computational fluid dynamics (CFD) technique to solve the full-term Navier-Stoke equation for the fluid in SFD [13]. Zhou [19] used the CFX, a commercial software, to analyze pressure distribution in SFD and found that the difference between the simulating and experimental results of the equivalent damping mainly comes from the failure to accurately simulate the cavitation phenomenon in the squeeze oil film damper in the model. Cui [20] used the ANSYS to established the two-phase model of an open SFD with a central feeding groove. The simulation showed that pressure distribution in the low-pressure zone and vapor volume fraction were sensitive to position of inlet hole. Furthermore, the variation frequency of oil force is closely related to the amount of the inlet holes. Krinner developed two novel methods to analyze the efficient time integration of mechanical systems with elasto-hydrodynamic lubricated joints[21, 22].In general, CFD provides an accurate way to investigate the cavitation in damper. Up to now, theoretical and experimental studies that directly simulate the cavitation in ERSFD are very limited.

In this paper, we establish two kinds of three-dimensional unsteady models of sealed and open-end ERSFDs based on the vapor-liquid two-phase flow theory and moving grid method. Then, the instantaneous pressure distribution characteristics of damper under full oil film and two-phase flow models are studied. Next, the influence of seal, different precession radius and precession frequencies on cavitation distribution and vapor volume fraction are analyzed. The results reveal the cavitation process in ERSFD and its effects on the pressure and oil force, which provide the foundation for the engineering.

2   NUMERICAL MODEL

2.1   Parameters of model

Because the elastic deformation is quite small under normal working conditions, this paper mainly considers the effect of cavitation in ERSFD on the flow field. The three-dimensional flow field model of ERSFD with four-boss elastic ring is established as shown in Figure 1 (a). The elastic ring divides the oil chamber of the damper into inner and outer oil films connected by oil holes, and each oil film is divided into the same four sections by the boss of the elastic ring. The two inlet/outlet holes are set at 18.5° and 198.5°respectively as shown in Figure 1 (b). The geometric parameters of the model are showed in Table 1. For the open-end model, the oil enters the inner oil film from the 1# and 2# inlet holes then flows out from both axial ends. Conversely, For the end sealed model, the oil enters in the damper from 1# inlet hole and flows from the 2# outlet hole.

The inlet hole is set as the pressure-inlet boundary where the pressure is set to p_{in} =0.1Mpa,while the outlet is set to the pressure-outlet boundary where the pressure is atmospheric pressure which is set to zero. Then, the outer wall of the inner oil chamber is set to the stationary boundary. Both axial ends and edges of the inner oil film are set to the deforming boundary condition. Next, the boundary condition of the inner wall of the inner oil chamber is set to the rigid body whose precession radius and frequency of the inner ring are defined by the user defined function (UDF). The mesh method of the dynamic mesh adopts the Smoothing Methods model which ensures that the number of mesh nodes and connectivity remain unchanged when the mesh is deformed by movement. In order to ensure that the grid will not break during the movement, a Diffusing grid change model with a diffusion coefficient, Ψ=0.5, is used to control the grid deformation by the diffusion equation.

According to Ref[17],the Reynolds number of the inner layer of the damper is defined by Eq (1) :

$$\text{Re} = \frac{\rho \omega c^2}{\mu}$$

(1)

where the density of the oil is ρ = 884.44kg/m3, the procession velocity of journal is ω = 2ϖf, the precession frequency is f = 120 Hz. Furthermore, the processional radius is ε= 0.2mm, the clearance of oil chamber c = 0.5mm, and the oil viscosity is μ = 0.023 Pa • s. Re is quite less than 2300, so the model is set to laminar flow according to Ref (17).

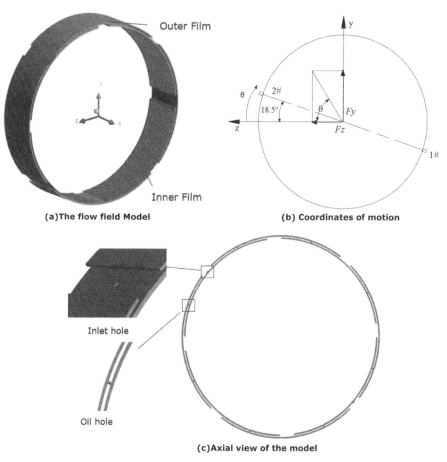

(a)The flow field Model

(b) Coordinates of motion

(c)Axial view of the model

Figure 1. Numerical model of ERSFD.

Table1. Parameters of model.

Parameters	Value
Outer radius of outer oil film, R_1	41.5mm
Inner radius of inner oil film, R_2	40mm
Clearance of oil chamber, c	0.5mm
Length of damper, L	20mm
Radius of inlet hole	1mm
Radius of oil hole	0.5mm
Radian of boss	0.045ϖ

2.2 Cavitation model

The Mixture two-phase flow model which simulates two-phase flows with different velocities in each phase is adopted in Fluent. The method simulates two-phase flow by solving the continuous equation, momentum equation, volume fraction equation of vapor phase, and relative velocity equation, so it has good convergence and high efficiency. According to the Ref[20], the governing equation of this model is Eq (2)-(5):

$$\frac{\partial(\rho_m)}{\partial t} + \nabla \cdot (\rho_m \nu_m) = 0 \tag{2}$$

$$\frac{\partial(\rho_m \nu_m)}{\partial t} + \nabla \cdot (\rho_m \nu_m \nu_m) = -\nabla p + \nabla \cdot \left[\mu_m\left(\nabla \nu_m + \nu_{mm}^{\mathrm{T}}\right)\right] + \rho_m g + F \tag{3}$$

$$\frac{\partial(\alpha_v \rho_v)}{\partial t} + \nabla \cdot (\alpha_v \rho_v \nu_v) = R_l - R_v \tag{4}$$

$$\frac{\partial((1-\alpha_v)\rho_l)}{\partial t} + \nabla \cdot ((1-\alpha_v)\rho_l \nu_l)) = R_v - R_l \tag{5}$$

where g is the acceleration of gravity, ρ_m is the density of the mixed phase, ρ_v is the density of the gas phase, and ρ_l is the density of the liquid phase. u_m is the average velocity of the mixed phase, u_l is the liquid phase velocity, u_v is the velocity of liquid phase, α_v is the volume fraction of the gas phase, F is the body force, R_l is the gas phase formation rate, and R_v is the gas phase condensation rate. Determining the mathematical model of mass exchange between the vapor and liquid phases is the foundation to analyze the flow of vapor, that is, determining R_l and R_v. Therefore, the Zwart-Gerber-Belamri (Z-G-B) model, which is more accurate and suitable for the cavitation calculation of the thin-film flow field, is used to determine the mass exchange of R_l and R_v without capturing the influence of non-condensable gas in the flow field. The calculation parameters are set as follows: bubble diameter is 1µm, nucleation site volume fraction is 0.0005, evaporation coefficient is 50, condensation coefficient is 0.01, vaporization pressure is 0.04 Mpa, and SIMPLE method is selected as pressure-velocity coupling in solution methods

3 RESULTS AND DISCUSSION

3.1 Comparison of oil pressure between full film model and cavitation model

Figure 2 shows that the inner oil pressures calculated by the full film or cavity model are bigger than the outer oil pressures respectively. However, there is not obvious difference between the inner and outer pressures calculated by full-film model and cavitation model in the non-squeezed zone. That is the inner and outer oil films are connected by oil holes, the outer pressure is mainly affected by the inner oil film connected to it. To further investigate the effect of cavitation, negative pressure area ration Ψ which is the ratio of the area under negative pressure S- to the total inner film area S, the equivalent outer and inner oil forces are calculated by:

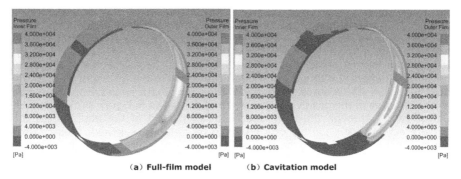

(a) Full-film model   (b) Cavitation model

Figure 2. Pressure of full film model and cavitation model.

$$\Psi = S_- / S \tag{6}$$

$$\begin{bmatrix} F_y \\ F_z \end{bmatrix} = \int_{\theta_i}^{\theta_j} \left(\int_{-\frac{L}{2}}^{\frac{L}{2}} P \, dx \right) \begin{bmatrix} sin\theta \\ cos\theta \end{bmatrix} R d\theta \ (i = 1,2,3,4.j = 1,2,3,4) \tag{7}$$

where θ_i and θ_j is the circumferential coordinates as shown in Figure 1 (b). Next, the phase ϑ of oil force can be calculated by

$$\vartheta = \arctan \left(F_y / F_z \right) \tag{8}$$

From Figure 1 (a), it can be found that a negative pressure zone will appear immediately following the positive pressure zone. Although the oil chamber is divided into different segments by the elastic ring boss, the oil film pressure distribution still shows good continuity, which is consistent with the results of the numerical model analysis in Ref[4]. Figure 2 (b) reveals that the negative pressure area in the cavitation model is

537

larger than that of the full oil film model, which agree with the quantitative results that the negative pressure area ration of inner film calculated by full-film model Ψ_f is 50.4% while the negative area ration under the influence of cavitation is Ψ_c is 61.21%. The reason is that the flow filed near the oil hole is not smooth and continuous because of the segmentation effect at boss position and the bigger pressure from oil hole.Moreover, the phase of the oil force is obviously changed by the effect of cavitation from ϑ_f,0.088, calculated by full film model to ϑ_c,-0.8636. In addition, because the outer oil film is not in direct contact with the journal, and its pressure value is less than the inner pressure, the inner pressure is mainly analyzed in the subsequent analysis.

3.2   Cavitation process

For the calculation parameters used in this paper ($\rho = 884.44kg/m3$, $f = 120$ Hz, $\varepsilon = 0.2mm$, c = 0.5mm, $\mu = 0.023$ Pa • s), cavitation does not exist throughout the all period. Due to the periodic precession of the rotor, the cavitation phenomenon repeatedly appears in the four oil chambers. Figure 3 (a, c, e, g) show the pressure during a cavitation process, and Figure 3 (b, d, f, e) show the distribution of the relative vapor volume fraction. Firstly, there is a negative pressure zone in the left quarter of the oil chamber when there is no cavitation, such as Figure 3 (a). Because the journal precession can be decomposed into tangential and radical movements, where the tangential motion causes a positive pressure zone in the inner oil cavity in the precession direction. Thus, next to the negative pressure area is the peak area of the positive pressure. Then, as shown in Figure 3 (c), when the journal passes through the boss, the supporting effect of the boss causes the squeeze action of the journal precession to be greatly weakened, thus causing a negative region adjacent to the peak of the positive pressure region. The negative pressure region is formed in advance, so that the cavitation region is as shown in Figure 3 (d). At the same time, the tangential squeezing effect is weakened by the division of the boss, and the positive pressure area originally formed by the tangential movements is reduced. Therefore, the original positive pressure area becomes a negative pressure area, thereby generating a cavitation area. However, due to the positive pressure effect of the inlet hole and the high-pressure backflow effect of the outer oil film, no obvious cavitation area was formed between the two cavitation areas. we can find that the values of pressure near the lowest oil hole keeps lower than around area at 0.51s when the inner pressure is increasing under the squeeze of journal. Next, the pressure around this area decrease to negative because of cavitation at 0.53s while the vapor volume fraction near the hole is lower than around area. Then, the pressure near the oil hole is bigger than the around area at 0.55s when the journal process to next chamber occurs cavitation, even the pressure should be zero or negative because there is no squeeze effect. At the same time, the vapor volume fraction near the hole is bigger than around area. This is because the inner oil film pressure and vapor volume fraction of the are affected by the outer film by the oil hole, so the pressure and gas volume fraction of the inner oil film will be affected through the oil holes and have this delay phenomenon. The delay phenomenon is able to influence the phase of the oil fore ϑ.

538

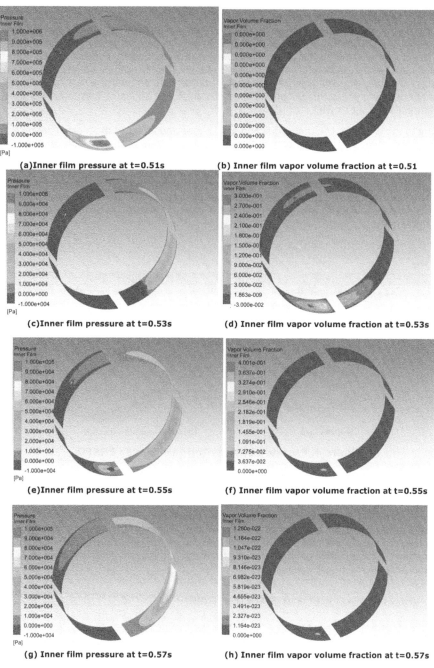

(a)Inner film pressure at t=0.51s

(b) Inner film vapor volume fraction at t=0.51

(c)Inner film pressure at t=0.53s

(d) Inner film vapor volume fraction at t=0.53s

(e)Inner film pressure at t=0.55s

(f) Inner film vapor volume fraction at t=0.55s

(g) Inner film pressure at t=0.57s

(h) Inner film vapor volume fraction at t=0.57s

Figure 3. Evolution of cavitation.

539

3.3 Influence of the seal

Figure 4 reveals the pressure and vapor volume fraction of a sealed damper which is another typical damper in engineering. The parameters of numerical model on sealed damper is the same as the open-end model, but the boundary conditions need to be changed. Firstly, Oil hole 1 # is set as the pressure inlet condition where the pressure is p_{in}=0.1Mpa, while oil hole 2# is set as the outlet hole boundary where the pressure is atmospheric pressure. Secondly, the both axial end zones of the outer and inner chambers are set as wall.

Due to the end seal, the tangential and axial squeezing effect of the journal precession is not as obvious as that in open end model, but the effect of radial squeezing and the oil holes are strengthened. By comparing with section 3.1, it can be found that the inner negative pressure area of sealed damper, Ψ_s=51.45%, is about 10% smaller than that of open-end damper. Like the open-end damper, the pressure distribution and cavitation distribution are affected by the oil holes as shown in Figure 4. The delay phenomenon similar to the open end damper shown in Figure 2 is able be found in seal damper too.

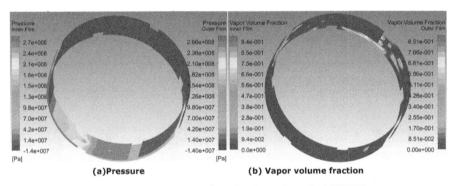

(a)Pressure   (b) Vapor volume fraction

Figure 4. Pressure and cavitation of sealed ERSFD.

3.4 Influence of the precession radius and frequencies

Vapor volume fraction φ, the ratio of vapor volume to the total volume of the oil cavity, is used to analyze the effect of the precession radius during a cavitation process. Calculating the mean value of vapor volume fraction φ_i at every point in the flow area, the vapor volume fraction φ can be obtained as

$$\varphi = \sum_{i=1}^{n} \varphi_i / n \qquad (9)$$

Figure 5 shows the vapor volume fraction at the beginning time step, λ, of a cavitation process at different precession radius and the two subsequent time points. We can find that the vapor volume ratio decreases with time increase, which corresponds to the cavitation process reproduced in Figure 3. Secondly, it can be found that the larger the precession radius is, the larger the vapor volume fraction is when cavitation occurs, and as time increases, this gap gradually decreases. The phenomenon indicates that the increase of the precession radius could increase the cavitation in the oil film.

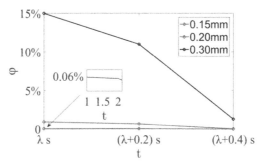

Fig 5. Vapor volume ratio under different precession radius.

Figure 6 depicts the cavitation ratio of the damper at different precession frequencies. We can find that the larger the precession frequency is, the larger the cavitation ratio is when cavitation occurs, and the cavitation ratio is decreasing with time running. This shows that the increase of the precession frequency will increase the cavitation in the oil cavity

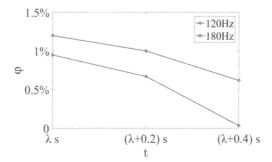

Fig 6. Vapor volume ratio under different precession frequencies.

4   CONCLUSION

In this paper, the CFD method is used to establish the three-dimensional flow field model of ERSFD. Then, the numerical model is used to investigate the cavitation evolution in ERSFD and its impact on oil pressure. The influences of precession radius, frequencies and end seal on cavitation of ERSFD are analyzed.

Firstly, the segmentation of boss breaks the continuity of the oil film, so the cavitation in ERSFD changes periodically with the segmented oil film, unlike the cavitation in SFD, which will be the same throughout the whole oil film. The distribution of cavitation is affected by oil inlet holes and oil holes. Thus, the phase of oil force is changed more than 0.8 radians under the influence of cavitation.

Secondly, the oil film pressure of the sealed ERSFD is quite bigger than that of open-end damper. Because of sealing effect, thus the inner negative pressure area of sealed

541

damper is 10% smaller than that of open-end damper. The cavitation mainly distribution and vapor volume fraction are more complicated than the open sealed damper.

Finally, the cavity ratio φ of the ERSFD increases with the increase of the precession radius and precession frequencies. Because the adjust effect of outer oil film and the oil holes, there is a "delay" phenomenon in cavitation damper. In addition, the effect of temperature is not considered in this paper. Due to the division of the oil chamber by the elastic ring, the flow field shape of ERSFD is more complex than that of SFD, so more experiments will be needed to study it in the future.

REFERENCES

[1] Zhang, W. and Q. Ding, elastic ring deformation and pedestal contact status analysis of elastic ring squeeze film damper. Journal of sound and vibration, 2015. 346(2015): p. 314–327.

[2] Siew, C.C., M. Hill and R. Holmes, Evaluation of various fluid-film models for use in the analysis of squeeze film dampers with a central groove. Tribology International, 2002. 35(8): p. 533–547.

[3] Han, B. and Q. Ding, Forced responses analysis of a rotor system with squeeze film damper during flight maneuvers using finite element method. Mechanism and Machine Theory, 2018. 122: p. 233–251.

[4] Han, Z., Q. Ding and W. Zhang, Dynamical analysis of an elastic ring squeeze film damper-rotor system. Mechanism and Machine Theory, 2019. 131: p. 406–419.

[5] Brewe, D.E., Theoretical Modeling of the Vapor Cavitation in Dynamically Loaded Journal Bearings. Journal of Tribology, 1986. 108(4): p. 628–637.

[6] Feng, N.S. and E.J. hahn, Density and Viscosity Models for Two-Phase Homogeneous Hydrodynamic Damper Fluids. A S L E Transactions, 1986. 29(3): p. 361–369.

[7] Diaz, S. and L. San andre S, A model for squeeze film dampers operating with air entrainment and validation with experiments. Journal of tribology, 2000. 123 (1): p. 125–133.

[8] Nikolajsen, J.L., Viscosity and Density Models for Aerated Oil in Fluid-Film Bearings©. Tribology Transactions, 1999. 42(1): p. 186–191.

[9] Diaz, S.E. and L.A. san andres, Measurements of Pressure in a Squeeze Film Damper with an air/oil Bubbly Mixture. Tribology Transactions, 1998. 41(2): p. 282–288.

[10] Diaz, S.E. and L.A. san andre S, Reduction of the Dynamic Load Capacity in a Squeeze Film Damper Operating with a Bubbly Lubricant. Journal of Engineering for Gas Turbines and Power, 1999. 121(4): p. 703–709.

[11] Zeidan, F. And J. Vance, Cavitation and Air Entrainment Effects on the Response of Squeeze Film Supported Rotors. Journal of Tribology, 1990. 112(2): p. 347–353.

[12] Zeidan, F.Y. and J.M. Vance, Cavitation Regimes in Squeeze Film Dampers and Their Effect on the Pressure Distribution. Tribology Transactions, 1990. 33(3): p. 447–453.

[13] Xing, C., M.J. Braun and H. Li, A Three-Dimensional Navier-Stokes–Based Numerical Model for Squeeze-Film Dampers. Part 1—effects of gaseous cavitation on pressure distribution and damping coefficients without consideration of inertia. Tribology Transactions, 2009. 52(5): p. 680–694.

[14] Xing, C., M.J. braun and H. Li, damping and added mass coefficients for a squeeze film damper using the full 3-d navier–stokes equation. Tribology international, 2010. 43(3): p. 654–666.

[15] Fan, T., S. Hamzehlouia and K. Behdinan, the effect of lubricant inertia on fluid cavitation for high-speed squeeze film dampers. Journal of vibroengineering, 2017. 19(8): p. 6122–6134.

[16] Gehannin, J., M. Arghir and O. Bonneau, evaluation of rayleigh–plesset equation based cavitation models for squeeze film dampers. Journal of tribology, 2009. 131(2).

[17] Tichy, J.A. the effect of fluid inertia in squeeze film damper bearings: a heuristic and physical description. In asme 1983 international gas turbine conference and exhibit. 1983.

[18] San andre S, L. and J.M. vance, effects of fluid inertia and turbulence on the force coefficients for squeeze film dampers. Journal of engineering for gas turbines and power, 1986. 108(2): p. 332–339.

[19] Zhou, H.L., et al., effects of oil supply conditions on equivalent damping and circumferential position damping of squeeze film damper. Journal of mechanical engineering, 2018. 54(6): p. 215–223.

[20] Cui, Y., et al., numerical simulation on cavitation flow field characteristics of squeeze film damper based on two-phase flow model. Journal of aerospace power, 2019. 34(8): p. 1781–1787.

[21] Krinner, A., T. Schindler and D. j. Rixen, Time integration of mechanical systems with elastohydrodynamic lubricated joints using Quasi-Newton method and projection formulations. Int. J. Numer. Meth. Engng, 2017(110): p. 523–548.

[22] Krinner, A., Multibody systems with lubricated contacts. 2018, Technical University of Munich.

12th International Conference on Vibrations in Rotating Machinery -
Institution of Mechanical Engineers, ISBN 978-0-367-67742-8

Experimental research on vibration reduction of turbine blades with underplatform dampers under rotating state

Y.N. Wu, H.J. Xuan

High-speed Rotating Machinery Laboratory, College of Energy Engineering, Zhejiang University, Hangzhou, China

ABSTRACT

Underplatform dampers are installed under the platform to reduce the blades' vibration by dry friction damping. In this paper, the vibration experiments of the turbine blades without and with underplatform dampers were carried out under rotating state. The experimental method of liquid jet excitation was used to simulate the airflow excitation of the turbine blades under working condition. Besides, the vibration signals of the turbine blades were measured by strain gages. The experiment results showed that the vibration strain of the blades with the underplatform dampers were about 30% lower than that without the underplatform dampers.

1 INTRODUCTION

The ongoing demand in aerospace industries for more efficient gas turbine engines has driven the design of many components to their structural mechanical limits. Turbine blades undergo complex loads during operation, which carry static stresses and dynamic stresses. Static stresses are caused by centrifugal loads, thermal loads and static fluid pressures. Mechanical vibrations, caused by additional dynamic loads of different origins, lead to dynamic stresses (1). On the basis of the static stress level, sustained and high dynamic stresses may lead to high cycle fatigue (HCF) and eventual failure of the blades (2). In particular, the wide operating speed range of aero-engines together with the high modal density of bladed disks makes it impossible to avoid all critical resonances during operation (3). Reducing the vibration amplitude at those resonances is therefore crucial. And the passive systems, based on friction damping (4) have been the most widely used approach over the years (5). Dry friction can provide damping in various locations on a blade, such as the shrouds, roots and blade tips. Therefore, there are many forms of dry friction damping, such as underplatform dampers (UPD), tip shrouds and damper wires, which are shown in Figure 1.

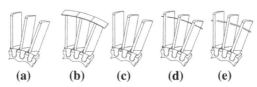

(a)   (b)   (c)   (d)   (e)

Figure 1. Common forms of dry friction damping (6): (a) roots joints, (b) tip shrouds, (c) underplatform dampers, (d) damper wires, (e) damper pins.

For gas turbine blades, the length of the blade is usually relatively short. In order to ensure their aerodynamic performance, underplatform dampers are usually selected for blade vibration reduction. Underplatform dampers consist of a metal device which sits in a groove on the underside of the platform between adjacent blades, and it is kept in place and loaded by the centrifugal force. When the blades vibrate, the relative motion between the adjacent platforms and the damper lead to friction at the contact interface, which in turn provide energy dissipation and damping to the system (7).

1980s, Griffin (8) made an attempt to model the dynamics of blades with underplatform dampers. He identified the damper stiffness as a key parameter in damper optimization. During the past 40 years, numerous numerical techniques have been developed to design and simulation calculations with underplatform dampers. Extensive papers on numerical modeling of underplatform dampers can be found. Cameron et al.(9), Wang et al (10), Sanliturk et al. (11), Petrov(12), He et al.(13) provided thorough reviews of the available literatures and methods solving nonlinear vibrations of mechanical systems with friction.

In order to obtain the damping effect of the damper with blade vibrations in the experimental research, many scholars have developed a variety of test methods, which can be divided into experiments performed on non-rotating test rigs and on rotating test rigs. The tests performed on non-rotating test rigs used two to three beamlike blade models where dampers are loaded by static gravity force to represent the centrifugal force acting on the damper (e.g. Sanliturk et al. (14), Szwedowicz et al. (15), Pešek (16), Pesareci (17), Gastaldi et al. (18)). These bench tests are difficult to include the effects of centrifugal loads (strain gradients, untwist, blade/rotor interaction, etc.), a critical driver of blade behavior. In order to get closer to the actual working conditions of the turbine blades, rotating tests are required which carried out on the rotating rigs with bladed disks in the vacuum condition. The beamlike blade models are reported by Stanbridge et al. (19) and Goetting et al. (20). The damping performance of a thin-walled damper, mounted under the platforms of two rotating, freestanding high pressure turbine blades, was investigated numerically and experimentally by Szwedowicz et al.(21). Sever (22) presented a methodology and results from an experimental investigation of forced vibration response for a rotating bladed disk with fitted underplatform "cottage-roof" friction dampers.

In this paper, turbine blades with wedge-shaped underplatform dampers are subjected to a rotating test, in which the blades were excited by liquid jet excitation。 The vibration responses of turbine blades under forced vibration are obtained, and the damping effect is obtained by comparing the vibration responses of blades with and without damping.

2   TEST RIG DESCRIPTION

We have modified the vertical high-speed rotating test rig, and established a test device that can be used for vibration damping test of different underplatform dampers under the rotating state. The schematic of the device shows in Figure 2, which consists of three parts.

1) Rotating rotor in a vacuum chamber. The rotor included two groups of blades. Each group of blades consisted of 5 adjacent blades. The middle blade was marked as the measuring blade, and strain gauges were attached to it. The blade rotation radius is around 300 mm. During the test, the chamber was vacuum with vacuum pump operation.

2) The driving of the rotating rotor. The rotor was driven by a motor and the speed was increased by the belt. Under the condition of good vacuum in the chamber, the test speed could reach 24000 r/min.
3) The excitation device for blades. The vibration excitation device mainly included fuel nozzles which were uniformly distributed in the circumferential direction and oil collector in the vacuum chamber, and an oil station for supplying and returning oil outside the vacuum chamber.

The photo of the high-speed rotating test rig shows in Figure 3.

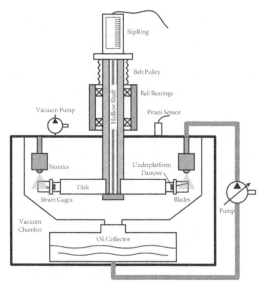

Figure 2. Schematic of the rotating test rig.

Figure 3. The photo of the high-speed rotating test rig.

3 EXCITATION AND MEASUREMENT OF THE BLADES

The experiments were performed with liquid jet to excite blades. Considering the vacuum situation of the experimental chamber, the vacuum- suitable spray liquid was selected as the aviation lubricants pattern. The aviation lubricating oil was firstly pressurized by oil pumps and then flowed into the vacuum chamber controlled by the valve. Uniformly distributed fuel nozzles are arranged above rotating blades. The liquid is ejected from fuel nozzles to form conical atomized liquid columns, and the rotating blades are regularly excited by the liquid columns, which causes the blades to forcibly vibrate. The excitation frequency is determined by the rotor speed and the number of nozzles, which means that when the rotor speed is slowly increased or decreased, the excitation frequency will also change accordingly, that is, the "sweep" can be achieved.

The pressure of the lubricating oil is controlled by the oil station outside the vacuum chamber. The outlet pressure of the oil station can reach a maximum of 0.9MPa. The total flow is determined by the number of nozzles and oil pressure. Because the vacuum chamber is sealed during the experiment, it is not convenient to measure the oil pressure at the nozzle, so the oil pressure at the nozzle has been calibrated before the test. Before the test, it is necessary to check that the relative distances of all nozzles and blades are consistent to ensure stable excitation frequencies and the same excitation force for each time.

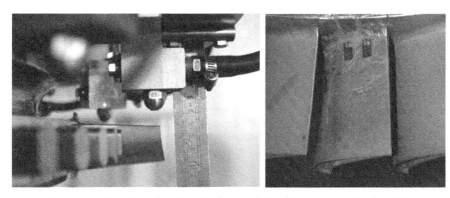

Figure 4. Liquid excitation device and strain gauges attachment.

The vibration responses of blade were measured by strain gauges which were attached to the surface of measuring blades where the dynamic stresses were relatively high above the platform (see Figure 4). A set of experimental blades consisted of 5 adjacent turbine blades. The underplatform dampers were installed below the platform of blades (Figure 5). The middle blade was identified as a measuring blade. Besides, the lead wires were pasted on the disk and finally passed through the hollow shaft. The other end of the lead wires were connected to the slip ring, and the signal transmitted by the slip ring was connected to the dynamic strain indicator.

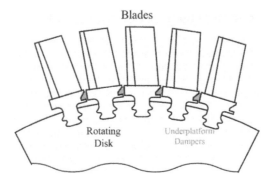

Blades

Rotating
Disk

Underplatform
Dampers

Figure 5. Installation of blades and underplatform dampers.

4 EXPERIMENTAL PROCESS

The experiment without underplatform dampers and the experiment with underplatform dampers were performed separately. Each test was repeated twice to ensure reproducibility. The data from repeated tests were averaged in subsequent test data processing. Finally, the strain amplitudes of blade vibration responses were compared under the same excitation conditions.

The experimental process is shown in Figure 6. In the test, the rotor first slowly increased speed. When the set rotation speed of the liquid-jet excitation (n1) was reached, the oil pump was turned on, and the blades were subjected to liquid-jet excitation and maintained. After the rotor continued to slowly rise to the target speed (n2), the rotor started to reduce speed after maintaining the target speed for a period of time. The acceleration of the speed increase and decrease was the same. Stop the liquid jet excitation when the rotor speed drops to n1, then stop the test after the rotating speed drops to zero. In order to prevent the zero drift of the strain gauges from seriously affecting the data, the collection of the strain data is performed slightly longer than the time of the liquid jet excitation time (b).

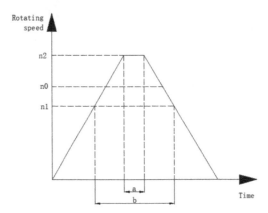

Figure 6. The experimental process.

5   EXPERIMENTAL RESULT

The strain data obtained from the test is shown in the blue line in Figure 7. It can be seen that the blade's strain generally changes with the rotation speed. This part of the strain is mainly caused by centrifugal force and belongs to the static strain (yellow line). When the speed is close to the resonance speed, the vibration amplitude increases significantly. This part of the strain is a dynamic strain (red line). In order to better observe the amplitude, the sliding strain is used to obtain the static strain, and the total strain is subtracted from the static strain to obtain the dynamic strain.

Fast Fourier Transform (FFT) is performed on the dynamic strain, and a three-dimensional waterfall chart is drawn (Figure 8). In order to be able to see peak changes clearly, the part with amplitude less than 10μɛ has been hidden in the figure. The peak value of the vibration amplitude of the blade changes with time. In the process of increasing and decreasing rotating speed, the strain amplitude increases sharply when the rotation speed passes near the resonance speed, and the peak strain amplitudes are taken as an important evaluation index.

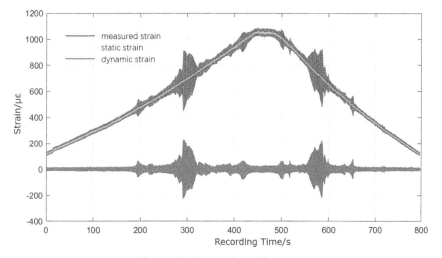

Figure 7. Strain data diagram.

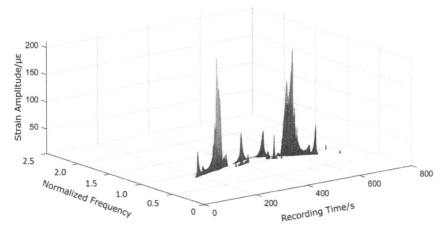

Figure 8. Three-dimensional waterfall chart.

Figure 9 shows the comparison of the vibration frequencies of the blades with and without underplatform dampers, in which we found that the resonance frequency of the blade with underplatform dampers is slightly higher than that without underplatform dampers. The two measured blades' resonance frequencies respectively increase by 3.11% and 2.93%.

The peak strain amplitude of the blade with underplatform dampers is lower than that without underplatform dampers (Figure 10). The peak vibration amplitudes of the two measured blades are respectively reduced by 31.10% and 34.39%.

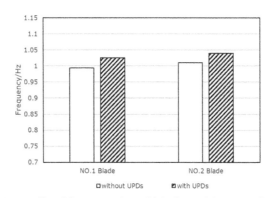

Figure 9. Normalized frequencies of blades without and with UPDs.

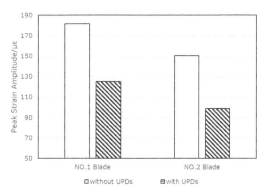

Figure 10. Peak strain amplitudes of blades without and with UPDs.

6   CONCLUSIONS

Experiments were designed and performed to study the vibration characteristics of the blades under the excitation of liquid spray in the rotating state. The vibration responses of the turbine blades with and without underplatform dampers were obtained, and the changes in resonance frequencies and peak strain amplitudes were compared.

From the obtained experimental results, the resonance frequency of the blades with underplatform dampers increased slightly and the amplitude decreased significantly. It can be considered that the underplatform dampers can reduce the blade vibration under the rotating state, especially the amplitude around the resonance peak. This can effectively reduce the stress of the blades due to vibration under working conditions, and improve the high cycle fatigue life of the blades.

REFERENCES

[1] Krack, M., Salles, L., & Thouverez, F. (2017) Vibration prediction of bladed disks coupled by friction joints. *Archives of Computational Methods in Engineering*. *24* (3), 589–636. doi: 10.1007/s11831-016-9183-2

[2] B. A. Cowles. (1996) High cycle fatigue in aircraft gas turbine - an industry prospective. *International Journal of Fracture*. 80, 147–163. doi: 10.1007/BF00012667.

[3] Pesaresi, L., Salles, L., Jones, A., Green, J. S., & Schwingshackl, C. W. (2017) Modelling the nonlinear behaviour of an underplatform damper test rig for turbine applications. *Mechanical Systems and Signal Processing*, *85*, 662–679. doi: 10.1016/j.ymssp.2016.09.007.

[4] Gaul, L., Nitsche, R. (2001) The role of friction in mechanical joints. *Applied Mechanics Reviews*. *54*(2), 93. doi: 10.1115/1.3097294.

[5] Griffin, J. H. (1989) A review of friction damping of turbine blade vibration. *International Journal of Turbo & Jet Engines*. 7,3–4. doi: 10.1515/TJJ.1990.7.3-4.297.

[6] Schurzig D.(2016) Development of a numerically efficient model for the dynamics of revolute clearance joints in adjustable stator cascades. Dissertation. *University of Hannover, Germany*.

[7] Sanliturk, K. Y., Ewins, D. J., & Stanbridge, A. B. (2001) Underplatform Dampers for Turbine Blades: Theoretical Modeling. Analysis, and Comparison With

Experimental Data. *Journal of Engineering for Gas Turbines and Power. 123(4)*, 919. doi: 10.1115/1.1385830.

[8] Griffin, J. H. (1980) Friction damping of resonant stresses in gas turbine engine airfoils. *Trans.ASME J.Eng.Power*. 102(2), 329–333. doi: 10.1115/1.3230256.

[9] Cameron, T. M., Griffin, J. H., Kielb, R. E., & Hoosac, T. M. (1990) An integrated approach for friction damper design. *Journal of Vibration and Acoustics, 112*(2), 175. doi:10.1115/1.2930110.

[10] Wang, J. H., Shieh, W. L. (1991) The influence of a variable friction coefficient on the dynamic behavior of a blade with a friction damper. *Journal of Sound & Vibration. 149*(1), 137–145. doi: 10.1016/0022-460X(91)90916-8.

[11] Csaba, G. (1999) Modelling of a Microslip Friction Damper Subjected to Translation and Rotation. *Asme International Gas Turbine & Aeroengine Congress & Exhibition*. American Society of Mechanical Engineers. doi: 10.1115/99-GT-149.

[12] Petrov, E. P., Ewins, D. J. (2003) Analytical Formulation of Friction Interface Elements for Analysis of Nonlinear Multi-Harmonic Vibrations of Bladed Discs. *ASME J. Turbomach*. 125, 364–371. doi: 10.1115/GT2002-30325.

[13] He, B., Ouyang, H., Ren, X., & He, S. (2017) Dynamic Response of a Simplified Turbine Blade Model with Under-Platform Dry Friction Dampers Considering Normal Load Variation. *Applied Sciences*. 7,228. doi: 10.3390/app7030228

[14] Sanliturk, K. Y., Stanbridge, A. B., and Ewins, D. J. (1995) Friction Dampers: Measurements, Modelling and Application to Blade Vibration Control. *Proceeding of Design Engineering Technology Conferences*. 84–2 (3), 1377–1382.

[15] Szwedowicz, J., Kissel, M., Ravindra, B., and Keller, R. (2001) Estimation of Contact Stiffness and its Role in the Design of a Friction Damper. ASME Paper No. 2001-GT-0290. doi: 10.1115/2001-GT-0290.

[16] Pešek, L., Hajžman, M., Půst, L., Zeman, V., Byrtus, M., & Brůha, J. (2015) Experimental and numerical investigation of friction element dissipative effects in blade shrouding. *Nonlinear Dynamics*. 79(3), 1711–1726. doi: 10.1007/s11071-014-1769-3.

[17] Pesaresi, L., Salles, L., Jones, A., Green, J. S., & Schwingshackl, C. W. (2017) Modelling the nonlinear behaviour of an underplatform damper test rig for turbine applications. *Mechanical Systems and Signal Processing*. 85, 662–679. doi: 10.1016/j.ymssp.2016.09.007.

[18] Gastaldi, C., & Berruti, T. (2018) Experimental Verification of the Dynamic Model of Turbine Blades Coupled by a Sealing Strip. *Applied Sciences*. 8(11), 2174. doi: 10.3390/app8112174.

[19] Stanbridge, A. B., Sever, I. A., & Ewins, D. J. (2002) Vibration measurements in a rotating blisk test rig using an LDV. *Fifth International Conference on Vibration Measurements by Laser Techniques: Advances and Applications*. International Society for Optics and Photonics. doi: 10.1117/12.468173.

[20] Florian G., Sextro, W., Panning, L., & Popp, A. K. (2004) Systematic Mistuning of Bladed Disk Assemblies With Friction Contacts. *Asme Turbo Expo: Power for Land, Sea, & Air*. doi: 10.1115/GT2004-53310.

[21] Szwedowicz, J., Gibert, C., Sommer, T. P., & Kellerer, R. (2008) Numerical and Experimental Damping Assessment of a Thin-Walled Friction Damper in the Rotating Setup With High Pressure Turbine Blades. *Asme Turbo Expo: Power for Land, Sea, & Air*. American Society of Mechanical Engineers. doi: 10.1115/1.2771240.

[22] Sever, I. A., Petrov, E. P., & Ewins, D. J. (2008) Experimental and numerical investigation of rotating bladed disk forced response using underplatform friction dampers. *Journal of Engineering for Gas Turbines and Power. 130*(4), 042503. doi: 10.1115/1.2903845.

12th International Conference on Vibrations in Rotating Machinery -
Institution of Mechanical Engineers, ISBN 978-0-367-67742-8

Rotor dynamics analysis of different bearing system configurations for a 30 kW high-speed turbocompressor

G. Zywica, P. Zych, M. Bogulicz

Department of Turbine Dynamics and Diagnostics, Institute of Fluid Flow Machinery, Polish Academy of Sciences, Poland

ABSTRACT

This paper focuses on the selection of a bearing system for a newly designed energetic turbocompressor. Two alternative concepts of rotor support (consisting of two or three bearings) are analysed. Because the rotor must pass through several resonant speeds before it reaches the nominal speed, an in-depth dynamic analysis was carried out. The research presented in this paper proves that an increase in the number of bearings used to support the shaft does not necessarily have a positive impact on the dynamic performance of the rotor. Ultimately, one version of the rotor-bearing system has been approved for implementation.

1 INTRODUCTION

The development of high-speed turbomachines requires the use of innovative solutions for bearing systems. In addition to the high rotational speed, the bearings of these machines must withstand high static and dynamic loads that act on the shaft as well as high thermal loads. This article discusses the research whose aim was to select a bearing system for the rotor of a newly designed energetic turbocompressor. The turbine's and compressor's discs, as well as the generator rotor, were mounted on the shaft of the turbocompressor with a rated speed of 100 krpm. The turbocompressor has been designed for use in a cogeneration system with an external combustion chamber, where it will generate up to 30 kW of electric power. Such systems may be an interesting alternative for gas microturbines and cogeneration systems that operate on the basis of the organic Rankine cycle (1).

A very high rotational speed, which often exceeds 100 krpm, can be a source of difficulty during the design process of rotor-bearings systems for micro-power turbomachines. The high rotational speed of the rotors is beneficial for the fluid-flow and allows a high efficiency to be achieved when small-diameter rotor discs are used. However, it is accompanied by unfavourable dynamic phenomena and the need to use advanced bearing systems. This is why much attention has been paid to the dynamic analysis of high-speed rotors in the scientific literature. This topic is explored, among other things, in publications on the design of rotors of automotive turbochargers, which are characterised by a high level of subharmonic vibrations and exhibit a tendency of self-excited vibration (2). These machines are very often equipped with floating-ring slide bearings which stabilise the whirling modes. In small turbochargers, ball bearings and ball bearings with dampers are also used (2,3). As for large turbochargers mounted in locomotives and marine engines, they can be equipped with preloaded 3-lobe bearings, 3-lobe bearings with a damper and taper land bearings (3). The dynamic problems of rotating systems get worse as the length of their rotors increases and the longer the rotors, the higher the number of eigenmodes within a range of operating speeds. Therefore, much attention is paid to the design of rotors of power machines, where the generator rotor must also be supported by bearings. Such a rotor can be driven via a coupling (4) or a gear (5), but it is typical of micro-power machines that their generator and rotor disc share the same shaft (6,7). This makes it possible to achieve high efficiency with a more compact and reliable design. Attempts have also been made to use the rotors of microturbines in a vertical configuration (8). In the case of ORC vapour microturbines, the use of a single shaft for the turbine and the generator makes it possible to build hermetically sealed oil-free turbogenerators (9). In order to reduce the vibration level of high-speed rotating machines, much attention must also be paid to balancing their rotors (10,11). In a case like this one, the loads caused by the unbalance are much higher than those caused by the magnetic field (11). Precise balancing of a rotor not only reduces the dynamic forces acting on the rotor and bearings but also reduces the impact of the machine on its surrounding environment (12). Too high unbalance of the rotor can also occur during operation, for example as a result of the deposition of impurities on the rotating elements (13). Due to the possible occurrence of various dynamical and operational problems, the development of high-speed turbomachines is very often accompanied by their testing under laboratory conditions. The tests of the dynamic performance of a microturbine rotor on a test rig is e.g. presented in article (14). The rotor was supported by ceramic rolling bearings and its maximum rotational speed (30 krpm) was greater than the critical speed of lateral and torsional vibrations.

Based on the available literature and our past experience, it can be said that selecting an appropriate configuration of the rotor and bearing system is crucial in the process of designing new high-speed turbomachines. The following part of the article describes the process of selecting the subassemblies and bearing system of a 30 kW turbocharger, which will be used to produce electric energy in a combined heat and power cogeneration system. Several alternative solutions for the rotating system are presented, then two of them were selected and analysed in detail. Numerical models were developed and static and dynamic analyses were carried out, which made it possible to compare the properties of the two systems. The characteristics of the rotors were obtained, including shaft deformations, eigenfrequencies and vibrations depending on the speed. This allowed us to choose a rotating system for further implementation.

2 SELECTION OF ROTOR CONFIGURATION AND BEARING TYPE

The principle of operation of the fluid-flow machine that is being designed is based on the Brayton cycle. The use of the so-called external combustion chamber is the fundamental difference between a classic gas turbine and the turbine that is currently under development. In such a system, the gas turbine is driven by hot air which is supplied under pressure (using a compressor) to the external combustion chamber with a heat exchanger. The air heated by the combustion gases in the heat exchanger is used as the working medium. Thanks to this solution, the turbine is not driven directly by combustion gases, which have a negative effect on it and shorten its service life. An important advantage of such a system is also the fact that almost all types of fuels (including renewable and non-renewable solid fuels) can be burnt in the combustion chamber. In classic gas turbines, only liquid and gas fuels that meet specific requirements can be used. When the fuel is burnt in an external chamber, its quality has no negative effect neither on the service life of the turbine nor on its operation.

During the preliminary analyses, several configurations of the rotor–bearings system were considered, in which the particular subassemblies and the bearings were positioned differently. The compressor, turbine and generator can operate on a common shaft or on two separate shafts, connected by a coupling. The three selected configurations of the rotating system are shown in Figure 1. In the first configuration, the generator is placed between the compressor and the turbine and all the subassemblies share the same shaft which is supported by two bearings. In the second case, the generator shaft is coupled to the turbocompressor shaft. Each shaft has its own bearing system. In the third configuration, the generator, compressor and turbine share the same shaft, which is supported by two bearings (located on both sides of the generator). The compressor and turbine are mounted on the overhanging part of the shaft.

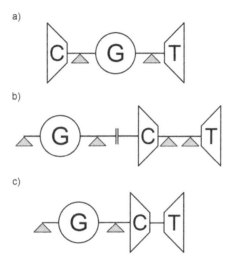

Figure 1. Alternative configurations of the turbocompressor-generator rotor-bearings system (G – generator, C – compressor, T – turbine).

Each of the systems presented has certain advantages and disadvantages in terms of their design and operation. The system depicted in Figure 1c was chosen for further analysis. The main advantages of this system are as follows: the favourable temperature distribution over the length of the shaft, a compact design which makes it possible to design a more rigid rotor, the lack of bearings located in the high-temperature zone (between the compressor and the turbine) and the absence of a need to use a coupling. Special attention must be paid to the favourable temperature distribution between the particular subassemblies. The highest temperature occurs in the gas turbine, where the air at the inlet has a temperature of 850°C. At the compressor outlet, the air has a temperature of approximately 200°C. The generator, whose maximum permissible operating temperature is 150°C, is the coolest part of the machine. It will be possible to maintain a large temperature gradient between the turbine and the generator. A similar arrangement of the subassemblies of a gas turbine was used in a solution described in article (15). However, an internal combustion chamber was used, from which the combustion gases went directly to the fluid-flow system of the gas turbine.

The turbocompressor rotor must be supported by bearings which will guarantee stable operation at very high rotational speeds and high temperatures. Their long operational life and low friction losses are also very important. That is why the possibilities of using different types of bearings (including aerodynamic and aerostatic gas bearings, foil bearings, oil-lubricated slide bearings and rolling element bearings) were explored. At the design stage of the prototype, the use of magnetic bearings was not considered since this solution is rather technically complicated and expensive. Aerodynamic gas bearings are characterised by low capacity and limited operating life. Although aerostatic gas bearings have higher capacities compared to aerodynamic ones, they must be constantly supplied with compressed gas. As for foil bearings, their alignment precision is low and the starting torque is quite high. Due to the fact that foil bearings are still under development, they are not available as off-the-shelf products. The floating ring bearings that are widely used in automotive turbochargers have the considerable disadvantage that they must be fed with large quantities of oil under pressure. This is obviously not a problem in combustion engines since they have oil lubrication systems. However, in the machine that is being designed, this would pose a problem because the use of the additional system, rather complex, would decrease its overall efficiency. Therefore, in the first version of the prototype, we decided to use high-precision angular rolling bearings. Such bearings have ceramic rolling elements and can therefore withstand very high rotational speeds. In addition, they ensure highly precise movements and have very low friction losses. Bearings of this type are widely available on the market since they are produced by several manufacturers.

3 NUMERICAL ANALYSIS

3.1 FEM model of the rotor

In order to obtain static and dynamic characteristics of the designed rotating system, numerical analyses were carried out. The calculations were made using the MADYN 2000 software, which is a modern tool capable of analysing the dynamic performance of rotating machines, including the rotors of turbocompressors with different bearing types (16).

During the preliminary design stage of the complete rotor, it turned out that the overhanging part of the shaft was quite long due to the need to have space for the intake system of the compressor and taking into account the dimensions of the compressor's and turbine's rotor discs. As this could have a negative impact on the dynamic performance of the rotor, an additional configuration of the rotating system has been proposed. Compared to the basic configuration shown in Figure 1c, the only difference in this new configuration was an auxiliary bearing placed between the compressor and the turbine. This modification was introduced to provide additional support for the overhang and flexible part of the shaft and to reduce the excessive vibrations that could occur in certain speed ranges. Since there will be a high temperature in this section of the machine (which can reach 400°C), it will not be possible to use a rolling bearing. An assumption was therefore made that a foil bearing (made of materials resistant to high temperatures) would be mounted between the compressor and the turbine. As the shaft will be precisely positioned using two rolling bearings placed on the two sides of the generator, it will be possible to reduce the assembly preload of the foil bearing. This will reduce the moment of friction, which will make it easier to start the machine and minimise power losses. The two analysed variants of the rotor are shown in Figure 2. Since the foil bearing must have a certain length, there was a need to lengthen the shaft in this case.

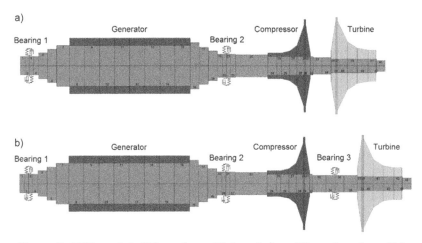

a)

Generator   Compressor   Turbine

Bearing 1   Bearing 2

b)

Generator   Compressor   Turbine

Bearing 1   Bearing 2   Bearing 3

Figure 2. FEM model of the rotor with two (a) and three bearings (b).

The two models of the rotor had 42 and 45 Timoshenko-type beam finite elements respectively, with six degrees of freedom at each node. The forces and moments acting on particular subassemblies as well as the properties of the materials were taken into account in the models. It was assumed that the following materials would be used: steel 42CrMo4 – for the shaft, aluminium alloy 7075 – for the compressor rotor disc and nickel-based alloy IN-738 – for the turbine rotor disc. The dimensions and mass of the rotor of the 2-pole synchronous generator were also considered in the FEM models. The mass and length of the complete rotating systems were as follows: 4.287 kg and 404 mm – for the two-bearing rotor; 4.347 kg and 434 mm – for the three-bearing rotor. The distance between the rolling bearings was the same in both variants and was equal to 218 mm. The stiffness of the rolling bearings was determined on the basis of empirical equations, where the dimensions of the bearings and rolling elements, as well as the assembly preload of the bearings, were taken into account. The stiffness and damping of the foil bearing were calculated using computer programs developed in the IMP PAN. In the calculations, the gas lubrication film and the support structure (the top foil and bump foils) were taken into account. The calculations took into account the bearing characteristics determined for the nominal speed.

3.2   Static analysis

A static analysis was carried out to verify the distribution of displacements, forces and stresses caused by gravitation and static loads that occur during the operation of the machine. The highest static displacement of the shaft supported by two bearings occurred at the free end of the shaft and was 10 μm. The use of the additional bearing made it possible to reduce this displacement to 6.7 μm (despite the fact that the shaft length was greater). In addition, the additional foil bearing changed the distribution of loads acting on the two rolling bearings. The static load of bearing 1 increased from 4 N to 10 N and that of bearing 2 decreased from 40 N to 22 N. The static load of the additional bearing was approximately 12 N. The reactions of the bearing supports are visible in Figure 3. This figure also shows the bending moments and shear forces in the shafts. It is clearly seen that the increase in the number of bearings resulted in a change of forces and moments acting on the rotor. The maximum bending moment decreased from 1.8 Nm to 0.7 Nm. The values of shear forces also decreased. The forces acting on the shaft have caused the appearance of low stresses, which can be considered as negligible given the mechanical properties of the material. The maximum reduced stresses did not exceed several megapascals. It can be said that both rotors were designed properly in terms of their strength.

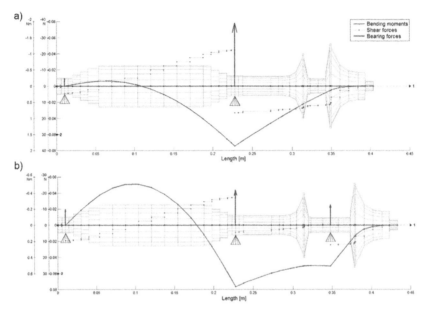

Figure 3. Results of the static analysis of the rotor with two (a) and three bearings (b).

3.3 Dynamic analysis

As part of the dynamic analysis of the rotors, a computational modal analysis and a forced vibration analysis were performed. Eigenfrequencies and eigenmodes were determined, taking into account the stiffness of the bearing supports and gyroscopic effects. The eigenfrequencies, determined in the range from 0 to 5,000 Hz, are shown in Table 1. They have been determined for the nominal speed of 100 krpm. The eigenfrequencies are marked by colours, which indicate the type of vibrations (**bending**, **axial** and **torsional** vibrations). In addition, the frequency values which correspond to bending vibrations occurring at backward whirls are italicised. The same eigenfrequencies are shown in graphical form in Figure 4 (in the range of 0–4000 Hz). The nominal rotational frequency of the turbocompressor rotor, which is equal to 1666.7 Hz, is also shown on the graphs.

Table 1. Natural frequencies of the rotor with two and three bearings.

Eigenmode No.	Frequency [Hz]		Eigenmode No.	Frequency [Hz]	
	two bearings	three bearings		two bearings	three bearings
1	110.0	109.4	9	1223.7	1074.8
2	144.7	153.4	10	1405.2	1156.8
3	222.4	273.7	11	2772.6	2291.6
4	433.0	429.9	12	3461.5	2859.2
5	502.0	493.6	13	4145.1	3272.6
6	506.9	498.1	14	4157.5	3633.3
7	634.0	559.4	15	4863.3	4027.3
8	939.5	745.3	-	-	-

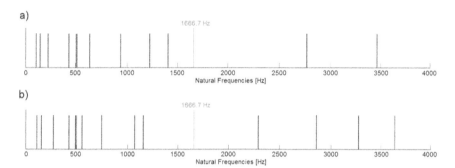

Figure 4. Natural frequencies of the rotor with two (a) and three bearings (b) (legend: bending, axial and torsional vibrations).

In both analysed cases, there were 15 eigenmodes in the range of 0–5000 Hz but they occurred at different frequencies. In general, the introduction of the additional bearing and the elongation of the shaft decreased the eigenfrequencies. This is clearly seen in Figure 4, where eigenmodes No. 9 and 10 have moved towards lower frequencies and increased their distance from the nominal rotational frequency. In addition, the frequency of eigenmode No. 11 decreased by 481 Hz and moved towards the rotational frequency. Besides bending eigenmodes, there were also two axial eigenmodes and three torsional eigenmodes in the analysed frequency range. Some bending eigenmodes occurred at backward whirls, so they did not affect the vibrations of the rotor supported by the rolling bearings.

In order to check the possibility of the excitation of lateral vibrations during the operation of the rotor-bearings system, a forced vibration analysis was carried out. The excitation force was the result of the residual unbalance, whose maximum value was assumed on the basis of the ISO 1940-1 standard. The calculations have been made for the following arrangements of the excitation forces:

a) one unbalance on the turbine disc,
b) two unbalances (rotated by 180°) on the turbine's and compressor's discs,
c) two unbalances on the turbine disc and the generator,
d) two unbalances (rotated by 180°) on the turbine disc and the generator.

The several arrangements of the unbalance were used to check the possibility of the excitation of different eigenmodes. The modal analysis had indicated that in the frequency range analysed, there was a rather high number of eigenmodes and that almost each of them presented a risk for the stable functioning of the rotor.

The results of the forced vibration analysis of the two- and three-bearing rotor are shown in Figures 5 and 6. Due to the page limitation of the article, only the selected characteristics are presented, that is to say, those obtained for the unbalance arrangement where the two additional masses were placed on the rotor discs of the turbine and the compressor (and the second mass was rotated by 180° relative to the first mass). This unbalance arrangement can be regarded as the most unfavourable in terms of the dynamic performance of the rotor. This arrangement resulted in the highest vibration levels at the critical speeds that were the closest to the nominal rotational speed. These vibration levels occurred at a speed of 68.6 krpm in the two-bearing system and at a speed of 161.5 krpm in the three-bearing system. The maximum vibration amplitude was 32 µm in the first case. In the second case, the highest vibration amplitude was 134 µm, but it occurred outside the range of operating speeds. In order to excite the vibrations of this type in the rotating system, an excitation force with a frequency that is a multiple of the rotational frequency would have to occur. Below the nominal speed, in the three-bearing system, the highest vibration amplitude was 48 µm and occurred at a speed of 59.7 krpm. The vibration amplitude in the operating speed range was higher compared to the basic system with two bearings. The increase in the vibration level of the rotor with an additional bearing resulted from the increase in its length. The reduction in the stiffness of the shaft was too large to be compensated by an additional support point (i.e the compliant foil bearing). Therefore, the vibration level was higher in some rotational speed ranges.

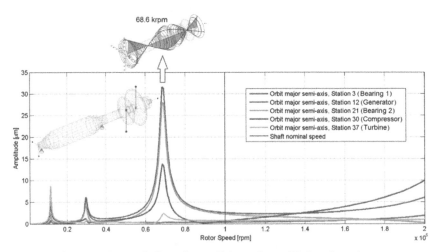

Figure 5. Forced vibrations of the rotor with two bearings.

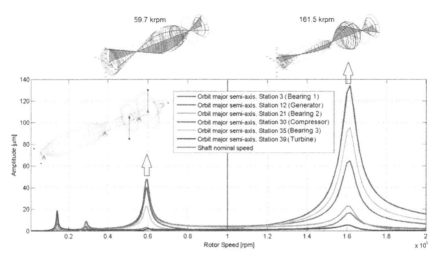

Figure 6. Forced vibrations of the rotor with three bearings.

Apart from the resonant speeds mentioned above, the vibration level was rather low. Even though there were certain increases in the vibration level at low speeds (in the range of 10–30 krpm), the vibration amplitude did not exceed 20 μm. This was due to the low dynamic excitation. If dynamic excitation originates from the unbalance, it increases proportionally to the square of the rotational speed. In the remaining unbalance arrangements, which are not discussed in detail in this article, the vibrations of the rotors determined by calculation were of low level.

It should also be noted that not all eigenmodes, whose frequencies are given in Table 1 and shown in Figure 4, will have a significant impact on the dynamic performance of the rotors analysed. This stems from the fact that some eigenmodes can only occur during the reverse precession, which should not happen in the case of rotors supported by rolling bearings.

5   CONCLUSIONS

The article discusses analyses that have sought to select the configuration of the rotating system and bearings for the rotor of a 30 kW compressor with a nominal rotational speed of 100 krpm. The system, in which the generator's, compressor's and turbine's rotors were placed one after another on a common shaft supported by two high-precision rolling bearings, was selected for further considerations. Such a configuration of the rotating system prevents the appearance of high thermal gradients within the rotor and facilitates management of the heat throughout the machine. After preparing the rotor design concept, it turned out that in the assumed configuration, the overhanging part of the shaft was quite long. Therefore, in addition to the initially selected configuration of the rotating system, another configuration was analysed (in which an additional bearing was placed between the compressor and the turbine). As there will be a very high temperature in this section of the machine, a decision has been made to use a gas foil bearing.

Based on the calculations made for the two alternative variants of the rotating system, the following conclusions can be drawn:

a) In order to use an additional bearing between the compressor and the turbine, it was necessary to lengthen the shaft. However, the static displacements of the shaft end were lower by about 30%.
b) Regardless of the number of bearings, in the frequency range from 0 to 5000 Hz, the two systems analysed had 15 eigenmodes that could be excited during the operation of the machine. The use of an additional bearing and the elongation of the shaft reduced the eigenfrequencies.
c) The forced vibration analysis has shown that in the speed range of 80-120 krpm, the two rotors analysed should work properly in terms of their dynamic performance. The critical speed, which was closest to the nominal speed, was at 68.6 krpm for the two-bearing system and at 59.7 krpm – for the three-bearing system. At lower speeds, the vibrations of the rotors were very low.

Based on the analyses carried out to date, we have decided to continue the development of the two-bearing turbocompressor. Since this solution is simple and less costly, it will be possible to accelerate work on the first prototype. Work on a heat-resistant gas foil bearing is also planned. In the future, it will be used either in the machine which is currently under development or in another fluid-flow machine. A final decision on whether to use a foil bearing will be made after first experimental tests.

The research presented in this article may be of interest to scientists and engineers who deal with innovative turbomachines and seek the best bearing systems for newly designed rotating machines. The results obtained can be useful when designing various types of high-speed turbomachines such as turbocompressors, gas/vapour micro-turbines and turbine expanders.

REFERENCES

[1] Kosowski, K., Tucki, K., Piwowarski, M., Stępień, R., Orynycz, O., Włodarski, W., "Thermodynamic Cycle Concepts for High-Efficiency Power Plants. Part B: Prosumer and Distributed Power Industry", *Sustainability*, 11, 2019, 2647.
[2] Kirk, R.G., Alsaeed, A.A., Gunter, E.J., Stability analysis of a high-speed automotive turbocharger, *Tribology Transactions*, 50, 2007, pp. 427–434.
[3] Chen, W.J., "Rotordynamics and bearing design of turbochargers", *Mechanical Systems and Signal Processing*, 29, 2012, pp. 77–89.
[4] Feng, S., Geng, H., Yu, L., "Rotordynamics analysis of a quill-shaft coupling-rotor-bearing system", *Proceedings of the Institution of Mechanical Engineers, Part C: Journal of Mechanical Engineering Science*, 229(8), 2015, pp. 1385–1398.
[5] Zywica, G., Brenkacz, L., Baginski, P., "Interactions in the rotor-bearings-support structure system of the multi-stage ORC Microturbine", *Journal of Vibration Engineering & Technologies*, 6, 2018, pp. 369–377.
[6] Arroyo, A., McLorn, M., Fabian, M., White, M., Sayma, A. I., "Rotor-dynamics of different shaft configurations for a 6 kW micro gas turbine for concentrated solar power", *Proceedings of the ASME Turbo Expo 2016*, June 13–17, Seoul, South Korea, GT2016–56479.
[7] Brenkacz, L., Zywica, G., Bogulicz, M., "Selection of the bearing system for a 1 kW ORC microturbine", *In: Cavalca K., Weber H. (eds) Proceedings of the 10th International Conference on Rotor Dynamics – IFToMM 2018. Mechanisms and Machine Science*, 60, 2019, pp. 223–235.
[8] Efimov, N.N., Papin, V.V., Bezuglov, R.V., "Determination of rotor surfacing time for the vertical microturbine with axial gas-dynamic bearings", *Procedia Engineering*, 150, 2016, pp. 294–299.

[9] Brenkacz, L., Zywica, G., Bogulicz, M., "Selection of the oil-free bearing system for a 30 kW ORC microturbine", *Journal of Vibroengineering*, 21(2), 2019, pp. 318–330.

[10] Lee, A.S., Ha, J.W., "Maximum unbalance responses of gear-coupled two-shaft rotor-bearing system", *Proceedings of the ASME Turbo Expo 2013*, June 16–19, Atlanta, USA, GT2003–38146.

[11] Hong, D.-K., Joo, D.-S., Woo, B.-C., Koo, D.-H., Ahn, C.-W, "Unbalance Response Analysis and Experimental Validation of an Ultra High Speed Motor-Generator for Microturbine Generators Considering Balancing", *Sensors*, 14 2014, pp. 16117–16127.

[12] Kaczmarczyk, T.Z., Zywica, G., Ihnatowicz, E., "Vibroacoustic diagnostics of a radial microturbine and a scroll expander operating in the organic Rankine cycle installation", *Journal of Vibroengineering*, 18(6), 2016, pp. 4130–4147.

[13] Zywica, G., Kaczmarczyk, T.Z., "Experimental evaluation of the dynamic properties of an energy microturbine with defects in the rotating system", *Eksploatacja i Niezawodnosc – Maintenance and Reliability*, 21(4), 2019, pp. 670–678.

[14] Zhang, D., Xie, Y., Feng, Z., "An investigation on dynamic characteristics of a high speed rotor with complex structure for microturbine test rig", *Proceedings of ASME Turbo Expo 2008*, June 9–13, Berlin, Germany, GT2008–50411.

[15] Vick, M., Young, T., Kelly, M., Tuttle, S., Hinnant, K., "A simple recuperated ceramic microturbine: design concept, cycle analysis, and recuperator component prototype tests", *Proceedings of ASME Turbo Expo 2016*, June 13–17, Seoul, South Korea, GT2016–57780.

[16] Schmied, J., "Application of MADYN 2000 to rotor dynamic problems of industrial machinery", In: Santos, I. (Ed.), *Proceedings of 13th SIRM: The 13th International Conference on Dynamics of Rotating Machinery*, 2019, Technical University of Denmark, Copenhagen, Denmark, pp. 490–500.

Digital twin of induction motors for torsional vibration analysis of electrical drive trains

T.P. Holopainen, J. Roivainen, T. Ryyppö

ABB Motors and Generators, Finland

ABSTRACT

Torsional vibrations must be considered in design of all high-power drive trains including an induction motor. Magnetic field in the air gap of induction motor generates additional stiffness and damping between the rotor and stator. The inclusion of these magnetic effects is limited by the availability of portable motor models. The aim of this paper is to introduce a portable digital twin of induction motor for electrical drive trains. This digital twin is based on the linearization of common equivalent-circuit models of induction motors. The calculation results for a motor-driven reciprocating compressor demonstrate the accuracy and advantages of developed digital twin.

1   INTRODUCTION

Electric power is transmitted from energy sources to various mechanical machines. An electric motor constitutes a part of this power flow by converting the electric energy to mechanical one. These electro-mechanical power drive trains are dynamical systems and sensitive for harmful vibrations. An essential part of the electrical drive-train design is a torsional vibration analysis. This analysis requires inertia and stiffness data of all drive train components together with loading and damping parameters. In most industrial cases, these components are designed and manufactured by separate producers. Usually, these producers are well-acquainted with their own components, but only superficially familiar with other components. This poses a remarkable challenge to the system integrator to successfully take the responsibility of drive-train vibrations.

An excellent example of these challenging systems is a motor-driven reciprocating-compressor. Increased requirements of calculation accuracy have resulted in the inclusion of magnetic stiffness and damping in torsional analyses. Due to this need, simple formulas, based on the motor characteristic data, have been presented for the evaluation of these magnetic parameters (1-3). These formulas are based on the equivalent circuit models, developed to describe steady-state and transient motor behaviour, and used also for the evaluation of magnetic stiffness and damping (4-10). The accuracy of equivalent circuit models can be improved by calculating the parameters by numerical methods using the actual geometry of active parts (11-13).

Application of these simple electro-mechanical motor models has been insignificant. A remarkable shortcoming has been the poor portability. It has been difficult to use the motor model as a part of standard torsional analysis without an in-depth knowledge of electromagnetics of motors. However, the motor is only one component of a power drive train with well-defined input and output parameters. The input is described by the supply voltage and shaft torque, and the output by the supply current and shaft speed. Only the details and generation of this "black box" requests the expertise of electrical engineers. In this paper this black box is referred as digital twin.

The main aim of this paper is to introduce a digital twin of induction motors for torsional vibration analysis of electrical drive trains. The aim is to describe the derivation of the digital twin, the accuracy of this representation, and the application to torsional analysis.

This paper is based strongly on the methods and findings of two previous papers (12-13). This paper starts by describing the requirements for a digital twin of electric motors used in torsional analyses. Next, the main equations of the equivalent circuit model are presented with a special focus on linearization. These equations form the basis for digital twins. After that, a digital twin of an example motor is derived, and the characteristics are described. Next, this digital twin is used to calculate the characteristics and forced response of a motor-compressor system. The results demonstrate the usefulness of digital-twin approach in torsional analyses.

2   DIGITAL TWIN IN TORSIONAL ANALYSIS

The electro-mechanical power drive train consists of separate components. The power flow is represented by voltage-current and torque-speed pairs between the components. The system components are separated by these simple interfaces. Thus, in principle, the components can be modelled separately, and connected by well-defined interface nodes. Figure 1 shows an example of a motor-driven reciprocating-compressor system.

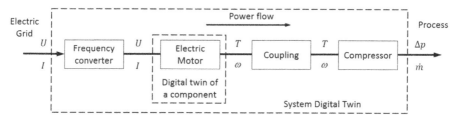

Figure 1. – Digital twin of a motor-driven reciprocating-compressor system.

The requirements for the digital twin of electric motors can be outlined as follows:

a) Accurate enough representation of actual system in various operating conditions
b) Automatic generation of model parameters for actual motors based on catalogue codes or serial number
c) Standardized interface based on input and output parameters
d) Protection of sensitive product information
e) Suitability for web-based systems with Application Programming Interface (API)

3   DIGITAL TWIN OF INDUCTION MOTORS

3.1   Equivalent-circuit model of electromagnetics
A single-cage equivalent-circuit model for an induction motor written in a reference frame that rotates at an arbitrary angular speed ω_k is (12)

$$
\begin{aligned}
-u_s^k &= R_{s0} - i_s^k + -\psi_{s,t}^k + \mathrm{j}\omega_k - \psi_s^k \\
0 &= R_{r0} - i_r^k + -\psi_{r,t}^k + \mathrm{j}(\omega_k - \omega) - \psi_r^k \\
T_e &= \tfrac{3}{2}p\mathrm{Im}\left(-\psi_s^{k*} - i_s^k\right)
\end{aligned} \tag{1}
$$

where $-u_s^k$, $-i_s^k$ and $-\psi_s^k$ are the space vectors of stator voltage, stator current and stator flux linkage, $-i_r^k$ and $-\psi_r^k$ are the space vectors of rotor current and rotor flux linkage, R_{s0} is the stator resistance, R_{r0} is the rotor resistance, ω is the angular speed of the rotor, ω_k the angular speed of the frame of reference with respect to the stator, p is the number of pole pairs, T_e is the electromagnetic torque, j is the imaginary unit, an asterisk denotes complex conjugation, the subscript t the time derivative, and subscript 0 refers to the parameter values of non-linear large-signal model.

The linear relation between the flux linkages and currents is

$$
\begin{aligned}
-\psi_{s0}^k &= L_{s0} - i_s^k + L_{m0} - i_r^k \\
-\psi_{r0}^k &= L_{m0} - i_s^k + L_{r0} - i_r^k
\end{aligned} \tag{2}
$$

where L_{s0} and L_{r0} are the self-inductances of the stator and rotor windings and L_{m0} is the mutual inductance between them.

The system of equations (Eq. 1) is non-linear when used for torsional vibrations because of the product of angular speed and rotor flux linkage in the second equation, and the product of stator flux linkage and stator current in the third equation. However, typical vibrations occur as oscillations around an equilibrium point determined by the current steady-state operation. Further, it is reasonable to assume that the small oscillations of the system around the equilibrium point are linear.

The system can be linearized at an operation point. Using the reference frame rotating at the synchronous speed, i.e. $\omega_k = \omega_s$, the linearization yields a linear small-signal model

$$
\begin{aligned}
R_s\Delta - i_s^s + L_s\Delta - i_{s,t}^s + \mathrm{j}\omega_s L_s\Delta - i_s^s + L_m\Delta - i_{r,t}^s + \mathrm{j}\omega_s L_m\Delta - i_r^s &= 0 \\
R_r\Delta - i_r^s + L_m\Delta - i_{s,t}^s + \mathrm{j}(\omega_s - \omega_0)L_m\Delta - i_s^s + L_r\Delta - i_{r,t}^s + \mathrm{j}(\omega_s - \omega_0)L_r\Delta - i_r^s &= \mathrm{j} - \psi_{r0}^s\Delta\omega \\
\Delta T_e = \tfrac{3}{2}p\mathrm{Im}\left(-\psi_{s0}^{s*}\Delta - i_s^s + -i_{s0}^s\Delta - \psi_s^{s*}\right) &
\end{aligned} \tag{3}
$$

where the currents have been chosen as the free variables, Δ denotes a small variation from the steady state value, i.e. a small signal variable, and $-i_{s0}^s$, $-\psi_{s0}^s$, $-\psi_{r0}^s$ and ω_0 are the steady-state stator current, stator flux-linkage, rotor flux-linkage and angular speed of large-signal model.

It can be added that the equivalent-circuit models of Eqs. 1 and 3 can be improved by including the skin effect of rotor bars. This skin effect may affect significantly the electromagnetic response of the motor to torsional oscillations. The modelling of this skin effect can be carried out by increasing the number of rotor branches, i.e. by using double- or triple-cage models (12).

3.2 Magnetic stiffness and damping

The angular speed ω in Eqs. 1 and 3 is given in electrical radians per second following the research tradition of electrical machines. The relation between these angular speeds and positions are

$$\omega = p\Omega_\mathrm{m}$$
$$\gamma = p\gamma_\mathrm{m} \tag{4}$$

where Ω_m is angular speed in mechanical or physical radians per second and γ_m is angular position in mechanical radians. Further, the electromagnetic stiffness and damping can be calculated by assuming harmonic variation of all small signal variables like

$$\Delta \underline{\tilde{i}}_\mathrm{s}^\mathrm{s} = \hat{i}_\mathrm{s1}^\mathrm{s} e^{j\omega_\mathrm{d}t} + \hat{i}_\mathrm{s2}^\mathrm{s} e^{-j\omega_\mathrm{d}t} \tag{5}$$

where $\hat{i}_\mathrm{s1}^\mathrm{s}$ and $\hat{i}_\mathrm{s2}^\mathrm{s}$ are the complex amplitudes of the stator current variation, and ω_d is the oscillation frequency. This yields for the Frequency Response Function (FRF) between the torque and angular position amplitudes as function of oscillation frequency

$$\hat{\underline{T}}_\mathrm{e}/\hat{\underline{\gamma}}_\mathrm{m}(\omega_\mathrm{d}) = -\frac{3j}{2}p\left(\underline{\Psi}_\mathrm{s0}^\mathrm{s*}\hat{\underline{i}}_\mathrm{s1}^\mathrm{s} + \hat{\underline{\Psi}}_\mathrm{s2}^\mathrm{s*}\underline{i}_\mathrm{s0}^\mathrm{s} - \underline{\Psi}_\mathrm{s0}^\mathrm{s}\hat{\underline{i}}_\mathrm{s2}^\mathrm{s*} - \hat{\underline{\Psi}}_\mathrm{s1}^\mathrm{s}\underline{i}_\mathrm{s0}^\mathrm{s*}\right)/\hat{\underline{\gamma}}_\mathrm{m} \tag{6}$$

The electromagnetic stiffness and damping coefficients are obtained from the real and imaginary parts of this FRF

$$k_\mathrm{em}(\omega_\mathrm{d}) = -\mathrm{Re}\left(\hat{\underline{T}}_\mathrm{e}/\hat{\underline{\gamma}}_\mathrm{m}\right)$$
$$C_\mathrm{em}(\omega_\mathrm{d}) = -\frac{1}{\omega_\mathrm{d}}\mathrm{Im}\left(\hat{\underline{T}}_\mathrm{e}/\hat{\underline{\gamma}}_\mathrm{m}\right) \tag{7}$$

3.3   Digital twin of induction motor

The equations of motion for a rigid rotor are

$$T_\mathrm{e} = J\Omega_{\mathrm{m},t} + T_\mathrm{m}$$
$$\Omega_\mathrm{m} = \gamma_{\mathrm{m},t} \tag{8}$$

where J is rotational inertia, and T_m is the load torque. By substituting the voltage and torque equations (Eq. 1) together with the flux linkages (Eq. 2) into the equations of motion (Eq. 8) yields in the stator reference frame ($\omega_\mathrm{k} = 0$)

$$-u_\mathrm{s}^0 = R_\mathrm{s0} - i_\mathrm{s}^0 + L_\mathrm{s0} - i_{\mathrm{s},t}^0 + L_\mathrm{m0} - i_{\mathrm{r},t}^0$$
$$0 = R_\mathrm{r0} - i_\mathrm{r}^0 + L_\mathrm{m0} - i_{\mathrm{s},t}^0 + L_\mathrm{r0} - i_{\mathrm{r},t}^0 - jp\Omega_\mathrm{m}\left(L_\mathrm{m0} - i_\mathrm{s}^0 + L_\mathrm{r0} - i_\mathrm{r}^0\right)$$
$$\tfrac{3}{2}pL_\mathrm{m0}\mathrm{Im}\left(-i_\mathrm{r}^{0*} - i_\mathrm{s}^0\right) = J\Omega_{\mathrm{m},t} + T_\mathrm{m}$$
$$\Omega_\mathrm{m} = \gamma_{\mathrm{m},t} \tag{9}$$

where the superscript 0 refers to the stator reference frame. Equation 9 can be presented in matrix form as

$$\mathbf{u} = \mathbf{R}_0\mathbf{x} + \mathbf{L}_0\mathbf{x}_{,t} \tag{10}$$

where the complex valued vectors and matrices are now defined as

$$\mathbf{u} = \left\{\begin{array}{cccc} -u_\mathrm{s}^0 & 0 & -T_\mathrm{m} & 0 \end{array}\right\}^\mathrm{T} \quad \mathbf{x} = \left\{\begin{array}{cccc} -i_\mathrm{s}^0 & -i_\mathrm{r}^0 & \Omega_\mathrm{m} & \gamma_\mathrm{m} \end{array}\right\}^\mathrm{T} \tag{11}$$

567

$$\mathbf{R}_0 = \begin{bmatrix} R_{s0} & 0 & 0 & 0 \\ -jp_mL_{m0} & R_r - jp_mL_{r0} & 0 & 0 \\ -j\frac{3}{4}pL_{m0} - i_r^{0*} & j\frac{3}{4}pL_{m0} - i_s^{0*} & 0 & 0 \\ 0 & 0 & -J & 0 \end{bmatrix} \quad \mathbf{L}_0 = \begin{bmatrix} L_{s0} & L_{m0} & 0 & 0 \\ L_{m0} & L_{r0} & 0 & 0 \\ 0 & 0 & J & 0 \\ 0 & 0 & 0 & J \end{bmatrix}$$

The solution for this <u>non-linear</u> first order differential equation (Eq. 10) can be sought by multiplying the equation by the inverse of the matrix $\mathbf{L}_0$

$$\mathbf{x}_{,t} = -\mathbf{L}_0^{-1}\mathbf{R}_0\mathbf{x} + \mathbf{L}_0^{-1}\mathbf{u} \tag{12}$$

The flexural shaft line can be modelled by adding stiffness and inertia elements to the system. As an example, the shaft of an induction motor is modelled by a two-inertia system consisting of two discs connected by a massless shaft. This yields for the vectors and matrices

$$\mathbf{u} = \{ -u_s^0 \quad 0 \quad 0 \quad -T_{se} \quad 0 \quad 0 \}^T \quad \mathbf{x} = \{ -i_s^0 \quad -i_r^0 \quad \Omega_{m1} \quad \Omega_{m2} \quad \gamma_{m1} \quad \gamma_{m2} \}^T \tag{13}$$

$$\mathbf{R}_0 = \begin{bmatrix} R_{s0} & 0 & 0 & 0 & 0 & 0 \\ -jp\Omega_mL_{m0} & R_r - jp\Omega_mL_{r0} & 0 & 0 & 0 & 0 \\ -j\frac{3}{4}pL_{m0} - i_r^{0*} & j\frac{3}{4}pL_{m0} - i_s^{0*} & c_1 & 0 & 0 & 0 \\ 0 & 0 & 0 & c_2 & k_{12} & -k_{12} \\ 0 & 0 & 0 & -J_1 & 0 & 0 \\ 0 & 0 & 0 & 0 & -J_2 & 0 \end{bmatrix}$$

$$\mathbf{L}_0 = \begin{bmatrix} L_{s0} & L_{m0} & 0 & 0 & 0 & 0 \\ L_{m0} & L_{r0} & 0 & 0 & 0 & 0 \\ 0 & 0 & J_1 & 0 & 0 & 0 \\ 0 & 0 & 0 & J_2 & 0 & 0 \\ 0 & 0 & 0 & 0 & J_1 & 0 \\ 0 & 0 & 0 & 0 & 0 & J_2 \end{bmatrix}$$

where index 1 refers to the rotor core and 2 to the shaft end, c_i is the viscous damping, k_{12} is the torsional stiffness between the elements, and T_{se} is the shaft-end torque. The mechanical inertia, stiffness and damping can be defined as effective values. Thus, a non-linear digital twin has been developed with the time dependent supply voltage $(-u_s^0)$ and load torque in the shaft-end (T_{se}) as input variables. The output variables are the supply current $\left(-i_s^0\right)$ and the shaft end angular speed $(_{m2})$.

Similarly, the combined voltage (Eq. 1) and motion (Eq. 8) equations can be linearized and presented in matrix form as

$$\Delta\mathbf{u} = \mathbf{R}\Delta\mathbf{x} + \mathbf{L}\Delta\mathbf{x}_{,t} \tag{14}$$

where the real valued vectors and matrices are now defined as

$$\Delta\mathbf{u} = \{ 0 \quad 0 \quad 0 \quad 0 \quad -\Delta T_m \quad 0 \}^T \Delta\mathbf{x} = \{ \Delta i_{sx}^s \quad \Delta i_{sy}^s \quad \Delta i_{rx}^s \quad \Delta i_{ry}^s \quad \Delta_m \quad \Delta\gamma_m \}^T \tag{15}$$

$$\mathbf{L} = \begin{bmatrix} L_s & 0 & L_m & 0 & 0 & 0 \\ 0 & L_s & 0 & L_m & 0 & 0 \\ L_m & 0 & L_r & 0 & 0 & 0 \\ 0 & L_m & 0 & L_r & 0 & 0 \\ 0 & 0 & 0 & 0 & J & 0 \\ 0 & 0 & 0 & 0 & 0 & J \end{bmatrix}$$

where the slip is defined as $s = (\omega_s - p_m)/\omega_s$, and the stator and rotor currents as $-\bar{i}_s^s = i_{sx}^s + ji_{sy}^s$ and $-\bar{i}_r^s = i_{rx}^s + ji_{ry}^s$. The solution of this <u>linear</u> first order differential equation (Eq. 14) can be sought by multiplying the equation by the inverse of the matrix $\mathbf{L}$

$$\Delta\mathbf{x}_{,t} = -\mathbf{L}^{-1}\mathbf{R}\Delta\mathbf{x} + \mathbf{L}^{-1}\Delta\mathbf{u} \tag{16}$$

The flexible rotor model can be included into the linear equations (Eq. 14) as was made for the non-linear equations above. This yields a linear digital twin at a steady-state operation point with the time dependent load torque in the shaft end (ΔT_{se}) as input variable, and the shaft end angular speed ($\Delta\Omega_{m2}$) as output variable. It can be mentioned that the eigenvalues and -modes can be easily defined for this linear system.

3.4 Identification of model parameters

The resistance and inductance parameters for the equivalent-circuit model were calculated by the time-harmonic FE analysis (14). The effective parameters (R_{s0}, L_{s0}, ...) were obtained for the rated operating point with steady-state currents from Eq. 1. The small-signal linearized parameters (R_s, L_s, ...) were obtained from Eq. 3.

4 CALCULATION EXAMPLE

A 3.7 MW induction motor was used in all the calculation examples. The main parameters of this motor are shown in Table 1.

Table 1. Rated parameters of the example motor.

Parameter	Value	Unit
Power	3725	kW
Frequency	60	Hz
Speed	895.3	rpm
Number of poles	8	
Connection	star	
Voltage	4000	V
Current	620	A
Rated torque	39.73	kNm
Breakdown torque	82.86	kNm
Moment of inertia	219.9	kgm^2

4.1 Parameters of digital twin

The rotor model consists of two disks with mass moments of inertia J_1 and J_2 connected by a massless shaft with a torsional stiffness k_{12}. The index 1 refers to the rotor core and 2 to the shaft drive-end. The torsional natural frequencies of the rotor were first calculated with a refined model for two cases with different boundary conditions, i.e., a) fixed in the drive end and free in the non-drive end, and b) free-free. The respective natural frequencies were 34.2 Hz and 175.8 Hz. The parameters of this simple model were fitted by adjusting the natural frequencies and the mass moment of inertia equal to those of the refined model. This approach yielded the parameters: $J_1 = 211.4 \text{ kgm}^2$, $J_2 = 8.3 \text{ kgm}^2$ and $k_{12} = 9.775 \text{ MNm/rad}$. It was assumed that the mechanical damping was negligible, and thus, the viscous damping parameters were set $c_1 = c_2 = 0 \text{ Nms/rad}$.

The electrical parameters were calculated by time-harmonic FE-analysis with a 2D-model of the core region (14). Table 2 shows the results for the no-load and rated operation at 60 Hz supply frequency. Similar parameters can be calculated for a two-cage and three-cage models. While the single-cage model has 5 parameters, the two- and three-cages models have 9 and 13 parameters, respectively.

Table 2. Electrical parameters of single cage model in rated and no-load operation.

	Non-linear			Linearized		Unit
	No-load	Rated		No-load	Rated	
R_{s0}	23.457	24.492	R_s	23.486	23.342	mΩ
R_{r0}	19.480	19.450	R_r	18.900	18.668	mΩ
L_{s0}	30.470	29.386	L_s	13.119	12.686	mH
L_{r0}	30.030	29.004	L_r	12.981	12.507	mH
L_{m0}	28.904	27.921	L_m	11.963	11.579	mH

Table 2 shows that the loading of the machine does not affect the parameters significantly. The effects of loading and speed are extensively studied in (13).

4.2 Magnetic stiffness and damping

Figure 2 Shows the magnetic torsional stiffness and damping in the rated operating condition calculated by linearized equivalent-circuit models and by FEM. The FE results obtained by the time-stepping method for each oscillation frequency separately are assumed to be the most accurate and will be used here and later as reference values. The results of Hauptmann et al. (3) are calculated by an analytic formula based on the rated torque, rated slip, supply frequency and breakdown torque. The single cage model is obtained by linearizing the system equations (Eq. 3) but using the values of large-signal model for resistances and inductances (R_{s0}, L_{s0}, ...). The single- and double-cage models with the epithet "linear RL" use the same system equations, but the linearized resistance and inductance (R_s, L_s, ...) are applied.

Figure 2 Shows that the linearized resistance and inductance parameters improve the accuracy of the model. Similarly, the double-cage model increases the accuracy significantly. The triple-cage model gives very similar results (not shown) than the double-cage model.

4.2 Eigenvalues of motor digital twin

The natural frequencies and damping factors, derived from the eigenvalues of the system matrix, were calculated for the digital twins of the motor using different cage models. The digital twin models are denoted by DTwin N, where N refers to the cage number. Table 3 shows the results together with the reference values obtained by FEM with time-stepping method. In this case a sinusoidal torque pulse (0...360°) was given to the rotor and the Frequency Response Function (FRF) between the speed and torque was derived. The parameters of the torque pulse were: $\Delta t = 0.005\text{s}$ and $T_{max} = 200$ kNm. The simulation time was 10 s and the modal parameters were estimated by using the peak picking method of real part of FRF.

Table 3 shows that due to the magnetic stiffness the frequency of rigid body mode is about 9.4 Hz and damping ratio 10 %. The prediction capability of digital twins is good. It was not possible to identify the electrical mode from FEM results using purely mechanical FRF between the rotor speed and torque. All the models yield the same natural frequency for the first elastic mode. However, FEM gives for the damping ratio 2.75 %. This is against the expectations, but the origin for this high damping value is not known.

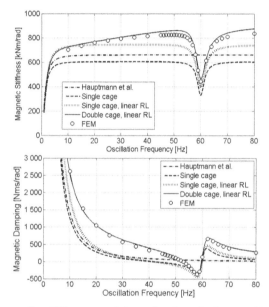

Figure 2. Magnetic stiffness and damping in rated operating condition.

Table 3. Natural frequencies and damping factors of motor digital twin in rated condition.

	Mode type					
	Rigid body		Electrical		First elastic	
Model	f [Hz]	ζ [%]	f [Hz]	ζ [%]	f [Hz]	ζ [%]
DTwin 1	9.417	_8.59	59.96	3.19	175.8	0.00
DTwin 2	9.437	10.19	59.77	3.66	175.8	0.00
DTwin 3	9.483	10.98	59.80	3.67	175.8	0.00
FEM	9.43_	_9.76	-	-	176.3	2.75

4.3 Motor driven reciprocating compressor

The example motor drives a compressor. The drive train consists of following components: motor, flexible coupling, flywheel and reciprocating four-cylinder compressor. This compressor can be used in direct-on-line operation with constant speed 895.3 rpm or in variable speed operation (450 – 900 rpm) supplied by a frequency converter. The mechanical drive train was generated by connecting the mock-up digital twins of all components. The number of inertias of these components was: motor 2, coupling 2, flywheel 1, and compressor 11. Viscous damping was added to the compressor cylinder locations. The structural damping induced by the flexible coupling was neglected due to missing modelling capabilities.

Table 4 shows the natural frequency and damping ratio for the lowest modes. For reference, the values are calculated also without magnetic effects. The natural frequencies and damping factors of the digital twin with triple-cage model were obtained from the eigenvalues of the system matrix. The first mode without magnetic effects is the rigid body mode. The main deformation of the second mode occurs in the flexible coupling. The third mode is an internal mode of the coupling, and in the fifth mode the flywheel and the compressor line are in the opposite phase without angular displacement of the motor.

Table 4. Natural frequency and damping ratio for five lowest modes of motor-compressor system at rated operation.

Model	Mode type									
	Rigid body		First elastic		Coup. local		Electric		Comp. local	
	f [Hz]	ζ [%]	f [Hz]	ζ [%]	f [Hz]	ζ [%]	f [Hz]	ζ [%]	f [Hz]	ζ [%]
DTwin 3 w/o EM	0.00	0.00	6.35	1.55	37.3	0.01	-	-	130.3	0.92
DTwin 3	4.18	8.81	10.16	8.00	37.3	0.02	59.80	3.67	130.3	0.92
FEM	4.19	8.59	10.06	5.63	37.3	-	-	-	-	-

Table 4 shows that the electromagnetic interaction increases clearly the natural frequency and damping ratio of the first two modes. The effect on modes 3 and 4 is negligible. This is logical due to the modal amplitudes of the modes 1 and 2 in contrast to the amplitudes of the modes 3 and 4. The prediction capability of digital twin with triple cage model is good.

Advantages of electromechanical digital twin are manifested when forced response analyses are needed. This is the case particularly with motor-driven reciprocating compressors. These compressors have large fluctuation of torque. Figure 3 shows, as an example, the coupling torque in the rated operating condition. The excitation torques are given separately for each throw using 24 lowest orders. In this case, the amplitude of coupling torque is 5.27 kNm that is 14.7 % of the average torque.

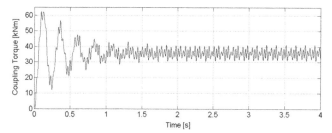

Figure 3. Coupling torque of reciprocating compressor train in rated operating condition.

The potential of reduced models comes up when large number of cases must be calculated. This is the case with variable speed applications. Figure 4 shows the vibratory torque amplitude in four cross sections of the example compressor train. An effective viscous damping value for the structural damping of the coupling was added by equating the dissipations of these damping models at the frequency 37.3 Hz. This frequency is the same as the natural frequency of the local coupling mode.

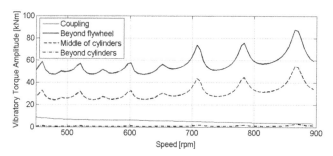

Figure 4. Vibratory torque amplitude in motor-driven compressor train.

5   DISCUSSION & FURTHER WORK

A digital twin of induction motors for torsional analysis of electrical drive trains was presented. This electromechanical digital twin represents the system dynamics between the electric power supply and shaft-end torque and speed. This digital twin consists of two parts: one non-linear system for the determination of slip and steady-state currents, and one linearized system for vibration studies.

The obtained results indicate that the accuracy can be significantly increased by using the double- or triple-cage model for the electromagnetics, and by using the resistances and inductances identified by the small-signal FE-model in steady-state condition.

The computational size of the presented digital twin is very small. However, the accuracy is relatively high due to the model parameters identified by FEM. In principle, the parameters of this digital twin can be generated automatically for actual motors. Thus, it suits well to be used with Application Programming Interface (API).

Currently, the main challenge is related to the standardisation of API protocols between component providers.

6   CONCLUSIONS

The obtained results show that the torsional vibration behaviour of an induction motor can be modelled accurately by a simple digital twin. This electromechanical digital twin represents the dynamics between the inputs, i.e. supply voltage and load torque, and outputs, i.e. supply current and shaft speed. The accuracy of this digital twin is achieved by linearization of system equations and by identification of system parameters with FEM. The calculation example of a motor-driven reciprocating-compressor train demonstrate the potential applications of this digital twin approach.

REFERENCES

[1]   Anon., "Guideline and recommended practice for control of torsional vibrations in direct-driven separable reciprocating compressors", *Gas Machinery Research Council, ACI Services*, Rev. 0, June 15, 2015.

[2]   Knop, G., "The importance of motor dynamics in reciprocating compressor drives", *Proc. 8th Conf. EFRC*, Düsseldorf, Germany, Sep 27- 282012.

[3]   Hauptmann, E., Howes, B., Eckert, B., "The influence on torsional vibration analysis of electromagnetic effects across an induction motor air gap", *Proc. of Gas Machinery Research Council*, Albuquerque, New Mexico, Oct. 6-9, 2013.

[4]   Concordia, C., "Induction motor damping and synchronizing torques", *AIEE Trans. Power Apparatus and Systems*, 71, 1952, pp. 364–366.

[5]   Jordan, H., Müller, J., Seinsch, H.O., "Über elektromagnetische und mechanische Ausgleichsvorgänge bei Drehstromantrieben", *Wiss. Ber. AEGTELEFUNKEN*, 52, 1979, pp. 263–270.

[6]   Jordan, H., Müller, J., Seinsch, H.O., "Über das Verhalten von Drehstromasynchronmotoren in drehelastischen Antrieben", *Wiss. Ber. AEGTELEFUNKEN*, 53, 1980, pp. 102–110.

[7]   Shaltout, A.A., "Analysis of torsional torques in starting of large squirrel cage induction motors", *IEEE Trans. Energy Convers.*, 9, 1994, pp. 135–142.

[8]   Leedy, A.W., "Simulink/MATLAB dynamic induction motor model for use as a teaching and research tool", *Int. Journal of Soft Computing and Engineering*, 3, 2013, pp. 102–107.

[9]   Brunelli, M., Fusi, A., Grasso, F., Pasteur, F., Ussi, A., "Torsional vibration analysis of reciprocating compressor trains driven by induction motors", *Proc. 9th Int. Conf. on Compressors and their Systems*, London, UK, Sept. 7-9, 2015.

[10] Fusi, A., Grasso, F., Sambataro, A., Baylon, A., Pugi, L., "Electro-mechanical modelling of a reciprocating compression train driven by induction motor", *Torsional Vibration Symposium 2017*, Salzburg, Austria, May 17-19, 2017.

[11] Holopainen, T.P., Repo, A.-K., Järvinen, J., "Electromechanical interaction in torsional vibrations of drive train systems including an electrical machine", *Proc. of the 8th International Conference on Rotor Dynamics*. Seoul, Korea, Sept. 12-15, 2010.

[12] Arkkio, A., Holopainen, T.P., "Space-vector models for torsional vibration of cage induction motors", *IEEE Transactions on Industry Applications*, 52, 2016, pp. 2988–2995.

[13] Holopainen, T.P., Arkkio, A., "Simple electromagnetic motor model for torsional analysis of variable speed drives with an induction motor", *Technische Mechanik*, 37, 2017, pp. 347–357.

[14] Repo, A.-K., Niemenmaa, A., Arkkio, A., "Estimating circuit models for a deep-bar induction motor using time harmonic finite element analysis", *Proc. 17th Int. Conf. on Electrical Machines*, Chania, Greece, Sept. 2-5, 2006.

12th International Conference on Vibrations in Rotating Machinery -
Institution of Mechanical Engineers, ISBN 978-0-367-67742-8

A reduced semi-analytical gas foil bearing model for transient run-up simulations

P. Zeise, M. Mahner, M. Bauer, M. Rieken, B. Schweizer

Institute of Applied Dynamics, Technical University of Darmstadt, Germany

ABSTRACT

A reduced semi-analytical model for gas foil bearings is presented, which allows time-efficient run-up simulations, while maintaining the main physical properties of the gas foil bearing. Hence, it may be used to optimise the rotor/bearing system over the complete speed range. The nonlinear force/displacement relationship of the bearing is approximated with hysteresis curves, generated in a pre-processing step by means of numerical simulations or measurements. Instability effects caused by the gas film cannot be predicted with the presented model, however, the nonlinear response of the rotor-bearing system may be predicted accurately, if the rotor speed is below the threshold speed of instability.

1   INTRODUCTION

In literature, different design variants for gas foil journal bearings have been presented and investigated. Foil bearings usually consist of top and bump foils. Alternatively, beam or wing foils may be used to support the top foil, see Ref. (1, 2); metal mesh (Ref. (3)) or spring (Ref. (4)) structures have also been analysed in literature.

The principle of operation of gas foil bearings is very similar to classical hydrodynamic oil film bearings. When the shaft starts to rotate, gas is dragged into the gap between the journal and the top foil and pressure is generated in the gas film. The gas gap is a function of the pressure field in the fluid film due to the elasticity of the foils. While the top foil provides a smooth surface for the fluid flow, the bump/beam foil yields the necessary elasticity and support against the housing. Dissipation in gas foil bearings is mainly generated by dry friction at the contact points between the foils and at the contact points with the housing. Viscous damping effects are usually of minor importance in gas foil bearings. As a consequence, the here presented model only takes dry friction into account.

A preload can be applied to bearings in various ways; for instance it may be created by defining wedges in the bearing with shims or with a lobed inner housing profile, see Refs. (5, 6). Or it may be induced by the deformation of the foil structure during the bearing assembly, see Ref. (7). The influence of preload on the static behaviour of the bearing is examined in Refs. (7, 8); the effect on the rotordynamic behaviour is discussed in Refs. (5, 6).

Furthermore, there exist single pad bearings (circular bearings) and three-pad bearings. In Ref. (9), single pad bump type bearings and preloaded three pad bump type bearings have been considered. In this work, the single pad bearing revealed a higher load capacity. On the other hand, preloaded three-pad bearings often show an enhanced rotordynamic performance at high rotor speeds (improved stability behaviour).

To make use of both advantages, several hybrid bearings have been presented, e.g. applying external pressure (see Ref. (10)) or using actuators to actively adapt the pad curvature (see Ref. (11)). The thermal behaviour of gas bearings is investigated in Ref. (12, 13) for instance. The influence of the top foil coating is investigated in Ref. (14).

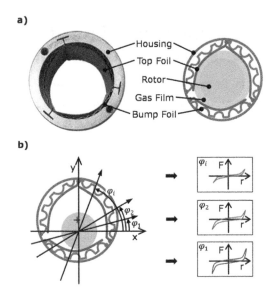

a)

Housing
Top Foil
Rotor
Gas Film
Bump Foil

b)

φ_i

φ_2

φ_1

Figure 1. A) Geometry of a three pad gas foil journal bearing b) Generation of the 2D hysteresis field.

In this manuscript, preloaded three-pad bearings with bump foils are considered, see Figure 1a). Using a preload has different advantages and disadvantages. Stiffness/ friction of the bearing is increased, if the preload is increased, which usually also increases the threshold speed for gas film induced instabilities (whirl/whip instabilities). However, the friction torque during the starting process of a rotor run-up will be increased. It should be mentioned that the friction torque in the air-borne region may not notably be affected by an increased preload, see Ref. (15). Calculation of gas foil bearings requires complex multi-physical models, see Refs. (7, 16). Such detailed quasi-static models are usually applied for optimisation of the bearing with respect to bearing load, thermal properties, etc. For transient rotordynamic simulations, very detailed thermo-elasto-hydrodynamic bearing models will entail very large simulation times or may even not be used for transient calculations. To analyse and optimise the rotor/bearing system over the complete speed range, reduced and very time-efficient models are necessary and useful.

In literature, different reduced bearing models have been developed. In the models presented in Refs. (17-19), the damping effects of the gas film are neglected and an infinite stiffness is assumed. The stiffness of the foil structure is modelled by a one-dimensional radial force function; dissipation is taken into account with a structural damping approach. In the current manuscript, precise nonlinear two-dimensional force fields are used by considering the radial and the circumferential dependency of the bearing forces. The model in this paper does not apply a structural damping approach; dissipation is modelled by a nonlinear physical friction model. It is mentioned in Ref. (18) that neglecting the fluid film stiffness may only be justified – for the bearing type analysed by the authors – if the rotor speed is high enough. It should be stressed that a different bearing design is analysed in the paper at hand. Here, foil bearings with a larger mechanical preload are considered. As has been shown in Refs. (7, 20), stiffness of the fluid film may be neglected compared to the structural stiffness of the foils in this case. Furthermore, due to the large mechanical preload, fluid film induced bearing instabilities may be suppressed in the operational speed range. Note that these positive features of highly preloaded bearing come at a price: The friction torque during the starting phase may be increased significantly if highly preloaded bearings are used, see Ref. (15).

More detailed bearing models in connection with transient simulations have been used in Refs. (21, 22). In Ref. (21), for instance, a transient bearing model has been presented, where the fluid film is taken into account. The stiffness of the foil structure and its influence on the gas gap is modelled by a linear force model; dissipation is considered with the help of a structural damping approach. Another detailed transient bearing model that takes the fluid film and an elastic fluid gap into account is presented in Ref. (22). The authors use a more detailed model for the foil structure: The foil model is based on a linear structural model, where dissipation is taken into account by a modal damping approach. Also, a comparison with test rig results is carried out.

A reduced hysteresis model for gas foil bearings has been presented in Ref. (20). Here, an extension and generalisation of this approach is discussed. The reduced gas-foil bearing model of Ref. (20) is based on a 2-dimensional look-up table approach. Therefore, hysteresis curves have to be generated in a pre-processing step either by measurements or by nonlinear finite-element simulations. To generate the hysteresis curves, the rotor shaft of a gas foil bearing is moved in radial r-direction and the radial and tangential bearing force components are determined as a function of the radial displacement. This procedure is carried out for different directions φ. As a result, two force fields in the r, φ-plane are obtained, namely a radial force field $F_r(r, \dot{r}, \varphi, \dot{\varphi}) - e_r$ and a tangential force field $F_\varphi(r, \dot{r}, \varphi, \dot{\varphi}) - e_\varphi$. Here, we consider gas foil bearings with a preload as in Ref. (7). Stiffness, friction and damping of gas foil bearings are generally determined by the structural components (top foil and bump foil) and by the gas film. For the considered bearing design, stiffness and damping properties of the gas film may be neglected in comparison to the structural components, if the preload is large enough. In this case, the fluid film may be considered as rather stiff compared with the foil structure, as outlined in Ref. (20). Moreover, if the preload is large enough, gas film induced instabilities (whirl/whip phenomena) are expected to occur at higher rotor speeds, i.e. above the nominal rotor speed. Hence, the bearing is mainly characterised by the stiffness of the top and bump foil and the friction between top and bump foil as well as between bump foil and housing. Note that friction between top foil and shaft is comparatively low due to the gas film, see Ref. (7).

The here considered bearing model only requires data from radial hysteresis curves. This reduces the effort for the preprocessing step (generation of the hysteresis splines) significantly, especially if the input data are generated by experimental hysteresis measurements. In order to enable arbitrary rotor motions during transient simulations, a special method based on a projection approach is applied. Numerical simulations with a detailed nonlinear finite-element reference model indicate that the presented reduced model, which only requires radial hysteresis curves, may also yield accurate results for arbitrary shaft motions.

Compared to the model derived in Ref. (20), the here presented extended approach contains two modifications. The first modification lies in the calculation of the tangential elastic bearing forces. In the original version of Ref. (20), the radial and the tangential elastic force fields are both directly determined by measurements/simulations. In the new extended version, only the radial elastic force component is measured/simulated, while the tangential elastic force component is calculated with the help of a force potential function. This modified force calculation has different advantages. The main benefit is that the extended approach will always yield conservative elastic forces fields. In Section 4.2, a run-up simulation is presented, which shows the advantages of the new approach. While the original force calculation based on Ref. (20) produces artificial self-excited oscillations, the new approach shows correct results without any non-physical vibration effects. A second modification lies in a refined generation of the hysteresis curves. While the model discussed in Ref. (20) is based on centred hysteresis curves, the extended approach is based on off-centred hysteresis curves, where the hysteresis curves are generated with respect to the static equilibrium position of the rotor.

2   REDUCED BEARING MODEL APPROACH

2.1   Generation of hysteresis curves

In Figure 1a), the here considered three-pad bearing is illustrated. The centre of the space-fixed x,y-system is located in the static equilibrium position of the bearing. Furthermore, a polar coordinate system is defined. The radial and tangential unit vectors are given by $\underline{e}_r = \begin{pmatrix} \cos\varphi \\ \sin\varphi \end{pmatrix}$ and $\underline{e}_\varphi = \begin{pmatrix} -\sin\varphi \\ \cos\varphi \end{pmatrix}$.

Here, φ denotes the angle between the x-axis and the displacement vector of the rotor journal. In polar coordinates, the velocity vector of the journal centre reads $\underline{v} = \dot{r}\underline{e}_r + r\dot{\varphi}\underline{e}_\varphi$. Ignoring the fluid film, which is assumed to be very stiff, the resulting bearing force can be expressed as a function of the rotor position and the rotor velocity. Representation of the bearing force in polar coordinates yields

$$\underline{F} = F_r(r,\dot{r},\varphi,\dot{\varphi})\underline{e}_r + F_\varphi(r,\dot{r},\varphi,\dot{\varphi})\underline{e}_\varphi \tag{1}$$

A key point of the presented bearing model lies in the fact that only hysteresis curves resulting from purely radial rotor displacements are required as input data. Thus, we consider a set of mere radial rotor displacements ($\varphi=\varphi_c=$ const.), which are prescribed by a harmonic oscillation with a fixed amplitude for instance, see Figure 1b). After a certain number of oscillation cycles, closed hysteresis curves for F_r and F_φ are obtained, which will be used as input data for the reduced bearing model, see Figure 2a). It should be noted that the shape of the hysteresis curves F_r and F_φ will usually depend not only on the angle φ_c, but also on the amplitude of the prescribed oscillation.

The hysteresis curves for F_r and F_φ can be separated into an upper and a lower curve, (F_r^u and F_r^l, F_φ^u and F_φ^l), see Figure 2a). The sign of $\dot{r}$ defines whether the upper or lower curve has to be used:

$$F_r^u(r,\varphi_c) = F_r(r,\dot{r}>0,\varphi,\dot{\varphi}=0), \tag{2}$$

$$F_r^l(r,\varphi_c) = F_r(r,\dot{r}<0,\varphi,\dot{\varphi}=0), \tag{3}$$

$$F_\varphi^u(r,\varphi_c) = F_\varphi(r,\dot{r}>0,\varphi,\dot{\varphi}=0), \tag{4}$$

$$F_\varphi^l(r,\varphi_c) = F_\varphi(r,\dot{r}<0,\varphi,\dot{\varphi}=0). \tag{5}$$

2.2   Implementation of hysteresis curves

For the reason of a clear representation, the here considered bearing model is explained in two steps. In the first step, the implementation of the hysteresis curves for a purely radial rotor movement (constant angle φ_c) is presented. In a second step, arbitrary rotor movements are considered.

a)

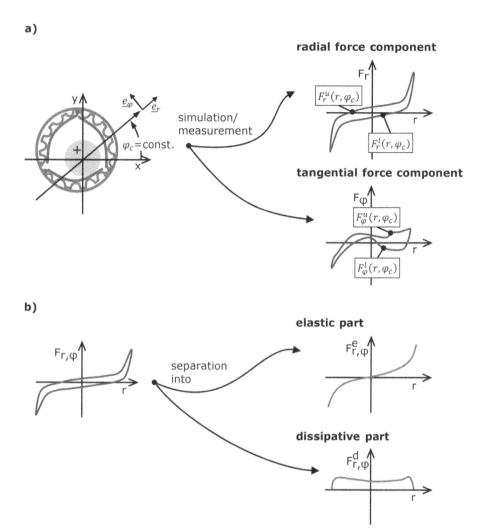

radial force component

$F_r^u(r, \varphi_c)$

$F_r^l(r, \varphi_c)$

tangential force component

$F_\varphi^u(r, \varphi_c)$

$F_\varphi^l(r, \varphi_c)$

simulation/measurement

φ_c=const.

b)

elastic part

dissipative part

$F_{r,\varphi}$

separation into

$F_{r,\varphi}^e$

$F_{r,\varphi}^d$

Figure 2. a) Generation of radial and tangential force components b) Separation into elastic and dissipative force parts.

2.2.1 Implementation for purely radial rotor motion ($r\neq const., \varphi=\varphi_c= const.$)

Specifying the rotor position, i.e. defining the radial coordinate r and the constant angle φ_c, the radial force component F_r and the tangential force component F_φ take the value of either the upper or the lower curve of the corresponding hysteresis curve (determined by the sign of $\dot{r}$). If the sign of $\dot{r}$ changes, an instantaneous jump from the upper to the lower curve (or vice versa) occurs. In the framework of our model approach, the hysteresis curves for the radial and tangential force components are separated into curves representing the centre line (F_r^e and F_φ^e) and curves defining the width of the hysteresis curves (F_r^d and F_φ^d), see Figure 2b). Physically, the centre line curves F_r^e and F_φ^e characterise the nonlinear elastic behaviour, while F_r^d and F_φ^d represent the dissipative effects. Hence, the bearing force components can be separated into the subsequent two parts

$$F_r(r, \dot{r}, \varphi_c, \dot{\varphi} = 0) = F_r^e + F_r^d, \tag{6}$$

$$F_\varphi(r, \dot{r}, \varphi_c, \dot{\varphi} = 0) = F_\varphi^e + F_\varphi^d. \tag{7}$$

The elastic and dissipative force parts are given by

$$F_r^e(r, \varphi_c) = \frac{F_r^u(r, \varphi_c) + F_r^l(r, \varphi_c)}{2}, \tag{8}$$

$$F_\varphi^e(, \varphi_c) = \frac{F_\varphi^u(r, \varphi_c) + F_\varphi^l(r, \varphi_c)}{2}, \tag{9}$$

$$F_r^d(r, \dot{r}, \varphi_c) = \frac{F_r^u(r, \varphi_c) - F_r^l(r, \varphi_c)}{2} \cdot \text{sign}(\dot{r}), \tag{10}$$

$${}_\varphi^d(\dot{r}, \varphi_c) = \frac{F_\varphi^u(r, \varphi_c) - F_\varphi^l(r, \varphi_c)}{2} \cdot \text{sign}(\dot{r}), \tag{11}$$

For a fixed angle φ_c, the elastic force parts F_r^e and F_φ^e only depend on the radial displacement r, while the dissipative force parts F_r^d and F_φ^d also depend on the sign of $\dot{r}$. To avoid numerical problems, the sign function is replaced by the polynomial

$$\text{sign}(\dot{r}) \approx \frac{1}{2\varepsilon^3} \dot{r}^3 + \frac{3}{2\varepsilon} \dot{r} \quad (-\varepsilon \leq \dot{r} \leq \varepsilon), \tag{12}$$

where ε characterises a small, user-defined constant (e.g. $\varepsilon = 10^{-8}$ m/s).

2.2.2 **Implementation for arbitrary rotor motion (r≠const.,φ≠const.)**
The here presented bearing model only uses radial hysteresis curves as input data. In order to use these data also for arbitrary motions, approximation formulas for F_r and F_φ have to be found on the basis of the force hystereses for purely radial journal motions.

2.2.2.1 ELASTIC FORCE PART
To determine the elastic force part for arbitrary rotor motions, two modifications are carried out. Firstly, the constant angle φ_c is simply replaced by a variable angle φ, which is possible since the elastic force parts only depend on the rotor position (r, φ) and not on $\dot{r}$ and $\dot{\varphi}$. Secondly, one has to ensure that the elastic force part remains conservative. A direct application of the components $F_r^e(r, \varphi)$ and $F_\varphi^e(r, \varphi)$ may generally not exactly yield (only approximatively) a conservative elastic force field.

The reason for the fact that the elastic force components $F_r^e(r, \varphi)$ and $F_\varphi^e(r, \varphi)$ will generally not represent a conservative elastic force field can mainly be traced back to the (rather intuitive) separation of the hysteresis curves into an elastic and dissipative part according to Figure 2b). In connection with hysteresis-type forces, it is very common to interpret the centre line of the hysteresis as elastic force part and the width as dissipative force part. This separation seems to be very reasonable and useful from the practical point of view, but it still remains an assumption. While this kind of force separation can be applied to one-dimensional hysteresis curves without further consideration, an application to two-dimensional hysteresis fields requires thorough attention as $F_r^e(r, \varphi)$ and $F_\varphi^e(r, \varphi)$ will generally not (exactly) represent a non-rotational force field.

If the force field is not conservative, non-physical and destabilising effects may occur. Therefore, we have to ensure that the elastic force part has a potential U_E, i.e. we have to guarantee that the following equation is fulfilled

$$-\nabla U_E = \underline{F}^e = F_r^e \underline{e}_r + F_\varphi^e \underline{e}_\varphi. \tag{13}$$

Since the components F_r^e and F_φ^e are obtained through simulation/measurement, equation (13) is usually not satisfied.

To ensure that F_r^e and F_φ^e remain conservative, the following procedure is applied. Firstly, the radial force component is calculated by $F_r^e(r,\varphi) = \frac{1}{2}\left(F_r^u(r,\varphi) + F_r^l(r,\varphi)\right)$, where the spline functions $F_r^u(r,\varphi)$ and $F_r^l(r,\varphi)$ are generated from measured or simulated data. The radial elastic force spline $F_r^e(r,\varphi)$ is analytically integrated in radial direction, which directly gives the elastic potential $U_E(r,\varphi) = -\int_0^r F_r^e(\bar{r},\varphi)d\bar{r}$. Then, the partial derivative with respect to φ is calculated, which directly yields an analytical expression for the tangential component, namely $F_\varphi^{e,con}(r,\varphi) = -\frac{1}{r}\frac{\partial U_E(r,\varphi)}{\partial \varphi}$. Hence, $F_\varphi^{e,con}(r,\varphi)$ is used instead of the measured/simulated spline $F_\varphi^e(r,\varphi)$ so that conservativity is guaranteed. In Section 4.2, a numerical example is given, which clearly shows the benefit of this procedure.

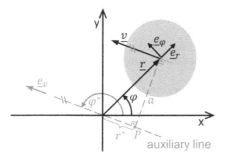

Figure 3. Geometric relations for arbitrary rotor movements.

2.2.2.2 DISSIPATIVE FORCE PART

The components of the dissipative force part F_r^d and F_φ^d are functions of the rotor position and the rotor velocity. Their definition according to Eq. (10) and Eq. (11) is only valid for purely radial motions and therefore needs to be generalised for arbitrary rotor motions. The basic idea for the approximated calculation of the dissipative force parts is shown in Figure 3. The rotor position is represented by the vector $\underline{r}$ (distance r, polar angle φ). The corresponding velocity is $\underline{v}$. Next, an auxiliary line through the origin is constructed, which orientation is determined by the current velocity vector $\underline{v}$. $\underline{e}_v$ denotes the related unit vector. φ^* represents the corresponding polar angle that can be calculated by $\varphi^* = \varphi + \arctan\left(\frac{r\dot{\varphi}}{\dot{r}}\right)$. Then, the position vector $\underline{r}$ is projected onto the auxiliary line (projection point P). The velocity vector $\underline{v}$ remains unchanged during the projection. The projected distance between P and the origin is given by $r^* = r\cos(\varphi^* - \varphi)$. The distance between the rotor and P reads $a = r\sin(\varphi^* - \varphi)$. To approximate the dissipative force part, we assume that a is sufficiently small. Based on this assumption and with the help of the unchanged velocity $\underline{v}$, one may approximately replace the current rotor position by the projection point P for calculating the dissipative force parts. Hence, the dissipative force part can be computed by using the projected position defined by r^* and φ^* and by making use of corresponding radial and tangential hysteresis curves F_r and F_φ. Then, we obtain the approximated radial and tangential force terms

$$F_{r^*}^d(r,\dot{r},\varphi,\dot{\varphi}) = \frac{F_r^u(r^*,\varphi^*) - F_r^l(r^*,\varphi^*)}{2} \tag{14}$$

$$F_{\varphi^*}^d(r,\dot{r},\varphi,\dot{\varphi}) = \frac{F_\varphi^u(r^*,\varphi^*) - F_\varphi^l(r^*,\varphi^*)}{2} \tag{15}$$

It should be noted that the sign function is no longer required, because the information of the direction of the motion is included in φ^*. Furthermore, it should be stressed that the approximated dissipative force parts according to Eqs. (14) and (15) are force components with respect to a polar coordinate system with the unit vectors $\underline{e}_{r^*} = \begin{pmatrix} \cos \varphi^* \\ \sin \varphi^* \end{pmatrix}$ and $\underline{e}_{\varphi^*} = \begin{pmatrix} -\sin \varphi^* \\ \cos \varphi^* \end{pmatrix}$.

Finally, the resultant bearing force, which consists of the elastic and the dissipative part, can then be computed by

$$\underline{F} = \underline{F}^e + \underline{F}^d \tag{16}$$

with

$$\underline{F}^e = F_r^e(r,\varphi)\underline{e}_r + F_\varphi^{e,con}(r,\varphi)\underline{e}_\varphi, \tag{17}$$

$$\underline{F}^d = F_{r^*}^d(r,\dot{r},\varphi,\dot{\varphi})\underline{e}_{r^*} + F_{\varphi^*}^d(r,\dot{r},\varphi,\dot{\varphi})\underline{e}_{\varphi^*}. \tag{18}$$

The accuracy of the approximation of the dissipative force part depends on the distance a. The smaller a, the better the approximation is.

2.2.3 Summary of the implementation

Firstly, hysteresis curves for the force components F_r and F_φ are computed for fixed angles $\varphi = \varphi_c$, i.e. $F_r(r,\dot{r},0°,0)$, $F_\varphi(r,\dot{r},0°,0)$; $F_r(r,\dot{r},15°,0)$, $F_\varphi(r,\dot{r},15°,0)$,..., $F_r(r,\dot{r},345°,0)$, $F_\varphi(r,\dot{r},345°,0)$. Secondly, the related force functions F_r^u, F_φ^u, F_r^l and F_φ^l are calculated and represented by Akima-Splines, for instance. In tangential direction, a linear interpolation approach is applied.

3   MODEL VERIFICATION

To validate the reduced bearing model of Section 2, a finite-element reference model is used (see Ref. (7)). The nonlinear FE model consists of a rigid rotor journal, the top foil and the bump foil and the rigid circular bearing housing. Since the considered bearing shows a symmetry with respect to the x,y-plane, a 2-dimensional model will be sufficient. The bump and top foil are modelled as geometrically nonlinear beams. The contact between journal and top foil is assumed to be frictionless, whereas the contacts between the bumps and the top foil are modelled as stick-slip contacts. Stick-slip contact is also assumed between bump foil and housing. The model contains approximately 9000 elements. Contact in normal direction is enforced with a penalty approach. The motion of the journal is kinematical prescribed according to

$$x_{\text{journal}} = -6\mu m \sin(-_1 t) - 4\mu m \sin(-4_1 t), \tag{19}$$

$$y_{\text{journal}} = -6\mu m \cos(-_1 t) - 4\mu m \cos(-4_1 t) - 30\mu m, \tag{20}$$

The corresponding journal orbit is shown in Figure 4a). The reaction forces F_{Rx} and F_{Ry} in x- and y-direction acting at the rotor journal are plotted in Figure 4b). The FE reference solution (ref_FEM) is compared with results obtained with the reduced bearing model. We consider the following three reduced bearing variants.

- The first reduced variant (red_var1) is the original model of Ref. (20), where centred hysteresis curves have been used.
- The second variant (red_var2) is based on non-centred hysteresis curves, where the origin of the coordinate system for the spline generation has been moved -30 µm in negative vertical direction (i.e. into the centre of the cycloid).
- The third variant (red_var3) also uses non-centred hysteresis curves; additionally, the tangential elastic forces are calculated by the potential function, see Eq. (13).

It should be mentioned that the spline data for the centred and non-centred hysteresis curves, which are required for the reduced bearing model, have been numerically generated with the FE reference model.

As can be seen, the non-centred bearing models (red_var2 and red_var3) show a good agreement with the FE reference model; the variants red_var2 and red_var3 almost show the same results. Little larger differences are observed with the centred bearing model (red_var1). It should finally be mentioned that the simulation time for the FE reference model is approx. 66 min (one cycle), while the simulation time for the reduced models is less than 1 s.

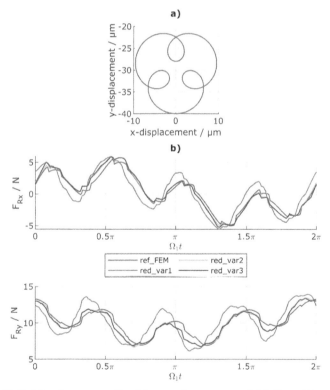

Figure 4. A) Orbit of a prescribed journal movement b) Resulting forces in x- and y- direction of the reference model and the reduced models.

4   RUN-UP SIMULATIONS

4.1   Run-up simulation 1 (heavy rotor)

For the first run-up simulation, a heavy rotor with a mass of approx. 2.3 kg is considered. The mass moments of inertia are approx. $I_x = I_y = 7200$ kg mm^2, $I_z = 330$ kg mm^2. The imbalances at the compressor and turbine side are $U_c = U_t = 2.5$ g mm, see Figure 5a). The centre of mass is located close to the compressor bearing (B_C) so that different static equilibrium positions have to be used for the left and right bearing (i.e. different force splines). Bump foil bearings with a larger preload are used, namely the heavily preloaded bearing of Ref. (7). The rotor is accelerated from 0 Hz to 1500 Hz in 20 s. Run-ups are simulated with the reduced bearing models red_var2 and red_var3. Figure 5b) shows the y-displacement of the rotor at the turbine and at the compressor bearing. Both model approaches show almost identical results. Increased amplitudes in the time period from 2 s to 6 s are observed. A FFT analysis is shown in Figure 5c) and Figure 5d). Subsynchronous oscillations with $\frac{1}{2}\Omega$, $\frac{1}{3}\Omega$ and $\frac{1}{4}\Omega$ (Ω =rotor speed) are detected resulting from the nonlinear bearing characteristic.

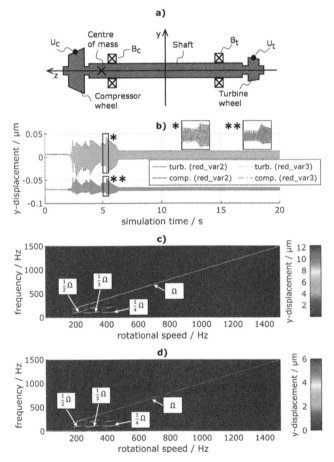

Figure 5. a) Sketch of the heavy rotor b) Rotor y-displacement at turbine and compressor bearing c) FFT at the turbine sided bearing (red_var3) d) FFT at the compressor sided bearing (red_var3).

4.2 Run-up simulation 2 (reduced rotor weight)

The rotor mass is now reduced to approx. 800 g. The mass moments of inertia are $I_x=I_y=2500$ kg mm^2, $I_z=120$ kg mm^2. The imbalances are $U_c=U_t=0.8$ g mm. The centre of mass is located in the middle between the turbine bearing and the compressor bearing, see Figure 6a). Now, bump type foil bearings with a lower preload are used, namely the lightly preloaded bearing of Ref. (7). Two reduced bearing models are compared: red_var2 and red_var3, see Section 3. The two reduced bearing models only differ in the calculation of the tangential component of the elastic force part, see Section 2. The rotor is accelerated from 0 Hz to 2000 Hz in 20 s.

Figure 6b) shows the rotor y-displacement at the compressor bearing for both reduced bearing models. The curves of both models are similar up to 13 s. Then, a sudden increase of the amplitude can be observed for the red_var2 model. The FFT analysis depicted in Figure 6c) reveals artificial oscillations with a frequency of approximately 70 Hz. These self-excited oscillations result from the nonphysical bearing model (note that conservative elastic force fields are usually not enforced with red_var2).

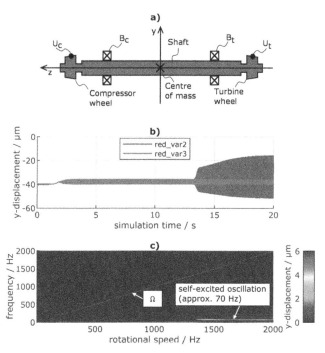

Figure 6. a) Sketch of the rotor with reduced weight b) Run-up simulations with bearing models red_var2 and red_var3 c) FFT of red_var2.

5 CONCLUSIONS

An enhanced reduced model for gas foil bearings is discussed in this paper, which allows very time-efficient run-up simulations. Although effects resulting from the gas film are neglected in this approach (especially gas film induced instabilities), the model may predict rather precisely the nonlinear stiffness and friction properties of the foil sandwich. Hence, the model may be interesting for rotordynamic studies below the threshold speed of instability. Compared to a previously presented model, the current approach contains two modifications. While the old model is based in centred hysteresis curves, the new approach uses non-centred hysteresis curves. A comparison with a detailed FE reference solution shows the improved accuracy of the non-centred hysteresis model. The second modification concerns an improved calculation of the elastic part of the bearing forces, where a special approach incorporating a force potential function is used. A numerical example clearly shows the benefit of this improvement: while the old model may predict physically unrealistic self-excited oscillations, the improved model shows correct results without artificial oscillations. The improved determination of the elastic force part may especially be interesting for the case, where measured (and not simulated) hysteresis curves are used as model input data. Measured data may contain larger data errors so that the detection of unrealistic self-excited oscillations might be more probable with the old bearing model.

REFERENCES

[1] Swanson, E.E. and P. Shawn O'Meara, *The wing foil: a novel compliant radial foil bearing design*. Journal of Engineering for Gas Turbines and Power, 2018. **140**(8).

[2] Shalash, K. and J. Schiffmann. *Comparative Evaluation of Foil Bearings With Different Compliant Structures for Improved Manufacturability*. in *ASME Turbo Expo 2017: Turbomachinery Technical Conference and Exposition*. 2017. American Society of Mechanical Engineers Digital Collection.

[3] San Andrés, L. and T.A. Chirathadam, *Performance characteristics of metal mesh foil bearings: predictions versus measurements*. Journal of Engineering for Gas Turbines and Power, 2013. **135**(12): p. 122503.

[4] Feng, K., et al., *Structural characterization of a novel gas foil bearing with nested compression springs: analytical modeling and experimental measurement*. Journal of Engineering for Gas Turbines and Power, 2016. **138**(1): p. 012504.

[5] Sim, K., et al., *Effects of mechanical preloads on the rotordynamic performance of a rotor supported on three-pad gas foil journal bearings*. Journal of Engineering for Gas Turbines and Power, 2014. **136**(12): p. 122503.

[6] Kim, T.H. and L.S. Andres, *Effects of a mechanical preload on the dynamic force response of gas foil bearings: measurements and model predictions*. Tribology Transactions, 2009. **52**(4): p. 569–580.

[7] Mahner, M., et al., *Numerical and Experimental Investigations on Preload Effects in Air Foil Journal Bearings*. Journal of Engineering for Gas Turbines and Power, 2018. **140**(3): p. 032505.

[8] Radil, K., S. Howard, and B. Dykas, *The role of radial clearance on the performance of foil air bearings*. Tribology transactions, 2002. **45**(4): p. 485–490.

[9] Kim, D., *Parametric studies on static and dynamic performance of air foil bearings with different top foil geometries and bump stiffness distributions*. Journal of tribology, 2007. **129**(2): p. 354–364.

[10] Kim, D. and D. Lee, *Design of three-pad hybrid air foil bearing and experimental investigation on static performance at zero running speed*. Journal of engineering for gas turbines and power, 2010. **132**(12).

[11] Sadri, H., H. Schlums, and M. Sinapius, *Investigation of Structural Conformity in a Three-Pad Adaptive Air Foil Bearing With Regard to Active Control of Radial Clearance*. Journal of Tribology, 2019. **141**(8).

[12] Bruckner, R.J. and B.J. Puleo. *Compliant Foil Journal Bearing Performance at Alternate Pressures and Temperatures*. in *ASME Turbo Expo 2008: Power for Land, Sea, and Air*. 2008. American Society of Mechanical Engineers.

[13] Lehn, A., M. Mahner, and B. Schweizer, *A Contribution to the Thermal Modeling of Bump Type Air Foil Bearings: Analysis of the Thermal Resistance of Bump Foils*. Journal of Tribology, 2017. **139**(6).

[14] San Andrés, L. and W. Jung. *Evaluation of Coated Top Foil Bearings: Dry Friction, Drag Torque, and Dynamic Force Coefficients*. in *ASME Turbo Expo 2018: Turbo-machinery Technical Conference and Exposition*. 2018. American Society of Mechanical Engineers.

[15] Mahner, M., et al., *An experimental investigation on the influence of an assembly preload on the hysteresis, the drag torque, the lift-off speed and the thermal behavior of three-pad air foil journal bearings*. Tribology International, 2019. **137**: p. 113–126.

[16] Li, C., J. Du, and Y. Yao, *Study of load carrying mechanism of a novel three-pad gas foil bearing with multiple sliding beams*. Mechanical Systems and Signal Processing, 2020. **135**: p. 106372.

[17] San Andres, L. and T.H. Kim, *Forced nonlinear response of gas foil bearing supported rotors*. Tribology International, 2008. **41**(8): p. 704–715.

[18] Balducchi, F., M. Arghir, and R. Gauthier, *Experimental analysis of the unbalance response of rigid rotors supported on aerodynamic foil bearings*. Journal of Vibration and Acoustics, 2015. **137**(6).

[19] Guo, Z., et al., *Measurement and prediction of nonlinear dynamics of a gas foil bearing supported rigid rotor system*. Measurement, 2018. **121**: p. 205–217.

[20] Zeise, P., et al. *A Reduced Model for Air Foil Journal Bearings for Time-Efficient Run-Up Simulations*. in *Turbo Expo: Power for Land, Sea, and Air*. 2019. American Society of Mechanical Engineers.

[21] Larsen, J.S. and I.F. Santos, *On the nonlinear steady-state response of rigid rotors supported by air foil bearings—Theory and experiments*. Journal of Sound and Vibration, 2015. **346**: p. 284–297.

[22] Bonello, P. and M.B. Hassan, *An experimental and theoretical analysis of a foil-air bearing rotor system*. Journal of Sound and Vibration, 2018. **413**: p. 395–420.

Stable turbocharger bearings

M.S. Ibrahim, A.S. Dimitri, H.N. Bayoumi, A. El-Shafei

Faculty of Engineering, Cairo University, Egypt

ABSTRACT

Turbochargers are increasingly used with internal combustion engines to increase power output, reduce fuel consumption and reduce emissions. These turbochargers are compact and run at speeds anywhere between 140,000 rpm to 220,000 rpm. This paper introduces the concepts of stable low-cost bearings for use with turbochargers. Based on years of analysis and testing, it is clear that misaligned bearings provide superior stability characteristics. Internally Skewed Bearings are introduced to be used in high speed applications. It is shown numerically that these bearings have superior stability characteristics and are suitable for turbocharger bearings because of their low manufacturing costs.

1   INTRODUCTION

Automotive Turbochargers usually use the Floating Ring bearing due to its ability to work under a wide range of rotation speeds, in addition to its low production cost. However, Floating Ring bearings like other circular hydrodynamic bearings show self-excited vibrations which directly appear in the form of very high noise in automotive turbochargers [1]. This noise by time became undesirable because automotive manufacturers are seeking low vibration, no noise turbochargers.

Instabilities in high speed flexible rotors, like turbochargers, have been described and elaborated in the literature, but few options are available for reducing these instabilities. Most of these options are extremely high in cost to be manufactured [1]. On the other hand, it was reported in the literature that the angular misalignment between shaft and bearing axis considerably improves the onset of instability leading to a stable bearing operation. We intend to exploit this feature in this paper.

Newkirk and Lewis [2] observed experimentally cases in which the instability occurred at six times the critical speed. Kato et al. [3] demonstrated experimentally the effect of offset misalignment to improve the journal bearing stability. In their experiments, El-Shafei et al. [4] applied an angular misalignment to a rotor system coupling, and it was demonstrated that the onset of oil whip instability was significantly improved. El-Shafei [5] [6] speculated that this angular misalignment causes oil flow disturbances in the axial direction, and presented a new class of journal bearings where the bearing axis is intentionally skewed to improve bearing stability. Moreover, El-Shafei et.al. [7] were able to demonstrate experimentally the success of skewed journal bearings in significantly improving the oil whip instability. Muszunska [8] interpreted that loading the bearings improves the rotor bearing system stability by affecting the bearing stiffness.

As investigated by Yu and Adams [9], they referred to the need of 16 damping and 16 stiffness coefficients in order to study misaligned journal bearings. Nikolakopoulos and Papadopoulos [10] calculated linear and nonlinear stiffness and damping coefficients for the misaligned journal bearing using an analytical developed model. They emphasised that the calculated linear [11] and non-linear [12] 32 coefficients are important in evaluating stability. Ahmed and El-Shafei [13] used the finite difference method in order to study the effect of misalignment in journal bearings, showing that shaft misalignment affects the values of force and moment stiffness and damping coefficients. Saqr et al. [14] developed a computer code to solve for the coefficients of the internally skewed bearing and many other bearings using the finite element method and applied this bearing type to a rigid rotor achieving improved stability.

Eling et al. [15] described a 60,000 rpm test rig to investigate stability of turbocharger bearings. Moreover, Eling et al. [16] described another 240,000 rpm test rig for the same investigation at high speeds. We at Cairo University have built similar test rigs to investigate several of the bearing concepts introduced by El-Shafei [5] through funding from the Science and Technology Fund (STDF) in Egypt. The test rigs have been built, and the testing on the 60,000 rpm test rig is on-going, and confirms the analysis. This paper presents the numerical results for using the internally skewed plain journal bearing on the 240,000 rpm test rig. The results illustrate the elimination of oil whip throughout the speed range up to 240,000 rpm with only a 0.25° skewness in the vertical direction.

2   INTERNALLY SKEWED BEARING MODEL

Hydrodynamic pressure distribution in internally skewed bearing is computed by solving Reynolds' equation assuming a Newtonian iso-viscous incompressible fluid in laminar flow. Because such a complex geometry is difficult to be analytically solved, a numerical finite element method [14] is used to solve this equation. Hydrodynamic pressure is changing along the bearing axis, because of the non-uniformity of the bearing clearance such that it starts with a certain value and continues to increase or decrease according to the sign of the skewness angle.

2.1   Mathematical model

The bearing geometry is described in Cartesian co-ordinate system as depicted in Figure 1, where, x is the direction of rotor weight in most of applications. An angular skewness of the bearing can be either horizontally or vertically skewed as illustrated in Figure 2. Thus, 'Ψ_x' and 'Ψ_y' are bearing angles of skewness around y and x axes, respectively. Assuming rotation around bearing mid plane ($z - L/2$), where L is the axial length of the pad. These angles can be positive or negative values.

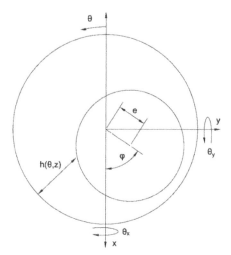

Figure 1. Schematic representation of the bearing nomenclature.

589

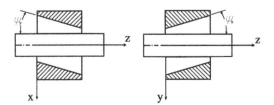

Figure 2. Schematic representation of skewness angles.

Assuming that, bearing radial clearance 'c', 'e' is the journal eccentricity and 'φ' is the bearing attitude angle. Therefore, the film thickness 'h' can be computed from:

$$h_0 = c + e\cos(\theta - \varphi) + z(\psi_y \cos\theta - \theta\psi_x \sin\theta)$$

or

$$h_0 = c + (e_x + z\theta_y)\cos\theta + (e_y - z\theta_x)\sin\theta$$

2.2 Pressure distribution and boundary conditions

Upon perturbation due to small amplitude whirling around the static equilibrium position, film thickness and pressure will be the superposition of static and dynamic terms, a first-order Taylor expansion of which yields the following linear approximation[17]:

$$h = h_0 + (\Delta x + z\Delta\theta_y)\cos\theta + (\Delta y - z\Delta\theta_x)\sin\theta$$

$$p = p_0 + \sum_{k=1}^{8} p_{x_k}\Delta x_k$$

Where, $p_{x_k} = \frac{\partial p}{\partial x_k}$ and $x_k = x, y, \theta_x, \theta_y$

Boundary conditions must be imposed in order to solve perturbed equations, there are two types of boundary conditions

(a) Geometry boundary condition in which the pressure at the periphery of the bearing is atmospheric:

$$p|_{\theta=0,2} = p|_{z=0,L} = 0$$

(b) Internal boundary condition caused by cavitation where fluid film ruptures and reforms. Reynolds boundary condition is applied, in which pressure and its gradient are set to zero maintaining continuity of flow. Reynolds boundary condition is written as:

$$p|_{\theta=\theta^*} = \frac{\partial p}{\partial\theta}\Big|_{\theta=\theta^*} = 0$$

Where, θ is the onset of cavitation in the circumferential direction. Moreover, film reformation is assumed to be at the location of maximum film thickness, i.e. $\theta = \Psi$, and pressure is set to be atmospheric:

$$p|_{\theta=\varphi} = 0$$

2.3 Resulting hydrodynamic forces and moments

When hydrodynamic pressures are calculated, nonlinear hydrodynamic forces and moments {F} are obtained from the summation of pressure integrated over the bearing area 'A' bounded by the leading and trailing edges of the pressurized film. They can be linearized for small whirling amplitudes.

$$\{F\} = \iint_A \left(p_0 + \sum_{k=1}^{8} p_{x_k} \Delta x_k \right) \begin{Bmatrix} -\cos\theta \\ -\sin\theta \\ z\sin\theta \\ -z\cos\theta \end{Bmatrix} R \, d\theta \, dz$$

Integration of static pressure yields static forces and moments {F₀} which balance externally applied loads, whereas integration of perturbed pressures results in perturbed forces and moments with respect to small displacements and velocities [18] [19]. The latter by definition yields stiffness and damping coefficients of the lubricant film necessary to determine the dynamic behaviour of a rotating journal around equilibrium considering the effect of skewness. They comprise 8 force-translation coefficients, 8 moment-rotation, 8 moment-translation coefficients and 8 force-rotation coefficients.

$$\begin{Bmatrix} F_x \\ F_y \\ M_{\theta_x} \\ M_{\theta_y} \end{Bmatrix} = \begin{Bmatrix} F_{x_0} \\ F_{y_0} \\ M_{\theta_{x0}} \\ M_{\theta_{y0}} \end{Bmatrix} + [K] \begin{Bmatrix} \lambda x \\ \lambda y \\ \lambda\theta_x \\ \lambda\theta_y \end{Bmatrix} + [C] \begin{Bmatrix} \lambda\dot{x} \\ \lambda\dot{y} \\ \lambda\dot{\theta}_x \\ \lambda\dot{\theta}_y \end{Bmatrix}$$

Where,

$$\{F_0\} = \iint_A p_0 \begin{Bmatrix} -\cos\theta \\ -\sin\theta \\ \left(z - \frac{L}{2}\right)\sin\theta \\ -\left(z - \frac{L}{2}\right)\cos\theta \end{Bmatrix} R \, d\theta \, dz = \begin{Bmatrix} F_{x_0} \\ F_{y_0} \\ M_{\theta_{x0}} \\ M_{\theta_{y0}} \end{Bmatrix}$$

3 ROTOR BEARING MODEL

3.1 Flexible rotor parameters

The rotor is modeled as a Timoshenko beam, including rotary inertia nad shear deformation [20]. Gyroscopic effects are included [14]. The bearings are solved using the bearings code described in section 2 above. Eling [16] created a numerical model of the rotor-bearing system depicted in Figure 3, using floating ring bearings as rotor supports. This flexible rotor is designed to run at speeds up to 240,000 rpm. In this study, the same rotor is modelled in plain journal bearings and internally skewed bearings with a 0.25 degrees skewness angle to discuss the change in dynamic coefficients after the introduction of the bearing skewness.

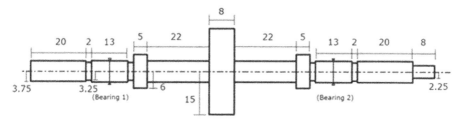

Figure 3. Schematic layout of the flexible rotor (all dimensions in mm).

Table 1. Rotor material parameters.

Material	Density (kg/m^3)	Elastic Modulus (GPa)	Shear Modulus (GPa)
Steel	7850	210	80

3.2   Bearing parameters

The rotor composed of a main disk located at the centre distance between both bearings. Both bearings are lubricated with a hydraulic oil of SAE 46 grade, this oil is assumed to be with a constant viscosity parameter estimated from an average operating temperature, all parameters in both bearings are almost the same except for the loading parameter as shown in Table 2.

Table 2. Bearing's parameters.

Bearing	Load (N)	Radial Clearance (μm)	Diameter (mm)	Length (mm)
Left bearing (non-drive end)	0.4994	20	7.5	4
Right bearing (drive end)	0.4817	20	7.5	4

The load, as shown in Table 2, is slightly different in both bearings, however we can approximate that both bearings are almost the same, so we can only model one bearing. The bearing clearance influence was studied to determine the value and configuration which achieves the best stability response and possibility of being manufactured [14]. The optimum radial clearance is found to be 20 μm. Internal skewness angle significantly affects the instability onset speed of the rotor system. Not only the angle value but also the configuration of this angle, whether positive or negative, were investigated. Tremendous iterations and combinations are conducted to select the angle of skewness, in addition to the configuration of both bearings. In this study, a 0.25° vertical angle of skewness shows the best stability response when it is modelled.

4   RESULTS

The Campbell diagram depicted in Figure 4, shows that there are two main critical speeds: one at nearly 78,000 rpm and the other at 218,000 rpm. The Campbell diagrams for both the plain journal bearings model and that with internally skewed bearings are observed to be the same. The mode shapes are shown in Figure 5.

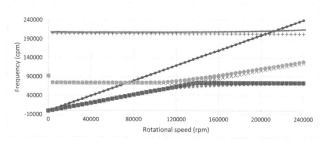

Figure 4. Flexible rotor Campbell diagram.

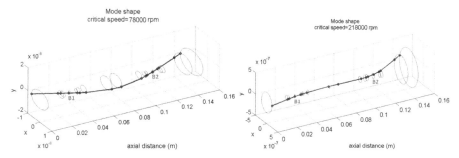

Figure 5. Flexible rotor Mode shapes.

In order to clearly check the stability response of the rotor model, the journal bearing is modelled in three models: (i) linearized short bearing model, (ii) non-linear short bearing model and (iii) numerical finite element model of the journal bearing.

Using the first model, which is the **linearized** plain short journal bearing model, the stability map for that flexible rotor model is depicted in Figure 6. It is obvious that there are two early unstable whirl modes, which can be also seen in the Campbell diagram below the 1X line as depicted in Figure 4. These modes correspond to the oil whirl modes due to journal bearing oil film flow. In fact, oil whirl modes are slightly observed in the rotor vibration, their vibration amplitudes depend mainly on the amount of rotor unbalance. The onset of these whirl modes are much earlier than predicted by the second model, the **non-linear** short journal bearing model. The waterfalls depicted in Figure 7 show that the actual oil whirl onset is nearly at 100K rpm which is higher than that 20K rpm whirling mode predicted by the linearized short bearing model, possibly due to lack of excitation.

The linearized short journal bearing model and the non-linear model predicted nearly the same critical speed which is 78K rpm, however there is a slight difference in oil whip prediction; the linearized model predicted that the oil whip onset is to be at 183K rpm (Figure 6) while the non-linear model shows that at its all four analysing positions (drive (DE) and non-drive end (NDE)), the oil whip instability onset is at 175K rpm (Figure 7).

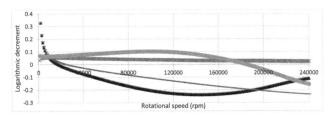

Figure 6. Flexible rotor in linearized short journal bearings stability map.

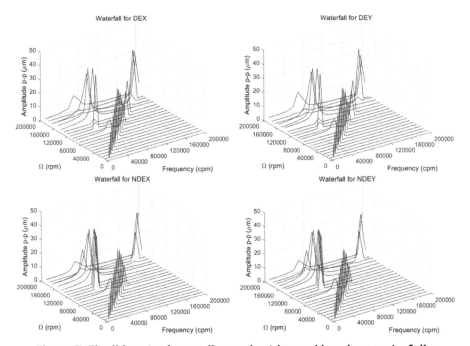

Figure 7. Flexible rotor in non-linear short journal bearings waterfalls.

Using the third bearing model analysis, a finite element model for plain journal bearings, and by neglecting the oil whirl modes, it becomes clear that the fourth forward mode goes unstable at a speed around 183K rpm as depicted in Figure 8. This is the oil whip instability at which the rotor vibration matches its critical speed causing severe vibration which has been frequently observed [4].

594

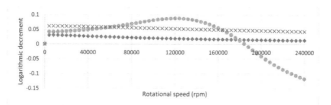

Figure 8. Flexible rotor in finite element journal bearings model stability map – showing only oil whip instability.

The stability map for the flexible rotor with bearing 0.25 degrees vertically skewed bearings is shown in Figure 9. It is shown that the oil whip severe instability is totally vanished all over the running speed range. The bearing skewness successfully eliminated the bearing oil whip undesired instability. Here also the oil whirl modes are removed from the plot.

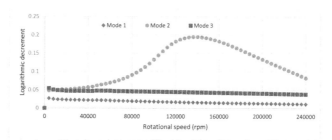

Figure 9. Flexible rotor in finite element vertically skewed journal bearings model stability map – showing only oil whip instability- at 40 °C.

Applying a vertical skew to a flexible rotor shows that skewness has extremely significant effect on the dynamic coefficients. Most importantly, direct force-translation stiffness and damping coefficients are severely affected by increasing the amount of skewness [13] [14]. In addition, skewness gives rise to force-rotation and moment-translation coefficients whose values are comparable to force-translation coefficients which in turn alters the behaviour of a rotor-bearing system. Moment-rotation coefficients were less significant [13] [14].

This bearing skewness is also showing promising response at extremely high temperatures when investigated. According to ISO 3104, The 46 grade oil viscosity at 100 °C is 6.8 mm^2/s then modelling the vertical skewed journal bearing at this very low viscosity value, Figure 10 shows a slight effect of the low viscosity on the instability onset which is relatively reduced to 232,000 rpm. However, a considerable increase in stable range remains in function.

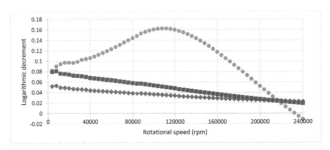

Figure 10. Flexible rotor in finite element vertically skewed journal bearings model stability map – showing only oil whip instability- at 100 oC.

The same rotor is modelled in floating ring bearings, at 40°C, in order to check the superiority of the new design of the vertically skewed journal bearing to be used in high speed turbochargers. The stability map depicted in Figure 11 shows that the onset instability speed of the rotor in the floating ring bearings will be 208000 rpm.

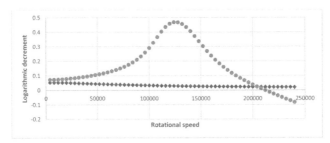

Figure 11. Flexible rotor in finite element floating ring bearings model stability map – showing only oil whip instability.

5   CONCLUSION

Flexible rotor stability supported on skewed journal bearings is investigated in this paper. The skewed journal bearing works effectively in high speed applications eliminating the oil whip instability. It can replace floating ring bearings in turbochargers. This is achieved because the direct force-translation stiffness and damping coefficients are severely affected by increasing the amount of skewness. Bearing angle configuration is determined throughout simulations so there is an optimum configuration achieving better stability response, In addition to its low cost of manufacture. Therefore, in applications at high speeds, where skewed bearings demonstrate superior dynamic performance, neither load carrying capacity nor power losses shall hinder the introduction of skewed bearings as a potential solution to instability problems.

Finally, test rigs are re-designed to be capable with testing more than one bearing type and are manufactured, so then it will be more interesting to compare the experimental results at these high speeds with the numerical ones presented in that paper.

REFERENCES

[1] H. Nguyen-Schäfer, *Rotordynamics of Automotive Turbochargers*, vol. 9783642275. Berlin, Heidelberg: Springer Berlin Heidelberg, 2012.

[2] Newkirk, B.L., and Lewis, J.F., 1956, "Oil Film Whirl-An Investigation on Disturbances Due to Oil Films in Journal Bearings", Trans. ASME, 78, pp. 21–27.

[3] Kato, T., Matsuoka, H., and Hori, Y., 1993, "Seismic Response of Linearly Stable, Misaligned Multirotor System", Tribol. Trans., 36_2, pp. 311–315.

[4] A. El-Shafei, S. H. Tawfick, M. S. Raafat, and G. M. Aziz, "Some experiments on oil whirl and oil whip," *J. Eng. Gas Turbines Power*, vol. 129, no. 1, pp. 144–153, 2007.

[5] A. El-Shafei, "Methods of Controlling the Instability in Fluid Film Bearngs," US Patent 7,836,601, 2010.

[6] A. El-Shafei, "Developments in Fluid Film Bearing Technology", IUTAM Symposium on Emerging Trends in Rotor Dynamics, pp. 201–215, Springer Netherlands, 2011.

[7] A. El-Shafei, A., S.H. Tawfick, and M.O.A.Mokhtar, "Experimental Investigation of the Effect of Angular Misalignment on the Instability of Plain Journal Bearings", Proceedings ASME/STLE 2009 International Joint Tribology Conference, 171–173, Memphis, TN, October, paper IJTC2009-15057, 2009.

[8] Muszunska, A., and Bently, D. E., "Fluid Generated Instabilities of Rotors", ORBIT, 10_1, pp. 6–14, 1989.

[9] H. Yu and M. L. Adams, "The linear model for rotor-dynamic properties of journal bearings and seals with combined radial and misalignment motions," J. Sound Vib., vol. 131, no. 3, pp. 367–378, Jun. 1989.

[10] P. G. Nikolakopoulos and C. A. Papadopoulos, "Non-linearities in misaligned journal bearings," Tribol. Int., vol. 27, no. 4, pp. 243–257, 1994.

[11] P. G. Nikolakopoulos and C. A. Papadopoulos, "Dynamic Stability of Linear Misaligned Journal Bearings Via Lyapunov's Direct Method," *Tribol. Trans.*, vol. 40, no. 1, pp. 138–146, Jan. 1997.

[12] P. G. Nikolakopoulos and C. A. Papadopoulos, "Lyapunov's Stability of Nonlinear Misaligned Journal Bearings," *J. Eng. Gas Turbines Power*, vol. 117, no. 3, pp. 576–581, Jul. 1995.

[13] A. M. Ahmed and A. El-Shafei, "Effect of Misalignment on the Characteristics of Journal Bearings," *J. Eng. Gas Turbines Power*, vol. 130, no. 4, pp. 1–8, Jul. 2008.

[14] T. Saqr, H. Bayoumi, and A. El-Shafei, "Stability Characteristics of Internally Skewed Plain Journal Bearings", to appear, 2020.

[15] R. Eling, M. Wierik, and R. A. J. van Ostayen, "Multiphysical modeling comprehensiveness to model a high speed Laval rotor on journal bearings," 14th EDF - Pprime Work. Influ. Des. Mater. J. thrust Bear. Perform., pp. 1–14, 2015.

[16] R. Eling, M. te Wierik, R. van Ostayen, and D. Rixen, "Rotordynamic and friction loss measurements on a high speed laval rotor supported by floating ring bearings," *Lubricants*, vol. 5, no. 1, pp. 1–14, 2017.

[17] J. W. Lund and K. K. Thomsen, "Calculation Method and Data for the Dynamic Coefficients of Oil-Lubricated Journal Bearings.," pp. 1–28, 1978.

[18] El-Shafei, A., "Modeling Fluid Inertia Forces of Short Journal Bearings for Rotordynamic Applications" ASME J. Vib. Acoust, Vol. 117, No. 4, pp. 462–469, 1995.

[19] El-Shafei, A., "Insights Into the Static and Dynamic Characteristics of Journal Bearings," Proceedings of the Fourth IFToMM International Conference on Rotordynamics, pp. 307–315, 1994.

[20] H. D. Nelson, "Finite Rotating Shaft Element Using Timoshenko Beam Theory.," *Am. Soc. Mech. Eng.*, vol. 102, no. 79-WA/DE-5, 1979.

12th International Conference on Vibrations in Rotating Machinery -
Institution of Mechanical Engineers, ISBN 978-0-367-67742-8

AMrotor - A MATLAB® toolbox for the simulation of rotating machinery

J. Maierhofer, M. Kreutz, T. Mulser, T. Thümmel, D.J. Rixen

Technical University of Munich, Chair of Applied Mechanics, Germany

ABSTRACT

During the last years, a MATLAB® toolbox for simulation of rotating machines has been developed at the Chair of Applied Mechanics of the Technical University of Munich. For the geometry of a rotor, a 2D silhouette can be given by a simple point description. Then, a mesh of beam elements (e.g 1D Timoshenko) is created which are assembled into a MCK-model using the finite element method. In the next step, different components like bearings, external forces and loads (e.g. unbalance) are added to the system. It is also possible to add time-variant loads and nonlinear or even active components, e.g. magnetic bearings. Different types of analysis can then be performed, like modal analysis, Campbell diagrams or time integration. The whole toolbox is programmed in an object oriented way. The code is meant to be a research code which focuses more on easy architecture than on execution performance. The goal is to enable easy implementation of new components with own methods and testing them.

This contribution will give a short overview of the structure behind the toolbox and its capabilities will be highlighted. Using a real application of a test rig with active magnetic bearings, the code is demonstrated and shown to give good results. For industry, the toolbox can be a very useful tool to make small simulations for estimating properties of simple rotating machines.

1   INTRODUCTION

The value of a model has always been measured by its fidelity in relation to reality. A good model is able to predict a certain behavior under the given excitation as correctly as possible. The model should be as simple as possible, but as complex as necessary. In the field of rotor dynamics, there have already been many development steps from the simplest rigid body models to the current standard, the finite element approach. To make the models usable, software is an indispensable tool. In general a distinction has to be made between finite element analysis (FEA) software that is specific for rotor dynamics and software that is a general purpose FEA software having modules for rotor dynamic problems. A further criterion is the costs and availability of the software package.

1.1   Available software for rotor dynamics

Commonly used rotor dynamics software modules come with *Ansys*, *Nastran* or *Comsol*. The advantage of these software tools is their capability of modeling a rotor with 3D elements that allow complex geometries. With the provided user interface, performing a simulation is quite straight forward. On the other hand, general codes come with the drawback that they are often not specific enough for rotor dynamics and therefore unhandy to use. The major drawback is that the code is closed and user-specific elements are hard or impossible to implement.

Examples for commercial, rotor dynamics specific software are *Dyrobes* or *XLRotor*. There are commercial software frameworks specific for rotor dynamics that are based on the MATLAB® ecosystem, like DYNROT and MADYN2000. This very powerful software allows all kinds of standard analysis types over a graphical user interface. Those software are intended for use by industry. The drawbacks are the high costs and that the code is closed.

Looking for open source software there is a vacuum of rotor dynamics software. Friswell et al. offer a set of scripts written in MATLAB® to accompany the book [1] This scripts are handy for learning and teaching but they are not suited for larger projects. The open source script set *VibronRotor* is based on the Octave platform and offers a good basis for linear system analyses but lacks the functionality of time integration. In the authors view, the software structure is not suitable for more extensive experimental implementations. A very new open source software code package is *ROSS* based on the python ecosystem. The ROSS development is supported by Petrobras, Universidade Federal do Rio de Janeiro (UFRJ) and Agência Nacional de Petróleo, Gás Natural e Biocombustíveis (ANP). The framework is fully object oriented and offers modern plotting options. The user interaction is fully handled with python scripts. The first public available commit is from Dec 2018. At the moment this code lacks the possibility of active components like active magnetic bearings. [10] The authors see a big future potential in this framework.

1.2 Introduction to AMrotor

The toolbox developed at the Chair of Applied Mechanics at the Technical University of Munich is called **AMrotor**. AMrotor is a FEA software that is highly specialized for the requirements of rotor system simulations and is based on the MATLAB® platform. As of state of today, the code does not use any MATLAB® Toolboxes which could allow it to run on Octave in near future. The code will be available under an open source license and is designed to be as flexible as possible in terms of research and teaching. Therefore, the code is not trimmed for speed nor has a graphical user interface. All simulations are performed using script files. Simple structures and clarity have the highest priority to have low barriers for beginners to implement their concepts in a fast way. The code is construced along the object oriented programming paradigm. The code is regularly verified and validated for simple systems, but it is not certified and should thus not be used as a certified code for designing industrial or critical systems.

2 MATHEMATICAL DESCRIPTION

The following section gives an overview of the used mathematical framework for the simulation of rotor systems with the toolbox AMrotor. The physics of rotor dynamics is comprehensively treated by Gasch [2], Genta [3] and Tıwarı [11] to name just a few.

2.1 Definition of coordinate systems

At first, the coordinate system and directions have to be defined. In section 2.1 a Laval (Jeffcott) rotor with the associated frame of reference is shown. The coordinate system is assumed to be inertial. The z-axis is orientated in the rotor axis and the positive direction of rotation is around the positive z-axis. The $x-y$ plane is located in the plane of the non-tilted disc.

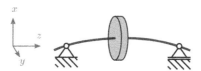

Figure 1. Definition of the coordinate system.

2.2 Rotor FEM - Timoshenko beams

AMrotor uses Timoshenko beam elements to discretize the shaft. Compared to Euler-Bernoulli beam elements, they include shear stresses and thus have less rigidifying effects in case the beam is not very thin. In addition to shear deformations, Timoshenko beams include rotational inertia terms. In Figure 2 a single beam element is shown with its degrees of freedom (dof) on 2 nodes.

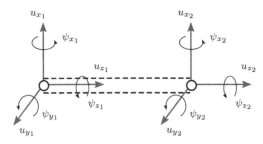

Figure 2. Timoshenko beam element with marked dofs.

The displacements and rotations of a node n is described by q_n. The dofs of all nodes of the system are concatenated in a global dof-vector q_{global}:

$$q_n = \begin{pmatrix} u_x & u_y & u_z & \psi_x & \psi_y & \psi_z \end{pmatrix}^T \qquad q_{global} = \begin{pmatrix} q_1 \\ q_2 \\ \dots \\ q_n \end{pmatrix} \qquad (1)$$

Assuming the z-axis is located on the neutral axis and that the neutral axis passes through the shear center of the cross sections, the dofs can be divided into axial, torsional and bending uncoupled subsets with local matrices which are then assembled into the global matrix with so called localisation matrices.

The axial and torsional matrices are written as axial mass M_A and stiffness K_A, where ρ is the material density, A is the cross-section area of the beam (i.e. rotor), E is the Young's modulus and l is the length of the element. The torsional behavior is obtained similarly with I_p the polar moment of inertia of a circular cross section and G the shear modulus. [11]

$$M_A = \frac{\rho A l}{6} \begin{pmatrix} 2 & 1 \\ 1 & 2 \end{pmatrix} \qquad K_A = \frac{EA}{l} \begin{pmatrix} 1 & -1 \\ -1 & 1 \end{pmatrix} \qquad (2)$$

$$M_T = \frac{\rho I_p l}{6} \begin{pmatrix} 2 & 1 \\ 1 & 2 \end{pmatrix} \qquad K_A = \frac{GI_p}{l} \begin{pmatrix} 1 & -1 \\ -1 & 1 \end{pmatrix} \qquad (3)$$

The bending behavior of the beam element is described by the flexural mass M_B, stiffness K_B and finally the gyroscopic matrix G_B. For their formulation, two additional parameters are necessary. The shear correction factor and the ratio between the shear and flexural behavior Ψ.

The full equations for the matrices would go beyond the scope of this article and they can be read in [11] or other literature. Instead we want to show a schematic matrix that shows the element matrices and how they are populated.

In order for the simulation to reflect reality well, damping should be included. The code offers a classical Rayleigh-Damping approach as often used in engineering practice.

$$C = a_1 * K + a_2 * M \qquad (4)$$

The damping is computed from the stiffness- and mass matrix of the finite element beam model and added before any further components are added to the system matrices.

symmetric (elementwise) 12 x 12 matrix

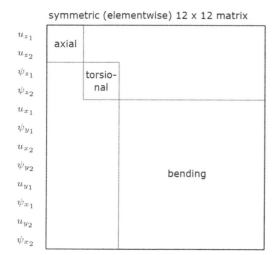

Entry in each component matrix of equation:

$$M\ddot{q} + \Omega G\dot{q} + Kq = f$$

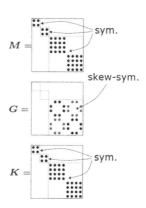

Figure 3. Population scheme for the element matrix.

All in all, this results in the second order differential equation eq. (5) with the gyroscopic and damping matrices combined in the matrix D.

$$M\ddot{q} + \underbrace{(G + C)}_{D} Kq = h \qquad (5)$$

2.3 Rotor components

Different components are implemented either with linear coefficients that are added to the system matrices via a localisation matrix or with a nonlinear force law that is evaluated for every time step. The local matrices $M_{comp}, C_{comp}, K_{comp}$ are 6x6 matrices consistent with the ordering of . Three examples are used here to show the capability of the framework.

Disc Significant jumps in the diameter of the cross-section (as occurring adisc is attached to the shaft) are modeled as ainfinite thin disc component ($K_{Disc} = 0, M_{Disc}, G_{Disc}$) attached to one node of the shaft. Modeling them as large diameter beam elements would lead to a significant increase of stiffness in that local area which is not true as in reality the stress flow is nearly unperturbed by the thin disc. This results in a nearly constant bending stiffness along the z-axis even when discs are attached. Using this strategy, also more geometrically complex components can be approximated by a disc.

$$
M_{Disc} = \begin{pmatrix} m & & & & \\ & m & & & 0 \\ & & m & & \\ & & & J_x & \\ & 0 & & & J_x \\ & & & & & J_z \end{pmatrix} \qquad G_{Disc} = \begin{pmatrix} 0 & & & & & \\ & 0 & & & & 0 \\ & & 0 & & & \\ & & & 0 & \Omega J_p & 0 \\ & & & -\Omega J_p & 0 & 0 \\ & 0 & 0 & 0 & 0 & 0 \end{pmatrix} \qquad (6)
$$

The parameters of the disc can be obtained by various methods. Either, an analytical calculation can be made for moderately complicated geometries with specific boundary conditions, or the data can be read from the CAD-system. The component can be modeled as completely rigidly connected to one node or with an elasticity in relation to the shaft node. This is done for example in [5] where the stiffness of an impeller flange is determined with static analyses from 3D finite element software.

Support Bearings To model the effect of bearings, simple uncoupled, linear spring-damper elements can be added in the axial, torsional and lateral directions at every node of the mesh. Again, they are written in a local matrix form and then added to the global system. More advanced bearing models up to full roller bearing multi body systems can be implemented by the user. Some advanced models of high speed ball bearings were already implemented in this environment by Wagner [12].

Active Magnetic Bearings To perform a modal analysis and other system analysis calculations, a linearized version of the system is considered for the left hand side (lhs) of the equation of motion (eom). The magnetic force of an active magnetic bearing can be linearized around its center position using two linear factors [8]. One linear term is expressed by the stiffness coefficient (k_s) for the position u_x of the shaft in the bearing, the other linear term beeing the stiffness coefficient (k_i) relating the influence of the current from the controller. The controller is a simple PID-controller in continous time-domain. Written for one direction:

$$
i(t) = k_P u_x(t) + k_D \dot{u}_x(t) + k_I \int u_x(t) dt \qquad (7)
$$

Including the controller into the force and additionally neglecting the integrational part for the linearized system gives the lhs coefficients:

$$
K_{AMB} = \begin{pmatrix} k_s + k_i k_P & & & & \\ & k_s + k_i k_P & & 0 & \\ & & 0 & & \\ & & & 0 & \\ & 0 & & & 0 \end{pmatrix} \qquad C_{AMP} = \begin{pmatrix} k_i + k_D & & & & \\ & k_i + k_D & & 0 & \\ & & 0 & & \\ & & & 0 & \\ & 0 & & & 0 \end{pmatrix} \qquad (8)
$$

During a full time integration, the non-linear force law with the discretized controller are fully accounted for on the right hand side (rhs). The force for the magnetic bearing can then be taken from any arbitrary function $f(x, i)$ in dependency of the position x and the actual current i. The function f can be obtained from a polynomial representation, full co-simulations or interpolated measurement data. The current i is evaluated for every time step according to the controller algorithm, using the actual position and the position before.

$$h_{AMB} = \begin{pmatrix} g(u_x(t), u_x(t - \Delta t), t, \Delta t) \\ g(u_x(t), u_x(t - \Delta t), t, \Delta t) \\ 0 \\ 0 \\ 0 \\ 0 \end{pmatrix} \tag{9}$$

2.4 Rotor loads

All loads are formulated as inhomogeneous parts of the ode (i.e. the right hand side). The loads vector is formulated in the inertial coordinate system with the same global coordinates as the system matrices explained in the previous section. The load vector is generally dependent on the state-space vector and is updated for every time step during time integration. To approximate the actual load step, the load is calculated for example using the previous state-space vector. All loads are fully neglected during system analysis (homogeneous) computations. In this case the user has to provide alinearized matrix descriptions of the components for the lhs to obtain meaningful system analyses results. Rotor faults are also formulated as rotor loads. An overview is given in [9].

Unbalance One very common load to rotor systems is the unbalance. It is defined by its z-position, phase angle φ and magnitude ($U = \varepsilon\, m$).

$$J_{unbalance} = U\Omega^2 \begin{pmatrix} \cos(\Omega t + \varphi) \\ \sin(\Omega t + \varphi) \\ 0 \\ 0 \\ 0 \\ 0 \end{pmatrix} + U\dot{\Omega} \begin{pmatrix} \sin(\Omega t + \varphi) \\ -\cos(\Omega t + \varphi) \\ 0 \\ 0 \\ 0 \\ 0 \end{pmatrix} \tag{10}$$

For a constant rotational speed, the angular acceleration terms vanish.

Forward Whirling Chirp A more sophisticated load is the whirling chirp. This describes a force whose direction is forward whirling in the fixed frame of reference around a node of the rotor. Its whirling frequency is not constant but varying with a chirp frequency. The cosinus chirp signal is defined as a linear ascending frequency $f(t) = f_0 + kt$ with $k = (f_1 - f_0)/t_1$. The amplitude then writes [7]:

$$x(t) = \cos(2\pi \int f(\tau)d\tau) = \cos(2\pi \int_0^t (f_0 + k\tau)d\tau) = \cos(2\pi(f_0 + k/2t)t) \tag{11}$$

Where k is the rise of the linear function with f_0 as start frequency, f_1 as end frequency and t_1 as reference time. The resulting load vector h containing the forward whirl chirp is then:

$$h_{whirlchirp} = \hat{F} \begin{pmatrix} \cos(2\pi(f_0 + k/2t)t) \\ \sin(2\pi(f_0 + k/2t)t) \\ 0 \\ 0 \\ 0 \\ 0 \end{pmatrix} \tag{12}$$

This excitation is used to generate time series results which are similar to real measurement where this kind of excitation can be applied to a rotor using an active magnetic bearing. Later, modal analyses are performed using this data.

2.5   System analysis

To analyze the system with different methods, it is transformed into a first order state space system. The gyroscopic matrix is part of , see eq. (5).

Modal analysis The modal analysis is conducted in the homogeneous state space domain, i.e. no load is applied to the system. The state space is formulated in the following way:

$$A\dot{y} + B\dot{y} = 0 \quad \begin{pmatrix} M & 0 \\ 0 & K \end{pmatrix}\begin{pmatrix} \ddot{q} \\ \dot{q} \end{pmatrix} + \begin{pmatrix} D & K \\ -K & 0 \end{pmatrix}\begin{pmatrix} \ddot{q} \\ q \end{pmatrix} = 0 \tag{13}$$

This leads to a generalized eigenvalue problem which can be solved directly by common algorithms.

$$-BV = AV\lambda \tag{14}$$

The eigenvalue is described with λ, the eigenvector V is given in the state space. To visualize a mode shape, only the second half of the vector is of interest.

Time integration For the time integration, the system is brought to the following state space form.

$$\dot{y} + Ay + Bu = \quad \begin{pmatrix} 0 & 1 \\ -M^{-1}K & -M^{-1}D \end{pmatrix}\begin{pmatrix} q \\ \dot{q} \end{pmatrix} + \begin{pmatrix} 0 & 0 \\ 0 & -M^{-1} \end{pmatrix}\begin{pmatrix} 0 \\ h \end{pmatrix} \tag{15}$$

Kreutz examines the differences for integration schemes in [4]. As a rule of thumb, an explicit Runge-Kutta (4,5) algorithm can be used. When the system is stiff, algorithms based on the numerical differentiation formulas will be appropriate.

3   SOFTWARE STRUCTURE

The following section will describe the structure of the AMrotor toolbox and the way to interact with it. This is not meant as detailed documentation but more as description of the overall concept behind it.

3.1   Object-oriented approach

The code is written along the object oriented programming paradigm. Although this is not completely strict. The simulation itself can be conducted in the main file as procedural, interpreted code.

3.2   Class hierarchy

To form the classes and their relation for the code, it was natural to follow the concept of presenting physical components as object. The diagram in Figure 4 shows the idea of the class hierarchy in a simplified manner. The central object is the *Rotorsystem* which consists of individual parts, like the *Rotor* itself, multiple *Components* and *Loads*. The *Components* class is an abstract class as they come in many different forms. As an example two are listed here. Many more different types like different bearings (including active magnetic bearings with PID-controller) and seals are implemented. This is the place where new components can be implemented for research purposes. They inherit useful information from their parent class *Component*.

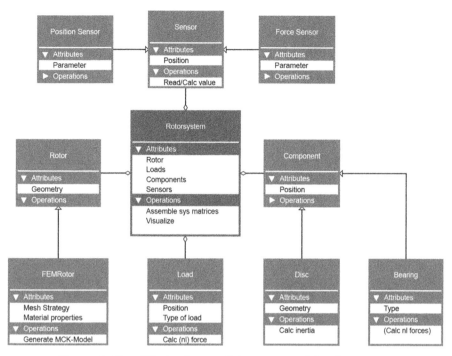

Figure 4. Simplified scheme of the core module of AMrotor.

The most important method of the *Rotorsystem* is the ability to assemble the full system matrices. Thereby a distinction is made between linear and nonlinear components. Linear ones can therefore be assembled via localisation matrices on the left-hand-side of the equation (i.e. in the MCK-model). Nonlinear or controlled components are then added to the right-hand-side and can be dependent of the complete state vector. They behave finally like loads. Loads can be defined in different directions as constant, time variant or depending on the state vector. Finally, *Sensors* are added to the whole system. Sensors can be of various type like position, velocity, acceleration or force. They are used to read or calculate the output results of the simulation at specific points on the rotor.

As shown in Figure 5, several *Experiments* with the *Rotorsystem*-object can be carried out after setting up the rotorsystem. The experiments are divided in two groups of analysis types. Firstly, the system analysis type that is used to investigate the properties of the left-hand-side (lhs) of the MCK-model (i.e. the homogeneous ode). This results in experiments like the modal analysis, frf-calculations from and to arbitrary points. Of course nonlinear components that are added to the right-hand-side (rhs) are not considered directly in this experiments. Therefore, the user can add a linearized version (around any equilibrium points) of the component to the rotorsystem which is then considered. A warning is given to the user when no linearized matrix version of the component is found. Secondly, the group of time-integration experiments use the full ode and applies different time integration schemes to it.

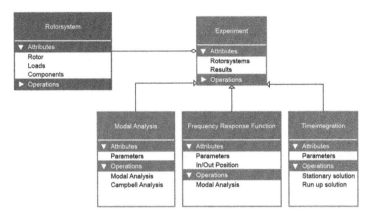

Figure 5. Simplified scheme of the experiment module of AMrotor.

3.3   Visualization and data output

All the experiments that are conducted with the rotor system store their results in the object of the experiment. The Figure 6 shows the way to plot the results. It is possible to have one result plotted in different ways. Therefore, it is again useful to divide the visualization code from the rest of the code and only hand the objects with relevant information to the graphs. There are a wide range of graphs available. For simplicity reasons only a few are shown here. The obtained datasets (from each sensor object) can also be outputed to standard universal (.unv) file to exchange with other software for further processing.

3.4   Simulation and configuration files

The user parameters of the simulation toolbox is set with two files. The first file is a *model* file. The file follows regular MATLAB® syntax. Here, the properties for each component are set in structs. Multiple components of one type can be set as array of structs. The second file is the *main simulation* file. There, the rotorsystem object is built from individual components. Each component receives its properties for the constructor from the associated struct. The setup of the components to the rotorsystem happens automatically by reading the configuration file and iterating through all available structs. Next, an experiment object is set up and filled with the parameters. After handing in the rotorsystem object, the experiment can be calculated. The same strategy is applied for the graphs and the data output. Because of this code structure, it is possible to set up different experiments from one rotorsystem object, that can be computed in parallel without any difficulty. That can be very useful for doing optimization or parameter studies.

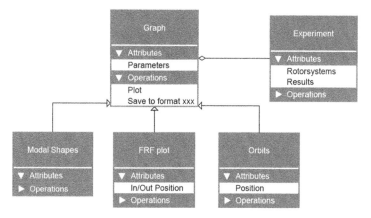

Figure 6. Simplified scheme of the graph module of AMrotor.

4 EXAMPLE

The application of the AMrotor toolbox is shown for an academic test rig which is used at the Chair of Applied Mechanics at TUM. The test rig was already used in [6] to perform an experimental and operational modal analysis with only the magnetic bearings as excitation and measuring device.

4.1 Experimental test rig

In Figure 7 the design of the magnetic bearing test rig (MBTR) is shown.

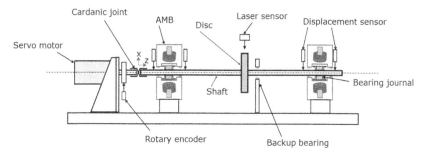

Figure 7. Scheme of the magnetic bearing test rig (MBTR).

The MBTR consists of a rotor (shaft, mass disc and bearing journals) mounted in two active magnetic bearings (AMB) and is driven by a servo motor. The servo motor and the rotor are connected via a cardanic joint. For controlling purposes, each AMB includes four eddy-current sensors. The disc can be extended by further masses to vary the unbalance. The deflection of the rotor at the position of the disc is measured by a laser sensor.

4.2 Steps to simulate the system

After setting up a model file for the presented system, the object for the rotor system can be constructed and a graph object for the 3D visualization is used to generate the output in Figure 8.

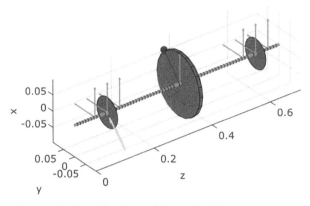

Figure 8. Visualization of the object *Rotorsystem*.

The discretized rotor mesh is shown in grey, while the disc components are shown in blue. They represent the additional added parts (bearing journal) for the active magnetic bearings and the disc in the middle. The radius and the thickness are only for the visualization. The simulation itself uses only the given inertia parameter. The AMB component is shown in orange as rings around the rotor shaft. The purple sphere represents an unbalance at the position of the disc. The physical parameters are not directly visualized as it is difficult to represent the proportions in an appropriate way. The yellow arrow symbolizes a load vector, like the described chirp force. Finally, the green quiver arrows show the positions of the displacement sensors. These are the locations and directions for which the user will obtain displacement results after the simulation.

Modal analysis The purpose of a system analysis is the detection of the system-inherent behavior without external excitation. The result of the system analysis is the determination of the natural frequencies and the associated mode shapes. The pictures figure. 9a and figure. 9b show visualizations of results of the system analysis. In figure. 9a, the first forward mode shape of the rotor is plotted. In figure. 9b, a Campbell diagram with the eigenfrequencies (ω) relative to the angular frequency (Ω) is shown. The shown diagram is filtered to only display frequencies arising from lateral modes. The first mode (blue) is correlated to the first bending mode, shown in figure. 9a. Modes two (orange) and three (gray) are closely related to rigid body modes coming from the relatively soft bearing stiffness with high damping values. The fourth mode (i.e. the green plotted mode around 105) shows a strong dependency of the rotational speed.

Time simulation The time simulation considers the active behavior of the magnetic bearings with a PID-controller for the position for each of the two bearings. The system is excited by the unbalance as depicted in Figure 8 and the whirling chirp excitation with a frequency range of 0 Hz to 250 Hz. Two different rotational speeds 0 and 250 are considered and displayed in section 4.2. An ODE15s algorithm is used to calculate the problem. This algorithm is used for stiff problem, which typically have multiple decaying solution components with greatly differing decaying speed. This is given for the shown system as the shaft is damped very weak and the magnetic bearings bring very high damping due to the controller. Therefore the vibration decay in the system can be very different. The simulation time was 1 s with a sample time of 1. The initial conditions are that the rotor is spinning with constant angular velocity and at $t = 0$ the two excitation force suddenly are present.

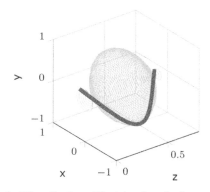

(a) Visualization of the lateral mode shape for the first lateral mode 15 Hz

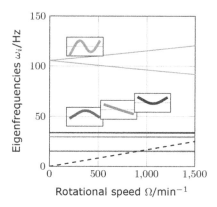

(b) Campbell diagram for the system

Figure 9. System analysis results.

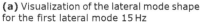
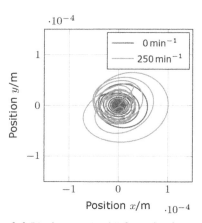

(a) Displacement orbit from the disc

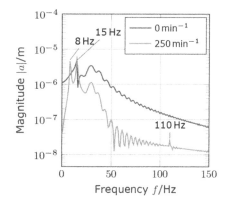

(b) Fourier transformation of the displacement in x-direction

Figure 10. Time integration results for two different rotational speeds.

Figure 10a shows the displacement orbit from the laser proximity measurements at the disc. The blue curve is only generated by the whirling excitation with increasing frequency. It is obvious that the unbalance starts to dominate the green curve at $250\,\text{min}^{-1}$ right from the beginning. It starts at 0 deflection and is pulled by the unbalance in an angle of $\pi/4$. The excitation force is visible as small curls in the curve. The Fourier transformation of the time series signal shows the frequency content (Figure 10b). Here, the dof at which the excitation force is applied is taken as driving point (left magnetic bearing). Comparing to the Campbell diagram, the green curve reveals the additional frequency peak at the rotational speed. It is not extremely sharp because of the transition effects in the beginning of the simulation. Because the excitation force has very low energy content in the higher frequency ranges, only the forward whirling eigenmotion at around 110 is excited which is clearly seen from the Campbell diagram.

609

5 CONCLUSION

In summary, a toolbox was outlined that can be used in research as well as for teaching in the field of rotor dynamics. One of the key elements is the 1D Timoshenko beam formulation that is generated by the user's input geometry. A second building block are the components that can be coupled to the lhs or rhs of the equation depending on the needs of the formulation. Different system analysis steps as well as time integration experiments are possible.

The application of the toolbox was shown with an example simulation of an academic test rig. This test-rig has two active magnetic bearings which are formulated in a linearized manner in order to perform investigations such as system analyses like modal analysis. Due to the rotational speed, different mode shapes are determined and shown in a Campbell diagram to see critical rotational speeds. Writing the magnetic bearing to the rhs allowed a transient time series simulation which is then examined using a Fourier transformation. Here, the excitation amplitude of the different eigenmodes can be observed.

Future aspects that are currently under development include the implementation of non-symmetric rotor elements like rectangular sections. Additionally, using coupling mechanisms will allow to consider the flexible housing dynamics described for instance by its frequency response functions.

REFERENCES

[1] Friswell, M. I., Penny, J. E. T., and Garvey, S. D. *Dynamics of Rotating Machines*. Cambridge University Press.

[2] Gasch, R., Nordmann, R., and Pfützner, H. *Rotordynamik*. Springer Berlin Heidelberg. <http://dx.doi.org/doi:%252010.1007/3-540-33884-5>.

[3] Genta, G. *Dynamics of Rotating Systems*. Springer US, 2005. <http://dx.doi.org/doi:%252010.1007/0-387-28687-x>.

[4] Kreutz, M. J., Maierhofer, J., Thümmel, T., and Rixen, D. Comparison of different time integration schemes and application to a rotor system with magnetic bearings in matlab. In *Proceedings of VIRM12 - Vibrations in Rotating Machinery*.

[5] Krügel, S., Maierhofer, J., Thümmel, T., and Rixen, D. "Rotor Model Reduction for Wireless Sensor Node Based Monitoring Systems". en. In: *Proceedings of 13th SIRM: The 13th International Conference on Dynamics of Rotating Machinery*, 978-87-7475-568-5. Technical University of Denmark. doi: 10.13140/RG.2.2.33454.05444.

[6] Maierhofer, J., Gille, M., Thümmel, T., and Rixen, D. "Using the Dynamics of Active Magnetic Bearings to perform an experimental Modal Analysis of a Rotor System". en. In *Proceedings of 13th SIRM: The 13th International Conference on Dynamics of Rotating Machinery*, Number 978-87-7475-568-5. Technical University of Denmark. doi: 10.13140/RG.2.2.26743.16809.

[7] MATLAB. *9.8.0.1323502 (R2020a)*. Natick", address = Natick", address = "Massachusetts: The MathWorks Inc., 2020.

[8] Schweitzer, G. & E. H. Maslen (2009). *Magnetic Bearings*. Springer Berlin Heidelberg. <http://dx.doi.org/doi:%252010.1007/978-3-642-00497-1>.

[9] Thuemmel, T., M. Rossner, C. Wagner, J. Maierhofer, & D. Rixen (2018, aug). Rotor orbits at operation speed and model-based diagnosis of multiple errors. In *Mechanisms and Machine Science*, pp. 222–237. Springer International Publishing. <http://dx.doi.org/doi:%252010.1007/978-3-319-99268-6_16>.

[10] Timbo, R., R. Martins, J. Mota, G. Buchmann, & F. Rangel (2020). Ross-rotordynamic open source software. *Journal of Open Source Software*.

[11] Tiwari (2017, nov). *Rotor Systems*. CRC Press. <http://dx.doi.org/doi:%252010.1201/9781315230962>.

[12] Wagner, C., A. Krinner, T. ThÃ¼mmel, & D. Rixen (2017, jun). Full dynamic ball bearing model with elastic outer ring for high speed applications. *Lubricants 5* (2), 17. <http://dx.doi.org/doi:%252010.3390/lubricants5020017>.

12th International Conference on Vibrations in Rotating Machinery -
Institution of Mechanical Engineers, ISBN 978-0-367-67742-8

A review of important nonlinear phenomena in rotor vibration

M.L. Adams

Mechanical & Aerospace Engineering, Case School of Engineering, Case Western Reserve University, Cleveland, Ohio, USA

ABSTRACT

For routine rotor vibration analyses and associated field troubleshooting measurement data analyses, linear vibration models are typically employed, justifiably so. However, an awareness and a basic understanding of inherent nonlinearities is a valuable asset to those routinely employing the linear models. Reviewed herein are nonlinear rotor vibration phenomena and topics on which the author has worked, both in laboratory research and in troubleshooting operating machinery in the field. Phenomena and topics reviewed include (1) response to very large rotor unbalance, (2) self-excited oil-whip hysteresis loop with seismic or depth charge events, (3) wrist-pin bearing damage in a reciprocating compressor, (4) rotor impact-on-bearing coefficient-of-restitution, (5) self-excited pad flutter in tilting-pad journal bearings, and (6) application of deterministic chaos theory in analyzing nonlinear rotor vibrations.

1   INTRODUCTION

All real machine vibratory systems contain some nonlinearities. Fortunately, those nonlinearities are usually relatively small and thus can justifiably be ignored in routine machinery design analyses. However, evidence of small-effect nonlinearities is still typically exhibited in measured rotor vibration signals as exampled in Figure 1. If the stiffness and damping coefficients are all constant, the governing differential equations are mathematically linear. Then, if excited by a harmonic forcing function, the differential equation(s) of motion, e.g., Eqn. (1) for Figure 2 model, has a steady-state response (particular solution) containing only the excitation frequency Ω, as expressed in Eqn. (2) for the Figure 2 one-degree-of-freedom model.

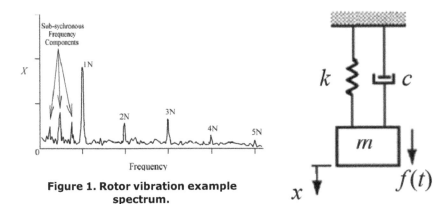

Figure 1. Rotor vibration example spectrum.

Figure 2. One degree of freedom.

$$m\ddot{x} + c\dot{x} + kx = F_0 \sin \Omega t \qquad (1)$$

$$x(t) = X_0 \sin(\Omega t + \phi) \qquad (2)$$

However, if the stiffness and/or damping coefficients are functions of the motion, i.e., $k=k(x)$ and $c=c(x)$, then the equation(s) of motion becomes nonlinear and the steady state response $x(t)$ can also contain harmonics of the forcing frequency, as exampled in Figure 1 for relatively small nonlinear effects. However, when the nonlinear effects become relatively large, a linear analysis simulation model can readily predict quite fallacious vibratory responses, as exampled in the next section.

2   VERY LARGE ROTOR UNBALANCE

The photographs in Figure 3 were first presented by Adams and McCloskey (1), being from two catastrophic failures in the 1970s of large 600MW steam turbine-generator units, Kinan No.3 (Japan) and Porsheville (France). A computational approach to simulate the rotor vibration response caused by very large rotor unbalance, e.g., detachment of last-stage low-pressure (LP) turbine blade, was given by Adams (2).

(a) Low pressure (LP) turbine casing   (b) LP turbine last-stage blades

(c) Brushless exciter shaft   (d) Generator shaft

Figure 3. Catastrophically failed 600MW steam turbine-generator unit, circa 1970s.

Clearly seen from Figure 3(a) is the LP turbine casing hole caused by the full-speed detachment of last stage LP turbine blades, the determined root cause of the resulting very large rotor unbalance. Correspondingly, time-transient simulations were computed for the LP turbine (Figure 4) to parametrically study rotor unbalance from small tip portion detachment up to that of the outer 1/3 of a last stage blade detachment.

The simulated steady state rotor vibration orbits from that study were first presented by Adams (2) and subsequently by Adams (7). One set of simulations modeled the actual fixed sleeve journal bearings (journal diameter 16 in., 40 cm). A second set of simulations employed 4-pad pivoted-pad journal bearings identically sized as the actual fixed sleeve journal bearings. The difference between the two sets of simulations is dramatic, as exampled in Figures 5 and 7.

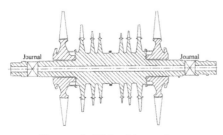

Figure 4. LP turbine rotor.

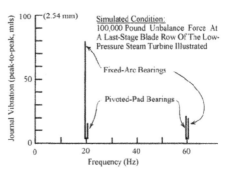

Figure 6. FFT of journal vibration (p-to-p).

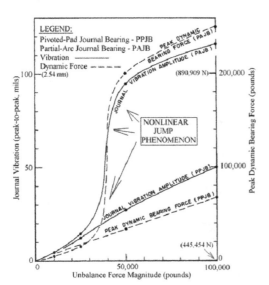

Figure 5. Vibration and peak dynamic force, comparing fixed-sleeve and PPJB.

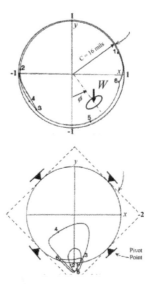

Figure 7. Journal-to-bearing orbits, timing marks each 1/2 revolution.

A spectrum comparing the two simulation sets is shown in Figure 6, and the steady state journal-relative-to-bearing orbits in Figure 7 (top orbit for fixed sleeve, bottom for PPJB). The journal-to-bearing orbit with fixed sleeve journal bearings exhibits a large 1/3N dominating subharmonic component, but the journal orbit with pivoted-pad journal bearings (PPJB) does not.

A heuristic view of this is that a truly linear system has closed "energy valves" that prevents any vibration energy from "leaking" between the various harmonics of a harmonically forced linear system. Correspondingly, with a relatively small degree of nonlinearity present, those "valves" leak a little, giving rise to a response spectrum as that of Figure 1, where a small amount of energy may get to various harmonics, (e.g.,. 1/2, 1/3, 1/4..., 2, 3, 4...). In the case of the LP turbine rotor-bearing system covered in Figures 4, through 7, it has a very lightly damped harmonic at 1,150 cpm, (close to 1/3 of 3,600 rpm operating speed) when running at 3,600 rpm. Actually, oil whip self-excited subharmonic rotor vibration occasionally shows up in the field at 3,600-rpm operation with this unit. The simulation results in Figures 5 show that as the unbalance is increased and correspondingly the nonlinearity also increased, a result the theorists call a "nonlinear jump phenomenon" occurs. Again heuristically speaking, as the postulated rotor unbalance is increased, the journal-to-bearing vibration amplitude correspondingly increases, which in turn increases the nonlinearity, which in turn further "opens up" the "energy valves" and so on, giving the observed jump phenomenon with fixed-sleeve journal bearings . As Figures 5 and 7(a) show, the 1/3 subharmonic participation dominates the large unbalance vibration and bearing dynamic force for fixed sleeve journal bearings.

Pivoted-pad journal bearings (PPJB) have long been employed to eliminate the oil-whip phenomenon for numerous types of rotating machinery. Thus unsurprisingly, the simulation results employing PPJBs does not yield the nonlinear jump to a dominate 1/3 subharmonic. The FFT of the two configurations, Figure 6, most compactly shows this important comparative difference.

3   JOURNAL BEARING OIL-WHIP HYSTERESIS LOOP

The nonlinear hysteresis loop associated with the journal bearing phenomenon oil whip was for a long time an interesting topic for the academics. However, in the seismically active region of Japan a team headed up by Professor Y. Hori (3, 4) at the University of Tokyo brought the practical importance of this topic to the wider engineering community. This work by Hori motivated the author to research the topic further, Adams, et al. (5).

A generic illustration of the journal bearing hysteresis loop is shown in Figure 8. It imbeds the classical oil-whip phenomenon within an expanded view that shows two stable vibration solutions at speeds below the small-perturbation oil-whip threshold speed ω_{th} (i.e., a Hopf bifurcation) and one unstable solution which is a boundary between the two stable solutions. Adams et al. (5) demonstrate this through computation simulations of a rigid symmetric rotor supported in two identical 360° journal bearings which they treat as a 2-DOF point mass rotor supported in one hydrodynamic fluid film journal bearing. They simulate hysteresis loop examples covering a wide range of parameter combinations covering a static load range from nearly zero to high loads. They confirm their simulation generated hysteresis loop examples using a specially configured laboratory table top test rig.

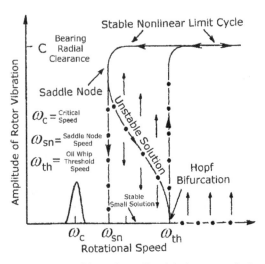

Figure 8. Journal bearing oil-whip hysteresis loop.

At vanishingly small radial static loads, a hysteresis loop does not occur and the oil-whip threshold speed ω_{th} is near twice the critical speed ω_c, as is well known. At progressively higher bearing static loads the hysteresis loop progressively opens up wider, with the oil-whip threshold speed ω_{th} occurring at progressively higher speeds, as also well known. But the lower speed limit ω_{sn} of the hysteresis loop gets progressively lower, asymptotically approaching the lower limit of 1.72 times the critical speed ω_c. The quite practical importance of this is that given a large static bearing load and a substantially large dynamic disturbance such as a major earthquake, the large amplitude stable oil-whip nonlinear limit cycle vibration can occur at a rotor speed less than twice the critical (resonance) speed ω_c. So while increased bearing static load raises the expected oil-whip threshold speed, it also lowers the speed above which a large amplitude oil-whip limit cycle vibration can occur. One can see a similarity between the hysteresis loop and the nonlinear jump phenomenon covered in the previous section. However, the two phenomena are not the same thing since the journal bearing hysteresis loop phenomenon is self-excited and has its own frequency, being initiated only by a large bump or transient ground motion disturbance such as a major earthquake.

Khonsari and Chang (6) show the existence of a position boundary encircling the static equilibrium point which delineates between initial positions which die out and initial positions which grow to a stable large amplitude limit cycle orbit. The author believes they located an unstable solution locus that exists within its hysteresis loop. Though they do not state this, their contribution is none-the-less valuable by confirming existence of an unstable solution boundary between the two stable solutions. The complete initial-condition boundary provided by the unstable solution would have to be defined in the appropriate dimensioned phase space. Utilizing the unstable-solution locus in the position/velocity space provides evaluation of dynamic disturbance intensities needed to cause the stable nonlinear limit cycle to occur. For example, how strong an earthquake does it takes to cause that to happen for a given land-based system, or how strong of a depth charge does it take for that to happen on a critical safety related pump in a nuclear powered submarine.

4   RECIPROCATING COMPRESSOR WRIST-PIN BEARING

The piston and connecting rod sub-assembly shown in Figure 9 is from a single-piston reciprocating compressor designed for use in both a home refrigerator and a window air conditioner. After some design improvements to this compressor design were implemented into production of these two refrigerant products, the refrigerator units started to show more than 4 times the rate of within-warranty compressor failures than the window air conditioner. This resulted in a multi-million dollar annual loss on the refrigerator product, thereby quickly getting attention of upper management.

A study of several of the failed compressors clearly revealed that it was the wrist pin bearing that was failing. The wrist pin sleeve bearing is press fitted into the piston and surrounds the wrist pin which is press fitted into the connecting rod, Figure 9. That the refrigerator model's compressor failure rate was over 4 times that of the air conditioner compressor mystified the manufacturer's top compressor engineers, because the wrist pin bearing peak load in the air conditioner was approximately 25% higher than in the refrigerator. The then existing wrist pin bearing radial load versus crank angle is illustrated for both applications in Figure 10.

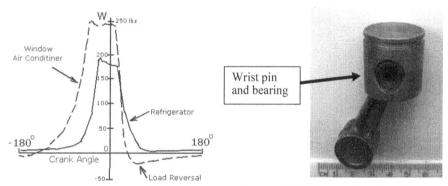

Figure 9. Piston and connecting rod of a small reciprocating refrigeration compressor.

Figure 10. Wrist pin bearing load curves versus crank angle.

In an attempt to uncover the root cause for the relatively large warranty failure rate in the refrigerator application, many different analyses and tests were conducted, sort of a "fishing expedition". One of the many analyses eventually pursued was computation of the wrist pin bearing's minimum film thickness within the 0-360° crank cycle. Then a recently new young employee (circa 1972) at the manufacture's corporate research center, the author was assigned that task. The author pointed out to his supervisor that the nonlinear dynamical orbit of the wrist pin relative to the bearing had to be simulated to predict the transient minimum oil film thickness, numerically resolving the Reynolds lubrication (partial differential) equation at each integration time step. Both the air conditioner and refrigerator time-varying loads (Figure 10) were inputs to the author's specially written simulation computer code, which employed a time-transient marching algorithm. Amazingly, the computed wrist pin orbits relative to the bearing, not the minimum film thickness, provided the answer to the bearing-failure root cause.

Figure 11 shows simulated wrist pin orbits from this analysis employing the Figure 10 load curves. They clearly showed the root cause of the refrigerator compressor's higher warranty failure rate. The load curves illustrated in Figure 10 show that the loading function for the air conditioner goes slightly negative and the one for the refrigerator does not, a feature that was not previously noted by the senior investigators prior to the author's simulation. That is, in the air conditioner application, just a slight amount of load reversal causes the wrist pin to substantially separate away from the oil-feed hole that channels oil from the rod bearing through a connecting hole in the rod. In contrast, for the refrigerator there was no load reversal and thus the wrist pin did not lift off the oil-feed hole as its oscillatory trajectory clearly shows. Extensive subsequent endurance tests completely confirmed what the computed nonlinear dynamical orbits implied. That is, the refrigerator wrist pin continuously rubbed on the bearing over the oil hole and thereby did not cyclically separate from the oil hole to allow sufficient lubricant (oil mixed with refrigerant) in for the next squeeze film action of each succeeding load cycle. Oil was mixed with the refrigerant because the refrigerant loop was hermetically sealed. With the root cause uncovered by the author's nonlinear dynamic analysis, modifications were then implemented to insure that the refrigerator compressor bearing load curve included a little load reversal. The high compressor failure rate ceased, once those units still "in the pipeline" cleared the retailers' stocks. Not a bad way for a young engineer to start a new job.

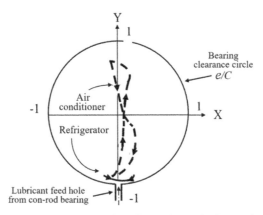

Figure 11. Wrist pin orbital trajectories relative to bearing.

5   SHAFT-ON-BEARING IMPACT

Impact is a quite nonlinear dynamic phenomenon. In Adams (7) rotor-stator rub-impacting is treated from the point of view as a result of excessive rotor vibration, and its identifying symptoms are treated. To properly simulate rotor-bearing hard point impact dynamic motion, there is the need for the impact coefficient-of-restitution value. The coefficient-of-restitution comes from experimentally determined information. A strictly theoretical hard point impact modeling approach is at the limits of what modern solid mechanics analysis tools can provide. Because the energy loss in such hard point impacts is the stress-strain hysteresis loop energy within the resulting complicated internal propagating waves within the two impacting hard bodies. Adams, Afshari and Adams (8) configured a quite elaborate experimental setup employing orthogonal x-y laser vibrometers to directly measure velocities through controlled bearing-shaft impacts. Their test apparatus is summarily

618

illustrated in Figure 12 and detailed in Adams (9). Test results obtained with this apparatus are presented in Figure 13, and cover wide ranges of journal speed and impact velocity.

As clear from Figure 13, test results obtained with this apparatus exhibit a fairly close repeatable grouping of test points, which is significant considering the non-triviality of capturing impact velocities, even with modern sensors and data reduction methods. These results indicate virtually no influence of journal sliding velocity, which probably reflects the quite small relative impact time during bearing-journal contact. These results approach a maximum restitution coefficient of about 0.8 at impact velocities vanishingly small, and asymptotically approach a value of about 0.5 as impact velocity increases.

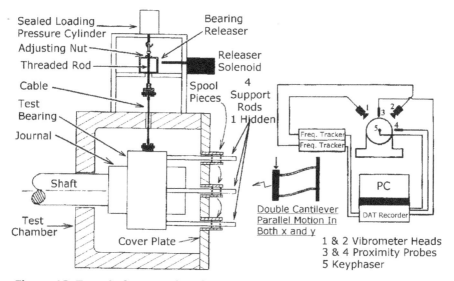

Figure 12. Test rig for rotor-bearing restitution coefficient measurement.

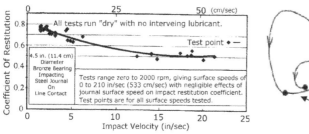

Figure 13. Coefficient-of-restitution values from bearing-on-journal impact velocity measurements.

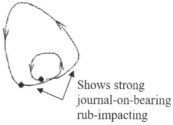

Figure 14. Measured orbit (filtered), Adams (7).

Figure 14 is a field measured journal orbit at a very lightly damped 1/2-speed harmonic encountered during a forced emergency coast down triggered by sudden very high rotor vibration. Reliable simulation of such impact events necessitates accurate coefficient-of-restitution data.

6 PAD FLUTTER IN UNLOADED BEARING TILTING-PADS

Figure 15(a) illustrates leading edge fatigue-crack damage on the statically unloaded top pads of a large tilting-pad journal bearing. Awareness of this problem first arose with routine bearing inspections during scheduled outages of large power plant steam turbine generator units employing large 4-pad tilting-pad journal bearings, Figure 15 (b). Around that same time the author was simulating nonlinear rotor dynamical characteristics of a vertical-centerline canned-motor primary coolant pump, employing water lubricated tilting pad journal bearings, for a pressurized-water nuclear reactor. In that work, the unanticipated discovery was made of a previously unrecognized dynamical phenomenon of tilting-pad journal bearings, namely self-excited pad flutter of the statically unloaded pad. This discovery had a low priority to other issues with the pump investigation, i.e., depth-charge initiated rotor response. A few years later after he became a professor the author renewed research interest on this self-excited pad-flutter phenomenon to thoroughly study it.

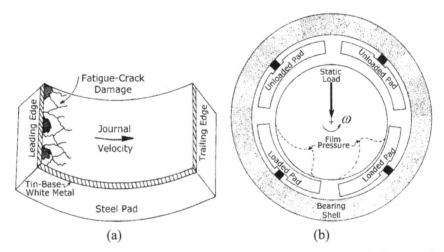

(a) (b)

Figure 15. Four-pad tilting-pad journal bearing loaded between bottom pads.

Adams and Payandeh (10) present extensive nonlinear dynamical simulations for a single-pad 2-DOF model with simultaneous radial and pitching pad motions. Figure 16 illustrates this model and provides a 4-frame time sequence of a typical case of unloaded pad flutter. They present a tabulation of the results for several different pad configurations and operating parameters which provide a broad coverage of design and operating conditions that will or will not promote pad flutter. Their work provides some general conclusions on the pad flutter phenomenon. Namely, when a pad's operating pivot clearance (C') is larger than the concentric clearance (C), a stable static equilibrium pad position may not exist, and if not, self-excited subsynchronous pad vibration will occur. The self-excited motion continuously seeks to find an instantaneous film pressure distribution that produces a null force and moment on the pad. The base frequency of this vibration is somewhat near 0.5 times the rotational speed, somewhat like classical journal bearing oil whip. In fact, if one observes the stationary journal centerline from a reference frame fixed in the fluttering pad, the journal appears to be undergoing a closed-orbit self-excited whirling. Therefore, this pad flutter phenomenon is really a camouflaged version of the classical journal bearing oil-whip phenomenon.

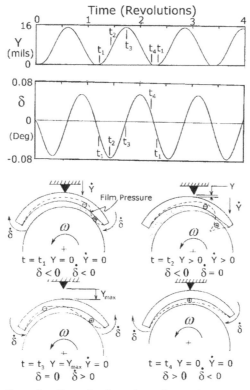

Figure 16. Simulation of self-excited motion of a statically unloaded tilting pad.

7   CHOAS ANAYSIS OF NONLINEAR ROTOR DYNAMICS

Using time-transient simulations, Adams and Abu-Mahfouz (11) explored several nonlinear rotor dynamical systems applying the chaos signal processing tools developed in the recent era, for example as described by Moon (12). Terms identifying the chaos signal processing tools include (a) bifurcation diagrams, (b) fractal dimensions of Poincare maps, along with time based rotor orbits. One practical example Adams and Abu-Mahfouz present is for a 3-pad tilting-pad journal bearing with static radial load directed exactly into a pivot location with a moderate rotor unbalance, Figure 17. As shown, for the 3-pad bearing with no bearing preload (C'/C=1), the static journal orbit is chaotic since there are two equally possible static equilibrium journal positions, one on either side of the loaded pivot point. Also shown of practical value is that with the addition of a moderate bearing preload (C'/C=0.85), the vicious chaotic orbital motion disappears.

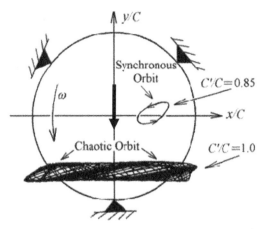

Figure 17. Rotor orbit with static load into pivot point of 3-tilting-pad bearing.

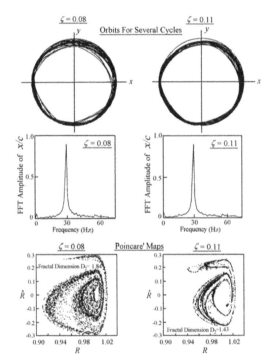

Figure 18. Poincare map of chaotic response reveals small loss in damping at an off-resonance condition with full-bore rub simulation.

$$R \equiv \frac{\sqrt{(x^2 + y^2)}}{C} \qquad \dot{R} = \frac{dR}{dt} \qquad C = \text{Radial Clearance}$$

Figure 18 shows the detection of a very small reduction of damping at an off-resonance condition. Neither the orbit trajectories nor the orbit FFTs reflect the small damping difference between 11% and 8% of critical damping. However, the fractal dimension and appearance of a Poinare map clearly do. It is only fair to point out that unlike an FFT, construction of a Poincare map requires coordinate data from several successive cycles of motion, one cycle for each dot.

CLOSING INSIGHT

Electrical engineering science has long been considerably imbedded in the use of linear-model analyses, justifiably so. In sharp contrast, chemical engineering science has long recognized the need for nonlinear-model analyses, such as for various constituent mixing processes. In mechanical engineering science linear models usually get the job done, however not always. Thus, the mechanical engineer may be the most likely one not to recognize when nonlinear effects are important. Vibrations in rotating machinery is such a mechanical engineering problem category.

REFERENCES

[1] Adams, M. L. and McCloskey, T. H., "Large Unbalance Vibration in Steam Turbine-Generator Sets," Proc., 3rd IMechE International Conference on Vibration in Rotating Machinery, York, 1984.
[2] Adams, M. L., "Non-Linear Dynamics of Flexible Multi-Bearing Rotors," Journal of Sound and Vibration, Vol. 71(1), 1980.
[3] Hori, Y, "Anti-Earthquake Considerations in Rotor Dynamics", Keynote address paper, Proc., 4th IMechE International Conference on Vibration in Rotating Machinery, ISBN 0 85298676 9, pp. 1–8, Edinburgh, 1988.
[4] Hori, Y. and Kato, T., "Earthquake Induced Instability of a Rotor Supported by Oil Film Bearing", ASME Journal of Vibrations and Acoustics, 112, 160–165, 1990.
[5] Adams, M. L., Adams, M. L. & Guo, J. S., "Simulations and Experiments of the Non-Linear Hysteresis Loop for Rotor-Bearing Instability", Proc., 6th International Conference on Vibration in Rotating Machinery, IMechE, Oxford University, 1996.
[6] Konsari, M. M. and Chang, Y. T., "Stability Boundary of Nonlinear Orbits within Clearance Circle of Journal Bearings", ASME Journal of Vibration and Acoustics, 115, 303–307, 1993.
[7] Adams, M. L., "Rotating Machinery Vibration - From Analysis to Troubleshooting", 2nd ed., CRC Press, Taylor & Francis, 2010, ISBN 978-1-4398-0717-0.
[8] Adams, M. L., Afshari, F. and Adams, M. L., "An Experiment to Measure the Restitution Coefficient for Rotor-Stator Impacts", Proc.,7th IMechE International Conference on Vibrations in Rotating Machinery, Nottingham University, 2000, ISBN 1 86058 273 7.
[9] Adams, M. L. "Rotating Machinery Research & Development Test Rigs", CRC Press, Taylor & Francis, 2017. ISBN-13: 978-1-138-03238-5.
[10] Adams, M. L. and Payandeh, S., "Self-Excited Vibration of Statically Unloaded Pads in Tilting-Pad Journal Bearings," ASME Journal of Lubrication Technology, 105(3), 1983.

[11] Adams, M. L. and Abu-Mahfouz, I., "Exploratory Research on Chaos Concepts as Diagnostic Tools for Assessing Rotating Machinery Vibration Signatures", Proc. IFTOMM 4th International Conference on Rotor Dynamics, Chicago, September 1994.

[12] Moon, F. C. "Chaotic Vibrations: An Introduction for Applied Scientists and Engineers". 309 p., New York: Wiley-Interscience, 2004.

12th International Conference on Vibrations in Rotating Machinery -
Institution of Mechanical Engineers, ISBN 978-0-367-67742-8

An experimental assessment of torsional and package vibration in an industrial engine-compressor system

B. Halkon[1], I. Cheong[2], G. Visser[2], P. Walker[1], S. Oberst[1]

[1]Centre for Audio, Acoustics and Vibration, University of Technology Sydney, Australia
[2]ALS Industrial, Queensland, Australia

ABSTRACT

An experimental field vibration measurement campaign was conducted on an engine-compressor system. Torsional vibrations were measured using both a strain-gauge based technique at the engine-compressor coupling and a rotational laser vibrometer at the torsional vibration damper. Package vibration measurements were simultaneously captured using a number of accelerometers mounted at various locations on the engine and compressor casings. Findings from the study include the observation that the coupling/damper dominant order 1.5 torsional vibration level was higher at idle (c14.1 Hz) than at full speed (c19.1 Hz) and that this is likely the result of the coincidence of the first torsional natural frequency (c19-20 Hz); vibration remained within limits. The package vibration observed was in general within limits and displayed the expected behaviour when shaft speeds coincided with structural resonances. Increasing of system load was observed to result in package vibration level increase in the engine but reduction in the compressor and this is suspected to be as a result of the effect of increased damping. Induced cylinder misfire scenarios were shown to lead to higher vibration levels. To the authors' knowledge, this is the first time that angular displacement, vibratory torque and package vibration have been simultaneously measured, analysed and reported in an industrial context/scenario. It is hoped that this contribution might, therefore, serve as a practical guide to vibration engineers that wish to embark on similar campaigns.

1   INTRODUCTION

Experimental vibration analysis has long been accepted as the most effective means of determining the health of rotating machinery (1), (2). Vibrations in rotating machinery are inevitable but their consequences extend from loss of efficiency through to safety-critical failures. Organisations relying on production processes powered by rotating machinery systems invest heavily in systems and services to monitor vibrational performance. Vibration velocity is generally regarded as the optimum measurement parameter, whether translational or rotational, but often assessment criteria might be specified as displacement or acceleration as well or instead with this choice often influenced by historical factors such as transducer availability, reliability and value proposition or data acquisition or processing options.

Direct shaft measurements are generally more challenging than those from nearby non-rotating components, often requiring significant production downtime and risk management for setup and installation. Slip rings or telemetry required to power transducers and/or transfer signals to non-rotating acquisition systems require larger installation package space and regular maintenance (3). Inevitable aspects such as shaft run-out or eccentricity and whole-body vibration present challenges with proximity probe type solutions and with tachos or encoders which, in addition to the measurement of torsional vibration, are generally used to post-process fixed sampled frequency data as engine order spectra or, indeed, for resampling into the angle domain. Even more the contemporary laser vibrometry (4), insensitive to shaft shape variation and translational vibration, is not the ideal rotating machinery transducer due to speckle noise and vibration component cross-sensitivity challenges (5).

This study has, for the first time, employed a range of rotating machinery applicable transducers and processing techniques simultaneously to better understand the links between torsional and package vibration characteristics for different operating conditions in an industrial context. Comparisons are made between torsional vibration measurements made directly from the torsional damper, using a laser vibrometer, and from the coupling and gearbox input shaft, using a torsional strain gauge configuration. Package vibrations, simultaneously captured at a series of locations on both the engine and compressor using a combination of uni- and tri-axial accelerometers mounted, are examined. Results are assessed, in the context of relevant industry standards/OEM guidelines, to understand the contributing factors towards possible excessive torsional vibration failure and to identify package vibration measurement locations that exhibit correlation to high torsional vibrations.

2 METHODOLOGY

2.1 Machinery arrangement and instrumentation configuration

The system consists of a 12-cylinder gas engine and an oil-flooded screw compressor, the two sub-systems being coupled by a shrink disc flexible rubber coupling with a "2:1" gearbox (116:59 teeth) at the compressor input (Figure 1). The compressor male and female rotors consist of 6 and 4 lobes respectively, thereby meshing at a 1.5:1 ratio. Anticipated dominant vibration frequencies of the system are therefore at engine half orders, 1.97x (1x male rotor), 1.31x (1x female rotor), 7.86x (1x lobe pass frequency, LPF) and 15.73x (2x LPF). The engine is mounted onto a foundation via adjustable chocks at six locations with the compressor being similarly mounted at four locations. Compressor loading is adjusted by varying the position of a slide valve from fully open (0% - unloaded) through fully closed (c100% - full load).

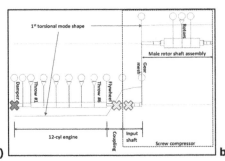

Figure 1. Machinery arrangement a) physically and b) schematically, also showing 1st torsional mode shape, laser vibrometer (green cross) and strain gauge (yellow cross) measurement locations.

A previously commissioned torsional system dynamic analysis, utilising the model shown in Figure 1b, determined the first torsional mode shape as identified. The twist node occurs in the flexible rubber coupling while the antinode appears at the front of the engine near the harmonic damper. The torsional vibration measurement locations (Figure 1b and Figure 2) were therefore chosen to be as sensitive as possible to the drivetrain predicted response, i.e. strain gauges installed at the region of highest stress, i.e. at the coupling nodes (Figure 2a), and laser vibrometer measurement taken at the region of greatest twist angle, i.e. at the damper anti-node (Figure 2b).

Figure 2. Torsional vib. instrumentation a) strain gauges at the coupling spacer and gearbox input shafts and b) laser vibrometer, beams highlighted.

2.2   Instrumentation configuration

Torsional strain at the coupling was monitored using foil type strain gauges (HBM 350 Ω, Darmstadt Germany; (6)), which were wired in a torsional full bridge, temperature-compensated configuration. The signals from the torsional strain gauges were conditioned using a torque telemetry system (Binsfeld TT10k, MI; (7)) configured with a 500 Hz low-pass filter and transmitter/receiver gains of 2000. Bridges were zeroed and shunt calibrated prior to data acquisition commencement.

Torsional vibration at the damper was directly monitored using a rotational laser vibrometer (Polytec RLV-5500, Waldbronn, Germany; (8)). The damper surface was treated with a 25 mm strip of retro-reflective tape (Figure 2b) to ensure reliable optical signal level. Angular velocity (100 °/s/V), displacement (1 °/V) and RPM (1000 min^{-1}/V) outputs were captured with the RPM and tracking filters set to "Slow".

Package vibrations were recorded using 50 mV/g tri-axial accelerometers (Kistler, NY (9)) and 100 mV/g uni-axial accelerometers (CTC, NY (10)) at locations and in orientations as set out in Table 1. All sensors were magnet mounted to the surface of the selected locations with a frequency response of up to at least 1 kHz expected (11). The following axis convention was used: x – axial, parallel to shaft rotation axis; y – horizontal, perpendicular to shaft axis, z – vertical.

Table 1 . Accelerometer identification and mounting locations.

ID	Label	Description	Type	Orientation
A1	E1H+V+A	Engine NDE Block, Above Mount - Male	Triaxial	X, Y, Z
A2	E2H+V	Engine DE Block, Above Mount - Male	Triaxial	Y, Z
A3	E3H+V+A	Engine DE – Shaft Centreline	Triaxial	X, Y, Z
A4	E4H	Engine mid-span block – Male	Uni-axial	Y
A5	E5H	Engine mid-span block – female	Uni-axial	Y
A6	E6H	Engine NDE Block, Above Mount - Female	Uni-axial	Y
A7	E7H	Engine DE Block, Above Mount - Female	Uni-axial	Y
A8	C1H+V+A	Compressor DE – Shaft Centreline	Triaxial	X, Y, Z
A9	C2H+V+A	Compressor Mid-span Casing Top	Triaxial	X, Y, Z
A10	C3H+V+A	Compressor Suction	Triaxial	X, Y, Z
A11	C4H	Compressor Mid-span Casing Female	Uni-axial	Y
A12	C5H	Compressor Mid-span Casing Male	Uni-axial	Y
A13	C6A	Compressor Male Rotor – Axial	Uni-axial	X

Dynamic torque, torsional vibration, package vibration and tachometer data (from the gearbox input shaft) were simultaneously collected via a multi-channel acquisition system (National Instruments cDAQ-9178/NI-9234/NI-9232; (12)). Data were collected in "raw", voltage format with a sampling frequency of 2.048 kHz, yielding a maximum observable frequency of 800 Hz, and recorded to disc in .tdms file format. Data post-processing was subsequently implemented in MATLAB.

2.3 Data post-processing
Torsional strain data were converted to torque, T, through the following equation:

$$T = \mu \varepsilon \mathbf{J\,G\,R} \tag{1a}$$

where μ = 2, ε = torsional strain, $\mathbf{J}$ = polar moment of inertia, $\mathbf{R}$ = radius and $\mathbf{G}$ = shear modulus (typically taken as 79 GPa for steel). A two second moving median filter was used to extract the static torque component from the resulting dynamic torque signal with this converted to steady state engine power, P (in kW), as follows:

$$P = T \times RPM / 9548.8 \tag{1b}$$

The dynamic torque component was extracted using a 2 Hz, 10th order Butterworth high-pass filter. Where data were captured during steady-state conditions, frequency response characteristics were investigated by applying a Fast Fourier Transform (FFT) algorithm to four second segments of data, yielding 0.25 Hz spectral resolution, with a Hann window utilised to minimise spectral leakage. Where possible and ideally during sufficiently slow-speed run-ups, order maps, with a minimum resolution of 0.25, were generated showing vibratory torque as a function of both engine speed and frequency.

Since the precise integration filter characteristics were uncertain, instead of using the available LDV output, angular *displacement* data were determined in post-processing from the LDV angular *velocity* output. A cumulative trapezoidal numerical integration technique, coupled with a 10th order Butterworth (2 Hz) high-pass filter to remove any integration drift ("edge effects" were ignored), was implemented. Again, for (pseudo) steady-state data, frequency spectra and order maps were similarly generated.

Accelerometer signals were 10th order Butterworth (1-800 Hz) bandpass filtered with the resulting signal integrated for velocity, again using a cumulative trapezoidal technique plus 2 Hz high-pass filter and edge-effects removal. Overall package vibration was assessed by computing the running RMS over a 5 sec period with the trend(s) plotted to assess severity against the relevant standard limits (13), (14). Waterfall plots with two averages per spectrum with 0.5 Hz resolution were generated to assess the dynamic package vibration at various timeframes during operation. Operational Deflection Shapes (ODS) were generated from both time and frequency domain data using ModalView (ABSignal; (15)). The former enables investigation of the package vibration response to transient events, e.g. load changes; the latter allows investigation of specific dominant vibration frequencies under steady-state conditions.

2.4   Vibration assessment criteria

Torsional vibration assessment criteria, presented in Table 2, were derived from the aforementioned torsional system dynamic analysis (vibratory torque amplitude of the various shafts in the drivetrain) and from the engine manufacturer service documentation (angular displacement amplitude at the harmonic damper housing).

Table 2 . Torsional vibration acceptability limits, a) vibratory torque and b) angular displacement.

a)		
System	**Location/Condition**	**Nm (pk-pk)**
Engine	Front drive	13219
	Crankshaft	38626
Compressor	Input shaft	6193
	Male rotor shaft	2106
Coupling	Nominal	7002
	Warm	4201

b)	
Order (of engine speed)	**Deg (pk-pk)**
0.5	2.5
1.0	0.7
1.5+	0.3

Package vibration assessment criteria, set out in Table 3, were determined from a combination of the engine manufacturer, ISO 10816-6 (13) and API 619 (14).

Table 3 . Package vibration acceptability limits, velocity-based.

Source/standard	Component	mm/s (RMS)	
		Low Alarm	High Alarm
Engine manufacturer	Engine	18.0 (Rough)	26.9 (Very rough)
ISO 10816-6 Class III	Engine	11.2 (Zone B/C)	17.8 (Zone C/D)
API 619	Compressor	8.0 (Concern)	12.0 (Alarm)

3   RESULTS

3.1   Engine performance

The engine performance data shown in Figure 3 are typical for the system under test under normal running conditions. Figure 3a shows the performance for 0% slide valve position, i.e. unloaded, initially with the engine running at low idle (~845 rpm) and subsequently at high idle (~1185 rpm). Figure 3b shows the same initially but, following a brief period at high idle, the slide valve position is progressively increased in 20% increments to full load. The engine achieved a power output of approximately 560 kW during high idle (full speed, no load) conditions. An approximately proportional relationship between valve position and power output can be seen in Figure 3b where, at the maximum slide valve position (full load), the engine achieves a power output of approximately 1070 kW, closely matching its nominal power rating.

3.2   Torsional vibration performance

3.2.1   *Vibratory torque*

Vibratory torque data at the coupling spacer and gearbox input shafts were interrogated at the following running conditions: low idle, high idle and at full speed under full load. Figure 4a&b show example time waveforms and corresponding FFTs for the measured data at the coupling shaft only at low idle (a) and under full load conditions (b). The high idle, unloaded signals were of similar form to those observed at full load albeit with reduced levels and are therefore not shown here for the sake of brevity. Similarly, the corresponding signals for the gearbox input shaft, which only show different levels again, are not shown. The FFT representation indicates a dominant frequency at order 1.5 with associated half order harmonics. At low idle, the order 1.5 peak was particularly pronounced whereas at high idle through to full load this dominance was reduced with the order 1 peak becoming increasingly significant as shown in the order map of Figure 4c and as highlighted by the arrow.

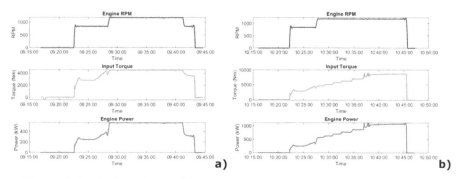

Figure 3. Typical engine performance a) low and high idle and b) low and high idle, up to full load.

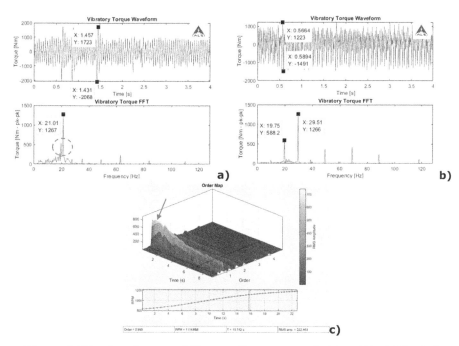

Figure 4. Vibratory torque at the coupling spacer a) at low idle, b) for full load and c) order map (0.25 res.) for no-load run-up; note order 2.5 peak.

The presence of an additional peak at about 19-20 Hz in the low idle FFT (see high-lighted region in Figure 4a), suggests that there may be a torsional resonance within this range. The presence of the first torsional natural frequency coinciding within 19-20 Hz would explain the higher vibratory torque response at low idle versus at high idle. This is shown in Table 4 which captures the vibratory torque response for both the coupling spacer and gearbox input shafts at the three running conditions (gearbox input under full load N/A due to loss of data). Despite this, with respect to Table 2a, the system was operating within vibratory torque limits in all cases.

3.2.2 *Angular displacement*

Specific regions in the data were interrogated to extract the torsional vibration in the form of angular displacement ("position") at the damper housing, under the following conditions: low idle and high idle; various slide valve positions up to (almost) full

load; and inducing up to two cold cylinders in the engine followed by varying the slide valve positions. Example torsional vibration FFT data at the harmonic damper are presented in Figure 5. Upon reviewing all data, it was determined that, during normal operating conditions, the angular displacement responses at orders 0.5 and 1.0 were below the limits. The response at order 1.5 was again dominant and approached the acceptability limit of 0.3 deg pk-pk under some conditions, as shown in Figure 6.

Table 4 . Vibratory torque response at coupling spacer and gearbox input.

Speed (RPM)	Condition	Torque (Nm pk-pk)	
		Coupling spacer	Gearbox input
845	Low idle	3791	4485
1185	High idle	1734	1861
1185	Full load	2714	/

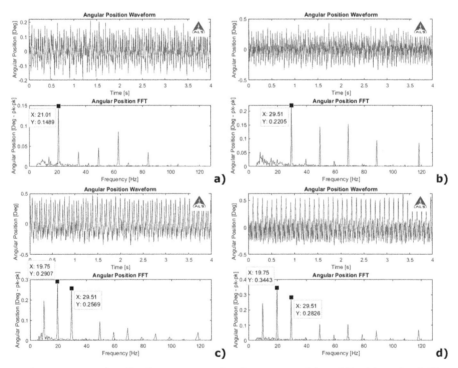

Figure 5. Angular displacement at the damper at a) low idle, b) under full load, c) with two cold cyls. at high idle and d) with one cold cyl. at full load.

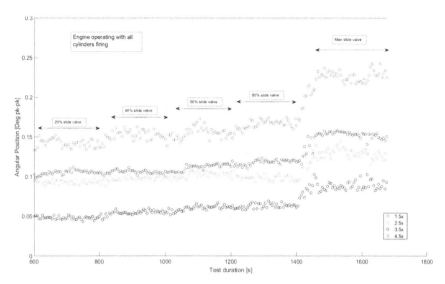

Figure 6. Order peak values under normal operating conditions.

The effect of cold cylinders can also be seen in Figure 5c&d with Table 5 summarising the angular displacement response under all running conditions. A significant increase in order 1.5 vibratory angular displacement of approximately 75% can be observed as one cold cylinder was induced. When a second cold cylinder was induced, an increase of approximately 220% results. In addition to the order 1.5 changes, the increased signifi- cance of the order 0.5 and 1 peaks is evident as shown in Figure 5c&d, with the order 1 peak becoming dominant in both cases. While running with two cold cylinders under loaded conditions was not performed following a risk assessment, based on these find- ings it is likely that the vibratory angular displacement with two cold cylinders under load would result in levels beyond the acceptability limits at 1.5, if not also at other orders.

Table 5 . Angular displacement response at damper housing at order 1.5.

Speed (RPM)	Condition	Deg (pk-pk)
845	Normal, unloaded (low idle)	0.19
1185	Normal, unloaded (high dle)	0.12
1185	Normal, 20% slide valve	0.14
1185	Normal, 40% slide valve	0.15
1185	Normal, 60% slide valve	0.16
1185	Normal, 80% slide valve	0.17
1185	Normal, 98% slide valve	0.23
1185	2 cold cylinders, unloaded	0.26
1185	1 cold cylinder, unloaded	0.21
1185	1 cold cylinder, max slide valve	0.29

633

3.3 Package vibration performance

3.3.1 *Overall vibration*

Example velocity running RMS example trends plotted against the relevant acceptability limits during operating conditions such as that shown in Figure 3b are presented in Figure 7. It can be observed that, in general, the vibration of the engine increases along with compressor loading while the opposite is true for the compressor, with the lowest vibration levels seen during full load. A summary of the overall package vibration under the steady-state, full load operating condition is presented in Table 6. The system was operating within limits at all locations.

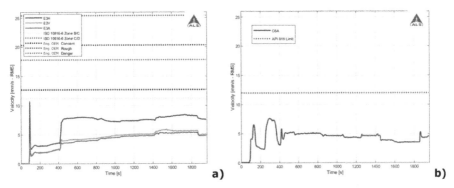

Figure 7. Example package vibration RMS at a) Engine drive end and b) Compressor male rotor.

Table 6. Overall package vibration velocities under full load.

Location	mm/s (RMS)		
	Axial	Horizontal	Vertical
A1	2.5	5.7	3.6
A2	-	4.8	2.3
A3	5.3	8.4	5.6
A4	-	7.0	-
A5	-	9.5	-
A6	-	4.9	-
A7	-	5.6	-
A8	5.6	4.2	3.9
A9	3.6	3.1	2.9
A10	4.7	6.5	4.7
A11	-	2.3	-
A12	-	2.0	-
A13	-	3.5	-

3.3.2 Waterfall analysis

A summary of the natural frequencies identified from waterfall plots, examples of which are shown in Figure 8 again for the engine and compressor, is presented in Table 7. The natural frequencies were identified by looking for signs of amplitude amplification during the run-up phase. The excitation source and dominant direction for those natural frequencies were also identified. From the table summary, all of the natural frequencies coinciding with a dominant excitation source showed sufficient separation margin of greater than 10%.

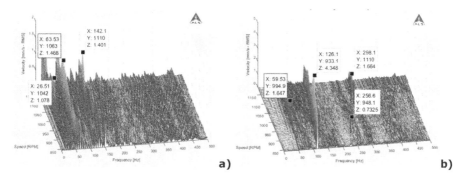

a) b)

Figure 8. Example package vibration waterfalls for no-load run-up at a) Eng. Drive End – Shaft Centerline (axial) and b) Comp. Male Rotor – Axial.

Table 7. Overall package vibration velocities under full load.

Sens.	Suspected natural frequencies (Hz)			Excit. source @ 1185 rpm		Sep. margin (%)	Comments
	X	Y	Z	(order)	(Hz)		
A3	26.5	26.5	26.5	1.5	29.6	10.5	Significant amplif. in horizontal
	63	-	-	3.5	69.1	8.9	Minor
	142	-	142	7.5	148.1	4.1	Minor
	-	-	195	10.5	207.4	6.0	Minor
A8	-	40	-	2.5	49.4	19.0	Minor
	59	-	-	3.5	69.1	14.6	Minor, possibly from engine frame
	-	-	80	4.5	88.9	10.0	Minor
	126	126	-	7.86	155.2	18.8	Significant amplif. in axial
	251	251	251	15.73	310.7	19.2	Minor
A13	59	-	-	3.5	69.1	14.6	Minor, possibly from engine frame
	126	-	-	7.86	155.2	18.8	Significant amplif. in axial
	256	-	-	15.73	310.7	17.6	Minor
	298	-	-	15.73	310.7	4.1	Minor

3.3.3 *Effect of increasing load*

In general and as expected, the engine vibration at dominant frequencies increases with increasing valve position. The engine vibration showed an increase in half order harmonics, particularly at orders 1.5, 2.5, 3.5, 4.5 and 6 (Figure 9a). The trend at those frequencies was also seen in the torsional vibration data, measured both by the strain gauges and the laser vibrometer, the latter as shown in Figure 6 and summarised in Table 5. It can be deduced that the engine vibration is directly correlated to the amount of loading imparted by the driven machinery. Meanwhile, the compressor dominant frequencies were found to decrease with increasing valve position, particularly at the lobe pass frequency (LPF) and harmonics as shown in in Figure 9b, again consistently with the torsional vibration. It can be hypothesized that loading of the screw compressor creates a damping effect on LPF related vibration.

3.3.4 *Effect of cold cylinders*

Investigation of the effect of inducing cold cylinders on the package vibration showed that the most sensitive location to detect an increase in torsional vibration in the engine through the monitoring of package vibration, was at the non-drive end of the

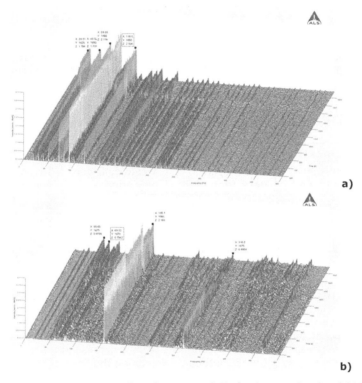

Figure 9. Example package vibration waterfalls for increasing load (20-98% slide valve), a) Eng. Non-Drive End and b) Comp. Male Rotor.

engine (i.e. at the front of the engine) in the horizontal direction. A similar observation was made of the compressor vibration spectra at the drive end – horizontal (gearbox input). This is to be expected since this shaft is physically coupled to the engine crankshaft. All other locations across the compressor casing did not show any significant changes upon inducing of cold cylinders, indicating that they are less sensitive to the torsional vibrations of the crankshaft system.

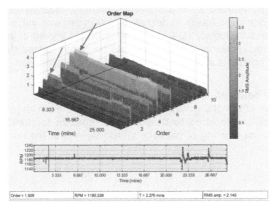

Figure 10. Example package vib. order map (0.1 res.) for cold cyls. at Eng. Non-Dr. End; note incr. at orders 1.5 and 4.5 upon inducing two cold cyls.

Table 8 . Example package vib. vels. for cold cyls. at Eng. Non Dr. End.

Order	mm/s (RMS)			
	Normal, unloaded	2 cold cylinders, unloaded	1 cold cylinder, unloaded	1 cold cylinder, full load
1.5	1.3	2.2	1.6	2.0
3.5	2.3	2.3	1.6	1.8
4.5	3.0	3.6	3.2	3.3

As shown in Figure 10, significant differences in the order 1.5 (c29.5 Hz), 3.5 (c69 Hz) and 4.5 (c88 Hz) frequencies occur upon inducing a cold cylinder. The results are summarized in Table 8 with the observation also in general agreement with the corresponding trend in torsional vibration data presented previously. It is postulated that the non-drive end of the engine is most sensitive to changes in torsional vibrations because of the presence of a torsional antinode or point of maximum vibratory twist at the damper housing as set out in section 2. Monitoring the trend of these vibrations could be an appropriate method of correlating package vibration to torsional vibration.

4   CONCLUSIONS

This paper has comprehensively reported an industrial vibration measurement campaign on an engine-compressor package, employing simultaneous measurement of torsional and package vibration. Significant insights into the preparation of the instrumentation, data acquisition and processing, in the context of the expected nature of the vibratory performance and documented limits of the system, have been detailed. The interaction of driven frequencies and structural resonances has been described in the context of an order 1.5 dominant peak and the first torsional natural frequency that were shown to lead to higher vibration at low vs. at high idle. The assessment and correlation of torsional and package vibration frequencies and levels, for increasing load and for induced cold cylinders, was presented. Finally, recommendations are made for the location (non-drive end of the engine and drive end of the equipment) and interpretation of *package* vibration measurements for the evaluation of *torsional* vibration. It is proposed that, in place of costly and time-consuming vibratory torque measurements, package vibration and/or rotational LDV measurements can reliably assess torsional vibration in an industrially valid context.

REFERENCES

[1] Downham, E., Woods, R., "The rationale of monitoring vibration on rotating machinery in continuously operating process plant", *Trans. ASME J. of Engineering for Industry* 71-Vibr-96, 1971.

[2] Mitchell, J. S., "Vibration analysis - its evolution and use in machinery health monitoring", *Soc. Env. Eng. Symp. Machine Health Monitoring*, London, 1975.

[3] Tawadros, P, Awadallah, M, Walker, PD, Zhang, N, "Using a low-cost bluetooth torque sensor for vehicle jerk and transient torque measurement", *Proc. IMechE, Part D: J. Auto Eng.*, 2019.

[4] Rothberg, S.J., et al., "An international review of laser Doppler vibrometry: making light work of vibration measurement", *Opt. Lasers in Eng*. 99, 2017.

[5] Halkon, B.J., Rothberg, S.J., "A practical guide to laser Doppler vibrometry measurements directly from rotating surfaces", *Proc. Eleventh Int. Conf. on Vibs. Rot. Mach., Manc., UK*, 2016, pp. 215–230, pp. 9-16.

[6] HBM, "Pre-wired strain gauges, model 350 CLY4-3L-3M" www.hbm.com/en/4550/pre-wired-strain-gauges-fast-safe-and-convenient [accessed 13-03-20].

[7] Binsfield Eng., Inc. "TorqueTrak 10K Torque Telemetry System" binsfeld.com/wp-content/uploads/2019/08/TT-10K_8695003_E.pdf [accessed 21-02-20].

[8] Polytec GmbH, "Rotational Laser Vibrometer RLV-5500 Operating Instructions" (41278-Man-RotVib-RLV5500-1116-05en), 2016.

[9] Kistler Instrument. Corp., "Ceramic Shear Triaxial Accelerometer type 8763B" www.kistler.com/?type=669&fid=10&model=download [accessed 13-03-20].

[10] CTC, Inc., "AC102 - Multi-Purpose Accelerometer" www.ctconline.com/ctc_100_mv_g_standard_size_accelerometers.aspx?prd=AC102 [accessed 13-03-20].

[11] Døssing, O., "Structural Testing, Part 1 – Mechanical Mobility Measurements", Brüel & Kjær, Denmark, 1988.

[12] Nat. Instruments, "NI cDAQ^{TM}-9178 Specifications, NI CompactDAQ Eight-Slot USB Chassis" www.ni.com/pdf/manuals/374046a.pdf [accessed 24-02-20].

[13] ISO 10816-6:1995(en), "Mechanical vibration — Evaluation of machine vibration by measurements on non-rotating parts — Part 6: Reciprocating machines with power ratings above 100 kW", 1995.

[14] API STD 619, "Rotary-Type Positive-Displacement Compressors for Petroleum, Petrochemical, and Natural Gas Industries", Fifth Edition, 2010.

[15] ABSignal, Inc., "ModalVIEW – modal testing & analysis tool" absignal.com/product/index.php [last accessed 01-03-20].

Author Index